Regression for Categorical Data

This book introduces basic and advanced concepts of modern categorical regression with a focus on the structuring constituents of regression. Meant for statisticians, applied researchers, and students, it includes many topics not normally included in books on categorical data analysis, including recent developments in flexible and high-dimensional regression.

In addition to standard methods such as logit and probit models and their extensions to multivariate settings, the book presents more recent developments in regularized regression with a focus on the selection of predictors; tools for flexible nonparametric regression that yield fits that are closer to the data; advanced models for count data; nonstandard tree-based ensemble methods; and tools for the handling of both nominal and ordered categorical predictors. Issues of prediction are explicitly considered in a chapter that introduces standard and newer classification techniques.

Software including an R package that contains datasets and code for most of the examples is available from http://www.stat.uni-muenchen.de/~tutz/catdata.

Dr. Gerhard Tutz is a Professor of Statistics in the Department of Statistics at Ludwig-Maximilians University, Munich. He was formerly a Professor at the Technical University Berlin. He is the author or co-author of nine books and more than 100 papers.

CAMBRIDGE SERIES IN STATISTICAL AND PROBABILISTIC MATHEMATICS

This series of high-quality upper-division textbooks and expository monographs covers all aspects of stochastic applicable mathematics. The topics range from pure and applied statistics to probability theory, operations research, optimization, and mathematical programming. The books contain clear presentations of new developments in the field and also of the state of the art in classical methods. While emphasizing rigorous treatment of theoretical methods, the books also contain applications and discussions of new techniques made possible by advances in computational practice.

A complete list of books in the series can be found at http://www.cambridge.org/uk/series/sSeries.asp?code=CSPM.

Recent titles include the following:

8. *A User's Guide to Measure Theoretic Probability*, by David Pollard
9. *The Estimation and Tracking of Frequency*, by B. G. Quinn and E. J. Hannan
10. *Data Analysis and Graphics Using R*, by John Maindonald and John Braun
11. *Statistical Models*, by A. C. Davison
12. *Semiparametric Regression*, by David Ruppert, M. P. Wand, and R. J. Carroll
13. *Exercises in Probability*, by Loic Chaumont and Marc Yor
14. *Statistical Analysis of Stochastic Processes in Time*, by J. K. Lindsey
15. *Measure Theory and Filtering*, by Lakhdar Aggoun and Robert Elliott
16. *Essentials of Statistical Inference*, by G. A. Young and R. L. Smith
17. *Elements of Distribution Theory*, by Thomas A. Severini
18. *Statistical Mechanics of Disordered Systems*, by Anton Bovier
19. *The Coordinate-Free Approach to Linear Models*, by Michael J. Wichura
20. *Random Graph Dynamics*, by Rick Durrett
21. *Networks*, by Peter Whittle
22. *Saddlepoint Approximations with Applications*, by Ronald W. Butler
23. *Applied Asymptotics*, by A. R. Brazzale, A. C. Davison, and N. Reid
24. *Random Networks for Communication*, by Massimo Franceschetti and Ronald Meester
25. *Design of Comparative Experiments*, by R. A. Bailey
26. *Symmetry Studies*, by Marlos A. G. Viana
27. *Model Selection and Model Averaging*, by Gerda Claeskens and Nils Lid Hjort
28. *Bayesian Nonparametrics*, edited by Nils Lid Hjort et al.
29. *From Finite Sample to Asymptotic Methods in Statistics*, by Pranab K. Sen, Julio M. Singer, and Antonio C. Pedrosa de Lima
30. *Brownian Motion*, by Peter Mörters and Yuval Peres
31. *Probability*, by Rick Durrett

Regression for Categorical Data

GERHARD TUTZ
Ludwig-Maximilians Universität

CAMBRIDGE
UNIVERSITY PRESS

CAMBRIDGE
UNIVERSITY PRESS

32 Avenue of the Americas, New York NY 10013-2473, USA

Cambridge University Press is part of the University of Cambridge.

It furthers the University's mission by disseminating knowledge in the pursuit of education, learning and research at the highest international levels of excellence.

www.cambridge.org
Information on this title: www.cambridge.org/9781107009653

First published 2012

A catalogue record for this publication is available from the British Library

Library of Congress Cataloguing in Publication data

Tutz, Gerhard.
Regression for categorical data / Gerhard Tutz.
p. cm. – (Cambridge series in statistical and probabilistic mathematics)
ISBN 978-1-107-00965-3 (hardback)
1. Regression analysis. 2. Categories (Mathematics) I. Title. II. Series.
QA278.2.T88 2011
519.5′36–dc22 2011000390

ISBN 978-1-107-00965-3 Hardback

Contents

Preface

The focus of this book is on applied structured regression modeling for categorical data. Therefore, it is concerned with the traditional problems of regression analysis: finding a parsimonious but adequate model for the relationship between response and explanatory variables, quantifying the relationship, selecting the influential variables, and predicting the response given explanatory variables.

The objective of the book is to introduce basic and advanced concepts of categorical regressions with the focus on the structuring constituents of regressions. The term "categorical" is understood in a wider sense, including also count data. Unlike other texts on categorical data analysis, a classical analysis of contingency tables in terms of association analysis is considered only briefly. For most contingency tables that will be considered as examples, one or more of the involved variables will be treated as the response. With the focus on regression modeling, the generalized linear model is used as a unifying framework whenever possible. In particular, parametric models are treated within this framework.

In addition to standard methods like the logit and probit models and their extensions to multivariate settings, more recent developments in flexible and high-dimensional regressions are included. Flexible or non-parametric regressions allow the weakening of the assumptions on the structuring of the predictor and yield fits that are closer to the data. High-dimensional regression has been driven by the advance of quantitative genetics with its thousands of measurements. The challenge, for example in gene expression data, is in the dimensions of the datasets. The data to be analyzed have the unusual feature that the number of variables is much higher than the number of cases. Flexible regression as well as high-dimensional regression problems call for regularization methods. Therefore, a major topic in this book is the use of regularization techniques to structure predictors.

Special topics that distinguish it from other texts on categorical data analysis include the following:

- Non-parametric regressions that let the data determine the shape of the functional relationships with weak assumptions on the underlying structure.
- Selection of predictors by regularized estimation procedures that allow one to apply categorical regression to higher dimensional modeling problems.
- The focus on regression includes alternative models like the hurdle model and zero-inflated regression models for count data, which are beyond generalized linear models.
- Non-standard tree-based ensemble methods that provide excellent tools for prediction.
- Issues of prediction are explicitly considered in a chapter that introduces standard and newer classification techniques, including the prediction of ordered categorical responses.
- The handling of categorical predictors, nominal as well as ordered ones. Regularization provides tools to select predictors and to determine which categories should be collapsed.

The present book is based on courses on the modeling of categorical data that I gave at Technical University Berlin and my home university, Ludwig-Maximilians Universität München. The students came from different fields – statistics, computer science, economics, business, sociology – but most of them were statistics students. The book can be used as a text for such courses that include students from interface disciplines. Another audience that might find the text helpful is applied researchers and working data analysts from fields where quantitative analysis is indispensable, for example, biostatisticians, econometricians, and social scientists. The book is written from the perspective of an applied statistician, and the focus is on basic concepts and applications rather than formal mathematical theory. Since categorical data analysis is such a wide field, not all approaches can be covered. For topics that are neglected, for example, exact tests and correlation models, an excellent source is always Alan Agresti's book *Categorical Data Analysis* (Wiley, 2002).

Most of the basic methods for categorical data analysis are available in statistical packages like SAS and SPSS or the free package R (R Development Core Team, 2010). Software including an R package that contains most of the datasets and code for the examples is available from http://www.stat.uni-muenchen.de/~tutz/catdata. Some references to R packages are given in the text, but code is available in the package only. When using the package one should be familiar with R. One of the many tutorials available at the R site might help. Also, introductory books that explicitly treat the use of R with some applications to categorical data like Everitt and Hothorn (2006) and Faraway (2006) could be helpful.

I had much help with computational issues in the examples; thanks to Jan Gertheiss, Sebastian Petry, Felix Heinzl, Andreas Groll, Gunther Schauberger, Sarah Maierhofer, Wolfgang Pößnecker, and Lorenz Uhlmann. I also want to thank Barbara Nishnik and Johanna Brandt for their skillful typing and Elise Oranges for all the corrections in grammar – thanks for all the commas. It was a pleasure to work with Lauren Cowles from Cambridge University Press.

Gerhard Tutz

Chapter 1

Introduction

Categorical data play an important role in many statistical analyses. They appear whenever the outcomes of one or more categorical variables are observed. A categorical variable can be seen as a variable for which the possible values form a set of categories, which can be finite or, in the case of count data, infinite. These categories can be records of answers (yes/no) in a questionnaire, diagnoses like normal/abnormal resulting from a medical examination, or choices of brands in consumer behavior. Data of this type are common in all sciences that use quantitative research tools, for example, social sciences, economics, biology, genetics, and medicine, but also engineering and agriculture.

In some applications all of the observed variables are categorical and the resulting data can be summarized in contingency tables that contain the counts for combinations of possible outcomes. In other applications categorical data are collected together with continuous variables and one may want to investigate the dependence of one or more categorical variables on continuous and/or categorical variables.

The focus of this book is on regression modeling for categorical data. This distinguishes between explanatory variables or predictors and dependent variables. The main objectives are to find a parsimonious model for the dependence, quantify the effects, and potentially predict the outcome when explanatory variables are given. Therefore, the basic problems are the same as for normally distributed response variables. However, due to the nature of categorical data, the solutions differ. For example, it is highly advisable to use a transformation function to link the linear or non-linear predictor to the mean response, to ensure that the mean is from an admissible range. Whenever possible we will embed the modeling approaches into the framework of generalized linear models. Generalized linear models serve as a background model for a major part of the text. They are considered separately in Chapter 3.

In the following we first give some examples to illustrate the regression approach to categorical data analysis. Then we give an overview on the content of this book, followed by an overview on the constituents of structured regression.

1.1 Categorical Data: Examples and Basic Concepts

1.1.1 Some Examples

The mother of categorical data analysis is the (2×2)-contingency table. In the following example data may be given in that simple form.

Example 1.1: Duration of Unemployment

The contingency table in Table 2.3 shows data from a study on the duration of employment. Duration of unemployment is given in two categories, short-term unemployment (less than 6 months) and long-term employment (more than 6 months). Subjects are classified with respect to gender and duration of unemployment. It is quite natural to consider gender as the explanatory variable and duration as the response variable.

TABLE 1.1: Cross-classification of gender and duration of unemployment.

Gender	Duration		Total
	$\leq$ 6 months	> 6 months	
male	403	167	570
female	238	175	413

□

A simple example with two influential variables, one continuous and the other categorical, is the following.

Example 1.2: Car in Household

In a sample of $n = 6071$ German households (German socio-economic household panel) various characteristics of households have been collected. Here the response of interest is if a household has at least one car ($y = 1$) or not ($y = 01$). Covariates that may be considered influential are income of household in Euros and type of household: (1) one person in household, (2) more than one person with children, (3) more than one person without children). In Figure 1.1 the relative frequencies for having a car are shown for households within intervals of length 50. The picture shows that the link between the probability of owning a car and income is certainly non-linear. □

In many applications the response variable has more than two outcomes, for example, when a customer has to choose between different brands or when the transport mode is chosen. In some applications the response may take ordered response categories.

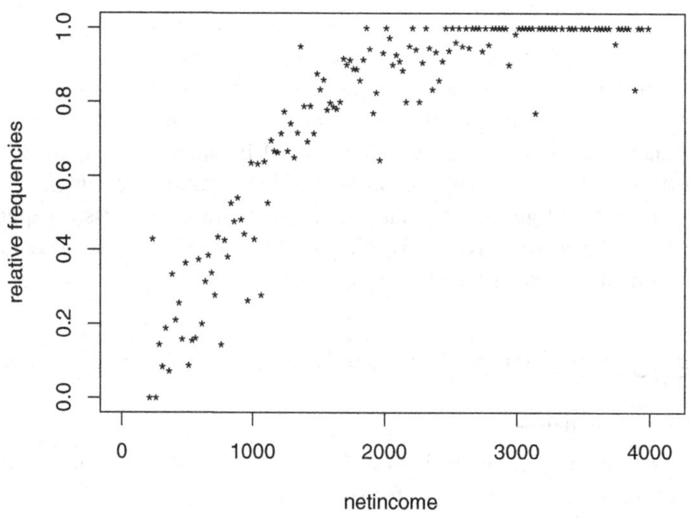

FIGURE 1.1: Car data, relative frequencies within intervals of length 50, plotted against net income in Euros.

Example 1.3: Travel Mode

Greene (2003) investigated the choice of travel mode of $n = 840$ passengers in Australia. The available travel modes were air, train, bus, and car. Econometricians want to know what determines the choice and study the influence of potential predictor variables as, for example, travel time in vehicle, cost, or household income. □

Example 1.4: Knee Injuries

In a clinical study focusing on the healing of sports-related injuries of the knee, $n = 127$ patients were treated. By random design, one of two therapies was chosen. In the treatment group an anti-inflammatory spray was used, while in the placebo group a spray without active ingredients was used. After 3, 7, and 10 days of treatment with the spray, the mobility of the knee was investigated in a standardized experiment during which the knee was actively moved by the patient. The pain Y occurring during the movement was assessed on a five-point scale ranging from 1 for no pain to 5 for severe pain. In addition to treatment, the covariate age was measured. A summary of the outcomes for the measurements after 10 days of treatment is given in Table 1.2. The data were provided by Kurt Ulm (IMSE Munich, Germany). □

TABLE 1.2: Cross-classification of pain and treatment for knee data.

	no pain 1	2	3	4	severe pain 5	
Placebo	17	8	14	20	4	63
Treatment	19	26	11	6	2	64

A specific form of categorical data occurs when the response is given in the form of counts, as in the following examples.

Example 1.5: Insolvent Companies in Berlin

The number of insolvent firms is an indicator of the economic climate; in particular, the dependence on time is of special interest. Table 1.3 shows the number of insolvent companies in Berlin from 1994 to 1996. □

TABLE 1.3: Number of insolvent companies in Berlin.

	Month											
	Jan.	Feb.	March	April	May	June	July	Aug.	Sep.	Oct.	Nov.	Dec.
1994	69	70	93	55	73	68	49	97	97	67	72	77
1995	80	80	108	70	81	89	80	88	93	80	78	83
1996	88	123	108	92	84	89	116	97	102	108	84	73

Example 1.6: Number of Children

There is ongoing research on the birthrates in Western countries. By use of microdata one can try to find the determinants that are responsible for the number of children a woman has during her lifetime. Here we will consider data from the German General Social Survey Allbus, which contains data on all aspects of life in Germany. Interesting predictors, among others, are age, level, and duration of education. □

In some applications the focus is not on the identification and interpretation of the dependence of a response variable on explanatory variables, but on prediction. For categorical responses prediction is also known as classification or pattern recognition. One wants to allocate a new observation into the class it stems from with high accuracy.

Example 1.7: Credit Risk

The aim of credit scoring systems is to identify risk clients. Based on a set of predictors, one wants to distinguish between risk and non-risk clients. A sample of 1000 consumers credit scores collected at a German bank contains 20 predictors, among them duration of credit in months, amount of credit, and payment performance in previous credits. The dataset was published in Fahrmeir and Hamerle (1984), and it is also available from the UCI Machine Learning Repository. □

1.1.2 Classification of Variables

The examples illustrate that variables in categorical data analysis come in different types. In the following some classifications of variables are given.

Scale Levels: Nominal and Ordinal Variables

Variables for which the response categories are qualitative without ordering are called *nominal.* Examples are gender (male/female), choice of brand (brand $A, \dots,$ brand K), color of hair, and nationality. When numbers $1, \dots, k$ are assigned to the categories, they have to be understood as mere labels. Any one-to-one mapping will do. Statistical analysis should not depend on the ordering, or, more technically, it should be *permutation invariant.*

Frequently the categories of a categorical variable are ordered. Examples are severeness of symptoms (none, mild, moderate, marked) and degree of agreement in questionnaires (strongly disagree, mildly disagree,. . . ,strongly agree). Variables of this type are measured on an ordinal scale level and are often simply called *ordinal.* With reference to the finite number of categories, they are also called *ordered categorical* variables. Statistical analysis may or may not use the ordering. Typically methods that use the ordering of categories allow for more parsimonious modeling, and, since they are using more of the information content in the data, they should be preferred. It should be noted that for ordinal variables there is no distance between categories available. Therefore, when numbers $1, \dots, k$ are assigned to the categories, only the ordering of these labels may be used, but not the number itself, because it cannot be assumed that the distances are equally spaced.

Variables that are measured on *metric* scale levels (*interval* or *ratio* scale variables) represent measurements for which distances are also meaningful. Examples are duration (seconds, minutes, hours), weight, length, and also number of automobiles in household $(0, 1, 2, \dots)$. Frequently metric variables are also called *quantitative*, in contrast to nominal variables, which are called *qualitative.* Ordinal variables are somewhat in between. Ordered categorical variables with few categories are sometimes considered as qualitative, although the ordering has some quantitative aspect.

A careful definition and reflection of scale levels is found in particular in the psychology literature. Measuring intelligence is no easy task, so psychologists needed to develop some foundation for their measurements and developed an elaborated mathematical theory of measurement (see, in particular, Krantz et al., 1971).

Discrete and Continuous Variables

The distinction between discrete and continuous variables is completely unrelated to the concept of scale levels. It refers only to the number of values a variable can take. A *discrete* variable has a finite number of possible values or values that can at least be listed. Thus count data like the number of accidents with possible values from $0, 1, \ldots$ are considered discrete. The possible values of a *continuous* variable form an interval, although, in practice, due to the limitations of measuring instruments, not all of the possible values are observed.

Within the scope of this book discrete data like counts are considered as categorical. In particular, when the mean of a discrete response variable is small it is essential to recognize the discrete nature of the data.

1.2 Organization of This Book

The chapters may be grouped into five different units. After a brief review of basic issues in structured regression and classical normal distribution regression within this chapter, in the first unit, consisting of Chapters 2 through 7, the *parametric modeling* of univariate categorical response variables is discussed. In Chapter 2 the basic regression model for binary response, the logit or logistic regression model, is described. Chapter 3 introduces the class of generalized linear models (GLMs) into which the logit model as well as many other models in this book may be embedded. In Chapters 4 and 5 the modeling of binary response data is investigated more closely, including inferential issues but also the structuring of ordered categorical predictors, alternative link functions, and the modeling of overdispersion. Chapter 6 extends the approaches to high-dimensional predictors. The focus is on appropriate regularization methods that allow one to select predictor variables in cases where simple fitting methods fail. Chapter 7 deals with count data as a special case of discrete response.

Chapters 8 and 9 constitute the second unit of the book. They deal with parametric *multinomial response models.* Chapter 8 focuses on unordered multinomial responses, and Chapter 9 discusses models that make use of the order information of the response variable.

The third unit is devoted to *flexible non-linear regression*, also called *non-parametric regression.* Here the data determine the shape of the functional form with much weaker assumptions on the underlying structure. Non-linear smooth regression is the subject of Chapter 10. The modeling approaches are presented as extensions of generalized linear models. One section is devoted to functional data, which are characterized by high-dimensional but structured regressors that often have the form of a continuous signal. Tree-based modeling approaches, which provide an alternative to additive and smooth models, are discussed in Chapter 11. The method is strictly non-parametric and conceptually very simple. By binary recursive partitioning the feature space is partitioned into a set of rectangles, and on each rectangle a simple model is fitted. Instead of obtaining parameter estimates, one obtains a binary tree that visualizes the partitioning of the feature space.

Chapter 12 is devoted to the more traditional topic of *contingency analysis*. The main instrument is the log-linear model, which assumes a Poisson distribution, a multinomial distribution, or a product-multinomial distribution. For Poisson-distributed response there is a strong connection to count data as discussed in Chapter 7, but now all predictors are categorical. When the underlying distribution is multinomial, log-linear models and in particular graphical models are used to investigate the association structure between the categorical variables.

In the fifth unit *multivariate regression models* are examined. Multivariate responses occur if several responses together with explanatory variables are measured on one unit. In particular, repeated measurements that occur in longitudinal studies are an important case. The challenge is to link the responses to the explanatory variables and to account for the correlation between responses. In Chapter 13, after a brief overview, conditional and marginal models are outlined. Subject-specific modeling in the form of random effects models is considered in Chapter 14.

The last unit, Chapter 15, examines *prediction issues*. For categorical data the problem is strongly related to the common classification problem, where one wants to find the true class from which a new observation stems. Classification problems are basically diagnostic problems with applications in medicine when one wants to identify the type of the disease, in pattern recognition when one aims at recognition of handwritten characters, or in economics when one wants to identify risk clients in credit scoring. In the last decade, in particular, the analysis of genetic data has become an interesting field of application for classification techniques.

1.3 Basic Components of Structured Regression

In the following the structuring components of regression are considered from a general point of view but with special emphasis on categorical responses. This section deals with the various assumptions made for the structuring of the independent and the dependent variables.

1.3.1 Structured Univariate Regression

Regression methods are concerned with two types of variables, the explanatory (or independent) variables $\boldsymbol{x}$ and the dependent variables y. The collection of methods that are referred to as regression methods have several objectives:

- Modeling of the response y given $\boldsymbol{x}$ such that the underlying structure of the influence of $\boldsymbol{x}$ on y is found.
- Quantification of the influence of $\boldsymbol{x}$ on y.
- Prediction of y given an observation $\boldsymbol{x}$.

In regression the response variable y is also called the *regressand*, the *dependent variable*, and the *endogeneous variable*. Alternative names for the independent variables $\boldsymbol{x}$ are *regressors, explanatory variables, exogeneous variables, predictor variables*, and *covariates*.

Regression modeling uses several structural components. In particular, it is useful to distinguish between the random component, which usually is specified by some distributional assumption, and the components, which specify the structuring of the covariates $\boldsymbol{x}$. More specifically, in a structured regression the mean μ (or any other parameter) of the dependent variable y is modeled as a function in $\boldsymbol{x}$ in the form

$$\mu = h(\eta(\boldsymbol{x})),$$

where h is a transformation and $\eta(\boldsymbol{x})$ is a structured term. A very simple form is used in classical linear regression, where one assumes

$$\mu = \beta_0 + x_1\beta_1 + \cdots + x_p\beta_p = \beta_0 + \boldsymbol{x}^T\boldsymbol{\beta}$$

with the parameter vector $\boldsymbol{\beta}^T = (\beta_1, \ldots, \beta_p)$ and the vector of covariates $\boldsymbol{x}^T = (x_1, \ldots, x_p)$. Thus, classical linear regression assumes that the mean μ is directly linked to a linear predictor $\eta(\boldsymbol{x}) = \beta_0 + \boldsymbol{x}^T\boldsymbol{\beta}$. Covariates determine the mean response by a linear term, and the link h is the identity function. The distributional part in classical linear regression follows from assuming a normal distribution for $y|\boldsymbol{x}$.

In binary regression, when the response takes a value of 0 or 1, the mean corresponds to the probability $P(y = 1|\boldsymbol{x})$. Then the identity link h is a questionable choice since the probabilities are between 0 and 1. A transformation h that maps $\eta(\boldsymbol{x})$ into the interval $[0, 1]$ typically yields more appropriate models.

In the following, we consider ways of structuring the dependence between the mean and the covariates, with the focus on discrete response data. To keep the structuring parts separated, we will begin with the structural assumption on the response, which usually corresponds to assuming a specific distributional form, and then consider the structuring of the influential term and finish by considering the link between these two components.

Structuring the Dependent Variable

A common way of modeling the variability of the dependent variable y is to assume a distribution that is appropriate for the data. For binary data with $y \in \{0, 1\}$, the distribution is determined by $\pi = P(y = 1)$. As special case of the binomial distribution it is abbreviated by $B(1, \pi)$. For count data $y \in \{0, 1, 2, \dots\}$, the Poisson distribution $P(\lambda)$ with mass function $f(x) = \lambda^x e^{-\lambda}/x!$, $x = 0, 1, \dots$ is often a good choice. An alternative is the negative binomial distribution, which is more flexible than the Poisson distribution. If y is continuous, a common assumption is the normal distribution. However, it is less appropriate if the response is some duration for which $y \geq 0$ has to hold. Then, for example, a Gamma-distribution $\Gamma(\nu, \alpha)$ that has positive support might be more appropriate. In summary, the choice of the distributional model mainly depends on the kind of response that is to be modeled. Figures 1.2 and 1.3 show several discrete and continuous distributions, which may be assumed. Each panel shows two distributions that can be thought of as referring to two distinct values of covariates. For the normal distribution model where only the mean depends on covariates, the distributions referring to different values of covariates are simply shifted versions of each other. This is quite different for response distributions like the Poisson or the Bernoulli distribution. Here the change of the mean, caused by different values of covariates, also changes the shape of the distribution. This phenomenon is not restricted to discrete distributions but is typically found when responses are discrete.

Sometimes the assumption of a specific distribution, even if it reflects the type of data collected, is too strong to explain the variability in responses satisfactorily. In practice, one often finds that count data and relative frequencies are more variable than is to be expected under the Poisson and the binomial distributions. The data show *overdispersion*. Consequently, the structuring of the responses should be weakened by taking overdispersion into account.

One step further, one may even drop the assumption of a specific distribution. Instead of assuming a binomial or a Poisson distribution, one only postulates that the link between the mean and a structured term, which contains the explanatory variables, is correctly specified. In addition, one can specify how the variance of the response depends on explanatory variables. The essential point is that the assumptions on the response are very weak, within quasi-likelihood approaches structuring of the response in the form of distributional assumptions is not necessary.

Structuring the Influential Term

It is tempting to postulate no structure at all by allowing $\eta(\boldsymbol{x})$ to be any function. What works in the unidimensional case has severe drawbacks if $\boldsymbol{x}^T = (x_1, \dots, x_p)$ contains many variables. It is hard to explain how a covariate x_j determines the response if no structure is assumed. Moreover, estimation becomes difficult and less robust. Thus often it is necessary to assume some structure to obtain an approximation to the underlying functional form that works in practice. Structural assumptions on the predictor can be strict or more flexible, with the degree of flexibility depending on the scaling of the predictor.

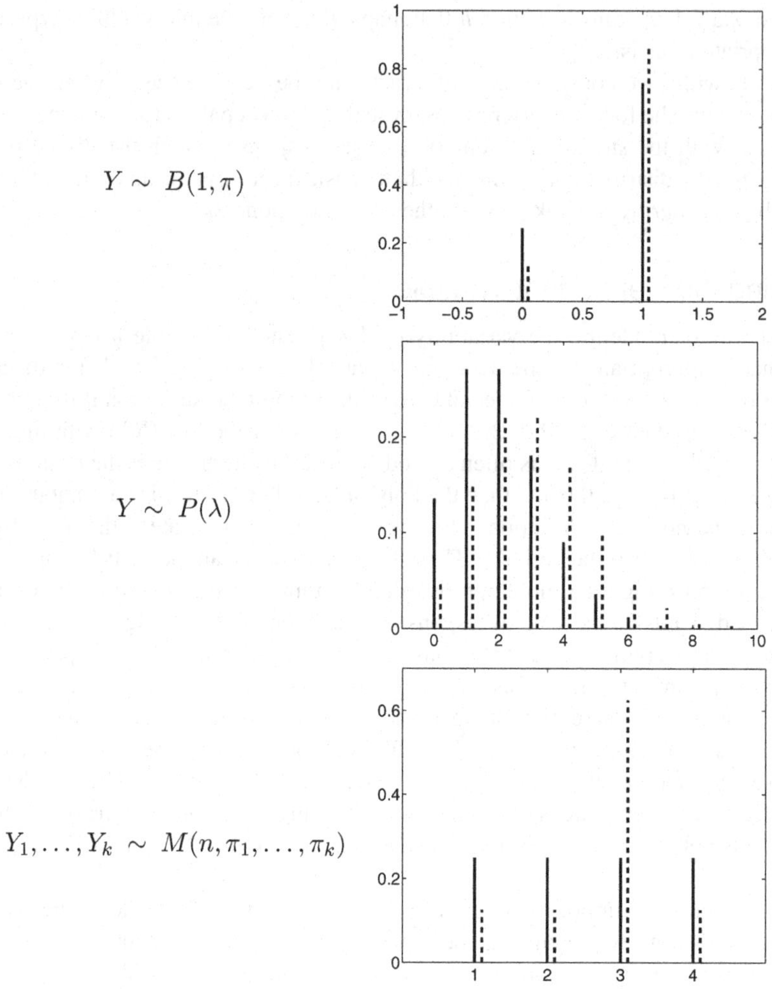

FIGURE 1.2: Binomial, Poisson, and multinomial distributions. Each panel shows two different distributions.

Linear Predictor

The most common form is the linear structure

$$\eta(\boldsymbol{x}) = \beta_0 + \boldsymbol{x}^T \boldsymbol{\beta},$$

which is very robust and allows simple interpretation of the parameters. Often it is necessary to include some interaction terms, for example, by assuming

$$\begin{aligned}\eta(\boldsymbol{x}) &= \beta_0 + x_1\beta_1 + \cdots + x_p\beta_p + x_1x_2\beta_{12} + x_1x_3\beta_{13} + \cdots + x_1x_2x_3\beta_{123}\\ &= \boldsymbol{z}^T\boldsymbol{\beta}.\end{aligned}$$

By considering $\boldsymbol{z}^T = (1, x_1, \ldots, x_p, x_1x_2, \ldots, x_1x_2x_3, \ldots)$ as variables, one retains the linear structure. For estimating and testing (not for interpreting) it is only essential that the structure is linear in the parameters. When explanatory variables are quantitative, interpreting the parameters is straightforward, especially in the linear model without interaction terms.

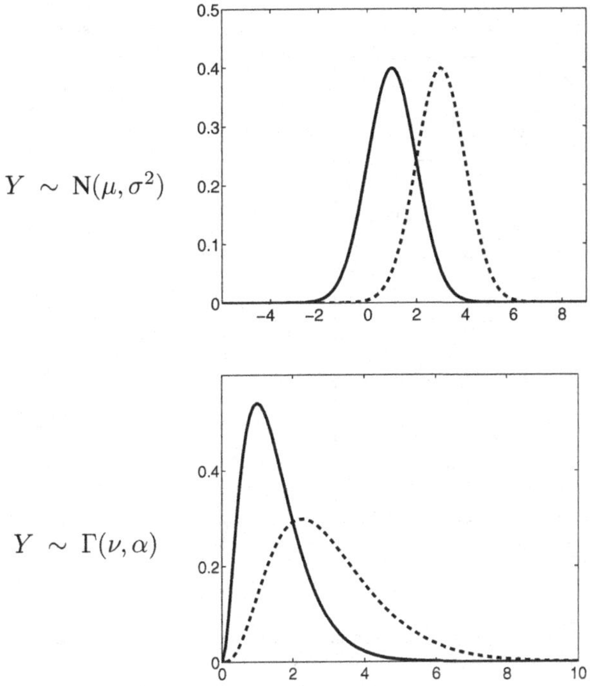

FIGURE 1.3: Normal and Gamma-distributions.

Categorical Explanatory Variables

Categorical explanatory variables, also called factors, take values from a finite set $1, \ldots, k$, with the numbers representing the factor levels. They cannot be used directly within the linear predictor because one would falsely assume fixed ordering of the categories with the distances between categories being meaningful. That is not the case for nominal variables, not even for ordered categorical variables. Therefore, specific structuring is needed for factors. Common structuring uses dummy variables and again yields a linear predictor. The coding scheme depends on the intended use and on the scaling of the variable. Several coding schemes and corresponding interpretations of effects are given in detail in Section 1.4.1. The handling of ordered categorical predictors is also considered in Section 4.4.3.

When a categorical variable has many categories, the question arises of which categories can be distinguished with respect to the response. Should categories be collapsed, and if so, which ones? The answer depends on the scale level. While for nominal variables, for which categories have no ordering, any fusion categories seems sensible, for ordinal predictors collapsing means fusing adjacent categories. Figure 1.4 shows a simple application. It shows the effect of the urban district and the year of construction on the rent per square meter in Munich. Urban district is a nominal variable that has 25 categories, year of construction is an ordered predictor, where categories are defined by decades. The coefficient paths in Figure 1.4 show how, depending on a tuning parameter, urban districts and decades are combined. It turns out that only 10 districts are really different, and the year of construction can be combined into 8 distinct categories (see also Section 6.5).

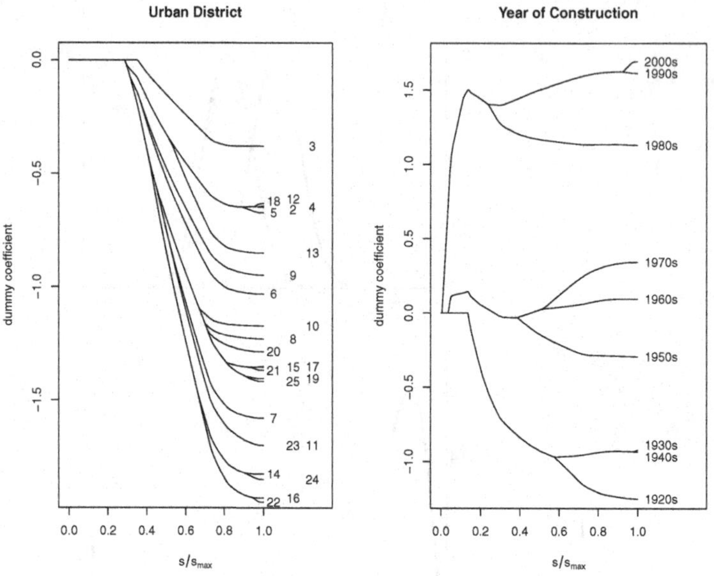

FIGURE 1.4: Effects of urban district and year of construction (in decades) on rent per square meter.

Additive Predictor

For quantitative explanatory variables, a less restrictive assumption is

$$\eta(\boldsymbol{x}) = f_{(1)}(x_1) + \cdots + f_{(p)}(x_p),$$

where $f_{(j)}(x_j)$ are unspecified functions. Thus one retains the additive form, which still allows simple interpretation of the functions $f_{(j)}$ by plotting estimates but the approach is much less restrictive than in the linear predictor. An extension is the inclusion of unspecified interactions, for example, by allowing

$$\eta(\boldsymbol{x}) = f_{(1)}(x_1) + \cdots + f_{(p)(x_p)} + f_{(13)}(x_1, x_3),$$

where $f_{(13)}(x_1, x_3)$ is a function depending on x_1 and x_3.

For categorical variables no function is needed because only discrete values occur. Thus, when, in addition to quantitative variables, $x_1, \ldots, x_p$, categorical covariates are available, they are included in an additional linear term, $\boldsymbol{z}^T\boldsymbol{\gamma}$, which is built from dummy variables. Then one uses the *partial linear predictor*

$$\eta(\boldsymbol{x}) = f_{(1)}(x_1) + \cdots + f_{(p)}(x_p) + \boldsymbol{\gamma}.$$

Additive Structure with Effect Modifiers

If the effect of a covariate, say gender (x_1), depends on age (x_2) instead of postulating an interaction model of the form $\eta = \beta_0 + x_1\beta_1 + x_2\beta_2 + x_1x_2\beta_{12}$, a more flexible model is given by

$$\eta = \beta_2(x_2) + x_1\beta_{12}(x_2),$$

where $\beta_2(x_2)$ is the smooth effect of age and $\beta_{12}(x_2)$ is the effect of gender (x_1), which is allowed to vary over age. Both functions $\beta_2(.)$ and $\beta_{12}(.)$ are unspecified and the data determine their actual form.

Tree-Based Methods

An alternative way to model interactions of covariates is the recursive partitioning of the predictor space into sets of rectangles. The most popular method, called CART for classification and regression trees, constructs for metric predictors partitions of the form $\{x_1 \leq c_1\} \cap \cdots \cap \{x_m \leq c_m\}$, where $c_1, \ldots, c_m$ are split-points from the regions of the variables $x_1, \ldots, x_m$. The splits are constructed successively beginning with a first split, producing, for example, $\{x_1 \leq c_1\}$, $\{x_1 > c_1\}$. Then these regions are split further. The recursive construction scheme allows us to present the resulting partition in a tree. A simple example is the tree given in Figure 1.5, where two variables are successively split by using split-points $c_1, \ldots, c_4$. The first split means that the dichotomization into $\{x_1 \leq c_1\}$ and $\{x_1 > c_1\}$ is of major importance for the prediction of the outcome. Finer prediction rules are obtained by using additional splits, for example, the split of the region $\{x_1 \leq c_1\}$ into $\{x_2 \leq c_2\}$ and $\{x_2 > c_2\}$. The big advantage of trees is that they are easy to interpret and the visualization makes it easy to communicate the underlying structure to practitioners.

The Link between Covariates and Response

Classical linear regression assumes $\mu = \eta(\boldsymbol{x})$ with $\eta(\boldsymbol{x}) = \boldsymbol{x}^T\boldsymbol{\beta}$. For binary regression models, the more general form $\mu = h(\eta(\boldsymbol{x}))$ is usually more appropriate, since h may be chosen such that μ takes values in the unit interval $[0, 1]$. Typically h is chosen as a distribution function, for example, the logistic distribution function $h(\eta) = \exp(\eta)/(1 + \exp(\eta))$ or the normal distribution function. The corresponding models are the so-called logit and probit models. However, any distribution function that is strictly monotone may be used as a response function (see Section 5.1).

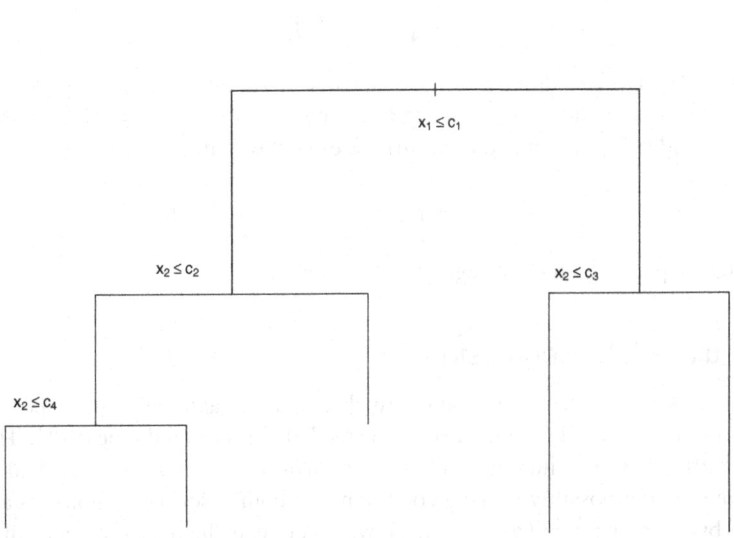

FIGURE 1.5: Example tree for two variables.

In many applications h is considered as known. Then, of course, there is the danger of misspecification. It is often more appropriate to consider alternative transformation functions and choose the one that yields the best fit. Alternatively, one can estimate the transformation itself, and therefore let the data determine the form of the transformation function (Section 5.2).

1.3.2 Structured Multicategorical Regression

When the response is restricted to a fixed set of possible values, the so-called response categories, the typical assumption for the distribution is the multinomial distribution. When $1, \dots, k$ denote the response categories of variable Y, the multinomial distribution specifies the probabilities $\pi_1(\boldsymbol{x}), \dots, \pi_k(\boldsymbol{x})$, where $\pi_r(\boldsymbol{x}) = P(Y = r|\boldsymbol{x})$.

The simple structure of univariate regression models is no longer appropriate because the response is multivariate. One has to model the dependence of all the probabilities $\pi_1(\boldsymbol{x}), \dots, \pi_k(\boldsymbol{x})$ on the explanatory variables. This may be accomplished by a multivariate model that has the basic structure

$$\pi_r(\boldsymbol{x}) = h_r(\eta_1(\boldsymbol{x}), \dots, \eta_k(\boldsymbol{x})), r = 1, \dots, k-1,$$

where $h_r, r = 1, \dots, k-1$, are transformation functions that are specific for the category. Since probabilities sum up to one, it is sufficient to specify $k-1$ of the k components. By using the $(k-1)$-dimensional vectors $\boldsymbol{\pi}(\boldsymbol{x})^T = (\pi_1(\boldsymbol{x}), \dots, \pi_{k-1}(\boldsymbol{x})), \boldsymbol{\eta}(\boldsymbol{x}))^T = (\eta_1(\boldsymbol{x}), \dots, \eta_{k-1}(\boldsymbol{x}))$, models have the closed form

$$\boldsymbol{\pi}(\boldsymbol{x}) = h(\boldsymbol{\eta}(\boldsymbol{x})).$$

The choice of the transformation function depends on the scale level of the response. If the response is nominal, for example, when modeling the choice of different brands or the choice of transport mode, other response functions are more appropriate than in the ordinal case, when the response is given on a rating scale with categories like very good, good, fair, poor, and very poor (see Chapter 8 for nominal and Chapter 9 for ordinal responses).

The structuring of the predictor functions is in analogy to univariate responses. Strict linear structures assume

$$\eta_r(\boldsymbol{x}) = \boldsymbol{x}^T \boldsymbol{\beta}_r,$$

where the parameter vector depends on the category r. By defining an appropriate design matrix, one obtains a multivariate generalized linear model form $\boldsymbol{\pi}(\boldsymbol{x}) = h(\boldsymbol{X}\boldsymbol{\beta})$. More flexible predictors use additive or partial additive structures of the form

$$\eta_r(\boldsymbol{x}) = f_{(r1)}(x_1) + \dots + f_{(rp)}(x_p),$$

with functions depending on the category.

1.3.3 Multivariate Regression

In many studies several response variables are observed for each unit. The responses may refer to different variables or to the same measurements that are observed repeatedly. The latter case is found in particular in longitudinal studies where measurements on an individual are observed at several times under possibly varying conditions. In both cases the response is a multivariate represented by a vector $\boldsymbol{y}_i^T = (y_{i1}, \dots, y_{im})$, which collects the measurements on unit i. Since measurements taken on one unit or cluster tend to be more similar, one has to assume that in general the measurements are correlated.

Structuring the Dependent Variables

The structuring of the vector of dependent variables by assuming an appropriate distribution is not as straightforward as in classical multivariate regression, where a multivariate normal distribution is assumed. Multivariate, normally distributed responses have been extensively investigated for a long time. They are simply structured with a clear separation of the mean and correlation structures, which are sufficient to define the distribution.

When the marginals, that is, single responses y_{it}, are discrete it is harder to find a sparse representation of the total response vector. Although the form of the distribution may have a simple form, the number of parameters can be extremely high. If, for example, the marginals are binary with $y_{it} \in \{0, 1\}$, the total distribution is multinomial but determined by 2^m probabilities for the various combinations of outcomes. With $m = 10$ measurements, the number of parameters is 1024. Simple measures like the mean of the marginals and the correlations between components do not describe the distribution sufficiently.

One strategy that is used in marginal modeling is to model the mean structure of the marginals, which uses univariate regression models, and in addition specify an association structure between components that does not have to be the correct association structure. An alternative approach uses random effects. One assumes that the components are uncorrelated given a fixed but unobserved latent variable, which is shared by the measurements within one unit or cluster. In both cases one basically uses parameterizations of the discrete marginal distributions.

Structuring the Influential Term

The structuring of the influential term is more complex than in univariate response models because covariates may vary across measurements within one cluster. For example, in longitudinal studies, where the components refer to repeated measurements across time, the covariates that are to be included may also vary across time. Although the specification is more complex, in principle, the same form of predictors as in univariate regression applies. One can use linear terms or more flexible additive terms. However, new structuring elements are useful, and two of them are the following.

In random effects models one models explicitly the heterogeneity of clustered responses by assuming cluster-specific random effects. For example, the binary response of observation t on cluster i with covariate vector $\boldsymbol{x}_{it}$ at measurement t may be modeled by

$$P(y_{it} = 1|\boldsymbol{x}_{it}, b_i) = h(\eta_{it}), \quad \eta_{it} = b_i + \boldsymbol{x}_{it}^T\boldsymbol{\beta}.$$

While $\boldsymbol{\beta}$ is a fixed effect that is common to all clusters, each cluster has its own cluster-specific random effect b_i, which models the heterogeneity across clusters. More generally, one can assume that not only the intercept but also the slopes of the variables are cluster-specific. Then one has to specify which components have category-specific effects. Moreover, one has to specify a distribution for these category-specific effects. Thus, some additional structuring and therefore decision making by the modeler is needed.

Another effect structure that can be useful in repeated measurements is the variation of effect strength across time, which can be modeled by letting the parameter depend on time:

$$\eta_{it} = b_i + \boldsymbol{x}_{it}^T\boldsymbol{\beta}_t.$$

Marginal modeling approaches are given in detail in Chapter 13, and random effects models are found in Chapter 14.

1.3.4 Statistical Modeling

Statistical modelling refers to the process of using models to extract information from data. In statistics that means, in particular, to separate the systematic effects or underlying patterns from the random effects. A model, together with its estimation method, can be seen as a measuring instrument. Like magnifying glasses and telescopes serve as instruments to uncover structures not seen to the unarmed eye, a model allows one to detect patterns that are in the data but not seen without the instrument. What is seen depends on the instrument. Only those patterns are found for which the instrument is sensitive. For example, linear models allow one to detect linear structures. If the underlying pattern is non-linear, they fail and the results can be very misleading. Or, effect modifiers are detected only if the model allows for them. In that sense the model determines what is found. More flexible models allow one to see more complex structures, at least if reliable estimation methods are found. In the same way as a telescope depends on basic conditions like the available amount of light, statistical models depend on basic conditions like the sample size and the strength of the underlying effects. Weak patterns typically can be detected only if much information is available.

The use and choice of models is guided by different and partly contradictory objectives. Models should be simple but should account for the complexity of the underlying structure. Simplicity is strongly connected to interpretability. Users of statistical models mostly prefer interpretable models over black boxes. In addition, the detection of interpretable and simple patterns and therefore understanding is the essential task of science. In science, models often serve to understand and test subject-matter hypotheses about underlying processes.

The use of models, parametric or more flexible, is based on the assumption that a stochastic data model has generated the data. The statistician is expected to use models that closely approximate the data driving the model, quantify the effects within the model, and account for the estimation error. Ideally, the analysis also accounts for the closeness of the model to the data-generating model, for example, in the form of goodness-of-fit tests.

A quite different objective that may determine the choice of the model is the exactness of the prediction. One wants to use that model that will give the best results in terms of prediction error when used on future data. Then black box machines, which do not try to uncover latent structures, also apply and may show excellent prediction results. Ensemble methods like random forests or neural networks work in that way. Breiman (2001b) calls them *algorithmic models* in contrast to *data models*, which assume that a stochastic data-generating model is behind the data. Algorithmic models are less models than an approach to find good prediction rules by designing algorithms that link the predictors to the responses. Especially in the machine learning community, where the handling of data is guided by a more pragmatic view, algorithms with excellent prediction properties in the form of black boxes are abundant.

What type of model is to be preferred depends mainly on the objective of the scientific question. If prediction is the focus, intelligently designed algorithms may serve the purpose well. If the focus is on understanding and interpretation, data models are to be preferred. The choice of the model depends on the structures that are of interest to the user and circumstances like sample size and strength of the parameters. When effects are weak and the sample size is small, simple parametric models will often be more stable than models that allow for complex patterns. As a model does not fit all datasets, for a single dataset different models that uncover different structures may be appropriate. Typically there is not a single best model for a set of data.

In the following chapters we will predominantly consider tools for data models; that is alternative models, estimation procedures, and diagnostic tools will be discussed. In the last chapter, where prediction is the main issue, algorithmic models/methods will also be included. Specification of models is treated throughout the book in various forms, ranging from classical test procedures to examine parameters to regularization techniques that allow one to select variables or link functions.

There is a rich literature on model selection. Burnham and Anderson (2002) give an extensive account of the information-theoretic approach to model selection; see also Claeskens and Hjort (2008) for a survey on the research in the field. The alternative modeling cultures, data modeling versus algorithmic modeling, was treated in a stimulating article by Breiman (2001b). His strong opinion on stochastic data models is worth reading, in particular together with the included and critical discussion.

1.4 Classical Linear Regression

Since the linear regression model is helpful as a background model, in this section a brief overview of classical linear regression is given. The section may be skipped if one feels familiar with the model. It is by no means a substitute for a thorough introduction to Gaussian response models, but a reminder of the basic concepts. Parametric regression models including linear models are discussed in detail in many statistics books, for example, Cook and Weisberg (1982), Ryan (1997), Harrell (2001), and Fahrmeir et al. (2011).

The basic multiple linear regression model is often given in the form

$$y = \beta_0 + x_1\beta_1 + \cdots + x_p\beta_p + \varepsilon,$$

where ε is a noise variable that fulfills $\mathrm{E}(\varepsilon) = 0$. Thus, for given $x_1, \ldots, x_p$ the expectation of the response variable $\mu = \mathrm{E}(y|x_1, \ldots, x_p)$ is specified as a linear combination of the explanatory variables $x_1, \ldots, x_p$:

$$\mu = \beta_0 + x_1\beta_1 + \cdots + x_p\beta_p = \beta_0 + \boldsymbol{x}^T\boldsymbol{\beta},$$

with the parameter vector $\boldsymbol{\beta}^T = (\beta_1, \ldots, \beta_p)$ and vector of covariates $\boldsymbol{x}^T = (x_1, \ldots, x_p)$.

1.4.1 Interpretation and Coding of Covariates

When interpreting the parameters of linear models it is useful to distinguish between quantitative (metrically scaled) covariates and categorical covariates, also called factors, which are measured on a nominal or ordinal scale level.

Quantitative Explanatory Variables

For a quantitative covariate like age, the linear model has the form

$$\mathrm{E}(y|x) = \beta_0 + x\beta_A = \beta_0 + \text{Age} * \beta_A.$$

From $\mathrm{E}(y|x+1) - \mathrm{E}(y|x) = \beta_A$ it is immediately seen that β_A reflects the change of the mean response if the covariate is increased by one unit. If the response is income in dollars and the covariate age is given in years, the units of β_A are dollars per year and β_A is the change of expected income resulting from increasing the age by one year.

Binary Explanatory Variables

A binary covariate like gender (G) has to be coded, for example, as a (0-1)-variable in the form

$$x_G = \begin{cases} 1 & \text{male} \\ 0 & \text{female.} \end{cases}$$

Then the interpretation is the same as for quantitative variables. If the response is income in dollars, the parameter β_G in

$$\mathrm{E}(y|x) = \beta_0 + x_G \beta_G$$

represents the change in mean income if x_G is increased by one unit. Since x_G has only two values, β_G is equivalent to the increase or decrease of the mean response resulting from the transition from $x_G = 0$ to $x_G = 1$. Thus β_G is the difference in mean income between men and women.

Multicategorical Explanatory Variables or Factors

Let a covariate have k possible categories and categories just reflect labels. This means that the covariate is measured on a nominal scale level. Often covariates of this type are called factors or factor variables. A simple example is in the medical profession P, where the possible categories gynaecologist, dermatologist, and so on, are coded as numbers $1, \dots, k$. When modeling a response like income, a linear term $P * \beta_P$ will produce nonsense since it assumes a linear relationship between the response and the values of $P \in \{1, \dots, k\}$, but the coding for profession by numbers is arbitrary. To obtain parameters that are meaningful and have simple interpretations one defines dummy variables. One possibility is dummy or (0-1)-coding.

(0-1)-Coding of $P \in \{1, \dots, k\}$

$$x_{P(j)} = \begin{cases} 1 & \text{if } P = j \\ 0 & \text{otherwise} \end{cases}$$

When an intercept is in the model only $k-1$ variables can be used. Otherwise, one would have too many parameters that would not be identifiable. Therefore, one dummy variable is omitted and the corresponding category is considered the *reference category*. When one chooses k as the reference category the linear predictor is determined by the first $k-1$ dummy variables:

$$\mathrm{E}(y|P) = \beta_0 + x_{P(1)}\beta_{P(1)} + \dots + x_{P(k-1)}\beta_{P(k-1)}. \tag{1.1}$$

Interpretation of the parameters follows directly from considering the response for different values of P:

$$\mathrm{E}(y|P = i) = \beta_0 + \beta_{P(i)}, \quad i = 1, \dots k-1, \quad \mathrm{E}(y|P = k) = \beta_0.$$

β_0 is the mean for the reference category k and $\beta_{P(i)}$ is the increase or decrease of the mean response in comparison to the reference category k. Of course any category can be used as the reference. Thus, for a categorial variable, $k-1$ functionally independent dummy variables are introduced. A particular choice of the reference category determines the set of variables, or in the terminology of analysis of variance, the set of contrasts. The use of all k dummy variables results in overparameterization because they are not functionally independent. The sum over all k dummy variables yields 1, and therefore β_0 would not be identifiable.

An alternative coding scheme is effect coding, where categories are treated in a symmetric way.

Effect Coding of $P \in \{1, \ldots, k\}$

$$x_{P(j)} = \begin{cases} 1 & \text{if } P = j \\ -1 & \text{if } P = k, \quad j = 1, \ldots, k-1 \\ 0 & \text{otherwise} \end{cases}$$

The linear predictor (1.1) now yields

$$\begin{aligned} E(y|P = i) &= \beta_0 + \beta_{P(i)}\ , \ i = 1, \ldots, k-1, \\ E(y|P = k) &= \beta_0 - \beta_{P(1)} - \ldots - \beta_{P(k-1)}. \end{aligned}$$

It is easily seen that

$$\beta_0 = \frac{1}{k} \sum_{j=1}^{k} E(y|P = j)$$

is the average response across the categories and

$$\beta_{P(j)} = E(y|P = j) - \beta_0$$

is the deviation of category j from the average response level given by β_0. Although only $k-1$ dummy variables are used, interpretation does not refer to a particular category. There is no reference category as in the case of (0-1)-coding. For the simple example of four categories one obtains

		$x_{P(1)}$	$x_{P(2)}$	$x_{P(3)}$
	1	1	0	0
P	2	0	1	0 .
	3	0	0	1
	4	-1	-1	-1

Alternative coding schemes are helpful, for example, if the categories are ordered. Then a split that distinguishes between categories below and above a certain level often reflects the ordering in a better way.

Split-Coding of $P \in \{1, \ldots, k\}$

$$x_{P(j)} = \begin{cases} 1 & \text{if } P > j \\ 0 & \text{otherwise} \qquad j = 1, \ldots, k-1 \end{cases}$$

With split-coding, the model $E(y|P) = \beta_0 + x_{P(1)}\beta_{P(1)} + \ldots + x_{P(k-1)}\beta_{P(k-1)}$ yields

$$\begin{aligned} &E(y|P = 1) = \beta_0, \\ &E(y|P = i) = \beta_0 + \beta_{P(1)} + \cdots + \beta_{P(i-1)}\ , \ i = 2, \ldots, k \end{aligned}$$

and therefore the coefficients

$$\beta_0 = \mathrm{E}(y|P=1),$$
$$\beta_{P(i)} = \mathrm{E}(y|P=i+1) - \mathrm{E}(y|P=i)\ ,\ i=1,\dots,k-1.$$

Thus the coefficients $\beta_{P(i)}, i = 1, \dots, k-1$ represent the difference in expectation when the factor level increases from i to $i+1$. They may be seen as the stepwise change over categories given in the fixed order $1, \dots, k$, with 1 serving as the reference category where the process starts. In contrast to (0-1)-coding and effect coding, split-coding uses the ordering of categories: $\beta_{P(1)} + \dots + \beta_{P(i-1)}$ can be interpreted as the change in expectation for the transition from category 1 to category i with intermediate categories $2, 3, \dots, i-1$.

For further coding schemes see, for example, Chambers and Hastie (1992). The structuring of the linear predictor is examined in more detail in Chapter 4, where models with binary responses are considered.

1.4.2 Linear Regression in Matrix Notation

Let the observations be given by $(y_i, x_{i1}, \dots, x_{ip})$ for $i = 1, \dots, n$, where y_i is the response and $x_{i1}, \dots, x_{ip}$ are the given covariates. The model takes the form

$$y_i = \beta_0 + x_{i1}\beta_1 + \dots + x_{ip}\beta_p + \varepsilon_i,\ i = 1, \dots n.$$

With $\boldsymbol{x}_i^T = (1, x_{i1}, \dots x_{ip}),\ \boldsymbol{\beta}^T = (\beta_0, \beta_1, \dots \beta_p)$ it may be written more compact as

$$y_i = \boldsymbol{x}_i^T \boldsymbol{\beta} + \varepsilon_i,$$

where $\boldsymbol{x}_i$ has length $\tilde{p} = p + 1$. In matrix notation one obtains

$$\begin{bmatrix} y_1 \\ \vdots \\ \vdots \\ y_n \end{bmatrix} = \begin{bmatrix} 1 & x_{11} & \dots & x_{1p} \\ 1 & x_{21} & \dots & x_{2p} \\ \vdots & & & \\ 1 & x_{n1} & \dots & x_{np} \end{bmatrix} \begin{bmatrix} \beta_0 \\ \beta_1 \\ \vdots \\ \beta_p \end{bmatrix} + \begin{bmatrix} \varepsilon_1 \\ \vdots \\ \vdots \\ \varepsilon_n \end{bmatrix}$$

or simply

$$\boldsymbol{y} = \boldsymbol{X}\boldsymbol{\beta} + \boldsymbol{\varepsilon},$$

where

$\boldsymbol{y}^T = (y_1, \dots, y_n)$ is the vector of responses;

$\boldsymbol{X}$ is the design matrix, which is composed from the explanatory variables;

$\boldsymbol{\beta}^T = (\beta_0, \dots, \beta_p)$ is the $(p+1)$-dimensional parameter vector;

$\boldsymbol{\varepsilon}^T = (\varepsilon_1, \dots, \varepsilon_n)$ is the vector of errors.

Common assumptions are that the errors have expectation zero, $\mathrm{E}(\varepsilon_i) = 0,\ i = 1, \dots, n$; the variance of each error component is given by $\mathrm{var}(\varepsilon_i) = \sigma^2,\ i = 1, \dots, n$, called *homoscedasticity* or *homogeneity*; and the error components from different observations are uncorrelated, $\mathrm{cov}(\varepsilon_i, \varepsilon_j) = 0,\ i \neq j$.

Multiple Linear Regression

$$y_i = \boldsymbol{x}_i^T\boldsymbol{\beta} + \varepsilon_i \quad \text{or} \quad \boldsymbol{y} = \boldsymbol{X}\boldsymbol{\beta} + \boldsymbol{\varepsilon}$$

Assumptions:

$$\mathrm{E}(\varepsilon_i) = 0, \quad \mathrm{var}(\varepsilon_i) = \sigma^2, \quad \mathrm{cov}(\varepsilon_i, \varepsilon_j) = 0, i \neq j$$

In matrix notation the assumption of homogeneous variances may be condensed into $\mathrm{E}(\boldsymbol{\varepsilon}) = \boldsymbol{0}$, $\mathrm{cov}(\boldsymbol{\varepsilon}) = \sigma^2\boldsymbol{I}$. If one assumes in addition that responses are normally distributed, one postulates $\boldsymbol{\varepsilon} \sim N(\boldsymbol{0}, \sigma^2\boldsymbol{I})$. It is easy to show that the assumptions carry over to the observable variables y_i in the form

$$\mathrm{E}(\boldsymbol{y}|\boldsymbol{X}) = \boldsymbol{X}\boldsymbol{\beta}, \quad \boldsymbol{y} \sim N(\boldsymbol{X}\boldsymbol{\beta}, \sigma^2\boldsymbol{I}).$$

The last representation may be seen as a structured representation of the classical linear model, which gives the distributional and systematic components separately without using the noise variable ε. With $\boldsymbol{\mu}$ denoting the mean response vector with components $\mu_i = \boldsymbol{x}_i^T\boldsymbol{\beta}$ one has the following form.

Multiple Linear Regression with Normally Distributed Errors

$$\boldsymbol{\mu} = \boldsymbol{X}\boldsymbol{\beta}, \quad \boldsymbol{y} \sim N(\boldsymbol{X}\boldsymbol{\beta}, \sigma^2\boldsymbol{I})$$

The separation of the structural and distributional components is an essential feature that forms the basis for the extension to more general models considered in Chapter 3.

1.4.3 Estimation

Least-Squares Estimation

A simple criterion to obtain estimates of the unknown parameter $\boldsymbol{\beta}$ is the least-squares criterion, which minimizes

$$Q(\boldsymbol{\beta}) = \sum_{i=1}^{n}(y_i - \boldsymbol{x}_i^T\boldsymbol{\beta})^2.$$

Thus the parameter $\hat{\boldsymbol{\beta}}$, which minimizes $Q(\boldsymbol{\beta})$, is the parameter that minimizes the squared distance between the actual observation y_i and the predicted value $\hat{y}_i = \boldsymbol{x}_i^T\hat{\boldsymbol{\beta}}$. The choice of the squared distance has the advantage that an explicit form of the estimate is available. It means that large discrepancies between y_i and $\hat{y}_i$ are taken more seriously than for the Euclidean distance $|y_i - \hat{y}_i|$, which would be an alternative criterion to minimize. Simple calculation shows that the derivative of $Q(\boldsymbol{\beta})$ has the form

$$\frac{\partial Q(\boldsymbol{\beta})}{\partial \beta_s} = \sum_{i=1}^{n} 2(y_i - \boldsymbol{x}_i^T\boldsymbol{\beta})x_{is},$$

$s = 0, \ldots, p$, where $x_{i0} = 1$. A minimum can be expected if the derivative equals zero. Thus the least-squares estimate has to fulfill the equation $\sum_i x_{is}(y_i - \boldsymbol{x}_i^T\hat{\boldsymbol{\beta}}) = 0$. In vector notation one obtains the form

$$\sum_{i=1}^{n} \boldsymbol{x}_i y_i = \sum_{i} \boldsymbol{x}_i \boldsymbol{x}_i^T \hat{\boldsymbol{\beta}},$$

which may be written in the form of the normal equation $\boldsymbol{X}^T\boldsymbol{y} = \boldsymbol{X}^T\boldsymbol{X}\hat{\boldsymbol{\beta}}$. Assuming that the inverse of $\boldsymbol{X}^T\boldsymbol{X}$ exists, an explicit solution is given by

$$\hat{\boldsymbol{\beta}} = (\boldsymbol{X}^T\boldsymbol{X})^{-1}\boldsymbol{X}^T\boldsymbol{y}.$$

Maximum Likelihood Estimation

The least-squares estimate is strongly connected to the maximum likelihood (ML) estimate if one assumes that the error is normally distributed. The normal distribution contains a quadratic term. Thus it is not surprising that the ML estimate is equivalent to minimizing squared distances. If one assumes $\varepsilon_i \sim N(0, \sigma^2)$ or, equivalently, $y_i \sim N(\boldsymbol{x}_i^T\boldsymbol{\beta}, \sigma^2)$, the conditional likelihood (given $\boldsymbol{x}_1, \dots, \boldsymbol{x}_n$) is given by

$$L(\boldsymbol{\beta}, \sigma^2) = \prod_{i=1}^{n} \frac{1}{\sqrt{2\pi\sigma^2}} \exp(-(y_i - \boldsymbol{x}_i^T\boldsymbol{\beta})^2/(2\sigma^2)).$$

The corresponding log-likelihood has the form

$$\begin{aligned} l(\boldsymbol{\beta}, \sigma^2) &= -\frac{1}{2\sigma^2}\sum_{i=1}^{n}(y_i - \boldsymbol{x}_i^T\boldsymbol{\beta})^2 - \frac{n}{2}\log(2\pi) - \log(\sigma^2) \\ &= -\frac{1}{2\sigma^2}Q(\boldsymbol{\beta}) - \frac{n}{2}\log(2\pi) - n\log(\sigma^2). \end{aligned}$$

As far as $\boldsymbol{\beta}$ is concerned, maximization of the log-likelihood is equivalent to minimizing the squared distances $Q(\boldsymbol{\beta})$. Simple derivation shows that maximization of $l(\boldsymbol{\beta}, \sigma^2)$ with respect to σ^2 yields

$$\hat{\sigma}^2_{ML} = \frac{1}{n}\sum_{i=1}^{n}(y_i - \boldsymbol{x}_i^T\hat{\boldsymbol{\beta}})^2.$$

It is noteworthy that the maximum likelihood estimate $\hat{\boldsymbol{\beta}}$ (which is equivalent to the least-squares estimate) does not depend on σ^2. Thus the parameter $\boldsymbol{\beta}$ is estimated without reference to the variability of the response.

Properties of Estimates

A disadvantage of the ML estimate $\hat{\sigma}^2_{ML}$ is that it underestimates the variance σ^2. An unbiased estimate is given by

$$\hat{\sigma}^2 = \frac{1}{n-(p+1)}\sum_{i=1}^{n}(y_i - \boldsymbol{x}_i^T\hat{\boldsymbol{\beta}})^2,$$

where the correction in the denominator reflects the number of estimated parameters in $\hat{\boldsymbol{\beta}}$, which is $p+1$ since an intercept is included. The essential properties of estimates are given in the so-called Gauss-Markov theorem. Assuming for all observations $\mathrm{E}(\varepsilon_i) = 0$, $\mathrm{var}(\varepsilon_i) = \sigma^2$, $\mathrm{cov}(\varepsilon_i, \varepsilon_j) = 0, i \neq j$, one obtains

(1) $\hat{\boldsymbol{\beta}}$ and $\hat{\sigma}^2$ are unbiased, that is, $\mathrm{E}(\hat{\boldsymbol{\beta}}) = \boldsymbol{\beta}, \mathrm{E}(\hat{\sigma}^2) = \sigma^2$.

(2) $\text{cov}(\hat{\boldsymbol{\beta}}) = \sigma^2(\boldsymbol{X}^T\boldsymbol{X})^{-1}$.

(3) $\hat{\boldsymbol{\beta}}$ is the best linear unbiased estimate of $\boldsymbol{\beta}$. This means that, for any vector, $\boldsymbol{c}\,\text{var}(\boldsymbol{c}^T\hat{\boldsymbol{\beta}}) \leq \text{var}(\boldsymbol{c}^T\tilde{\boldsymbol{\beta}})$ holds where $\tilde{\boldsymbol{\beta}}$ is an unbiased estimator of $\boldsymbol{\beta}$, which has the form $\tilde{\boldsymbol{\beta}} = \boldsymbol{A}\boldsymbol{y} + \boldsymbol{d}$ for some matrix $\boldsymbol{A}$ and vector $\boldsymbol{d}$.

Estimators in Linear Multiple Regression

Least-squares estimate

$$\hat{\boldsymbol{\beta}} = (\boldsymbol{X}^T\boldsymbol{X})^{-1}\boldsymbol{X}^T\boldsymbol{y}$$

Unbiased estimate of σ^2

$$\hat{\sigma}^2 = \frac{1}{n-p-1}\sum_{i=1}^{n}(y_i - \boldsymbol{x}_i^T\boldsymbol{\beta})^2$$

1.4.4 Residuals and Hat Matrix

For single observations the discrepancy between the actual observation and the fitted value $\hat{y}_i = \boldsymbol{x}_i^T\hat{\boldsymbol{\beta}}$ is given by the simple residual

$$r_i = y_i - \boldsymbol{x}_i^T\hat{\boldsymbol{\beta}}.$$

It is a preliminary indicator for ill-fitting observations, that is, observations that have large residuals. Since the classical linear model assumes that the variance is the same for all observations, one might suspect that the residuals also have the same variance. However, because $\hat{\boldsymbol{\beta}}$ depends on all of the observations, they do not. Thus, for the diagnosis of an ill-fitting value, one has to take the variability of the estimate into account. For the derivation of the variance a helpful tool is the hat matrix. Consider the vector of residuals given by

$$\boldsymbol{r} = \boldsymbol{y} - \hat{\boldsymbol{y}} = \boldsymbol{y} - \boldsymbol{H}\boldsymbol{y} = (\boldsymbol{I} - \boldsymbol{H})\boldsymbol{y},$$

where $\boldsymbol{H}$ is the projection matrix $\boldsymbol{H} = \boldsymbol{X}(\boldsymbol{X}^T\boldsymbol{X})^{-1}\boldsymbol{X}^T$. The matrix $\boldsymbol{H}$ is called the hat matrix because one has $\hat{\boldsymbol{y}} = \boldsymbol{H}\boldsymbol{y}$; thus $\boldsymbol{H}$ maps $\hat{\boldsymbol{y}}$ into $\boldsymbol{y}$. $\boldsymbol{H}$ is a projection matrix because it is symmetric and idempotent, that is, $\boldsymbol{H}^2 = \boldsymbol{H}$. It represents the projection of the observed values into the space spanned by $\boldsymbol{H}$. The decomposition

$$\boldsymbol{y} = \hat{\boldsymbol{y}} + \boldsymbol{y} - \hat{\boldsymbol{y}} = \boldsymbol{H}\boldsymbol{y} + (\boldsymbol{I} - \boldsymbol{H})\boldsymbol{y}$$

is orthogonal because $\hat{\boldsymbol{y}}^T(\boldsymbol{y} - \hat{\boldsymbol{y}}) = \boldsymbol{y}^T\boldsymbol{H}(\boldsymbol{I} - \boldsymbol{H})\boldsymbol{y} = \boldsymbol{y}^T(\boldsymbol{H} - \boldsymbol{H})\boldsymbol{y} = 0$. The covariance of $\boldsymbol{r}$ is easily derived by

$$\text{cov}(\boldsymbol{r}) = (\boldsymbol{I} - \boldsymbol{H})\,\text{cov}(\boldsymbol{y})(\boldsymbol{I} - \boldsymbol{H}) = \sigma^2(\boldsymbol{I} - \boldsymbol{H} - \boldsymbol{H} + \boldsymbol{H}^2) = \sigma^2(\boldsymbol{I} - \boldsymbol{H}).$$

Therefore one obtains with the diagonal elements from $\boldsymbol{H} = (h_{ij})$ the variance $\text{var}(r_i) = \sigma^2(1 - h_{ii})$. Scaling to the same variance produces the form

$$\tilde{r}_i = \frac{r_i}{\sqrt{1 - h_{ii}}},$$

with $\text{var}(\tilde{r}_i) = \sigma^2$. If, in addition, one divides by the estimated variance $\hat{\sigma}^2 = (\boldsymbol{r}^T\boldsymbol{r})/(n-p-1)$, where $p+1$ is the length of $\boldsymbol{x}_i$, one obtains the *studentized residual*

$$r_i^* = \frac{\tilde{r}_i}{\hat{\sigma}} = \frac{y_i - \hat{\mu}_i}{\hat{\sigma}\sqrt{1-h_{ii}}},$$

which behaves much like a Student's t random variable except for the fact that the numerator and denominator are not independent.

The hat matrix itself is a helpful tool in diagnosis. From $\hat{\boldsymbol{y}} = \boldsymbol{H}\boldsymbol{y}$ it is seen that the element h_{ij} of the hat matrix $\boldsymbol{H} = (h_{ij})$ shows the amount of leverage or influence exerted on $\hat{y}_i$ by y_j. Since $\boldsymbol{H}$ depends only on $\boldsymbol{x}$, this influence is due to the "design" and not to the dependent variable. The most interesting influence is that of y_i on the fitted value $\hat{y}_i$, which is reflected by the diagonal element h_{ii}. For the projection matrix $\boldsymbol{H}$ one has

$$\text{rank}(\boldsymbol{H}) = \sum_{i=1}^{n} h_{ii} = p+1$$

and $0 \leq h_{ii} \leq 1$. Therefore, $(p+1)/n$ is the average size of a diagonal element. As a rule of thumb, an $\boldsymbol{x}$-point for which $h_{ii} > 2(p+1)/n$ holds is considered a high-leverage point (e.g., Hoaglin and Welsch, 1978).

Case Deletion as Diagnostic Tool

In case deletion let the deletion of observation i be denoted by subscript (i). Thus $\boldsymbol{X}_{(i)}$ denotes the matrix that is obtained from $\boldsymbol{X}$ by omitting the ith row; in $\boldsymbol{\mu}_{(i)}, \boldsymbol{y}_{(i)}$ the ith observation component is also omitted. Let $\hat{\boldsymbol{\beta}}_{(i)}$ denote the least-squared estimate resulting from the reduced dataset. The essential connection between the full dataset and the reduced set is given by

$$(\boldsymbol{X}_{(i)}^T\boldsymbol{X}_{(i)})^{-1} = (\boldsymbol{X}^T\boldsymbol{X})^{-1} + (\boldsymbol{X}^T\boldsymbol{X})^{-1}\boldsymbol{x}_i\boldsymbol{x}_i^T(\boldsymbol{X}^T\boldsymbol{X})^{-1}/(1-h_{ii}), \tag{1.2}$$

where $h_{ii} = \boldsymbol{x}_i^T(\boldsymbol{X}^T\boldsymbol{X})^{-1}\boldsymbol{x}_i$ is the diagonal element of $\boldsymbol{H}$. One obtains after some computation

$$\hat{\boldsymbol{\beta}}_{(i)} = \hat{\boldsymbol{\beta}} - (\boldsymbol{X}^T\boldsymbol{X})^{-1}\boldsymbol{x}_i r_i/(1-h_{ii}).$$

Thus the change in $\boldsymbol{\beta}$ that results if the ith observation is omitted may be measured by

$$\Delta_i\hat{\boldsymbol{\beta}} = \hat{\boldsymbol{\beta}} - \hat{\boldsymbol{\beta}}_{(i)} = (\boldsymbol{X}^T\boldsymbol{X})^{-1}\boldsymbol{x}_i r_i/(1-h_{ii}).$$

Again, the diagonal element of the hat matrix plays an important role. Large values of h_{ii} yield values $\hat{\boldsymbol{\beta}}_{(i)}$, which are distinctly different from $\hat{\boldsymbol{\beta}}$.

The simple *deletion residual* is given by

$$r_{(i)} = y_i - \hat{\mu}_{(i)},$$

where $\hat{\mu}_{(i)} = \boldsymbol{x}_i^T\boldsymbol{\beta}_{(i)}$. It measures the deviation of y_i from the value predicted by the model fitted to the remaining points and therefore reflects the accuracy of the prediction. From $\hat{\mu}_{(i)} = \boldsymbol{x}_i^T\boldsymbol{\beta}_{(i)}$ one obtains $\text{var}(r_{(i)}) = \sigma^2 + \sigma^2\boldsymbol{x}_i^T(\boldsymbol{X}_{(i)}^T\boldsymbol{X}_{(i)})^{-1}\boldsymbol{x}_i = \sigma^2(1+h_{(i)})$, where $h_{(i)} = \boldsymbol{x}_i^T(\boldsymbol{X}_{(i)}^T\boldsymbol{X}_{(i)})^{-1}\boldsymbol{x}_i$. It follows easily from equation (1.2) that $h_{(i)}$ is given by $h_{(i)} = h_{ii}/(1-h_{ii})$. One obtains the standardized value

$$\tilde{r}_{(i)} = \frac{r_{(i)}}{\sqrt{1+h_{(i)}}},$$

which has variance σ^2. With $\hat{\sigma}^2_{(i)} = \boldsymbol{r}^T_{(i)}\boldsymbol{r}_{(i)}/(n-p-1)$ one obtains the *studentized* version

$$r^*_{(i)} = \frac{\tilde{r}_{(i)}}{\hat{\sigma}_{(i)}} = \frac{y_i - \hat{\mu}_{(i)}}{\hat{\sigma}_{(i)}\sqrt{1+h_{(i)}}} = \frac{y_i - \hat{\mu}_i}{\hat{\sigma}_{(i)}\sqrt{1-h_{ii}}}.$$

The last transformation follows from $\hat{\mu}_{(i)} = \boldsymbol{x}_i^T\hat{\boldsymbol{\beta}}_{(i)} = \boldsymbol{x}_i^T(\hat{\boldsymbol{\beta}} - (\boldsymbol{X}^T\boldsymbol{X})^{-1}\boldsymbol{x}_i r_i/(1-h_{ii})) = \hat{\mu}_i - \boldsymbol{x}_i^T(\boldsymbol{X}^T\boldsymbol{X})^{-1}\boldsymbol{x}_i r_i/(1-h_{ii}) = \hat{\mu}_i - h_{ii}r_i/(1-h_{ii}) = \hat{\mu}_i - r_i h_{(i)}$. Therefore one has $y_i - \hat{\mu}_{(i)} = (y_i - \hat{\mu}_i)(1+h_{(i)})$.

The studentized deletion residual is related to the studentized residual by $r^*_{(i)} = r_i\hat{\sigma}/\hat{\sigma}_{(i)}$. It represents a standardization of the scaled residual $(y_i - \hat{\mu}_i)/\sqrt{1-h_i}$, which has variance $\hat{\sigma}^2$ by an estimate of σ^2, which does not depend on the ith observation. Therefore, when normality holds, the standardized case deletion residual is distributed as Student's t with $(n-p)$ degrees of freedom. Cook and Weisberg (1982) refer to r_i^* as the studentized residuals with internal studentization, in contrast to external studentization for $r^*_{(i)}$. The r_i^*'s are also called cross-validatory or jackknife residuals. Rawlings et al. (1998) used the term studentized residuals for r_i^*. For more details on residuals see Cook and Weisberg (1982).

Residuals

Simple residual

$$r_i = y_i - \hat{\mu}_i = \boldsymbol{y}_i - \boldsymbol{x}_i^T\hat{\boldsymbol{\beta}}$$

Studentized residual

$$r_i^* = \frac{y_i - \hat{\mu}_i}{\hat{\sigma}\sqrt{1-h_{ii}}}$$

Case deletion residual

$$r_{(i)} = y_i - \hat{\mu}_{(i)} = y_i - \boldsymbol{x}_i^T\hat{\boldsymbol{\beta}}_{(i)}$$

Studentized case deletion residual

$$r^*_{(i)} = \frac{y_i - \hat{\mu}_i}{\hat{\sigma}_{(i)}\sqrt{1-h_{ii}}}$$

1.4.5 Decomposition of Variance and Coefficient of Determination

The sum of squared deviations from the mean may be partitioned in the following way:

$$\sum_{i=1}^n (y_i - \bar{y})^2 = \sum_{i=1}^n (\hat{\mu}_i - \bar{y})^2 + \sum_{i=1}^n (\hat{\mu}_i - y_i)^2, \tag{1.3}$$

where $\bar{y} = \sum_i y_i/n$ is the mean over the responses. The partitioning has the form SST $=$ SSR $+$ SSE, where

SST = $\sum_{i=1}^n (y_i - \bar{y})^2$ is the *total sum of squares*, which represents the total variation in y that is to be explained by x-variables;

$\text{SSR} = \sum_{i=1}^{n}(\hat{\mu}_i - \bar{y})^2$ (R for regression) is the *regression sum of squares* built from the squared deviations of the fitted values around the mean;

$\text{SSE} = \sum_{i=1}^{n}(\hat{\mu}_i - y_i)^2$ (E for error) is the sum of the squared residuals, also called the *error sum of squares*.

The partitioning (C.3) may also be seen from a geometric view. The fitted model based on the least-squares estimate $\hat{\boldsymbol{\beta}}$ is given by

$$\hat{\boldsymbol{\mu}} = \boldsymbol{X}\hat{\boldsymbol{\beta}} = \boldsymbol{X}(\boldsymbol{X}^T\boldsymbol{X})^{-1}\boldsymbol{X}^T\boldsymbol{y} = \boldsymbol{H}\boldsymbol{y},$$

which represents a projection of $\boldsymbol{y}$ into the space span($\boldsymbol{X}$), which is spanned by the columns of $\boldsymbol{X}$. Since span($\boldsymbol{X}$) contains the vector $\mathbf{1}$ and projections are linear operators, one obtains with $P_{\boldsymbol{X}}$ denoting the projection into span($\boldsymbol{X}$) and $\bar{\boldsymbol{y}}^T = (\bar{y}, \ldots, \bar{y})$ the orthogonal decomposition

$$\boldsymbol{y} - \bar{\boldsymbol{y}} = \hat{\boldsymbol{\mu}} - \bar{\boldsymbol{y}} + \boldsymbol{y} - \hat{\boldsymbol{\mu}},$$

where $\hat{\boldsymbol{\mu}} - \bar{\boldsymbol{y}} = \boldsymbol{H}\boldsymbol{y} - \bar{\boldsymbol{y}}$ is the projection of $\boldsymbol{y} - \bar{\boldsymbol{y}}$ into span($\boldsymbol{X}$) and $\boldsymbol{y} - \hat{\boldsymbol{\mu}} = \boldsymbol{y} - \boldsymbol{H}\boldsymbol{y}$ is from the orthogonal complement of span($\boldsymbol{X}$) such that $(\boldsymbol{y} - \hat{\boldsymbol{\mu}})^T(\hat{\boldsymbol{\mu}} - \bar{\boldsymbol{y}}) = 0$.

The *coefficient of determination* is defined by

$$R^2 = \frac{\text{SSR}}{\text{SST}} = \frac{\sum_i(\hat{\mu}_i - \bar{y})^2}{\sum_i(y_i - \bar{y})^2} = 1 - \frac{\sum_i(\hat{\mu}_i - y_i)^2}{\sum_i(y_i - \bar{y}_i)^2}.$$

Thus R^2 gives the proportion of variation explained by the regression model and therefore is a measure for the adequacy of the linear regression model.

From the definition it is seen that R^2 is not defined for the trivial case where $y_i = \bar{y}$, $i = 1, \ldots, n$, which is excluded here. Extreme values of R^2 are

$$R^2 = 0 \Leftrightarrow \hat{\mu}_i = \bar{y}, \quad \text{that is, a horizontal line is fitted;}$$

$$R^2 = 1 \Leftrightarrow \hat{\mu}_i = y_i, \quad \text{that is, all observations are on a line with slope unequal to 0.}$$

Although R^2 is often considered as a measure of goodness-of-fit, it hardly reflects goodness-of-fit in the sense that a high value of R^2 tells us that the underlying model is a linear regression model. Though built from residuals, R^2 compares the residuals of the model that specifies a linear effect of variables $\boldsymbol{x}$ and the residuals of the simple intercept model (the null model). Thus it measures the additional explanatory values of variable vector $\boldsymbol{x}$ within the linear model. It cannot be used to decide whether a model shows appropriate fit. Rather, R^2 and its generalizations measure the strength of association between covariates and the response variable. If R^2 is large, the model should be useful since some aspect of the association between the response and covariates is captured by the linear model. On the other hand, if R^2 is close to zero, that does not mean that the model has a bad fit. On the contrary, if a horizontal line is fitted ($R^2 = 0$), the data may be very close to the fitted data, since any positive value $\sum_i(\hat{\mu}_i - y_i)^2$ is possible. $R^2 = 0$ just means that there is no linear association between the response and the linear predictor beyond the horizontal line. R^2 tells how much of the variation is explained by the included variables within the linear approach. It is a *relative measure* that reflects the improvement by the inclusion of predictors as compared to the simple model, where only the constant term is included.

Now we make some additional remarks to avoid misrepresentation. That R^2 is not a tool to decide if the linear model is true or not may be easily seen from considering an underlying linear model. Let a finite number of observations be drawn from the range of $\boldsymbol{x}$-values. Now, in addition to the sample of size n, let n_0 observations be drawn at a fixed design point $\boldsymbol{x}_0$. Then, for $n_0 \to \infty$, it follows that $\bar{y} \to \mu_0 = \mathrm{E}(y|x_0)$ and $\hat{\mu}_i \to \mu_0$ for $i \gg n$ such that $R^2 \to 0$. This means that although the linear model is true, R^2 approaches zero and therefore cannot be a measure for the truth of the model. On the other hand, if a non-linear model is the underlying model and observations are only drawn at two distinct design points, R^2 will approach 1 since two points may always be fitted by a line. The use of R^2 *is restricted to linear models*. There are examples where R^2 can be larger than 1 if a non-linear function is fitted by least squares (see Exercise 1.4).

1.4.6 Testing in Multiple Linear Regression

The most important tests are for the hypotheses

$$H_0: \beta_1 = \cdots = \beta_p = 0 \qquad H_1 : \beta_i \neq 0 \text{ for at least one variable} \tag{1.4}$$

and

$$H_0: \boldsymbol{\beta}_i = 0 \qquad H_1 : \boldsymbol{\beta}_i \neq 0. \tag{1.5}$$

The first null hypothesis asks if there is any explanatory value of the covariates, whereas the latter concerns the question if one specific variable may be omitted – given that all other variables are included. The test statistics may be easily derived as special cases of linear hypotheses (see next section). For normally distributed responses one obtains for $H_0: \beta_j = 0$

$$t = \frac{\hat{\beta}_j}{\hat{cov}(\hat{\beta}_j)} \quad \sim t(n-p-1),$$

where $\hat{cov}(\hat{\beta}_j) = \hat{\sigma}\sqrt{a_{jj}}$ with a_{jj} denoting the jth diagonal element of $(\boldsymbol{X}^T\boldsymbol{X})^{-1}$ and H_0 is rejected if $|t| > t_{1-\alpha/2}(n-p-1)$. For $H_0: \beta_1 = \ldots = \beta_p = 0$ and normally distributed responses one obtains

$$F = \frac{n-p-1}{p}\,\frac{R^2}{1-R^2} = \frac{(\mathrm{SST}-\mathrm{SSE})/p}{\mathrm{SSE}\,/(n-p-1)} \overset{H_0}{\sim} F(p, n-p-1)$$

and H_0 is rejected if $F > F_{1-\alpha}(p, n-p-1)$. The F-test for the global hypothesis $H_0: \beta_1 = \cdots = \beta_p = 0$ is often given within an analysis-of-variance (ANOVA) framework. Consider again the partitioning of the total sum of squares:

$$\mathrm{SST} = \mathrm{SSR} + \mathrm{SSE},$$

$$(\boldsymbol{y}-\bar{\boldsymbol{y}})^T(\boldsymbol{y}-\bar{\boldsymbol{y}}) = (\hat{\boldsymbol{\mu}}-\bar{\boldsymbol{y}})^T(\hat{\boldsymbol{\mu}}-\bar{\boldsymbol{y}}) + (\boldsymbol{y}-\hat{\boldsymbol{\mu}})^T(\boldsymbol{y}-\hat{\boldsymbol{\mu}}).$$

If the regression model holds, the error sum of squares has a scaled χ^2-distribution, SSE $\sim \sigma^2\chi^2(n-p-1)$. The degrees of freedom follow from considering the residuals $\boldsymbol{r} = \boldsymbol{y} - \boldsymbol{X}\hat{\boldsymbol{\beta}} = \boldsymbol{y} - \boldsymbol{H}\boldsymbol{y} = (\boldsymbol{I}-\boldsymbol{H})\boldsymbol{y}$, which represent an orthogonal projection by use of the projection matrix $\boldsymbol{I}-\boldsymbol{H}$. Since SSE is equivalent to the squared residuals, SSE $= \boldsymbol{r}^T\boldsymbol{r}$, and $\boldsymbol{I}-\boldsymbol{H}$ has rank $n-p-1$, one obtains $\sigma^2\chi^2(n-p-1)$. If the regression model holds and in addition $\beta_1 = \cdots = \beta_p = 0$, then one obtains for SSE and SSR the χ^2-distributions

$$\mathrm{SST} \sim \sigma^2\chi^2(n-1), \;\; \mathrm{SSR} \sim \sigma^2\chi^2(p).$$

In addition, in this case SSR and SSE are independent. The corresponding means squares are given by

$$\mathrm{MSE} = \mathrm{SSE}\,/(n-p-1), \mathrm{MSR} = \mathrm{SSR}\,/p.$$

It should be noted that while SSE and SSR sum up to SST, the sum of MSE and MSR does not give the average over all terms.

TABLE 1.4: ANOVA table for multiple linear regression.

Source of variation	SS	df	MS
Regression	$\text{SSR} = \sum_{i=1}^{n}(\hat{y}_i - \bar{y}_i)^2$	p	$MSR = \frac{\text{SSR}}{p}$
Error	$\text{SSE} = \sum_{i=1}^{n}(\hat{\mu}_i - y_i)^2$	$n - (p+1)$	$MSE = \frac{\text{SSE}}{n-p-1}$

Submodels and the Testing of Linear Hypotheses

A general framework for testing all kinds of interesting hypotheses is the testing of linear hypotheses given by

$$H_0 : \boldsymbol{C\beta} = \boldsymbol{\xi} \qquad H_1 : \boldsymbol{C\beta} \neq \boldsymbol{\xi}.$$

The simple null hypothesis $H_0 : \beta_1 = \beta_2 = \cdots = \beta_p, = 0$ turns into

$$H_0 : \begin{pmatrix} 0 & 1 & & \\ \vdots & & 1 & \\ \vdots & & & \ddots & \\ 0 & & & & 1 \end{pmatrix} \begin{pmatrix} \beta_0 \\ \beta_1 \\ \vdots \\ \beta_p \end{pmatrix} = \begin{pmatrix} 0 \\ 0 \\ \vdots \\ 0 \end{pmatrix}.$$

The hypothesis $H_0 : \beta_i = 0$ is given by $H_0 : (0 \ldots 1 \ldots 0)\boldsymbol{\beta} = 0$. Comparisons of covariates of the form $H_0 : \beta_i = \beta_j$ are given by

$$H_0 : (0, \ldots, 1, \ldots, -1, \ldots 0)\boldsymbol{\beta} = 0,$$

where 1 corresponds to β_i and -1 corresponds to β_j. Hypotheses like $H_0 : \beta_1 = \cdots = \beta_p = 0$ or $H_0 : \beta_i = 0$ are linear hypotheses, and the corresponding models may be seen as submodels of the multiple regression model. They are submodels because the parameter space is more restricted than in the original multiple regression model.

Let the more general $\tilde{M}$ be a submodel of M ($\tilde{M} \subset M$), where M is the unrestricted multiple regression model and $\tilde{M}$ is restricted to a linear subspace of dimension $(p+1) - s$, that is, $\text{rank}(\boldsymbol{C}) = s$. For example, if the restricted model contains only the intercept, one has $\text{rank}(\boldsymbol{C}) = 1$ and the restricted model specifies a subspace of dimension one. Let $\hat{\boldsymbol{\beta}}$ denote the usual least-squares estimate for the multiple regression model and $\tilde{\boldsymbol{\beta}}$ be the restricted estimate that minimizes

$$(\boldsymbol{y} - \boldsymbol{X\beta})^T(\boldsymbol{y} - \boldsymbol{X\beta})$$

under the restriction $\boldsymbol{C\beta} = \boldsymbol{\xi}$. Using Lagrange multipliers yields

$$\tilde{\boldsymbol{\beta}} = \hat{\boldsymbol{\beta}} - (\boldsymbol{X}^T\boldsymbol{X})^{-1}\boldsymbol{C}^T[\boldsymbol{C}(\boldsymbol{X}^T\boldsymbol{X})^{-1}\boldsymbol{C}^T]^{-1}[\boldsymbol{C}\hat{\boldsymbol{\beta}} - \boldsymbol{\xi}]. \tag{1.6}$$

One obtains two discrepancies, namely, the discrepancy between M and the data and the discrepancy between $\tilde{M}$ and the data. As a discrepancy measure one may use a residual or error sums of squares:

$$\text{SSE}(M) = \sum_{i=1}^{n}(\boldsymbol{y}_i - \boldsymbol{x}_i^T\hat{\boldsymbol{\beta}})^2 = (\boldsymbol{y} - \boldsymbol{X}\hat{\boldsymbol{\beta}})^T(\boldsymbol{y} - \boldsymbol{X}\hat{\boldsymbol{\beta}}),$$

$$\text{SSE}(\tilde{M}) = \sum_{i=1}^{n}(\boldsymbol{y}_i - \boldsymbol{x}_i^T\tilde{\boldsymbol{\beta}})^2 = (\boldsymbol{y} - \boldsymbol{X}\tilde{\boldsymbol{\beta}})^T(\boldsymbol{y} - \boldsymbol{X}\tilde{\boldsymbol{\beta}}).$$

Since $\tilde{M}$ is a more restricted model, $\mathrm{SSE}(\tilde{M})$ tends to be greater than $\mathrm{SSE}(M)$. One may decompose the discrepancy $\mathrm{SSE}(\tilde{M})$ by considering

$$\mathrm{SSE}(\tilde{M}) = \mathrm{SSE}(M) + \mathrm{SSE}(\tilde{M}|M), \tag{1.7}$$

where $\mathrm{SSE}(\tilde{M}|M) = \mathrm{SSE}(\tilde{M}) - \mathrm{SSE}(M)$ is the increase of residuals that results from using the more restrictive model $\tilde{M}$ instead of M. It may also be seen as the amount of variation explained by M but not by $\tilde{M}$. The notation refers to the interpretation as a conditional discrepancy; $\mathrm{SSE}(\tilde{M}|M)$ is the discrepancy of $\tilde{M}$ within model M, that is, the additional discrepancy between data and model. This results from fitting $\tilde{M}$ instead of the less restrictive model M. The decomposition (1.7) may be used for testing the fit of $\tilde{M}$ given that M is an accepted model. This corresponds to testing $H_0 : \boldsymbol{C\beta} = \boldsymbol{\xi}$ (corresponding to $\tilde{M}$) within the multiple regression model (corresponding to M).

An important property of the decomposition (1.7) is that it is based on orthogonal components. Behind (1.7) is the trivial decomposition

$$\boldsymbol{y} - \boldsymbol{X}\tilde{\boldsymbol{\beta}} = (\boldsymbol{X}\hat{\boldsymbol{\beta}} - \boldsymbol{X}\tilde{\boldsymbol{\beta}}) + (\boldsymbol{y} - \boldsymbol{X}\hat{\boldsymbol{\beta}}), \tag{1.8}$$

where $\boldsymbol{y} - \boldsymbol{X}\hat{\boldsymbol{\beta}}$ is orthogonal to $\boldsymbol{X}\hat{\boldsymbol{\beta}} - \boldsymbol{X}\tilde{\boldsymbol{\beta}}$, i.e. $(\boldsymbol{y} - \boldsymbol{X}\hat{\boldsymbol{\beta}})^T(\boldsymbol{X}\hat{\boldsymbol{\beta}} - \boldsymbol{X}\tilde{\boldsymbol{\beta}}) = 0$. Decomposition (1.7) follows from (1.8) by considering $\mathrm{SSE}(\tilde{M}) = (\boldsymbol{y} - \boldsymbol{X}\tilde{\boldsymbol{\beta}})^T(\boldsymbol{y} - \boldsymbol{X}\tilde{\boldsymbol{\beta}})$, $\mathrm{SSE}(M) = (\boldsymbol{y} - \boldsymbol{X}\hat{\boldsymbol{\beta}})^T(\boldsymbol{y} - \boldsymbol{X}\hat{\boldsymbol{\beta}})$. From (1.8) and (1.6) the explicit form of $\mathrm{SSE}(\tilde{M}|M)$ follows as

$$\mathrm{SSE}(\tilde{M}|M) = (\boldsymbol{C}\hat{\boldsymbol{\beta}} - \boldsymbol{\xi})^T[\boldsymbol{C}(\boldsymbol{X}^T\boldsymbol{X})^{-1}\boldsymbol{C}^T]^{-1}(\boldsymbol{C}\hat{\boldsymbol{\beta}} - \boldsymbol{\xi}).$$

If M holds, $\mathrm{SSE}(M)$ is χ^2-distributed with $\mathrm{SSE}(M)/\sigma^2 \sim \chi^2(n-p-1)$; if $\tilde{M}$ holds (H_0 is true), $\mathrm{SSE}(\tilde{M}|M)/\sigma^2 \sim \chi^2(s)$ and $\mathrm{SSE}(M)$ and $\mathrm{SSE}(\tilde{M}|M)$ are independent. One obtains

$$\begin{array}{lclcl}
\mathrm{SSE}(\tilde{M}) & = & \mathrm{SSE}(\tilde{M}|M) & + & \mathrm{SSE}(M) \\
\sigma^2\chi^2(n-p-1+s) & & \sigma^2\chi^2(s) & & \sigma^2\chi(n-p-1) \\
\text{if } \tilde{M} \text{ holds} & & \text{if } \tilde{M} \text{ holds} & & \text{if } M \text{ holds}
\end{array}$$

Thus, if $\tilde{M}$ holds,

$$F = \frac{\left(\mathrm{SSE}(\tilde{M}) - \mathrm{SSE}(M)\right)/s}{\mathrm{SSE}(M)/(n-p-1)} \sim F(s,\, n-p-1),$$

which may be used as test statistics for $H_0 : \ \boldsymbol{C\beta} = \boldsymbol{\xi}$. H_0 is rejected if F is larger than the $(1-\alpha)$-quantile $F_{1-\alpha}(s,\, n-p-1)$.

1.5 Exercises

1.1 Consider a linear model that specifies the rent to pay as a function of the size of the flat and the city (with data for 10 cities available). Let the model be given as

$$\mathrm{E}(y|\text{size}, C = i) = \beta_0 + \text{size} * \beta_s + \beta_{C(i)}, \quad i = 1, \ldots, 10,$$

where $\beta_{C(i)}$ represents the effect of the city. Since the parameters $\beta_{C(1)}, \ldots, \beta_{C(10)}$ are not identifiable, one has to specify an additional constraint.

(a) Give the model with dummy variables by using the symmetric side constraint $\sum_i \beta_{C(i)} = 0$.

(b) Give the model with dummy variables by specifying a reference category.

(d) Specify $\boldsymbol{C}$ and $\boldsymbol{\xi}$ of the linear hypothesis $H_0 : \boldsymbol{C\beta} = \boldsymbol{\xi}$ if you want to test if rent does not vary over cities.

(e) What is the meaning if the hypothesis $H_0 : \beta_{C(j)} = 0$ for fixed j holds?

(f) Find the transformation that transforms parameters with a reference category into parameters with a symmetric side constraint, and vice versa for a general number of cities k.

1.2 The R Package *catdata* provides the dataset *rent*.

(a) Use descriptive tools to learn about the data.

(b) Fit a linear regression model with response *rent* (net rent in Euro) and explanatory variables *size* (size in square meters) and *rooms* (number of rooms). Discuss the results.

(c) Fit a linear regression model with response *rent* and the single explanatory variable *rooms*. Compare with the results from (b) and explain why the coefficients differ even in sign.

1.3 The dataset *rent* from R Package *catdata* contains various explanatory variables.

(a) Use the available explanatory variables when fitting a linear regression model with the response *rent*. Include polynomial terms and dummy variables if necessary. Evaluate if explanatory variables can be excluded.

(b) Fit a linear model with the response *rentm* (rent per square meter) by using the available explanatory variables. Discuss the effects and compare to the results from (a).

1.4 Kockelkorn (2000) considers the model $y_i = \mu(x_i) + \varepsilon_i$ with $\mu(x) = x^\beta$ if $x \geq 0$ and $\mu(x) = -(-x)^\beta$ if $x < 0$. For some $z > 0$ let observations (y_i, x_i) be given by $\{(0,1), (0,-1), (-z^3, -z), (z^3, z)\}$.

(a) Compute the value β that minimizes the least-squares criterion.

(b) Compute R^2 as a function of z and investigate what values R^2 takes.

Chapter 2

Binary Regression: The Logit Model

Categorical regression has the same objectives as metric regression. It aims at an economic representation of the link between covariables considered as the independent variables and the response as the dependent variable. Moreover, one wants to evaluate the influence of the independent variables regarding their strength and the way they exert their influence. Predicting new observations can be based on adequate modeling of the response pattern.

Categorical regression modeling differs from classical normal regression in several ways. The most crucial difference is that the dependent variable y follows a quite different distribution. A categorical response variable can take only a limited number of values, in contrast to normally distributed variables, in which any value might be observed. In the simplest case of binary regression the response takes only two values, usually coded as $y = 0$ and $y = 1$. One consequence is that the scatterplots look different. Figure 2.1 shows data from the household panel described in Example 1.2. The outcomes "car in household" ($y = 1$) and "no car in household" ($y = 0$) are plotted against net income (in Euros). It is seen that for low income the responses $y = 0$ occur more often, whereas for higher income $y = 1$ is observed more often. However, the structural connection between the response and the covariate is hardly seen from this representation. Therefore, in Figure 2.1 the relative frequencies for owning a car are shown for households within intervals of length 50. The picture shows that a linear connection is certainly not the best choice. This leads to the second difference, which concerns the link between the covariates and the mean of the response. Although the covariates might enter the model as a linear term in the same way as in classical regression models, the link between response and linearly structured explanatory variables usually has to be modified to avoid improper models. Thus at least two structuring elements have to be modified: the distribution of the response and the link between the response and the explanatory variables.

In the following, first distributions of y are considered and then methods for structuring the link between the response and the independent variables are outlined. The chapter introduces the concept of models for binary responses. Estimation and inference are considered in Chapter 4, and extensions are given in Chapter 5.

2.1 Distribution Models for Binary Responses and Basic Concepts

2.1.1 Single Binary Variables

Let the binary response y be coded by $y = 1$ and $y = 0$. In Example 1.2, the two outcomes refer to “car in household” and “no car in household”. Often $y = 1$ is considered as success

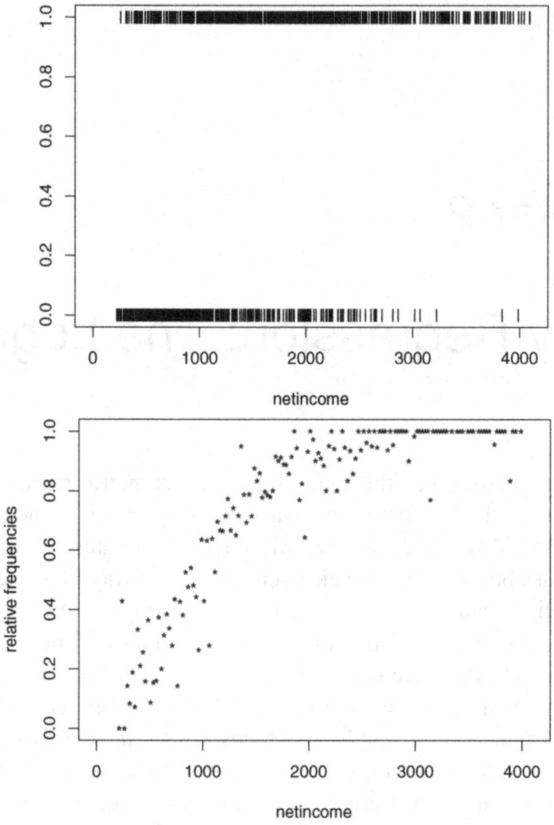

FIGURE 2.1: Car data: Upper panel shows the raw data with $y = 1$ for "car in household" and $y = 0$ for "no car in household" plotted against net income in Euros. Lower panel shows the relative frequencies within intervals of length 50, plotted against net income.

and $y = 0$ as failure, a convention that will be used in the following. The distribution of the simple binary random variable $y \in \{0, 1\}$ is completely characterized by the probability

$$\pi = P(y = 1).$$

The probability for $y = 0$ is then given by $P(y = 0) = 1 - \pi$. The mean of y is simply computed by

$$\mathrm{E}(y) = 1 \times \pi + 0 \times (1 - \pi) = \pi.$$

Therefore, the response probability π represents the mean of the binary distribution. The variance is computed equally simple by

$$\mathrm{var}(y) = \mathrm{E}(y - \mathrm{E}(y))^2 = (1 - \pi)^2 \pi + (0 - \pi)^2 (1 - \pi) = \pi(1 - \pi).$$

It is seen that the variance is completely determined by π and depends on π with minimal value zero at $\pi = 0$ and $\pi = 1$ and maximal value at $\pi = 1/2$. This is in accordance with intuition; if $\pi = 0$, only $y = 0$ can be observed; consequently, the variance is zero since there is no variability in the responses. The same holds for $\pi = 1$, where only $y = 1$ can be observed.

2.1.2 The Binomial Distribution

In many applications a binary variable is observed repeatedly and the focus is on the number of successes (occurrence of $y = 1$). The classical example is the flipping of a coin n times and then counting the number of trials where heads came up. More interestingly, the trials may refer to a standardized treatment of n persons and the outcome is the number of persons for whom treatment was successful. The same data structure is found if in Example 1.2 income is measured in categories, where categories refer to intervals of length 50. Considering the households within one interval as having the same response distribution, one has repeated trials with n denoting the number of households within a specific interval.

The basic assumptions underlying the binomial distribution are that the random variables $y_1, \dots, y_n$ with fixed n are binary, $y_i \in \{0, 1\}$ with the same response probability $\pi = P(y_i = 1), i = 1, \dots, n$, and are independent. Then the number of successes $y = y_1 + \dots + y_n$ in n trials is called a *binomial random variable*, $y \sim \mathrm{B}(n, \pi)$, and has the distribution function

$$P(y = r) = \binom{n}{r} \pi^r (1 - \pi)^{n-r}, \quad r = 1, \dots, n.$$

The mean and variances of a binomial variable are easily calculated. One obtains

$$\mathrm{E}(y) = n\pi, \ \mathrm{var}(y) = n\pi(1 - \pi).$$

The random variable y counts the number of successes in n trials. Often it is useful to look at the relative frequencies or proportions y/n rather than at the number of successes y. The form of the distribution remains the same; only the support changes since for n trials y takes values from $\{0, 1, \dots, n\}$, whereas y/n takes values from $\{0, 1/n, \dots, 1\}$. The probability function of y/n is given by

$$P(y/n = z) = \binom{n}{nz} \pi^{nz} (1 - \pi)^{n-nz},$$

where $z \in \{0, 1/n, \dots (n - 1)/n, 1\}$. The distribution of y/n is called a *scaled binomial distribution*, frequently abbreviated by $y/n \sim \mathrm{B}(n, \pi)/n$. One obtains

$$\mathrm{E}(y/n) = \pi, \ \mathrm{var}(y/n) = \pi(1 - \pi)/n.$$

It is seen that y/n has mean π. Therefore, the relative frequency is a natural and unbiased estimate for the underlying π. The variance of the estimate y/n depends on π and n with large values for π close to 0.5 and small values if π approaches 0 or 1. Generally the variance decreases with increasing n. It is noteworthy that in Figure 2.1 the estimates vary in their degree of trustworthiness since the sample sizes and underlying probabilities vary across intervals.

Odds, Logits, and Odds Ratios

A binary response variable is distinctly different from a continuous response variable. While the essential characteristics of a continuous response variable are the mean $\mu = \mathrm{E}(y)$, the variance $\sigma^2 = \mathrm{var}(y)$, skewness, and other measures, which frequently may vary independently, in a binary response variable all these characteristics are determined by only one value, which often is chosen as the probability of $y = 1$. Instead of using π as an indicator of the random behavior of the response, it is often useful to consider some transformation of π. Of particular importance are *odds* and *log-odds* (also called *logits*), which are functions of π.

Odds	Log Odds or Logits
$\gamma(\pi) = \frac{\pi}{1-\pi}$	$\text{logit}\ (\pi) = \log(\gamma(\pi)) = \log(\frac{\pi}{1-\pi})$

The odds $\gamma(\pi) = \pi/(1-\pi)$ are a directed measure that compares the probability of the occurrence of $y = 1$ and the probability of the occurrence of $y = 0$. If $y = 1$ is considered a "success" and $y = 0$ a "failure", a value $\gamma = 1/4$ means that failure is four times as likely as a success. While γ compares $y = 1$ to $y = 0$, the inverse compares $y = 0$ to $y = 1$, yielding

$$\gamma = \frac{P(y=1)}{P(y=0)}, \qquad \frac{1}{\gamma} = \frac{P(y=0)}{P(y=1)}.$$

Therefore, if odds are considered as functions of π, one obtains

$$\gamma(1-\pi) = \frac{1-\pi}{\pi} = \frac{1}{\gamma(\pi)}$$

because $\gamma(1-\pi)$ corresponds to comparing $y = 0$ to $y = 1$. Odds fulfill

$$\gamma(\pi)\gamma(1-\pi) = 1.$$

For the log-odds the relation is additive:

$$\text{logit}(1-\pi) = \log\left(\frac{1-\pi}{\pi}\right) = -\ \text{logit}(\pi),$$

and therefore

$$\text{logit}(\pi) + \text{logit}(1-\pi) = 0.$$

Comparing Two Groups

The concept of odds and log-odds may be illustrated by the simple case of comparing two groups on a binary response. The groups can be two treatment groups – a treatment group and a control group – in a clinical trial or correspond to two populations, for example, men and women. The data are usually collected in a (2×2)-contingency table with the underlying probabilities given in the following table.

	y: 1	y: 0	
1	π_1	$1-\pi_1$	1
2	π_2	$1-\pi_2$	1

In the table $\pi_t = P(y = 1|T = t)$, $t = 1, 2$, denotes the probability of response $y = 1$, corresponding to success. The probability for failure is then determined by $P(y = 0|T = t) = 1 - P(y = 1|T = t)$. A comparison of the two groups may be based on various measures. One may use the *difference of success probabilities*:

$$d_{12} = \pi_1 - \pi_2,$$

which implies the difference of failures, since $P(y = 0|T = 1) - P(y = 0|T = 2) = \pi_2 - \pi_1$. Groups are equivalent with respect to the response if the difference is zero. An alternative measure is the proportion, frequently called the *relative risk*:

$$r_{12} = \pi_1/\pi_2,$$

which has an easy interpretation. A relative risk of 2 means that the probability of success in group 1 is twice the probability of success in group 2. There is no difference between groups when the relative risk is 1.

Another quite attractive measure compares the odds rather than the probabilities. The *odds ratio* between groups 1 and 2 is

$$\gamma_{12} = \frac{\pi_1/(1-\pi_1)}{\pi_2/(1-\pi_2)} = \frac{\gamma(\pi_1)}{\gamma(\pi_2)}.$$

The ratio of the odds $\gamma(\pi_1)$ and $\gamma(\pi_2)$ compares odds instead of ratios of probabilities of success. The odds ratio γ_{21} between groups 2 and 1, $\gamma_{21} = \gamma(\pi_2)/\gamma(\pi_1)$, is simply the inverse, $\gamma_{21} = 1/\gamma_{12}$. Groups are equivalent as far as the response is concerned if $\gamma_{21} = \gamma_{12} = 1$. The odds ratio also measures the association between rows and columns, which is the association between the grouping variable and the response.

Rather than measuring association by odds ratios, one can use the log-transformed odds ratios

$$\log(\gamma_{12}) = \log\left(\frac{\gamma(\pi_1)}{\gamma(\pi_2)}\right).$$

While odds ratios can equal any non-negative number, log-odds ratios have the advantage that they are not restricted at all. A value of 0 means that the groups are equivalent.

Regression modeling in the case of two groups means modeling the response y as a function of a binary predictor. The models for π_t have to distinguish between $T = 1$ and $T = 2$. A model that uses the probabilities itself has the form

$$\pi_1 = \beta_0 + \beta, \qquad \pi_2 = \beta_0,$$

and the effect β is equivalent to the difference in probabilities. This means in particular that the effect is restricted to the interval $[-1, 1]$. A more attractive model is the linear model for the log-odds which has the form

$$\log(\pi_1/(1-\pi_1)) = \beta_0 + \beta, \qquad \log(\pi_2/(1-\pi_2)) = \beta_0, \tag{2.1}$$

and one easily derives that β is equal to the log-odds ratio and $\exp(\beta)$ is equal to the odds ratio:

$$\beta = \log(\gamma_{12}) = \log\left(\frac{\gamma(\pi_1)}{\gamma(\pi_2)}\right), \quad \exp(\beta) = \gamma_{21} = \frac{\gamma(\pi_2)}{\gamma(\pi_1)}.$$

Therefore, parameters correspond to common measures of association. The value $\beta = 0$, which means that no effect is present, corresponds to the case of no association, where the odds ratio is 1 and the log odds ratio is 0. Since model (2.1) parameterizes the logits, it is called the *logit model*, a model that will be considered extensively in the following. Model (2.1) is just the simplest version of a logit model where only one binary predictor is included.

2.2 Linking Response and Explanatory Variables

2.2.1 Deficiencies of Linear Models

Let $(y_i, \boldsymbol{x}_i), i = 1, \dots, n$, be observations with binary response $y_i \in \{0, 1\}$ on a covariate vector $\boldsymbol{x}_i$. One could try to link y_i and $\boldsymbol{x}_i$ by using the classical regression model

$$y_i = \beta_0 + \boldsymbol{x}_i^T\boldsymbol{\beta} + \varepsilon_i, \tag{2.2}$$

where y_i is the response given $\boldsymbol{x}_i$ and ε_i represents the noise with $\mathrm{E}(\varepsilon_i) = 0$. Several problems arise from this approach. First, the structural component $\mathrm{E}(y_i) = \beta_0 + \boldsymbol{x}_i^T\boldsymbol{\beta}$, which specifies the assumed dependence of the mean of y on the covariates $\boldsymbol{x}_i$, is given by

$$\pi_i = \beta_0 + \boldsymbol{x}_i^T\boldsymbol{\beta},$$

where $\pi_i = P(y_i = 1|\boldsymbol{x}_i)$. Thus the range where a model of this type may hold is restricted severely. Since $\pi_i \in [0, 1]$ even for the simplest model with univariate predictor $\pi_i = \beta_0 + x_i\beta$, it is easy to find regressor values x_i such that $\pi_i > 1$ or $\pi_i < 0$ if $\beta \neq 0$. The second problem concerns the random component of the model. From $\pi_i = \beta_0 + \boldsymbol{x}_i^T\boldsymbol{\beta}$ it follows that ε_i takes only two values, $\varepsilon_i \in \{1 - \pi_i, -\pi_i\}$. Moreover, the variance of ε_i is given by $\mathrm{var}(\varepsilon_i) = \mathrm{var}(y_i) = \pi_i(1 - \pi_i)$, which is the usual variance of a binary variable. Therefore the model is heteroscedastic, with the variance being directly connected to the mean. In summary, the usual linear regression model that assumes homogenous variances, continuous noise, and linear influence for all values of covariates is unsatisfactory in several ways. Nevertheless, as is shown in the next section, one may obtain a binary regression model by using the continuous regression model as a background model.

2.2.2 Modeling Binary Responses

In the following we show two ways to obtain more appropriate models for binary responses. The first gives some motivation for a model by considering underlying continuous responses that yield a rather general family of models. The second is based on conditional normal distribution models that yield a more specific model.

Binary Responses as Dichotomized Latent Variables

Binary regression models can be motivated by assuming that a linear regression model holds for a continuous variable that underlies the binary response. In many applications one can imagine that an underlying continuous variable steers the decision process that results in a categorical outcome. For example, the decision to buy a car certainly depends on the wealth of the household. If the wealth exceeds a certain threshold, a car is bought ($y_i = 1$); if not, one observes $y_i = 0$. In a bioassay, where $y_i = 1$ stands for death of an animal depending on the dosage of some poison, the underlying variable that determines the response may represent the damage on vital bodily functions.

To formalize the concept, let us assume that the model $\tilde{y}_i = \gamma_0 + \boldsymbol{x}_i^T\boldsymbol{\beta} + \varepsilon_i$ holds for a latent response $\tilde{y}_i$. Moreover, let $-\varepsilon_i$ have the distribution function F, which does not depend on $\boldsymbol{x}_i$. The essential concept is to consider y_i as a dichotomized version of the latent variable $\tilde{y}_i$ with the link between the observable variable y_i and the latent variable $\tilde{y}_i$ given by

$$y_i = 1 \quad \text{if} \quad \tilde{y}_i \geq \theta, \tag{2.3}$$

where θ is some unknown threshold. One obtains

$$\begin{aligned}\pi_i = \pi(\boldsymbol{x}_i) &= P(y_i = 1|\boldsymbol{x}_i) = P(\tilde{y}_i \geq \theta) = P(\gamma_0 + \boldsymbol{x}_i^T\boldsymbol{\beta} + \varepsilon_i \geq \theta)\\ &= P(-\varepsilon_i \leq \gamma_0 - \theta + \boldsymbol{x}_i^T\boldsymbol{\beta}) = F(\beta_0 + \boldsymbol{x}_i^T\boldsymbol{\beta}),\end{aligned}$$

where $\beta_0 = \gamma_0 - \theta$. The resulting model has the simple form

$$\pi(\boldsymbol{x}_i) = F(\beta_0 + \boldsymbol{x}_i^T\boldsymbol{\beta}). \tag{2.4}$$

In this model the mean of the response is still determined by a term that is linear in $\boldsymbol{x}_i$ but the connection between π_i and the linear term $\eta_i = \beta_0 + \boldsymbol{x}_i^T\boldsymbol{\beta}$ involves some transformation that is determined by the distribution function F.

The basic assumption behind the derivation of model (2.4) is that the observable variable y_i is a coarser, even binary, version of the latent variable $\tilde{y}_i$. It is important that the derivation from latent variables be seen as a mere motivation for the binary response models. The resulting model and parameters may be interpreted without reference to any latent variable. This is also seen from the handling of the threshold θ. Since it is never observed and is not identifiable, it vanishes in the parameter $\beta_0 = \gamma_0 - \theta$. A point in favor of the derivation from the latent model is that $\pi_i = F(\beta_0 + \boldsymbol{x}_i^T\boldsymbol{\beta})$ is always from the admissible range $\pi_i \in [0, 1]$ because F is a distribution function for which $F(\eta) \in [0, 1]$ always holds. A simple example is the *probit model*,

$$\pi(\boldsymbol{x}_i) = \Phi(\beta_0 + \boldsymbol{x}_i^T\boldsymbol{\beta}),$$

where Φ is the distribution function of the standardized normal distribution $\mathrm{N}(0, 1)$. The more widely used model is the *logit model*,

$$\pi(\boldsymbol{x}_i) = F(\beta_0 + \boldsymbol{x}_i^T\boldsymbol{\beta}),$$

where F is the logistic distribution function $F(\eta) = \exp(\eta)/(1 + \exp(\eta))$. An alternative derivation of the logit model is given in the following.

Modeling the Common Distribution of a Binary and a Continuous Distribution

The motivation by latent variables yields the very general class of models (2.4), where F may be any continuous distribution function. Models of this type will be considered in Chapter 5. Here we want to motivate the choice of a special distribution by considering a model for the total distribution of $y, \boldsymbol{x}$.

Let us consider the bivariate random variable $(y, \boldsymbol{x})$, where $y \in \{0, 1\}$ is a binary variable and $\boldsymbol{x}$ is a vector of continuous random variables. The most common continuous distribution is the normal distribution. But, rather than assuming $\boldsymbol{x}$ to be normally distributed in the total population, let us assume that $\boldsymbol{x}$ is conditionally normally distributed within the groups that are determined by $y = 0$ and $y = 1$. One assumes

$$\boldsymbol{x}|y = r \sim \mathrm{N}(\boldsymbol{\mu}_r, \boldsymbol{\Sigma}),$$

where the covariance matrix $\boldsymbol{\Sigma}$ is the same for all conditions $y = r$. Simple derivations based on Bayes' theorem shows that $y|\boldsymbol{x}$ is determined by

$$\pi(\boldsymbol{x}) = P(y = 1|\boldsymbol{x}) = \frac{\exp(\beta_0 + \boldsymbol{x}^T\boldsymbol{\beta})}{1 + \exp(\beta_0 + \boldsymbol{x}^T\boldsymbol{\beta})}, \tag{2.5}$$

where

$$\beta_0 = -\frac{1}{2}\boldsymbol{\mu}_1'\boldsymbol{\Sigma}^{-1}\boldsymbol{\mu}_1 + \frac{1}{2}\boldsymbol{\mu}_0\boldsymbol{\Sigma}^{-1}\boldsymbol{\mu}_0 + \log\left(\frac{P(y = 1)}{P(y = 0)}\right),$$

$$\boldsymbol{\beta} = \boldsymbol{\Sigma}^{-1}(\boldsymbol{\mu}_1 - \boldsymbol{\mu}_0).$$

It is seen that model (2.5) is of the general form (2.4), with F being the logistic distribution function $F(\eta) = \exp(\eta)/(1 + \exp(\eta))$. A more general form is considered in Exercise 2.1.

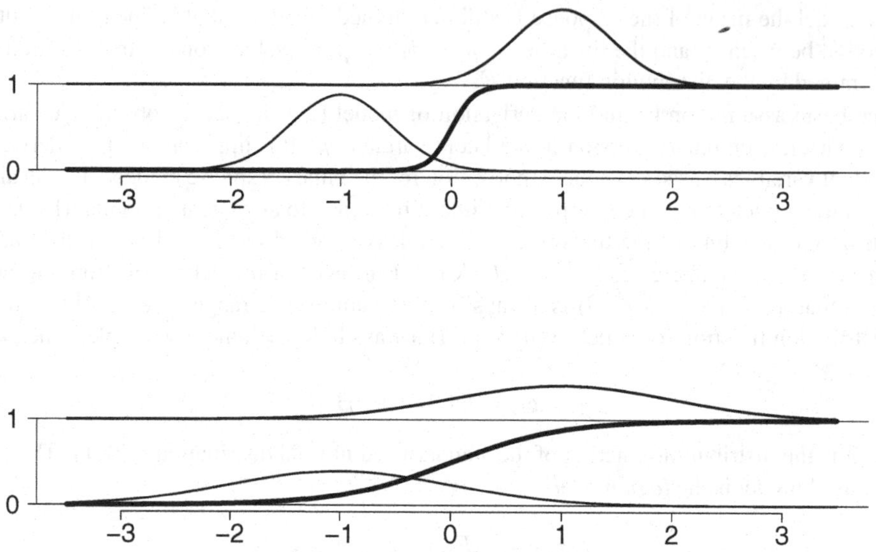

FIGURE 2.2: Logistic regression model resulting from conditionally normally distributed $x|y = i \sim \mathrm{N}(\pm 1, \sigma^2)$, $\sigma^2 = 0.16$ (upper panel), $\sigma^2 = 1$ (lower panel).

To illustrate the functional form that is specified by the logistic model, let us consider the simple case of a one-dimensional variable x. The derivation from the normal distribution is illustrated in Figure 2.2. Let x be normally distributed with $x|y = 0 \sim \mathrm{N}(-1, \sigma^2), x|y = 1 \sim \mathrm{N}(1, \sigma^2)$; then one obtains $\beta = 2/\sigma^2$. It is seen from Figure 2.2 how the response is determined by the variance within the subpopulations. The resulting logistic response function is rather flat for large $\sigma(\sigma = 1)$ and distinctly steeper for smaller $\sigma(\sigma = 0.4)$. Thus, if the groups determined by $y = 0$ and $y = 1$ are well separated by the covariate x (small σ), the response $\pi(x)$ varies strongly with x.

Basic Form of Binary Regression Models

In structured regression models one often distinguishes between two components, the *structural component* and the *random component.* A general form of the structural component is $\mu = h(\eta(\boldsymbol{x}))$, where μ denotes the mean, h is a transformation function, and $\eta(\boldsymbol{x})$ denotes a structured predictor. In the case of a binary variable, the structural component that specifies the mean response has exactly this form, given by

$$\pi_i = F(\beta_0 + \boldsymbol{x}_i^T \boldsymbol{\beta}), \tag{2.6}$$

with the linear predictor $\eta(\boldsymbol{x}_i) = \beta_0 + \boldsymbol{x}_i^T \boldsymbol{\beta}$ and the distribution function F. For a binary response y_i, the *random component* that describes the variation of the response is already specified by (2.6). The distribution of y_i is determined uniquely if π_i is fixed. In particular, the variance is a simple function of π_i, since $\mathrm{var}(y_i) = \pi_i(1 - \pi_i)$.

2.3 The Logit Model

In the following we consider the logit model, which is the most widely used model in binary regression. Alternative models will be considered in Chapter 5.

2.3.1 Model Representations

In the previous section the logit model was derived from the assumption that covariates are normally distributed given the response categories. Collecting the predictors in $\boldsymbol{x}$, where an intercept is included, the basic form is given by

$$\pi(\boldsymbol{x}) = F(\eta(\boldsymbol{x})),$$

with $F(\eta) = \exp(\eta)/(1 + \exp(\eta))$ and the linear predictor $\eta(\boldsymbol{x}) = \boldsymbol{x}^T\boldsymbol{\beta}$. Simple derivation shows that the following equivalent representations of the model hold.

Binary Logit Regression Model

$$\pi(\boldsymbol{x}) = \frac{\exp(\boldsymbol{x}^T\boldsymbol{\beta})}{1 + \exp(\boldsymbol{x}^T\boldsymbol{\beta})}$$

or

$$\frac{\pi(\boldsymbol{x})}{1 - \pi(\boldsymbol{x})} = \exp(\boldsymbol{x}^T\boldsymbol{\beta})$$

or

$$\log\left(\frac{\pi(\boldsymbol{x})}{1 - \pi(\boldsymbol{x})}\right) = \boldsymbol{x}^T\boldsymbol{\beta}$$

The different forms of the model focus on different aspects of the dependence of the response on the covariates. The first form shows how the response probability $\pi(\boldsymbol{x})$ is determined by the covariates $\boldsymbol{x}$. The second form shows the dependence of the odds $\pi(\boldsymbol{x})/(1 - \pi(\boldsymbol{x}))$ on the covariates. The third form shows that in the case of the logit model, the logits log $(\pi(\boldsymbol{x}))/(1 - \pi(\boldsymbol{x}))$ depend linearly on the covariates.

2.3.2 Logit Model with Continuous Predictor

In the following we consider first properties of the model for the simple case of univariate predictors. The logistic regression model for a one-dimensional covariate x postulates a monotone association between $\pi(x) = P(y = 1|x)$ and a covariate x of the form

$$\pi(x) = \frac{\exp(\beta_0 + x\beta)}{1 + \exp(\beta_0 + x\beta)}, \tag{2.7}$$

where β_0, β are unknown parameters. The function that determines the form of the response is the logistic distribution function $F(\eta) = \exp(\eta)/(1 + \exp(\eta))$. The functional form of F is given in Figure 2.3, where it is seen that F is strictly monotone with $F(0) = 0.5$. In addition, F is symmetric, that is, $F(\eta) = 1 - F(-\eta)$ holds for all η. F transforms the linear predictor $\beta_0 + x\beta$ such that $\pi(x)$ is between 0 and 1 and $\pi(x)$ is monotone in x.

From $\pi(x) = F(\beta_0 + x\beta)$ one obtains that $\pi(x) = 0.5$ if $\beta_0 + x\beta = 0$ or, equivalently, if $x = -\beta_0/\beta$. Thus $x = -\beta_0/\beta$ may be used as an anchor point on the x-scale where

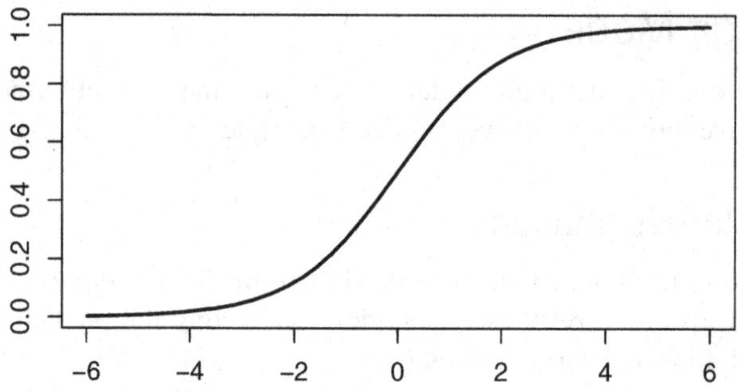

FIGURE 2.3: Logistic distribution function $F(\eta) = \exp(\eta)/(1 + \exp(\eta))$.

$\pi(x) = 0.5$. The slope of the function $\pi(x)$ is essentially determined by β. If β is large and positive, the probability increases strongly for increasing x. If β is negative, it decreases with the decrease being stronger if β is very small. Figure 2.4 shows the function $\pi(x)$ for several values of β.

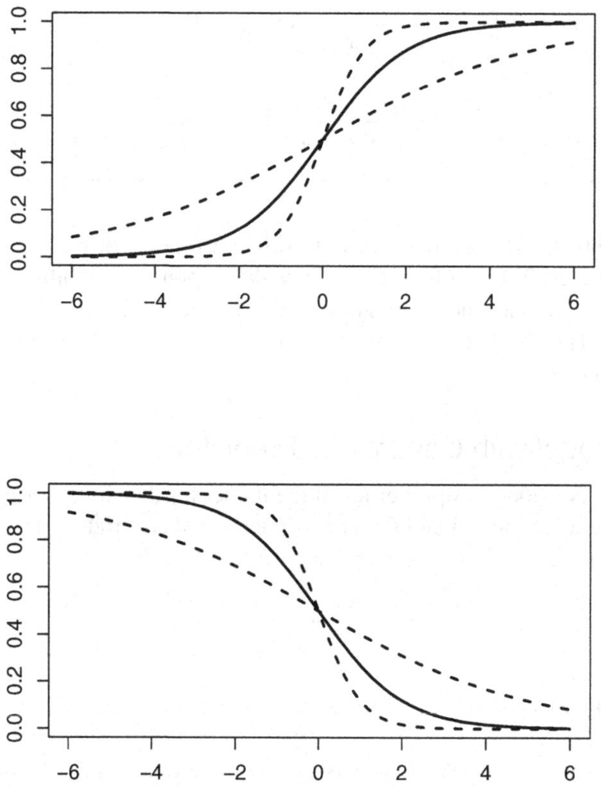

FIGURE 2.4: Response curve $\pi(x) = F(\beta_0 + \beta x)$ for $\beta_0 = 0$ and $\beta = 0.4$, $\beta = 1$, $\beta = 2$ (top) and for $\beta_0 = 0$ and $\beta = -0.4$, $\beta = -1$, $\beta = -2$ (bottom).

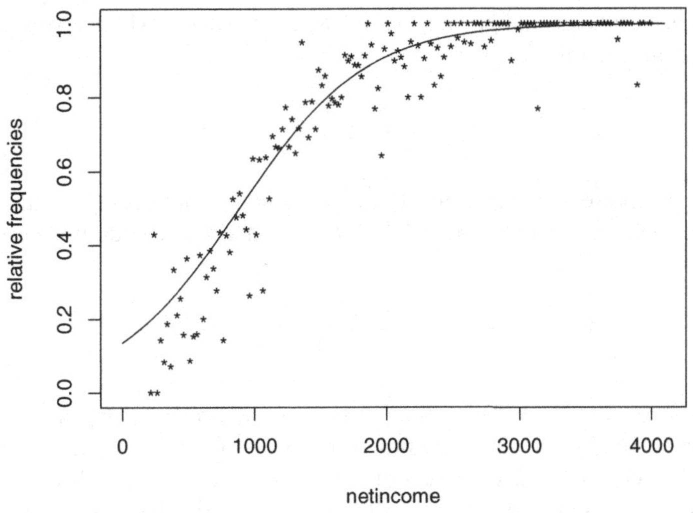

FIGURE 2.5: Car in household against income in Euros.

Example 2.1: Car in Household

In Figure 2.5, the logit model is fitted to the household data, where $\pi(x) = P(y=1|x)$ is the probability that at least one car is owned. The estimates are $\hat{\beta}_0 = -1.851, \hat{\beta} = 0.00209$. The model suggests a distinct increase of the probability with increasing net income. Figure 2.5 also shows the relative frequencies in intervals of length 50. It is seen that the fit is quite good for higher income but less satisfactory for low income. □

It is often useful to work with the alternative representations of the model:

$$\frac{\pi(x)}{1-\pi(x)} = \exp(\beta_0 + x\beta) \tag{2.8}$$

and

$$\log(\frac{\pi(x)}{1-\pi(x)}) = \beta_0 + x\beta. \tag{2.9}$$

Representation (2.8) is based on the odds $\pi(x)/(1-\pi(x)) = P(y=1|x)/P(y=0|x)$, whereas the left-hand side of (2.9) represent the log-odds or logits for a given x. Thus (2.8) may be seen as the odds representation of the model while (2.9) gives the logit representation. Since it is easier to think in odds than in log-odds, most program packages give an estimate of e^{β} in addition to an estimate of β.

When interpreting parameters for the logit model one has to take into account the logit transformation. For the linear model $\mathrm{E}(y|x) = \beta_0 + x\beta$, the parameter β is simply the change in mean response if the x-value increases by one unit, that is, $\beta = \mathrm{E}(y|x+1) - \mathrm{E}(y|x)$. For the logit model, the corresponding change is measured in logits. Let

$$\text{logit}(x) = \log(\gamma(x)) = \log\left(\frac{\pi(x)}{1-\pi(x)}\right)$$

denote the logit or log-odds at value x. Then, from $\text{logit}(x) = \beta_0 + x\beta$, it follows that β is given by

$$\beta = \text{logit}(x+1) - \text{logit}(x).$$

Thus β measures the change in logits if x is increased by one unit. Alternatively, one might use the form (2.8) of the logit model:

$$\pi(x)/(1-\pi(x)) = \exp(\beta_0 + x\beta) = \mathrm{e}^{\beta_o}(\mathrm{e}^{\beta})^x.$$

It is seen that e^{β_o} represents the odds at value $x = 0$ and the odds change by the factor e^{β} if x increases by one unit. With $\gamma(x) = \pi(x)/(1-\pi(x))$ denoting the odds at a covariate value x, one obtains

$$e^{\beta} = \frac{\pi(x+1)/(1-\pi(x+1))}{\pi(x)/(1-\pi(x))} = \frac{\gamma(x+1)}{\gamma(x)}. \tag{2.10}$$

Thus e^{β} is the factor by which the odds $\gamma(x)$ increase or decrease (depending on $\beta > 0$ or $\beta < 0$) if x is increased by one unit. As a proportion between $\gamma(x+1)$ and $\gamma(x)$, the term e^{β} is an *odds ratio*. The odds ratio is a measure of the dependence of y on x. It reflects how strong the odds change if x increases by one unit. It is noteworthy that the logit model assumes that the odds ratios given by (2.10) do not depend on x. Therefore, interpretation of β or e^{β} is very simple.

Let us consider the car data for several scalings of net income. The first model uses income in Euros as the predictor. Alternatively, one might center income around some fixed value, say 1500, such that the variable x corresponds to *net income* $-$ 1500. As a third version we consider the covariate (*net income* $-1500)/1000$ such that one unit corresponds to 1000 Euros. The estimates for all three models are given in Table 2.1. For the first model, $\mathrm{e}^{\hat{\beta}_0} = 0.157$ corresponds to the odds that there is a car in the household (instead of no car) for zero income. For income centered around 1500, $\mathrm{e}^{\hat{\beta}_0} = 3.6$ refers to the odds at income 1500 Euros. In the first two models the change in odds by increasing income by one Euro is determined by the factor $\mathrm{e}^{\hat{\beta}} = 1.002$, which does not help much in terms of interpretation. For a rescaled covariate, measured in 1000 Euros, the factor is $\mathrm{e}^{\hat{\beta}} = 8.064$, which gives some intuition for the increase in odds when income is increased by one unit, which means 1000 Euros.

TABLE 2.1: Parameter for logit model with predictor income.

Income in Euros	
Parameter	Odds
$\hat{\beta}_0 = -1.851$	$\mathrm{e}^{\hat{\beta}_0} = 0.157$
$\hat{\beta} = 0.00209$	$\mathrm{e}^{\hat{\beta}} = 1.002$
Income in Euros, centered at Euro 1500	
Parameter	Odds
$\hat{\beta}_0 = 1.281$	$\mathrm{e}^{\hat{\beta}_0} = 3.600$
$\hat{\beta} = 0.00209$	$\mathrm{e}^{\hat{\beta}} = 1.002$
Income in thousands of Euros, centered at Euro 1500	
Parameter	Odds
$\hat{\beta}_0 = 1.281$	$\mathrm{e}^{\hat{\beta}_0} = 3.600$
$\hat{\beta} = 2.088$	$\mathrm{e}^{\hat{\beta}} = 8.069$

An alternative way to get some understanding of the parameter values has been suggested by Cox and Snell (1989). It may be shown that $1/\beta$ is approximately the distance between the

75% point and the 50% point of the estimated logistic curve. The distance between the 95% point and the 50% point is approximately $3/\beta$. From $1/\hat{\beta} = 478$ for the first model of the car data (and $1/\hat{\beta} = 0.478$ for the scaling in 1000 Euros) one gets some intuition for the increase in the logistic curve without having to plot it. As a general rule one might keep the following in mind:

Change in x by $\pm 1/\beta$:	Change in π from $\pi = 0.5$ to 0.5 ± 0.23,
Change in x by $\pm 3/\beta$:	Change in π from $\pi = 0.5$ to 0.5 ± 0.45.

The value $\pi = 0.5$ itself occurs for the x-value $-\beta_0/\beta$.

Multivariate Predictor

When several continuous covariates are available the linear predictor may have the form

$$\eta(\boldsymbol{x}) = \beta_0 + x_1\beta_1 + \ldots + x_m\beta_m.$$

The interpretation is the same as for univariate predictors; however, one should be aware that the other variables are present. From

$$\text{logit}(\boldsymbol{x}) = \log\frac{\pi(\boldsymbol{x})}{1-\pi(\boldsymbol{x})} = \beta_0 + x_1\beta_1 + \ldots + x_m\beta_m$$

and

$$\gamma(x) = \frac{\pi(\boldsymbol{x})}{1-\pi(x)} = \mathrm{e}^{\beta_0}(\mathrm{e}^{\beta_1})^{x_1}\ldots(\mathrm{e}^{\beta_m})^{x_m}$$

one derives immediately that β_j corresponds to the *additive change in logits* when variable β_j is increased by one unit *while all other variables are kept fixed*,

$$\beta_j = \text{logit}(x_1,\ldots,x_j+1,\ldots,x_m) - \text{logit}(x_1,\ldots,x_m),$$

and e^{β_j} corresponds to the multiplicative change in odds when x_j is increased by one unit,

$$\mathrm{e}^{\beta_i} = \frac{\gamma(x_1,\ldots,x_j+1,\ldots,x_m)}{\gamma(x_1,\ldots,x_m)}.$$

It is essential that the effects of predictors measured by β_j or e^{β_j} assume that all other values are kept fixed. A strong assumption is implied here, namely that the effect of a covariate is the same, whatever values the other variables take. If that is not the case, one has to include interaction effects between predictors (see Chapter 4).

Example 2.2: Vasoconstriction

A classical example in logistic regression is the vasoconstriction data that were used by Finney (1947). The data (Figure 2.2) were obtained in a carefully controlled study in human physiology where a reflex "vasoconstriction" may occur in the skin of the digits after taking a single deep breath. The response y is the occurrence ($y = 1$) or non-occurrence ($y = 0$) of vasoconstriction in the skin of the digits of one subject after he or she inhaled a certain volume of air at a certain rate. The responses of three subjects are available. The first contributed 9 responses, the second contributed 8 responses, and the third contributed 22 responses.

Although the data represent repeated measurements, usually independent observations were assumed. The effect of the volume of air and inspiration rate on the occurrence of vasoconstriction may be based on the binary logit model

$$\text{logit}(\boldsymbol{x}) = \log(\frac{\pi(\boldsymbol{x})}{1-\pi(\boldsymbol{x})}) = \beta_0 + \text{volume}\beta_1 + \text{rate}\beta_2.$$

Then interpretation of the parameters refers to the multiplicative change in odds when the variables are increased by one unit. Alternatively, one can apply the logit model with log-transformed variables,

$$\text{logit}(\boldsymbol{x}) = \log(\frac{\pi(\boldsymbol{x})}{1-\pi(\boldsymbol{x})}) = \beta_0 + \log(\text{volume})\beta_1 + \log(\text{rate})\beta_2,$$

which is equivalent to

$$\frac{\pi(\boldsymbol{x})}{1-\pi(\boldsymbol{x})} = e^{\beta_0}\,\text{volume}^{\beta_1}\text{rate}^{\beta_2}.$$

Then the effect of the covariates volume and rate is multiplicative on the odds. When rate changes by the factor c the odds change by the factor c^{β_1}. The maximum likelihood estimates for the log-transformed covariates are $\beta_0 = -2.875$, $\beta_1 = 5.179$, $\beta_2 = 4.562$. The logit model for the original variables yields $\beta_0 = -9.5296$, $\beta_1 = 3.8822$, $\beta_2 = 2.6491$. In both models the covariates are highly significant. □

TABLE 2.2: Vasoconstriction data.

Index	Volume	Rate	Y	Index	Volume	Rate	Y
1	3.70	0.825	1	20	1.80	1.800	1
2	3.50	1.090	1	21	0.40	2.000	0
3	1.25	2.500	1	22	0.95	1.360	0
4	0.75	1.500	1	23	1.35	1.350	0
5	0.80	3.200	1	24	1.50	1.360	0
6	0.70	3.500	1	25	1.60	1.780	1
7	0.60	0.750	0	26	0.60	1.500	0
8	1.10	1.700	0	27	1.80	1.500	1
9	0.90	0.750	0	28	0.95	1.900	0
10	0.90	0.450	0	29	1.90	0.950	1
11	0.80	0.570	0	30	1.60	0.400	0
12	0.55	2.750	0	31	2.70	0.750	1
13	0.60	3.000	0	32	2.35	0.030	0
14	1.40	2.330	1	33	1.10	1.830	0
15	0.75	3.750	1	34	1.10	2.200	1
16	2.30	1.640	1	35	1.20	2.000	1
17	3.20	1.600	1	36	0.80	3.330	1
18	0.85	1.415	1	37	0.95	1.900	0
19	1.70	1.060	0	38	0.75	1.900	0
				39	1.30	1.625	1

2.3.3 Logit Model with Binary Predictor

When both the predictor and the response variable are binary the usual representation of data is in a (2×2)-contingency table, as in the next example.

Example 2.3: Duration of Unemployment

A simple example of a dichotomous covariate is given by the contingency table in Table 2.3, which shows data from a study on the duration of unemployment. The duration of unemployment is given in two categories, short-term unemployment (less than 6 months) and long-term employment (more than 6 months). Subjects are classified with respect to gender and duration of unemployment. Gender is considered as the explanatory variable and duration as the response variable. □

When a binary variable is used as an explanatory variable it has to be coded. Thus, instead of $G = 1$ for males and $G = 2$ females, one uses a dummy variable. There are two forms in common use, (0-1)-coding and effect coding.

TABLE 2.3: Cross-classification of gender and duration of unemployment.

		Duration				
		≤ 6 Months	> 6 Months	Marginals	Odds	Log-Odds
Gender	male	403	167	570	2.413	0.881
	female	238	175	413	1.360	0.307

Logit Model with (0-1)-Coding of Covariates

Let x_G denote the dummy variable for G in (0-1)-coding, given by $x_G = 1$ for males and $x_G = 0$ for females. With y denoting the dichotomous response, specified by $y = 1$ for short-term unemployment and $y = 0$ for long-term unemployment and $\pi(x_G) = P(y = 1|x_G))$, the corresponding logit model has the form

$$\log\left(\frac{\pi(x_G)}{1-\pi(x_G)}\right) = \beta_0 + x_G\beta \quad \frac{\pi(x_G)}{1-\pi(x_G)} = e^{\beta_0}\left(e^{\beta}\right)^{x_G}. \tag{2.11}$$

It is immediately seen that β_0 is given by

$$\beta_0 = \log\left(\frac{\pi(x_G = 0)}{1-\pi(x_G = 0)}\right) = \text{logit}(x_G = 0),$$

which corresponds to the logits in the reference category, where $x_G = 0$. The effect of the covariate takes the form

$$\begin{aligned}\beta &= \log\left(\frac{\pi(x_G = 1)}{1-\pi(x_G = 1)}\right) - \log\left(\frac{\pi(x_G = 0)}{1-\pi(x_G = 0)}\right)\\ &= \text{logit}(x_G = 1) - \text{logit}(x_G = 0),\end{aligned}$$

which is the *additive* change in logits for the transition from $x_G = 0$ to $x_G = 1$. A simpler interpretation holds for the transformed parameters

$$e^{\beta_0} = \frac{\pi(x_G = 0)}{1-\pi(x_G = 0)} = \gamma(x_G = 0),$$

which corresponds to the odds in the reference category $x_G = 0$, and

$$e^{\beta} = \frac{\pi(x_G = 1)/(1-\pi(x_G = 1)}{\pi(x_G = 0)/(1-\pi(x_G = 0))} = \frac{\gamma(x_G = 1)}{\gamma(x_G = 0)} = \gamma(1|0),$$

which corresponds to the odds ratio between $x_G = 1$ and $x_G = 0$. It may also be seen as the factor by which the logits change if $x_G = 0$ is replaced by $x_G = 1$.

Logit Model with (0-1)-Coding

$$\beta_0 = \text{logit}(x_G = 0) \qquad e^{\beta_0} = \gamma(x_G = 0)$$

$$\beta = \text{logit}(x_G = 1) - \text{logit}(x_G = 0) \qquad e^{\beta} = \frac{\gamma(x_G = 1)}{\gamma(x_G = 0)}$$

Example 2.4: Duration of Unemployment

For the contingency table in Example 2.3 one obtains for the logit model with gender given in (0-1)-coding

$$\hat{\beta}_0 = 0.307, \quad e^{\hat{\beta}_0} = 1.360, \quad \hat{\beta} = 0.574, \quad e^{\hat{\beta}} = 1.774.$$

It is usually easier to interpret e^{β} rather than β itself. While β refers to the change in logits, e^{β} gives the odd ratio. Thus $e^{\hat{\beta}} = 1.774$ means that the odds of short-time unemployment for men are almost twice the odds for women. □

Logit Model with Effect Coding

An alternative form of the logit model is given by

$$\log\left(\frac{P(y=1|G=i)}{1-P(y=1|G=i)}\right) = \beta_0 + \beta_i.$$

Here β_i is the effect of the factor G if $G = i$. But since only two logits are involved, namely, for $G = 1$ and $G = 2$, one needs a restriction to make the parameters $\beta_0, \beta_1, \beta_2$ identifiable. The symmetric constraint $\beta_1 + \beta_2 = 0$ (or $\beta_2 = -\beta_1$) is equivalent to the model

$$\log\left(\frac{\pi(x_G)}{1-\pi(x_G)}\right) = \beta_0 + x_G\beta,$$

where $\beta = \beta_1$ and x_G is given in effect coding, that is, it is given by $x_G = 1$ for males and $x_G = -1$ for females. With $\gamma(x_G) = \pi(x_G)/(1-\pi(x_G))$ one obtains $\gamma(x_G = 1) = e^{\beta_0} e^{\beta}, \gamma(x_G = -1) = e^{\beta_0} e^{-\beta}$. Simple computation shows that

$$\beta_0 = \frac{1}{2}(\text{logit}(x_G = 1) + \text{logit}(x_G = -1))$$

is the *arithmetic* mean over the logits of categories $G = 1$ and $G = 2$. Therefore,

$$e^{\beta_0} = (\gamma(x_G = 1)\gamma(x_G = -1))^{1/2}$$

is the *geometric* mean over the odds $\gamma(x_G = 1)$ and $\gamma(x_G = -1)$, representing some sort of "baseline" odds. For β one obtains

$$\beta = \frac{1}{2}(\text{logit}(x_G = 1) - \text{logit}(x_G = -1)),$$

which is half the change in logits for the transition from $x_G = -1$ to $x_G = 1$. Thus

$$e^{\beta} = (\gamma(x_G = 1)/\gamma(x_G = -1))^{\frac{1}{2}} = \gamma(x_G = 1|x_G = -1)^{1/2}$$

is the square root of the odds ratio. Since $\gamma(x_G = 1) = e^{\beta_0} e^{\beta}$, the term e^{β} is the factor that modifies the "baseline" odds e^{β_0} to obtain $\gamma(x_G = 1)$. To obtain $\gamma(x_G = -1)$ one has to use the factor $e^{-\beta}$.

Logit Model with Effect Coding

$$\beta_0 = \frac{1}{2}(\text{logit}(x_G = 1) + \text{logit}(x_G = -1)) \qquad e^{\beta_0} = (\gamma(x_G = 1)\gamma(x_G = -1))^{1/2}$$

$$\beta = \frac{1}{2}(\text{logit}(x_G = 1) - \text{logit}(x_G = -1)) \qquad e^{\beta} = \gamma(x_G = 1|x_G = -1)^{\frac{1}{2}}$$

Example 2.5: Duration of Unemployment
For the data in Example 2.3 one obtains for the logit model with gender given in effect coding

$$\hat{\beta}_0 = 0.594, \quad e^{\hat{\beta}_0} = 1.811, \qquad \hat{\beta} = 0.287, \quad e^{\hat{\beta}} = 1.332.$$

As in (0-1)-coding it is easier to interpret e^{β} rather than β itself. While $\hat{\beta}_0$ is the average (arithmetic mean) over logits, $e^{\hat{\beta}_0} = 1.811$ is the geometric mean over the odds. One obtains the odds in the male population by applying the factor $e^{\hat{\beta}} = 1.332$, which shows that the odds are better in the male population. Multiplication with $e^{-\hat{\beta}} = 1/1.332$ yields the odds in the female population. □

2.3.4 Logit Model with Categorical Predictor

In the more general case, categorical covariates have more than just two outcomes. As an example let us again consider the duration of unemployment given in two categories, short-term unemployment (less than 6 months) and long-term employment (more than 6 months), but now depending on education level. In Table 2.4 subjects are classified with respect to education level and duration of unemployment. Education level is considered the explanatory variable and duration the response variable. Interpretation of the parameters is similar to the binary covariate case. In the following we will again distinguish between the the restrictions yielding (0-1)-coding and effect coding.

TABLE 2.4: Cross-classification of level of education and duration of unemployment.

		Duration		
		≤ 6 Months	> 6 Months	
No specific training	1	202	96	298
Low level training	2	307	162	469
High level training	3	87	66	153
University degree	4	45	18	63

Logit Model with (0-1)-Coding

Consider a categorical covariable or factor A with categories $A \in \{1, \ldots, I\}$ and let $\pi(i) = P(y = 1|A = i)$ denote the response probability. Then a general form of the logit model is given by

$$\log(\frac{\pi(i)}{1-\pi(i)}) = \beta_0 + \beta_i \quad \text{or} \quad \frac{\pi(i)}{1-\pi(i)} = e^{\beta_0}\, e^{\beta_i}. \tag{2.12}$$

Since one has only I logits, $\log(\pi(i)/(1-\pi(i))), i = 1, ..., I$, but $I+1$ parameters $\beta_0, \beta_1, ..., \beta_I$, a constraint for the parameters is necessary. One possibility is to set $\beta_I = 0$, thereby defining I as the reference category. Then the intercept $\beta_0 = \log(\pi(I)/(1-\pi(I)))$ is the logit for the reference category I and

$$\beta_i = \log\left(\frac{\pi(i)}{1-\pi(i)}\right) - \log\left(\frac{\pi(I)}{1-\pi(I)}\right) = \log \gamma(i|I)$$

is the additive change in logits if $A = I$ is replaced by $A = i$, which is equivalent to the logarithm of the odds ratio of $A = i$ to $A = I$. Alternatively, one may consider the exponentials

$$e^{\beta_0} = \frac{\pi(I)}{1-\pi(I),} \qquad\qquad e_i^{\beta} = \gamma(i|I).$$

The latter is the odds ratio of $A = i$ to $A = I$. The model (2.12) may be given in the form of a regression model by using the dummy variables $x_{A(i)} = 1$ if $A = i$, and $x_{A(i)} = 0$ otherwise. Then one has a model with multiple predictor $x_{A(1)}, \ldots, x_{A(I-1)}$

$$\log \frac{\pi(i)}{1-\pi(i)} = \beta_0 + x_{A(1)}\beta_1 + \cdots + x_{A(I-1)}\beta_{I-1}.$$

Example 2.6: Duration of Unemployment with Predictor Education Level

For the data in Table 2.4 one obtains the odds and the parameters for the logit model in (0-1)-coding as given in Table 2.5. □

TABLE 2.5: Odds and parameters for logit model with predictor education level.

Level	Short Term	Long Term	Odds e^{β_i}	β_i	Standard Error
1	202	96	$\gamma(1/4) = 0.843$	−0.170	0.30
2	307	162	$\gamma(2/4) = 0.761$	−0.273	0.29
3	87	66	$\gamma(3/4) = 0.529$	−0.637	0.32
4	45	18	$\gamma(4/4) = 1$	0	0

Parameters with (0-1)-Coding

$$\beta_0 = \log\left(\frac{\pi(I)}{1-\pi(I)}\right) \qquad e^{\beta_0} = \frac{\pi(I)}{1-\pi(I)}$$
$$\beta_i = \log(\gamma(i|I)) \qquad e_i^{\beta} = \gamma(i|I)$$

Logit Model with Effect Coding

An alternative restriction on the parameters in model (2.12)) is the symmetric restriction $\beta_1 + ... + \beta_I = 0$. By considering $\beta_I = -\beta_1 - ... - \beta_{I-1}$ one obtains

$$\log\left(\frac{\pi(i)}{1-\pi(i)}\right) = \beta_0 + \beta_i\ , \quad i = 1, ..., I-1,$$
$$\log\left(\frac{\pi(I)}{1-\pi(I)}\right) = \beta_0 - \beta_1 - ... - \beta_{I-1}.$$

This is equivalent to the form

$$\log\left(\frac{\pi(i)}{1-\pi(i)}\right) = \beta_0 + x_{A(1)}\beta_1 + ... + x_{A(I-1)}\beta_{I-1},$$

where the $x_{A(i)}$'s are given in effect coding, that is, $x_{A(i)} = 1$ if $A = i$; $x_{A(i)} = -1$ if $A = k$; and $x_{A(i)} = 0$ otherwise. One obtains for the parameters

$$\beta_0 = \frac{1}{I}\sum_{i=1}^{I} \log\left(\frac{\pi(i)}{1-\pi(i)}\right) = \frac{1}{I}\sum_{i=1}^{I} \log(\gamma(i)),$$
$$\beta_i = \log\left(\frac{\pi(i)}{1-\pi(i)}\right) - \beta_0 = \log(\gamma(i)) - \beta_0.$$

Therefore β_0 corresponds to the arithmetic mean of logits across all categories of A and is some sort of a baseline logit, whereas β_i represents the (additive) deviation of category i from this baseline. For e^{β_0} one obtains

$$e^{\beta_0} = (\gamma(1) \cdot \ldots \cdot \gamma(I))^{1/I},$$

which is the geometric mean over the odds. Since $\gamma(i) = e^{\beta_0} e^{\beta_i}$, e^{β_i} represents the factor that transforms the baseline odds into the odds of population i.

Parameters with Effect Coding

$$\beta_0 = \frac{1}{I}\sum_{i=1}^{I} \log(\gamma(i)) \qquad e^{\beta_0} = (\gamma(1) \cdot \ldots \cdot \gamma(I))^{1/I}$$
$$\beta_i = \log(\gamma(i)) - \beta_0 \qquad e^{\beta_i} = \gamma(i)\, e^{-\beta_0}$$

Logit Model with Several Categorical Predictors

If more than one predictor is included, the interpretation of parameters is pretty much the same. Let us consider two predictors, $A \in \{1, \ldots, I\}$ and $B \in \{1, \ldots, J\}$, and the *main effect model* that contains dummies but no interactions. With $\pi(A = i, B = j) = P(y = 1|A = i, B = j)$ the model with the reference category I for factor A and J for factor B is given by

$$\log(\frac{\pi(A = i, B = j)}{1 - \pi(A = i, B = j)})$$
$$= \beta_0 + x_{A(1)}\beta_{A(1)} + \ldots + x_{A(I-1)}\beta_{A(I-1)} + x_{B(1)}\beta_{B(1)} + \ldots + x_{B(J-1)}\beta_{B(J-1)},$$

where $x_{A(i)}, x_{B(j)}$ are 0-1 dummy variables. It is easily derived (Exercise 2.2) that the parameters have the form given in the following box.

Parameters with (0-1)-Coding

$$e^{\beta_0} = \frac{\pi(A = I, B = J)}{1 - \pi(A = I, B = J)} = \gamma(A = I, B = J),$$
$$\text{odds for} \quad A = I, B = J$$
$$e^{\beta_{A(i)}} = \frac{\pi(A = i, B = j)/(1 - \pi(A = i, B = j))}{\pi(A = I, B = j)/(1 - \pi(A = I, B = j))} = \frac{\gamma(A = i, B = j)}{\gamma(A = I, B = j)},$$
$$\text{odds ratio compares} \quad A = i \quad \text{to} \quad A = I, \quad \text{any} \quad B = j$$
$$e^{\beta_{B(j)}} = \frac{\pi(A = i, B = j)/(1 - \pi(A = i, B = j))}{\pi(A = i, B = J)/(1 - \pi(A = i, B = J))} = \frac{\gamma(A = i, B = j)}{\gamma(A = i, B = J)},$$
$$\text{odds ratio compares} \quad B = j \quad \text{to} \quad B = J, \quad \text{any} \quad A = i$$

The exponential of β_0 corresponds to the *odds* for the reference category ($A = I, B = J$), the exponential of $\beta_{A(i)}$ corresponds to the *odds ratio* between $A = i$, and the reference category $A = I$ for any category of B. For $\beta_{B(j)}$, the corresponding odds ratio does not depend on A. It should be noted that this simple interpretation no longer holds if interactions are included.

2.3.5 Logit Model with Linear Predictor

In the general case one has several covariates, some of which are continuous and some of which are categorical. Then one may specify a logit model $\text{logit}(\boldsymbol{x}) = \eta(\boldsymbol{x})$ with the linear predictor given by

$$\eta(\boldsymbol{x}) = \boldsymbol{x}^T\boldsymbol{\beta} = \beta_0 + x_1\beta_1 + \ldots + x_p\beta_p,$$

which yields

$$\log(\frac{\pi(\boldsymbol{x})}{1-\pi(\boldsymbol{x})}) = \beta_0 + x_1\beta_1 + \ldots + x_p\beta_p.$$

The x-values in the linear predictor do not have to be the original variables. When variables are categorical the x-values represent dummy variables; when variables are continuous they can represent transformations of the original variables. Components within the linear predictor can be

$x_i, x_i^2, x_i^3, \ldots,$	main effects or polynomial terms built from variable x_i;
$x_{A(i)}, \ldots, x_{A(I-1)},$	main effects (dummy variables) as transformations of categorical variable $A \in \{1, \ldots, I\}$;
$x_{A(i)} \cdot x_{B(j)},$ $i = 1, \ldots, I-1,$ $j = 1, \ldots, J-1;$	interactions (products of dummy variables) between two factor $A \in \{1, \ldots, I\}, B \in \{1, \ldots, J\}$;
$x_i x_{B(j)}, j = 1, \ldots, J,$	interactions between metrical variable x_i and categorical variable $B \in \{1, \ldots, J\}$.

One may also include products of more than two variables (or dummies), which means interactions of a higher order. The structuring of the linear predictor is discussed more extensively in Section 4.4.

2.4 The Origins of the Logistic Function and the Logit Model

In this chapter and in some of the following the logistic function occurs quite often. Therefore some remarks on the origins of the function seem warranted. We will follow Cramer (2003), who gives a careful account of the historical development.

A simple model for the growth of populations assumes that the growth rate $dw(t)/dt$ is directly proportional to the size $w(t)$ at time point t. The corresponding differential equation $dw(t)/dt = \beta w(t)$ has the solution $w(t) = c\exp(\beta t)$, where c is a constant. This exponential growth model model means unopposed growth and is certainly not realistic over a wide range of time.

An alternative model that has been considered by the Belgian astronomer Quetelet (1795–1874) and the mathematician Verhulst (1804–1849) may be derived from a differential equation with an extra term. It is assumed that the rate is given by

$$dw(t)/dt = \tilde{\beta} w(t)(S - w(t)),$$

where S denotes the upper limit or saturation level. The growth rate is now determined by the size $w(t)$ but also by the term $(S - w(t))$, which has the opposite effect, that is, the large size decelerates growth. By transforming the equation to

$$dF(t)/dt = \beta F(t)(1 - F(t)),$$

where $F(t) = w(t)/S$ and $\beta = S\tilde{\beta}$, one obtains that the solution has the form of the logistic function

$$F(t) = \exp(\alpha + \beta t)/(1 + \exp(\alpha + \beta t))$$

with constant α. Therefore, the logistic function results as the solution of a rather simple growth model. It is widely used in modeling population size but also in marketing when the objective is the modeling of market penetration of new products.

The logit model came up much later than the probit model. Cramer (2003) traces the roots of the probit model back to Fechner (1801–1887), who is still well known to psychologists because of Fechner's law, which relates the physical strength of a stimulus and its strength as perceived by humans. It seems to have been reinvented several times until Bliss (1934) introduced the term *probit* for probability unit. The probit model used to be the classical model of bioassays. Berkson (1994) introduced as an alternative the logit model, which was easier to estimate. The logit model was not well received, but after the ideological conflict had abated the logit model was widely adopted in the late 1950s.

2.5 Exercises

2.1 Consider the bivariate random variable $(y, \boldsymbol{x})$, where $y \in \{0, 1\}$ is a binary variable and $\boldsymbol{x}$ is a vector of continuous random variables. Assume that $\boldsymbol{x}$ is conditionally normally distributed within the groups determined by $y = 0$ and $y = 1$, that is, $\boldsymbol{x}|y = r \sim \mathrm{N}(\boldsymbol{\mu}_r, \boldsymbol{\Sigma}_r)$, where the covariance matrix $\boldsymbol{\Sigma}_r$ depends on the category $y = r$. Use Bayes' theorem to derive the conditional distribution $y|\boldsymbol{x}$ and show that one obtains a logit model that is determined by a linear predictor, but x-variables themselves are included in quadratic form.

2.2 Consider a binary logit model with two factors, $A \in \{1, \dots, I\}$ and $B \in \{1, \dots, J\}$, and linear predictor $\eta = \beta_0 + x_{A(1)}\beta_{A(1)} + \dots + x_{A(I-1)}\beta_{A(I-1)} + x_{B(1)}\beta_{B(1)} + \dots + x_{B(J-1)}\beta_{B(J-1)}$. Show that the parameters can be represented as log odds and log odds ratios.

2.3 A logit model is used to model the probability of a car in household depending on the factor "type of household" (1: household includes more than one person and children, 2: household includes more than one person without children, 3: one-person household). The logit model uses two (0-1)-dummy variables for type 1 and type 2.

(a) Write down the model and interpret the parameters, which were estimated as $\beta_0 = -0.35$, $\beta_1 = 2.37$, $\beta_2 = 1.72$.

(b) Compute the parameters if category 1 is chosen as the reference category.

(c) Let the model now be given in effect coding with parameters $\tilde{\beta}_1, \tilde{\beta}_1$. Find the parameters $\tilde{\beta}_1, \tilde{\beta}_1$ and interpret.

(d) Give a general form of how parameters for a factor $A \in \{1, \dots, k\}$ in (0-1)-coding β_j can be transformed into parameters in effect coding $\tilde{\beta}_j$.

2.4 A logit model for the data from the previous exercise was fitted, but now including a linear effect of income. One obtains parameter estimates $\beta_0 = -2.06$, $\beta_1 = 1.34$, $\beta_2 = 0.93$, and $\beta_{income} = 0.0016$.

(a) Write down the model and interpret the parameters.

(b) Why do these parameters differ from the parameters in the previous exercise?

2.5 In a treatment study one wants to investigate the effect of age (x_1) in years and dose (x_2) in milligrams (mg) on side effects. The response is headache ($y = 1$: headache; $y = 1$: none). One fits a logit model with the predictor $\eta = \beta_0 + x_1\beta_1 + x_2\beta_2$.

(a) One obtains $\beta_0 = -3.4$, $\beta_1 = 0.02$, $\beta_2 = 0.15$. Interpret the parameters.

(b) Plot the logits of headache as a function of dose for a patient of age 40 and a patient of age 60.

(c) Plot the (approximate) probability of headache as a function of dose for a patient of age 40 and a patient of age 60.

(d) Give the probability of headache for a patient of age 40 when the dose is 5.

Assume now that an interaction effect has to be included and one has the predictor $\eta = \beta_0 + x_1\beta_1 + x_2\beta_2 + x_1x_2\beta_{12}$ with values $\beta_0 = -3.8$, $\beta_1 = 0.02$, $\beta_2 = 0.15, \beta_{12} = 0.005$.

(a) Interpret the effect of dose if age is fixed at 40 (60).

(b) Plot the logits of headache as a function of dose for a patient of age 40 and a patient of age 60.

(c) Plot the (approximate) probability of headache as a function of dose for a patient of age 40 and a patient of age 60.

(d) How can the effect of dose be interpreted if age is fixed at 40 (60)?

Chapter 3

Generalized Linear Models

In this chapter we embed the logistic regression model as well as the classical regression model into the framework of generalized linear models. Generalized linear models (GLMs), which have been proposed by Nelder and Wedderburn (1972), may be seen as a framework for handling several response distributions, some categorical and some continuous, in a unified way. Many of the binary response models considered in later chapters can be seen as generalized linear models, and the same holds for part of the count data models in Chapter 7.

The chapter may be read as a general introduction to generalized linear models; continuous response models are treated as well as categorical response models. Therefore, parts of the chapter can be skipped if the reader is interested in categorical data only. Basic concepts like the deviance are introduced in a general form, but specific forms that are needed in categorical data analysis will also be given in the chapters where the models are considered. Nevertheless, the GLM is useful as a background model for categorical data modeling, and since McCullagh and Nelder's (1983) book everybody working with regression models should be familiar with the basic concept.

3.1 Basic Structure

A generalized linear model is composed from several components. The random component specifies the distribution of the conditional response y_i given $\boldsymbol{x}_i$, whereas the systematic component specifies the link between the expected response and the covariates.

(1) *Random component and distributional assumptions*

Given $\boldsymbol{x}_i$, the y_i's are (conditionally) independent observations from a simple exponential family. This family has a probability density function or mass function of the form

$$f(y_i|\theta_i, \phi_i) = \exp\left\{\frac{y_i\theta_i - b(\theta_i)}{\phi_i} + c(y_i, \phi_i)\right\}, \tag{3.1}$$

where

θ_i	is the natural parameter of the family,
ϕ_i	is a scale or dispersion parameter, and
$b(.)$ and $c(.)$	are specific functions corresponding to the type of the family.

As will be outlined later, several distributions like the binomial, normal, or Poisson distribution are members of the simple exponential family.

(2) *Systematic component*

The systematic component is determined by two structuring components, the linear term and the link between the response and the covariates. The linear part that gives the GLM its name specifies that the variables $\boldsymbol{x}_i$ enter the model in linear form by forming the linear predictor

$$\eta_i = \boldsymbol{x}_i^T \boldsymbol{\beta},$$

where $\boldsymbol{\beta}$ is an unknown parameter vector of dimension p. The relation between the linear part and the conditional expectation $\mu_i = \mathrm{E}(y_i|\boldsymbol{x}_i)$ is determined by the transformation

$$\mu_i = h(\eta_i) = h(\boldsymbol{x}_i^T \boldsymbol{\beta}), \tag{3.2}$$

or, equivalently, by

$$g(\mu_i) = \eta_i = \boldsymbol{x}_i^T \boldsymbol{\beta}, \tag{3.3}$$

where

h	is a known one-to-one *response function*,
g	is the so-called *link function*, that is, the inverse of h.

Equations (3.2) and (3.3) reflect equivalent ways to specify how the mean of the response variable is linked to the linear predictor. The response function h in (3.2) shows how the linear predictor has to be transformed to determine the expected mean. Equation (3.3) shows for which transformation of the mean the model becomes linear. A simple example is the logistic model, where the mean μ_i corresponds to the probability of success π_i. In this case one has the two forms

$$\pi_i = \frac{\exp(\boldsymbol{x}_i^T \boldsymbol{\beta})}{1 + \exp(\boldsymbol{x}_i^T \boldsymbol{\beta})},$$

yielding the response function $h(\eta_i) = \exp(\eta_i)/(1 + \exp(\eta_i))$, and

$$\log(\frac{\pi_i}{1 - \pi_i}) = \boldsymbol{x}_i^T \boldsymbol{\beta}, \tag{3.4}$$

where the link function $g = h^{-1}$ is specified by the logit transformation $g(\pi) = \log(\pi/(1-\pi))$.

Based on the latter form, which corresponds to (3.3), it is seen that a GLM is a linear model for the transformed mean where additionally it is assumed that the response has a distribution in the simple exponential family. A specific generalized linear model is determined by

- the type of the exponential family that specifies the distribution of $y_i|\boldsymbol{x}_i$;
- the form of the linear predictor, that is, the selection and coding of covariates;
- the response or link function.

Before considering the various models that fit into this framework, let us make some remarks on simple exponential families: In simple exponential families the natural parameter is linked to the mean of the distribution. Thus the parameter θ_i may be seen as $\theta_i = \theta(\mu_i)$, where θ is considered as a transformation of the mean. Parametrization of specific distributions most often uses different names and also different sets of parameters; for example, λ_i is often used in the case of the Poisson distribution and the exponential distribution. These parameters determine uniquely the mean μ_i and therefore the natural parameter θ_i.

3.2 Generalized Linear Models for Continuous Responses

3.2.1 Normal Linear Regression

The normal linear regression model is usually given with an error term in the form

$$y_i = \boldsymbol{x}_i^T \boldsymbol{\beta} + \epsilon_i$$

with normal error, $\epsilon_i \sim \mathrm{N}(0, \sigma^2)$. Alternatively, the model may be specified in GLM terminology by

$$y_i | \boldsymbol{x}_i \sim \mathrm{N}(\mu_i, \sigma^2) \qquad \text{and} \qquad \mu_i = \eta_i = \boldsymbol{x}_i^T \boldsymbol{\beta}.$$

The form separates the distribution from the systematic component. While $y_i|\boldsymbol{x}_i \sim \mathrm{N}(\mu_i, \sigma^2)$ assumes that the response is normal with the variance not depending on the observation, the link between the mean and the predictor is provided by assuming $\mu_i = \eta_i = \boldsymbol{x}_i^T \boldsymbol{\beta}$. Thus, the classical linear model uses the identity as a link function. It is easily seen that the normal distribution is within the exponential family by considering

$$f(y) = \exp\left\{-\frac{1}{2}(\frac{y-\mu}{\sigma})^2 - \log(\sqrt{2\pi}\,\sigma)\right\} = \exp\left\{\frac{y\mu - \mu^2/2}{\sigma^2} - \frac{y^2}{2\sigma^2} - \log(\sqrt{2\pi}\,\sigma)\right\}.$$

Therefore, the natural parameter and the function b are given by

$$\theta(\mu) = \mu, \quad b(\theta) = \theta^2/2 = \mu^2/2, \quad \phi = \sigma^2.$$

The separation of random and systematic components makes it easy to allow for alternative links between the mean and the predictors. For example, if the response is income or reaction time, the responses are expected to be positive. Then, a more appropriate link that at least ensures that means are positive is

$$\mu = \exp(\eta) = \exp(\boldsymbol{x}^T \boldsymbol{\beta}).$$

Of course, the influence of the covariates and consequently the interpretation of the parameters differ from those in the classical linear model. In contrast to the linear model,

$$\mu = \boldsymbol{x}^T \boldsymbol{\beta} = x_1\beta_1 + \ldots + x_p\beta_p,$$

where the change of x_j by one unit means an *additive* effect of β_j on the expectation. The modified link

$$\mu = \exp(x_1\beta_1 + \ldots + x_p\beta_p) = e^{x_1\beta_1} \cdot \ldots \cdot e^{x_p\beta_p}$$

specifies that the change of x_j by one unit has a *multiplicative* effect on μ by the factor e^{β_j}, since $e^{(x_j+1)\beta_j} = e^{x_j\beta_j}e^{\beta_j}$. In Figure 3.1 the normal regression model is illustrated for one explanatory variable. The left picture shows the linear model and the right picture the log-link model. The straight line and the curve show the means as functions of x; the densities of the response are shown only at three distinct x-values.

3.2.2 Exponential Distribution

In cases where responses are strictly non-negative, for example, in the analysis of duration time or survival, the normal distribution model is rarely adequate. A classical distribution that is often used when time is the response variable is the exponential distribution

$$f(y) = \lambda e^{-\lambda y} = \exp(-\lambda y + \log(\lambda)), \quad y \geq 0.$$

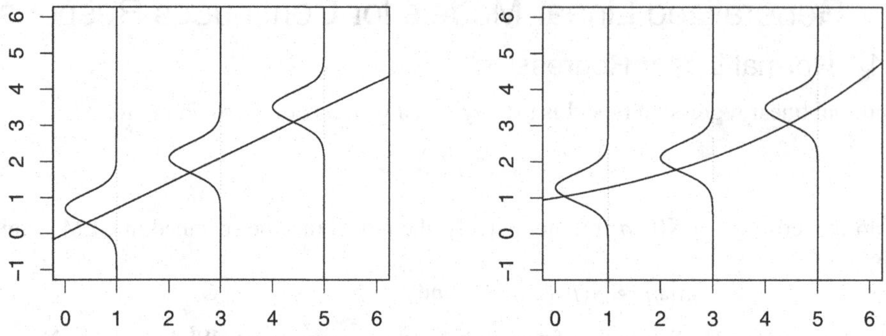

FIGURE 3.1: Normal regression with identity link (left) and with log-link (right).

With $\theta = -\lambda, \phi = 1$, and $b(\theta) = -\log(-\theta)$, the exponential distribution is of the simple exponential family type. Since the expectation and variance of the exponential distribution are given by $1/\lambda$ and $1/\lambda^2$, it is seen that in contrast to the normal distribution the variance increases with increasing expectation. Thus, although there is a fixed link between the expectation and the variance, the distribution model captures an essential property that is often found in real datasets. The so-called *canonical link*, which fulfills $\theta(\mu) = \eta$, is given by

$$g(\mu) = -\frac{1}{\mu} \quad \text{or} \quad h(\eta) = -\frac{1}{\eta}.$$

Since $\mu > 0$, the linear predictor is restricted to $\eta = \boldsymbol{x}^T\boldsymbol{\beta} < 0$, which implies severe restrictions on $\boldsymbol{\beta}$. Therefore, often a more adequate link function is given by the log-link

$$g(\mu) = \log(\mu) \quad \text{or} \quad h(\eta) = \exp(\eta),$$

yielding $\mu = \exp(\eta) = \exp(\boldsymbol{x}^T\boldsymbol{\beta})$.

3.2.3 Gamma-Distributed Responses

Since the exponential distribution is a one-parameter distribution, its flexibility is rather restricted. A more flexible distribution model for non-negative responses like duration or insurance claims is the Γ-distribution. With $\mu > 0$ denoting the expectation and $\nu > 0$ the shape parameter, the Gamma-distribution has the form

$$\begin{aligned} f(y) &= \frac{1}{\Gamma(\nu)}\left(\frac{\nu}{\mu}\right)^{\nu} y^{\nu-1}\exp\left(-\frac{\nu}{\mu}\,y\right) \\ &= \exp\left(\frac{-(1/\mu)y - \log(\mu)}{1/\nu} + \nu\log(\nu) + (\nu-1)\log(y) - \log(\Gamma(\nu))\right). \end{aligned}$$

In exponential family parameterization one obtains the dispersion parameter $\phi = 1/\nu$ and $\theta(\mu) = -1/\mu, b(\theta) = -\log(-\theta)$. In contrast to the exponential distribution, the dispersion parameter is not fixed. While it is $\phi = 1$ for the exponential distribution, it is an additional parameter in the Γ-distribution. As is seen from Figure 3.2, the parameter ν is a shape parameter. For $0 < \nu < 1$, $f(y)$ decreases monotonically, whereas for $\nu > 1$ the density has a mode at $y = \mu - \mu/\nu$ and is positively skewed. Usually the Γ-distribution is abbreviated by $\Gamma(\nu, \alpha)$, where $\alpha = \nu/\mu$. When using the expectation as a parameter, as we did in the specification of the density, we will write $\Gamma(\nu, \frac{\nu}{\mu})$.

The variance of the Gamma-distribution is given by $\text{var}(y) = \nu/\alpha^2 = \mu^2/\nu$. Thus the variance depends strongly on the expectation, an effect that is often found in practice. The dependence may be characterized by the coefficient of variation. The coefficient of variation, given by $c = \sigma/\mu$, is a specific measure of variation that scales the standard deviation by the expectation. For Gamma-distributions, the coefficient of variation for the ith observation is given by $\sigma_i/\mu_i = \mu_i/(\sqrt{\nu}\mu_i) = 1/\sqrt{\nu}$. Since it does not depend on the observation, one may set $c = \sigma_i/\mu_i$. Therefore, the assumption of a Gamma-distribution implies that the coefficient of variation is held constant across observations. It is implicitly assumed that large means are linked to large variances. This is in contrast to the assumption that is often used for normal distributions, when variances are assumed to be constant over observations.

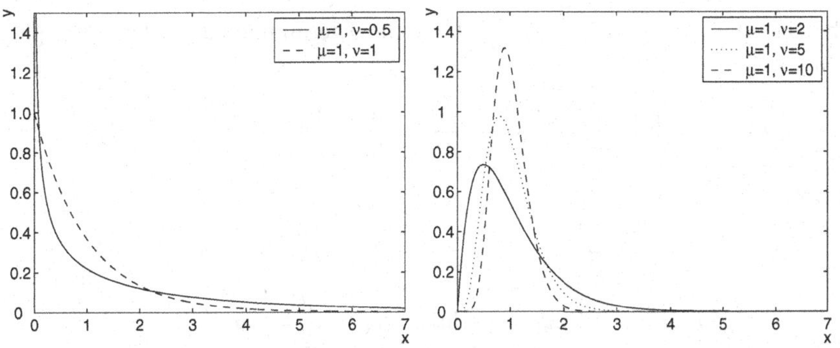

FIGURE 3.2: Gamma-distributions for several μ,ν.

The canonical link for the Gamma-disribution is the same as for the exponential distribution. Figure 3.3 shows the exponential and the Gamma regression model for the log-link function. We can see how the shifting of the mean along the logarithmic function changes the form of the distribution. In contrast to the normal model, where densities are simply shifted, for Gamma-distributed responses, the form of the densities depends on the mean. Moreover, Figure 3.3 shows that densities are positive only for positive x-values. For the normal model shown in Figure 3.1 the log-link ensures that the mean is positive, but nevertheless the model also allows negative values. Thus, for a strictly positive-valued response the normal model is often not a good choice, but, of course, the adequacy of the model depends on the values of x that are modeled and the variance of the response.

3.2.4 Inverse Gaussian Distribution

An alternative distribution with a strictly non-negative response, which can be used to model responses like duration, is the inverse Gaussian-distribution. In its usual form it is given by the density

$$f(y) = \left(\frac{\lambda}{2\pi y^3}\right)^{1/2} \exp\left\{-\frac{\lambda}{2\mu^2 y}(y-\mu)^2\right\}, \qquad y > 0,$$

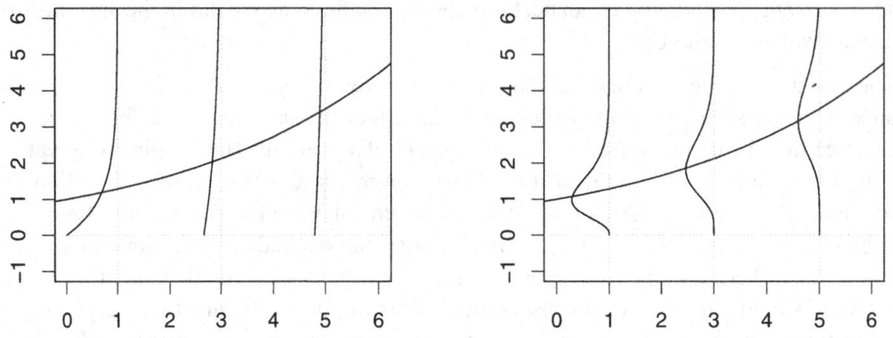

FIGURE 3.3: Exponential (left) and Gamma-distributed (right) regression model with log-link.

with abbreviation $IG(\mu, \lambda)$, where $\mu, \lambda > 0$ are the determining parameters. Straightforward derivation yields

$$f(y) = \exp\left\{\frac{y(-1/(2\mu^2)) + 1/\mu}{1/\lambda} - \frac{\lambda}{2y} - \frac{1}{2}\log(\lambda 2\pi) - \frac{3}{2}\log(y)\right\},$$

and therefore

$$\theta = -\frac{1}{2\mu^2}, \quad b(\theta) = -1/\mu = -\sqrt{-2\theta}, \quad \phi = 1/\lambda,$$
$$c(y, \phi) = -1/(2y\phi) - \frac{1}{2}\log(2\pi/\phi) - \frac{3}{2}\log(y).$$

The canonical link function, for which $\theta(\mu) = \eta$ holds, is given by

$$g(\mu) = -\frac{1}{2\mu^2} \qquad \text{or} \qquad h(\eta) = -\frac{1}{\sqrt{2\eta}},$$

which implies the severe restriction $\eta = \boldsymbol{x}^T\boldsymbol{\beta} > 0$. A link function without these problems is the log-link function $g(\mu) = \log(\mu)$ and thus $ih(\eta) = \exp(\eta)$.

The inverse Gaussian distribution has several interesting properties, including that the ML estimates of the mean μ and the dispersion $1/\lambda$, given by

$$\hat{\mu} = \frac{1}{n}\sum_{i=1}^{n} y_i, \quad \frac{1}{\hat{\lambda}} = \frac{1}{n}\sum\left(\frac{1}{y_i} - \frac{1}{\bar{y}}\right),$$

are independent. This is similar to the normal distribution, for which the sample mean and the sample variance are independent. Based on the independence, Tweedie (1957) suggested an analog of the analysis of variance for nested designs (see also Folks and Chhikara, 1978).

3.3 GLMs for Discrete Responses

3.3.1 Models for Binary Data

The simplest case of a discrete response is when only "success" or "failure" is measured with the outcome $y \in \{0, 1\}$. The Bernoulli distribution has for $y \in \{0, 1\}$ the probability mass function

$$f(y) = \pi^y(1-\pi)^{1-y} = \exp\left\{y \log\left(\frac{\pi}{1-\pi}\right) + \log(1-\pi)\right\},$$

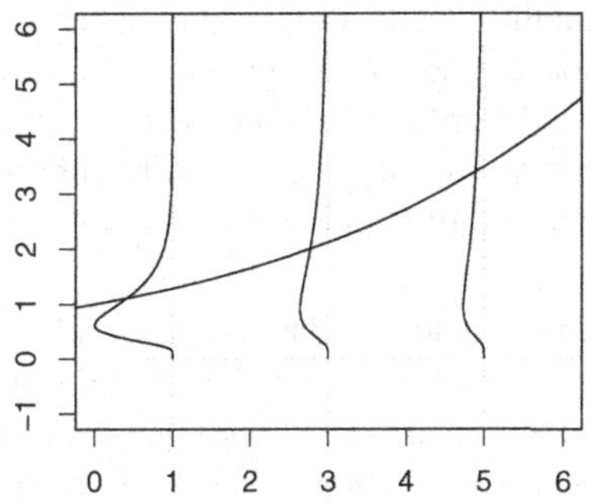

FIGURE 3.4: Inverse Gaussian-distributed model with $\lambda = 3$ and log-link.

where $\pi = P(y = 1)$ is the probability for "success." With $\mu = \pi$ it is an exponential family with $\theta(\pi) = \log(\pi/(1-\pi))$, $b(\theta) = \log(1 + \exp(\theta)) = -\log(1-\pi)$, $\phi = 1$.

The classical link that yields the logit model is

$$\pi = \frac{\exp(\eta)}{1 + \exp(\eta)}, \quad g(\pi) = \log\left(\frac{\pi}{1-\pi}\right)$$

(see Chapter 2). Alternatively, any strictly monotone distribution function F like the normal distribution or extreme value distributions may be used as a response function, yielding $\pi = F(\boldsymbol{x}_i^T \boldsymbol{\beta})$, with the response and link functions given by $h(\eta) = F(\eta)$, $g(\pi) = F^{-1}(\pi)$.

3.3.2 Models for Binomial Data

If experiments that distinguish only between "success" and "failure" are repeated independently, it is natural to consider the number of successes or the proportion as the response variable. For m trials one obtains the binomially distributed response $\tilde{y} \in \{0, \ldots, m\}$. The probability function has the parameters m and the probability π of success in one trial. For $\tilde{y} \in \{0, \ldots, m\}$ it has the form

$$f(\tilde{y}) = \binom{m}{\tilde{y}} \pi^{\tilde{y}} (1-\pi)^{m-\tilde{y}} = \exp\left\{ \frac{\frac{\tilde{y}}{m} \log\left(\frac{\pi}{1-\pi}\right) + \log(1-\pi)}{1/m} + \log\binom{m}{\tilde{y}} \right\}.$$

By considering the proportion of successes $y = \tilde{y}/m$ instead of the number of successes $\tilde{y}$, one obtains an exponential family with the same specifications as for binary responses: $\mu = \mathrm{E}(\tilde{y}/m) = \pi$, $\theta(\pi) = \log(\pi/(1-\pi))$, and $b(\theta) = (1 + \exp(\theta)) = -\log(1-\pi)$. Only the dispersion parameter is different, given as $\phi = 1/m$. The distribution of y has the usual binomial form

$$f(y) = \binom{m}{my} \pi^{my} (1-\pi)^{m-my} = \binom{m}{\tilde{y}} \pi^{\tilde{y}} (1-\pi)^{m-\tilde{y}},$$

but with values $y \in \{0, 1/m, \ldots, 1\}$. Because the support is different from the usual binomial distribution, it is called the *scaled binomial distribution*. It consists of a simple rescaling of the number of successes to proportions and therefore changes the support.

For the binomial distribution the specification of the dispersion parameter differs from that for the other distributions considered here. With indices one has for observation $y_i = \tilde{y}_i/m_i$ the dispersion parameter $\phi_i = 1/m_i$, where m_i is the number of replications. Because m_i is fixed, the dispersion is fixed (and known) but may depend on the observations since the number of replications may vary across observations. In contrast to the other distributions, the dispersion depends on i.

An alternative way of looking at binomial data is by considering them as grouped observations, that is, grouping of replications (see Section 3.5). For the special case $m = 1$ there is no difference between the binomial and the rescaled binomial distributions. Of course, the binary case may be treated as a special case of the binomial case. Consequently, the link and response functions are treated in the same way as in the binary case.

3.3.3 Poisson Model for Count Data

Discrete responses often take the form of counts, for example, the number of insurance claims or case numbers in epidemiology. Contingency tables may be seen as counts that occur as entries in the cells of the table. A simple distribution for count data is the Poisson distribution, which for integer values $y \in \{0, 1, \ldots\}$ and parameter $\lambda > 0$ has the form

$$f(y) = \frac{\lambda^y}{y!} e^{-\lambda} = \exp\{y \log(\lambda) - \lambda - \log(y!)\}.$$

With expectation $\mu = \lambda$ the parameters of the exponential family are given by $\theta(\mu) = \log(\mu)$, $b(\theta) = \exp(\theta) = \mu, \phi = 1$. A sensible choice of the link function should account for the restriction $\lambda > 0$. Thus a widely used link function is the log-link yielding

$$\log(\lambda) = \boldsymbol{x}^T \boldsymbol{\beta} \quad \text{or} \quad \lambda = \exp(\boldsymbol{x}^T \boldsymbol{\beta}), \text{ respectively.}$$

The distribution is shown for three distinct x-values in Figure 3.5. It is seen that differing means imply different shapes of the distribution. While the distribution of the response is skewed for low means, it is nearly symmetric for large values of the mean.

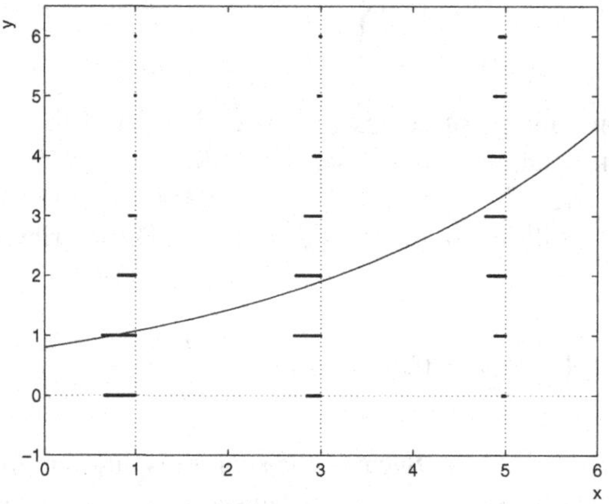

FIGURE 3.5: Poisson regression with log-link.

3.3.4 Negative Binomial Distribution

An alternative distribution for count data is the negative binomial distribution, which has mass function

$$f(\tilde{y}) = \frac{\Gamma(\tilde{y}+\nu)}{\Gamma(\tilde{y}+1)\Gamma(\nu)} \left(\frac{\nu}{\tilde{\mu}+\nu}\right)^{\nu} \left(\frac{\tilde{\mu}}{\tilde{\mu}+\nu}\right)^{y}, \quad y = 0, 1, \ldots, \tag{3.5}$$

where $\nu, \tilde{\mu} > 0$ are parameters. We will use the abbreviation $\mathrm{NB}(\nu, \mu)$. The distribution may be motivated in several ways. It may be seen as a mixture of Poisson distributions in the so-called Gamma-Poisson model. The model assumes that the parameter λ of the Poisson distribution is itself a random variable that is Gamma-distributed with $\lambda \sim \Gamma(\nu, \frac{\nu}{\tilde{\mu}})$ with shape parameter ν and expectation $\tilde{\mu}$. Given λ, it is assumed that $\tilde{y}$ is Poisson-distributed, $\tilde{y}|\lambda \sim \mathrm{P}(\lambda)$. Then the marginal distribution of $\tilde{y}$ is given by (3.5). Since it is often more appropriate to assume that the total counts result from heterogeneous sources with individual parameters, the negative binomial model is an attractive alternative to the Poisson model. From the variance of the Gamma-distribution $\tilde{\mu}^2/\nu$, it is seen that for $\nu \to \infty$ the mixture of Poisson distributions shrinks to just one Poisson distribution and one obtains the Poisson model as the limiting case. The expectation and variance of the negative binomial are given by

$$\mathrm{E}(\tilde{y}) = \tilde{\mu}, \quad \mathrm{var}(\tilde{y}) = \tilde{\mu} + \tilde{\mu}^2/\nu.$$

Thus, for $\nu \to \infty$, one obtains $\mathrm{E}(\tilde{y}) = \mathrm{var}(\tilde{y})$, which is in accordance with the Poisson distribution. The parameter ν may be seen as an additional dispersion parameter that yields a larger variation for small values. Thus it is more appropriate to consider $1/\nu$ as an indicator for the amount of variation.

For integer-valued ν the negative binomial is also considered in the form

$$f(y) = \binom{\nu+y-1}{\nu-1} \pi^{\nu}(1-\pi)^{y} \quad y = 0, 1, \ldots, \tag{3.6}$$

where $\pi = \nu/(\tilde{\mu}+\nu) \in (0,1)$ may be seen as an alternative parameter with a simple interpretation. If independent Bernoulli variables with probability of occurrence π are considered, then the negative binomial distribution (3.6) reflects the probability for the number of trials that are necessary in addition to ν to obtain ν hits. The most familiar case is $\nu = 1$, where one considers the number of trials (plus one) that are necessary until the first hit occurs. The corresponding geometric distribution is a standard distribution, for example, in fertility studies where the number of trials until conception is modeled.

Within the exponential family framework one obtains with $\pi = \nu/(\tilde{\mu}+\nu)$ from (3.5)

$$f(\tilde{y}) = \exp\left\{[\log(\pi) + (\tilde{y}/\nu)\log(1-\pi)]/(1/\nu) + \log\left(\frac{\Gamma(\tilde{y}+\nu)}{\Gamma(\tilde{y}+1)\Gamma(\nu)}\right)\right\}.$$

For fixed ν one has a simple exponential family for the scaled response $y = \tilde{y}/\nu$ and the dispersion $\phi = 1/\nu$. Since $\tilde{y}/\nu$ is considered as the response, one has expectation $\mu = \mathrm{E}(y) = \tilde{\mu}/\nu$ and therefore $\theta(\mu) = \log(1-\pi) = \log(\mu/(\mu+1))$ and $b(\theta) = -\log(1-\exp(\theta))$. The canonical link model that fulfills $\theta(\mu) = \eta$ is given by

$$\log\left(\frac{\mu}{\mu+1}\right) = \eta \quad \text{or} \quad \mu = \frac{\exp(\eta)}{1-\exp(\eta)}.$$

The canonical link may cause problems because, for $\eta \to 0$, one has $\mu \to \infty$. For the log-link, $\log(\mu) = \eta$ or $\mu = \exp(\eta)$; however, the predictor η is not restricted.

The negative binomial response $y = \tilde{y}/\nu$ is scaled by the specified parameter ν. Thus, when treated within the framework of GLMs, the parameter has to be fixed in advance.

3.4 Further Concepts

3.4.1 Means and Variances

The distribution of the responses is assumed to be in the exponential family $f(y_i|\theta_i, \phi_i) = \exp\{(y_i\theta_i - b(\theta_i))/\phi_i + c(y_i, \phi_i)\}$. In the previous sections examples have been given for the dependence of the natural parameters θ_i on μ_i and the parameters that characterize the distribution. For example, for the Bernoulli distribution one obtains $\theta_i = \theta(\mu_i)$ in the form $\theta_i = \log(\mu_i/(1-\mu_i))$, and since $\mu_i = \pi_i$, one has $\theta_i = \log(\pi_i/(1-\pi_i))$.

In general, in the exponential families the mean is directly related to the function $b(\theta_i)$ in the form

$$\mu_i = b'(\theta_i) = \partial b(\theta_i)/\partial\theta, \tag{3.7}$$

and for the variances one obtains

$$\sigma_i^2 = \text{var}(y_i) = \phi_i b''(\theta_i) = \phi_i \partial^2 b(\theta_i)/\partial\theta^2. \tag{3.8}$$

Thus the variances are composed from the dispersion parameter ϕ_i and the so-called *variance function* $b''(\theta_i)$. As is seen from (3.7) and (3.8), in GLMs there is a strict link between the mean μ_i and the variance since both are based on derivatives of $b(\theta)$. Because θ_i depends on the mean through the functional form $\theta_i = \theta(\mu_i)$, the variance function is a function of the mean, that is, $v(\mu_i) = \partial^2 b(\theta_i)/\partial\theta^2$, and the variance can be written as $\sigma_i^2 = \phi_i v(\mu_i)$. For the normal distribution one obtains $v(\mu_i) = 1$, and for the Poisson $v(\mu_i) = \mu_i$ (see Table 3.1).

The link between the mean and variance includes the dispersion parameter ϕ_i. However, the latter is not always an additional parameter. It is fixed for the exponential, Bernoulli, binomial, and Poisson distributions. Only for the normal, Gamma, negative binomial, and inverse Gaussian is it a parameter that may be chosen data-dependently. In all these cases the dispersion has the general form

$$\phi_i = \phi a_i,$$

where a_i is known with $a_i = 1/m_i$ for the binomial distribution and $a_i = 1$ otherwise. The parameter ϕ is the actual dispersion that is known ($\phi = 1$ for exponential, Bernoulli, binomial, Poisson) or an additional parameter. The only case where $a_i \neq 1$ is the binomial distribution, which may be considered as replications of Bernoulli variables.

Means and Variances

$$\mu_i = b'(\theta_i)$$
$$\sigma_i^2 = \phi_i b''(\theta_i) = \phi_i v(\mu_i)$$

TABLE 3.1: Exponential family of distributions.

$$f(y_i|\theta_i,\phi_i) = \exp\left\{\frac{y_i\theta_i - b(\theta_i)}{\phi_i} + c(y_i,\phi_i)\right\},\ \phi_i = \phi a_i$$

(a) Components of the exponential family

Distribution	Notation	μ_i	$\theta(\mu_i)$	$b(\theta_i)$	ϕ	a_i
Normal	$N(\mu_i,\sigma^2)$	μ_i	μ_i	$\theta_i^2/2$	σ^2	1
Exponential	$E(\lambda_i)$	$1/\lambda_i$	$-1/\mu_i$	$-\log(-\theta_i)$	1	1
Gamma	$\Gamma(\nu,\frac{\nu}{\mu_i})$	μ_i	$-1/\mu_i$	$-\log(-\theta_i)$	$\frac{1}{\nu}$	1
Inverse Gaussian	$IG(\mu_i,\lambda)$		$-1/(2\mu_i^2)$	$-(-2\theta_i)^{1/2}$	$1/\lambda$	1
Bernoulli	$B(1,\pi_i)$	π_i	$\log(\frac{\mu_i}{1-\mu_i})$	$\log(1+\exp(\theta_i))$	1	1
Binomial (rescaled)	$B(m_i,\pi_i)/m_i$	π_i	$\log(\frac{\mu_i}{1-\mu_i})$	$\log(1+\exp(\theta_i))$	1	$\frac{1}{m_i}$
Poisson	$P(\lambda_i)$	λ_i	$\log(\mu_i)$	$\exp(\theta_i)$	1	1
Negative binomial (rescaled)	$NB(\nu,\frac{\nu(1-\pi_i)}{\pi_i})/\nu$	$\frac{\nu(1-\pi_i)}{\pi_i}/\nu$	$\log(\frac{\mu_i}{\mu_i+1})$	$-\log(1-e^{\theta})$	$\frac{1}{\nu}$	1

(b) Expectation and variance

Distribution	$\mu_i = b'(\theta_i)$	var. fct. $b''(\theta_i)$	variance $\phi_i\, b''(\theta_i)$
Normal	$\mu_i = \theta_i$	1	σ^2
Exponential	$\mu_i = -\frac{1}{\theta_i}$	μ_i^2	μ_i^2
Gamma	$\mu_i = -\frac{1}{\theta_i}$	μ_i^2	$\frac{\mu_i^2}{\nu}$
Inverse Gaussian	$\mu_i = (-2\theta)^{-1/2}$	μ_i^3	μ_i^3/λ
Bernoulli	$\mu_i = \frac{\exp(\theta_i)}{1+\exp(\theta_i)}$	$\pi_i(1-\pi_i)$	$\pi_i(1-\pi_i)$
Binomial	$\mu_i = \frac{\exp(\theta_i)}{1+\exp(\theta_i)}$	$\pi_i(1-\pi_i)$	$\frac{1}{m_i}\pi_i(1-\pi_i)$
Poisson	$\lambda_i = \exp(\theta_i)$	λ_i	λ_i
Negative binomial	$\mu_i = \frac{\exp(\theta_i)}{1-\exp(\theta_i)}$	$\mu(1+\mu)$	$\frac{\mu(1+\mu)}{\nu}$

3.4.2 Canonical Link

The choice of the link function depends on the distribution of the response. For example, if y is non-negative, a link function is appropriate that specifies non-negative means without restricting the parameters. For each distribution within the simple exponential family there is one link function that has some technical advantages, the so-called canonical link. It links the linear predictor directly to the canonical parameter in the form

$$\theta_i = \boldsymbol{x}_i^T\boldsymbol{\beta}.$$

Since θ_i is determined as a function $\theta(\mu_i)$, the canonical link g may be derived from the general form $g(\mu_i) = \boldsymbol{x}_i^T\boldsymbol{\beta}$ as the transformation that transforms μ_i to θ_i. In Table 3.1 the canonical links are given, and one obtains, for example,

$$\begin{aligned} g(\mu) &= \mu && \text{for the normal distribution,}\\ g(\mu) &= \log(\pi/(1-\pi)) && \text{for the Bernoulli distribution,}\\ g(\mu) &= -1/\mu && \text{for the Gamma-distribution.}\end{aligned}$$

The last example shows that the canonical link might not always be the best choice because $-1/\mu_i = \boldsymbol{x}_i^T\boldsymbol{\beta}$ or $\mu_i = -1/\boldsymbol{x}_i^T\boldsymbol{\beta}$ implies severe restrictions on $\boldsymbol{\beta}$ arising from the restriction that μ_i has to be non-negative.

3.4.3 Extensions Including Offsets

When modeling insurance claims or numbers of cases y_i within a time interval, the time interval may depend on i and therefore may vary across observations. With Δ_i denoting the underlying time interval for observation y_i, one may assume Poisson-distributed responses $y_i \sim \mathrm{P}(\Delta_i\lambda_i)$, where λ_i is the underlying intensity for one unit of time, which may be any time unit like minutes, days, or months. A sensible approach to modeling will not specify the expectation of y_i, which is $\Delta_i\lambda_i$, but the intensity λ_i in dependence on covariates and include the time intervals as known constants. The model

$$\lambda_i = \exp(\boldsymbol{x}_i^T\boldsymbol{\beta})$$

yields for the expectation $\mu_i = \mathrm{E}(y_i)$

$$\mu_i = \Delta_i\lambda_i = \exp(\log(\Delta_i) + \boldsymbol{x}_i^T\boldsymbol{\beta}).$$

Since y_i follows a Poisson distribution one has a GLM but with a known additive constant in the predictor. Constants of this type are called *offsets*; they are not estimated but considered as fixed and known. In the special case where all observations are based on the same length of the time interval, that is, $\Delta_i = \Delta$, the offset $\log(\Delta_i)$ is omitted because it cannot be distinguished from the intercept within $\boldsymbol{x}_i^T\boldsymbol{\beta}$. For more examples see Section 7.4.

3.5 Modeling of Grouped Data

In the previous sections observations have been given in the *ungrouped* form $(y_i, \boldsymbol{x}_i)$, $i = 1, \dots, n$. Often, for example, if covariates are categorial or in experimental studies, several of the covariate values $\boldsymbol{x}_1, \dots, \boldsymbol{x}_n$ will be identical. Thus the responses for fixed covariate vectors may be considered as replications with identical mean. By relabeling the data one obtains the form

$$(y_{ij}, \boldsymbol{x}_i), \quad i = 1, \dots, N, \quad j = 1, \dots, n_i,$$

where observations $y_{i1}, \dots, y_{in_i}$ have a fixed covariate vector $\boldsymbol{x}_i$ with n_i denoting the sample size at covariate value $\boldsymbol{x}_i$, yielding the total sum of observations $n = n_1 + \dots + n_N$. Since means depend on covariates, one has

$$\mu_i = \mu_{ij} = \mathrm{E}(y_{ij}) = h(\boldsymbol{x}_i^T\boldsymbol{\beta}), \quad j = 1, \dots, n_i,$$

and also the natural parameter $\theta_i = \theta(\mu_i)$ depends on i only. Let the dispersion parameter $\phi_i = \phi$ be constant over replications $y_{ij}, j = 1, \dots, n_i$. Then one obtains for the mean over individual responses at covariate value $\boldsymbol{x}_i$, $\bar{y}_i = \frac{1}{n_i}\sum_{j=1}^{n_i} y_{ij}$, the density or mass function

$$f(\bar{y}_i) = \exp\left\{\frac{(\bar{y}_i\theta_i - b(\theta_i))}{\phi/n_i} + \bar{c}(y_{i1}, \dots, y_{in_i}, \phi)\right\}, \tag{3.9}$$

where $\bar{c}(\ldots)$ is a modified normalizing function (Exercise 3.3). Thus, the mean has an exponential family distribution with the same natural parameter θ_i and function $b(\theta_i)$ as for ungrouped data. However, the dispersion parameter has changed. With the value

$$\phi_i = \phi/n_i$$

it reflects the reduced dispersion due to considering the mean $\bar{y}_i$ across n_i observations. For grouped data one has

$$\mu_i = \mathrm{E}(\bar{y}_i) = \mathrm{E}(y_{ij}) = b'(\theta_i)$$

$$\sigma_i^2 = \mathrm{var}(\bar{y}_i) = \frac{1}{n_i}\,\mathrm{var}(y_{ij}) = \frac{\phi}{n_i} b''(\theta_i).$$

The variance function $v(\mu_i) = b''(\theta(\mu_i))$ is the same as for ungrouped observations, but the variance is $\mathrm{var}(\bar{y}_i) = (\phi/n_i)v(\mu_i)$.

Consequently, one obtains for grouped data, considering $\bar{y}_1, \ldots, \bar{y}_N$ as responses, a GLM with the dispersion given by $\phi_i = \phi/n_i$. This is exactly what happens in the transition from the Bernoulli response to the scaled binomial response. The binomial response may be seen as grouped data from Bernoulli responses with local sample size given by n_i.

In the grouped case we will use N as the number of grouped observations with local sample sizes $n_1, \ldots, n_N$. For ungrouped data the number of observations is n. In both cases the dispersion parameter is denoted by ϕ_i.

In general, the mean y_i over replications $y_{ij}, \ldots, y_{in_i}$ does not have the same type of distribution as the original observations y_{ij}. For example, in Poisson or binomially distributed responses the mean is not integer-valued and thus is not Poisson or binomially distributed, respectively. However, for these distributions the sum $\tilde{y}_i = \sum_{j=1}^{n_i} y_{ij}$ has the same type of distribution as the replications.

3.6 Maximum Likelihood Estimation

For GLMs the most widely used method of estimation is maximum likelihood. The basic principle is to construct the likelihood of the unknown parameters for the sample data, where the likelihood represents the joint probability or probability density of the observed data, considered as a function of the unknown parameters. Maximum likelihood (ML) estimation for all GLMs has a common form. This is due to the assumption that the responses come from an exponential family. The essential feature of the simple exponential family with density $f(y_i|\theta_i, \phi_i) = \exp\{(y_i\theta_i - b(\theta_i))/\phi_i + c(y_i, \phi_i)\}$ is that the mean and variance are given by

$$\mathrm{E}(y_i) = \partial b(\theta_i)/\partial\theta, \quad \mathrm{var}(y_i) = \phi_i \partial^2(\theta_i)/\partial\theta^2,$$

where the parameterization is in the canonical parameter θ_i. As will be seen, the likelihood and its logarithm, the log-likelihood, are determined by the assumed mean and variance.

Log-Likelihood and Score Function

From the exponential family one obtains for independent observations $y_1, \ldots, y_n$ the log-likelihood

$$l(\boldsymbol{\beta}) = \sum_{i=1}^{n} l_i(\theta_i) = \sum_{i=1}^{n} (y_i\theta_i - b(\theta_i))/\phi_i,$$

where the term $c(y_i, \phi_i)$ is omitted because it does not depend on θ_i and therefore not on $\boldsymbol{\beta}$. For the maximization of the log-likelihood one computes the derivation $\boldsymbol{s}(\boldsymbol{\beta}) = \partial l(\boldsymbol{\beta})/\partial\boldsymbol{\beta}$, which is called the score function. For the computation it is useful to consider the parameters as resulting from transformations in the form $\theta_i = \theta(\mu_i), \mu_i = h(\eta_i), \eta_i = \boldsymbol{x}_i^T\boldsymbol{\beta}$. One has the transformation structure

$$\eta_i \quad \underset{g=h^{-1}}{\overset{h}{\rightleftarrows}} \quad \mu_i \quad \underset{\mu=\theta^{-1}}{\overset{\theta}{\rightleftarrows}} \quad \theta_i$$

yielding $\theta_i = \theta(\mu_i) = \theta(h(\eta_i))$. Then the score function $\boldsymbol{s}(\boldsymbol{\beta}) = \partial l(\boldsymbol{\beta})/\partial\boldsymbol{\beta}$ is given by

$$\boldsymbol{s}(\boldsymbol{\beta}) = \frac{\partial l(\boldsymbol{\beta})}{\partial\boldsymbol{\beta}} = \sum_{i=1}^{n} \frac{\partial l_i(\theta_i)}{\partial\theta}\frac{\partial\theta(\mu_i)}{\partial\mu}\frac{\partial h(\eta_i)}{\partial\eta}\frac{\partial\eta_i}{\partial\boldsymbol{\beta}}.$$

With $\mu_i = \mu(\theta_i)$ denoting the transformation of θ_i into μ_i one obtains

$$\frac{\partial l_i}{\partial\theta} = (y_i - b'(\theta_i))/\phi_i = (y_i - \mu_i)/\phi_i,$$
$$\frac{\partial\theta(\mu_i)}{\partial\mu} = \left(\frac{\partial\mu(\theta_i)}{\partial\theta}\right)^{-1} = \left(\frac{\partial^2 b(\theta_i)}{\partial\theta^2}\right)^{-1} = \phi_i/\operatorname{var}(y_i),$$
$$\frac{\partial\eta_i}{\partial\boldsymbol{\beta}} = \boldsymbol{x}_i,$$

and therefore the score function

$$\boldsymbol{s}(\boldsymbol{\beta}) = \sum_{i=1}^{n} s_i(\boldsymbol{\beta}) = \sum_{i=1}^{n} \boldsymbol{x}_i \frac{\partial h(\eta_i)}{\partial\eta}\frac{(y_i - \mu_i)}{\operatorname{var}(y_i)}.$$

With $\sigma_i^2 = \phi_i v(\mu_i) = \operatorname{var}(y_i)$, the estimation equation $\boldsymbol{s}(\hat{\boldsymbol{\beta}}) = \mathbf{0}$ has the form

$$\sum_{i=1}^{n} \boldsymbol{x}_i \frac{\partial h(\eta_i)}{\partial\eta}\frac{(y_i - \mu_i)}{\phi_i v(\mu_i)} = \mathbf{0}. \tag{3.10}$$

In (3.10) the response (or link) function is found in the specification of the mean $\mu_i = h(\boldsymbol{x}_i^T\boldsymbol{\beta})$ and in the derivative $\partial h(\eta_i)/\partial\eta$, whereas from higher moments of the distribution of y_i only the variance $\sigma_i^2 = \phi_i v(\mu_i)$ is needed. Since $\phi_i = \phi a_i$, the dispersion parameter ϕ may be canceled out and the estimate $\hat{\boldsymbol{\beta}}$ does not depend on ϕ.

For the canonical link the estimation equation simplifies. Since $\theta_i = \eta_i = \boldsymbol{x}_i^T\boldsymbol{\beta}$, the score function reduces to $\boldsymbol{s}(\boldsymbol{\beta}) = \sum_i (\partial l_i/\partial\theta)(\partial\eta_i/\partial\boldsymbol{\beta})$ and one obtains

$$\boldsymbol{s}(\boldsymbol{\beta}) = \sum_{i=1}^{n} \boldsymbol{x}_i(y_i - \mu_i)/\phi_i.$$

In particular, one has $\partial h(\eta_i)/\partial\eta = \operatorname{var}(y_i)/\phi_i$ if the canonical link is used.

In matrix notation the score function is given by

$$s(\boldsymbol{\beta}) = \boldsymbol{X}^T\boldsymbol{D}\,\boldsymbol{\Sigma}^{-1}(\boldsymbol{y} - \boldsymbol{\mu}),$$

where $\boldsymbol{X}^T = (\boldsymbol{x}_1, \ldots, \boldsymbol{x}_n)$ is the design matrix, $\boldsymbol{D} = \operatorname{Diag}(\partial h(\eta_1)/\partial\eta, \ldots, \partial h(\eta_n)/\partial\eta)$ is the diagonal matrix of derivatives, $\boldsymbol{\Sigma} = \operatorname{Diag}(\sigma_1^2, \ldots, \sigma_n^2)$ is the covariance matrix, and $\boldsymbol{y}^T = (y_1, \ldots, y_n)$, $\boldsymbol{\mu}^T = (\mu_1, \ldots, \mu_n)$ are the vectors of observations and means. Sometimes it is useful to combine $\boldsymbol{D}$ and $\boldsymbol{\Sigma}$ into the weight matrix $\boldsymbol{W} = \boldsymbol{D}\,\boldsymbol{\Sigma}^{-1}\boldsymbol{D}^T$, yielding $s(\boldsymbol{\beta}) = \boldsymbol{X}^T\boldsymbol{W}\boldsymbol{D}^{-1}(\boldsymbol{y} - \boldsymbol{\mu})$ and $\boldsymbol{F}(\boldsymbol{\beta}) = \boldsymbol{X}^T\boldsymbol{W}\boldsymbol{X}$.

Information Matrix

In maximum likelihood theory the information matrix determines the asymptotic variance. The *observed information matrix* is given by

$$\boldsymbol{F}_{obs}(\boldsymbol{\beta}) = -\frac{\partial^2 l(\boldsymbol{\beta})}{\partial \boldsymbol{\beta} \partial \boldsymbol{\beta}^T} = \left(-\frac{\partial^2 l(\boldsymbol{\beta})}{\partial \beta_i \partial \beta_j} \right)_{i,j}.$$

Its explicit form shows that it depends on the observations and therefore is random. The *(expected) information* or *Fisher matrix*, which is not random, is given by

$$\boldsymbol{F}(\boldsymbol{\beta}) = \mathrm{E}(\boldsymbol{F}_{obs}(\boldsymbol{\beta})).$$

For the derivation it is essential that $\mathrm{E}(\boldsymbol{s}(\boldsymbol{\beta})) = \boldsymbol{0}$ and that $\mathrm{E}(-\partial^2 l_i / \partial \boldsymbol{\beta} \partial \boldsymbol{\beta}^T) = \mathrm{E}((\partial l_i / \partial \boldsymbol{\beta})$ $(\partial l_i / \partial \boldsymbol{\beta}^T))$, which holds under general assumptions (see, for example, Cox and Hinkley, 1974). Thus one obtains

$$\begin{aligned}
\boldsymbol{F}(\boldsymbol{\beta}) &= \mathrm{E}\left(\sum_{i=1}^{n} \boldsymbol{s}_i(\boldsymbol{\beta}) \boldsymbol{s}_i(\boldsymbol{\beta})^T \right) = \mathrm{E}\left(\sum_{i=1}^{n} \boldsymbol{x}_i \boldsymbol{x}_i^T \left(\frac{\partial h(\eta_i)}{\partial \eta} \right)^2 \frac{(y_i - \mu_i)^2}{\mathrm{var}(y_i)^2} \right) \\
&= \sum_{i=1}^{n} \boldsymbol{x}_i \boldsymbol{x}_i^T \left(\frac{\partial h(\eta_i)}{\partial \eta} \right)^2 / \sigma_i^2,
\end{aligned}$$

where $\sigma_i^2 = \mathrm{var}(y_i)$. By using the design matrix $\boldsymbol{X}$ one obtains the information matrix $\boldsymbol{F}(\boldsymbol{\beta})$ in the form

$$\boldsymbol{F}(\boldsymbol{\beta}) = \boldsymbol{X}^T \boldsymbol{W} \boldsymbol{X},$$

where $\boldsymbol{W} = \mathrm{Diag}\left(\left(\frac{\partial h(\eta_1)}{\eta} \right)^2 / \sigma_1^2, \dots, \left(\frac{\partial h(\eta_n)}{\partial \eta} \right)^2 / \sigma_n^2 \right)$ is a diagonal weight matrix that has the matrix form $\boldsymbol{W} = \boldsymbol{D}\, \boldsymbol{\Sigma}^{-1} \boldsymbol{D}^T$.

For the canonical link the corresponding simpler form is

$$\boldsymbol{F}(\boldsymbol{\beta}) = \sum_{i=1}^{n} \boldsymbol{x}_i \boldsymbol{x}_i^T \sigma_i^2 / \phi_i^2 = \boldsymbol{X}^T \boldsymbol{W} \boldsymbol{X},$$

with weight matrix $\boldsymbol{W} = (\sigma_1^2/\phi_1^2, \dots, \sigma_n^2/\phi_n^2)$ and the observed information is identical to the information matrix, $\boldsymbol{F}_{obs}(\boldsymbol{\beta}) = \boldsymbol{F}(\boldsymbol{\beta})$. It is immediately seen that for the normal distribution model with a (canonical) identity link one has with $\phi_i = \sigma_i^2 = \sigma^2$ the familiar form

$$\boldsymbol{F}(\boldsymbol{\beta}) = \sum_{i=1}^{n} \boldsymbol{x}_i \boldsymbol{x}_i^T / \sigma^2 = \boldsymbol{X}^T \boldsymbol{X} / \sigma^2.$$

In this case it is well known that the covariance of the estimator $\hat{\boldsymbol{\beta}}$ is given by $\mathrm{cov}(\hat{\boldsymbol{\beta}}) = \sigma^2 (\boldsymbol{X}^T \boldsymbol{X})^{-1} = \boldsymbol{F}(\boldsymbol{\beta})^{-1}$. For GLMs the result holds only asymptotically ($n \to \infty$). With

$$\widehat{\mathrm{cov}}(\hat{\boldsymbol{\beta}}) \approx (\boldsymbol{X}' \hat{\boldsymbol{W}} \boldsymbol{X})^{-1},$$

where $\hat{\boldsymbol{W}}$ means the evaluation of $\boldsymbol{W}$ at $\hat{\boldsymbol{\beta}}$, that is, $\partial h(\eta_i)/\partial \eta$ is replaced by $\partial h(\hat{\eta}_i)/\partial \eta$, $\hat{\eta}_i = \boldsymbol{x}_i^T \hat{\boldsymbol{\beta}}$, and $\sigma_i^2 = \phi_i v(\hat{\mu}_i)$, $\hat{\mu}_i = h(\hat{\eta}_i)$.

It should be noted that in the grouped observations case the form of the likelihood, score function, and Fisher matrix are the same; only the summation index n has to be replaced by N.

Log-Likelihood

$$l(\boldsymbol{\beta}) = \sum_{i=1}^{n}(y_i\theta_i - b(\theta_i))/\phi_i$$

Score Function

$$\begin{aligned} \boldsymbol{s}(\boldsymbol{\beta}) = \frac{\partial l(\boldsymbol{\beta})}{\partial \boldsymbol{\beta}} &= \sum_{i=1}^{n} \boldsymbol{x}_i \frac{\partial h(\eta_i)}{\partial \eta}\frac{(y_i - \mu_i)}{\sigma_i^2} \\ &= \boldsymbol{X}^T\boldsymbol{D}\boldsymbol{\Sigma}^{-1}(\boldsymbol{y}-\boldsymbol{\mu}) = \boldsymbol{X}^T\boldsymbol{W}\boldsymbol{D}^{-1}(\boldsymbol{y}-\boldsymbol{\mu}) \end{aligned}$$

Information Matrix

$$\begin{aligned} \boldsymbol{F}(\boldsymbol{\beta}) = \mathrm{E}\left(-\frac{\partial^2 l(\boldsymbol{\beta})}{\partial\boldsymbol{\beta}\partial\boldsymbol{\beta}^T}\right) &= \sum_{i=1}^{n}\boldsymbol{x}_i\boldsymbol{x}_i^T\left(\frac{\partial h(\eta_i)}{\partial\eta}\right)^2/\sigma_i^2 \\ &= \boldsymbol{X}^T\boldsymbol{D}\boldsymbol{\Sigma}^{-1}\boldsymbol{D}\boldsymbol{X} = \boldsymbol{X}^T\boldsymbol{W}\boldsymbol{X} \end{aligned}$$

If ϕ is an unknown (normal, Gamma-distribution), the moments estimate is

$$\hat{\phi} = \frac{1}{n-p}\sum_{i=1}^{n}\frac{(y_i-\hat{\mu}_i)^2}{v(\hat{\mu}_i)}.$$

For the normal model $\hat{\phi}$ reduces to the usual unbiased and consistent estimates $\hat{\phi} = \hat{\sigma}^2 = \sum_i (y_i - \hat{\mu}_i)^2/(n-p)$. For the Gamma-distribution one obtains

$$\hat{\phi} = \frac{1}{\hat{\nu}} = \frac{1}{n-p}\sum_{i=1}^{n}\left(\frac{y_i-\hat{\mu}_i}{\hat{\mu}_i}\right)^2.$$

In the approximation ϕ has to be replaced by $\hat{\phi}$ when computing $\boldsymbol{F}(\hat{\boldsymbol{\beta}})$. The likelihood, score function, and Fisher matrix are summarized in a box. In the matrix form derivations are collected in the matrix $\boldsymbol{D} = \mathrm{Diag}\,(\partial h(\eta_1)/\partial\eta, \dots, \partial h(\eta_n)/\partial\eta)$. By using $\boldsymbol{W}$ and $\boldsymbol{D}$, the dependence on $\boldsymbol{\beta}$ is suppressed, and actually one has $\boldsymbol{W} = \boldsymbol{W}(\boldsymbol{\beta}), \boldsymbol{D} = \boldsymbol{D}(\boldsymbol{\beta})$.

Under regularity conditions the ML estimate $\hat{\boldsymbol{\beta}}$ exists and is unique asymptotically ($n \to \infty$). It is consistent and the distribution may be approximated by a normal distribution with the covariance given by the inverse Fisher matrix. More precisely, under assumptions that ensure the convergence of $\boldsymbol{F}(\hat{\boldsymbol{\beta}})/n$ to a limit $\boldsymbol{F}_0(\hat{\boldsymbol{\beta}})$, one obtains for $\sqrt{n}(\hat{\boldsymbol{\beta}} - \boldsymbol{\beta})$ asymptotically a normal distribution $\mathrm{N}(\boldsymbol{0}, \boldsymbol{F}_0(\hat{\boldsymbol{\beta}})^{-1})$. For finite n one uses the approximation $\mathrm{cov}(\hat{\boldsymbol{\beta}}) \approx \boldsymbol{F}_0(\hat{\boldsymbol{\beta}})^{-1}/n$, which is approximated by $\boldsymbol{F}(\hat{\boldsymbol{\beta}})^{-1}$. For regularity conditions see Haberman (1977) and Fahrmeir and Kaufmann (1985). Bias correction by approximation of the first-order bias of ML estimates was investigated by Cordeiro and McCullagh (1991) and Firth (1993).

Approximation

$$\hat{\boldsymbol{\beta}} \overset{a}{\sim} \mathrm{N}\,(\boldsymbol{\beta}, \boldsymbol{F}(\hat{\boldsymbol{\beta}})^{-1}),$$

$$\boldsymbol{F}(\hat{\boldsymbol{\beta}}) = \boldsymbol{X}^T\hat{\boldsymbol{W}}\boldsymbol{X} = \textstyle\sum_{i=1}^{n}\boldsymbol{x}_i\boldsymbol{x}_i^T\left(\frac{\partial h(\hat{\eta}_i)}{\partial\eta}\right)^2/(\hat{\phi}v(\hat{\mu}_i))$$

The unifying concept of GLMs may be seen in the common form of the log-likelihood, the score function (which determines the estimation equation), and the information matrix (which determines the variances of estimators). Specific models result from specific choices of

- the link or response function, yielding the derivative matrix $\boldsymbol{D}$, which contains $\partial h(\eta_i)/\partial \eta$;
- the distribution, yielding the covariance matrix $\boldsymbol{\Sigma}$, which contains $\sigma_i^2 = \text{var}(y_i)$;
- the explanatory variables, which determine the design matrix $\boldsymbol{X}$.

In GLMs these constituents may be chosen freely. In principle, any link function can be combined with any distribution and any set of explanatory variables. Of course there are combinations of links and distributions that are more sensible than others.

3.7 Inference

Main questions in inference concern

- the adequacy of the model or goodness-of-fit of the model,
- the relevance of explanatory variables,
- the explanatory value of the model.

In the following these questions are considered in a different order. First the deviance is introduced, which measures the discrepancy between the observations and the fitted model. The deviance is a tool for various purposes. The relevance of the explanatory variables may be investigated by comparing the deviance of two models, the model that contains the variable in question and the model where this variable is omitted. Moreover, for grouped observations the deviance may be used as a goodness-of-fit statistic.

3.7.1 The Deviance

When fitting a GLM one wants some measure for the discrepancy between the fitted model and the observations. The deviance is a measure for the discrepancy that is based on the likelihood ratio statistic for comparing nested models. The nested models that are investigated are the GLM that is under investigation and the most general possible model. This so-called *saturated model* fits the data exactly by assuming as many parameters as observations.

Let $l(\boldsymbol{y}; \hat{\boldsymbol{\mu}}, \phi)$ denote the maximum of the log-likelihood of the model where $\boldsymbol{y}^T = (y_1, \dots, y_n)$ represents the data, $\hat{\boldsymbol{\mu}}^T = (\hat{\mu}_1, \dots, \hat{\mu}_n)$, $\hat{\mu}_i = h(\boldsymbol{x}_i^T \hat{\boldsymbol{\beta}})$ represent the fitted values based on the ML estimate $\hat{\boldsymbol{\beta}}$; and the dispersion of observations has the form $\phi_i = \phi a_i$ with known a_i. For the saturated model that matches the data exactly one has $\hat{\boldsymbol{\mu}} = \boldsymbol{y}$ and the log-likelihood is given by $l(\boldsymbol{y}; \boldsymbol{y}, \phi)$. With $\theta(\hat{\mu}_i), \theta(y_i)$ denoting the canonical parameters of the GLM under investigation and the saturated model, respectively, the *deviance* is given by

$$\begin{aligned} D(\boldsymbol{y}, \hat{\boldsymbol{\mu}}) &= -\phi 2\{l(\boldsymbol{y}; \hat{\boldsymbol{\mu}}, \phi) - l(\boldsymbol{y}; \boldsymbol{y}, \phi)\} \\ &= 2\sum_{i=1}^{n} \{y_i(\theta(y_i) - \theta(\hat{\mu}_i)) - (b(\theta(y_i)) - b(\theta(\hat{\mu}_i)))\}/a_i. \end{aligned}$$

$D(\boldsymbol{y}, \hat{\boldsymbol{\mu}})$ is known as deviance of the model under consideration while $D^+(\boldsymbol{y}, \hat{\boldsymbol{\mu}}) = D(\boldsymbol{y}, \hat{\boldsymbol{\mu}})/\phi$ is the so-called *scaled deviance*. The deviance is linked to the likelihood ratio statistic

TABLE 3.2: Deviances for several distributions.

Distribution	Deviance
Normal	$D(\boldsymbol{y}, \hat{\boldsymbol{\mu}}) = \sum_{i=1}^{n} (y_i - \hat{\mu}_i)^2$
Gamma	$D(\boldsymbol{y}, \hat{\boldsymbol{\mu}}) = 2\sum_{i=1}^{n} -\log\left(\frac{y_i}{\hat{\mu}_i}\right) + \left[\frac{(y_i - \hat{\mu}_i)}{\hat{\mu}_i}\right]$
Inverse Gaussian	$D(\boldsymbol{y}, \hat{\boldsymbol{\mu}}) = \sum_{i=1}^{n} (y_i - \hat{\mu}_i)^2/(\hat{\mu}_i^2 y_i)$
Bernoulli	$D(\boldsymbol{y}, \hat{\boldsymbol{\mu}}) = 2\sum_{i=1}^{n} y_i \log\left(\frac{y_i}{\hat{\mu}_i}\right) + (1 - y_i)\log\left(\frac{1 - y_i}{1 - \hat{\mu}_i}\right)$
Poisson	$D(\boldsymbol{y}, \hat{\boldsymbol{\mu}}) = 2\sum_{i=i}^{n} y_i \log\left(\frac{y_i}{\hat{\mu}_i}\right) - [(y_i - \hat{\mu}_i)]$

$\lambda = -2\{l(\boldsymbol{y}; \hat{\boldsymbol{\mu}}, \phi) - l(\boldsymbol{y}; \boldsymbol{y}, \phi)\}$, which compares the current model to the saturated model by $D(\boldsymbol{y}, \hat{\boldsymbol{\mu}}) = \phi\lambda$.

Simple derivation yields the deviances given in Table 3.2. For the normal model, the deviance is identical to the error or residual sum of squares SSE and the scaled deviance takes the form SSE $/\sigma^2$. For the Bernoulli distribution, one has $\theta(\mu_i) = \log(\mu_i/(1 - \mu_i))$ and one obtains $D(\boldsymbol{y}, \hat{\boldsymbol{\mu}}) = 2\sum_i d(y_i, \hat{\pi}_i)$, where $d(y_i, \hat{\pi}_i) = -\log(1 - |y_i - \hat{\pi}_i|)$ (for more details see Section 4.2). In the cases of the Poisson and the Gamma deviances, the last term given in brackets [...] can be omitted if the model includes a constant term because then the sum over the terms is zero.

The deviance as a measure of discrepancy between the observations and the fitted model may be used in an informal way to compare the fit of two models. For example, two models with the same predictor but differing link functions can be compared by considering which one has the smaller deviance. However, there is no simple way to interpret the difference between the deviances of these models. This is different if the difference of deviances is used for nested models, for example, to investigate the relevance of terms in the linear predictor. The comparison of models with and without the term in question allows one to make a decision based on significance tests with a known asymptotic distribution. The corresponding analysis of deviance (see Section 3.7.2) generalizes the analysis of variance, which is in common use for normal linear models.

For ungrouped data some care has to be taken in the interpretation as a goodness-of-fit measure. As an absolute measure of goodness-of-fit, which allows one to decide if the model has satisfactory fit or not, the deviance for ungrouped observations is appropriate only in special cases. For the interpretation of the value of the deviance it would be useful to have a benchmark in the form of an asymptotic distribution. Since the deviance may be derived as a likelihood ratio statistic, it is tempting to assume that the deviance is asymptotically χ^2-distributed. However, in general, the deviance *does not* have an asymptotic χ^2-distribution in the limit for $n \to \infty$. Standard asymptotic theory of likelihood ratio statistics for nested models assumes that the ranks of the design matrices that build the two models and therefore the degrees of freedom are fixed for increasing sample size. In the present case this theory does not apply because the degrees of freedom of the saturated model increase with n. This is already seen in the case of the normal distribution, where $D(\boldsymbol{y}, \hat{\boldsymbol{\mu}}) = (n - p)\hat{\sigma}^2 \sim \sigma^2\chi^2(n - p)$. For $n \to \infty$, the limiting distribution does not have a χ^2- distribution with fixed degrees of freedom. Similar effects occur for binary data.

This is different if one considers the deviance of the binomial distribution or the Poisson distribution. The binomial distribution may be seen as a replication version of the Bernoulli distribution. Thus, when the number of replications increases, $n_i \to \infty$, the proportion y_i is asymptotically normally distributed. For the Poisson distribution, asymptotic normality of the observations follows if $\mu_i \to \infty$ for each observation. In these cases the χ^2-distribution may be used as an approximation (see Section 3.8).

3.7.2 Analysis of Deviance and the Testing of Hypotheses

Let us consider the nested models $\tilde{M} \subset M$, where M is a given GLM with $\mu_i = h(\boldsymbol{x}_i^T\boldsymbol{\beta})$, and $\tilde{M}$ is a submodel that is characterized by the linear restriction $\boldsymbol{C\beta} = \boldsymbol{\xi}$, where $\boldsymbol{C}$ is a known $(s \times p)$-matrix with $\text{rank}(\boldsymbol{C}) = s \le p$ and $\boldsymbol{\xi}$ is an s-dimensional vector. This means that $\tilde{M}$ corresponds to the null hypothesis $H_0 : \boldsymbol{C\beta} = \boldsymbol{\xi}$, which specifies a simpler structure of the predictor.

Analysis of Deviance

With $\tilde{\boldsymbol{\mu}}^T = (\tilde{\mu}_1, \dots, \tilde{\mu}_n)$ denoting the fitted values for the restricted model $\tilde{M}$ and $\hat{\boldsymbol{\mu}}^T = (\hat{\mu}_1, \dots, \hat{\mu}_n)$ denoting the fit for model M, one obtains the corresponding deviances

$$D(M) = -\phi 2\{l(\boldsymbol{y}, \hat{\boldsymbol{\mu}}; \phi) - l(\boldsymbol{y}, \boldsymbol{y}; \phi)\},$$
$$D(\tilde{M}) = -\phi 2\{l(\boldsymbol{y}, \tilde{\boldsymbol{\mu}}; \phi) - l(\boldsymbol{y}, \boldsymbol{y}; \phi)\}.$$

The difference of deviances

$$D(\tilde{M}|M) = D(\tilde{M}) - D(M) = -2\phi\{l(\boldsymbol{y}, \tilde{\boldsymbol{\mu}}; \phi) - l(\boldsymbol{y}, \hat{\boldsymbol{\mu}}; \phi)\} \tag{3.11}$$

compares the fits of models $\tilde{M}$ and M. The difference of scaled deviances $D(\tilde{M}|M)/\phi$ is eqiuivalent to the likelihood ratio statistic for testing H_0. Similar to the partitioning of the sum of squares in linear regression, one may consider the partitioning of the deviance of the restricted model $\tilde{M}$ into

$$D(\tilde{M}) = D(\tilde{M}|M) + D(M).$$

$D(\tilde{M}|M)$ gives the increase in discrepancy between the data and the fit if model $\tilde{M}$ is fitted instead of the less restrictive model M. For normal distributions this corresponds to the partitioning of the sum of squares

$$\text{SSE}(\tilde{M}) = \text{SSE}(\tilde{M}|M) + \text{SSE}(M)$$

(see Section 1.4.6), which, for the special model where $\tilde{M}$ contains only an intercept, reduces to SST = SSR + SSE. In the normal case one obtains for SSE(M) a $\sigma^2\chi^2(n-p)$-distribution, if M holds (p denotes the dimension of the predictor in M). If $\tilde{M}$ holds, SSE$(\tilde{M}|M)$ and SSE(M) are independent with SSE$(\tilde{M}|M) \sim \sigma^2\chi^2(s)$ and SSE$(\tilde{M}) \sim \sigma^2\chi^2(n+s-p)$. For testing H_0 one uses the F-statistic

$$\frac{\{\text{SSE}(\tilde{M}) - \text{SSE}(M)\}/s}{\hat{\sigma}^2} \sim \text{F}(s, n-p),$$

where $\hat{\sigma}^2 = \text{SSE}(M)/(n-p)$. In the general case of GLMs one uses

$$\frac{D(\tilde{M}) - D(M)}{\phi} = \frac{D(\tilde{M}|M)}{\phi},$$

which under mild restrictions is asymptotically $\chi^2(s)$-distributed. This means that the difference

$$D(\tilde{M}) - D(M) = D(\tilde{M}|M)$$

has an asymptotically a $\phi\chi^2(s)$-distribution.

When using the χ^2-approximation the deviance has to be scaled by $1/\phi$. For the binomial $(\phi_i = 1/n_i, \phi = 1)$ Bernoulli, exponential, and Poisson $(\phi = 1)$ distributions one may use the difference $D(\tilde{M}) - D(M)$ directly, whereas for the normal, Gamma, and inverse Gaussian the dispersion parameter has to be estimated. In the normal regression case $\frac{1}{s}D(\tilde{M}|M)/\hat{\phi}$ has $\mathrm{F}(s, n-p)$-distribution. In the general case the approximation by the F-distribution may be used if $\hat{\phi}$ is consistent for ϕ, has approximately a scaled χ^2-distribution, and $D(\tilde{M}) - D(M)$ and $\hat{\phi}$ are approximately independent (Jorgenson, 1987). In analogy to the ANOVA table, in normal regression one obtains a table for the analysis of deviance (see Table 3.3).

TABLE 3.3: Analysis of deviance table.

	df	cond. deviance	df
$D(\tilde{M})$	$n-p+s$		
$D(M)$	$n-p$	$D(\tilde{M}\|M)$	s

It should be noted that only the difference of deviances $D(\tilde{M}|M)$ has asymptotically a $\phi\chi^2(s)$-distribution. The degrees of freedom of $D(\tilde{M})$ have the basic structure "number of observations minus number of fitted parameters." In $\tilde{M}$, by considering an additional s-dimensional restriction, the effective parameters in the model are reduced to $p - s$, yielding $df = n - (p - s) = n - p + s$. In the case of grouped data, the deviances $D(\tilde{M})$ and $D(M)$ themselves are asymptotically distributed with $D(\tilde{M}) \sim \chi^2(N-p+s)$, and $\chi^2(N-p)$, where N denotes the number of grouped observations (see Section 3.8). While $D(\tilde{M})$ and $D(M)$ are different for grouped and ungrouped data, the difference $D(\tilde{M}) - D(M)$ is the same.

Next we give a summary of results on distributions for the classical linear case and the deviances within the GLM framework. For the classical linear model one has

$$\begin{array}{ccccc} \mathrm{SSE}(\tilde{M}) & = & \mathrm{SSE}(\tilde{M}|M) & + & \mathrm{SSE}(M) \\ \sigma^2\chi^2(n-p+s) & & \sigma^2\chi^2(s) & & \sigma^2\chi^2(n-p) \\ \text{if } \tilde{M} \text{ holds} & & \text{if } \tilde{M} \text{ holds} & & \text{if } M \text{ holds} \end{array}$$

For grouped data within the GLM framework, one has the asymptotic distributions

$$\begin{array}{ccccc} D(\tilde{M}) & = & D(\tilde{M}|M) & + & D(M) \\ \phi\chi^2(N-p+s) & & \phi\chi^2(s) & & \phi\chi^2(N-p) \\ \text{if } \tilde{M} \text{ holds} & & \text{if } \tilde{M} \text{ holds} & & \text{if } M \text{ holds} \\ \text{grouped data} & & & & \text{grouped data} \end{array}$$

The approach may be used to test sequences of nested models,

$$M_1 \subset M_2 \subset \ldots \subset M_m,$$

by using the successive differences $(D(M_i) - D(M_{i+1}))/\phi$. The deviance of the most restrictive model is given as sum of these differences:

$$\begin{aligned} D(M_1) = & \; (D(M_1) - D(M_2)) + (D(M_2) - D(M_3)) \\ & + \ldots + (D(M_{m-1}) - D(M_m)) + D(M_m) \\ = & \; D(M_1|M_2) + \ldots + D(M_{m-1}|M_m) + D(M_m). \end{aligned}$$

Thus the discrepancy of the model M_1 is the sum of the "conditional" deviances $D(M_i|M_{i+1}) = D(M_i) - D(M_{i+1})$ and the discrepancy between the most general model M_m and the saturated model. However, when one starts from a model M_m and considers sequences of simpler models, one should be aware that different sequences of submodels are possible (see Section 4.4.2).

3.7.3 Alternative Test Statistics for Linear Hypotheses

The analysis of deviance tests if a model can be reduced to a model that has a simpler structure in the covariates. The simplified structure is specified by the null hypothesis H_0 of the pair of hypotheses

$$H_0 : \boldsymbol{C\beta} = \boldsymbol{\xi} \quad \text{against} \quad H_1 : \boldsymbol{C\beta} \neq \boldsymbol{\xi},$$

where $\text{rank}(\boldsymbol{C}) = s$. Alternative test statistics that can be used are the Wald test and the score statistic.

Wald Test

The *Wald statistic* has the form

$$w = (\boldsymbol{C}\hat{\boldsymbol{\beta}} - \boldsymbol{\xi})^T \left[\boldsymbol{C}\,\boldsymbol{F}^{-1}(\hat{\boldsymbol{\beta}})\,\boldsymbol{C}^T\right]^{-1} (\boldsymbol{C}\hat{\boldsymbol{\beta}} - \boldsymbol{\xi}).$$

It uses the weighted distance between the unrestricted estimate $\boldsymbol{C}\hat{\boldsymbol{\beta}}$ of $\boldsymbol{C\beta}$ and its hypothetical value $\boldsymbol{\xi}$ under H_0. The weight is derived from the distribution of the difference $(\boldsymbol{C}\hat{\boldsymbol{\beta}} - \boldsymbol{\xi})$, for which one obtains asymptotically $\text{cov}(\boldsymbol{C}\hat{\boldsymbol{\beta}} - \boldsymbol{\xi}) = \boldsymbol{C}\,\boldsymbol{F}^{-1}(\hat{\boldsymbol{\beta}})\,\boldsymbol{C}^T$. Therefore, w is the squared length of the standardized estimate $(\boldsymbol{C}\,\boldsymbol{F}^{-1}(\hat{\boldsymbol{\beta}})\,\boldsymbol{C}^T)^{-1/2}(\boldsymbol{C}\hat{\boldsymbol{\beta}} - \boldsymbol{\xi})$, and one obtains for w under H_0 an asymptotic $\chi^2(s)$-distribution.

An advantage of the Wald statistic is that it is based on the ML estimates of the full model. Therefore, it is not necessary to compute an additional fit under H_0. This is why most program packages give significance tests for single parameters in terms of the Wald statistic. When a single parameter is tested with $H_0 : \beta_j = 0$, the corresponding matrix $\boldsymbol{C}$ is $\boldsymbol{C} = (0, 0, \ldots, 1, \ldots, 0)$. Then the Wald statistic has the simple form

$$w = \frac{\beta_j^2}{\hat{a}_{jj}},$$

where $\hat{a}_{jj}$ is the jth diagonal element of the estimated inverse Fisher matrix $\boldsymbol{F}^{-1}$. Since w is asymptotically $\chi^2(1)$-distributed, one may also consider the square root,

$$z = \sqrt{w} = \frac{\beta_j}{\sqrt{\hat{a}_{jj}}},$$

which follows asymptotically a standard normal distribution. Thus, for single parameters, program packages usually give the standard error $\sqrt{\hat{a}_{jj}}$ and the p-value based on z.

Score Statistic

The score statistic is based on the following consideration: The score function $\boldsymbol{s}(\boldsymbol{\beta})$ for the unrestricted model is the zero vector if it is evaluated at the unrestricted ML estimate $\hat{\boldsymbol{\beta}}$. If, however, $\hat{\boldsymbol{\beta}}$ is replaced by the MLE $\tilde{\boldsymbol{\beta}}$ under H_0, $\boldsymbol{s}(\tilde{\boldsymbol{\beta}})$ will be significantly different from zero if H_0 is not true. Since the covariance of the score function is approximately the Fisher matrix, one uses the *score statistic*,

$$u = \boldsymbol{s}(\tilde{\boldsymbol{\beta}})^T\, \boldsymbol{F}^{-1}(\tilde{\boldsymbol{\beta}})\, \boldsymbol{s}(\tilde{\boldsymbol{\beta}}),$$

which is the squared weighted score function evaluated at $\tilde{\boldsymbol{\beta}}$.

An advantage of the Wald and score statistics is that they are properly defined for models with overdispersion since only the first and second moments are involved. All test statistics have the same asymptotic distribution. If they are differing strongly, that may be seen as a hint that the conditions for asymptotic results may not hold. A survey on asymptotics for test statistics was given by Fahrmeir (1987).

Test Statistics for Linear Hypotheses

$$H_0 : \boldsymbol{C}\boldsymbol{\beta} = \boldsymbol{\xi} \qquad H_1 : \boldsymbol{C}\boldsymbol{\beta} \neq \boldsymbol{\xi}$$

$$\text{with} \quad \text{rank}(\boldsymbol{C}) = s$$

Likelihood Ratio Statistic

$$\begin{aligned}\lambda &= -2\{l(\boldsymbol{y}; \tilde{\boldsymbol{\mu}}, \hat{\phi}) - l(\boldsymbol{y}, \hat{\boldsymbol{\mu}}, \hat{\phi})\} \\ &= (D(\boldsymbol{y}; \tilde{\boldsymbol{\mu}}, \hat{\phi}) - D(\boldsymbol{y}, \hat{\boldsymbol{\mu}}, \hat{\phi}))/\hat{\phi}\end{aligned}$$

Wald Statistic

$$w = (\boldsymbol{C}\hat{\boldsymbol{\beta}} - \boldsymbol{\xi})^T (\boldsymbol{C}\boldsymbol{F}^{-1}(\hat{\boldsymbol{\beta}})\boldsymbol{C}^T)^{-1} (\boldsymbol{C}\hat{\boldsymbol{\beta}} - \boldsymbol{\xi})$$

Score Statistic

$$u = \boldsymbol{s}(\tilde{\boldsymbol{\beta}})^T \boldsymbol{F}^{-1}(\tilde{\boldsymbol{\beta}}) \boldsymbol{s}(\tilde{\boldsymbol{\beta}})$$

Approximation

$$\lambda, w, u \sim \chi^2(s)$$

3.8 Goodness-of-Fit for Grouped Observations

It has already been mentioned that for grouped observations the deviance has an asymptotic χ^2-distribution. Hence, it may be used to test the model fit.

3.8.1 The Deviance for Grouped Observations

The analysis of deviance and alternative tests provide an instrument that helps to decide if a more parsimonious model $\tilde{M}$ may be chosen instead of the more general model M, where $\tilde{M} \subset M$. The test statistics may be seen as tools to investigate the fit of model $\tilde{M}$ *given model* M. However, they are of limited use for investigating if a model is appropriate for the given data, that is, the model fit compared to the data. The only possibility would be to choose M as the saturated model. But then the deviance has no fixed distribution in the limit.

A different situation occurs if replications are available. If, for a fixed covariate vector $\boldsymbol{x}_i$, independent replications $y_{i1}, \dots, y_{in_i}$ are observed, the mean across replications $\bar{y}_i = \sum_j y_{ij}/n_i$ again represents a GLM and the deviance for the means $\bar{y}_1, \dots, \bar{y}_N$ may be used. For grouped data with response $\bar{y}_i$, $i = 1, \dots, N$, the essential difference is that the scale parameter is given as $\phi_i = \phi/n_i$, where ϕ is the dispersion for the single observations. Since grouped observations $y_i, \dots y_{in_i}$ share the same predictor value $\boldsymbol{x}_i$, the log-likelihood is given by

$$l = \sum_{i=1}^{N}\sum_{j=1}^{n_i}(y_{ij}\theta_i - b(\theta_i))/\phi + c(y_{ij},\phi) = \sum_{i=1}^{N}\frac{\bar{y}_i\theta_i - b(\theta_i)}{\phi/n_i} + \sum_{i=1}^{N}\sum_{j=1}^{n_i}c(y_{ij},\phi).$$

Thus maximization with respect to $\boldsymbol{\beta}$ yields the same results if l is maximized in the ungrouped or grouped form. The deviance, however, changes for grouped data because the saturated model for data $(\bar{y}_i, x_i), i = 1, \dots, N$, means that only the N observations $\bar{y}, \dots, \bar{y}_N$ have to be fitted perfectly.

With $\bar{\boldsymbol{y}}^T = (\bar{y}_1, \dots, \bar{y}_N)$, $\hat{\boldsymbol{\mu}}^T = (\hat{\boldsymbol{\mu}}_1, \dots, \hat{\boldsymbol{\mu}}_N)$, and $\hat{\boldsymbol{\mu}}_i = h(\boldsymbol{x}_i^T\hat{\boldsymbol{\beta}})$, where $\boldsymbol{\mu}_i = \mathrm{E}(\bar{y}_i) = \mathrm{E}(y_{ij}), j = 1, \dots, n_i$, the deviance for grouped observations has the form

$$\begin{aligned} D(\bar{\boldsymbol{y}}, \hat{\boldsymbol{\mu}}) &= -\phi 2\{l(\bar{\boldsymbol{y}}; \hat{\boldsymbol{\mu}}, \phi) - l(\bar{\boldsymbol{y}}; \bar{\boldsymbol{y}}, \phi)\} \\ &= 2\sum_{i=1}^{N}\{\bar{y}_i(\theta(\bar{y}_i) - \theta(\hat{\boldsymbol{\mu}}_i)) - (b(\theta(\bar{y}_i)) - b(\theta(\hat{\boldsymbol{\mu}}_i)))\}n_i. \end{aligned}$$

The deviances for various distributions are given in Table 3.4. For ungrouped data with $N = n, n_i = 1$, one obtains the deviances as given in Table 3.2. The grouped deviance for Bernoulli variables is equivalent to the deviance of the binomial distribution. The reason is obvious because the binomial distribution implicitly assumes replications.

TABLE 3.4: Deviances for grouped observations.

Distribution	Deviance for grouped observations
Normal	$D(\bar{\boldsymbol{y}}, \hat{\mu}) = \sum_{i=1}^{N} n_i(\bar{y}_i - \hat{\mu}_i)^2$
Gamma	$D(\bar{\boldsymbol{y}}, \hat{\mu}) = 2\sum_{i=1}^{N} n_i\left(-\log\left(\frac{\bar{y}_i}{\hat{\mu}_i}\right) + \frac{\bar{y}_i - \hat{\mu}_i}{\hat{\mu}_i}\right)$
Inverse Gaussian	$D(\bar{\boldsymbol{y}}, \hat{\mu}) = \sum_{i=1}^{N} n_i(\bar{y}_i - \hat{\mu}_i)^2/(\hat{\mu}_i^2\bar{y}_i)$
Bernoulli	$D(\bar{\boldsymbol{y}}, \hat{\mu}) = 2\sum_{i=1}^{N}\left(n_i\bar{y}_i\log\left(\frac{\bar{y}_i}{\hat{\mu}_i}\right) + n_i(1-\bar{y}_i)\log\left(\frac{1-\bar{y}_i}{1-\hat{\mu}_i}\right)\right)$
Poisson	$D(\bar{\boldsymbol{y}}, \hat{\mu}) = 2\sum_{i=1}^{N} n_i\left(\bar{y}_i\log\left(\frac{\bar{y}_i}{\hat{\mu}_i}\right) - (\bar{y}_i - \hat{\mu}_i)\right)$

The advantage of replications or grouped data is that for this kind of data the scaled deviance or likelihood ratio statistic $D(\bar{\boldsymbol{y}}, \hat{\boldsymbol{\mu}})/\phi$ provides a goodness-of-fit statistic that may be used to test if the model is appropriate for the data. The maximal likelihood $l(\bar{\boldsymbol{y}}; \bar{\boldsymbol{y}}, \phi)$ is the likelihood of a model with N parameters, one per covariate value $\boldsymbol{x}_i$, where only the assumption of distribution with independent, identically distributed responses is made. Thus $D(\bar{\boldsymbol{y}}, \hat{\boldsymbol{\mu}})/\phi$ may be used to test the current model, implying the form of the linear predictors and a specific link function, against the distribution model. Under fixed cells asymptotics $(N \text{ fixed}, n_i \to \infty, n_i/n \to c_i, c_i > 0)$ and regularity conditions one obtains for $D(\bar{\boldsymbol{y}}, \hat{\boldsymbol{\mu}})/\phi$ a limiting χ^2-distribution with $N - p$ degrees of freedom, where p denotes the dimension of the predictor $\boldsymbol{x}_i$. If $D(\bar{y}, \hat{\boldsymbol{\mu}})/\phi$ is larger than the $1 - \alpha$ quantile of $\chi^2(N - p)$, the model is questionable as a tool for investigating the connection between covariates and responses.

It should be noted that the difference between likelihoods for different models and therefore the analysis of variance yields the same results for ungrouped and grouped modeling. For two models $\tilde{M} \subset M$ and corresponding fits $\hat{\mu}_i, \tilde{\mu}_i$, the difference $D(\bar{\boldsymbol{y}}, \tilde{\boldsymbol{\mu}}) - D(\bar{\boldsymbol{y}}, \hat{\boldsymbol{\mu}})$ for grouped data are the same as the difference $D(\boldsymbol{y}, \tilde{\boldsymbol{\mu}}) - D(\boldsymbol{y}, \hat{\boldsymbol{\mu}})$ for ungrouped observations. Therefore $(D(\bar{\boldsymbol{y}}, \tilde{\boldsymbol{\mu}}) - D(\bar{\boldsymbol{y}}, \hat{\boldsymbol{\mu}}))/\phi$ is asymptotically $\chi^2(s)$-distributed, where s is the difference between the number of parameters in M and $\tilde{M}$.

For the normal linear model alternative tests for the lack-of-fit are available. The partitioning of (ungrouped) least squares data $(y_{ij}, \boldsymbol{x}_i)$, $j = 1, \dots, n_i$, yields

$$\sum_{i=1}^{N}\sum_{j=1}^{n_i}(y_{ij} - \hat{\mu}_i)^2 = \sum_{i=1}^{N} n_i(\bar{y}_i - \hat{\mu}_i)^2 + \sum_{i=1}^{N}\sum_{j=1}^{n_i}(y_{ij} - \bar{y}_i)^2,$$

which has the form

$$D(\tilde{M}) = D(\tilde{M}|M) + D(M),$$

where $\tilde{M}$ stands for the linear model and M for a model where only $y_{ij} = \mu_i + \varepsilon_{ij}$ with $\varepsilon_{ij} \sim \mathrm{N}(0, \sigma^2)$ is assumed. Since in computing $D(M)$ no assumption on linearity is assumed, it is also called the *pure error sum of squares* and $D(\tilde{M}|M)$ the *lack of fit sum of squares*. By use of the mean squares one obtains for $\boldsymbol{H}_0 : \boldsymbol{\mu} = \boldsymbol{X\beta}$ against $H_1 : \boldsymbol{\mu} \neq \boldsymbol{X\beta}$ the F-statistic

$$F = \frac{\sum_{i=1}^{N} n_i(\bar{y}_i - \hat{\mu}_i)^2/(N-p)}{\sum_{i=1}^{N}\sum_{j=1}^{n_i}(y_{ij} - \bar{y}_i)^2/(n-N)} \sim \mathrm{F}(N-p, n-N).$$

Linearity is dismissed if $F > F_{1-\alpha}(N-p, n-N)$. When using the F-statistic, not all levels of covariates need to have repeated observations, only some of the n_i's have to be larger than 1. However, the test is still based on the assumptions that responses are normally distributed and have variance σ^2. Note that $D(\tilde{M}|M)$ is the deviance for grouped observations.

3.8.2 Pearson Statistic

An alternative measure for the discrepancy between the data and the model is the Pearson statistic:

$$\chi_P^2 = \sum_{i=1}^{N} \frac{(\bar{y}_i - \hat{\mu}_i)^2}{v(\hat{\mu}_i)/n_i}, \tag{3.12}$$

where $\bar{y}_i$ is the mean for grouped observations, $\hat{\mu}_i$ is the estimated mean, and $v(\hat{\mu}_i)$ is the corresponding variance function that is linked to the variance by $\mathrm{var}(y_i) = v(\mu_i)\phi/n_i$. If fixed cells asymptotics applies (N fixed, $n_i \to \infty$, $n_i/n \to c_i, c_i > 0$), χ_P^2 is asymptotically χ^2-distributed, that is, χ_P^2 has an approximately $\phi\chi^2(N-p)$-distribution. The dispersion parameter ϕ has to be known and fixed since the estimation of ϕ is based on this statistic. Replacing ϕ by the dispersion estimate from grouped observations $\hat{\phi}_N = \chi_p^2/(N-p)$ would yield the trivial result $\chi_p^2/\hat{\phi}_N = N - p$.

Goodness-of-Fit for Grouped Observations

Deviance

$$D = -\phi 2 \sum_{i=1}^{N} \{l(\bar{y}_i; \hat{\mu}_i, \phi) - l(\bar{y}_i; \bar{y}_i, \phi)\}$$

Pearsons χ^2

$$\chi^2 = \sum_{i=1}^{N} n_i \frac{(\bar{y}_i - \hat{\mu}_i)^2}{v(\hat{\mu}_i)}$$

Fixed cells asymptotic (N fixed, $n_i \to \infty$, $n_i/n \to c_i, c_i > 0$)

$$D, \chi^2 \text{ approximatively } \phi\chi^2(N-p)$$

3.9 Computation of Maximum Likelihood Estimates

Maximum likelihood estimates are obtained by solving the equation $\boldsymbol{s}(\hat{\boldsymbol{\beta}}) = \mathbf{0}$. In general, there is no closed form of the estimate available, and iterative procedures have to be applied. In matrix notation the score function is given by

$$s(\boldsymbol{\beta}) = \boldsymbol{X}^T \boldsymbol{D}\, \boldsymbol{\Sigma}^{-1}(\boldsymbol{y} - \boldsymbol{\mu}) = \boldsymbol{X}^T \boldsymbol{W} \boldsymbol{D}^{-1}(\boldsymbol{y} - \boldsymbol{\mu}),$$

where in $\boldsymbol{D}$, $\boldsymbol{\Sigma}$, and $\boldsymbol{W}$ the dependence on $\boldsymbol{\beta}$ is suppressed (see Section 3.6).

The *Newton-Raphson method* is an iterative method for solving non-linear equations. Starting with an initial guess $\boldsymbol{\beta}^{(0)}$, the solution is found by successive improvement. Let $\boldsymbol{\beta}^{(k)}$ denote the estimate in the kth step, where $k = 0$ is the initial estimate. If $s(\boldsymbol{\beta}^{(k)}) \neq 0$, one considers the linear Taylor approximation

$$s(\boldsymbol{\beta}) \approx s_{\text{lin}}(\boldsymbol{\beta}) = s(\hat{\boldsymbol{\beta}}^{(k)}) + \frac{\partial s(\hat{\boldsymbol{\beta}}^{(k)})}{\partial \boldsymbol{\beta}}(\boldsymbol{\beta} - \hat{\boldsymbol{\beta}}^{(k)}).$$

Instead of solving $s(\hat{\boldsymbol{\beta}}) = 0$ one solves $s_{\text{lin}}(\hat{\boldsymbol{\beta}}) = 0$, yielding

$$\hat{\boldsymbol{\beta}} = \hat{\boldsymbol{\beta}}^{(k)} - \left(\frac{\partial s(\hat{\boldsymbol{\beta}}^{(k)})}{\partial \boldsymbol{\beta}}\right)^{-1} s(\hat{\boldsymbol{\beta}}^{(k)}).$$

Since $\partial s(\boldsymbol{\beta})/\partial \boldsymbol{\beta} = \partial^2 l(\boldsymbol{\beta})/\partial \boldsymbol{\beta} \partial \boldsymbol{\beta}^T$, one obtains with the Hessian matrix $\boldsymbol{H}(\beta) = \partial^2 l(\boldsymbol{\beta})/\partial \boldsymbol{\beta} \partial \boldsymbol{\beta}^T$ the new estimate

$$\hat{\boldsymbol{\beta}}^{(k+1)} = \hat{\boldsymbol{\beta}}^{(k)} - \boldsymbol{H}(\hat{\boldsymbol{\beta}}^{(k)})^{-1} s(\hat{\boldsymbol{\beta}}^{(k)})$$

or, by using the observed information matrix $\boldsymbol{F}_{\text{obs}}(\boldsymbol{\beta}) = -H(\boldsymbol{\beta})$,

$$\hat{\boldsymbol{\beta}}^{(k+1)} = \hat{\boldsymbol{\beta}}^{(k)} + \boldsymbol{F}_{\text{obs}}(\hat{\boldsymbol{\beta}}^{(k)})^{-1} s(\hat{\boldsymbol{\beta}}^{(k)}).$$

Iterations are carried out until the changes between successive steps are smaller than a specified threshold ε. Iteration is stopped if

$$\| \hat{\boldsymbol{\beta}}^{(k+1)} - \hat{\boldsymbol{\beta}}^{(k)} \| / \| \hat{\boldsymbol{\beta}}^{(k)} \| < \varepsilon.$$

Convergence is usually fast, with the number of correct decimals in the approximation roughly doubling at each iteration.

An alternative method is the *Newton method with Fisher scoring*. The essential difference is that the observed information matrix $\boldsymbol{F}_{\text{obs}}$ is replaced by the expected information $\boldsymbol{F}(\boldsymbol{\beta}) = \mathrm{E}(\boldsymbol{F}_{\text{obs}}(\boldsymbol{\beta}))$ (or $\boldsymbol{H}(\beta)$ by $-\boldsymbol{F}(\boldsymbol{\beta})$), yielding

$$\hat{\boldsymbol{\beta}}^{(k+1)} = \hat{\boldsymbol{\beta}}^{(k)} + \boldsymbol{F}(\hat{\boldsymbol{\beta}}^{(k)})^{-1} s(\hat{\boldsymbol{\beta}}^{(k)}). \tag{3.13}$$

The iterative scheme (3.13) may alternatively be seen as an iterative weighted least-squares fitting procedure. Let pseudo- or working observations be given by

$$\tilde{\eta}_i(\hat{\boldsymbol{\beta}}) = \boldsymbol{x}_i^T \hat{\boldsymbol{\beta}} + \left(\frac{\partial h(\eta_i)}{\partial \eta}\right)^{-1} (y_i - \mu_i(\hat{\boldsymbol{\beta}}))$$

and $\tilde{\boldsymbol{\eta}}(\hat{\boldsymbol{\beta}})^T = (\tilde{\eta}_1(\hat{\boldsymbol{\beta}}), \dots, \tilde{\eta}_n(\hat{\boldsymbol{\beta}}))$ denote the vector of pseudo-observations given by $\tilde{\boldsymbol{\eta}}(\hat{\boldsymbol{\beta}}) = \boldsymbol{X}\hat{\boldsymbol{\beta}} + \boldsymbol{D}(\hat{\boldsymbol{\beta}})^{-1}(\boldsymbol{y} - \hat{\boldsymbol{\mu}})$. One obtains by simple substitution

$$\hat{\boldsymbol{\beta}}^{(k+1)} = (\boldsymbol{X}^T \boldsymbol{W}(\hat{\boldsymbol{\beta}}^{(k)})\boldsymbol{X})^{-1}\boldsymbol{X}^T \boldsymbol{W}(\hat{\boldsymbol{\beta}}^{(k)})\tilde{\boldsymbol{\eta}}(\hat{\boldsymbol{\beta}}^{(k)}).$$

Thus $\hat{\boldsymbol{\beta}}^{(k+1)}$ has the form of a weighted least-squares estimate for the working observations $(\tilde{\eta}_i^{(k)}, \boldsymbol{x}_i), i = 1, \dots, n$, with the weight $\boldsymbol{W}(\hat{\boldsymbol{\beta}}^{(k)})$ depending on the iteration.

For a canonical link one obtains $\boldsymbol{F}_n(\boldsymbol{\beta}) = \boldsymbol{X}^T \boldsymbol{W} \boldsymbol{X}$ with $\boldsymbol{W} = \boldsymbol{\phi}^{-1}\boldsymbol{\Sigma}\boldsymbol{\phi}^{-1}$, $\boldsymbol{\phi} = \mathrm{diag}(\phi_1, \dots, \phi_n)$ and score function $s(\boldsymbol{\beta}) = \boldsymbol{X}^T \boldsymbol{\phi}^{-1}(\boldsymbol{y} - \boldsymbol{\mu})$ and therefore

$$\hat{\boldsymbol{\beta}}^{(k+1)} = \hat{\boldsymbol{\beta}}^{(k)} + (\boldsymbol{X}^T \boldsymbol{W}\boldsymbol{X})^{-1}\boldsymbol{X}^T \boldsymbol{\phi}^{-1}(\boldsymbol{y} - \boldsymbol{\mu}),$$

which corresponds to least-squares fitting

$$\hat{\boldsymbol{\beta}}^{(k+1)} = (\boldsymbol{X}^T \boldsymbol{W}\boldsymbol{X})^{-1}\boldsymbol{X}^T \boldsymbol{W}\boldsymbol{\eta}(\hat{\boldsymbol{\beta}}^{(k)})$$

with $\tilde{\boldsymbol{\eta}}(\boldsymbol{\beta}) = \boldsymbol{X}\boldsymbol{\beta} + \boldsymbol{W}^{-1}\boldsymbol{\phi}^{-1}(\boldsymbol{y} - \boldsymbol{\mu})$. If $\boldsymbol{\phi} = \phi \boldsymbol{I}$, one obtains

$$\hat{\boldsymbol{\beta}}^{(k+1)} = \hat{\boldsymbol{\beta}}^{(k)} + \phi(\boldsymbol{X}^T \boldsymbol{\Sigma}\boldsymbol{X})^{-1}\boldsymbol{X}^T(\boldsymbol{y} - \boldsymbol{\mu}).$$

3.10 Hat Matrix for Generalized Linear Models

With weight matrix $\boldsymbol{W}(\boldsymbol{\beta}) = \boldsymbol{D}(\boldsymbol{\beta})\boldsymbol{\Sigma}(\boldsymbol{\beta})^{-1}\boldsymbol{D}(\boldsymbol{\beta})^T$, the iterative fitting procedure has the form

$$\hat{\boldsymbol{\beta}}^{(k+1)} = (\boldsymbol{X}^T \boldsymbol{W}(\hat{\boldsymbol{\beta}}^{(k)})\boldsymbol{X})^{-1}\boldsymbol{X}^T \boldsymbol{W}(\hat{\boldsymbol{\beta}}^{(k)})\tilde{\boldsymbol{\eta}}(\hat{\boldsymbol{\beta}}^{(k)}).$$

At convergence one obtains

$$\hat{\boldsymbol{\beta}} = (\boldsymbol{X}^T \boldsymbol{W}(\hat{\boldsymbol{\beta}})\boldsymbol{X})^{-1}\boldsymbol{X}^T \boldsymbol{W}(\hat{\boldsymbol{\beta}})\tilde{\boldsymbol{\eta}}(\hat{\boldsymbol{\beta}}).$$

Thus $\hat{\boldsymbol{\beta}}$ may be seen as the least-squares solution of the linear model

$$\boldsymbol{W}^{T/2}\tilde{\boldsymbol{\eta}}(\hat{\boldsymbol{\beta}}) = \boldsymbol{W}^{T/2}\boldsymbol{X}\boldsymbol{\beta} + \tilde{\varepsilon},$$

where in $\boldsymbol{W} = \boldsymbol{W}(\hat{\boldsymbol{\beta}})$ the dependence on $\hat{\boldsymbol{\beta}}$ is suppressed. The corresponding hat matrix has the form

$$\boldsymbol{H} = \boldsymbol{W}^{T/2}\boldsymbol{X}(\boldsymbol{X}^T \boldsymbol{W}\boldsymbol{X})^{-1}\boldsymbol{X}^T \boldsymbol{W}^{1/2}.$$

Since the matrix $\boldsymbol{H}$ is idempotent and symmetric, it may be seen as a projection matrix for which $\text{tr}(\boldsymbol{H}) = \text{rank}(\boldsymbol{H})$ holds. Moreover, one obtains for the diagonal elements of $\boldsymbol{H} = (h_{ij})$ $0 \leq h_{ii} \leq 1$ and $\text{tr}(\boldsymbol{H}) = p$ (if X has full rank).

It should be noted that, in contrast to the normal regression model, the hat matrix depends on $\hat{\boldsymbol{\beta}}$ because $\boldsymbol{W} = \boldsymbol{W}(\hat{\boldsymbol{\beta}})$. The equation $\boldsymbol{W}^{T/2}\hat{\boldsymbol{\eta}} = \boldsymbol{H}\boldsymbol{W}^{T/2}\tilde{\boldsymbol{\eta}}(\boldsymbol{\beta})$ shows how the hat matrix maps the adjusted variable $\tilde{\boldsymbol{\eta}}(\boldsymbol{\beta})$ into the fitted values $\hat{\boldsymbol{\eta}}$. Thus $\boldsymbol{H}$ may be seen as the matrix that maps the adjusted observation vector $\boldsymbol{W}^{T/2}\tilde{\boldsymbol{\eta}}$ into the vector of "fitted" values $\boldsymbol{W}^{T/2}\hat{\boldsymbol{\eta}}$, which is a mapping on the transformed predictor space.

For the linear model the hat matrix represents a simple projection having the form $\hat{\boldsymbol{\mu}} = \boldsymbol{H}\boldsymbol{y}$. In the case of generalized linear models, it may be shown that approximatively

$$\boldsymbol{\Sigma}^{-1/2}(\hat{\boldsymbol{\mu}} - \boldsymbol{\mu}) \simeq \boldsymbol{H}\boldsymbol{\Sigma}^{-1/2}(\boldsymbol{y} - \boldsymbol{\mu}) \tag{3.14}$$

holds, where $\boldsymbol{\Sigma} = \boldsymbol{\Sigma}(\hat{\boldsymbol{\beta}})$. Thus $\boldsymbol{H}$ may be seen as measure of the influence of $\boldsymbol{y}$ on $\hat{\boldsymbol{\mu}}$ in standardized units of changes. From (4.6) follows

$$\hat{\boldsymbol{\mu}} - \boldsymbol{\mu} \simeq \boldsymbol{\Sigma}^{1/2}\boldsymbol{H}\boldsymbol{\Sigma}^{-1/2}(\boldsymbol{y} - \boldsymbol{\mu}), \tag{3.15}$$

such that the influence in unstandardized units is given by the projection matrix $\boldsymbol{\Sigma}^{1/2}\boldsymbol{H}\boldsymbol{\Sigma}^{-1/2}$, which is idempotent but not symmetric. Note that for the normal regression model with an identity link one has $\boldsymbol{W} = \boldsymbol{I}/\sigma^2$, $\boldsymbol{H} = \boldsymbol{X}(\boldsymbol{X}^T\boldsymbol{X})^{-1}\boldsymbol{X}^T$, and (4.6) and (3.15) hold exactly. From (3.15) one derives the approximation

$$\text{cov}(\boldsymbol{y} - \hat{\boldsymbol{\mu}}) \simeq \boldsymbol{\Sigma}^{1/2}(\boldsymbol{I} - \boldsymbol{H})\boldsymbol{\Sigma}^{T/2}.$$

An alternative property is obtained from considering the estimating equation $\boldsymbol{X}^T\hat{\boldsymbol{W}}\hat{\boldsymbol{D}}^{-T}(\boldsymbol{y} - \hat{\boldsymbol{\mu}}) = \boldsymbol{0}$, which yields directly

$$\boldsymbol{P}_{\boldsymbol{X},\boldsymbol{W}}\hat{\boldsymbol{D}}^{-T}\hat{\boldsymbol{\mu}} = \boldsymbol{P}_{\boldsymbol{X},\boldsymbol{W}}\hat{\boldsymbol{D}}^{-T}\boldsymbol{y}, \tag{3.16}$$

with $\boldsymbol{P}_{\boldsymbol{X},\boldsymbol{W}} = \boldsymbol{X}(\boldsymbol{X}^T\hat{\boldsymbol{W}}(\hat{\boldsymbol{\beta}})\boldsymbol{X})^{-1}\boldsymbol{X}^T\hat{\boldsymbol{W}}$ being the projection on the subspace span $(\boldsymbol{Z})$ with respect to the matrix $\boldsymbol{W}$, which means an orthogonal projection with respect to the product $\boldsymbol{x}_1^T\boldsymbol{W}\boldsymbol{x}_2$, $\boldsymbol{x}_1, \boldsymbol{x}_2 \in \mathbb{R}^n$. $\boldsymbol{P}_{\boldsymbol{Z},\boldsymbol{W}}$ is idempotent but not symmetric. Thus the projections of the transformed values $\hat{\boldsymbol{D}}^{-T}\boldsymbol{y}$, $\hat{\boldsymbol{D}}^{-T}\hat{\boldsymbol{\mu}}$ on the η-space are identical. From (3.16) one obtains easily

$$\boldsymbol{H}\boldsymbol{W}^{T/2}\boldsymbol{D}^{-T}\hat{\boldsymbol{\mu}} = \boldsymbol{H}\boldsymbol{W}^{T/2}\boldsymbol{D}^{-T}\boldsymbol{y},$$

which yields

$$\boldsymbol{H}\boldsymbol{\Sigma}^{-1/2}\hat{\boldsymbol{\mu}} = \boldsymbol{H}\boldsymbol{\Sigma}^{-1/2}\boldsymbol{y},$$

meaning that the orthogonal projections (based on $\boldsymbol{H}$) of standardized values $\boldsymbol{\Sigma}^{-1/2}\hat{\boldsymbol{\mu}}$, $\boldsymbol{\Sigma}^{-1/2}\boldsymbol{y}$ are identical. With $\boldsymbol{\chi} = \boldsymbol{\Sigma}^{-1/2}(\boldsymbol{y} - \hat{\boldsymbol{\mu}})$ denoting the standardized residual, one has

$$\boldsymbol{H}\boldsymbol{\chi} = 0 \text{ and } (\boldsymbol{I} - \boldsymbol{H})\boldsymbol{\chi} = \boldsymbol{\chi}.$$

There is a strong connection to the χ^2-statistic since $\chi^2 = \boldsymbol{\chi}^T\boldsymbol{\chi}$. The matrix $\boldsymbol{H}$ has the form $\boldsymbol{H} = (\boldsymbol{W}^{T/2}\boldsymbol{X})(\boldsymbol{X}^T\boldsymbol{W}^{1/2}\boldsymbol{W}^{T/2}\boldsymbol{X})^{-1}(\boldsymbol{X}^T\boldsymbol{W}^{1/2})$, which shows that the projection is into the subspace that is spanned by the columns of $\boldsymbol{X}^T\boldsymbol{W}^{1/2}$. The essential difference from ordinary linear regression is that the hat matrix does depend not only on the design but also on the fit.

3.11 Quasi-Likelihood Modeling

In generalized linear models it is assumed that the true density of the responses follows an exponential family. However, in applications it is not too infrequently found that the implicitly specified variation of the responses is not consistent with the variation of the data. Approaches that use only the first two moments of the response distribution are based on so-called quasi-likelihood estimates (Wedderburn, 1974; McCullagh and Nelder, 1989). When using quasi-likelihood estimates, the exponential family assumption is dropped, and the mean and variance structures are separated. No full distributional assumptions are necessary. Under appropriate conditions, parameters can still be estimated consistently, and asymptotic inference is possible under appropriate modifications.

Quasi-likelihood approaches assume, like GLMs, that the mean and variance structures are correctly specified by

$$\mathrm{E}(y_i|\boldsymbol{x}_i) = \mu_i = h(\boldsymbol{x}^T\boldsymbol{\beta}), \qquad \mathrm{var}(y_i|\boldsymbol{x}_i) = \sigma_i^2(\mu_i) = \phi v(\mu_i), \tag{3.17}$$

where $v(\mu)$ is a variance function and ϕ is a dispersion parameter. The main difference from GLMs is that the mean and variance do not have to be specified by an exponential family. The usual maximum likelihood estimates for GLMs are obtained by setting the score function equal to zero:

$$\boldsymbol{s}(\hat{\boldsymbol{\beta}}) = \boldsymbol{X}^T\boldsymbol{D}(\hat{\boldsymbol{\beta}})\boldsymbol{\Sigma}^{-1}(\hat{\boldsymbol{\beta}})(\boldsymbol{y} - \boldsymbol{\mu}) = \boldsymbol{0}. \tag{3.18}$$

The ML estimation equation uses the link (in $\boldsymbol{D}(\hat{\boldsymbol{\beta}})$) and the variance function (in $\boldsymbol{\Sigma}^{-1}(\hat{\boldsymbol{\beta}})$), but no higher moments. The solution of (3.18) yields the ML estimates when the mean and variance correspond to an exponential family. The quasi-likelihood (QL) estimates are obtained when the specification of the mean and the variance is given by (3.17) without reference to an exponential family. One may understand

$$\boldsymbol{s}_Q(\boldsymbol{\beta}) = \boldsymbol{X}^T\boldsymbol{D}(\boldsymbol{\beta})\boldsymbol{\Sigma}^{-1}(\boldsymbol{\beta})(\boldsymbol{y} - \boldsymbol{\mu})$$

with the specifications (3.17) as a quasi-score function with the corresponding estimation equation $\boldsymbol{s}_Q(\hat{\boldsymbol{\beta}}) = \boldsymbol{0}$. It is also possible to construct a quasi-likelihood function $Q(\boldsymbol{\beta}, \phi)$ that has the derivative $\boldsymbol{s}_Q(\boldsymbol{\beta}) = \partial Q/\partial\boldsymbol{\beta}$ (Nelder and Pregibon, 1987; McCullagh and Nelder, 1989). It can be shown that the asymptotic properties are similar to those for GLMs. In particular, one obtains asymptotically a normal distribution with the covariance given in the form of a pseudo-Fisher matrix:

$$\boldsymbol{F}_Q(\boldsymbol{\beta}) = \sum_{i=1}^{n} \boldsymbol{x}_i\boldsymbol{x}_i^T \left(\frac{\partial h(\eta_i)}{\partial\eta}\right)^2 / \sigma_i^2 = \boldsymbol{X}^T\boldsymbol{D}\boldsymbol{\Sigma}^{-1}\boldsymbol{D}\boldsymbol{X} = \boldsymbol{X}^T\boldsymbol{W}\boldsymbol{X}$$

(see also Wedderburn, 1974; McCullagh and Nelder, 1989).

It should be noted that quasi-likelihood models weaken the distributional assumptions considerably. One obtains estimates of parameters without assuming a specific distribution. One just has to specify the mean and the variance and is free to select a variance function $v(\mu_i)$ that is not determined by a fixed distribution.

A major area of application is the modeling of overdispersion. For example, in count data the assumption of the Poisson distribution means that the variance depends on the mean in the form $\mathrm{var}(y_i) = \mu_i = \exp(\boldsymbol{x}_i^T\boldsymbol{\beta})$. In quasi-likelihood approaches one might assume that the variance is $\mathrm{var}(y_i) = \phi\mu_i = \phi\exp(\boldsymbol{x}_i^T\boldsymbol{\beta})$, with an additional unknown dispersion parameter ϕ. Since the Poisson distribution holds for $\phi = 1$ only, one does not assume the Poisson model to hold (see Sections 5.3 and 7.5).

More flexible models allow the variance function to depend on additional parameters (e.g., Nelder and Pregibon, 1987). Then the variance function has the form $\phi v(\mu; \alpha)$, where α is an additional unknown parameter. An example is $v(\mu; \alpha) = \mu^\alpha$. For fixed α, the quasi-likelihood estimate $\hat{\boldsymbol{\beta}}$ (and an estimate $\hat{\phi}$) is obtained by solving the corresponding estimation equation. Estimation of α, given $\hat{\boldsymbol{\beta}}$ and $\hat{\phi}$, can be carried out by the method of moments. Cycling between the two steps until convergence gives a joint estimation procedure. Asymptotic results for $\hat{\boldsymbol{\beta}}$ remain valid if α is replaced by a consistent estimate $\hat{\alpha}$.

Within quasi-likelihood approaches one can also model that the dispersion parameter depends on covariates. Then one assumes $\mu = h(\boldsymbol{x}^T\boldsymbol{\beta})$, $\phi = \phi(\boldsymbol{z}^T\gamma)$, $\text{var}(y) = \phi v(\mu)$, where $\boldsymbol{z}$ is a vector of covariates affecting the dispersion parameter. Cycling between the estimating equation for $\boldsymbol{\beta}$ and an estimation equation for $\boldsymbol{\gamma}$ yields estimates for both parameters. Alternatively, a joint estimation of parameters can be obtained by the general techniques for fitting likelihood and quasi-likelihood models described by Gay and Welsch, 1988. Consistent estimation of β and α requires that not only the mean but also the dispersion parameter be correctly specified (for details see Pregibon, 1984; Nelder and Pregibon, 1987; Efron, 1986; McCullagh and Nelder, 1989; Nelder, 1992).

3.12 Further Reading

Surveys and Books. A source book on generalized linear models is McCullagh and Nelder (1989). Multivariate extensions are covered in Fahrmeir and Tutz (2001). Shorter introductions were given by Dobson (1989) and Firth (1991). A Bayesian perspective on generalized linear models is outlined in Dey et al. (2000). More recently, a general treatment of regression, including GLMs was given by Fahrmeir et al. (2011).

Quasi-likelihood. Quasi-likelihood estimates were considered by Wedderburn (1974), McCullagh (1983), and McCullagh and Nelder (1989). Efficiency of quasi-likelihood estimates was investigated by Firth (1987). A rigorous mathematical treatment was given by Heyde (1997). Asymptotic properties were also studied by Xia et al. (2008).

R packages. GLMs can be fitted by use of the model fitting functions *glm* from the *MASS* package. Many tools for diagnostics and inferences are available.

3.13 Exercises

3.1 Let independent observations $y_{i1}, \dots, y_{in_i}$ have a fixed covariate vector $\boldsymbol{x}_i$ with n_i denoting the sample size at covariate value $\boldsymbol{x}_i$. The model to be examined has the form $\mu_i = \mu_{ij} = \text{E}(y_{ij}) = h(\boldsymbol{x}_i^T\boldsymbol{\beta})$.

(a) Let observations be binary with $y_{ij} \in \{0, 1\}$. Show that the distribution of the mean $\bar{y}_i = \frac{1}{n_i}\sum_{j=1}^{n_i} y_{ij}$ also has the form of a simple exponential family. Compare the canonical parameter and the dispersion parameter with the values of the exponential family of the original binary response.

(b) Show that the mean $\bar{y}_i$ always has the form of a simple exponential family if the mean of y_{ij} has the form of a simple exponential family.

3.2 Let observations $(y_i, \boldsymbol{x}_i), i = 1, \dots, n$, with binary response $y_i \in \{0, 1\}$ be given. The used models are the logit model $P(y_i = 1|\boldsymbol{x}_i) = \Phi(\boldsymbol{x}_i^T\boldsymbol{\beta})$, with F denoting the logistic distribution function, and the probit model $P(y_i = 1|\boldsymbol{x}_i) = \Phi(\boldsymbol{x}_i^T\boldsymbol{\beta})$, with Φ denoting the standard normal distribution function. Give the log-likelihood function and derive the score function $\boldsymbol{s}(\boldsymbol{\beta})$, the matrix of derivatives $\boldsymbol{D}(\boldsymbol{\beta})$, and the variance matrix $\boldsymbol{\Sigma}(\boldsymbol{\beta})$ for both models. In addition, give the observed and expected information matrices.

3.3 Consider a GLM with Poisson-distributed responses.

(a) Derive the Fisher matrix for the canonical link function.

(b) Show that the asymptotic distribution for grouped data in N groups has approximate covariance $\text{cov}(\hat{\boldsymbol{\beta}}) \approx (\boldsymbol{X}^T \text{diag}(\hat{\boldsymbol{\mu}})\boldsymbol{X})^{-1}$, where $\log(\hat{\boldsymbol{\mu}}) = \boldsymbol{X}\boldsymbol{\beta}$.

3.4 Let the independent observations $y_{i1}, \dots, y_{in_i}$, observed at predictor value $\boldsymbol{x}_i$, follow a Poisson distribution, $y_{ij} \sim \text{P}(\lambda_{ij})$. Then one obtains for the sum $\tilde{y}_i = \sum_{j=1}^{n_i} y_{ij} \sim \text{P}(\tilde{\lambda}_i)$, where $\tilde{\lambda}_i = \sum_j \lambda_{ij}$. Discuss modeling strategies for the observations. Consider in particular models for single variables y_{ij}, for accumulated counts $\tilde{y}_i$, and for average counts $\tilde{y}_i / n_i$.

3.5 Let $(y_i, \boldsymbol{x}_i)$ denote independent observations. A linear model with log-transformed responses is given by $\log(y_i) = \boldsymbol{x}_i^T \boldsymbol{\beta} + \epsilon_i$, $\epsilon_i \sim \text{N}(0, \sigma^2)$. Compare the model to the GLM

$$y_i | \boldsymbol{x}_i \sim \text{N}(\mu_i, \sigma^2) \quad \text{and} \quad \mu_i = \exp(\boldsymbol{x}_i^T \boldsymbol{\beta}),$$

and explain the difference between the two models.

3.6 The R Package *catdat* provides the data set *rent.*

(a) Find a GLM with response *rent* (net rent in Euro) and the explanatory variables *size* (size in square meters) and *rooms* (number of rooms) that fits the data well. Try several distribution functions like Gaussian and Gamma and try alternative links.

(b) Discuss strategies to select a model from the models fitted in (a).

Chapter 4

Modeling of Binary Data

In Chapter 2 in particular, the logit model is considered as one specific binary regression model. In this section we will discuss modeling issues for the more general binary regression model

$$P(y_i = 1|\boldsymbol{x}_i) = \pi(\boldsymbol{x}_i) = h(\boldsymbol{x}_i^T \boldsymbol{\beta}),$$

where the response function h is a fully specified function, which in the case of the logit model is the logistic distribution function $h(\eta) = \exp(\eta)/(1 + \exp(\eta))$. A general parametric binary regression model is determined by the *link function* (the inverse of the response function) and the *linear predictor*. While the link function determines the functional form of the response probabilities, the linear predictor determines which variables are included and in what form they determine the response. In particular, when categorical and metric variables are present, the linear predictor can, in addition to simple linear terms, contain polynomial versions of continuous variables, dummy variables, and interaction effects. Therefore some care should be taken when specifying constituents of the model like the linear predictor. Statistical regression modeling always entails a series of decisions concerning the structuring of the dependence of the response on the predictor. Various aspects are important when making these decisions, among them are the following:

- Discrepancy between data and model. Does the fit of the model support the inferences drawn from the model?
- Relevance of variables and form of the linear predictor. Which variables should be included and how?
- Explanatory power of the covariates.
- Prognostic power of the model.
- Choice of link function. Which link function fits the data well and has a simple interpretation.

Figure 4.1 illustrates aspects of regression modeling and the corresponding evaluation instruments. Since the evaluation of a model starts with parameter estimation, it is at the top of the panel. When estimates have been obtained one can deal with problems concerning the appropriateness of the model, the specification of the predictor, and the obtained explanatory value. Some of the tools that can be used to cope with these problems are given at the bottom of the panel. It should be noted that these aspects are not independent. A model should represent an appropriate approximation of the data when one investigates if the linear predictor may be

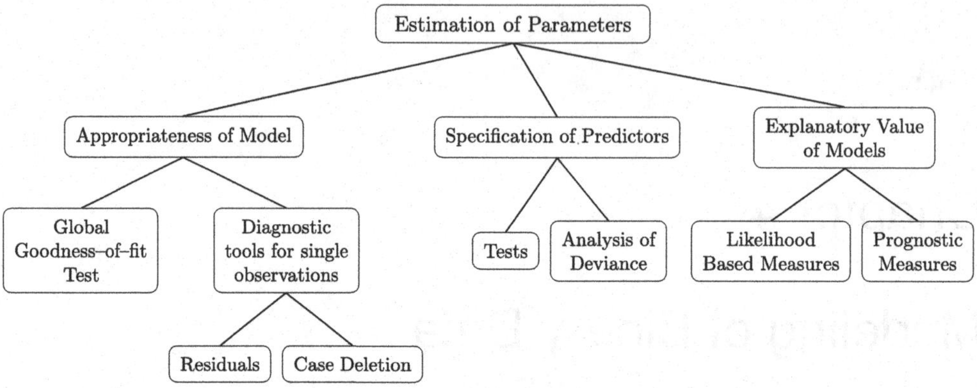

FIGURE 4.1: Aspects of regression modeling.

simplified. On the other hand, the specification determines the goodness-of-fit of the model. There is also a strong link between the specification of the linear predictor and the explanatory value of the covariates. Only the focus is different. While the specification of the linear predictor aims at finding an adequate form of the covariates and at reducing the variables to a set of variables that is really needed, the explanatory value of a model aims at quantifying the effect of the covariates within the model.

Since the estimation of parameters plays such a prominent role in modeling, the first section of this chapter is devoted to estimation, in particular maximum likelihood (ML) estimation. Since most of the tools considered in this chapter are likelihood-based, we will defer alternative estimation concepts to later sections. In this chapter the focus is on tools; therefore, only part of the model structures will be discussed. For example, the specification of the link function and weaker distributional assumptions will be considered later, in Chapter 5.

We will consider here in particular tools for the following aspects:

- Discrepancy between data and fit: global goodness-of-fit of a model (Section 4.2)
- Diagnostic tools for single observations (Section 4.3)
- Specification of the linear predictor (Section 4.4)
- Explanatory value of covariates (Section 4.6)

4.1 Maximum Likelihood Estimation

When a model is assumed to represent a useful relationship between an observed response variable and several explanatory variables, the first step in inference is the estimation of the unknown parameters. In the linear logit model

$$P(y_i = 1|\boldsymbol{x}_i) = \frac{\exp(\boldsymbol{x}_i^T\boldsymbol{\beta})}{1+\exp(\boldsymbol{x}_i^T\boldsymbol{\beta})},$$

as well as in the more general model $\pi(\boldsymbol{x}_i) = h(\boldsymbol{x}_i^T\boldsymbol{\beta})$, the parameters to be estimated are the regression parameters $\boldsymbol{\beta}$. In the following we consider the more general model, where h is a fully specified function. The logit model uses $h(\eta) = \exp(\eta)/(1+\exp(\eta))$. Alternative response functions like the normal distribution function, which yields the probit model, will be considered in Chapter 5.

The most widely used general method of estimation in binary regression models is maximum likelihood. The basic principle is to construct the likelihood of the unknown parameters for the sample data. The likelihood represents the joint probability or probability density of the observed data, considered as a function of the unknown parameters.

In the following we will distinguish between the case of single binary responses and the more general case of scaled binomials (or proportions) $\bar{y}_i$. A proportion $\bar{y}_i$ is computed from n_i independent binary observations observed at the same measurement point $\boldsymbol{x}_i$.

Single Binary Responses

For independent observations $(y_i, \boldsymbol{x}_i), i = 1, \ldots, n$, with $y_i \in \{0, 1\}$, the *likelihood* for the conditional responses $y_i|\boldsymbol{x}_i$ is given by

$$L(\boldsymbol{\beta}) = \prod_{i=1}^{n} \pi(\boldsymbol{x}_i)^{y_i}(1 - \pi(\boldsymbol{x}_i))^{1-y_i}.$$

Each term in the product represents the probability that y_i is observed since $\pi(\boldsymbol{x}_i)^{y_i}(1 - \pi(\boldsymbol{x}_i))^{1-y_i}$ simplifies to $\pi(\boldsymbol{x}_i)$ if $y_i = 1$ and to $1 - \pi(\boldsymbol{x}_i)$ if $y_i = 0$. The product is built since the observations $y_1, \ldots, y_n$, given $\boldsymbol{x}_1, \ldots, \boldsymbol{x}_n$, are considered as independent. The maximum likelihood estimates of $\boldsymbol{\beta}$ are those values $\hat{\boldsymbol{\beta}}$ that maximize the likelihood. It is usually more convenient to maximize the log-likelihood rather than the likelihood itself. Since the logarithm is a strictly monotone transformation, the obtained values $\hat{\boldsymbol{\beta}}$ will be the same. Therefore, the *log-likelihood*

$$l(\boldsymbol{\beta}) = \log(L(\boldsymbol{\beta})) = \sum_{i=1}^{n} l_i(\boldsymbol{\beta}) = \sum_{i=1}^{n} y_i \log(\pi(\boldsymbol{x}_i)) + (1 - y_i)\log(1 - \pi(\boldsymbol{x}_i))$$

is used. The value $\hat{\boldsymbol{\beta}}$ that maximizes $l(\boldsymbol{\beta})$ can be obtained by solving the system of equations $\partial l(\boldsymbol{\beta})/\partial \boldsymbol{\beta} = \mathbf{0}$. It is common to consider the derivatives that yield the equations as functions of $\boldsymbol{\beta}$. One considers the so-called *score function*:

$$\boldsymbol{s}(\boldsymbol{\beta}) = \partial l(\boldsymbol{\beta})/\partial \boldsymbol{\beta} = (\partial l(\boldsymbol{\beta})/\partial \beta_1, \ldots, \partial l(\boldsymbol{\beta})/\partial \beta_p)^T,$$

which has the form

$$\boldsymbol{s}(\boldsymbol{\beta}) = \sum_{i=1}^{n} \boldsymbol{s}_i(\boldsymbol{\beta}) = \sum_{i=1}^{n} \boldsymbol{x}_i \frac{\partial h(\eta_i)}{\partial \eta} \frac{(y_i - \pi(\boldsymbol{x}_i))}{\sigma_i^2},$$

where $\eta_i = \boldsymbol{x}_i^T \boldsymbol{\beta}$ and $\sigma_i^2 = var(y_i) = \pi(\boldsymbol{x}_i)(1 - \pi(\boldsymbol{x}_i))$. The maximum likelihood (ML) estimate is then found by solving $\boldsymbol{s}(\hat{\boldsymbol{\beta}}) = \mathbf{0}$. The system of equations has to be solved iteratively (see Section 3.9).

For the logit model one obtains by simple calculation

$$\frac{\partial l(\boldsymbol{\beta})}{\partial \beta_j} = \sum_{i=1}^{n} x_{ij}(y_i - \pi(\boldsymbol{x}_i)),$$

yielding the score function $\boldsymbol{s}(\boldsymbol{\beta}) = \sum_{i=1}^{n} \boldsymbol{x}_i(y_i - \pi(\boldsymbol{x}_i))$. Therefore, the likelihood equations to be solved are

$$\sum_{i=1}^{n} x_{ij} y_i = \sum_{i=1}^{n} x_{ij} \pi(\boldsymbol{x}_i)),$$

which equate the sufficient statistics $\sum_{i=1}^{n} x_{ij} y_i$, $j = 1 \ldots, p$ for $\boldsymbol{\beta}$ to their expected values (Exercise 4.3).

In maximum likelihood theory, the asymptotic variance of the estimator is determined by the *information* or *Fisher matrix* $\boldsymbol{F}(\boldsymbol{\beta}) = E(-\frac{\partial^2 l(\boldsymbol{\beta})}{\partial \boldsymbol{\beta} \partial \boldsymbol{\beta}^T})$. As derived in Chapter 3, $\boldsymbol{F}(\boldsymbol{\beta})$ is given by $\boldsymbol{F}(\boldsymbol{\beta}) = \boldsymbol{X}^T \boldsymbol{W} \boldsymbol{X}$, where $\boldsymbol{X}$ with $\boldsymbol{X}^T = (\boldsymbol{x}_1, \ldots, \boldsymbol{x}_n)$ is the design matrix and $\boldsymbol{W} = \text{Diag}\,((\frac{\partial h(\eta_1)}{\partial \eta})^2/\sigma_1^2, \ldots, (\frac{\partial h(\eta_n)}{\partial \eta})^2/\sigma_n^2)$ is a diagonal weight matrix.

For the logit model one obtains the simpler form $\boldsymbol{F}(\boldsymbol{\beta}) = \sum_{i=1}^{n} \boldsymbol{x}_i \boldsymbol{x}_i^T \sigma_i^2 = \boldsymbol{X}^T \boldsymbol{W} \boldsymbol{X}$ with weight matrix $\boldsymbol{W} = (\sigma_1^2, \ldots, \sigma_n^2)$. An approximation to the covariance of $\hat{\boldsymbol{\beta}}$ is given by

$$\hat{cov}(\hat{\boldsymbol{\beta}}) \approx \boldsymbol{F}(\boldsymbol{\beta})^{-1} = (\boldsymbol{X}^T \boldsymbol{W} \boldsymbol{X})^{-1},$$

where in practice $\boldsymbol{W}$ is replaced by $\hat{\boldsymbol{W}}$, which denotes the evaluation of $\boldsymbol{W}$ at $\hat{\boldsymbol{\beta}}$.

Grouped Data: Estimation for Binomially Distributed Responses

In many applications, for given values of the covariates, several independent binary responses are observed. Since $P(y = 1|\,\boldsymbol{x}) = \pi(\boldsymbol{x})$ is assumed to depend on $\boldsymbol{x}$ only, the mean is assumed to be the same for all the binary observations collected at this value. More formally, let $y_{i1}, \ldots, y_{in_i}$ denote the independent dichotomous responses collected at value $\boldsymbol{x}_i$ and let $\boldsymbol{x}_1, \ldots, \boldsymbol{x}_N$ denote the distinct values of covariates or measurement points where responses are observed. The model has the form

$$P(y_{ij} = 1|\boldsymbol{x}_i) = h(\boldsymbol{x}_i^T \boldsymbol{\beta}), j = 1, \ldots, n_i,$$

or simpler $\pi(\boldsymbol{x}_i) = h(\boldsymbol{x}_i^T \boldsymbol{\beta})$, where $\pi(\boldsymbol{x}_i) = P(y_{ij} = 1|\boldsymbol{x}_i), j = 1, \ldots, n_i$.

For the purpose of estimation one may use the original binary variables $y_{i1}, \ldots, y_{in_i}$, $i = 1, \ldots, N$, $(y_{ij} \sim B(1, \pi(\boldsymbol{x}_i))$ or, equivalently, the binomial distribution of $y_{i1} + \cdots + y_{in_i} \sim B(n_i, \pi(\boldsymbol{x}_i))$. For the collection of binary variables the likelihood has the form

$$\begin{aligned} L(\boldsymbol{\beta}) &= \prod_{i=1}^{N} \prod_{j=1}^{n_i} \pi(\boldsymbol{x}_i)^{y_{ij}} (1 - \pi(\boldsymbol{x}_i))^{1-y_{ij}} \\ &= \prod_{i=1}^{N} \pi(\boldsymbol{x}_i)^{y_i} (1 - \pi(\boldsymbol{x}_i))^{n_i - y_i}, \end{aligned} \tag{4.1}$$

where $y_i = y_{i1} + \cdots + y_{in_i}$ denotes the number of successes. The likelihood for the number of successes $y_i \sim B(n_i, \pi(\boldsymbol{x}_i))$ has the form

$$L_{bin}(\boldsymbol{\beta}) = \prod_{i=1}^{N} \binom{n_i}{y_i} \pi(\boldsymbol{x}_i)^{y_i} (1 - \pi(\boldsymbol{x}_i))^{n_i - y_i}.$$

The binary observations likelihood $L(\boldsymbol{\beta})$ and the binomial likelihood L_{bin} differ in the binomial factor, which is irrelevant in maximization because it does not depend on $\boldsymbol{\beta}$. Therefore, the relevant part of the log-likelihood (omitting the constant) is

$$\begin{aligned} l(\boldsymbol{\beta}) = \log(L(\boldsymbol{\beta})) &= \sum_{i=1}^{N} y_i \log(\pi(\boldsymbol{x}_i)) + (n_i - y_i) \log(1 - \pi(\boldsymbol{x}_i)) \\ &= \sum_{i=1}^{N} n_i \{\bar{y}_i \log\left(\frac{\pi(\boldsymbol{x}_i)}{1 - \pi(\boldsymbol{x}_i)}\right) + \log(1 - \pi(\boldsymbol{x}_i))\}, \end{aligned} \tag{4.2}$$

where $\bar{y}_i = y_i/n_i$ denotes the relative frequency. The score function is given by $\boldsymbol{s}(\boldsymbol{\beta}) = \sum_{i=1}^{n} \boldsymbol{x}_i(\partial h(\eta_i)/\partial\eta)(\bar{y}_i - \pi(\boldsymbol{x}_i))/\text{var}(\bar{y}_i)$, and the Fisher matrix, which is needed for the standard errors, is

$$\boldsymbol{F}(\boldsymbol{\beta}) = E(-\partial l^2(\boldsymbol{\beta})/\partial\boldsymbol{\beta}\partial\boldsymbol{\beta}^T) = \sum_{i=1}^{N} n_i \frac{(\partial h(\boldsymbol{x}_i^T\boldsymbol{\beta})/\partial\eta)^2}{h(\boldsymbol{x}_i^T\boldsymbol{\beta})(1 - h(\boldsymbol{x}_i^T\boldsymbol{\beta}))} \boldsymbol{x}_i\boldsymbol{x}_i^T.$$

For the logit model one obtains the score function $\boldsymbol{s}(\boldsymbol{\beta}) = \sum_{i=1}^{n} n_i\boldsymbol{x}_i(\bar{y}_i - \pi(\boldsymbol{x}_i))$ and the Fisher matrix

$$\boldsymbol{F}(\boldsymbol{\beta}) = \sum_{i=1}^{N} n_i h(\boldsymbol{x}_i^T\boldsymbol{\beta})(1 - h(\boldsymbol{x}_i^T\boldsymbol{\beta}))\boldsymbol{x}_i\boldsymbol{x}_i^T.$$

An approximation to the covariance of $\hat{\boldsymbol{\beta}}$ is again given by the inverse information matrix.

Asymptotic Properties

Under regularity assumptions, ML estimators for binary regression models have several favorable asymptotic properties ($n \to \infty$, where $n = n_1 + \cdots + n_N$ in the binomial case). The ML estimator exists and is unique asymptotically; it is consistent and asymptotically normally distributed with $\hat{\boldsymbol{\beta}} \sim N(\boldsymbol{\beta}, \boldsymbol{F}(\hat{\boldsymbol{\beta}})^{-1})$. Moreover, it is asymptotically efficient compared to a wide class of other estimates. General results specifying the necessary regularity conditions were given by Haberman (1977) and Fahrmeir and Kaufmann (1985).

Existence of Maximum Likelihood Estimates

For a finite sample size it may happen that ML estimates do not exist. It is easy to construct data structures for which the estimates tend to infinity. For a univariate predictor, let the data be separated such that all data with a predictor below a fixed value c have response 0 and above that value response 1. Then the best fit is obtained when the parameter for the predictor becomes infinitely large. The general case was investigated by Albert and Anderson (1984) and Santner and Duffy (1986). They call a dataset *completely separated* if there exists a vector $\boldsymbol{\theta}$ such that

$$\boldsymbol{x}_i^T\boldsymbol{\theta} > 0 \text{ if } y_i = 1 \text{ and } \boldsymbol{x}_i^T\boldsymbol{\theta} < 0 \text{ if } y_i = 0$$

hold for $i = 1, \ldots, n$, where $\boldsymbol{x}_i$ contains an intercept. It is called *quasi-completely separated* if there exists a vector $\boldsymbol{\theta}$ such that

$$\boldsymbol{x}_i^T\boldsymbol{\theta} \geq 0 \text{ if } y_i = 1 \text{ and } \boldsymbol{x}_i^T\boldsymbol{\theta} \leq 0 \text{ if } y_i = 0$$

holds for $i = 1, \ldots, n$. A dataset is said to have *overlap* if there is no complete separation and no quasi-complete separation. They show that the ML estimate exists if and only if the dataset has overlap. Thus, from a geometric point of view, ML estimates exist if there is no hyperplane that separates the 0 and 1 responses. Christmann and Rousseeuw (2001) showed how to compute the smallest number of observations that need to be removed to make the ML estimate non-existent. This number of observations is called the regression depth and measures the amount of separation between 0 and 1 responses. For literature on alternative estimators, in particular robust procedures, see Section 4.7.

Estimation Conditioned on Predictor Values

ML estimation as considered in the previous section is conditional on the $\boldsymbol{x}$-values. Although both variables y and $\boldsymbol{x}$ can be random, estimating the parameters of a model for $P(y = 1|\boldsymbol{x})$

does not depend on the marginal distribution of $\boldsymbol{x}$. Therefore, the ML estimates have the same form in cases where one has a total sample of *iid* observations $(y_i, \boldsymbol{x}_i)$ or a sample of responses conditional on $\boldsymbol{x}$-values.

In applications one also finds samples conditional on the response. In such a stratified sample one observes $\boldsymbol{x}$-values given $y = 1$ and $\boldsymbol{x}$-values given $y = 0$. A common case is case-control studies in biomedicine, where $y = 1$ refers to cases and $y = 0$ are controls. In both populations the potential risk factors $\boldsymbol{x}$ are observed given the population. In econometrics, this type of sampling is often called *choice-based sampling*, referring to the sampling of characteristics of a choice maker given his or her choice was made.

The association between response y and predictor $\boldsymbol{x}$ as captured in the logit model also can be inferred from samples that are conditional on responses. Let us consider the most simple case of one binary predictor. Therefore, one has $y \in \{0, 1\}$ and $x \in \{0, 1\}$. The coefficient β in the logit model $\text{logit}(P(y = 1|x)) = \beta_0 + x\beta$ is given by $\beta = \log(\gamma)$, where γ is the odds ratio, which contains the association between y and x. However, the odds ratio γ can be given in two forms:

$$\gamma = \frac{P(y=1|x=1)/P(y=0|x=1)}{P(y=1|x=0)/P(y=0|x=0)} = \frac{P(x=1|y=1)/P(x=0|y=1)}{P(x=1|y=0)/P(x=0|y=0)}.$$

The first form corresponds to the logit model $\text{logit}(P(y = 1|x)) = \beta_0 + x\beta$, the second form to the logit model $\text{logit}(P(x = 1|y)) = \tilde{\beta}_0 + y\tilde{\beta}$. For the latter model, which models response x given y, the parameter that contains the association between these variables is the same, $\tilde{\beta} = \log(\gamma) = \beta$. Therefore, ML estimation of the latter model, based on a sample given y, yields an estimate of the coefficient β of the original logit model $\text{logit}(P(y = 1|x)) = \beta_0 + x\beta$. Asymptotic properties hold for the transformed model $\text{logit}(P(x = 1|y)) = \tilde{\beta}_0 + y\tilde{\beta}$.

In general, the use of estimators for samples that are conditional on y may be motivated by the specific structure of the logit model. Therefore, we go back to the derivation of the binary logit model to assume that predictors are normally distributed given $y = r$ (Section 2.2.2). With $f(\boldsymbol{x}|r)$ denoting the density given $y = r$ and $p(r) = P(y = r)$ denoting the marginal probability, it follows from Bayes' theorem that

$$P(y = 1|\boldsymbol{x}) = \frac{\exp\{\log([p(1)f(\boldsymbol{x}|1)]/[p(0)f(\boldsymbol{x}|0)])\}}{1 + \exp\{\log([p(1)f(\boldsymbol{x}|1)]/[p(0)f(\boldsymbol{x}|0)])\}}.$$

Therefore, the linear logit model holds if

$$\log \frac{p(1)f(\boldsymbol{x}|1)}{p(0)f(\boldsymbol{x}|0)} = \beta_0 + \boldsymbol{x}^T\boldsymbol{\beta},$$

holds. The equivalent form,

$$\log \frac{f(\boldsymbol{x}|1)}{f(\boldsymbol{x}|0)} = \beta_0 - \log(p(1)/p(0)) + \boldsymbol{x}^T\boldsymbol{\beta},$$

shows that a logit model holds if $\log(f(\boldsymbol{x}|1)/f(\boldsymbol{x}|0))$ has a linear form that contains the term $\boldsymbol{x}^T\boldsymbol{\beta}$ and only the intercept depends on the marginal probabilities. The essential point is that the marginals determine only the intercept. Thus, for identical densities but different marginal probabilities, the coefficient $\boldsymbol{\beta}$, which measures the association between y and $\boldsymbol{x}$, is unchanged. The argument holds more generally when the linear term $\boldsymbol{x}^T\boldsymbol{\beta}$ is replaced by a function $\eta(\boldsymbol{x}^T, \boldsymbol{\beta})$; linearity in $\boldsymbol{x}$ is just the most prominent case. It is one of the strengths of the logit model that the parameter $\boldsymbol{\beta}$ does not depend on the marginal probabilities.

Nevertheless, the likelihood for a given y differs from the likelihood given predictors. By using $f(\boldsymbol{x}_i|y_i) = P(y_i|\boldsymbol{x}_i)f(\boldsymbol{x}_i)/p(y_i)$, one obtains for the log-likelihood conditional on y

$$l_{cond} = \sum_{i=1}^{n} \log(P(y_i|\boldsymbol{x}_i)) + \log(f(\boldsymbol{x}_i)) - \log(p(y_i)).$$

The first term on the right-hand side is equivalent to the conditional log-likelihood given $\boldsymbol{x}$-values. The second term corresponds to the marginal distribution of $\boldsymbol{x}$, which can be maximized by the empirical distribution. The third term refers to the marginal distribution of y, which is fixed by the sampling and changes the intercept. Of course it has to be shown that maximization of l_{cond}, which includes non-parametric maximization of the marginal distribution of $\boldsymbol{x}$, yields coefficient estimates $\boldsymbol{\beta}$ that have the properties of ML estimates. For more details of choice-based sampling see Prentice and Pyke (1979), Carroll et al. (1995), and Scott and Wild (1986).

4.2 Discrepancy between Data and Fit

4.2.1 The Deviance

Before drawing inferences from a fitted model it is advisable to critically assess the fit. Thus, when fitting a regression model one usually wants some measure for the discrepancy between the fitted model and the observations. It should be obvious that the sum of the squared residuals $\sum_i (y_i - \hat{\pi}_i)^2$ that is used in normal regression models is inappropriate because it is designed for a symmetric distribution like the normal and in addition assumes homogeneous variances. It does not account for the specific type of distribution (and noise) when responses are binary. If the unknown parameters are estimated by maximum likelihood, a common measure for the discrepancy between the data and the fitted model that is specific for the underlying distribution of responses is the deviance. The deviance may be seen as a comparison between the data fit and the perfect fit.

The deviance is strongly related to basic concepts of statistics. A basic test statistic that is used to evaluate nested models is the *likelihood ratio statistic*:

$$\lambda = -2 \log \frac{L(\text{submodel})}{L(\text{model})},$$

where L(model) represents the maximal likelihood when a model is fit and L(submodel) represents the maximal likelihood if a more restrictive model, a so-called submodel, is fit. By considering the submodel as the binary regression model and the model as the most general possible model (with perfect fit), one considers

$$\begin{aligned} \lambda &= -2\{\log L(\text{fitted submodel}) - \log L(\text{fitted model})\} \\ &= -2\{l(\text{fitted submodel}) - l(\text{fitted model})\}. \end{aligned}$$

The most general model that produces a perfect fit is often called the *saturated model.* It is assumed to have as many parameters as observations: therefore it is the perfect fit.

Deviance for Binary Response

Suppose that the data are given by $(y_i, \boldsymbol{x}_i), i = 1, \dots, n$, where $y_i \in \{0, 1\}$. Let $l(\boldsymbol{y}; \hat{\boldsymbol{\pi}})$ denote the log-likelihood for the fitted model with $\boldsymbol{y}^T = (y_1, \dots, y_n)$, $\hat{\boldsymbol{\pi}}^T = (\hat{\pi}_1, \dots, \hat{\pi}_n)$, $\hat{\pi}_i = \hat{\pi}(\boldsymbol{x}_i) = h(\boldsymbol{x}_i^T \hat{\boldsymbol{\beta}})$. The perfectly fitting model is represented by the likelihood $l(\boldsymbol{y}; \boldsymbol{y})$, where the fitted values are the observations themselves. Then the *deviance* for the binary responses is given by the difference of $l(\boldsymbol{y}, \boldsymbol{y})$ and $l(\boldsymbol{y}, \hat{\boldsymbol{\pi}})$:

$$\begin{aligned} D(\boldsymbol{y}, \hat{\boldsymbol{\pi}}) &= 2\{l(\boldsymbol{y}, \boldsymbol{y}) - l(\boldsymbol{y}, \hat{\boldsymbol{\pi}})\} \\ &= 2\left\{\sum_{i=1}^{n} y_i \log\left(\frac{y_i}{\hat{\pi}_i}\right) + (1-y_i)\log(\frac{1-y_i}{1-\hat{\pi}_i})\right\} \\ &= -2\sum_{i=1}^{n}\{y_i \log(\hat{\pi}_i) + (1-y_i)\log(1-\hat{\pi}_i)\}, \end{aligned} \quad (4.3)$$

where the convention $0 \cdot \infty = 0$, is used. Since $l(\boldsymbol{y}, \boldsymbol{y}) = 0$, the deviance reduces to $D(\boldsymbol{y}, \hat{\boldsymbol{\pi}}) = -2l(\boldsymbol{y}, \hat{\boldsymbol{\pi}})$. An alternative form of the deviance is

$$D(\boldsymbol{y}, \hat{\boldsymbol{\pi}}) = 2\sum_{i=1}^{n} d(y_i, \hat{\pi}_i), \quad (4.4)$$

where

$$d(y_i, \hat{\pi}_i) = \begin{cases} -\log(\hat{\pi}_i) & y_i = 1 \\ -\log(1-\hat{\pi}_i) & y_i = 0. \end{cases}$$

The simpler form,

$$d(y_i, \hat{\pi}_i) = -\log(1 - |y_i - \hat{\pi}_i|),$$

shows that for binary data the deviance implicitly uses the difference between observations and fitted values. From the latter form it is also seen that $D(\boldsymbol{y}, \hat{\boldsymbol{\pi}}) \geq 0$ and $D(\boldsymbol{y}, \hat{\boldsymbol{\pi}}) = 0$ only if $\hat{\boldsymbol{\pi}} = \boldsymbol{y}$.

In extreme cases the deviance degenerates as a measure of goodness-of-fit. Consider the simple case of a single sample and n independent Bernoulli variables. Let n_1 denote the number of "hits" ($y_i = 1$) and $n_2 = n - n_1$ the number of observations $y_i = 0$. Since $\hat{\pi}_i$ is the same for all observations, one obtains with $\hat{\pi} = n_1/n$

$$\begin{aligned} D(\boldsymbol{y}, \hat{\boldsymbol{\pi}}) &= -2(n_1 \log(\hat{\pi}) + n_2 \log(1-\hat{\pi})) \\ &= -2n\{\hat{\pi}\log(\hat{\pi}) + (1-\hat{\pi})\log(1-\hat{\pi})\}, \end{aligned}$$

which is completely determined by the relative frequency $\hat{\pi}$ and therefore does not reflect the goodness-of-fit. The asymptotic distribution depends heavily on π with a degenerate limit as $n \to \infty$.

Another concept that is related to the deviance is the Kullback-Leibler distance, which is a general directed measure for the distance between two distributions (Appendix D). For discrete distributions with support $\{z_1, \ldots, z_m\}$ and the vectors of probabilities for the two probability functions given by $\boldsymbol{\pi} = (\pi_1, \ldots, \pi_m)^T$ and $\boldsymbol{\pi}^* = (\pi_1^*, \ldots, \pi_m^*)^T$, it has the form

$$KL(\boldsymbol{\pi}, \boldsymbol{\pi}^*) = \sum_{i=1}^{m} \pi_i \log(\frac{\pi_i}{\pi_i^*}).$$

Considering the data as a degenerate mass function $(1 - y_i, y_i)$ with support $\{0, 1\}$, one obtains

$$D(\boldsymbol{y}, \hat{\boldsymbol{\pi}}) = 2\sum_{i=1}^{n} KL((1-y_i, y_i), (1-\hat{\pi}_i, \hat{\pi}_i)).$$

The maximum likelihood estimator may be motivated as a minimum distance estimator that minimizes the sums of the Kullback-Leibler distances considered in the last equation.

Deviance for Proportions

Let the data be given in the form $(y_i, \boldsymbol{x}_i), i = 1, \ldots, N$, where the y_i's are binomially distributed. Each response variable $y_i \sim B(n_i, \pi(\boldsymbol{x}_i))$ can be thought of as being composed from n_i binary variables. Therefore one should distinguish between the total number of (binary) observations $n = n_1 + \cdots + n_N$ and the number of observed binomials. The model-specific probabilities $\pi_i = \pi(\boldsymbol{x}_i)$ depend only on the measurement points $\boldsymbol{x}_1, \ldots, \boldsymbol{x}_N$; thus only N potentially differing probabilities are specified.

Let $\bar{\boldsymbol{y}}^T = (\bar{y}_1, \ldots, \bar{y}_N)$ denote the observations where $\bar{y}_i = y_i/n_i$ represents the relative frequencies that are linked to $y_i = y_{i1} + \cdots + y_{in_i} \sim B(n_i, \pi_i)$, the number of successes for the repeated trials at measurement point $\boldsymbol{x}_i$. Then the corresponding difference of log-likelihoods based on (4.2) is given by

$$D(\bar{\boldsymbol{y}}, \hat{\boldsymbol{\pi}}) = 2(l(\bar{\boldsymbol{y}}, \bar{\boldsymbol{y}}) - l(\bar{\boldsymbol{y}}, \hat{\boldsymbol{\pi}})) = 2\sum_{i=1}^{N} n_i\{\bar{y}_i \log(\frac{\bar{y}_i}{\hat{\pi}_i}) + (1 - \bar{y}_i)\log(\frac{1-\bar{y}_i}{1-\hat{\pi}_i})\}.$$

The essential difference between the deviance for single binary observations and the deviance for binomial distributions is in the definition of the saturated model. While in the first case observations $y_1, \ldots, y_n$ are fitted perfectly by the saturated model, in the latter case the means $\bar{y}_1, \ldots, \bar{y}_N$ are fitted perfectly. This has severe consequences when trying to use the deviance as a measure of the goodness-of-fit of the model.

Deviance as Goodness-of-Fit Statistic

Since the deviance may be derived as a likelihood ratio statistic, it is tempting to assume that the deviance is asymptotically χ^2-distributed. However, this does not hold for the binary variables deviance $D(\boldsymbol{y}, \hat{\boldsymbol{\pi}})$ if $n \to \infty$. The reason is that the degrees of freedom are not fixed; they increase with the sample size. Thus there is no benchmark (in the form of an approximate distribution) to which the absolute value of $D(\boldsymbol{y}, \hat{\boldsymbol{\mu}})$ may be compared. The case is different for binomial distributions. If N, the number of measurement points, is fixed, and $n_i \to \infty$ for $i = 1, \ldots, N$, then the degrees of freedom are fixed, and, if the model holds, $D(\bar{\boldsymbol{y}}, \hat{\boldsymbol{\pi}})$ is under weak conditions asymptotically χ^2-distributed with $D(\bar{\boldsymbol{y}}, \hat{\boldsymbol{\pi}}) \stackrel{(a)}{\sim} \chi^2(N - dim(\boldsymbol{x}))$, where $dim(\boldsymbol{x})$ denotes the length of the vector $\boldsymbol{x}$ that is equivalent to the number of estimated parameters. Thus, for n_i sufficiently large, the χ^2-distribution provides a benchmark to which the value of $D(\bar{\boldsymbol{y}}, \hat{\boldsymbol{\pi}})$ may be compared. Usually the model is considered to show an unsatisfactory fit, if $D(\bar{\boldsymbol{y}}, \hat{\boldsymbol{\pi}})$ is larger than the $1 - \alpha$ quantile of the χ^2-distribution for user-specified significance value α.

Nevertheless, since program packages automatically give the deviance, it is tempting to compare the value of the deviance to the (often absurdly high) degrees of freedom that are found for ungrouped binary data (see Example 4.1). Although the deviance does not provide a goodness-of-fit statistic for simple binary observations, it is useful in residual analysis and informal comparisons of link functions (see Sections 4.3 and 5.1).

TABLE 4.1: Main effects model for unemployment data.

	Estimate	Std. Error	z-value	Pr(>|z|)
Intercept	1.2140	0.2032	5.97	0.0000
Age	−0.0312	0.0060	−5.18	0.0000
Gender	0.6710	0.1393	4.82	0.0000

Example 4.1: Unemployment

In a study on the duration of unemployment with sample size $n = 982$ we distinguish between short-term unemployment (≤ 6 months) and long-term unemployment (> 6 months). Short-term unemployment is considered as success ($y = 1$). The response and the covariates are gender (1: male; 0: female) and age ranging from 16 to 61 years of age. From the estimates of coefficients in Table 4.1 it is seen that the probability of short-term unemployment is larger for males than for females and decreases with age. The deviance for ungrouped data, where the response is a 0-1 variable, is 1224.1 on 979 *df*. That deviance cannot be considered as a goodness-of-fit statistic. However, if data are grouped with the response represented by the binomial distribution given a fixed combination of gender and age, the deviance is 87.16 on 88 *df*. For the grouped case, 92 combinations of gender and age effects are possible; $N = 91$ of these combinations are found in the data; therefore, the *df* are $91 - 3$ since three parameters have been fitted. In the grouped case the deviance has an asymptotic distribution and one finds that the main effect model is acceptable. □

Example 4.2: Commodities in Household

In Table 4.2 the fit of two linear logit models is shown. In the first model the response is "car in household," and in the second model it is "personal computer (pc) in household." For both models the only covariate is net income and only one-person households are considered. Since the responses are strictly binary, the deviance cannot be considered as a goodness-of-fit statistic that is asymptotically χ^2-distributed. Thus the values of the deviance cannot be compared to the degrees of freedom. But since the number of observations and the linear predictor are the same for both models, one might compare the deviance of the two models in an informal way. The linear logit model seems to fit the data better when the response "pc in household" is considered rather than "car in household." However, it is not evaluated whether the fit is significantly better. □

TABLE 4.2: Effects of net income on alternative responses for linear logit model.

	$\hat{\beta}_0$	$\hat{\beta}$	Deviance	*df*
Car	−2.42	0.0019	1497.7	1294
PC	−3.88	0.0011	614.8	1294

4.2.2 Pearson Statistic

When proportions are considered and the n_i is large, the deviance yields a goodness-of-fit statistic. An alternative statistic in this setting is the *Pearson statistic*:

$$\chi_P^2 = \sum_{i=1}^{N} n_i \frac{(\bar{y}_i - \hat{\pi}_i)^2}{\hat{\pi}_i(1 - \hat{\pi}_i)},$$

where $\hat{\pi}_i = h(\boldsymbol{x}_i^T \hat{\boldsymbol{\beta}})$. If N is fixed and $n_i \to \infty$ for $i = 1, \dots, N$, then χ_P^2 is also asymptotically χ^2-distributed with $N - dim(\boldsymbol{x})$ degrees of freedom given the model holds.

The Pearson statistic is familiar from introductory texts on statistics. In the analysis of contingency tables it usually takes the form

$$\chi^2 = \sum_{\text{cells}} \frac{(\text{observed cell counts} - \text{expected cell counts})^2}{\text{expected cell counts}}, \tag{4.5}$$

where the expected cell counts are computed under the assumption that the model under investigation holds. Thus one considers the discrepancy between the actual counts and what is to be expected if the model holds. The denominator is a weight on the squared differences that takes differences less serious if the expected cell counts are large. In fact, this weighting scheme is a standardization to obtain the asymptotic χ^2-distribution. This is seen from showing that χ^2_P has the form (4.5). The fitting of the binary regression model $\pi_i = h(\boldsymbol{x}_i^T \boldsymbol{\beta})$ may be seen within the framework of contingency analysis with $\boldsymbol{x}_1, \dots, \boldsymbol{x}_N$ refering to the rows and $y = 0, y = 1$ refering to the columns. Then the probabilities within one row (corresponding to $\boldsymbol{x}_i$) are given by $\pi_i, 1 - \pi_i$; the cell counts are given by $n_i \bar{y}_i, n_i - n_i \bar{y}_i$; and the (estimated) expected cell counts are given by $n_i \hat{\pi}_i, n_i - n_i \hat{\pi}_i$. From (4.5) one obtains

$$\chi^2 = \sum_{i=1}^{N} \frac{(n_i \bar{y}_i - n_i \hat{\pi}_i)^2}{n_i \hat{\pi}_i} + \frac{(n_i \bar{y}_i - n_i \hat{\pi}_i)^2}{n_i - n_i \hat{\pi}_i},$$

which, after simple derivation, turns out to be equivalent to χ^2_P. The essential difference between the representation (4.5) and χ^2_P is that in the form (4.5) the sum is across all cells of the corresponding contingency table, while in χ^2_P the sum is only across the different values of $\boldsymbol{x}_i$, which correspond to rows.

The deviance and Pearson's χ^2 have the same asymptotic distribution. More concisely, one postulates for fixed N that $n = n_1 + \cdots + n_N \to \infty$ and $n_i/N \to \lambda_i$, where $\lambda_i \in (0, 1)$. To distinguish these assumptions from the common asymptotical conditions where only $n \to \infty$ is assumed, one uses the term *fixed cells asymptotics*, referring to the assumption that the number of rows N is fixed. If the values of the deviance D and χ^2_P differ strongly, this may be taken as a hint that the requirement of fixed cells asymptotics does not hold and both test statistics are not reliable (see also Section 8.6.2, where alternative asymptotic concepts that hold for the more general power-divergence family of goodness-of-fit statistics are briefly discussed).

Goodness-of-Fit Statistics for Proportions

Pearson Statistic

$$\chi^2_P = \sum_{i=1}^{N} n_i \frac{(\bar{y}_i - \hat{\pi}_i)^2}{\hat{\pi}_i (1 - \hat{\pi}_i)}$$

Deviance

$$D = 2 \sum_{i=1}^{N} n_i \left\{ \bar{y}_i \log\left(\frac{\bar{y}_i}{\hat{\pi}_i}\right) + (1 - \bar{y}_i) \log\left(\frac{1 - \bar{y}_i}{1 - \hat{\pi}_i}\right) \right\}$$

Approximation $(n_i/N \to \lambda_i)$

$$\chi^2_P, D \overset{(a)}{\sim} \chi^2(N - p)$$

A general problem with global or omnibus tests to assess the fit of a parametric model is that large values of the test statistic indicate lack-of-fit but do not show the particular reason for the lack-of-fit. It is only by comparing fits that one may get some insight about its nature. In Section 4.4 nested models are compared that differ in the predictor. In that case the difference of deviances has a fixed asymptotic distribution. In the case of different link functions, considered in Section 5.1, the models are not nested, but a comparison of the goodness-of-fit measures may show which link function has the best fit.

4.2.3 Goodness-of-Fit for Continuous Predictors

The deviance and Pearson's χ_P^2 may be used to assess the adequacy of the model when the response is binomial with not too small cell counts. They cannot be used when predictors are continuous since then cell counts would always be one and there would be no benchmark in the form of a χ^2-statistic available (see also Example 4.2). An easy way out seems to be to categorize all of the continuous variables. However, then one looses power and it works only in the case of one or two continuous variables. When more continuous variables are in the model it is not clear how to find categories that contain enough observations. Several methods have been proposed to assess the fit of a model when continuous variables are present. We briefly consider some of them in the following.

Hosmer-Lemeshow Test

Hosmer and Lemeshow (1980) proposed a Pearson-type test statistic where the categorizing is based on the fitted values. One orders the responses according to the fitted probabilities and then forms N equally sized groups. Thus the quantiles of the response values determine the groups. More precisely, the n/N observations with the smallest fitted probabilities form the first group, the next n/N observations form the second group, and so on. Hosmer and Lemeshow (1980) proposed to use $N = 10$ groups that are called "deciles of risk."

Let y_{ij} denote the jth observation in the ith group, where $i = 1, \ldots, N, j = 1, \ldots, n_i$. Then the average of the observations of the ith group, $\bar{y}_i = \sum_{j=1}^{n_i} y_{ij}/n_i$, is compared to the average of the fitted probabilities, $\hat{\pi}_i = \sum_{j=1}^{n_i} \hat{\pi}_{ij}/n_i$, where $\hat{\pi}_{ij}$ denotes the fitted value for observation y_{ij} in a Pearson-type statistic:

$$\chi_{HL}^2 = \sum_{i=1}^{N} n_i \frac{(\bar{y}_i - \hat{\pi}_i)^2}{\hat{\pi}_i(1 - \hat{\pi}_i)}.$$

The test statistic has a rather complicated distribution, but Hosmer and Lemeshow (1980) showed by simulations that the asymptotic distribution can be approximated by a χ^2-distribution with $df = N - 2$.

Since the grouping is based on the model that is assumed to hold, one cannot expect good power for the Hosmer-Lemeshow statistic, and indeed the test statistic has moderate power. More importantly, as for all global tests, a large value of the test statistic indicates lack-of-fit but does not show if, for example, the linear term has been misspecified.

A methodology that is similar to the Hosmer-Lemeshow approach but makes use of categorical variables in the model has been proposed by Pulkstenis and Robinson (2002). An adjusted Hosmer-Lemeshow test that uses an alternative standardization was proposed by Pigeon and Heyse (1999). Kuss (2002) showed that Hosmer-Lemeshow-type tests are numerically unstable.

Alternative Tests

Alternative test strategies may be based on the score test. Brown (1982) embedded the logit model into a family of models and proposed a test for the specific parameters that select the logit model. Tsiatis (1980) divided the space of the covariate into distinct regions and tested if a term that is constant in these regions may be omitted from the linear predictor.

An alternative approach for goodness-of-fit tests uses non-parametric regression techniques. Azzalini et al. (1989) proposed a pseudo-likelihood ratio test by using a kernel-smoothed estimate of the response probabilities. For binomial data, $y_i \sim B(n_i, \pi(x_i))$, a smooth nonparametric estimate of the response probability (for a unidimensional predictor) is

$$\hat{\pi}(x) = \sum_{i=1}^{N} y_i K(\frac{x - x_i}{h}) / \sum_{i=1}^{N} n_i K(\frac{x - x_i}{h}).$$

The hypotheses to be investigated are

$$H_0: \quad \pi(x) = \pi(x, \boldsymbol{\beta}) \text{ specified by a parametric model}$$
$$H_1: \quad \pi(x) \text{ is a smooth function.}$$

The corresponding pseudo-likelihood ratio statistic is

$$\sum_{i=1}^{N} y_i \log(\frac{\hat{\pi}(x_i)}{\pi(x, \hat{\boldsymbol{\beta}})}) + (n_i - y_i) \log(\frac{1 - \hat{\pi}(x_i)}{1 - \pi(x, \hat{\boldsymbol{\beta}})}),$$

where $\hat{\pi}(x)$ is the smoothed estimate based on a selected smoothing parameter h. Since the test statistic is not asymptotically χ^2, the null hypothesis behavior of the test statistic is examined by simulating data from the fitted parametric model.

Rather than estimating the probability by smoothing techniques, one may also use smooth estimates of the standardized residuals. LeCessie and van Houwelingen (1991) smooth the (ungrouped) Pearson residuals $r(x_i) = r_P(y_i, \hat{\pi}_i)$ (see next section) to obtain $\tilde{r}(x) = \sum_i r(x_i) K(\frac{x-x_i}{h}) / \sum_i K(\frac{x-x_i}{h})$. The test statistic has the form $T = \sum_i \tilde{r}(x_i) v(x_i)$, where $v(x_i) = \sum_i \{K(x - x_i)/h)\}^2 / \sum_i K((x - x_i)/h)^2$ is the inverse of the variance of the smoothed residuals. LeCessie and van Houwelingen (1991) derive the mean and the variance and demonstrate that the test statistic may be approximated by a normal distribution. The equivalence to a score test in a random effects model was shown by LeCessie and van Houwelingen (1995). The approaches may be extended to multivariate predictors by using multivariate kernels but are restricted to few dimensions since kernel-based estimates for higher dimensions suffer from the curse of dimensionality (see Section 10.1.4).

4.3 Diagnostic Checks

Goodness-of-fit tests provide only global measures of the fit of a model. They tell nothing about the reason for a bad fit. Regression diagnostics aims at identifying reasons for it. Diagnostic measures should in particular identify observations that are not well explained by the model as well as those that are influential for some aspect of the fit.

4.3.1 Residuals

The goodness-of-fit statistics from the preceding section provide global measures for the discrepancy between the data and the fit. In particular, if the fit is bad, one wants to know if all the observations contribute to the lack-of-fit or if the effect is due to just some observations. Residuals measure the agreement between single observations and their fitted values and help to identify poorly fitting observations that may have a strong impact on the overall fit of the model. In the following we consider responses from a binomial distribution.

For scaled binomial data the *Pearson residual* has the form

$$r_P(\bar{y}_i, \hat{\pi}_i) = \frac{\bar{y}_i - \hat{\pi}_i}{\sqrt{\hat{\pi}_i(1 - \hat{\pi}_i)/n_i}}.$$

It is just the raw residual scaled by the estimated standard deviation of $\bar{y}_i$. It is also the signed square root of the corresponding component of the Pearson statistics that has the form $\chi_P^2 = \sum_i r_P(\bar{y}_i, \hat{\pi}_i)^2$. For small n_i the distribution of $r_P(\bar{y}_i, \hat{\pi}_i)$ is rather skewed, an effect that is ameliorated by using the transformation to *Anscombe* residuals:

$$r_A(\bar{y}_i, \hat{\pi}_i) = \sqrt{n_i}\frac{t(\bar{y}_i) - [t(\hat{\pi}_i) + (\hat{\pi}_i(1-\hat{\pi}_i))^{-1/3}(2\hat{\pi}_i - 1)/6n_i]}{(\hat{\pi}_i(1-\hat{\pi}_i))^{1/6}},$$

where $t(u) = \int_0^u s^{-1/3}(1-s)^{-1/3}ds$ (see Pierce and Schafer, 1986). Ansombe residuals consider an approximation to

$$\frac{t(\bar{y}_i) - E(t(\bar{y}_i))}{\sqrt{var(t(\bar{y}_i))}}$$

by use of the delta method, which yields

$$\frac{t(\bar{y}_i) - t(E(t(\bar{y}_i)))}{t'(E(\bar{y}_i))\sqrt{var(\bar{y}_i)}}.$$

The Pearson residual cannot be expected to have unit variance because the variance of the residual has not been taken into account. The standardization in $r_P(\bar{y}_i; \hat{\pi}_i)$ just uses the estimated standard deviation of $\bar{y}_i$. As shown in Section 3.10, the variance of the residual vector may be approximated by $\boldsymbol{\Sigma}^{1/2}(\boldsymbol{I} - \boldsymbol{H})\boldsymbol{\Sigma}^{T/2}$, where the hat matrix is given by $\boldsymbol{H} = \boldsymbol{W}^{T/2}\boldsymbol{X}(\boldsymbol{X}^T\boldsymbol{W}\boldsymbol{X})^{-1}\boldsymbol{X}^T\boldsymbol{W}^{1/2}$. Therefore, when looking for ill-fitting observations, one prefers the *standardized Pearson residuals*:

$$r_{P,s}(\bar{y}_i, \hat{\pi}_i) = \frac{\bar{y}_i - \hat{\pi}_i}{\sqrt{(1-h_{ii})\hat{\pi}_i(1-\hat{\pi}_i)/n_i}},$$

where h_{ii} denotes the ith diagonal element of $\boldsymbol{H}$. The standardized Pearson residuals are simply the Pearson residuals divided by $\sqrt{1-h_{ii}}$.

Alternative residuals derive from the deviance. From the basic form of the deviance $D = \sum_i r_D(y_i, \hat{\pi}_i)^2$ one obtains

$$r_D(\bar{y}_i, \hat{\pi}_i) = sign(\bar{y}_i - \hat{\pi}_i)\sqrt{n_i\{\bar{y}_i \log\left(\frac{\bar{y}_i}{\hat{\pi}_i}\right) + (1-\bar{y}_i)\log\left(\frac{1-\bar{y}_i}{1-\hat{\pi}_i}\right)\}},$$

where $sign(\bar{y}_i - \hat{\pi}_i)$ is 1 when $\bar{y}_i \geq \hat{\pi}_i$ and is -1 when $\bar{y}_i < \hat{\pi}_i$. For the special case $n_i = 1$ it simplifies to

$$r_D(y_i, \hat{\pi}_i) = sign(y_i - \hat{\pi}_i)\sqrt{-\log(1 - |y_i - \hat{\pi}_i|)}.$$

A transformation that yields a better approximation to the normal distribution is the *adjusted residuals*:

$$r_D\hat{a}(\bar{y}_i, \hat{\pi}_i) = r_D(\bar{y}_i, \hat{\pi}_i) + (1 - 2\hat{\pi}_i)/\sqrt{n_i\hat{\pi}_i(1-\hat{\pi}_i) * 36}.$$

Standardized deviance residuals are obtained by dividing by $\sqrt{1-h_{ii}}$:

$$r_{D,s}(\bar{y}_i, \hat{\pi}_i) = \frac{r_D(\bar{y}_i, \hat{\pi}_i)}{\sqrt{1-h_{ii}}}.$$

Typically residuals are visualized in a graph. In an index plot the residuals are plotted against the observation number, or index. It shows which observations have large values and may be considered outliers. For finding systematic deviations from the model it is often more informative to plot the residuals against the fitted linear predictor. If one suspects that particular variables should be transformed before being included in the linear predictor, one may also plot residuals against the ordered fitted values of that explanatory variable.

An alternative graph compares the standardized residuals to the order statistic of an $N(0,1)$-sample. In this plot the residuals are ordered and plotted against the corresponding quantiles of a normal distribution. If the model is correct and residuals can be expected to be approximately normally distributed (depending on local sample size), the plot should show approximately a straight line as long as outliers are absent.

Example 4.3: Unemployment

In a study on the duration of unemployment with sample size $n = 982$ we distinguish between short-term unemployment (≤ 6 months) and long-term unemployment (> 6 months). For illustration, a linear logit model is fitted with the covariate age, ranging from 16 to 61 years of age. Figure 4.2 shows the fitted response function. It is seen that in particular for older unemployed persons, the fitted values tend to be larger than the observed proportions. The effect is also seen in the plot of residuals against fitted values in Figure 4.3. The quantile plot in Figure 4.3 shows a rather straight line but the slope is rather steep, indicating that the variability of deviances differs from that of a standard normal distribution. A less restrictive nonparametric fit of the data is considered in Example 10.1. □

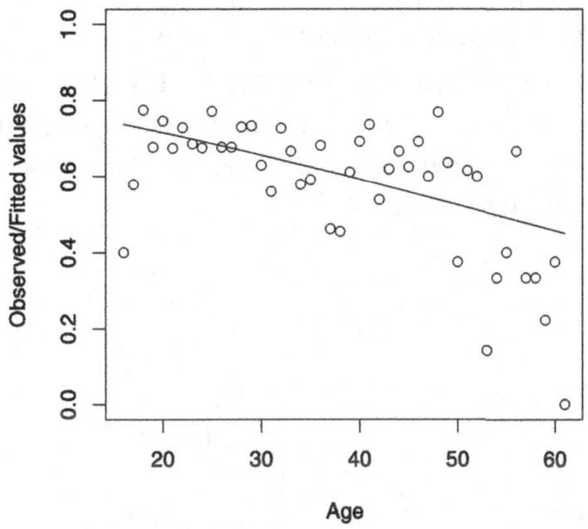

FIGURE 4.2: Observations and fitted probabilities for unemployment data plotted against age.

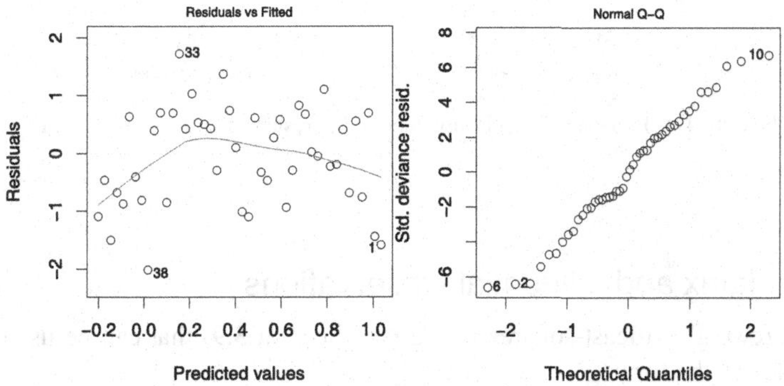

FIGURE 4.3: Deviance residuals for unemployment data plotted against fitted values (left) and quantile plot (right).

TABLE 4.3: Short- and long-term unemployment depending on age

Observ.		1	2	3	4	5	6	7	8	9	10	11	12	13	14	15	16
Age		16	17	18	19	20	21	22	23	24	25	26	27	28	29	30	31
	1	2	11	31	42	50	54	43	35	25	27	21	21	19	22	17	14
	0	3	8	9	20	17	26	16	16	12	8	10	10	7	8	10	11
Observ.		17	18	19	20	21	22	23	24	25	26	27	28	29	30	31	32
Age		32	33	34	35	36	37	38	39	40	41	42	43	44	45	46	47
	1	8	14	11	13	15	6	5	11	9	14	7	13	4	10	9	9
	0	3	7	8	9	7	7	6	7	4	5	6	8	2	6	4	6
Observ.		33	34	35	36	37	38	39	40	41	42	43	44	45	46		
Age		48	49	50	51	52	53	54	55	56	57	58	59	60	61		
	1	10	7	3	8	3	1	2	2	2	3	2	2	3	0		
	0	3	4	5	5	2	6	4	3	1	6	4	7	5	1		

Example 4.4: Food-Stamp Data

Data that have often been used in diagnostics of binary data are the food-stamp data from Künsch et al. (1989), which consist of $n = 150$ persons, 24 of whom participated in the federal food-stamp program. The response indicates participation, and the predictor variables represent the dichotomous variables tenancy (TEN), supplemental income (SUP), as well as the log-transformation of the monthly income, $\log(\text{monthly income} + 1)$ (LMI). Künsch et al. (1989) show that two values are poorly accounted for by the logistic model and are most influential for the ML fit. If these two observations and four other observations are left out, the data have no overlap and the ML estimate does not exist. Figure 4.4 shows quantile plots for Pearson and Anscombe residuals. It is seen that Anscombe residuals are much closer to the normal distribution than Pearson residuals. □

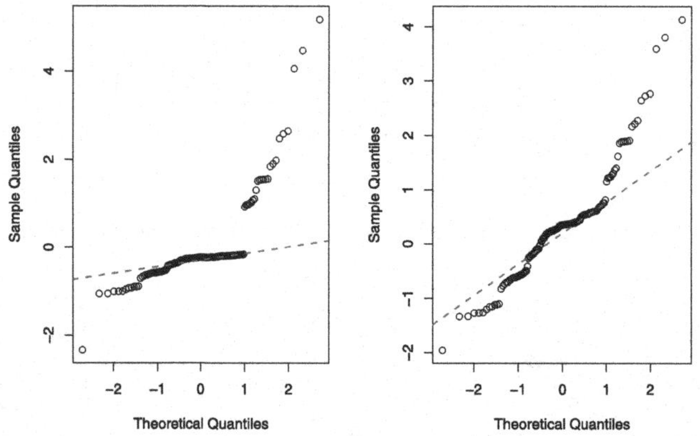

FIGURE 4.4: Pearson (left) and Anscombe (right) residuals for food-stamp data.

4.3.2 Hat Matrix and Influential Observations

The iteratively reweighted least-squares fitting (see Section 3.9) that can be used to compute the ML estimate has the form

$$\hat{\boldsymbol{\beta}}^{(k+1)} = (\boldsymbol{X}^T \boldsymbol{W}(\hat{\boldsymbol{\beta}}^{(k)})\boldsymbol{X})^{-1}\boldsymbol{X}^T \boldsymbol{W}(\hat{\boldsymbol{\beta}}^{(k)})\tilde{\eta}(\hat{\boldsymbol{\beta}}^{(k)})$$

with the adjusted variable $\tilde{\eta}(\hat{\beta}) = \hat{\eta} + D(\hat{\beta})^{-1}(y - \mu(\hat{\beta}))$, where $\hat{\eta} = X\hat{\beta}$ and the diagonal matrices $W = \text{Diag}\,((\partial h(\hat{\eta}_1)/\eta)^2/\sigma_1^2, \ldots, (\partial h(\hat{\eta}_n)/\partial\eta)^2/\sigma_n^2)$ and $D(\hat{\beta}) = (\partial h(\hat{\eta}_1)/\partial\eta, \ldots, \partial h(\hat{\eta}_n)/\partial\eta)$. At convergence one obtains

$$\hat{\beta} = (X^T W(\hat{\beta})X)^{-1} X^T W(\hat{\beta})\tilde{\eta}(\hat{\beta}).$$

Thus $\hat{\beta}$ may be seen as the least-squares solution of the linear model

$$W^{1/2}\tilde{\eta}(\hat{\beta}) = W^{1/2}X\beta + \tilde{\varepsilon},$$

where, in $W = W(\hat{\beta})$, the dependence on $\hat{\beta}$ is suppressed. The corresponding hat matrix has the form

$$H = W^{1/2}X(X^T WX)^{-1}X^T W^{1/2}.$$

Since the matrix H is idempotent and symmetric, it may be seen as a projection matrix for which $\text{tr}(H) = \text{rank}(H)$ holds. Moreover, one obtains for the diagonal elements of $H = (h_{ij})$ $0 \leq h_{ij} \leq 1$ and $\text{tr}(H) = p$ (if X has full rank).

The equation $W^{1/2}\hat{\eta} = HW^{1/2}\tilde{\eta}(\beta)$ shows how the hat matrix maps the adjusted variable $\tilde{\eta}(\beta)$ into the fitted values $\hat{\eta}$. Thus H may be seen as the matrix that maps the adjusted observation vector $W^{1/2}\tilde{\eta}$ into the vector of "fitted" values $W^{1/2}\hat{\eta}$, which is a mapping on the transformed predictor space. Moreover, it may be shown that approximately

$$\Sigma^{-1/2}(\hat{\mu} - \mu) \simeq H\Sigma^{-1/2}(y - \mu) \tag{4.6}$$

holds, where $\Sigma = \Sigma(\hat{\beta})$. Thus $H = (h_{ij})$ may be seen as a measure of the influence of y on $\hat{\mu}$ in standardized units of changes.

In summary, large values of h_{ii} should be useful in detecting influential observations. But it should be noted that, in contrast to the normal regression model, the hat matrix depends on $\hat{\beta}$ because $W = W(\hat{\beta})$. The essential difference from an ordinary linear regression is that the hat matrix does depend not only on the design but also on the fit.

A further property that is not too hard to derive is

$$H\Sigma^{-1/2}\hat{\mu} = H\Sigma^{-1/2}y,$$

meaning that the orthogonal projection (based on H) of standardized values $\Sigma^{-1/2}\hat{\mu}$, $\Sigma^{-1/2}y$ are identical. With $\chi = \Sigma^{-1/2}(y - \hat{\mu})$ denoting the standardized residual, one has

$$H\chi = 0 \text{ and } (I - H)\chi = \chi.$$

There is a strong connection to the Pearson χ_P^2 statistic since $\chi_P^2 = \chi^T\chi$. The matrix H has the form $H = (W^{T/2}X)(X^T W^{1/2} W^{T/2} X)^{-1}(X^T W^{1/2})$, which shows that the projection is into the subspace that is spanned by the columns of $X^T W^{1/2}$.

4.3.3 Case Deletion

A strategy to investigate the effect of single observations on the parameter estimates is to compare the estimate $\hat{\beta}$ with the estimate $\hat{\beta}_{(i)}$, obtained from fitting the model to the data without the ith observation. An overall measure that includes all the components of the vector of coefficients is due to Cook (1977). *Cook's distance* for observation i has the form

$$c_i = (\hat{\beta}_{(i)} - \hat{\beta})^T \text{cov}(\hat{\beta})^{-1}(\hat{\beta}_{(i)} - \hat{\beta}) = (\hat{\beta}_{(i)} - \hat{\beta})^T X^T WX(\hat{\beta}_{(i)} - \hat{\beta}).$$

It may be seen as a confidence interval displacement diagnostic resulting from the exclusion of the ith observation. It is derived from the asymptotic confidence region for $\hat{\boldsymbol{\beta}}$ given by the likelihood distance $-2\{l(\boldsymbol{\beta}) - l(\hat{\boldsymbol{\beta}})\} = c$. By approximation of $l(\boldsymbol{\beta})$ by a second-order Taylor approximation at $\hat{\boldsymbol{\beta}}$ one obtains

$$c_i = (\boldsymbol{\beta} - \hat{\boldsymbol{\beta}})^T \operatorname{cov}(\hat{\boldsymbol{\beta}})^{-1} (\boldsymbol{\beta} - \hat{\boldsymbol{\beta}}).$$

Cook's distance is obtained by replacing $\boldsymbol{\beta}$ by $\hat{\boldsymbol{\beta}}_{(i)}$ and using an estimate of $\operatorname{cov}(\hat{\boldsymbol{\beta}})^{-1}$ that is composed from previously defined elements.

Computation of $\hat{\boldsymbol{\beta}}_{(i)}$ requires an iterative procedure for each observation. This may be avoided by using approximations. One may approximate $\hat{\boldsymbol{\beta}}_{(i)}$ by a one-step estimate, which is obtained by performing one Fisher scoring step starting from $\hat{\boldsymbol{\beta}}$. The corresponding approximation has the form

$$c_{i,1} = h_{ii} r_{P,i}^2 / (1 - h_{ii})^2,$$

where $r_{P,i} = r_P(\bar{y}_i, \hat{\pi}_i)$ is the Pearson residual for the ith observation and h_{ii} is the ith diagonal element of the hat matrix $\boldsymbol{H}$. The approximation is based on the one-step approximation of $\hat{\boldsymbol{\beta}}_{(i)}$:

$$\hat{\boldsymbol{\beta}}_{(i)} \approx \hat{\boldsymbol{\beta}} - w_{ii}^{1/2} r_{P,i} (\boldsymbol{X}^T \boldsymbol{W} \boldsymbol{X})^{-1} \boldsymbol{x}_i / (1 - h_{ii}),$$

where $\boldsymbol{x}_i$ is the covariate vector of observation i and w_{ii} is the ith diagonal element of $\boldsymbol{W}$.

Large values of Cook's measure c_i or its approximation indicate that the ith observation is influential. Its presence determines the value of the parameter vector. A useful way of presenting this measure of influence is in an index plot.

Example 4.5: Unemployment

Cook's distances for unemployment data (Figure 4.5) show that observations 33, 38, 44, which correspond to ages 48, 53, 59, are influential. All three observations are rather far from the fit. □

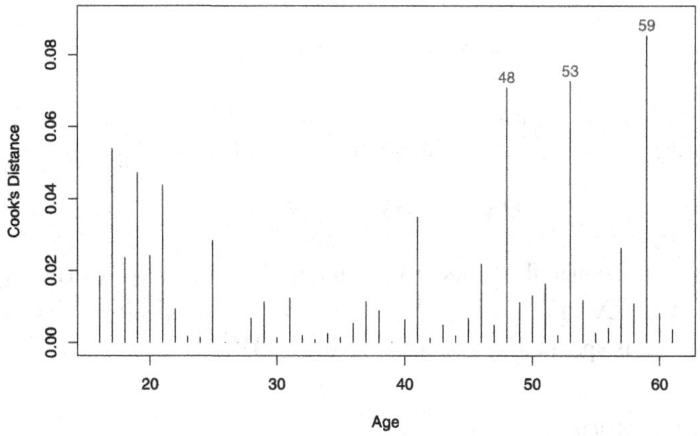

FIGURE 4.5: Cook distance for unemployment data.

Example 4.6: Exposure to Dust (Non-Smokers)

In a study on the effects of dust on bronchitis conducted in a German plant, the observed covariates were mean dust concentration at working place in mg/m^3 (dust), duration of exposure in years (years), and

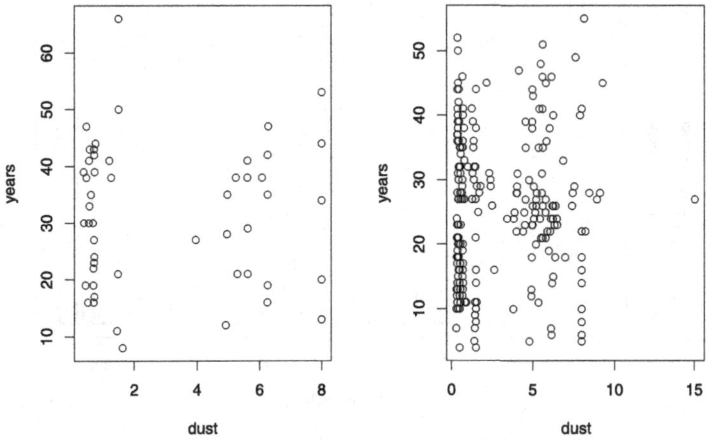

FIGURE 4.6: Dust exposure data, non-smokers with bronchitis (left panel), non-smokers without bronchitis (right panel).

smoking (1: yes; 0: no). Bronchitis is considered the binary response (1: present; 0: not present). The total sample was $n = 1246$. In previous analysis the focus has been on the estimation of threshold limiting values (Ulm, 1991; Küchenhoff and Ulm, 1997).

In the following we will first consider the subsample of non-smokers. Figure 4.6 shows the observations for workers with bronchitis and without bronchitis. It is seen that there is one person with an extreme value in the observation space. Since interaction terms and quadratic terms turn out not to contribute significantly to the model fit, in the following the main effect logit model is used. Table 4.4 shows the estimated coefficient for the main effects model. While years of exposure is highly significant, dust concentration is not significant in the subsample of non-smokers. Figure 4.7 shows Cook's distances for the subsample. Observations that show large values of Cook's distance are observations $730, 1175, 1210$, which have values $(1.63, 8), (8, 32), (8, 13)$ for (dust, years); all three observations correspond to persons with bronchitis. The observations are not extreme in the range of years, which is the influential variable. The one observation $(15.04, 27)$, which is very extreme in the observation space, corresponds to observation 1245, which is the last in the plot of Cook's distances and has a rather small value of Cook's distance. The fit without that observation, as given in Table 4.5, shows that the coefficient for concentration of dust

TABLE 4.4: Main effects model for dust exposure data (non-smokers).

	Estimate	Std. Error	z-Value	Pr(>|z|)
Intercept	−3.1570	0.4415	−7.15	0.0000
Dust	0.0053	0.0564	0.09	0.9248
Years	0.0532	0.0132	4.04	0.0001

TABLE 4.5: Main effects model for dust exposure data without one observation (non-smokers).

	Estimate	Std. Error	z-Value	Pr(>|z|)
Intercept	−3.1658	0.4419	−7.16	0.0000
Dust	0.0120	0.0580	0.21	0.8361
Years	0.0529	0.0131	4.03	0.0001

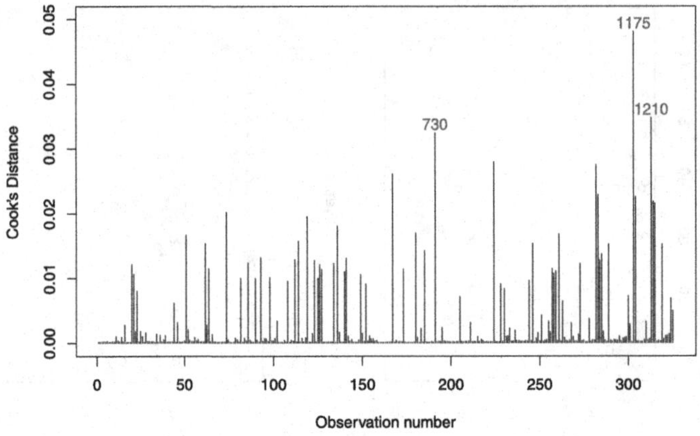

FIGURE 4.7: Cook distances for dust data (non-smokers).

has distinctly changed. However, the variable dust shows no significant effect, and therefore it is only a consequence that the Cook's distance is small. □

Example 4.7: Exposure to Dust

In the following the exposure data, including non-smokers, is used. It turns out that for the full dataset concentration of dust, years of exposure, and smoking are significantly influential (see Table 4.6). Again interaction effects between the covariates can be omitted. Figure 4.8 shows the observation space for all

TABLE 4.6: Main effects model for dust exposure data.

	Estimate	Std. Error	z-Value	Pr(>|z|)
(Intercept)	−3.0479	0.2486	−12.26	0.0000
Dust	0.0919	0.0232	3.95	0.0001
Years	0.0402	0.0062	6.47	0.0000
smoking	0.6768	0.1744	3.88	0.0001

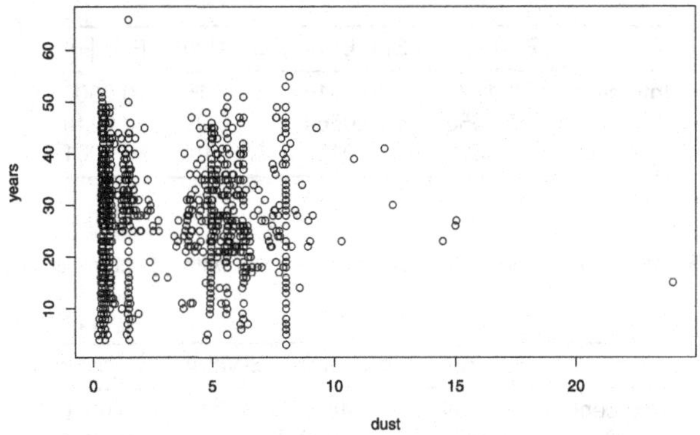

FIGURE 4.8: Dust exposure data.

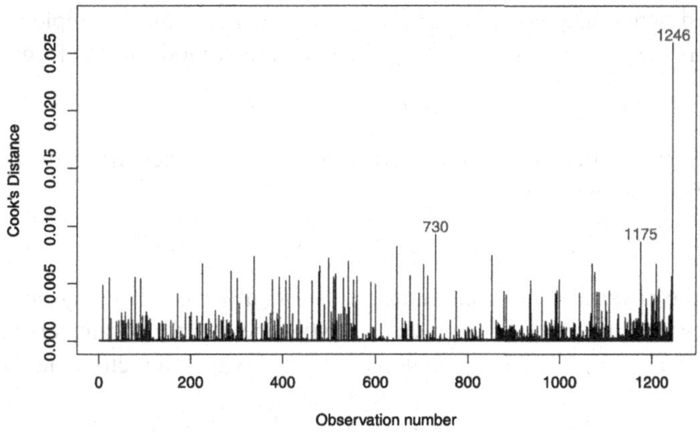

FIGURE 4.9: Cook distances for dust exposure data.

of the observations. It is seen that also in the full dataset one observation is positioned very extreme in the observation space. That observation (1246) is seen to have an extreme value of Cook's distance (Figure 4.9). When that extreme value is excluded, coefficient estimates for the variables years and smoking are similar to the estimates for the full dataset. However, coefficients differ for the variable concentration of dust by about 8% (see Table 4.7). Since observation 1246 is very far away from the data and the mean exposure is a variable that is not easy to measure exactly, one might suspect that the value of that variable is not trustworthy. One might consider the observation an outlier that should be omitted since it yields biased estimates. □

TABLE 4.7: Main effects model for dust exposure data without one observation.

	Estimate	Std. Error	z-Value	Pr(>|z|)
(Intercept)	−3.0620	0.2491	−12.29	0.0000
Dust	0.0992	0.0239	4.15	0.0000
Years	0.0398	0.0062	6.40	0.0000
Smoking	0.6816	0.1745	3.91	0.0001

4.4 Structuring the Linear Predictor

The structuring of the linear predictor, and in particular the coding of categorical predictors, have already been considered briefly in Section 1.4. In the following we again have a look at the linear predictor and introduce a notation scheme due to Wilkinson and Rogers (1973).

4.4.1 The Linear Predictor: Continuous Predictors, Factors, and Interactions

The parametric binary regression model

$$\pi(\boldsymbol{x}) = P(y = 1|\boldsymbol{x}) = h(\boldsymbol{x}^T\boldsymbol{\beta})$$

that is considered here contains the linear predictor $\eta = \boldsymbol{x}^T\boldsymbol{\beta}$. In the simplest case the linear predictor contains the p variables $x_1, \dots, x_p$ in a main effect model of the form

$$\eta = \beta_0 + x_1\beta_1 + \cdots + x_p\beta_p.$$

However, frequently one finds that some interaction terms are necessary in the predictor, yielding a linear predictor of the form

$$\eta = \beta_0 + x_1\beta_1 + \cdots + x_p\beta_p + x_1x_2\beta_{12} + x_1x_3\beta_{13} + \cdots .$$

When some of the covariates are continuous, polynomial terms also may be included in the linear predictor, which means that the predictor is still linear in the parameters but not linear in the variables. If, for example, x_1 is continuous, a more flexible predictor containing non-linear effects of x_1 is

$$\eta = \beta_0 + x_1\beta_1 + \cdots + x_p\beta_p + x_1^2\beta_1^{(2)} + x_1^3\beta_1^{(3)} \cdots .$$

In many applications the influential variables are factors that may take a finite number of values. As in classical regression, these categorical covariates are included in the predictor in the form of dummy variables. When a categorical variable or factor A takes values in $\{1, \dots, k\}$ the linear predictor has the form

$$\eta = \beta_0 + x_{A(1)}\beta_{A(1)} + \ldots + x_{A(k-1)}\beta_{A(k-1)},$$

where $x_{A(i)}$ are dummy variables. In (0-1)-coding one has $x_{A(i)} = 1$ if $A = i$ and $x_{A(i)} = 0$ otherwise, which implies that $A = k$ is used as a reference category. When an additional continuous covariate x is available, the predictor of the main effects model becomes

$$\eta = \beta_0 + x\beta_x + x_{A(1)}\beta_{A(1)} + \cdots + x_{A(k-1)}\beta_{A(k-1)}. \tag{4.7}$$

The inclusion of an interaction between the continuous predictor and the categorical variables means that all the products $xx_{A(i)}$ are included, yielding the more complicated predictor

$$\begin{aligned}\eta = \beta_0 + x\beta_x + x_{A(1)}\beta_{A(1)} + \cdots + x_{A(k-1)}\beta_{A(k-1)} + xx_{A(1)}\beta_{x,A(1)} + \cdots \\ + xx_{A(k-1)}\beta_{x,A(k-1)}.\end{aligned} \tag{4.8}$$

Interaction of this form means that the effect of one covariate is modified by the other. From the restructuring

$$\eta = \beta_0 + \cdots + x(\beta_x + x_{A(1)}\beta_{x,A(1)} + \cdots) + \cdots$$

one sees that the slope of x is modified by the categorical variable, yielding different slopes for different categories of A. The restructuring

$$\eta = \beta_0 + \cdots + x_{A(i)}(\beta_{A(i)} + x\beta_{x,A(i)}) + \cdots$$

shows how the effects of variable A are modified by the value of x. Interactions between two factors $A \in \{1, \dots, k_A\}$ and $B \in \{1, \dots, k_B\}$ may be modeled by including all products of dummy variables $x_{A(i)}x_{B(j)}$. For the interpretation of interactions that is familiar from linear modeling, one should have in mind that the the effects do not refer directly to the mean $\mathrm{E}(y|\boldsymbol{x})$ but to the transformed mean $g(\mathrm{E}(y|\boldsymbol{x}))$, which in the case of the logit model is the logits $\log(\pi(\boldsymbol{x}))/(1 - \pi(\boldsymbol{x}))$. In particular, for categorical predictors, interaction parameters have simple interpretations in terms of odds ratios (Exercise 4.4).

Since the form of the linear predictor is rather tedious to write down, it is helpful to use a notation that is due to Wilkinson and Rogers (1973). The notation uses operators to combine predictors in model formula terms. Let x, z denote continuous, metrically scaled variables and A, B, C denote factors.

Basic Model Term

The model terms X and A stand for themselves. However, the linear predictor that is built depends on the type of variable. For a continuous variable x, the model term X means that $x\beta_x$ is included. For the factor A, the term to be included is a set of dummy variables with corresponding weights $x_{A(i)}\beta_{A(i)}$.

The + Operator

The + operator in a model formula means that the corresponding terms are added in the algebraic expression of the predictor. Therefore $X + A$ means that the predictor has the form (4.7).

The Dot Operator

The dot operator is used to form products of the constituent terms. For example, the model term $X.Z$ refers to the algebraic expression $xz\beta_{x,z}$, whereas $X.A$ refers to the algebraic expression containing all products $xx_{A(i)}$, which is the interaction term in (4.8). Of course $X.Z$ is equivalent to $Z.X$. Polynomial terms of continuous variables my be built by $X.X$, referring to $x^2\beta$. The dot operator dominates the + operator, so that $A.B + X$ is equivalent to $(A.B) + X$. The notation of higher interaction terms is straightforward. $A.B.C$ denotes inclusion of the set of products $x_{A(i)}x_{B(j)}x_{C(k)}$. Of course it helps that the dot operator is commutative, that is, $(A.B).C$ is equivalent to $A.(B.C)$. It should be noted that sometimes different notations are used for the dot operator. For example, the statistical software R uses the notation $A : B$ for $A.B$.

The Crossing Operator

The crossing operator is helpful for including marginal effects. The model term $A * B$ is an abbreviation for $A+B+A.B$. Similarly, the term $A*B*C$ means that the main effects and all two factor interactions are included. It is abbreviated as $A+B+C+A.B+A.C+B.C+A.B.C$. Crossing is distributive, meaning that $A * (B + C)$ is equivalent to $A * B + A * C$ (and to $A + B + C + A.B + A.C$).

TABLE 4.8: Wilkinson-Rogers notation for linear predictions with metric covariates X, Z and factors A, B.

	Model Term	Linear Predictor	Linear Predictor
			$A = i, B = j$
Single	X	$x\beta_x$	$x\beta_x$
	A	$\sum_s x_{A(s)}\beta_{A(s)}$	$\beta_{A(i)}$
Addition	$X + Z$	$x\beta_x + z\beta_z$	$x\beta_x + z\beta_z$
	$A + B$	$\sum_s x_{A(s)}\beta_{A(s)} + \sum_s x_{B(s)}\beta_{B(s)}$	$\beta_{A(i)} + \beta_{B(j)}$
	$X + A$	$x\beta_x + \sum_s x_{A(s)}\beta_{A(s)}$	$x\beta_x + \beta_{A(i)}$
Interaction	$X.Z$	$xz\beta_{x,z}$	$xz\beta_{xz}$
	$A.B$	$\sum_{s,r} x_{A(s)}x_{B(r)}\beta_{AB(s,r)}$	$\beta_{AB(ij)}$
	$X.A$	$\sum_s xx_{A(s)}\beta_{xA(s)}$	$x\beta_{xA(i)}$
Hierarchical interaction	$X * Z$	$x\beta_x + z\beta_z + xz\beta_{xz}$	$x\beta_x + z\beta_z + xz\beta_{xz}$
	$A * B$	$\sum_s x_{A(s)}\beta_{A(s)} + \sum_r x_{B(r)}\beta_{B(r)}$ $+\sum_{s,r} x_{A(s)}x_{B(r)}\beta_{AB(s,r)}$	$\beta_{A(i)} + \beta_{B(j)} + \beta_{AB(ij)}$
	$X * A$	$x\beta_x + \sum_s x_{A(s)}\beta_{A(s)}$ $+\sum_s xx_{A(s)}\beta_{xA(s)}$	$x\beta_x + \beta_{A(i)} + x\beta_{xA(i)}$

Table 4.8 shows model terms in the notation of Wilkinson-Rogers together with the corresponding linear predictors for continuous covariates X, Z and factors A, B. It should be noted that the interactions built by these model terms are a specific form of parametric interaction. Alternative concepts of interactions may be derived (see also Section 10.3.3 for smooth interaction terms).

4.4.2 Testing Components of the Linear Predictor

Most interesting testing problems concerning the linear predictor are linear hypotheses of the form

$$H_0 : \boldsymbol{C\beta} = \boldsymbol{\xi} \text{ against } H_1 : \boldsymbol{C\beta} \neq \boldsymbol{\xi},$$

where $\boldsymbol{C}$ is a fixed matrix of full rank $s \leq p$ and $\boldsymbol{\xi}$ is a fixed vector. In the simplest case one tests if one parameter can be omitted by considering

$$H_0 : \beta_j = 0 \text{ against } H_1 : \beta_j \neq 0,$$

which has the form of a linear hypothesis with $\boldsymbol{C} = (0, \ldots, 1, \ldots, 0)$ and $\boldsymbol{\xi} = 0$. The general no-effects hypothesis

$$H_0 : \boldsymbol{\beta} = \boldsymbol{0} \text{ against } H_1 : \boldsymbol{\beta} \neq \boldsymbol{0}$$

corresponds to a linear hypothesis with $\boldsymbol{C}$ denoting the unit matrix and $\boldsymbol{\xi} = \boldsymbol{0}$. If one wants to test if a factor A has no effect, one has to test simultaneously if all the corresponding parameters are zero:

$$H_0 : \beta_{A(1)} = \cdots = \beta_{A(k-1)} = 0 \text{ against } H_1 : \beta_{A(j)} \neq 0 \text{ for one } j.$$

It is easily seen that in this case one also tests linear hypotheses. General test statistics for linear hypotheses have already been given in Section 3.7.2. Therefore, in the following the tests are considered only briefly.

Likelihood Ratio Statistic and the Analysis of Deviance

Let M denote the model with linear predictor $\eta = \boldsymbol{x}^T\boldsymbol{\beta}$ and $\tilde{M}$ denote the submodel that is constrained by $\boldsymbol{C\beta} = \boldsymbol{\xi}$. By using the notation of Section 4.2, the likelihood ratio statistic that compares models $\tilde{M}$ and M has the form

$$\lambda = -2\{l(\boldsymbol{y}, \tilde{\boldsymbol{\pi}}) - l(\boldsymbol{y}, \hat{\boldsymbol{\pi}})\},$$

where $\tilde{\boldsymbol{\pi}}^T = (\tilde{\pi}_1, \ldots, \tilde{\pi}_n)$, $\tilde{\pi}_i = h(\boldsymbol{x}_i^T\tilde{\boldsymbol{\beta}})$, denotes the fit of submodel $\tilde{M}$ and $\hat{\boldsymbol{\pi}}^T = (\hat{\pi}_1, \ldots, \hat{\pi}_n)$, $\hat{\pi}_i = h(\boldsymbol{x}_i^T\hat{\boldsymbol{\beta}})$, denotes the fit of model M. Since for a binary distribution the deviances of models $\tilde{M}$ and M are given by $-2l(\boldsymbol{y}, \tilde{\boldsymbol{\pi}})$ and $-2l(\boldsymbol{y}, \hat{\boldsymbol{\pi}})$, respectively, λ is equivalent to the difference between deviances:

$$\lambda = D(\boldsymbol{y}, \tilde{\boldsymbol{\pi}}) - D(\boldsymbol{y}, \hat{\boldsymbol{\pi}}) = D(\tilde{M}) - D(M)$$

and has under mild conditions an asymptotic χ^2-distribution with $s = rg(\boldsymbol{C})$ degrees of freedom. It is noteworthy that the difference of deviances yields the same value if computed from binomial responses or from the single binary variables that build the binomial response.

As shown in Section 3.7.2, a sequence of nested models $M_1 \subset M_2 \subset \cdots \subset M_m$ can be tested by considering the difference between successive models $D(M_i|M_{i+1}) = D(M_i) - D(M_{i+1})$ within the decomposition

$$D(M_1) = D(M_1|M_2) + \cdots + D(M_{m-1}|M_m) + D(M_m).$$

TABLE 4.9: Empirical odds and log-odds for duration of unemployment.

Gender	Education Level	Short Term	Long Term	Odds	Log-Odds
m	1	97	45	2.155	0.768
	2	216	81	2.667	0.989
	3	56	32	1.750	0.560
	4	34	9	3.778	1.330
w	1	105	51	2.059	0.722
	2	91	81	1.123	0.116
	3	31	34	0.912	-0.092
	4	11	9	1.222	0.201

TABLE 4.10: Hierarchies for level (L) and gender (G).

Model	Deviance	df (p-value)	Cond. Deviance	df (p-value)	Tested Effect
1	32.886	7 (0.000)			
			6.557	3 (0.087)	L(G ignored, L.G ignored)
1+L	26.329	4 (0.000)			
			18.808	1 (0.000)	G(L.G ignored, L taken into account)
1+L+G	7.521	3 (0.057)			
			7.521	3 (0.057)	L.G
L∗G	0	0			
1	32.886	7 (0.000)			
			17.959	1 (0.000)	G(L ignored, L.G ignored)
1+G	14.928	6 (0.021)			
			7.406	3 (0.060)	L(L.G ignored, G taken into account)
1+L+G	7.521	3 (0.057)			
			7.521	3 (0.057)	L.G
L∗G	0	0			

The result is usually presented in an analysis of deviance table that gives the deviances of models, their differences, and the corresponding degrees of freedom (see Example 4.8).

Example 4.8: Duration of Unemployment

With the response duration of unemployment (1: short-term unemployment, less than 6 months; 0: long-term unemployment) and the covariates gender (1: male; 0: female) and level of education (1: lowest, up to 4: highest, university degree), one obtains the data given in Table 4.9. Analysis is based on the grouped data structure given in this table. The saturated logit model is given by

$$\text{logit}(\pi(G, L)) = \beta_0 + x_G\beta_G + x_L\beta_L + x_G x_L\beta_{GL},$$

which can be abbreviated by $G * L$. The testing of $\beta_{GL}, \beta_G, \beta_L$ yields the analysis of deviance table given in Table 4.10. It contains the deviances and the differences of the two sequences of nested models $1 \subset 1+L \subset 1+L+G \subset G*L$ and $1 \subset 1+G \subset 1+L+G \subset G*L$. In both sequences the intercept model is the strongest and the saturated model is the weakest. Thus different paths between these models can be tested. Starting from the saturated model, the first transition seems possible, since $D = 7.52$ on 3 df, which corresponds to a p-value of 0.057. However, in the first sequence the next transition to model $1 + L$ yields the difference of deviances 18.81 on 1 df; therefore gender cannot be omitted. In the other sequence one considers the transition to model $1 + G$, obtaining the difference of deviances 7.41 on 3 df, which corresponds to a a p-value of 0.06. Although simplification seems possible, the model $1 + G$ does

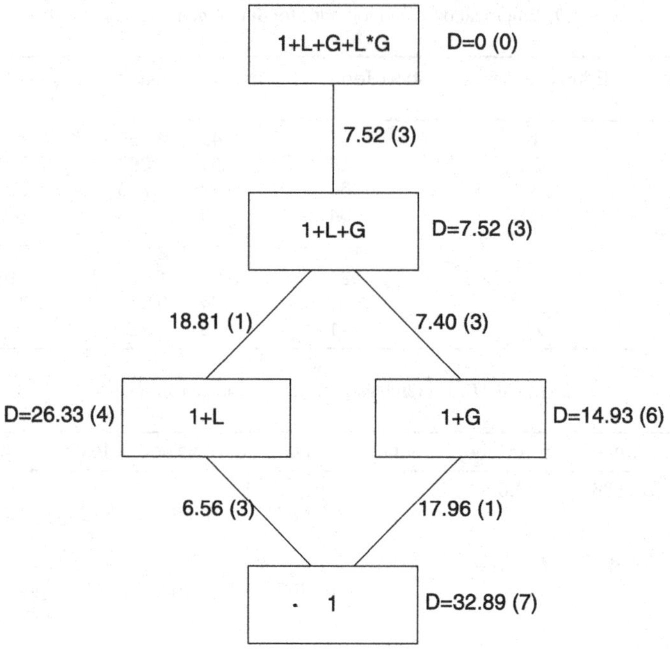

FIGURE 4.10: Analysis of deviance for unemployment data with factors level (L) and gender (G).

not fit will since the deviance 14.93 on 3 df is rather large, corresponding to a p-value of 0.02. Therefore, simplification beyond the main effect model does not seem to be warranted. □

Alternative Test Statistics

Alternatives to the likelihood ratio statistic are the Wald test and the score test for linear hypotheses given by

$$w = (\boldsymbol{C}\hat{\boldsymbol{\beta}} - \boldsymbol{\xi})^T[\boldsymbol{C}\boldsymbol{F}^{-1}(\hat{\boldsymbol{\beta}})\boldsymbol{C}^T]^{-1}(\boldsymbol{C}\hat{\boldsymbol{\beta}} - \boldsymbol{\xi}),$$

and

$$u = s^T(\tilde{\boldsymbol{\beta}})\boldsymbol{F}^{-1}(\tilde{\boldsymbol{\beta}})s(\tilde{\boldsymbol{\beta}}).$$

Asymptotically, all three test statistics, the likelihood ratio statistic, the Wald test, and the score test, have the same distribution $\lambda, w, u \overset{(a)}{\sim} \chi^2(\text{rank}\,\boldsymbol{C})$.

Contrasts and Quasi-Variances

Let us again consider the influence of a factor A that takes values in $\{1, \ldots, k\}$. The corresponding linear predictor without constraints has the form

$$\eta = \beta_0 + x_{A(1)}\beta_{A(1)} + \ldots + x_{A(k)}\beta_{A(k)}.$$

Since the parameters $\beta_{A(1)}, \ldots, \beta_{A(k)}$ are not identifiable, side constraints have to be used. For example, by choosing the reference category k one sets $\beta_{A(k)} = 0$. Among the full set of parameters $\beta_{A(1)}, \ldots, \beta_{A(k)}$ only contrasts, that is, linear combinations $\boldsymbol{c}^T\boldsymbol{\beta}$ with $\boldsymbol{c}^T = (c_1, \ldots, c_k)$,

TABLE 4.11: Estimates, standard errors, quasi-standard-errors and quasi-variances for duration of unemployment data.

Level	Estimate $\hat{\beta}_{A(i)}$	Standard Error	Quasi-Standard Error	Quasi-Variance
1	−0.170	0.305	0.124	0.015
2	−0.273	0.295	0.097	0.009
3	−0.637	0.323	0.163	0.026
4	0	0	0.278	0.077

$\sum_i c_i = 0$, $\boldsymbol{\beta}^T = (\beta_{A(1)}, \ldots, \beta_{A(k)})$, are identified. A simple contrast is $\beta_{A(i)} - \beta_{A(j)}$, which compares levels i and j. For illustration we consider again the effect of the education level on the binary response duration of unemployment (Example 2.6). Table 4.11 shows the estimates together with standard errors for reference category 4. The given standard errors refer to the contrasts $\hat{\beta}_{A(i)} - \hat{\beta}_{A(4)}$ because $\hat{\beta}_{A(4)} = 0$. The disadvantage of the presentation of standard errors for fixed reference category is that all other contrasts are not seen. An alternative presentation of standard errors uses so-called *quasi-variances*, which have been proposed by Firth and De Menezes (2004). Quasi-variances $q_1, \ldots, q_k$ are constructed such that the variance of a contrast, $\text{var}(\boldsymbol{c}^T\hat{\boldsymbol{\beta}})$, is approximately $\sum_{i=1}^k c_i^2 q_i$. The corresponding "quasi-standard-errors" are given by $q_1^{1/2}, \ldots, q_k^{1/2}$. Quasi-variances (or quasi-standard-errors) can be used to determine the variance of contrasts. Consider the quasi-variances given in Table 4.11. For example, the approximate standard error for $\hat{\beta}_{A(1)} - \hat{\beta}_{A(4)}$ can be computed from quasi-variances by $(0.015 + 0.077)^{1/2}$, or from quasi-standard-errors, based on Pythagorean calculation, by $(0.124^2 + 0.278^2)^{1/2}$, yielding 0.303, which is a rather good approximation of the standard error of $\hat{\beta}_{A(1)}$ given in Table 4.11. But with the quasi-variances given in Table 4.11 also standard errors for $\hat{\beta}_{A(1)} - \hat{\beta}_{A(2)}$ can be computed yielding $(0.015 + 0.009)^{1/2} = 0.155$, which is rather large when compared to $\hat{\beta}_{A(1)} - \hat{\beta}_{A(2)} = 0.103$. Quasi-variances are a helpful tool in reporting

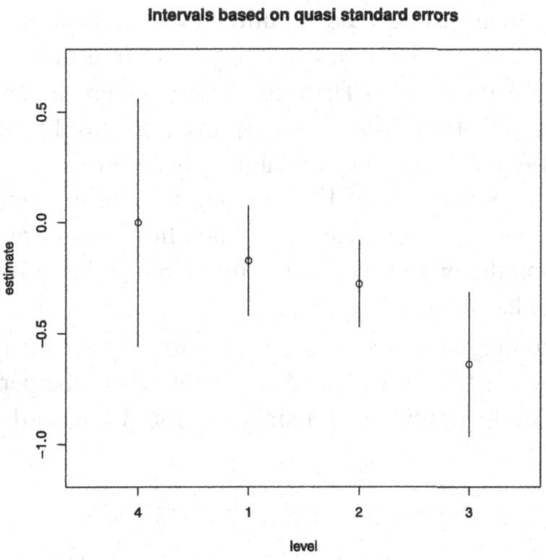

FIGURE 4.11: Quasi-standard-errors for duration of unemployment data.

standard errors for the level of factors. They are useful for investigating contrasts but should not be misinterpreted as standard errors for the parameters themselves. Firth and De Menezes (2004) give conditions under which the approximation is exact and investigate the accuracy in a variety of settings.

4.4.3 Ordered Categorical Predictors

The inclusion of categorical covariates in the linear predictor may strongly increase the dimension of the predictor space. Since a factor A that takes values in $\{1, \dots, k\}$ means that $k-1$ dummy variables have to be included, the number of observations within one factor level may become very small. In particular, if one wants to model interactions of factors, empty cells (for combinations of factors) will occur if the number of levels is large. The consequence frequently is that estimates of effects do not exist. In that case practioners typically join some factor categories to reduce the dimension of the predictor space, but at the cost of loosing information.

The coding in dummy variables for factors is based on the nominal scale of the factor. The categories $\{1, \dots, k\}$ represent simple labels for levels of the factor; the ordering of the categories is arbitrary and should not be used. In practice, however, the categories of a factor are often ordered, and the variable that constitutes the predictor is an ordered categorical variable. Then the ordering should be used to obtain a more sparse representation and therefore more stable estimates. One way to use the ordering is to define assigned scores to the categories of the factor. Although that "solution" to the dimensional problem is widespread, it is not satisfactory. By assigning scores and using a linear or quadratic model for the scores one definitely assumes a higher scale level for the factor. However, linear or quadratic terms are appropriate only when the influential variable is metrically scaled, albeit discrete. In particular, differences of values of the variable have to be meaningful. For ordered factors this is usually not the case. If the factor levels represent subjective judgements like "strong agreement," "slight agreement," and "strong disagreement," the assigned scores are typically useless. Interpretation depends on the assigned scores, which are to a certain extent arbitrary, and different sets of scores yield different effect strengths.

It is slightly different when the ordinal scale of the factor is due to an underlying continuous variable. For example, often variables like income brackets and categorized age are available and it is known how the categories have been built. For a categorized covariate age, available in categories $[20, 29), [30, 39), \dots, [60, 80)$, one may build mid-range scores and use them as a substitute for the true underlying age. Then one approximates the unobserved covariate age in the predictor. What works for covariate age may work less well when the predictor is categorized income. Since only categories are available, it is hard to know what values hide in the highest interval, because it has been learned that incomes can be extremely high. Then the mid-range score of the last category is a mere guess. In addition, the score of the highest category is at the boundary of the predictor space and therefore tends to be influential. Thus one has an influential observation but has to choose a score.

An alternative to assigning scores that takes the ordering seriously but does not assume knowledge of the true score is to use penalized estimates where the penalty explicitly uses the ordering of categories. Let the predictor of a single factor A with ordered categories $1, \dots, k$ be given in the form

$$\eta = \beta_0 + x_{A(2)}\beta_2 + \dots + x_{A(k)}\beta_k,$$

where $x_{A(2)}, \dots, x_{A(k)}$ denotes $k-1$ dummy variables in (0–1)-coding, therefore implicitly using category 1 as the reference category. However, instead of maximizing the usual log-likelihood $l(\boldsymbol{\beta})$, estimates of the parameter vector $\boldsymbol{\beta}^T = (\beta_0, \beta_2, \dots, \beta_k)$ are obtained by maximizing the *penalized log-likelihood*:

$$l_p(\boldsymbol{\beta}) = l(\boldsymbol{\beta}) - \frac{\lambda}{2} P(\boldsymbol{\beta}),$$

where the penalty term is given by

$$P(\boldsymbol{\beta}) = \sum_{j=2}^{k} (\beta_j - \beta_{j-1})^2$$

with $\beta_1 = 0$. Therefore, the differences of adjacent parameters are penalized with the strength of the penalty determined by λ. For $\lambda = 0$, maximization of $l_p(\boldsymbol{\beta})$ yields the usual maximum likelihood estimate whereas $\lambda \to \infty$ yields $\hat{\beta}_j = 0, j = 1, \dots, k$. For an appropriately chosen λ, the penalty restricts the variability of the parameters across the response categories, thereby assuming a kind of "smooth effect" of the ordinal categories on the dependent variable. The method is strongly related to the penalization techniques considered in Chapter 6.

In matrix form the penalty is given by $P(\boldsymbol{\beta}) = \boldsymbol{\beta}^T \boldsymbol{D}^T \boldsymbol{D} \boldsymbol{\beta} = \boldsymbol{\beta}^T \boldsymbol{K} \boldsymbol{\beta}$, where $\boldsymbol{K} = \boldsymbol{D}^T \boldsymbol{D}$ and $\boldsymbol{D}$ is the $((k-1) \times k)$-matrix:

$$\boldsymbol{D} = \begin{pmatrix} 0 & 1 & 0 & \dots & & 0 \\ 0 & -1 & 1 & & & \vdots \\ 0 & & -1 & 1 & & \\ \vdots & & \ddots & & \ddots & \\ 0 & 0 & \dots & 0 & -1 & 1 \end{pmatrix}. \tag{4.9}$$

The corresponding penalized score function is

$$\boldsymbol{s}_p(\boldsymbol{\beta}) = \sum_{i=1}^{n} \boldsymbol{x}_i \frac{\partial h(\eta_i)}{\partial \eta} (y_i - \mu_i)/\sigma_i^2 - \lambda \boldsymbol{K} \boldsymbol{\beta}.$$

The estimation equation $\boldsymbol{s}_p(\hat{\boldsymbol{\beta}}_p) = \boldsymbol{0}$ may be solved by iterative pseudo-Fisher scoring:

$$\hat{\boldsymbol{\beta}}_p^{(k+1)} = \hat{\boldsymbol{\beta}}_p^{(k)} + \boldsymbol{F}_p(\hat{\boldsymbol{\beta}}_p^{(k)})^{-1} \boldsymbol{s}_p(\hat{\boldsymbol{\beta}}_p^{(k)})^{-1},$$

where $\boldsymbol{F}_p(\boldsymbol{\beta}) = \boldsymbol{F}(\boldsymbol{\beta}) - \lambda \boldsymbol{K}$ and $\boldsymbol{F}(\boldsymbol{\beta}) = E(-\partial^2 l / \partial \boldsymbol{\beta} \partial \boldsymbol{\beta}^T)$. Approximate covariances are obtained by the sandwich matrix

$$\text{cov}(\hat{\boldsymbol{\beta}}_p) \approx (\boldsymbol{F}(\hat{\boldsymbol{\beta}}_p) + \lambda \boldsymbol{K})^{-1} \boldsymbol{F}(\hat{\boldsymbol{\beta}}_p) (\boldsymbol{F}(\hat{\boldsymbol{\beta}}_p) + \lambda \boldsymbol{K})^{-1}.$$

An advantage of the penalized estimate $\hat{\boldsymbol{\beta}}_p$ is that it will exist even in cases where the unpenalized ML estimate $\hat{\boldsymbol{\beta}}$ does not exist. Unconstrained estimates may not exist even in the simple case of a single factor. For the logit model, unconstrained estimates are given by $\hat{\beta}_0 = \log(p_1/(1-p_1)), \hat{\beta}_j = \log(p_j/(1-p_j)) - \hat{\beta}_0, j = 2, \dots, p$, with p_j denoting the observed relative frequencies in category j. Therefore, estimates do not exist if one of the relative frequencies is zero or one. In the more general case, where the intercept term is replaced by a linear term $\boldsymbol{x}^T \boldsymbol{\beta}_x$, which contains an additional vector of predictors, non-existence will occur more frequently.

Alternative Representation

Before demonstrating the advantages of penalized estimates, an alternative representation of the approach by use of split-coding is given. Let us separate the weights on predictors from the intercept by considering the decomposition $\boldsymbol{\beta}^T = (\beta_0, \boldsymbol{\beta}_c^T)$. In analogy the matrix $\boldsymbol{D}$ is decomposed into $\boldsymbol{D} = [\boldsymbol{0}|\boldsymbol{D}_c]$. Then the penalty may be given as

$$J(\boldsymbol{\beta}) = \boldsymbol{\beta}_c^T \boldsymbol{K}_c \boldsymbol{\beta}_c, \tag{4.10}$$

where $\boldsymbol{K}_c = \boldsymbol{D}_c^T \boldsymbol{D}_c$ and $\boldsymbol{\beta}_c^T = (\beta_2, \dots, \beta_k)$. The penalized maximization problem can be transformed into a problem with a simpler penalty by use of the inverse $\boldsymbol{D}_c^{-1}$, which is a lower triangular matrix with 1 on and below the diagonal. By using the parametrization $\tilde{\boldsymbol{\beta}} = \boldsymbol{D}_c \boldsymbol{\beta}_c$ with $\tilde{\boldsymbol{\beta}}^T = (\tilde{\beta}_1, \dots, \tilde{\beta}_{k-1})$ the penalty $\boldsymbol{\beta}_c^T \boldsymbol{K}_c \boldsymbol{\beta}_c$ turns into the simpler form

$$J(\boldsymbol{\beta}) = \tilde{\boldsymbol{\beta}}^T \tilde{\boldsymbol{\beta}} = \sum_{i=1}^{k-1} \tilde{\beta}_i^2,$$

which is a ridge penalty on the transformed parameters. The corresponding transformation of the design matrix is based on the decomposition $\boldsymbol{X} = [\boldsymbol{1}|\boldsymbol{X}_c]$. One obtains $\boldsymbol{X}\boldsymbol{\beta} = [\boldsymbol{1}\beta_0|\boldsymbol{X}_c \boldsymbol{D}_c^{-1}\tilde{\boldsymbol{\beta}}]$. Then one row of the design matrix $\boldsymbol{X}_c \boldsymbol{D}_c^{-1}$ is given by the vector $(\tilde{x}_{A(1)}, \dots, \tilde{x}_{A(k-1)}) = (x_{A(2)}, \dots, x_{A(k)})\boldsymbol{D}_c^{-1}$, where the dummies $\tilde{x}_{A(i)}$ are given in split-coding:

$$\tilde{x}_{A(i)} = \begin{cases} 1 & \text{if } A > i \\ 0 & \text{otherwise,} \end{cases}$$

which is a coding scheme that distinguishes between categories $\{1, \dots, i\}$ and $\{i+1, \dots, k\}$ (for split-coding compare Section 1.4.1). The transformation $\tilde{\boldsymbol{\beta}} = \boldsymbol{D}_c \boldsymbol{\beta}_c$ yields the parameters $\tilde{\beta}_1 = \beta_2$, $\tilde{\beta}_2 = \beta_3 - \beta_2$, $\tilde{\beta}_{k-1} = \beta_k - \beta_{k-1}$, which are used in the corresponding predictor $\eta = \beta_0 + \sum_{i=1}^{k-1} \tilde{x}_{A(i)} \tilde{\beta}_i$. The transformation shows that the smoothness penalty $\boldsymbol{\beta}^T \boldsymbol{K} \boldsymbol{\beta}$ can be represented as a ridge penalty for the parameters corresponding to split-coding, and thus smoothness across categories is transformed into penalizing transitions between groups of adjacent categories.

Example 4.9: Simulation

In a small simulation study the penalized regression approach is compared to pure dummy coding and a binary regression model with a linear predictor that takes the group labels as (metric) independent variable. The underlying model is the logit model $P(y=1) = \exp(\boldsymbol{x}^T\boldsymbol{\beta})/(1+\exp(\boldsymbol{x}^T\boldsymbol{\beta}))$. As true values of $\boldsymbol{\beta}$ we assume an approximately linear structure (Figure 4.12, top left) and an obviously non-linear coefficient vector (Figure 4.12, bottom left). In each of the 100 simulations $n = 330$ values of $\boldsymbol{x}$ were generated. Figure 4.12 shows the mean squared error for the estimation of $\boldsymbol{\beta}$. It is seen that the linear predictor performs well when the underlying structure is approximately linear but fails totally when the structure is non-linear. Simple dummy coding should adapt to both scenarios, but due to estimation uncertainty it performs worse than the linear predictor approach in the first scenario but better in the second scenario. The penalized estimate with smoothed dummies outperforms both approaches distinctly. It adapts well to the approximately linear and to the distinctly non-linear structure. □

So far we assumed a single independent variable, but models with several predictors are an obvious extension. Only the penalty matrix has to be modified. Now one uses a block-diagonal structure with the blocks given by the penalty matrix for a single ordered predictor. In the following example three ordinal predictors are used.

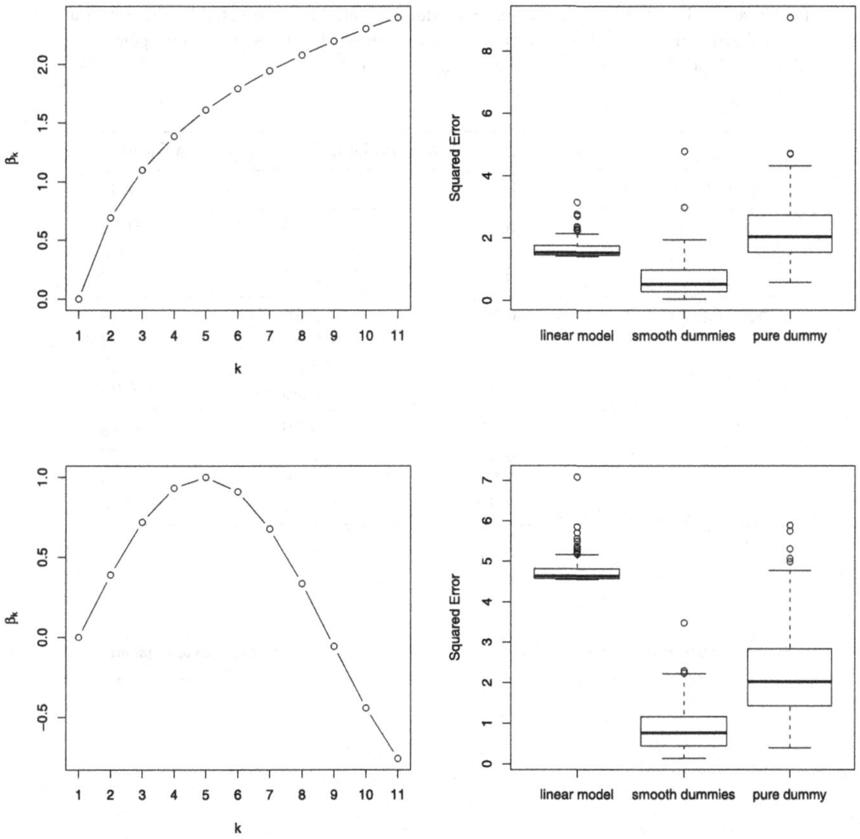

FIGURE 4.12: True coefficient vectors (left) and squared errors (right) for the considered methods after 100 simulation runs with $\sigma^2 = 2$.

Example 4.10: Choice of Coffee Brand

In this example the binary response is coffee brand, which is only separated into cheap coffee from a German discounter and real branded products. The ordinal predictors are monthly income (in four categories), social class (in five categories), and age group in five categories, below 25, 25–39, 40–49, 50–59, above 59). While one might consider some sort of midpoints for age group and monthly income, it is hardly imaginable for social class. Table 4.12 shows the estimated coefficients of corresponding dummy variables for a logit model with group labels as predictors. The comparison refers to linear modeling, pure dummy and smoothed ordered dummies. It can be seen that penalization yields much less variation in the coefficients for single variables when compared to pure dummy coding.

To investigate the methods' performance in terms of prediction accuracy, the data were split randomly into training ($n = 100$) and test ($m = 100$) data. The training data are used to fit the model, the test set for evaluation only. As a measure of prediction accuracy we take the sum of squared deviance residuals (SSDR) on the test set. It is remarkable that the pure dummy model cannot be fitted due to complete data separation in 68% of the splits. Figure 4.13 summarizes the results in terms of SSDR after 200 random splits. It is seen that smooth dummies are to be preferred, in particular because the fit of pure dummies exists only in 32% of the splits. □

As the previous example shows, ML estimates often do not exist when simple dummy coding is used. Penalized estimates have the advantage that estimates exist under weaker

TABLE 4.12: Coefficients of corresponding dummy variables, estimated by the use of a (generalized) linear model, i.e., logit model, with group labels as predictors, penalized regression types I and II ($\lambda = 10$ in each case), and a logit model based on pure dummy coding.

		Linear Model	Smooth Dummies	Pure Dummy
Intercept		−0.36	−0.81	−0.38
Income	2	0.02	−0.05	−0.13
	3	0.03	−0.02	0.29
	4	0.05	−0.04	0.17
Social Class	2	−0.28	−0.14	−0.92
	3	−0.56	−0.31	−1.39
	4	−0.84	−0.39	−1.28
	5	−1.12	−0.56	−1.96
Age Group	2	−0.10	0.06	0.79
	3	−0.20	−0.09	−0.29
	4	−0.30	0.05	0.84
	5	−0.40	−0.18	−0.04

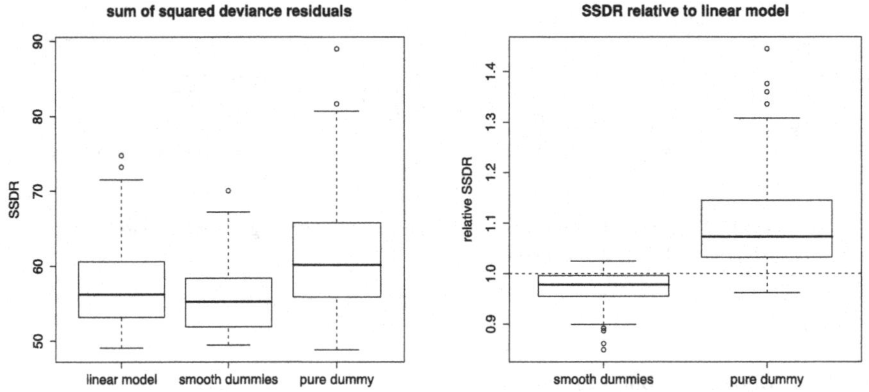

FIGURE 4.13: Performance (in terms of SSDR) of a (generalized) linear regression on the group labels, penalized regression for smooth dummy coefficients, and a pure dummy model (for the latter only the 68% successful estimates have been used) (left). Observed values for all considered methods; (right) SSDR values relative to linear model.

assumptions. It may be seen as an disadvantage that an additional tuning parameter has to be selected. However, this is easily done for example by cross-validation or minimization of information criteria like the AIC. For details and more simulations and applications, see Gertheiss and Tutz (2009b).

The ridge-type penalty ensures the existence of estimates and yields a smooth estimation of effects for ordered categorical predictors. An alternative strategy is to search for the categories that actually have different effects, or, in other words, to identify the categories that can be collapsed with respect to the dependent variable. In the case of ordered categorical predictors, collapsing naturally refers to adjacent categories. This selection and collapsing approach for ordered predictors is considered within selection procedures in Section 6.5.

4.5 Comparing Non-Nested Models

The analysis of deviance is a useful tool that allows one to distinguish between relevant and irrelevant terms in the linear predictor. The underlying strategy is based on a comparison of the nested models. One typically compares the model that contains the term in question to the model in which the term is omitted.

Analysis of deviance cannot be used if the models to be compared are non-nested, because differences of deviances have no standard distribution. If two models are not nested, one should distinguish between two cases. When the two models have the same number of parameters one can compute goodness-of-fit statistics, if available, and at least find out which model shows a better fit to the data. Although there is no benchmark for the comparison in the form of a standard distribution, it shows how strong the goodness-of-fit varies across models. However, if in addition the models to be compared have different numbers of parameters, goodness-of-fit will tend to favor the model that contains more parameters because the additional flexibility of the model will yield a better fit. Various criteria that take the number of parameters into account have been proposed. A widely used criterion is Akaike's information criterion which he named AIC, for "an information criterion" (Akaike, 1973). It is defined by

$$AIC = -2\log(L(\hat{\boldsymbol{\beta}})) + 2 \cdot (\text{number of fitted parameters}),$$

where $\log(L(\hat{\boldsymbol{\beta}}))$ denotes the log-likelihood of the fitted model evaluated at the ML estimate $\hat{\boldsymbol{\beta}}$. The second term may be seen as a penalty accounting for the number of fitted parameters. For binary response models with ungrouped data the deviance is given by $D(\boldsymbol{y}, \hat{\boldsymbol{\pi}}) = -2\log(L(\hat{\boldsymbol{\beta}}))$ and one obtains

$$AIC = D(\boldsymbol{y}, \hat{\boldsymbol{\pi}}) + 2 \cdot (\text{number of fitted parameters}).$$

Therefore, in the AIC criterion the number of fitted parameters is added to the deviance as a measure for the discrepancy between the data and the fit. The correction to the deviance may be derived by asymptotic reasoning. AIC has been shown to be an approximately unbiased estimate of the mean log-density of a new independent data set (see also Appendix D). A careful investigation of AIC and alternative model choice criteria was given by Burnham and Anderson (2002). While AIC is an information-theoretic measure based on Kullback-Leibler distance, the BIC (for Bayesian information criterion) has been derived in a Bayesian context by Schwarz (1978). It has the form

$$BIC = -2\log(L(\hat{\boldsymbol{\beta}})) + \log(n) \cdot (\text{number of fitted parameters}).$$

AIC and BIC are not directly comparable since the underlying targets differ. Based on its derivation as an unbiased estimate of the mean density of a new dataset with equal sample size, AIC is specific for the sample size at hand. In contrast, derivation of BIC assumes a true generating model, independent of sample size, although selection of the true model is obtained only asymptotically. For a comparison of these selection criteria including multimodel inference, see Burnham and Anderson (2004).

Akaike's Criterion

$$AIC = -2\log(L) + 2 \cdot (\text{number of fitted parameters})$$

Bayesian Information Criterion

$$BIC = -2\log(L) + \log(n) \cdot (\text{number of fitted parameters})$$

Although a comparison of non-nested models by testing is not straightforward, some methods were proposed in the econometric literature; see, for example, Vuong (1989) , who proposed likelihood ratio tests for model selection with tests on non-nested hypotheses. An alternative strategy is to compare the models in terms of prediction accuracy. When the sample is split several times into a learning dataset (for fitting of the model) and a test dataset (for evaluating prediction performance) one chooses the model that has the better performance (see Chapter 15).

4.6 Explanatory Value of Covariates

When using a regression model one is usually interested in describing the strength of the relation between the dependent variable and the covariates. In classical linear regression the most widely used measure is the squared multiple correlation coefficient R^2, also called the *coefficient of determination.* The strength of R^2 is its simple interpretation as the proportion of variation explained by the regression model. As a descriptive measure it may be derived from the partioning of squared residuals:

$$\sum_{i=1}^{n}(y_i - \bar{y})^2 = \sum_{i=1}^{n}(\hat{\mu}_i - \bar{y})^2 + \sum_{i=1}^{n}(\hat{\mu}_i - y_i)^2, \tag{4.11}$$

where $\bar{y}$ denotes the mean across observations $y_1, \ldots, y_n$ and $\hat{\mu}_i = \boldsymbol{x}_i^T \hat{\boldsymbol{\beta}}$ denotes the fitted values when $\hat{\boldsymbol{\beta}}$ is estimated by least squares and $\boldsymbol{x}_i$ contains an intercept. The term on the left-hand side is the total sum of squares (SST), the first term on the right-hand side corresponds to the sum of squares explained by regression (SSR), and the second term is the error sum of squares (SSE). The empirical *coefficient of determination* R^2 is defined by

$$R^2 = \frac{\text{SSR}}{\text{SST}} = \frac{\sum_i(\hat{\mu}_i - \bar{y})^2}{\sum_i(y_i - \bar{y})^2}.$$

With SST being the total variation without covariates, R^2 gives the proportion of variation explained by the covariates. If $\hat{\mu}_i$ is obtained by least-squares fitting, R^2 is equivalent to the squared correlation between the observations y_i and fitted values $\hat{\mu}_i$ (for more on R^2 see also Section 1.4.5).

The partitioning of squared residuals (4.11) may be seen as an empirical version of the decomposition of variance into "between variance" $\text{var}(\text{E}(y|\boldsymbol{x}))$ (explained by regression on $\boldsymbol{x}$) and "within variance" $\text{E}(\text{var}(y|\boldsymbol{x}))$ (or error):

$$\text{var}(y) = \text{var}(\text{E}(y|\boldsymbol{x})) + \text{E}(\text{var}(y|\boldsymbol{x})),$$

which holds without distributional assumptions for random variables $y, \boldsymbol{x}$. Thus R^2 is an empirical version of the true or theoretical proportion of explained variance:

$$R_T^2 = \text{var}(\text{E}(y|\boldsymbol{x}))/\,\text{var}(y) = \{\text{var}(y) - \text{E}(\text{var}(y|\boldsymbol{x}))\}/\,\text{var}(y),$$

which represents the population coefficient of determination. For the population measure R_T^2 one has a similar property as for the empirical coefficient of determination. It represents the proportion of explained variance but also equals the squared correlation between the random variables y and $E(y|\boldsymbol{x})$:

$$R_T^2 = \text{var}(\text{E}(y|\boldsymbol{x}))/\,\text{var}(y) = \text{cor}(y, \text{E}(y|\boldsymbol{x}))^2. \tag{4.12}$$

Therefore, in the linear model the empirical coefficient of determination R^2, obtained from least-squares fitting, as well as the empirical correlation coefficient are estimates of R_T^2.

The extension to GLMs has some pitfalls. Although the equivalence (4.12) holds more generally for GLMs, the empirical correlation coefficient between the data and values fitted by ML estimation is not the same as $R^2 = \text{SSR} / \text{SST}$. Moreover, in GLMs it is more difficult to interpret R_T^2 as a proportion of the explained variation. For example, when responses are binary, the mean and variance of the response are strictly linked. Therefore, $\text{var}(y|\boldsymbol{x})$ is not fixed but varies with $\boldsymbol{x}$. This is different from the case of a normally distributed vector $(y, \boldsymbol{x})$ for which $\text{var}(y|\boldsymbol{x})$ is a constant value and therefore the variability of $\text{E}(y|\boldsymbol{x})$ and $\text{var}(y|\boldsymbol{x})$, which determine the decomposition of $\text{var}(y)$, are strictly separated. In addition, the population measure R_T^2 can be severely restricted to small values, although a strong relation between the predictor and the response is present. Cox and Wermuth (1992) considered a linear regression model for binary responses with a single explanatory variable, $P(y = 1|x) = \pi + \beta(x - \mu_x)$, where x has mean μ_x and variance σ_x^2, and $\pi = P(y = 1)$. The value of $R_T^2 = \beta^2\sigma_x^2/(\pi(1-\pi))$ is primarily determined by the variance of x, and for a sensible choice of the variance $R_T^2 = 0.36$ is the largest value that can be achieved. Therefore, explained variation as a proportion of the explained variance is restricted to rather small values.

The different interpretations of the coefficient of determination in the linear model have led to various measures for non-linear models, most of them more descriptive in nature. Although there is no widely accepted direct analog to R^2 from least-squares regression, a number of R^2 for binary response models are in common use. We give some of them in the following.

4.6.1 Measures of Residual Variation

The coefficient of determination is a member of a more general class of measures that aim at the proportional decrease in variation obtained in going from a simple model to a more complex model that includes explanatory variables. The measures have the general form

$$R(M_0|M) = \frac{D(M_0) - D(M)}{D(M_0)}, \tag{4.13}$$

where $D(M)$ measures the (residual) variation of model M and $D(M_0)$ the variation of the simpler model M_0, which typically is the intercept model (compare Efron, 1978). The *proportional reduction in variation measure* (4.13) can be seen as a descriptive statistic or a population measure depending on the measure of variation that is used. For example, the empirical coefficient of determination R^2 is obtained by using $D(M_0) = SST$, $D(M) = SSE$; the population value is based on the corresponding distribution models and uses $D(M_0) = var(y)$, $D(M) = E(var(y|\boldsymbol{x}))$. However, in most applications measures of the form (4.13) are used as descriptive tools without reference to an underlying population value. The form (4.13) has more generally been referred to as a measure of proportional reduction in loss (or error) that is used to quantify reliability (Cooil and Rust, 1994).

Least squares as measures of variation are linked to the estimation in linear models with normally distributed responses. For generalized linear models the preferred estimation method is maximum likelihood. The corresponding variation measure is the deviance, which explicitly uses the underlying distribution. Let $l(\hat{\boldsymbol{\beta}})$ denote the maximal log-likelihood of the fitted model, $l(\hat{\beta}_0)$ denote the log-likelihood of the intercept model, and $l(\text{sat})$ be the log-likelihood of the saturated model. With $D(M) = D(\hat{\boldsymbol{\beta}}) = -2(l(\hat{\boldsymbol{\beta}}) - l(\text{sat}))$, $D(M_0) = D(\hat{\beta}_0) = -2(l(\hat{\beta}_0) - l(\text{sat}))$ one obtains the deviance-based measure

$$R_{dev}^2 = \frac{D(\hat{\beta}_0) - D(\hat{\boldsymbol{\beta}})}{D(\hat{\beta}_0)} = \frac{l(\hat{\boldsymbol{\beta}}) - l(\hat{\beta}_0)}{l(\text{sat}) - l(\hat{\beta}_0)}.$$

R^2_{dev} compares the reduction of the deviance when the regression model is fitted instead of the simple intercept model to the deviance of the intercept model.

For ungrouped binary observations one has $D(\hat{\boldsymbol{\beta}}) = -2l(\hat{\boldsymbol{\beta}})$ and the coefficient is equivalent to McFadden's (1974) *likelihood ratio index* (also called pseudo-R^2):

$$R^2_{dev} = \frac{l(\hat{\beta}_0) - l(\hat{\boldsymbol{\beta}})}{l(\hat{\beta}_0)}.$$

It is seen that $0 \le R^2_{dev} \le 1$ with

$$R^2_{dev} = 0 \quad \text{if} \quad l(\hat{\beta}_0) = l(\hat{\boldsymbol{\beta}}), \text{that is, if all other parameters have zero estimates:}$$
$$R^2_{dev} = 1 \quad \text{if} \quad D(\hat{\boldsymbol{\beta}}) = 0, \text{that is, if the model shows perfect fit, } \hat{\pi}_i = y_i.$$

R^2_{dev} is directly linked to the likelihood ratio statistic λ by $\lambda = R^2_{MF}(-2l(\hat{\beta}_0))$, which has asymptotic χ^2-distribution.

It should be noted that the deviance-based measure depends on the level of aggregation because the deviance for grouped observations differs from the deviance for individual observations. Deviances based on ungrouped data are to be preferred because if one fits a model that contains many predictors, the fit can be perfect for grouped data yielding $D(\hat{\boldsymbol{\beta}}) = 0$, although individual observations are not well explained. As an explanatory measure for future data, which will will come as individual data it is certainly insufficient. For the use of deviances for grouped observations see also Theil (1970) and Goodman (1971).

Although the deviance has some appeal when considering GLMs, alternative measures of variation can be used. One candidate is squared error with $D(M_0) = SST$, $D(M) = SSR$, yielding R^2_{SE} (e.g., Efron, 1978) . For binary observations, the use of the empirical variance $D(M_0) = n\bar{p}(1-\bar{p})$, where $\bar{p}$ is the proportion of ones in the total sample, and $D(M) = \sum_i \hat{\pi}_i(1-\hat{\pi}_i)$ yields Gini's concentration measure:

$$G = \frac{\sum_i \hat{\pi}_i^2 - n\bar{p}^2}{n\bar{p}(1-\bar{p})},$$

which was also discussed by Haberman (1982) (Exercise 4.7).

A prediction-oriented criterion that has been proposed uses the predictions based on $\bar{p}$ (without covariates) and $\hat{\pi}_i$ (with covariates). The variation measures D are defined by

$$D(M) = \sum_i L(y_i, \hat{\pi}_i), \quad D(M_0) = \sum_i L(y_i, \bar{p}),$$

where L is a modified (0–1) loss function with $L(y, \hat{\pi}) = 1$ if $|y - \hat{\pi}| > 0.5$, and $L(y, \hat{\pi}) = 0.5$ if $|y - \hat{\pi}| = 0.5$, and $L(y, \hat{\pi}) = 0$ if $|y - \hat{\pi}| < 0.5$. The corresponding measure R_{class} compares the number of misclassified observations ("non-hits") obtained without covariates to the number of misclassified observations obtained by using the model M, where the simple classification rule $\hat{y} = 1$ if $\hat{\pi}_i > 0.5$ and $\hat{y} = 0$ if $\hat{\pi}_i < 0.5$, with an adaptation for ties, is used. The measure uses the threshold 0.5, which corresponds to prior probabilities $P(y = 1) = P(y = 0)$ in terms of classification (see Chapter 15). The measure has a straightforward interpretation but depends on the fixed threshold. It is also equivalent to Goodman and Kruskal's λ (Goodman and Kruskal, 1954); see also van Houwelingen and Cessie (1990). A problem with measures like the number of misclassified errors is that they are a biased measure of the underlying true error rate in future samples. Since the sample is used to estimate a classification rule and to evaluate its performance, it underestimates the true error. The resulting error is also called a reclassification error. Better measures are based on cross-classification or leaving-one-out versions (for details see Section 15.3).

Measures for Explanatory Value of Covariates

Pseudo-R^2

$$R^2_{dev} = \frac{l(\hat{\beta}_0) - l(\hat{\boldsymbol{\beta}})}{l(\hat{\beta}_0)}.$$

Squared Error

$$R^2_{SE} = 1 - \frac{\sum_{i=1}^n (y_i - \hat{\pi}_i)^2}{\sum_{i=1}^n (y_i - \hat{y})^2}$$

Gini

$$G = \frac{\sum_i \hat{\pi}_i^2 - n\bar{p}^2}{n\bar{p}(1-\bar{p})}.$$

Reduction in Classification Error

$$R_{\text{class}} = \frac{\text{non-hits}(M_0) - \text{non-hits}(M)}{\text{non-hits}(M_0)},$$

Cox and Snell

$$R^2_{LR} = 1 - \left(\frac{L(\hat{\beta}_0)}{L(\hat{\boldsymbol{\beta}})}\right)^{2/n}$$

4.6.2 Alternative Likelihood-Based Measures

Since the coefficient of determination has several interpretations, alternative generalizations are possible. Cox and Snell (1989) considered

$$R^2_{LR} = 1 - \left(\frac{L(\hat{\beta}_0)}{L(\hat{\boldsymbol{\beta}})}\right)^{2/n},$$

where $L(\hat{\beta}_0), L(\hat{\boldsymbol{\beta}})$ denote the likelihoods of the two models. When $L(\hat{\beta}_0), L(\hat{\boldsymbol{\beta}})$ are the likelihoods of a normal response model, R^2_{LR} reduces to the standard R^2 of a classical linear regression. With a binary regression model R^2_{LR} cannot obtain a value of one even if the model predicts perfectly. Therefore, a correction has been suggested by Nagelkerke (1991). He proposed using $R^2_{corr} = R^2_{LR}/(1 - L(\hat{\beta}_0))^{2/n}$, which is a simple rescaling by using the maximal value $(1 - L(\hat{\beta}_0))^{2/n}$ that can be obtained by R^2_{LR}.

4.6.3 Correlation-Based Measures

In most program packages, summary measures that are based on the correlation between observations y_i and fit $\hat{\pi}_i$ are given. One candidate is the (squared) correlation between y_i and $\hat{\pi}_i$. Some motivation for the use of the squared correlation is that in the linear model and least-squares fitting the squared correlation is equivalent to the ratio of the explained variation SSR/SST, although the equivalence does not hold for GLMs. Zheng and Agresti (2000) prefer the correlation to its square because it has a familiar interpretation and works on the original scale of the observations.

Widely used association measures in logistic regression are rank-based measures. Concordance measures compare pairs of tupels $(y_i, \hat{\pi}_i)$ built from observation y_i and the corresponding prediction $\hat{\pi}_i$. Let a pair be given by $(y_i, \hat{\pi}_i), (y_j, \hat{\pi}_j)$, where $y_i < y_j$. The pair is called *concordant* if the same ordering holds for the predictions, $\hat{\pi}_i < \hat{\pi}_j$; it is *discordant* if $\hat{\pi}_i > \hat{\pi}_j$ holds. The pair is *tied* if $\hat{\pi}_i = \hat{\pi}_j$. If $y_i = y_j$ holds, the pair is also tied. If a pair is concordant, discordant, or tied, it can be expressed by using the sign function: $\text{sign}(d) = 1$ if $d > 0$, $\text{sign}(d) = 0$ if $d = 0$, and $\text{sign}(d) = -1$ if $d < 0$. Then the product $\text{sign}(y_i - y_j)\,\text{sign}(\hat{\pi}_i - \hat{\pi}_j)$ takes value 1 for concordant pairs and value -1 for discordant pairs. Several measures in common use are given in a separate box. The measures are also given in an alternative form since they can be expressed by using N_c for the number of concordant pairs, N_d for the number of discordant pairs, and N for the number of pairs with different observations $y_i \neq y_j$.

Rank-Based Measures

Kendall's τ_a

$$\begin{aligned}\tau_a &= \sum_{i<j} \text{sign}(y_i - y_j)\,\text{sign}(\hat{\pi}_i - \hat{\pi}_j)/(n(n-1)/2)\\ &= (N_c - N_d)/(n(n-1)/2),\end{aligned}$$

Somers' D

$$\begin{aligned}D_S &= \sum_{i<j} \text{sign}(y_i - y_j)\,\text{sign}(\hat{\pi}_i - \hat{\pi}_j)/\sum_{i<j} \text{sign}(y_i - y_j)^2\\ &= (N_c - N_d)/N,\end{aligned}$$

Goodman and Kruskal's γ

$$\begin{aligned}\gamma &= \sum_{i<j} \text{sign}(y_i - y_j)\,\text{sign}(\hat{\pi}_i - \hat{\pi}_j)/\sum_{i<j} \text{sign}(y_i - y_j)^2\,\text{sign}(\hat{\pi}_i - \hat{\pi}_j)^2\\ &= (N_c - N_d)/(N_c + N_d),\end{aligned}$$

A disadvantage of rank-based association measures is that in many applications they can not distinguish between different link functions. If the predicted values remain monotonic, link functions that yield quite different fits yield the same association value. The correlation coefficient used by Zheng and Agresti (2000) is able to distinguish between link functions because it uses the exact estimate, although it is more sensitive to outliers.

In general, association measures between observations and fit try to quantify how strong the link between y_i and $\hat{\pi}_i$ is. They may be used as descriptive statistics, but one should be aware of what they are measuring. While the correlation coefficient measures linear dependence, rank-based procedures measure the association between ordered values and therefore monotonicity. They reflect in particular goodness-of fit in the sample. One should be cautious with squared values of association measures, considered, for example, by Mittlböck and Schemper (1996). In particular, for squared rank-based measures it is unclear what they are measuring. Moreover, one should not take them at face value because as descriptive statistics they are random variables depending on the sample, a fact that is often ignored. A more sensible approach first defines a population measure and then finds ways to estimate it. For most useful measures there is an underlying population measure. For example, the empirical correlation coefficient has the theoretic analog $\text{cor}(y, E(y|\boldsymbol{x}))$. Also, Kendall's τ_a and Somers' D may be seen as estimates of an underlying measure (nicely described by Newson, 2002). The advantage of an underlying population value is that one can study the properties of the empirical measures as an estimator. That is a non-trivial task because one cannot expect $(y_i, \hat{\pi}_i)$ to be iid observations even when $(y_i, \boldsymbol{x}_i)$ are iid observations and the empirical measure will not be the best estimator for the population value. Zheng and Agresti (2000) used the population correlation coefficient $\text{cor}(y, E(y|\boldsymbol{x}))$ and investigate bias and MSE for several estimators and show that in particular the cross-validation estimator has poor performance.

In various studies measures have been compared. For example, Mittlböck and Schemper (1996) investigated the performance of most of the measures considered here. They demonstrated that the squared correlation coefficient, R^2_{SE}, and G are very similar over a wide range of values and are numerically consistent with the empirical coefficient of determination when a linear model is an appropriate approximation, which means for small values of R^2. Squared τ_a, γ, and R^2_{class} were found to yield quite different values. However, it cannot be expected that all the measures will behave in the same way since different population measures are behind. Numerical consistency or inconsistency is interesting, but in cases where the linear model is inappropriate nothing can be inferred on the accuracy of a descriptive statistic if what one is trying to estimate is not well defined.

Of course population values should have an intuitively clear interpretation. The Kullback-Leibler distance measure, which is behind the deviance, might not be very convincing for practioners. Simple measures with clear interpretations are measures that represent performance in classification. Behind the number of hits is the probability of a correct classification. Therefore, population measure and estimate refer to interesting quantities. But the number of hits refers to hits in the learning sample and therefore can be an overoptimistic estimate. To infer on the performance in future samples and find appropriate estimates, one should not rely on the empirical analog but find alternative estimators. Prediction-based measures and estimators are considered in more detail in Chapter 15.

General definitions of population measures that quantify the strength of the relation between a response variable and a vector of covariates have been given by Joe (1989), Osius (2004), Soofi et al. (2000). Van der Linde and Tutz (2008) considered R^2 measures derived from symmetric Kullback-Leibler discrepancies.

4.7 Further Reading

Robust Estimators. With the development of diagnostic tools for binary regression models (e.g., Pregibon, 1981; Landwehr et al., 1984; Fowlkes, 1987), estimates that are robust against outliers have been suggested. The resistant fitting procedure proposed by Pregibon (1982) is based on the downgrading of the influence of observations with high residuals. Copas (1988) considered the substantial bias of resistant fitting, which yields numerically larger coefficients, yielding a more extreme fit, closer to 0 or 1. He considered a bias-corrected version and proposed a misclassification model where transpositions between the possible outcomes 0 and 1 happen with a small probability. Carroll and Pederson (1993) studied an estimate that is closely related to Copas' misclassification estimate but which is consistent for the logistic model. Rousseeuw and Christmann (2003) considered estimates that are robust against separation and connected to the approach used by Tutz and Leitenstorfer (2006). An interesting approach to robust fitting by a forward search through the data was proposed by Atkinson and Riani (2000). An alternative form of robustification is the use of shrinkage estimators as considered in Chapter 6.

Weighted least-squares Estimator. Grizzle, Starmer, and Koch (1969) proposed an least-squares estimator for categorical responses. The so-called Grizzle-Starmer-Koch approach was very influential in the modeling of categorical response data. Although it allows for an explicit form of the estimate it has the disadvantage that it can not be used for continuous predictors. With the computational facilities available today ML estimates are widely preferred.

R packages. GLMs can be fitted by use of the model fitting functions *glm* from the *MASS* package. For the selection and smoothing of ordinal predictors one can use the package *ordPens*. Quasi-variances can be computed with *qvcalc*.

4.8 Exercises

4.1 Derive the likelihood of a binary response model $\pi(\boldsymbol{x}_i) = h(\boldsymbol{x}_i^T\boldsymbol{\beta})$ for independent observations $(y_i, \boldsymbol{x}_i), i = 1, \dots, n$, with $y_i \in \{0, 1\}$.

(a) Find the derivatives $\partial l(\boldsymbol{\beta})/\partial \beta_j$ for the components of $\boldsymbol{\beta}$ by elementary differentiation.

(b) Show that the resulting score function $\boldsymbol{s}(\boldsymbol{\beta}) = \partial l(\boldsymbol{\beta})/\partial \boldsymbol{\beta}$ is that of a generalized linear model.

(c) Derive the entries of the matrix of second derivatives $\boldsymbol{H}(\boldsymbol{\beta}) = \partial l^2(\boldsymbol{\beta})/\partial\boldsymbol{\beta}\partial\boldsymbol{\beta}^T$.

(d) Determine $\boldsymbol{F}(\boldsymbol{\beta}) = -\operatorname{E}\boldsymbol{H}(\boldsymbol{\beta})$ and show that it has the form of the Fisher matrix of a GLM.

4.2

(a) Derive the likelihood of a binary response model $P(y_{ij} = 1|\boldsymbol{x}_i) = \pi(\boldsymbol{x}_i) = h(\boldsymbol{x}_i^T\boldsymbol{\beta})$ for independent binomial observations $(y_i, \boldsymbol{x}_i), i = 1, \dots, N$, with $y_i = y_{i1} + \dots + y_{in_i} \sim B(n_i, \pi(\boldsymbol{x}_i))$.

(b) Show that the resulting likelihood equation for the logit model has the form $t = \operatorname{E}(T)$, where $T = \sum_i y_i\boldsymbol{x}_i$ is a sufficient statistic of $\boldsymbol{\beta}$ and t is the observed value.

(c) Derive the sufficient statistic for the logit model when only two populations are distinguished, namely males ($x = 1$) and females ($x = -1$) and interpret the components of the statistic.

4.3 The following table shows a comparison of a new agent and an active control with the binary response 1 for much improvement and 0 for improvement (compare Example 9.8). The model to be considered is the logit model with predictor $\eta = \beta_0 + x_A\beta$, where $x_A = 1$ represents the new agent and $x_A = 0$ represents the active control.

Drug	1	0
1: New agent	24	37
0: Active control	11	51

(a) Fit the model by using the underlying single binary observations, that is, use 24 binary responses with response 1 and $x_A = 1$, etc, and examine the significance of the parameter estimates.

(b) Fit the model by using the binomial distribution, where one considers only two populations, the first with 61 observations the second with 62 observations. Compare with the results from (a)

(c) Compute the null deviance D_0 (deviance for the model with intercept only) and the deviance for the model D_M for the single and grouped observations cases. How can the difference of deviances be used to examine the goodness-of-fit of the model?

4.4 Consider a binary logit model with two factors $A \in \{1, \dots, I\}$, $B \in \{1, \dots, J\}$. It can be given by $\log(\pi(A = i, B = j))/(1 - \pi(A = i, B = j)) = \beta_0 + \beta_{A(i)} + \beta_{B(j)} + \beta_{AB(ij)}$ with side constraints $\beta_{A(I)} = \beta_{B(J)} = \beta_{AB(Ij)} = \beta_{AB(iJ)} = 0$ for all i, j or with linear predictor $\eta(A = i, B = j) = \beta_0 + x_{A(1)}\beta_{A(1)} + \dots + x_{A(I-1)}\beta_{A(I-1)} + x_{B(1)}\beta_{B(1)} + \dots + x_{B(J-1)}\beta_{B(J-1)} + x_{A(1)}x_{B(1)}\beta_{AB(11)} + \dots + x_{A(I-1)}x_{B(J-1)}\beta_{AB(I-1,J-1)}$, where $x_{A(i)}, x_{B(j)}$ are dummy variables in 0–1 coding.

(a) Show how the parameters can be represented as functions of odds and odds ratios (one obtains, for example, $e^{\beta_{AB(ij)}} = \log(\gamma(i,j)/\gamma(i,J))/\gamma(I,j)/\gamma(I,J)$, where $\gamma(i,j) = \pi(A=i, B=j)/(1-\pi(A=i, B=j))$.

(b) Interpret the parameters in terms of odds ratios.

(c) Let the probability of being a regular reader of a specific journal depending on gender and age be given by the following table:

		Age	
		Young (1)	Old (0)
Gender	Male (1)	0.5	0.4
	Female (0)	0.8	0.6

Compute the parameters of the corresponding logit model and interpret them.

(d) Compute the parameter estimates of a saturated logit model for the data in Table 4.9 and interpret them.

4.5 The dataset *dust* that was used in Example 4.6 is available from package *catdata* (or at http://www.stat.uni-muenchen.de/sfb386/ under the name "Chronic bronchitis and dust concentration").

(a) Fit models for non-smokers and smokers separately that include quadratic terms of dust concentration and years of exposure. Decide what effects are needed. Consider alternatively models that use log-transformed predictors. Compare the fitted models.

(b) Fit an appropriate model that includes smoker status as a predictor.

4.6 Table 2.2 shows the vasoconstriction dataset.

(a) Compare the logit model with predictors volume and rate to the logit model with log-transformed predictors.

(b) Investigate if an interaction term is needed.

(c) Consider binary response models with alternative link functions and decide on an appropriate model.

4.7 Show that for binary observations the proportional reduction in variance $(D(M_0) - D(M))/D(M_0)$ is equivalent to Gini's concentration measure $G = (\sum_i \hat{\pi}_i^2 - n\bar{p}^2)/(n\bar{p}(1-\bar{p}))$ if one uses the empirical variance $D(M_0) = n\bar{p}(1-\bar{p})$, where $\bar{p}$ is the proportion of ones in the total sample, and $D(M) = \sum_i \hat{\pi}_i(1-\hat{\pi}_i)$.

Chapter 5

Alternative Binary Regression Models

In this chapter we will first consider alternatives to the logit link. Although the logit model has some advantages, which are discussed in Section 5.1.3, alternative link functions may be more appropriate in concrete applications. Moreover, we will consider extensions of the simple binary regression model that allow for overdispersion in the response and the conditional likelihood approach.

5.1 Alternative Links in Binary Regression

As in Chapter 4, we will consider models of the form $\pi(\boldsymbol{x}) = h(\boldsymbol{x}^T\boldsymbol{\beta})$, but in this section we will denote them by

$$\pi(\boldsymbol{x}) = F(\boldsymbol{x}^T\boldsymbol{\beta}).$$

The use of F for the response function refers to the derivation of the models from latent variables models, where F is the distribution function of the latent variable. Therefore, the response functions used here are strictly monotone distribution functions. The inverse functions F^{-1} correspond to the link. In the following, common choices for link and response functions are motivated.

5.1.1 Binary Response Models

Probit Model

A widely used model, particularly in economics, is the probit model, which is based on the standard normal distribution $\phi(\eta) = (2\pi)^{-1/2}\int_{-\infty}^{\eta} e^{-x^2/2}dx$.

> Probit Model
>
> $$\pi(\boldsymbol{x}) = \phi(\boldsymbol{x}^T\boldsymbol{\beta}), \quad \phi^{-1}(\pi(\boldsymbol{x})) = \boldsymbol{x}^T\boldsymbol{\beta}$$

In applications the probit model usually yields approximately the same results as the logit model. The goodness-of-fit is comparable, the same variables turn out to be relevant, and $p-$values are about the same, although the values of the estimates should not be compared directly (see Section 5.1.2). Very large sample sizes are needed to distinguish between the logit and the probit model. One may consider it as a drawback that the response function has no explicit form and that parameters do not have the same simple interpretation in terms of log-odds as in the logit model; nevertheless, the results are similar.

Complementary Log-Log Model and Log-Log Model

A distribution function that is distinctly different from the logistic distribution function is the *minimum extreme value* (or *Gompertz*) distribution $F(\eta) = 1 - \exp(-\exp(\eta))$. While the logistic distribution function is symmetric, the Gompertz distribution is asymmetric (see Figure 5.1 for the density distribution function). The model has the following representations.

Complementary Log-Log Model

$$\pi(\boldsymbol{x}) = 1 - \exp(-\exp(\boldsymbol{x}^T\boldsymbol{\beta})) \qquad \log(-\log(1 - \pi(\boldsymbol{x}))) = \boldsymbol{x}^T\boldsymbol{\beta} \quad (5.1)$$

The name complementary log-log model derives from the second form, where one sees that the link is log-log effecting the complementary probability $1 - \pi(\boldsymbol{x})$.

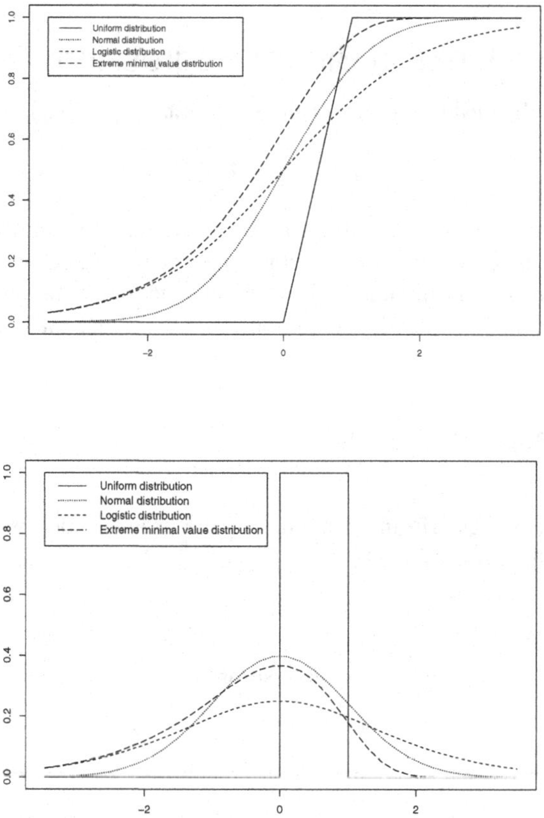

FIGURE 5.1: Response functions that correspond to distribution functions. Upper panel shows distribution functions of uniform distribution, normal distribution, logistic distribution, and minimum extreme value distribution; lower panel shows the corresponding densities.

A closely related model is the log-log model. The complementary log-log model (5.1) models $\pi(\boldsymbol{x}) = P(y = 1|\boldsymbol{x})$. Since the two possible values of $y, y = 1$ and $y = 0$, may be interchanged, one might use model (5.1) as well for the response $y = 0$, assuming $1 - \pi(\boldsymbol{x}) = 1 - \exp(-\exp(\boldsymbol{x}^T\boldsymbol{\beta}))$, which yields the log-log model.

Log-Log Model

$$\pi(\boldsymbol{x}) = \exp(-\exp(-\boldsymbol{x}^T\boldsymbol{\beta})) \qquad \log(-\log(\pi(\boldsymbol{x}))) = -\boldsymbol{x}^T\boldsymbol{\beta}$$

The use of $-\boldsymbol{x}^T\boldsymbol{\beta}$ rather than $\boldsymbol{x}^T\boldsymbol{\beta}$ (which is only important for the interpretation of parameters) has the advantage that the model has the form $\pi(\boldsymbol{x}) = F(\boldsymbol{x}^T\boldsymbol{\beta}))$, where F is the *maximum value* (or *Gumbel*) distribution $F(\eta) = \exp(-\exp(-\eta))$.

For symmetrical distributions like the logistic distribution it is just a matter of taste if one uses response y or the transformed response $\tilde{y} = 1 - y$. If $y = 1$ corresponds to success, $\tilde{y} = 1$ corresponds to failure. If $\boldsymbol{\beta}$ is the parameter vector of the logistic model when modeling success, $-\boldsymbol{\beta}$ is the parameter vector when modeling failure. Gomertz and Gumbel distributions are not symmetrical. They are connected in the following way: If a random variable ε follows the Gompertz distribution $F(\eta)$, the variable $-\varepsilon$ follows the Gumbel distribution $1 - F(-\eta)$, and vice versa. If $\boldsymbol{\beta}$ is the parameter vector of the Gompertz model for response y, $-\boldsymbol{\beta}$ is the parameter vector of the Gumbel model for response $\tilde{y}$. However, the Gumbel and the Gompertz models for fixed response y are not equivalent. Goodness-of-fit as well as parameter vectors will differ.

Exponential Model

Suppose that Z is a count variable taking values $0, 1, 2 \ldots$, which may refer to the number of cars in a household or in medicine to the number of symptoms. An often useful approximation to the distribution of Z is the Poisson distribution that has probability function $P(Z = z) = e^{-\lambda}\lambda^z/z!, z = 0, 1, 2, \ldots$, where λ represents the expectation of Z. If one is interested only in the dichotomization $Z = 0$ or $Z > 0$ (no car in household versus at least one car in household), one obtains for the dichotomous variable

$$y = \begin{cases} 1 & Z > 0 \\ 0 & Z = 0 \end{cases}$$

that

$$P(y = 1) = P(Z > 0) = 1 - P(Z = 0) = 1 - e^{-\lambda}.$$

By assuming that the expectation λ depends on the covariates in a linear way, $\lambda = \boldsymbol{x}^T\boldsymbol{\beta}$, one obtains the exponential model. Of course the complementary log-log model also can be motivated in this way, because the specification $\lambda = \exp(\boldsymbol{x}^T\boldsymbol{\beta})$ yields the complementary log-log model.

Exponential Distribution or Complementary Log Model

$$\pi(\boldsymbol{x}) = 1 - \exp(-\boldsymbol{x}^T\boldsymbol{\beta}) \qquad -\log(1-\pi(\boldsymbol{x})) = \boldsymbol{x}^T\boldsymbol{\beta}$$

Since the Poisson distribution is strongly connected to the exponential distribution, which is the waiting time until the next event in a Poisson process, it is not surprising that the exponential distribution model may also be motivated by a waiting time distribution model. Let a duration time (survival or sojourn time) T have exponential distribution $T \sim E(\lambda)$ with distribution function $F(x) = 1 - \exp(-\lambda x), \lambda > 0$. At time point τ one considers the dichotomization

$$y = \begin{cases} 1 & T < \tau \\ 0 & T \geq \tau, \end{cases}$$

which determines if the process T has ended or not until then; one obtains with parameterization $\lambda = \boldsymbol{x}^T\boldsymbol{\gamma}$

$$P(y=1) = P(T < \tau) = 1 - \exp(-\tau\boldsymbol{x}^T\boldsymbol{\gamma}) = 1 - \exp(-\boldsymbol{x}^T\boldsymbol{\beta}),$$

where $\boldsymbol{\beta} = \tau\boldsymbol{\gamma}$.

The motivation by dichotomizations of counts or waiting times shows that the model might be appropriate in many applications. Sometimes a problem is the non-convergence of estimates. It may arise because the link function is not differentiable everywhere. Since $F(x) = 1 - \exp(-\lambda x), x \geq 0$, the model actually has the form $\pi(\boldsymbol{x}) = 1 - \exp(-\boldsymbol{x}^T\boldsymbol{\beta})$ if $\boldsymbol{x}^T\boldsymbol{\beta} \geq 0$ and $\pi(\boldsymbol{x}) = 0$ if $\boldsymbol{x}^T\boldsymbol{\beta} < 0$. The simple form $\pi(\boldsymbol{x}) = 1 - \exp(-\boldsymbol{x}^T\boldsymbol{\beta})$ holds only if $\boldsymbol{x}^T\boldsymbol{\beta} \geq 0$, which implies severe restrictions on the parameter space. For discussions and applications see Wacholder (1986), Baumgarten et al. (1989), Guess and Crump (1978), Whittemore (1983), and Cornell and Speckman (1967). A nice overview on modeling with this link function is found in Piegorsch (1992).

If $\pi(x)$ is replaced by $1-\pi(x)$, one obtains the simple log-link or exponential model, which may be seen as a model that uses the distribution function $F(\eta) = \exp(\eta)$ for $\eta \leq 0$.

Exponential or Log-Link Model

$$\pi(\boldsymbol{x}) = \exp(\boldsymbol{x}^T\boldsymbol{\beta}) \qquad \log(\pi(\boldsymbol{x})) = \boldsymbol{x}^T\boldsymbol{\beta}$$

Cauchy Model

A model that is offered by some program packages is based on the Cauchy distribution. The "cauchit" link uses the (standard) Cauchy distribution function $F(\eta) = \tan^{-1}(\eta)/\pi + 1/2$, where $\tan^{-1} = \arctan$ is the inverse of the tangens and $\pi = 3.14159\ldots$. The Cauchy distribution is somewhat peculiar because it has no mean, variance, or higher moments defined, although the mode and median are well defined and are both equal to zero. It coincides with the Student's t-distribution with one degree of freedom. The cauchit link function $g(u) = \tan(\pi(u - 1/2))$ yields the following model.

Cauchy Model

$$\pi(\boldsymbol{x}) = \tan^{-1}(\boldsymbol{x}^T\boldsymbol{\beta})/\pi + 1/2 \qquad \tan(\pi(\pi(\boldsymbol{x}) - \frac{1}{2})) = \boldsymbol{x}^T\boldsymbol{\beta} \tag{5.2}$$

It should be noted that $\pi(\boldsymbol{x})$ denotes the probability whereas π in (5.2) it denotes the fixed and well-known number $\pi = 3.14159\ldots$. When compared to the normal distribution, the Cauchy distribution has heavier tails, thus allowing more extreme values than the normal distribution. For the modeling of binary responses it is attractive when observations occur for which the linear predictor is large in absolute value, indicating that the outcome is rather certain and yet the outcome is different. The model is more tolerant to these "outliers" than the logit or probit model. An early reference to the cauchit link model is Morgan and Smith (1993), where an example is given in which cauchit performs better than the probit link.

Identity Link Model

In a normal regression model the most widely used link is the identity link yielding $\mathrm{E}(y_i) = \boldsymbol{x}_i^T\boldsymbol{\beta}$. With some care it can also be used in binary regressions. However, the assumption $\pi(\boldsymbol{x}_i) = \boldsymbol{x}_i^T\boldsymbol{\beta}$ distinctly ignores that $\pi(\boldsymbol{x})$ is restricted to the unit interval. Nevertheless, it may be applied in cases where the covariate space is strongly limited in a way that the restriction to the unit interval holds. In Example 4.1 the logistic model yields an almost straight line and the logistic model and the linear model yield similar results (see Figure 4.3).

Identity Link Model

$$\pi(\boldsymbol{x}) = \boldsymbol{x}^T\boldsymbol{\beta}$$

5.1.2 Comparing Link Functions

The models considered in this section have the basic form $\pi(\boldsymbol{x}_i) = F(\boldsymbol{x}_i^T\boldsymbol{\beta})$, where F is a distribution function. Figure 5.1 shows the response functions for several of these models. At first sight the response functions seem to differ rather strongly and therefore should yield quite different discrepancies between data and fits. However, one should be aware that the distribution functions used as links are not comparable because they refer to different means and variances. For example, the standard normal distribution that is used in the normal model has mean zero and variance one while the logistic distribution (underlying the logistic regression model) has mean zero and variance π^2 (where $\pi = 3.14159\ldots$). Thus, it is no wonder that parameter estimates for the probit model and the logit model usually are quite different (although having about the same exploratory value). It is useful to consider again the derivation of binary regression models from latent regression models. In Section 2.2.2, it has been shown that all models of the form $\pi(\boldsymbol{x}_i) = F(\beta_0 + \boldsymbol{x}_i^T\boldsymbol{\beta})$ (with separated intercept) may be derived from an underlying continuous response model, $\tilde{y}_i = \gamma_0 + \boldsymbol{x}_i^T\boldsymbol{\beta} - \varepsilon_i$, where ε_i has the distribution function F and $y_i = 1$ if $\tilde{y}_i$ is above some threshold θ. One obtains

$$y_i = 1 \quad \Leftrightarrow \quad \tilde{y}_i = \gamma_0 + \boldsymbol{x}_i^T\boldsymbol{\beta} - \varepsilon_i \geq \theta \quad \Leftrightarrow \quad \varepsilon_i \leq \beta_0 + \boldsymbol{x}_i^T\boldsymbol{\beta},$$

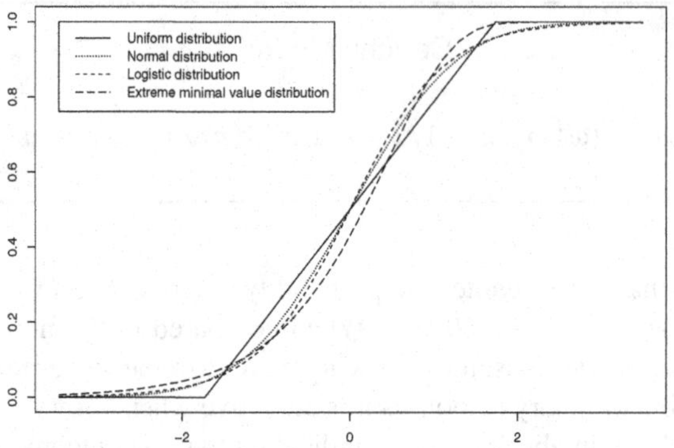

FIGURE 5.2: Response functions, standardized to mean zero and variances one, for several models.

where $\beta_0 = \gamma_0 - \theta$. If one wants to compare parameters of models with different link functions, one should at least assume that the distribution functions ε_i have the same mean and variance. Standardization of ε_i yields

$$y_i = 1 \Leftrightarrow \quad \frac{\varepsilon_i - E(\varepsilon_i)}{\sqrt{var(\varepsilon_i)}} \leq \frac{\gamma_0 - \theta - E(\varepsilon_i)}{\sqrt{var(\varepsilon_i)}} + x_{i1}\frac{\beta_1}{\sqrt{var(\varepsilon_i)}} + ... + x_{ip}\frac{\beta_p}{\sqrt{var(\varepsilon_i)}}$$

with the "standardized" parameters

$$\tilde{\beta}_0 = \frac{\beta_0 - E(\varepsilon_i)}{\sqrt{var(\varepsilon_i)}}, \quad \tilde{\beta}_i = \frac{\beta_i}{\sqrt{var(\varepsilon_i)}}.$$

With $F_{\text{stand}}(\eta)$ denoting the standardized distribution function (centered around zero with variance one), the models

$$\pi(\boldsymbol{x}) = F(\beta_0 + \boldsymbol{x}^T\boldsymbol{\beta}) \quad \text{and} \quad \pi(\boldsymbol{x}) = F_{\text{stand}}(\tilde{\beta}_0 + \boldsymbol{x}^T\tilde{\boldsymbol{\beta}})$$

are equivalent. It is seen from Figure 5.2 that the standardized response (distribution) functions are not so far apart. In particular, the normal and the logistic distribution functions are quite close. Therefore, it is quite natural that they yield similar goodness-of-fit, although parameter estimates for the unstandardized versions will differ.

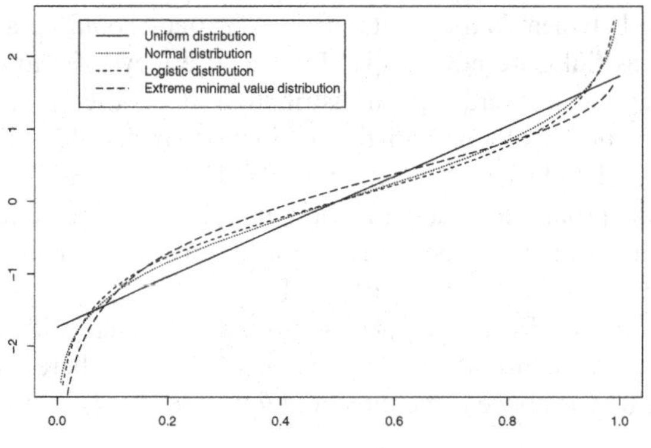

FIGURE 5.3: Link functions $F^{-1}(\pi)$ plotted against π.

Table 5.2 shows the fit of several binary response models for the response "car in household" with net income in the linear predictor. It is seen that the original estimates $\hat{\beta}$ are quite different but the standardized values are similar. In this dataset the Cauchy model showed the best fit, followed by the logit model.

TABLE 5.1: Means and variances of ε for several models.

Distrib.	Logit Logistic	Probit Normal	Complementary Log-Log Min. Extr. Value	Log-Log Max. Extr. Value	Compl. Exp Exponential
Mean	0	0	-0.577	0.577	1
Variance	$\pi^2/3$	1	$\pi^2/6$	$\pi^2/6$	1

TABLE 5.2: Estimates and standardized estimates for several link functions modeling the dependence of car ownership on net income.

	$\hat{\beta}_0$	$\tilde{\beta}_0$(standardized)	$\hat{\beta}$	$\tilde{\beta}$(standardized)	Deviance
Logit	-2.424	-1.334	0.0020	0.0011	1497.7
Probit	-1.399	-1.399	0.0011	0.0011	1505.0
C.log-log	-1.743	-0.909	0.0010	0.0008	1538.2
Cauchy	-2.655		0.0023		1479.7

5.1.3 Choice between Models and Advantages of Logit Models

For the choice of the link function, in particular, two aspects are relevant: goodness-of-fit and ease of interpretation. If one focusses on goodness-of-fit, one simply chooses the model that shows the best fit. Nevertheless, one always should be aware that only a selection of possible fits has been considered, so the best fit is only the best fit among the considered ones. In addition, differences of deviances have no standard distribution since there is no hierarchical order between the models. If responses are binomially distributed with not too small sample sizes n_i, the deviance can be compared to the χ^2-distribution showing if the models have a distinct lack-of-fit. An alternative approach to the selection of link functions is to estimate the fit in a non-parametric way. One can also fit parametric families of link functions that include the classical link functions as members. Then parameter estimates decide on the specific link function (for both approaches see Section 5.2).

If the deviances are not too different, a result that is usually found when comparing the logit model with the probit model, issues of interpretation become dominant. The logit model has several advantages that make it the most widely used model:

(1) The parameters are easily interpretable in terms of log-odds (for $\boldsymbol{\beta}$) or odds (for $\exp(\boldsymbol{\beta})$).

(2) If all of the covariates are categorical, the hypothesis $H_0 : \boldsymbol{\beta} = \mathbf{0}$ that none of the covariates has any exploratory value is equivalent to the statement that the response variable and the covariates are independent. For single predictors the hypothesis $H_0 : \beta_j = 0$ corresponds to the conditional independence of the response given the other variables (compare Chapter 12).

(3) The model is linked to the normal distribution, which plays a central role in statistics, by the derivation from the assumption $\boldsymbol{x}|y = i \sim N(\boldsymbol{\mu}_i, \boldsymbol{\Sigma})$; see Section 2.2.2.

(4) The logit link is the canonical link function, linking the natural parameter $\log(\pi/(1-\pi))$ directly to the linear predictor. One consequence is that the conditions for the existence of maximum likelihood estimates are weaker because the log-likelihood function is concave under weak conditions (see Fahrmeir and Kaufmann, 1985).

(5) The effects of variables may also be estimated if observations are drawn from the conditional distribution of $\boldsymbol{x}$ given y instead of the usual form of observing responses y given x. In econometrics this is called choice-based sampling (see Section 4.1).

Some motivation for the logit model may also be obtained from looking at the origins of the logistic distribution function (see Section 2.4).

5.2 The Missing Link

In Section 5.1 binary regression models with known link function were considered. Although various link functions may be used, there is always the danger that the true link function might not be among them or, more realistically, that none of the link functions provides a sufficiently good approximation to the data. This may be important since it has been demonstrated that misspecification of the link function can lead to substantial bias in the regression parameters (see Czado and Santner, 1992 , for binomial responses). Similar results have been demonstrated for single-index models by Horowitz and Härdle (1996).

Parametric Families of Link Functions

One approach to obtain more flexible models is to fit the parametric families of link functions. Several families have been proposed, including the link functions, that are in common use; see Prentice (1976), Pregibon (1980), Aranda-Ordaz (1983), Morgan (1985), Stukel (1988), Czado (1992), and Czado (1997). When using families of response functions, the common response function $F(\eta)$ is replaced by $F(\eta, \psi)$ with additional parameter ψ. As an example, let us consider a useful family allowing for right-tail modification of the logistic link (see Czado, 1997):

$$F(\eta, \psi) = \frac{\exp(h(\eta, \psi))}{1 + \exp(h(\eta, \psi))},$$

where

$$h(\eta, \psi) = \begin{cases} \frac{(\eta+1)^{\psi}-1}{\psi} & \eta > 0 \\ \eta & \text{otherwise.} \end{cases}$$

If $\psi = 1$, the logistic link results; for $\psi < 1$ ($\psi > 1$) the right tail is heavier (lighter) than for the logistic distribution. Families of this type have the advantage that the parameterization is orthogonal in a neighbourhood around $\boldsymbol{\beta} = \mathbf{0}$. When using families of response functions there is usually a parameter ψ^* that corresponds to the canonical link. In the case of the right-tail modification family one has $\psi^* = 1$.

A common approach to decide on the link is to treat it as a testing problem:

$$H_0 : \psi = \psi^* \quad \text{against} \quad H_1 : \psi \neq \psi^*.$$

If H_0 is not rejected, one keeps the canonical link; if H_0 is rejected, ψ and the regression parameters are estimated jointly to obtain the MLE $\hat{\boldsymbol{\delta}} = (\hat{\boldsymbol{\beta}}, \hat{\psi})$. Asymptotic theory, including strong consistency and asymptotic normal distributions was derived by Czado and Munk (2000). The use of the testing problem as a tool of model selection has been critized by various authors. In particular, Czado and Munk (2000) point out that for large sample sizes H_0 is frequently rejected in favor of H_1, although the mean space of both link functions is almost indistinguishable and therefore not scientifically relevant. They propose an alternative testing strategy that takes a measure of discrepancy between the response functions into account.

Non-Parametric Fitting of Link Functions

Families of link functions have the advantage that the link function is estimated by using parameters, and by fitting a larger family one avoids having to estimate separately non-nested models. A disadvantage is that the functions are still restricted to belong to the specified family. More flexible models are obtained by estimating link functions non-parametrically.

Especially in the economic literature, models of this type are known under the name single-index models. A *single-index model* has the form

$$\mu_i = h(\boldsymbol{x}_i^T \boldsymbol{\beta}),$$

where h is a smooth but *unspecified* function and $\mu_i = E(y_i|\boldsymbol{x}_i)$ denotes the conditional mean given covariates $\boldsymbol{x}_i$. The linear predictor $\boldsymbol{x}_i^T\boldsymbol{\beta}$ is known as the *single index*. The model may be seen as a special case of a *projection pursuit regression*, which assumes that μ_i has the additive form $h_1(\boldsymbol{x}_i^T\boldsymbol{\beta}_1) + \cdots + h_m(\boldsymbol{x}_i^T\boldsymbol{\beta}_m)$ with unknown functions $h_1, \ldots, h_m$, which transform the indices; see Friedman and Stützle (1981). Several approaches to the estimation of single-index models have been proposed in the literature (for references see the end of the chapter). However, in single-index models typically the response is assumed to be metrically scaled, and often a normal distribution is assumed. Moreover, single-index models do not assume that the function $h(.)$ is monotone. Therefore, the approaches are less helpful when one wants to estimate the unknown link function in generalized models. Although monotonicity is not needed when searching for a single index $\boldsymbol{x}_i^T\boldsymbol{\beta}$, it is useful for interpreting the parameters. If the link function $h(.)$ is not monotone, it is hard to interpret the parameters because a positive coefficient might increase or decrease the mean depending on the value of the other predictors. Thus, what might be helpful for dimension reduction is less helpful for fitting models that have easy interpretations.

Estimation of the unknown link function when the underlying distribution is from a simple exponential family was considered, for example, by Weisberg and Welsh (1994), Ruckstuhl and Welsh (1999), and Muggeo and Ferrara (2008). Weisberg and Welsh (1994) proposed estimating regression coefficients using the canonical link and then estimating the link via kernel smoothers given the estimated parameters. Then the parameters are re-estimated. Alternating between estimation of link and parameters yields consistent estimates.

The basic principle of alternating between these two estimates was also used by Yu and Ruppert (2002), but instead of kernel smoothers the unknown function is approximated by an expansion in basis functions. The approach may be outlined briefly as follows. For the unknown link function one uses the expansion

$$h(\eta_i) = \sum_{j=1}^{m} \alpha_j \phi_j(\eta_i),$$

where $\eta_i = \boldsymbol{x}_i^T\boldsymbol{\beta}$. To make the problem identifiable, $||\boldsymbol{\beta}|| = 1$ is postulated, where $||.||$ denotes the Euclidean norm. Moreover, $\boldsymbol{\beta}$ contains no intercept; it is included in $h(.)$. Splines are obtained by using the truncated power series basis functions $B_j(.)$ of degree q, which have also been used, for example, by Ruppert (2002). Thus the functions have the form $B_1(\eta) = 1, B_2(\eta) = \eta, B_{q+1}(\eta) = \eta^q$, $B_{q+j}(\eta) = |\eta - \tau_j|_+, j > 1$, where $\tau_1, \tau_2, \ldots$ are fixed knots. In a P-spline regression (see Section 10.1.3), usually a rather high number of equidistant knots is used (say $m = 20$ or 40) and the smoothness of the function estimate is controlled by an appropriate penalization. Ruppert (2002) suggested penalizing the squared coefficients that belong to the truncated powers, that is, $\sum_{j=q+2}^{m} \alpha_j^2$.

Let the response vector be given by $\mathbf{y}^T = (y_1, \ldots, y_n)$ and the design matrix by $\mathbf{X} = (\mathbf{x}_{(1)}, \ldots, \mathbf{x}_{(p)})$, where $\mathbf{x}_{(j)} = (x_{1j}, \ldots, x_{nj})^T$ denotes the observations of the jth covariate, $j = 1, \ldots, p$. Then, in the simplest case of normally distributed responses, an estimator of the single-index model is formulated as a minimizer of the penalized least-squares criterion:

$$Q(\boldsymbol{\alpha}, \boldsymbol{\beta}) = (\mathbf{y} - \boldsymbol{\phi}(\boldsymbol{\eta})\boldsymbol{\alpha})^T(\mathbf{y} - \boldsymbol{\phi}(\boldsymbol{\eta})\boldsymbol{\alpha}) + \lambda_P \boldsymbol{\alpha}^T \mathbf{P} \boldsymbol{\alpha}, \tag{5.3}$$

where $\boldsymbol{\eta} = \mathbf{X}\boldsymbol{\beta}$, $\boldsymbol{\phi}(\boldsymbol{\eta}) = (\phi_1(\boldsymbol{\eta}), \ldots, \phi_m(\boldsymbol{\eta})) = (\mathbf{1}, \boldsymbol{\eta}, \ldots, \boldsymbol{\eta}^q, (\boldsymbol{\eta} - \mathbf{1}\tau_1)_+^q, \ldots, (\boldsymbol{\eta} - \mathbf{1}\tau_m)_+^q)$, $\boldsymbol{\phi}_j(\boldsymbol{\eta}) = (\phi_j(\eta_1), \ldots, \phi_j(\eta_n))^T$, $\mathbf{P} = \text{diag}\{\mathbf{0}_{q+1}, \mathbf{1}_m\}$, and λ_P is a penalization parameter. Yu and Ruppert (2002) suggest solving (5.3) by using common non-linear least-squares routines while Leitenstorfer and Tutz (2011) and Tutz and Petry (2011) used boosting techniques.

In the case of GLMs, it is more appropriate to fit the model $\mu_i = h_0(h(\eta_i))$, where $h_0(.)$ is a fixed transformation function, which has to be chosen, and the inner function $h(.)$ is considered as unknown and has to be estimated. Typically, the choice of $h_0(.)$ depends on the distribution of the response. When the response is binary, a canonical choice is the logistic distribution function. The main advantage of specifying a fixed link function is that it may be selected such that the predictor is automatically mapped into the admissible range of the mean response. The expansion in basis functions is applied to the inner function $h(.)$.

Estimates are obtained by iteratively estimating the regression coefficients $\boldsymbol{\beta}$ and the parameters of the link function $\boldsymbol{\alpha}$, where $h(\eta_i) = \sum_{j=1}^{m} \alpha_j \phi_j(\eta_i) = \boldsymbol{\alpha}^T \boldsymbol{\Phi}_i$. In matrix notation, let $\hat{\boldsymbol{\beta}}^{(l)}$ and $\hat{\boldsymbol{\eta}}^{(l)} = \boldsymbol{X}\hat{\boldsymbol{\beta}}^{(l)}$ denote the parameter estimate and the fitted predictor in the lth step. Moreover, $\boldsymbol{\Phi}^{(l)} = (\boldsymbol{\Phi}_1^{(l)}, \ldots, \boldsymbol{\Phi}_n^{(l)})^T$ with $\boldsymbol{\Phi}_i^{(l)} = (\phi_1(\hat{\eta}_i^{(l)}), \ldots, \phi_m(\hat{\eta}_i^{(l)}))^T$ is the current design matrix for the basis functions. Then two steps are iterated:

Estimation of basis coefficients for a fixed predictor. For a fixed predictor $\hat{\boldsymbol{\eta}}^{(l-1)} = \mathbf{X}\hat{\boldsymbol{\beta}}^{(l-1)}$ the model $\boldsymbol{\mu} = h_0((\Phi^{(l-1)})^T \boldsymbol{\alpha}^{(l)})$ is fitted by one step of penalized Fisher scoring that uses the matrix of derivations $\hat{\mathbf{D}}^{(l-1)} = \text{diag}(\partial h_0(\hat{h}^{(l-1)}(\hat{\eta}_i^{(l-1)})/\partial h^{(l-1)}(\eta))$ evaluated at the estimate of the previous step, the diagonal matrix of variances evaluated at $h_0(\hat{h}^{(l-1)}(\eta_i^{(l-1)}))$, and a penalty matrix $\mathbf{P}_h$ that penalizes the second derivation of the estimated (approximated) response function (for penalties see Section 10.1.3).

Estimation of regression coefficients for a fixed response function. For $h(.)$ fixed, one fits the model $\boldsymbol{\mu} = h_0(h(\mathbf{X}\boldsymbol{\beta}^{(l)})$. Fisher scoring has the form

$$\hat{\boldsymbol{\beta}}^{(l)} = (\boldsymbol{X}^T \hat{\boldsymbol{D}}_{\eta}^{(l-1)} (\hat{\boldsymbol{\Sigma}}^{(l-1)})^{-1} \hat{\boldsymbol{D}}_{\eta}^{(l-1)} \boldsymbol{X})^{-1} \boldsymbol{X}^T \hat{\boldsymbol{D}}_{\eta}^{(l-1)} (\hat{\boldsymbol{\Sigma}}^{(l-1)})^{-1} (\boldsymbol{y} - \hat{\boldsymbol{\mu}}^{(l-1)}), \tag{5.4}$$

where $\hat{\boldsymbol{D}}_{\eta}^{(l-1)} = \text{diag}(\partial h_0(\hat{h}^{(l-1)}(\hat{\eta}_i^{(l-1)}))/\partial\eta)$ is the matrix of derivatives evaluated at the values of the previous iteration and $\hat{\boldsymbol{\Sigma}}^{(l-1)}$ is the variance from the previous step.

The second step can be modified to include a selection step that includes the most relevant predictor within a boosting procedure; see Section 6.4, where more details are given.

5.3 Overdispersion

In practice it is not too rarely found that models have large deviance although there seems to be no systematic lack-of-fit. Additional noise that is not accounted for may make the responses more variable than is to be expected under the assumed distribution model. The data show *overdispersion*. Although underdispersion, which signals lower variability than expected, is also found, it is much rarer. There are several strategies for dealing with overdispersion, but first we consider potential sources for the phenomenon in binary data.

5.3.1 Sources of Overdispersion

Correlated Observations

When considering binomial data $y_i = y_{i1} + \cdots + y_{in_i} \sim B(n_i, \pi_i)$ in prevous sections it was assumed that $y_{11}, \ldots, y_{1n_1}, y_{21}, \ldots, y_{Nn_N}$ are independent given the covariates. In particular, if measurements $y_{i1}, \ldots, y_{in_i}$ are collected at one unit, this assumption may be violated.

If one assumes that $y_{i1}, \ldots, y_{in_i}, y_{ij} \sim B(1, \pi_i)$ are correlated, one obtains

$$\text{var}(y_i) = var(\sum_{j=1}^{n_i} y_{ij}) = \sum_{j=1}^{n_i} \text{var}(y_{ij}) + \sum_{r \neq s} \text{cov}(y_{ir}, y_{is}).$$

By using $\text{var}(y_{ij}) = \pi_i(1 - \pi_i), \text{cov}(y_{ir}, y_{is}) = \rho(\text{var}(y_{ir}) \text{var}(y_{is}))^{1/2}$ with ρ denoting the correlation coefficient, one has

$$\text{var}(y_i) = n_i\pi_i(1 - \pi_i)[1 + (n_i - 1)\rho] = n_i\pi_i(1 - \pi_i)\phi_i$$

with dispersion parameter $\phi_i = 1 + (n_i - 1)\rho$. The resulting effects are

- for $n_i = 1$ one has no overdispersion, since $\phi_i = 1$;
- for a positive correlation, $\rho > 0$, the variability is larger than expected under the binomial probability model (provided $n_i > 1$);
- for a negative correlation, $\rho < 0$, underdispersion is found (assuming $n_i > 1$); however, since $\phi_i \geq 0$ has to hold, the negative correlation is restricted by $\rho \geq -1/(n_i - 1)$.

Unobserved Heterogeneity

A possible source of heterogeneity is that unobserved or unobservable variables induce extra variability in y. Let us assume that there is a latent variable D_i that selects a probability and the response y_i given the selected value (and observed covariates) has the usually assumed binomial distribution (see also Williams, 1982). To be more specific, the following is assumed.

(1) A latent variable $D_i \in [0, 1]$ with

$$\text{E}(D_i) = \pi_i, \quad \text{var}(D_i) = \delta\pi_i(1 - \pi_i), \delta \geq 0.$$

selects a value ϑ_i from $[0, 1]$.

(2) Given $D_i = \vartheta_i$, the response has the binomial distribution

$$y_i | D_i = \vartheta_i \sim B(n_i, \vartheta_i).$$

By using $\text{E}(\text{E}(Y|X)) = \text{E}(Y)$ and $\text{var}(Y) = \text{var}\,\text{E}(Y|X) + E(\text{var}(Y|X))$, one obtains for the marginal distribution of y_i

$$\text{E}(y_i) = n_i\pi_i, \quad \text{var}(y_i) = n_i\pi_i(1 - \pi_i)\phi_i,$$

where $\phi_i = 1 + (n_i - 1)\delta$. Thus the variable y_i has the usual mean to be expected under the binomial model, but the variances are inflated. If $\delta > 0$ and $n_i > 1$, the variability is larger than in the binomial model. The limiting case $\delta = 0$ means that the latent variable has zero variance. Then the latent variable has no effect and the binomial model holds.

It is noteworthy that the overdispersion resulting from correlated responses and the assumption of an underlying latent variable yields almost the same model for overdispersion. A difference is that the latent variable approach allows only overdispersion whereas the correlation model also allows for a restricted form of underdispersion.

There are several strategies for dealing with overdispersion data. One is the explicit modeling of heterogeneity and an example is the beta-binomial model, which is considered in the following. Alternatively, one can assume a normal distribution for the unobserved heterogeneity or a finite mixture; both modeling approaches will be treated in Chapter 14. A second strategy is to use generalized estimation functions, which are also considered in the next section.

5.3.2 Beta-Binomial Model

A candidate for the distribution of the latent variable $D_i \in [0, 1]$ is the beta distribution (see Appendix A). If one assumes $D_i \sim Beta(a_i, b_i)$ with parameters $a_i, b_i > 0$, the means and variances are given by

$$\mathrm{E}(D_i) = \pi_i = \frac{a_i}{a_i + b_i}, \qquad \mathrm{var}(D_i) = \pi_i(1 - \pi_i)/(a_i + b_i + 1).$$

Assuming the parametric model $\pi_i = h(\boldsymbol{x}_i^T\beta)$, one obtains $\mathrm{var}(D_i) = \delta_i\pi_i(1 - \pi_i)$ with the dispersion parameter $\delta_i = 1/(a_i + b_i + 1)$.

The marginal distribution of y_i is called the *beta-binomial distribution*; the explicit form is obtained by integrating with respect to the density of the beta distribution:

$$\begin{aligned} P(y_i; n_i, a_i, b_i) &= \int P(y_i|D_i = \vartheta_i)p(\vartheta_i)d\vartheta_i \\ &= \int \binom{n_i}{y_i}\vartheta_i^{y_i}(1-\vartheta_i)^{n_i-y_i}\frac{\Gamma(a_i+b_i)}{\Gamma(a_i)\Gamma(b_i)}\vartheta_i^{a_i-1}(1-\vartheta_i)^{b_i-1}d\vartheta_i \\ &= \binom{n_i}{y_i}\frac{(a_i+y_i-1)_{y_i}(b_i+n_i-y_i-1)_{n_i-y_i}}{(a_i+b_i+n_i-1)_{n_i}}, \end{aligned}$$

where $(k)_r = k(k-1)\ldots(k-r+1)$. An alternative form is

$$P(y_i; n_i, a_i, b_i) = \frac{B(a_i + y_i, b_i + n_i - y_i)}{B(a_i, b_i)},$$

$y_i \in \{0, 1 \ldots, n_i\}$, where $B(p, q) = \Gamma(p)\Gamma(q)/\Gamma(p + q)$ is the beta function.

The model contains the parameters $\boldsymbol{\beta}$ implicitly. Instead of a_i, b_i one may use the parameters $\pi_i = a_i/(a_i + b_i), \delta_i = 1/(a_i + b_i + 1)$, which yields $a_i = \pi_i(1 - \delta_i)/\delta_i, b_i = (1 - \pi_i)(1 - \delta_i)/\delta_i$ and turns the density into $B(\pi_i(1 - \delta_i)/\delta_i + y_i, (1 - \pi_i)(1 - \delta_i)/\delta_i + n_i - y_i)/B(\pi_i(1 - \delta_i)/\delta_i, (1 - \pi_i)(1 - \delta_i)/\delta_i)$. For simplicity it is often assumed that $\delta_i = \delta$ is the same for all observations. Then, assuming that the beta-binomial distribution holds, one specifies the mean by $\pi_i = h(\boldsymbol{x}_i^T\boldsymbol{\beta})$ with a fixed response function h. Thus the mean is given by $\mu_i = n_i\pi$ and the variance by $\mathrm{var}(y_i = n_i\pi(1 - \pi)[1 + (n_i - 1)\delta]$, with $\delta = 0$ representing the limiting case of a binomial distribution.

Beta-binomial models cannot be treated within the framework of generalized linear models. Crowder (1987) and Hinde and Démetrio (1998) gave algorithms for solving the maximum likelihood equations. The latter obtained the fit by iterating between estimates of β for fixed δ and estimates of δ for fixed $\boldsymbol{\beta}$. Prentice (1986) considered a more general model, where δ_i could also depend on covariates.

5.3.3 Generalized Estimation Functions and Quasi-Likelihood

Explicit parametric modeling of heterogeneity as used in the derivation of the beta-binomial model has several drawbacks. Numerical integration can be avoided only for specific distributions. More seriously, the assumption of a specific distribution function determines inferences, although it is hard to validate. To avoid the restrictive assumption of a specific latent variable one can use generalized estimation equations that are based on quasi-likelihood approaches (see also Section 3.11).

In *quasi-likelihood approaches* it is not necessary to specify a distribution for the responses. Much weaker, one only specifies the first two moments. For count data y_i with $y_i \in \{0, 1 \ldots, n_i\}$ one might assume that the means and variances are given by

$$\mathrm{E}(y_i) = n_i \pi_i = n_i h(\boldsymbol{x}_i^T \boldsymbol{\beta}), \quad \mathrm{var}(y_i) = n_i \pi_i (1 - \pi_i)\phi,$$

which corresponds to an overdispersed binomial distribution. For proportions $\bar{y}_i = p_i = y_i / n_i$ one has $\mathrm{E}(\bar{y}_i) = \pi_i = h(\boldsymbol{x}_i^T \beta)$, $\mathrm{var}(\bar{y}_i) = \pi_i(1 - \pi_i)\phi/n_i$.

The essential point in these equations is that the mean and variance are specified, with the variance having the simple form of an inflated binomial variance:

$$\mathrm{var}(y_i) = v(\pi_i)\phi,$$

where $v(\pi_i)$ is a known (or fully specified) variance function. This means that the functional form of the variance (depending on covariates) is assumed to be known. Only the scaling factor ϕ is unknown and has to be estimated. An estimate of ϕ is

$$\hat{\phi} = \frac{1}{N - p} \sum_{i=1}^{N} \frac{(y_i - \hat{\mu}_i)^2}{n_i v(\pi_i)} = \frac{1}{N - p} \sum_{i=1}^{N} \frac{(\bar{y}_i - \hat{\pi}_i)^2}{v(\pi_i)/n_i}.$$

For $v(\pi_i) = \pi_i(1 - \pi_i)$ one obtains $\hat{\phi} = \chi_P^2/(N - p)$, where χ_P^2 is Pearson's goodness-of-fit statistic.

The variance function $v(\pi_i) = \pi_i(1 - \pi_i)$ is very easy to handle. One fits the ordinary binary regression model and uses the ML estimate. To obtain the correct covariances matrix of $\hat{\beta}$ one multiplies the maximum likelihood covariance by $\hat{\phi}$ since the covariance is approximated by $\mathrm{cov}(\boldsymbol{\beta}) \approx \hat{\phi} \boldsymbol{F}(\boldsymbol{\beta})^{-1}$, where $\boldsymbol{F}(\hat{\boldsymbol{\beta}})$ is the Fisher matrix of the ordinary model. Maximum likelihood standard errors are multiplied by $\sqrt{\hat{\phi}}$ and t statistics are divided by $\sqrt{\hat{\phi}}$.

An alternative estimate of ϕ, based on the deviance, is $\tilde{\phi} = D/(N - p)$. It is comparable to $\hat{\phi}$ if all n_i's are of similar size. While $\hat{\phi}$ is also consistent for small local samples, this does not hold for $\tilde{\phi}$ (compare McCullagh and Nelder, 1989).

The estimation equation that is used to obtain an estimate of $\boldsymbol{\beta}$ has the form

$$\sum_{i=1}^{N} \boldsymbol{x}_i \frac{h'(\boldsymbol{x}_i^T \boldsymbol{\beta})}{\mathrm{var}(p_i)} (p_i - h(\boldsymbol{x}_i^T \boldsymbol{\beta})) = \mathbf{0}. \tag{5.5}$$

When the variance of p_i is the inflated binomial variance, specified by $\mathrm{var}(p_i) = \phi h(\boldsymbol{x}_i' \boldsymbol{\beta})(1 - h(\boldsymbol{x}_i' \boldsymbol{\beta}))/n_i$, the solution of equation (5.5) does not depend on ϕ. Of course, when $\phi = 1$ and

$v(\pi_i) = \pi_i(1-\pi_i)$ is assumed, equation (5.5) is equivalent to the ML estimation equation for the corresponding binomial model.

When using these estimates it is assumed that the mean and variance are correctly specified. More generally, one may consider equation (5.5) as a *generalized estimation function* under the assumption that only the mean is correctly specified, whereas the variance is considered a working covariance that does not have to be the variance of the data-generating model (e.g., Gourieroux et al., 1984). It may be shown that under regularity conditions one obtains an asymptotically normally distributed estimate $\hat{\boldsymbol{\beta}} \sim N(\boldsymbol{\beta}, \hat{\boldsymbol{F}}^{-1}\hat{\boldsymbol{V}}\hat{\boldsymbol{F}}^{-1})$, where the sandwich matrix $\hat{\boldsymbol{F}}^{-1}\hat{\boldsymbol{V}}\hat{\boldsymbol{F}}^{-1}$ is determined by

$$\hat{\boldsymbol{F}} = \sum_{i=1}^{N} \boldsymbol{x}_i \boldsymbol{x}_i^T \frac{h'(\boldsymbol{x}_i^T\hat{\boldsymbol{\beta}})^2}{\hat{\text{var}}(p_i)} \qquad \hat{\boldsymbol{V}} = \sum_{i=1}^{g} \boldsymbol{x}_i \boldsymbol{x}_i^T \frac{h'(\boldsymbol{x}_i^T\hat{\boldsymbol{\beta}})^2}{\hat{\text{var}}(p_i)^2}(p_i - h(\boldsymbol{x}_i^T\hat{\boldsymbol{\beta}}))^2$$

with $\hat{\text{var}}(p_i) = \hat{\phi} v(h(\boldsymbol{x}_i'\hat{\boldsymbol{\beta}}))$.

Although the inflated binomial variance is easy to handle, the assumption of correlated responses as well as the modeling by latent variables suggest that for count data $y_i \in \{0, 1 \ldots, n_i\}$ variances are given by $var(y_i) = n_i\pi_i(1-\pi_i)[1+(n_i-1)\delta]$, where δ is a dispersion parameter. The corresponding quasi-likelihood approach, which assumes that the mean and variance are correctly specified, is less easy to handle because the dispersion parameter does not cancel out. Williams (1982) proposed an algorithm that iterates between estimates of β for fixed δ and estimates of δ for fixed β.

Liang and McCullagh (1993) compared various approaches to the modeling of overdispersion; an overview can be found in Poortema (1999). Lambert and Roeder (1995) introduced a convexity plot that detects overdispersion, relative variance curves, and tests that help to understand the nature of the overdispersion. Relative variance curves and tests sometimes distinguish the source of the overdispersion better than score tests.

Example 5.1: Teratology

In a teratology experiment considered by Moore and Tsiatis (1991) and Liang and McCullagh (1993), 58 rats on iron-deficient diets were assigned to four groups (see Table 5.3). In the first group only placebo injections were given, and in the other groups iron supplements were given. The animals were impregnated and sacrificed after three weeks. The response was whether the fetus was dead ($y_{ij} = 1$) for each fetus in each rat's litter.

TABLE 5.3: Response counts of (litter size, number dead) for 58 litters of rats in low-iron teratology study.

Group 1: Untreated (low iron)
(10,1)(11,4)(12,9)(4,4)(10,10)(11,9)(9,9)(11,11)(10,10)(10,7)(12,12)
(10,9)(8,8)(11,9)(6,4)(9,7)(14,14)(12,7)(11,9)(13,8)(14,5)(10,10)
(12,10)(13,8)(10,10)(14,3)(13,13)(4,3)(8,8)(13,5)(12,12)
Group 2: Injections days 7 and 10
(10,1)(3,1)(13,1)(12,0)(14,4)(9,2)(13,2)(16,1)(11,0)(4,0)(1,0)(12,0)
Group 3: Injections days 0 and 7
(8,0)(11,1)(14,0)(14,1)(11,0)
Group 4: Injections weekly
(3,0)(13,0)(9,2)(17,2)(15,0)(2,0)(14,1)(8,0)(6,0)(17,0)

Source: Moore and Tsiatis (1991)

Since the observations for the ith litter $y_{i1}, \ldots y_{in_i}$ were measured on one female rat, one might suspect overdispersion. Let $y_{ij} \sim B(1, \pi_i)$ denote the response of the jth fetus in litter i and

$y_i = y_{i1} + \cdots + y_{in_i} \sim B(n_i, \pi_i)$ denote the number dead out of the n_i fetuses in litter i. As a structural component one assumes for $\pi_{ij} = \mathrm{E}(y_{ij})$ a logit model

$$\mathrm{logit}(\pi_{ij}) = \beta_0 + x_{G(2)}\beta_2 + x_{G(3)}\beta_3 + x_{G(4)}\beta_4,$$

where $x_{G(i)}$ is a (0-1)-dummy variable with $x_{G(i)} = 1$ if the observation is from group i and 0 otherwise. The naive approach assumes that all observations $y_{11}, y_{12}, \ldots$ are independent binary variables. One obtains a deviance of 173.45 on 54 degrees of freedom, which, however, should not be interpreted as a goodness-of-fit statistic. In Table 5.4 results are given for the approaches that yield the same estimates but differing standard errors. The independence model assumes that all binary observations are independent. The quasi-likelihood approach with an inflated binomial variance uses $var(y_i) = \phi n_i \pi_i (1 - \pi_i)$. In the weaker generalized estimation approach, independence was used as a working covariance. It is seen that standard errors are larger when overdispersion is taken into account. The simple independence model is certainly not appropriate. In Table 5.5 estimates are given for the beta-binomial model and for the mixed model approach, estimated by penalized quasi-likelihood and Gauss-Hermite quadrature with 14 quadrature points. The mixed model assumes that each unit (rat) has its own intercept, which follows a normal distribution. Therefore, heterogeneity across units is modeled quite flexibly (for details see Chapter 14). It is seen that the estimates for the beta-binomial model are slightly smaller than the estimated values in Table 5.4. For the mixed model approach estimates are distinctly larger, an effect that is frequently found in subject-specific models (Chapter 14). In addition, a discrete mixture model with two components was fitted. The model assumed that the response was a mixture of two components that have distinct intercepts (see Chapter 14). The two intercepts of the components, which are not given in the table, were -0.211 and 2.458. The estimated effects are quite close to the estimates for the mixed model with normally distributed random effects. □

TABLE 5.4: Estimates and standard errors for independence model, quasi-likelihood model, and generalized estimation functions fitted to teratology data with logit link.

	Estimates	Standard Errors		
		Independence Model	Quasi-Likelihood $var(y_i) = \phi n_i \pi_i (1 - \pi_i)$	GEE Independence
β_0	1.144	0.129	0.219	0.275
β_2	−3.323	0.331	0.560	0.440
β_3	−4.476	0.731	1.237	0.610
β_4	−4.130	0.476	0.806	0.576
$\hat{\phi}$		–	2.865	1.007

TABLE 5.5: Estimates for beta-binomial model and several mixture logit models fitted to teratology data

		Mixed Models		
	Beta-Binomial	Pen Quasi-Likelihood	Gauss-Hermite	Discrete Mixture
β_0	1.345 (0.244)	1.687 (0.306)	1.802 (0.362)	-
β_2	−3.087 (0.521)	−4.130 (0.614)	−4.515 (0.736)	−4.309(0.481)
β_3	−3.865 (0.863)	−5.274 (0.981)	−5.855 (1.189)	−5.509(0.824)
β_4	−3.919 (0.684)	−5.109 (0.747)	−5.594 (0.919)	−5.082(0.595)
		$\hat{\sigma} = 1.456$	$\hat{\sigma} = 1.533$	

5.4 Conditional Likelihood

In some applications the model involves parameters that are of minor interest to the investigator. These parameters are often called nuisance or incidental parameters. For example, in a clinical trial that compares the effect of a treatment (new drug or new therapy) with a current standard, let the response be 'success' or 'failure'. Suppose that data on the effect of treatment are available from g sources or strata. In a multi-center clinical trial the strata are the medical centers in which the trials are taken. One obtains for each stratum a 2 × 2 table, the table for the ith stratum has the form

	Success	Failure		Proportions
Treatment	y_{i1}	$n_{i1} - y_{i1}$	n_{i1}	$p_{i1} = y_{i1}/n_{i1}$
Control	y_{i2}	$n_{i2} - y_{i2}$	n_{i2}	$p_{i2} = y_{i2}/n_{i2}$
	$y_{i\cdot}$	$n_i - y_{i\cdot}$	n_i	

The main effect logit model, which models the effect of the stratum i and the treatment on the binary response ($y = 1$ for success and $y = 0$ for failure), has the form

$$\log\left(\frac{\pi(i, x_T)}{1 - \pi(i, x_T)}\right) = \beta_i + x_T\beta, \tag{5.6}$$

where $\pi(i, x_T)$ is the probability of success given the stratum i and the treatment group specified by x_T ($x_T = 1$ for treatment and $x_T = 0$ for control). The treatment effect is given by β while β_i represents the stratum effect.

The problem with the linear logistic model is that it contains g parameters, which have to be estimated on the basis of $2g$ observed binomial proportions. Thus for large g maximum likelihood estimates will hardly be efficient. Moreover, the parameter of interest is β, and the parameters $\beta_1, \ldots, \beta_{g-1}$ are nuisance parameters. When testing the hypothesis "no treatment effect" ($H_0 : \beta = 0$) one alternative is to condition on the success totals $y_{i1} + y_{i2} = y_{i\cdot}$, obtaining a hypergeometric distribution that does not depend on the nuisance parameters.

The Hypergeometric Distribution

Suppose that independently two random samples of size n_1, n_2 are drawn and it is observed whether attribute A occurs or not. One obtains a 2 × 2 table with fixed marginals n_1, n_2. The following table gives numbers of subjects who possess the attribute in question.

	A	$\bar{A}$	
Population 1	$y = y_{11}$	y_{12}	n_1
Population 2	y_{21}	y_{22}	n_2
	n_A	$n_{\bar{A}}$	n

From the marginals, n_1, n_2 and therefore $n = n_1 + n_2$ are fixed, while n_A and $n_{\bar{A}}$ are random. If also the marginals $n_A, n_{\bar{A}}$ are fixed, the observations in the table no longer follow a binomial distribution. The selection is the same as when drawing n_1 marbles from a box which contains n marbles, n_A of them possessing attribute A. Thus the distribution of $y = y_{11}$ is given by the *hypergeometric distribution*

$$P(y|\boldsymbol{n}, \boldsymbol{n}_A) = \frac{\binom{n_A}{y}\binom{n_{\bar{A}}}{n_1 - y}}{\binom{n}{n_1}} = \frac{\binom{n_1}{y}\binom{n_2}{n_A - y}}{\binom{n}{n_A}}, \tag{5.7}$$

where the range of possible values for y is given by the integers, satisfying $l = \max\{0, n_1 - (n - n_A)\} \leq y \leq \min\{n_1, n_A\} = u$. The notation $P(y|\boldsymbol{n}, \boldsymbol{n}_A)$ shows that the distribution of y is conditionally on the marginal counts $\boldsymbol{n}_A^T = (n_A, n_{\bar{A}})$ and $\boldsymbol{n}^T = (n_1, n_2)$. The second form in (5.7) results from the symmetry of the problem. Distribution (5.7) is abbreviated by $H(\boldsymbol{n}, \boldsymbol{n}_A)$. The hypergeometric distribution may also be derived directly from the independent binomial distributions $y_{11} \sim B(n_1, \pi)$, $y_{21} \sim B(n_2, \pi)$. Then the conditional distribution of $y = y_{11}$ conditionally on $y_{11} + y_{21} = n_A$ is given by (5.7).

The Non-Central Hypergeometric Distribution

In the more general case the response probability will be different for the two populations. By assuming independent binomial distributions

$$y_{11} \sim B(n_1, \pi_1),\ y_{21} \sim B(n_2, \pi_2),$$

one obtains for the conditional distribution of $y = y_{11}$, conditionally on $y_{11} + y_{21} = n_A$, the non-central hypergeometric distribution $H(\boldsymbol{n}, \boldsymbol{n}_A, \gamma)$, given by

$$P(y|\boldsymbol{n}, \boldsymbol{n}_A, \gamma) = \frac{\binom{n_1}{y}\binom{n_2}{n_A - y}\gamma^y}{P_o(\gamma)},$$

where $P_o(\gamma)$ is the polynomial in γ,

$$P_o(\gamma) = \sum_{j=l}^{u} \binom{n_1}{j}\binom{n_2}{n_A - j}\gamma^j,$$

and $\gamma = \{\pi_1/(1-\pi_1)\}/\{\pi_2/(1-\pi_2)\}$ is the odds ratio. The non-central hypergeometric distribution follows a simple exponential family with the (conditional) log likelihood given by

$$y\log(\gamma) - \log(P_o(\gamma)) + \log\left\{\binom{n_1}{y}\binom{n_2}{n_A - y}\right\}$$

This has the form $y\theta - b(\theta) + c(y)$ with $\theta = \log(\gamma)$ and $b(\theta) = \log(P_o(e^\theta))$, yielding

$$E(y) = b'(\theta) = P_1(e^\theta)/P_o(e^\theta),$$
$$var(y) = b''(\theta) = P_2(e^\theta)/P_o(e^\theta) - \{P_1(e^\theta)/P_o(e^\theta)\}^2,$$

where $P_1(e^\theta) = \partial P_o(\theta)/\partial\theta$, $P_2(e^\theta) = \partial^2 P_o(\theta)/\partial\theta^2$ given by

$$P_r(\gamma) = \sum_{j=l}^{n} \binom{n_1}{j}\binom{n_2}{n_A - j} j^r \gamma^j$$

(see McCullagh and Nelder, 1989).

When testing the nullhypothesis $H_0 : \beta = 0$ in model (5.6) conditioning on the success totals $y_{i1} + y_{i2} = y_{i\cdot}$ for each stratum yields the hypergeometric distribution

$$y_{i1}|\boldsymbol{y}_{i\cdot} \sim H(\boldsymbol{n}_{i\cdot}, \boldsymbol{y}_{i\cdot}, e^\beta),$$

where $\boldsymbol{n}_{i\cdot}^T = (n_{i1}, n_{i2}), \boldsymbol{y}_{i\cdot}^T = (y_{i1} + y_{i2}, n_i - y_{i1} - y_{i2})$ are the marginals under H_0. Then the conditional likelihood involves only one parameter, e^β. One obtains for the mean and variance of the hypergeometric distribution

$$\mathrm{E}(y_{i1}) = n_{i1}\frac{y_{i\cdot}}{n_i}, \quad \mathrm{var}(y_{i1}) = n_{i1}\frac{y_{i\cdot}}{n_i}\left(1 - \frac{y_{i\cdot}}{n_i}\right)\frac{n_i - n_{i1}}{n_i - 1}. \tag{5.8}$$

A statistic for the nullhypothesis may be based on the difference between observed counts and the expected values of the nullhypothesis. A test statistic that has been proposed by Mantel and Haenszel (1959), and in a similar form by Cochran (1954), is

$$T = \frac{\{\sum_{i=1}^{g}(y_{i1} - \mathrm{E}(y_{i1}))\}^2}{\sum_{i=1}^{g}\mathrm{var}(y_{i1})} \tag{5.9}$$

In the Mantel-Haenszel statistic a continuity correction is included, yielding

$$\chi^2_{MH} = \frac{\{|\sum_{i=1}^{g}(y_{i1} - \mathrm{E}(y_{i1}))| - 0.5\}^2}{\sum_{i=1}^{g}\mathrm{var}(y_{i1})}$$

with $\mathrm{E}(y_{i1})$ and $\mathrm{var}(y_{i1})$ derived from the hypergeometric distribution under H_0(5.8). By using the observed proportions $p_{i1} = y_{i1}/n_{i1}, p_{i2} = y_{i2}/n_{i2}$ it may be rewritten as

$$\chi^2_{MH} = \frac{\left\{|\sum_{i=1}^{g}\frac{n_{i1}n_{i2}}{n_i}(p_{i1} - p_{i2})| - 0.5\right\}^2}{\sum\frac{n_{i1}n_{i2}}{n_i - 1}\bar{p}_i(1 - \bar{p}_i)}$$

where $\bar{p}_i = (n_{i1}p_{i1} + n_{i2}p_{i2})/n_i$. Cochran (1954) proposed the statistic (5.9) with the variances replaced by variances derived from the two binomials for treatment and control. Under the nullhypothesis the tests have a large-sample χ^2-distribution. For more details on these tests see Agresti (2002). A general treatment of conditional likelihood approaches is given in McCullagh and Nelder (1989). An alternative approach to reduce the number of parameters is to assume that the stratum-specific parameters are random effects (see Chapter 14).

5.5 Further Reading

Single-index model. Several approaches to the estimation of single-index models have been proposed. One popular technique is based on an average derivative estimation, which exploits the fact that the average gradient of $h(\mathbf{x}_i'\boldsymbol{\beta})$ is proportional to $\boldsymbol{\beta}$ (see Powell et al., 1989; Hristache et al., 2001). An M-estimation that considers the unknown link function as an infinite-dimensional nuisance parameter was considered by Klein and Spady (1993) and Härdle et al. (1993). Weisberg and Welsh (1994) proposed an algorithm that alternates between the estimation of $\boldsymbol{\beta}$ and $h(.)$. Yu and Ruppert (2002) suggested using penalized regression splines. They reported more stable estimates compared to earlier approaches based on local regression (e.g., Carroll et al., 1997). Xia et al. (2002) proposed a symbiosis of sliced inverse regression average derivative estimation and local linear smoothing. Naik and Tsai (2001) proposed a model selection criterion for single-index models that selects variables and also smoothing parameters for the unknown link function.

R packages. Binary response models including quasi-likelihood models can be fitted with the function *glm*. Various link function can be specified in the family function. Generalized estimation functions are available in the library *gee*, function *gee*. The beta-binomial model can be fitted by the function *vglm*in the library *VGAM*. For the fitting procedures of mixed and finite mixture models see Chapter 14.

5.6 Exercises

5.1 The dataset *dust* is available from package *catdata* (or at http://www.stat.uni-muenchen.de/sfb386/ under the name "Chronic bronchitis and dust concentration").

(a) Fit models with different link functions for non-smokers and smokers separately.

(b) Compare the fitted models and discuss model selection.

5.2 The dataset *birth* is available from package *catdata.* Consider the binary response "Did a perineal tear occur" and explanatory variables weight, height, head circumference of child, and month of birth.

(a) Fit models with different link functions.

(b) Select an appropriate model.

5.3 The package *flexmix* contains the data set *betablocker*, which is from a 22-center clinical trial of beta-blockers for reducing mortality after myocardial infarction (see also Aitkin, 1999). In addition to centers, there is only one explanatory variable, treatment, coded as 0 for control and 1 for beta-blocker treatment.

(a) For the 44 binomial observations, fit the simple logit model, an appropriate quasi-likelihood model, and by using generalized estimation equations. Compare the effects and standard errors.

(b) Test if the treatment by beta-blockers has an effect by using test statistics based on conditional likelihood.

Chapter 6

Regularization and Variable Selection for Parametric Models

In several chapters we discussed parametric regression modeling for a moderate number of explanatory variables based on maximum likelihood methods. In some areas of application, however, the number of explanatory variables may be very high. For example, in genetics, where binary regression is a frequently used tool, the number of predictors may be even larger than the number of predictors. In this "$p > n$ problem" maximum likelihood and similar estimators are bound to fail. Typical data of this type are microarray data, where the expressions of thousands of predictors (genes) are observed and only some hundred samples are available. For example, the dataset considered by Golub et al. (1999a), which constitutes a milestone in the classification of cancer, consists of gene expression intensities for 7129 genes of 38 leukemia patients, from which 27 were diagnosed with acute lymphoblastic leukemia and the remaining patients acute myeloid leukemia.

In high-dimensional problems the reduction of the predictor space is the most important issue. A reduction technique with a long history is *stepwise variable selection*. However, stepwise variable selection as a discrete process is extremely variable. The results of a variable selection procedure may be determined by small changes in the data. The effect is often poor performance (see, e.g., Frank and Friedman, 1993). Moreover, it is challenging to investigate the sampling properties of stepwise variable selection procedures.

An alternative to stepwise subset selection is *regularization methods*. Ridge regression is a familiar regularization method that adds a simple penalty term to the log-likelihood and thereby shrinks estimates toward zero. In recent years several alternative regularization techniques based on penalties have been proposed, including methods that perform "smooth" variable selection. These methods select variables simultaneously via optimizing a penalized likelihood, and hence allow one to estimate standard errors. In the following we consider several penalty methods as well as boosting techniques, which are ensemble methods but also serve as regularization methods in structured regression.

Important aspects for regression modeling by regularization techniques are

- *existence of unique estimates* – this is where maximum likelihood estimates often fail;
- *prediction accuracy* – a model should be able to yield a decent prediction of the outcome;
- *sparseness and interpretation* – thea parsimonious model that contains the strongest effects is easier to interpret than a big model with hardly any structure.

We start with the conventional stepwise selection procedures and then consider regularized estimates. Most of the methods focus on variable selection, but regularization can also be helpful when one wants to know which categories of a categorical predictor should be distinguished.

6.1 Classical Subset Selection

In subset selection the predictor space is reduced by retaining only a subset of the variables. The main strategies are best subset selection, forward selection, backward selection, and a combination of the latter two methods.

Best subset selection aims at finding the best subset of predictors among all subsets of the variables $x_1, \ldots, x_p$. "Best" may be defined by minimizing some criterion like AIC or BIC. For binary or Poisson-distribution models, where estimates have to be computed iteratively, the full subset selection is extremely demanding if the number of variables is large.

Forward selection seeks a path through all possible subsets by sequentially adding one predictor into the model. The decision for a predictor may be based on test statistics. Let M denote the current model and M_r the model M with the additional variable x_r. Then, a test on the significance of variable x_r within model M_r is given by the difference of the deviances:

$$D(M|M_r) = D(M) - D(M_r).$$

One selects that variable x_{r_0} for which the corresponding p-value p_r is minimized, $r_0 = \arg\min_r p_r$, provided that the p-value is below some prechosen inclusion level α_{in}. The procedure aims to select variables rather than single terms or parameters. This means that, in a mixture of variables, some of them metric and some of them categorical, the differences of deviances $D(M|M_r)$ compare models of different sizes. While a metric covariate typically contributes only one term (or parameter) to the model, a categorical predictor, unless it is binary, will contribute more than one term (parameter) to the predictor. In cases where each variable corresponds to one term in the linear predictor, for example, in a main effect model with only metric predictors, minimization of the p-values is equivalent to maximizing $D(M|M_r)$. Then one implicitly chooses the variable that most improves the fit, since the deviance $D(M_{r_0})$ as a measure for the discrepancy between data and model is minimized. The procedure stops when no additional variable contributes significantly to the model M. One should be aware that the significance level α_{in} is a threshold rather than a significance level. Since many tests are performed, one has a multiple test problem and control of the multiple significance level is difficult.

Alternative test statistics that might be computationally easier to handle are the score test and the Wald test. Fahrmeir and Frost (1992) suggested the score test and computed the test statistic by efficient sweeps of the inverse information matrix. Since deviances use maximum likelihood for both models M and M_r, one might also run into problems with the existence of ML estimates. In contrast, the score test uses only the estimates of the restricted model; computation of the parameter estimates for the larger model is not required.

Backward selection starts with the full model and sequentially deletes predictors. The choice of the variable to delete is again typically based on test statistics. Let M denote the current model and $M_{\setminus r}$ the model M without predictor r. Then the deviance

$$D(M_{\setminus r}|M) = D(M_{\setminus r}) - D(M)$$

tests the significance of variable x_r within model M. One selects the variable r_0 for which the corresponding p-value p_r is maximal provided it is above some pre-chosen exclusive level α_{out}. The procedure stops when each predictor in the model has a p-value below the level α_{out}.

Computationally more efficient procedures may be obtained by using the Wald test (Fahrmeir and Frost, 1992). Backward selection strategies are restricted to cases where an estimate for the full model exists. This is often a problem when the number of predictors is large.

Forward and backward strategies may also be combined. After a new variable has been taken into the model (forward step) one investigates if one of the other variables in the model may now be deleted by performing a backward step. Both steps have to be controlled by inclusion and exclusion thresholds α_{in} and α_{out}, which, however, provide only local control of the model search.

Best subset selection and forward/backward strategies have several disadvantages. As already noted, subset selection is a discrete process, either a variable is in or out of the model, and therefore extremely variable. The instability of stepwise regression models was demonstrated for example by Breiman (1996b). Moreover, one should be very cautious with the interpretation of the found effects. Standard errors computed for the final model are not trustworthy because they simply ignore the model search. Taking the model search into account would yield much larger standard errors.

Subset selection has been studied extensively for normal distribution models; see, for example, Seeber (1977), Miller (1989), and Furnival and Wilson (1974). The latter gave an efficient algorithm that performs best subset selection up to 30 or 40 predictors. Lawless and Singhal (1978, 1987) developed efficient screening and all-subsets procedures for generalized linear models by use of likelihood ratio statistics. These methods were discussed within a more general framework by Fahrmeir and Frost (1992).

6.2 Regularization by Penalization

Regularization methods that are derived from maximum likelihood estimates are based on the *penalized log-likelihood*:

$$l_p(\boldsymbol{\beta}) = \sum_{i=1}^{n} l_i(\boldsymbol{\beta}) - \frac{\lambda}{2} J(\boldsymbol{\beta}),$$

where $l_i(\boldsymbol{\beta})$ is the usual log-likelihood contribution of the ith observation, λ is a tuning parameter, and $J(\boldsymbol{\beta})$ is a functional that penalizes the size of the parameters. By maximizing the penalized log-likelihood $l_p(\boldsymbol{\beta})$ one seeks estimates that are close to usual ML estimate but with regularized parameters. For example, the ridge penalty, which is one of the oldest penalization methods, uses the penalty $J(\boldsymbol{\beta}) = \sum_{j=1}^{p} \beta_j^2$. It penalizes the length of the parameter $\boldsymbol{\beta}$ and yields estimates that are shrunk toward zero.

There is a good reason for penalizing the length of the parameter. Segerstedt (1992) showed that under regularity assumptions the mean of the squared length of the ML estimate, $\mathrm{E}(\|\hat{\boldsymbol{\beta}}\|^2)$, is asymptotically $\|\boldsymbol{\beta}\|^2 + \mathrm{tr}(\boldsymbol{F}^{-1}(\boldsymbol{\beta}))$, where $\boldsymbol{F}^{-1}(\boldsymbol{\beta})$ denotes the Fisher matrix at the true value $\boldsymbol{\beta}$, which is an approximation to the covariance of $\hat{\boldsymbol{\beta}}$. Therefore, most common regularization techniques impose a penalty on the size of the regression coefficients, yielding shrunk estimates, which in particular have reduced variance. Shrinkage methods are in particular useful for obtaining estimates in applications where the use of the ML estimator involves problems. In a simple Gaussian linear regression the ML estimate is obtained by solving the estimation equation $(\boldsymbol{X}^T\boldsymbol{X})\boldsymbol{\beta} = \boldsymbol{X}^T\boldsymbol{y}$, which is easily solved if $\boldsymbol{X}^T\boldsymbol{X}$ is of full rank and the inverse $(\boldsymbol{X}^T\boldsymbol{X})^{-1}$ exists. In the case of collinearity, $\boldsymbol{X}^T\boldsymbol{X}$ is not of full rank, the ML estimate is not unique, and one has to determine which parts of $\boldsymbol{\beta}$ still may be estimated. Moreover, it has been shown that collinearity leads to poor performance of the estimators. Figure 6.1 illustrates the instability of ML estimates for strongly correlated data. It shows the estimates for two datasets that were drawn from the same underlying linear structure (same non-zero coefficients, $\beta_j = 5$, for the first six variables, zero coefficients for the other variables). It is seen that estimates take quite extreme values, since for strongly correlated predictors high positive values of estimated coefficients balance negative values. The effect is a high variability of estimates. For each drawing one obtains quite different estimates that are far from the true values. Shrinkage estimators like ridge regression estimators, however, yield much smaller values (open circles in Figure 6.1), which are distinctly closer to the true values. Shrinkage methods become important especially when many predictors are available and therefore the corresponding design matrix $\boldsymbol{X}$ is very large and usually contains redundant columns. Similar effects occur in a binary regression, where an additional problem occurs, since ML estimates do not exist if the data may be separated (see Section 4.1). When many predictors are available, the tendency that data structures occur in which the responses $y = 0$ and $y = 1$ are separated increases strongly.

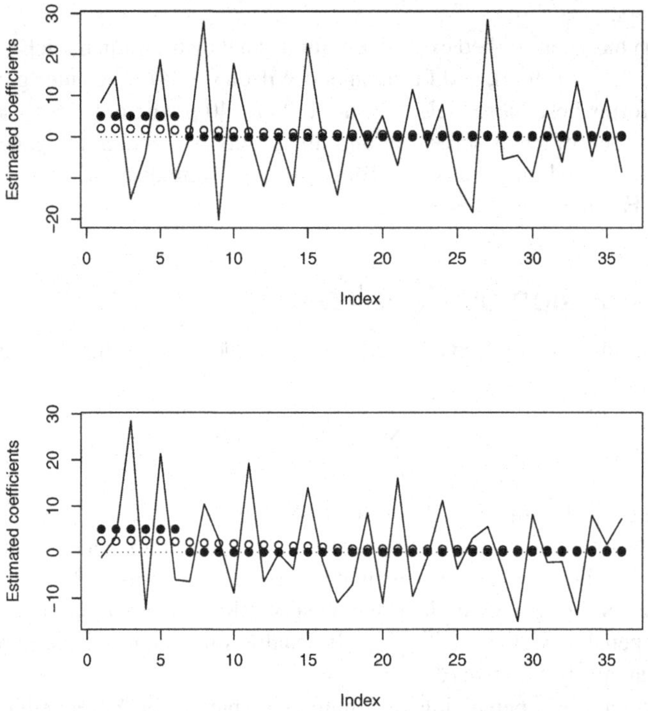

FIGURE 6.1: Maximum likelihood estimates for two datasets with correlated data; the first six variables are influential, and all other variables have zero coefficient.

Some shrinkage methods may also be seen as a continuous alternative to the selection of predictors. Since variables are retained or discarded, variable selection is a strictly discrete process that often exhibits high variance in prediction error. Shrinkage methods reduce the influence of variables in a much smoother way and therefore show less variability. Although shrinkage methods may be seen as providing alternative (more stable) methods to estimate the "true" regression coefficients, a more pragmatic view is that shrinkage methods yield regression models, true or not, that in particular in high-dimensional problems show better prediction error than usual maximum likelihood estimates. Therefore, the selection of the tuning parameter is often based on an estimate of the prediction error.

In the following let the predictor be given by $\eta_i = \boldsymbol{x}_i^T \boldsymbol{\beta}$, where $\boldsymbol{\beta}^T = (\beta_0, \beta_1, \ldots, \beta_p)$ and $\boldsymbol{x}_i$ contains a constant term. Frequently used penalties are of the *bridge penalty* type (Frank and Friedman, 1993):

$$J(\boldsymbol{\beta}) = \sum_{j=1}^{p} |\beta_j|^{\gamma}, \quad \gamma > 0. \tag{6.1}$$

For $\gamma = 2$ one obtains a *ridge regression* (Hoerl and Kennard, 1970) for $\gamma = 1$, the so-called *lasso* (Tibshirani, 1996). Alternatively, estimates are found by maximizing the log-likelihood

$$l(\boldsymbol{\beta}) = \sum_{i=1}^{n} l_i(\boldsymbol{\beta})$$

subject to the constraint

$$\sum_{j=1}^{p} |\beta_j|^{\gamma} \leq t. \tag{6.2}$$

For $\gamma \geq 1$, this approach is equivalent to maximizing the penalized likelihood with $\lambda \geq 0$ since the constraint area is convex (Fu, 1998). Maximization of l_p is usually referred to as a *penalized regression* whereas maximization of l subject to (6.2) is called a *constrained regression.* In the following we will consider penalties of the bridge penalty type and others.

6.2.1 Ridge Regression

Ridge regression as introduced by Hoerl and Kennard (1970) for linear models and extended to GLM type models by Nyquist (1991) is based on the penalty $J(\boldsymbol{\beta}) = \sum_{j=1}^{p} \beta_j^2$ yielding the penalized log-likelihood

$$l_p(\boldsymbol{\beta}) = \sum_{i=1}^{p} l_i(\boldsymbol{\beta}) - \frac{\lambda}{2} \sum_{j=1}^{p} \beta_j^2.$$

For deriving estimates it is useful to rewrite the penalty in the form

$$J(\boldsymbol{\beta}) = \sum_{j=1}^{p} \beta_j^2 = \boldsymbol{\beta}^T \boldsymbol{P} \boldsymbol{\beta},$$

where $\boldsymbol{P} = (p_{ij})$ differs from the $(p+1) \times (p+1)$ identity matrix only by having $p_{11} = 0$ instead of $p_{11} = 1$. The corresponding penalized score function $\boldsymbol{s}_p(\boldsymbol{\beta})$ is given by

$$\boldsymbol{s}_p(\boldsymbol{\beta}) = \sum_{i=1}^{n} \boldsymbol{x}_i \frac{\partial h(\eta_i)}{\partial \eta} (y_i - \mu_i)/\sigma_i^2 - \lambda \boldsymbol{P} \boldsymbol{\beta},$$

yielding the estimation equation

$$\boldsymbol{X}^T \boldsymbol{D}(\boldsymbol{\beta}) \boldsymbol{\Sigma}^{-1}(\boldsymbol{\beta})(\boldsymbol{y} - \boldsymbol{\mu}) - \lambda \boldsymbol{P} \boldsymbol{\beta} = \boldsymbol{0},$$

where $\boldsymbol{y}^T = (y_1, \ldots, y_n)$, $\boldsymbol{\mu}^T = (\mu_1, \ldots, \mu_n)$, $\boldsymbol{X}^T = (\boldsymbol{x}_1 \ldots \boldsymbol{x}_n)$, $\boldsymbol{D}(\boldsymbol{\beta}) = \text{diag}(\partial h(\eta_1)/\partial \eta,$ $\ldots \partial h(\eta_r)/\partial \eta)$, $\boldsymbol{\Sigma}(\boldsymbol{\beta}) = \text{diag}(\sigma_1^2, \ldots, \sigma_n^2)$, $\sigma_i^2 = \text{var}(y_i)$. For the normal distribution model one obtains with $\sigma_i^2 = \sigma^2$, $\boldsymbol{D} = \boldsymbol{I}$, $\tilde{\lambda} = \lambda \sigma^2$ the explicit solution

$$\hat{\boldsymbol{\beta}} = (\boldsymbol{X}^T \boldsymbol{X} + \tilde{\lambda} \boldsymbol{P})^{-1} \boldsymbol{X}^T \boldsymbol{y},$$

which is the maximum likelihood estimate except for the term $\tilde{\lambda} \boldsymbol{P}$. For the covariance one obtains $\text{cov}(\hat{\boldsymbol{\beta}}) = \sigma^2 (\boldsymbol{X}^T \boldsymbol{X} + \lambda \sigma^2 \boldsymbol{P})^{-1} \boldsymbol{X}^T \boldsymbol{X} (\boldsymbol{X}^T \boldsymbol{X} + \lambda \sigma^2 \boldsymbol{P})^{-1}$.

For generalized linear models iterative procedures, for example, Fisher scoring, have to be used. Fisher scoring for solving $\boldsymbol{s}_p(\hat{\boldsymbol{\beta}}) = \mathbf{0}$ has the form

$$\hat{\boldsymbol{\beta}}^{(k+1)} = \hat{\boldsymbol{\beta}}^{(k)} + \boldsymbol{F}_p(\hat{\boldsymbol{\beta}}^{(k)})^{-1}\boldsymbol{s}_p(\hat{\boldsymbol{\beta}}^{(k)}),$$

where $\boldsymbol{F}_p(\boldsymbol{\beta}) = \mathrm{E}(-\partial l_p/\partial\boldsymbol{\beta}\partial\boldsymbol{\beta}^T) = \boldsymbol{F}(\boldsymbol{\beta}) + \lambda\boldsymbol{P}$ with $\boldsymbol{F}(\boldsymbol{\beta})$ being the usual Fisher matrix and $\boldsymbol{W}(\boldsymbol{\beta}) = \mathrm{diag}((\partial h(\hat{\eta}_1)/\partial\eta)^2/\sigma_1^2, \ldots))$. By adding λ to the diagonal of $\boldsymbol{F}(\boldsymbol{\beta})$, the matrix $\boldsymbol{F}_p(\boldsymbol{\beta})$ becomes invertible even if $\boldsymbol{F}(\boldsymbol{\beta})$ is not.

The penalty term $(\lambda/2)\sum_j \beta_j^2$ contains only one tuning parameter λ, which determines the amount of shrinkage for all β_j. Since the parameter β_j depends on the scaling of the corresponding covariate x_j, solutions of $\boldsymbol{s}_p(\boldsymbol{\beta}) = \mathbf{0}$ are not equivariant under scaling of the covariates. Therefore, usually covariates are standardized before solving the estimation equation.

Ridge estimates have nice properties. Under weak conditions that hold for generalized linear models, estimates exist and are unique for $\lambda > 0$ (Fu, 1998). For small λ, the resulting estimate will be mildly biased and the covariance may be approximated by

$$\widehat{\mathrm{cov}}(\hat{\boldsymbol{\beta}}) = (\boldsymbol{X}^T\boldsymbol{W}(\hat{\boldsymbol{\beta}})\boldsymbol{X} + \lambda\boldsymbol{P})^{-1}(\boldsymbol{X}^T\boldsymbol{W}(\hat{\boldsymbol{\beta}})\boldsymbol{X})(\boldsymbol{X}^T\boldsymbol{W}(\hat{\boldsymbol{\beta}})\boldsymbol{X} + \lambda\boldsymbol{P})^{-1}.$$

Early attempts to generalized ridge regression were restricted to logistic regression; see Anderson and Blair (1982), Schaefer et al. (1984), and Duffy and Santner (1989). Ridge regression in generalized linear models has been investigated by Nyquist (1991), Segerstedt (1992), and LeCessie (1992).

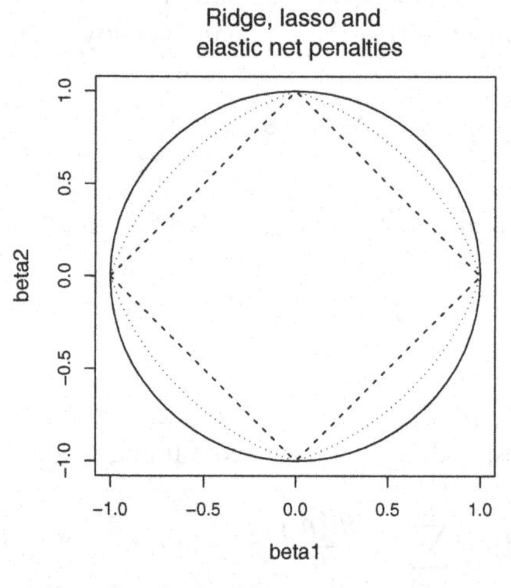

FIGURE 6.2: Constraint regions for the ridge penalty (circle), the lasso (diamond), and the elastic net (in between) for a two-dimensional predictor space.

Example 6.1: Heart Disease

The selection of variables by regularization will be illustrated by the use of the heart disease data that are available from the R package *glmpath* (see Park and Hastie, 2007). The data contain 462 observations on 9 variables and the binary response coronary heart disease. The explanatory variables are sbp (systolic blood pressure), tobacco (cumulative tobacco), ldl (low density lipoprotein cholesterol), adiposity, famhist (family history of heart disease), typea (type A behavior), obesity, alcohol (current alcohol consumption), and age (age at onset). Figure 6.3 shows the coefficient buildups for the ridge estimate based on 10-fold cross-validation (standardized explanatory variables, package *lqa*). Parameter estimates are not plotted against λ but against $||\boldsymbol{\beta}||/\max||\boldsymbol{\beta}||$, where $\max||\boldsymbol{\beta}||$ denotes the maximum value that $||\boldsymbol{\beta}||$ can take. Therefore, small values of $||\boldsymbol{\beta}||$ correspond to large values of λ and large values of $||\boldsymbol{\beta}||$ to small values of λ. For $\lambda = 0$ one obtains the ML estimates on the right-hand side. It is seen that, depending on the value of the smoothing parameter, the estimates are shrunk toward zero. □

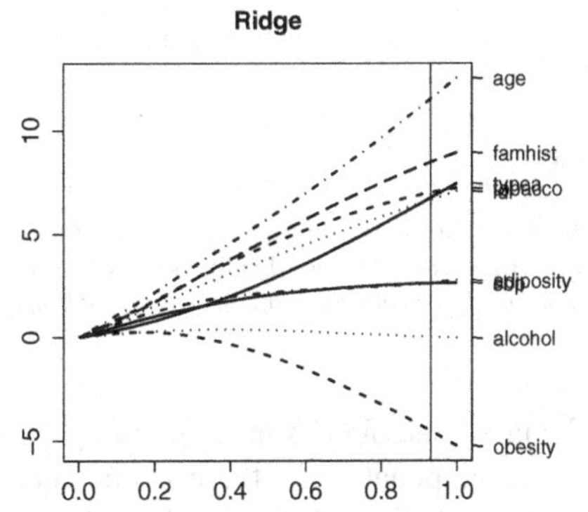

FIGURE 6.3: Coefficient paths for heart disease data when using ridge (package *lqa*; vertical line shows estimate selected by 10-fold cross-validation).

6.2.2 L_1-Penalty: The Lasso

Ridge regression often achieves better prediction performance than maximum likelihood based regression. However, ridge regression does not produce a parsimonious model, since all variables are retained. With a large number of predictors one often wants to determine a smaller subset that contains the strongest variables. Tibshirani (1996) proposed a new technique, called the lasso, for "least absolute shrinkage and selection operator", that shrinks some coefficients and sets others to 0. It tends to avoid the high variability of subset selection while producing a sparse model that shows good prediction performance. The lasso uses the L_1-penalty

$$J(\boldsymbol{\beta}) = \sum_{j=1}^{p} |\beta_j|, \tag{6.3}$$

which is a member of the bridge penalty family. In signal regression the L_1-penalization approach has also been called *basis pursuit* (Chen et al., 2001). It was originally proposed for the linear model in the constrained regression version, which means that the log-likelihood is maximized subject to the constraint

$$\sum_{j=1}^{p} |\beta_j| \le t$$

for some t. The constrained regression version is helpful for illustrating why lasso often produces coefficients that are exactly zero.

Figure 6.2 shows the constraint regions for the ridge penalty, the lasso, and the elastic net (the latter will be introduced in the next section). The constraint region for the ridge penalty is

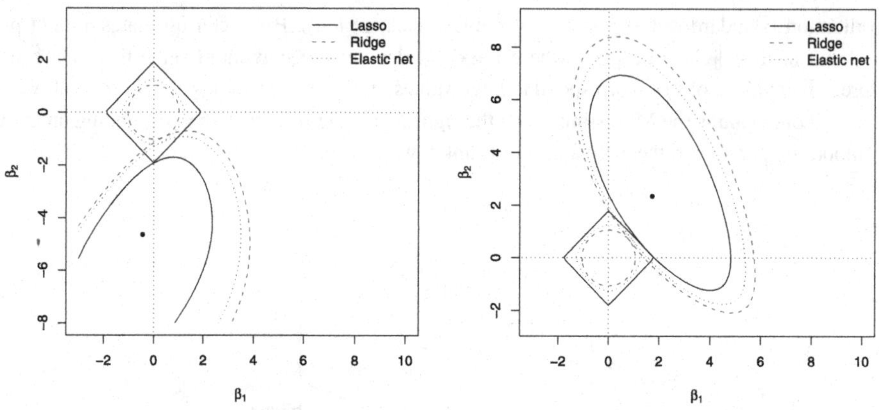

FIGURE 6.4: Constraint regions for the ridge penalty (circle), the lasso (diamond), and the elastic net (in between) together with the log-likelihood functions for negatively correlated predictors (left) and positively correlated predictors (right) for binary regression models.

the disk $\beta_1^2 + \beta_2^2 \leq t$; for the lasso one obtains the diamond $|\beta_1^2| + |\beta_2^2| \leq t$. Figure 6.4 shows the constraint regions for the ridge penalty and the lasso penalty in the two-dimensional case together with the contours of the likelihood function for a binary regression model. The left picture shows the contours of the log-likelihood function for a binary logit model with $n = 30$ and negatively correlated predictors ($\varrho = -0.87$), and the picture on the right shows the log-likelihood for $n = 30$ and positively correlated predictors ($\varrho = 0.87$). The point within the contours is the ML estimate for the specific dataset, and the penalized estimate is represented by the point where the contours touch the penalty region. Maximization of the likelihood subject to the constraint region yields an estimate that is closer to zero than the ML estimate. Of course, the amount of shrinkage depends on the size of the constraint region, which is determined by t or, equivalently, by λ. The advantage of the lasso over ridge regression is that the constraint region is not smooth. Since the diamond has distinct corners, if a solution occurs at a corner, then one parameter is set to zero. The same happens in higher dimensions, but the constraint regions are harder to visualize. For three dimensions (see Figure 6.5), if the penalty region touches at a corner, two parameters are set to zero; if it touches at a connection between two corners, one parameter is set to zero.

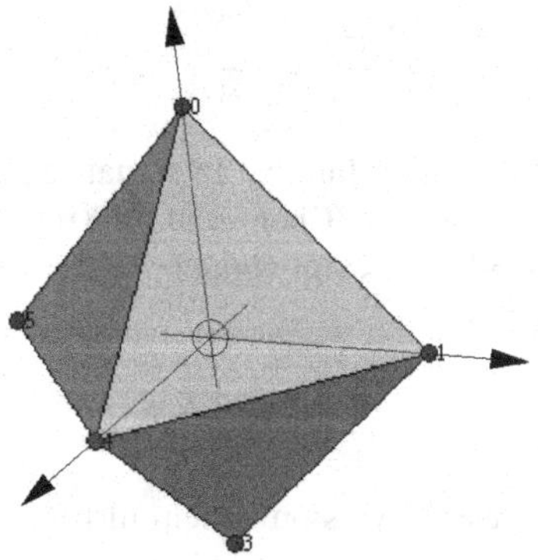

FIGURE 6.5: Constraint regions for the lasso (diamond) for a three-dimensional predictor space.

In contrast to linear regression, in binary regression the contours for a finite sample size are only approximately elliptical. An elliptical approximation is obtained by using a second-order Taylor approximation of the log-likelihood at the maximum likelihood estimate $\hat{\boldsymbol{\beta}}_{ML}$, obtaining

$$l(\boldsymbol{\beta}) \approx l(\hat{\boldsymbol{\beta}}_{ML}) + \frac{1}{2}(\boldsymbol{\beta} - \hat{\boldsymbol{\beta}}_{ML})^T \boldsymbol{F}(\hat{\boldsymbol{\beta}}_{ML})(\boldsymbol{\beta} - \hat{\boldsymbol{\beta}}_{ML}),$$

where $\boldsymbol{F}(\hat{\boldsymbol{\beta}}_{ML})$ is the Fisher matrix. For normal regression models the approximation is exact and one obtains (apart from constants)

$$l(\boldsymbol{\beta}) = l(\hat{\boldsymbol{\beta}}_{ML}) - \frac{1}{2\sigma^2}(\boldsymbol{\beta} - \hat{\boldsymbol{\beta}}_{ML})^T \boldsymbol{X}^T \boldsymbol{X}(\boldsymbol{\beta} - \hat{\boldsymbol{\beta}}_{ML}),$$

and therefore a quadratic function centered at $\hat{\boldsymbol{\beta}}_{ML}$. For a large sample size, the contours of the likelihood also show an almost elliptical form for binary models.

By using the L_1-penalty (6.3) the lasso does both, continuous shrinkage and automatic variable selection, simultaneously. Concerning prediction error, it has been shown that the performance is not better than ridge regression in any case. In a comparison of the lasso, ridge, and bridge regression it has been shown that neither of them uniformly dominates the other two (see Tibshirani, 1996; Fu, 1998). The big advantage of the lasso is its sparse representation, which makes it attractive for practitioners.

The implicit shrinkage of the lasso can be illustrated by the idealized case of orthonormal columns in the design matrix of a linear model. Then the lasso penalty $\tilde{\lambda} J(\boldsymbol{\beta})$ yields the *soft thresholding* rule

$$\hat{\beta}_j = S(\beta_j^{ML}, \tilde{\lambda}) = \text{sign}(\hat{\beta}_j^{ML})(|\hat{\beta}_j^{ML}| - \tilde{\lambda})_+,$$

where β_j^{ML} denotes the ML estimate of the jth component, $\tilde{\lambda} = \lambda\sigma^2$, and $(z)_+ = 1$, if $z \geq 0$, $(z)_+ = 0$, if $z < 0$. Figure 6.6 shows the estimated values as functions of the ML estimate. It is seen that estimates are set to zero if the ML estimate is below some threshold and are shrunk if the ML estimate is above the threshold. The term "soft thresholding" was built to distinguish it from hard thresholding, where estimates are set to zero if the ML estimate is below some threshold and retained if the ML estimate is above the threshold (also given in Figure 6.6). Moreover, estimates for SCAD are included, which will be considered in Section 6.2.5.

For the lasso in linear models, various computational procedures were proposed. Tibshirani (1996) used a combined quadratic programming method, Fu (1998) gave a modified Newton-Raphson algorithm and introduced the shooting algorithm, and Osborne et al. (2000) considered the lasso and its dual. Fan and Li (2001) proposed an alternative algorithm based on quadratic approximations. A fast implementation for large-scale logistic regression with the lasso has been presented by Genkin et al. (2004). Alternative approaches aim at the estimation of the entire path of the coefficient estimates as λ varies, to find estimates $\hat{\boldsymbol{\beta}}(\lambda), 0 < \lambda < \infty$. Efron et al. (2004) proposed *LARS*, which determines the exact piecewise linear coefficient paths for lasso in the linear case. However, in the case of GLMs, the paths are not piecewise linear. Park and Hastie (2007) proposed an efficient path algorithm for generalized linear models that uses the predictor-corrector method. Rosset (2004) suggested a general path-following algorithm that can be used for any loss function. Zhao and Yu (2004) proposed a boosted lasso, which includes backward steps. An extremely fast algorithm based on pathwise coordinate optimization was given by Friedman et al. (2007), Friedman et al. (2010). It uses coordinate descent methods, which have been proposed earlier (for example Fu, 1998) but were not fully appreciated at the time.

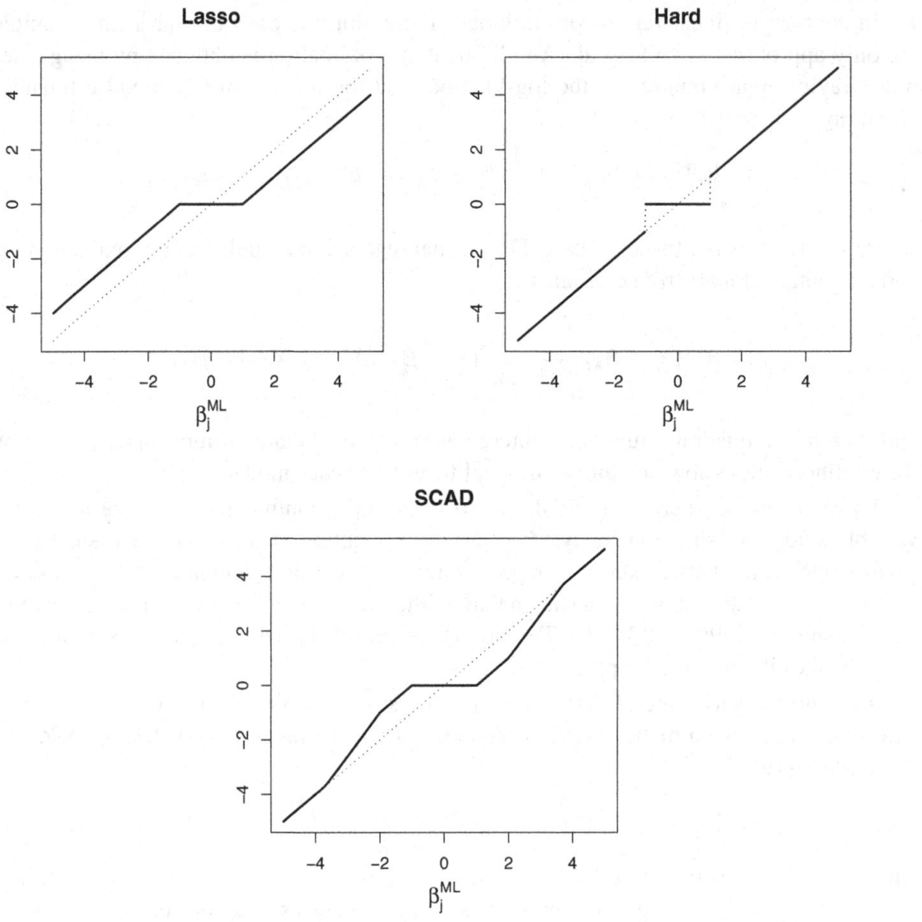

FIGURE 6.6: Estimates for lasso, hard thresholding, and SCAD when columns in the design matrix are orthonormal.

The Adaptive Lasso

Variable selection procedures are often discussed in terms of *oracle* properties, which refer to the identification of the right subset model. For parameter vector $\boldsymbol{\beta}$ let $A = \{j : \beta_j \neq 0\}$ denote the active set, where $|A| = p_0 < p$. For simplicity, let $\boldsymbol{\beta}$ be partitioned into $\boldsymbol{\beta}^T = (\boldsymbol{\beta}_1^T, \boldsymbol{\beta}_2^T)$, where $\boldsymbol{\beta}_1$ represents the active set and $\boldsymbol{\beta}_2 = \mathbf{0}$. Then oracle properties means that (1) estimates $\hat{\boldsymbol{\beta}}^T = (\hat{\boldsymbol{\beta}}_1^T, \hat{\boldsymbol{\beta}}_2^T)$ must asymptotically satisfy $\hat{\boldsymbol{\beta}}_1 \neq \mathbf{0}$ and $\hat{\boldsymbol{\beta}}_2 = \mathbf{0}$, and (2) the optimal estimation rate is obtained, so that the estimator performs as well as if the underlying model were known. A *selection procedure* is *consistent* if asymptotically the right subset model is found, $\lim_n P(A_n = A) = 1$, where A_n is the active set for n observations. Zou (2006) showed that lasso variable selection can be inconsistent and gave necessary conditions for consistency. He proposed an extended version of lasso, for which the penalty has the form

$$J(\boldsymbol{\beta}) = \sum_{j=1}^{p} w_j |\beta_j|, \tag{6.4}$$

where w_j are known weights. By using weights on coefficients the variables are not equally penalized, which adds some flexibility. He showed that for cleverly chosen data-dependent weights the adaptive lasso has oracle properties. One choice of weights is based on a root-n consistent estimator $\tilde{\boldsymbol{\beta}}$ of $\boldsymbol{\beta}$, for example, the ML estimate. Then weights are fixed by $w_j = 1/|\tilde{\beta}_j|^\gamma$, for fixed chosen $\gamma > 0$. The oracle properties that Zou (2006) derived for the adaptive lasso use that for growing sample size the weights for zero-coefficients get inflated, whereas the weights on non-zero-coefficients converge to a finite constant. Moreover, Zou (2006) showed that the adaptive lasso leads to near-minimax-optimal estimators.

Example 6.2: Heart Disease

Figure 6.7 shows the coefficient buildups for the lasso and the adaptive lasso (standardized explanatory variables; package *lqa*) plotted against $||\boldsymbol{\beta}||/\max||\boldsymbol{\beta}||$. The vertical line shows the regularization obtained when using 10-fold cross-validation. It is seen that, in contrast to the ridge, not all variables are found to be influential. Based on 10-fold cross-validation one concludes that the variables alcohol and adiposity can be omitted. Moreover, it is seen that for this dataset the adaptive lasso is very close to the simple lasso. □

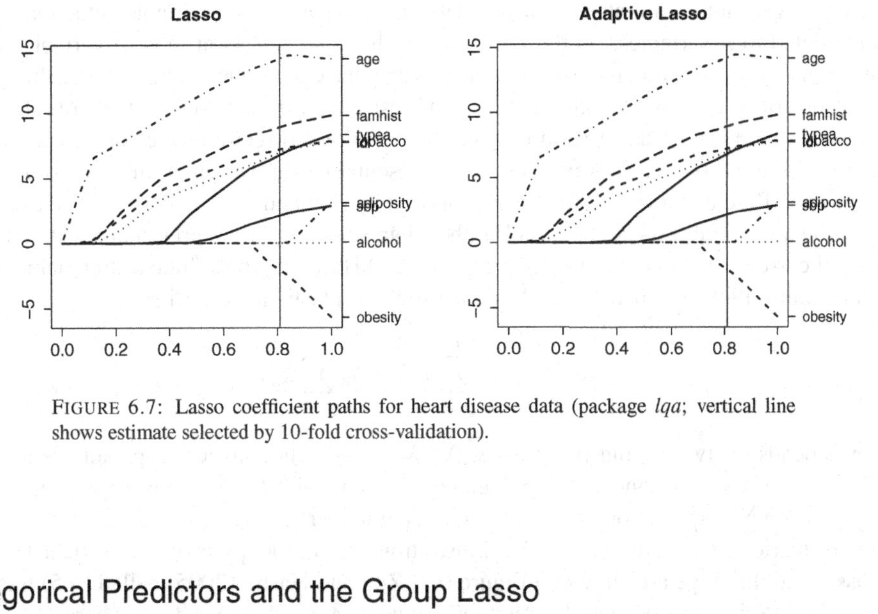

FIGURE 6.7: Lasso coefficient paths for heart disease data (package *lqa*; vertical line shows estimate selected by 10-fold cross-validation).

Categorical Predictors and the Group Lasso

The lasso as considered in the previous section selects individual predictors. That approach is sensible when all variables are of the same type, for example, if all the variables are continuous, or all are binary. For a mixture of predictors, some of them categorical and some of them binary, the penalty is unsatisfactory. If the categorical predictor (factor) is represented by dummy variables, the lasso penalty selects individual dummy variables instead of whole factors, and the solution depends on how the dummy variables are encoded. A sensible procedure should select whole factors or continuous variables. The group lasso proposed by Yuan and Lin (2006) can overcome these problems.

Let the p-dimensional predictor be structured as $\boldsymbol{x}_i^T = (\boldsymbol{x}_{i1}^T, \dots, \boldsymbol{x}_{i,G}^T)$, where $\boldsymbol{x}_{ij}$ corresponds to the jth *group of variables*. A group of variables may refer to the dummy variables of one factor, with df_j denoting the number of the variables in the jth group. A continuous variable that has a linear form within the predictor obviously has $df_j = 1$. A group of variables may also refer to interactions between factors or between factors and continuous variables, where df_j is the number of individual interaction terms. Correspondingly the parameter vector is partitioned into subvectors, $\boldsymbol{\beta}^T = (\boldsymbol{\beta}_1^T, \dots, \boldsymbol{\beta}_G^T)$. The *group lasso* uses the penalty

$$J(\boldsymbol{\beta}) = \sum_{j=1}^{G} \sqrt{df_j} \|\boldsymbol{\beta}_j\|_2,$$

where $\|\boldsymbol{\beta}_j\|_2 = (\beta_{j1}^2 + \dots + \beta_{j,df_j}^2)^{1/2}$ is the L_2-norm of the parameters of the jth group. The penalty encourages sparsity in the sense that either $\hat{\boldsymbol{\beta}}_j = \mathbf{0}$ or $\beta_{js} \neq 0$ for $s = 1, \dots, df_j$. For a geometrical interpretation of the penalty, see Yuan and Lin (2006). Meier et al. (2008) showed that under sparsity the resulting estimates are consistent even when the number of predictors is larger than the sample size. The penalty may be seen as a special case of the composite absolute penalty family proposed by Zhao et al. (2009).

6.2.3 The Elastic Net

Although the lasso has several advantages, it has some severe limitations, pointed out by Zou and Hastie (2005). If there are high correlations between predictors, it has been observed that ridge regression dominates the lasso (Tibshirani, 1996). As a variable selection method it is restricted to n variables. In the $p > n$ case, the lasso selects at most n variables before it saturates. Moreover, the lasso does not necessarily have a unique solution, since the penalty term is not strictly convex. A point of special interest concerns the variables that are selected by the lasso. If there is a group of variables among which the correlations are very high, the lasso tends to select only one of the variables as a representative. In particular, this way of selecting variables is different from the *elastic net*, proposed by Zou and Hastie (2005). The elastic net does automatic variables selection, and, rather than selecting one representative, it can select *groups* of correlated variables. According to Zou and Hastie it works "like a stretchable fishing net that retains all the big fish." The elastic net uses the elastic net criterion

$$J(\boldsymbol{\beta}) = \lambda_1 \sum_{j=1}^{p} |\beta_j| + \lambda_2 \sum_{j=1}^{p} \beta_j^2, \tag{6.5}$$

which depends on two tuning parameters, $\lambda_1, \lambda_2 > 0$. The elastic net penalty is a convex combination of the lasso and the ridge penalty. In constraint form it may be written as $(1 - \alpha)\sum_j |\beta_j| + \alpha \sum_j \beta_j^2 \leq t$ for some t and tuning parameter $\alpha = \lambda_2/(\lambda_1 + \lambda_2)$. With $\alpha \in [0, 1]$ the lasso and ridge are limiting cases. For illustration, the contour plots of the elastic net penalty, the lasso, and the ridge are shown in Figure 6.2. Zou and Hastie (2005) called (6.5) the naïve elastic net criterion and proposed a rescaled solution $\hat{\boldsymbol{\beta}} = (1 + \lambda_2)\hat{\boldsymbol{\beta}}_{net}$, where $\hat{\boldsymbol{\beta}}_{net}$ is the penalized least-squares solution of the naïve elastic net criterion. They give several reasons for choosing $(1 + \lambda_2)$ as scaling factor.

The interesting property of the elastic net is the *grouping effect*. A regression method exhibits the grouping effect if the regression coefficients of a group of highly correlated variables tend to be equal, up to a change of sign if negatively correlated. Zou and Hastie (2005) show that for penalized least-squares problems the coefficient paths of predictors x_i and x_j with sample correlation ϱ_{ij} are confined by

$$|\hat{\beta}_i - \hat{\beta}_j| / \sum_i |y_i| \leq \frac{1}{\lambda_2} \sqrt{2(1 - \varrho_{ij})},$$

where $\hat{\beta}_i$, $\hat{\beta}_j$ are naïve net solutions with parameters λ_1, λ_2. If x_i and x_j are highly correlated, ($\varrho_{ij} \to 1$), the coefficient paths of x_i and x_j are very close. Thus the elastic net shows the grouping the effect, which is important, for example, in genetics, where groups of genes that are relevant are to be selected ("grouped selection").

To illustrate why the grouping effect is useful we use the idealized example given by Zou and Hastie (2005). With Z_1 and Z_2 being two independent $\mathrm{U}(0, 20)$ variables, the response is generated as $\mathrm{N}(Z_1 + 0.1Z_2, 1)$. It is assumed that one observes only noisy versions of Z_1 and Z_2:

$$x_1 = Z_1 + \epsilon_1, \quad x_2 = -Z_1 + \epsilon_2, \quad x_3 = Z_1 + \epsilon_3,$$
$$x_4 = Z_2 + \epsilon_4, \quad x_5 = -Z_2 + \epsilon_5, \quad x_6 = Z_2 + \epsilon_6,$$

where ϵ_i are independent identically distributed $\mathrm{N}(0, 1/16)$. The variables x_1, x_2, and x_3 may be considered as forming one group and x_4, x_5, and x_6 as forming a second group. Figure 6.8 shows the coefficient buildups for the lasso and a method that shows the grouping effect for sample size $n = 100$. The method used is BlockBoost, which is described in the next section. It is seen that BlockBoost selects the variables x_1, x_2, and x_3, and the corresponding estimates are (up to sign) identical. The strong group consistency of x_1, x_2, and x_3 is distinctly identified. Lasso shows quite different coefficient buildups, selecting as strongly influential the variables x_1 and x_3 and, with rather weak effect, x_2. While the coefficient paths for BlockBoost reflect the high correlation of x_1, x_2 and x_3, the paths of the lasso are rather irregular. The elastic net behaves quite similar to BlockBoost (compare Zou and Hastie, 2005).

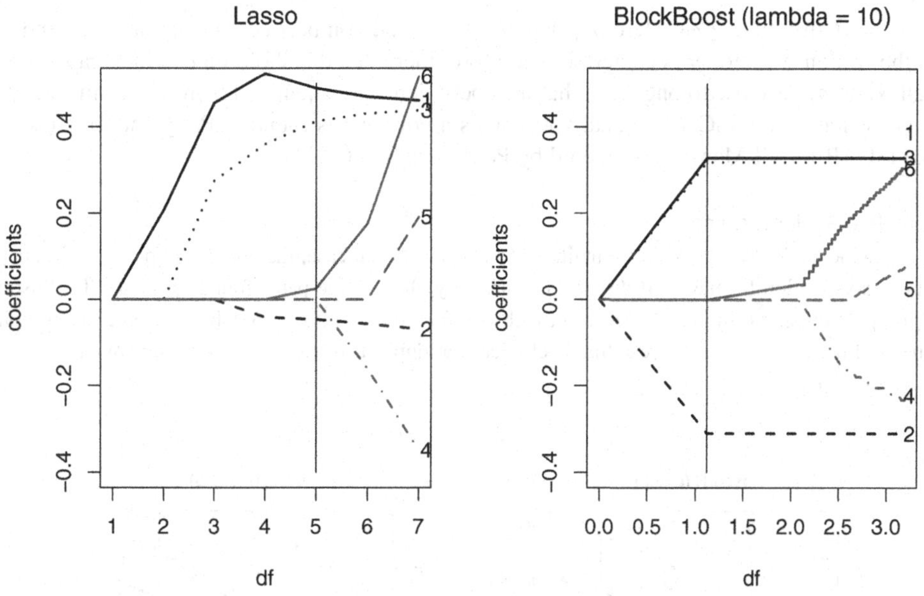

FIGURE 6.8: Coefficient buildups for lasso (left) and BlockBoost (right) of the hidden factors example. Vertical lines indicate the degrees of freedom corresponding to the tuning parameter(s) chosen by 10-fold cross-validation.

6.2.4 Alternative Estimators with Grouping Effect

The grouping effect of the elastic net gives similar coefficients to highly correlated variables. More recently, alternative penalties were proposed that aim at the grouping effect.

OSCAR

Bondell and Reich (2008) proposed a method called OSCAR, for Octagonal Shrinkage and Clustering Algorithm for Regression. The method shrinks, like the lasso, some coefficients to zero, but in addition yields the exact equality of some of the coefficients. The predictors with equal coefficients form clusters that are represented by a single coefficient. OSCAR in constrained form uses the restriction

$$\sum_j |\beta_j| + c \sum_{j<k} \max\{|\beta_j|, |\beta_k|\} \leq t$$

for some tuning parameters $t > 0, c \geq 0$. The parameter c controls the relative weighting of the L_1-norm and the pairwise L_∞-norm. With $c = 0$ the lasso is a special case. For two predictors, the constraint region forms an octagon. The vertices on the diagonals and on the axis encourage equality of coefficients (when a vertice on the diagonal is hit) and sparsity (when a vertice on the axis is hit). Varying c changes the angle formed in the octagon, yielding a diamond if $c = 0$ and a square if $c \to \infty$.

The exact grouping property derived by Bondell and Reich (2008) uses parameters λ and c from the penalized version $J(\boldsymbol{\beta}) = \lambda\{\sum_j |\beta_j| + c\sum_{j<k} \max\{|\beta_j|, |\beta_k|\}$. For signed coefficients, so that $\hat{\beta}_j \geq 0$ for all j, they show that for the linear model there exists a c_0 such that

$$0 \leq c_0 \leq 2\lambda^{-1}|\boldsymbol{y}|\sqrt{2(1-\varrho_{jk})}$$

and $\hat{\beta}_j = \hat{\beta}_k$ for all $c \geq c_0$. Here ϱ_{jk} denotes the correlation between the variables x_j and x_k, and the response vector $\boldsymbol{y}$ is centered. Therefore, there is a threshold on c, which can be very small when ϱ_{jk} is close to one, such that the coefficients are equal and therefore form a cluster. A representation of OSCAR's penalty region as a polytope is found in Petry and Tutz (2012). The OSCAR for GLMs was considered by Petry and Tutz (2011).

Example 6.3: Heart Disease

For elastic net and OSCAR, which contain the lasso as a special case, the coefficient paths based on 10-fold cross-validation for the heart disease data are very close to the paths found for lasso. To illustrate the grouping effect we show the coefficient paths for $c = 0.2$ and $c = 0.5$, which enforce the grouping property. Figure 6.9 shows the resulting coefficient buildups. It is seen that for strong smoothing some effects are set equal. □

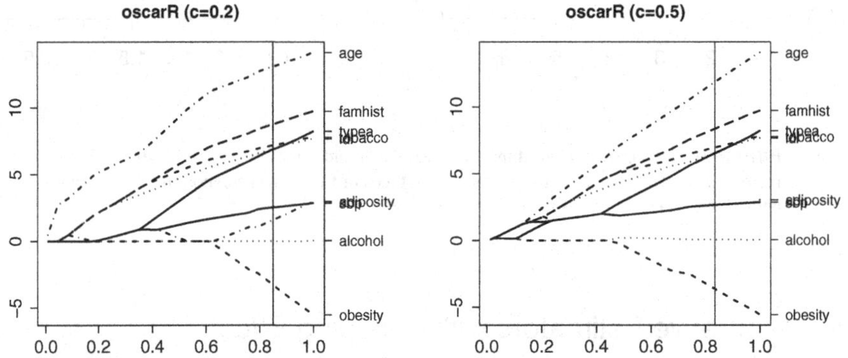

FIGURE 6.9: OSCAR coefficient paths for heart disease data (two values of c).

Correlation-Based Penalties

An alternative approach that aims at the grouping effect uses explicitly the correlation between pairs of predictors. The correlation-based penalty has the form

$$J_c(\boldsymbol{\beta}) = \sum_{i=1}^{p-1}\sum_{j>i}\left\{\frac{(\beta_i - \beta_j)^2}{1-\varrho_{ij}} + \frac{(\beta_i + \beta_j)^2}{1+\varrho_{ij}}\right\} = 2\sum_{i=1}^{p-1}\sum_{j>i}\frac{\beta_i^2 - 2\varrho_{ij}\beta_i\beta_j + \beta_j^2}{1-\varrho_{ij}^2}, \tag{6.6}$$

where ϱ_{ij} denotes the (empirical) correlation between the ith and the jth predictors. It is designed to focus on the grouping effect, that is, its highly correlated effects show comparable values of estimates ($|\hat{\beta}_i| \approx |\hat{\beta}_j|$) with the sign being determined by positive or negative correlation. For strong positive correlation ($\varrho_{ij} \to 1$) the first term becomes dominant, having the effect that estimates for β_i, β_j are similar ($\hat{\beta}_i \approx \hat{\beta}_j$). For strong negative correlation ($\varrho_{ij} \to -1$) the second term becomes dominant and $\hat{\beta}_i$ will be close to $-\hat{\beta}_j$. Consequently, for weakly correlated data the performance is quite close to the ridge penalty.

Figure 6.10 shows the two-dimensional contour plots for selected values of ϱ together with the constraint region for the ridge penalty and the lasso. It is seen that contours for the ridge and lasso are highly symmetric; $x_1 = 0$ is an axis of symmetry as well as $x_2 = 0$. In contrast, the constrained region for the correlation-based estimator is an ellipsoid that becomes narrower with increasing correlation. Spectral decomposition of $J_c(\beta)$ yields eigenvectors $(1, 1)$ and $(1, -1)$ with corresponding eigenvalues $\lambda/(1-\varrho)$ and $\lambda/(1+\varrho)$. Thus, for $\varrho > 0$, the first eigenvalue becomes dominant while for $\varrho < 0$ it is the second eigenvalue that determines the orientation of the ellipsoid. When computing the penalized least-squares criterion, the effect is that, for $\varrho > 0$, estimates are preferred for which the components $\hat{\beta}_1, \hat{\beta}_2$ are similar; for $\varrho < 0$, similarity of $\hat{\beta}_1$ and $-\hat{\beta}_2$ is preferred. This may be seen from the contour plots, since for $\varrho > 0$ the increase in $P_c(\beta)$ is slower when moving in the direction of the first eigenvector $(1, 1)$ than in the orthogonal direction $(1, -1)$. For $\varrho < 0$, the eigenvalue corresponding to $(1, -1)$ is larger, and therefore parameter values where β_1 is close to $-\beta_2$ are preferred. Thus the use of penalty P_c implies shrinkage, with the strength of shrinkage being determined by λ, but shrinkage differs from ridge shrinkage, which occurs for the special case $\varrho_{ij} = 0$.

Assume that $\lambda > 0$ and $\varrho_{ij}^2 \neq 1$ for $i \neq j$. Then $J_c(\boldsymbol{\beta})$ is strictly convex and the estimate exists and is unique. For linear models an explicit solution to the penalized least-squares problem is obtained, called the *correlation-based estimator*:

$$\hat{\boldsymbol{\beta}}_c = (\mathbf{X}^T\mathbf{X} + \lambda \boldsymbol{M})^{-1}\mathbf{X}^T\boldsymbol{y}, \tag{6.7}$$

where $\mathbf{X}^T = (\mathbf{x}_1 \ldots \mathbf{x}_n)$ is the design matrix, $\mathbf{y}$ collects the responses, $\mathbf{y}^T = (y_1, \ldots, y_n)$, and $\boldsymbol{M}$ is a matrix that is determined by the correlations $\varrho_{ij}, i, j = 1, \ldots, p$. The explicit form exploits that the correlation-based penalty (9.7) can be written as a quadratic form:

$$J_c(\boldsymbol{\beta}) = \boldsymbol{\beta}^T \boldsymbol{M} \boldsymbol{\beta}, \tag{6.8}$$

where $\boldsymbol{M} = (m_{ij})$ has entries $m_{ij} = 2\sum_{s\neq i} 1/(1-\varrho_{is}^2)$ if $i = j$, and $m_{ij} = -2\varrho_{ij}/(1-\varrho_{ij}^2)$ if $i \neq j$. For GLM-type models the ML estimate is obtained by penalized Fisher scoring (see Section 6.2.1). Although the correlation-based penalty enforces the grouping effect with good results in simulations, it does not enforce sparsity. Therefore, Tutz and Ulbricht (2009) proposed a specific form of blockwise boosting, called BlockBoost. To obtain the grouping effect of the correlation-based estimator combined with variable selection, a boosting procedure is used that updates in each step the coefficients of more than one variable. The procedure differs from common componentwise boosting, where just one variable is selected and the corresponding coefficient is adjusted. The algorithm is able to handle high-dimensional data and,

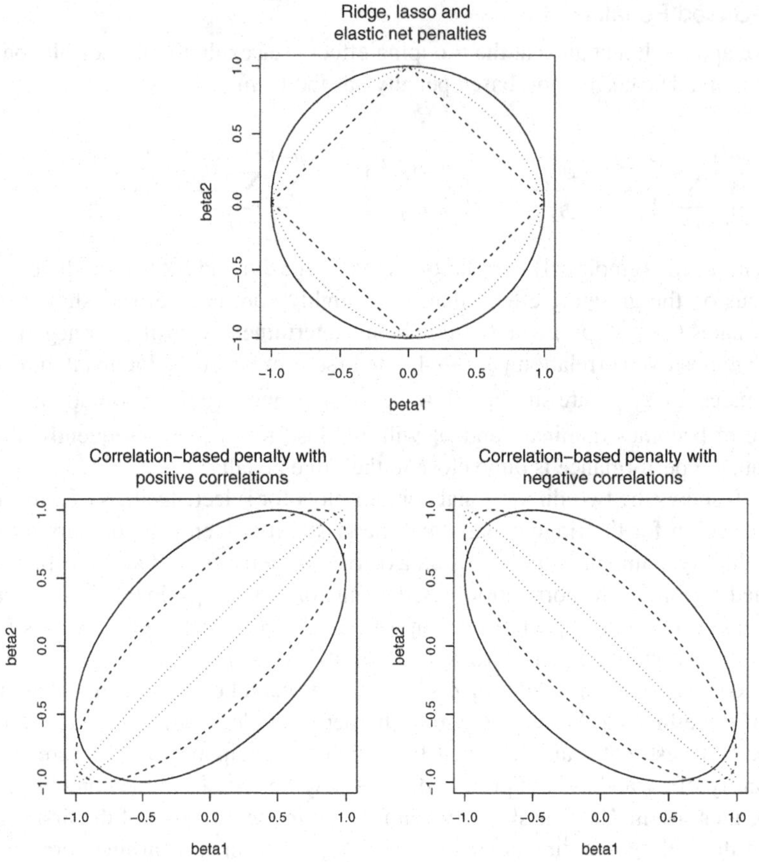

FIGURE 6.10: Top panel: Two-dimensional contour plots for ridge, lasso (dashed line), and elastic net with $\alpha = 0.5$ (dotted line). Lower panel left: Contour plots of correlation-based penalty for positive correlations: $\varrho = 0.5$ (solid line), $\varrho = 0.8$ (dashed line), and $\varrho = 0.99$ (dotted line). Lower panel right: Contour plots of correlation-based penalty for negative correlations: $\varrho = -0.5$ (solid line), $\varrho = -0.8$ (dashed line), and $\varrho = -0.99$ (dotted line).

depending on the tuning parameter, enforces sparsity. The grouping effect is demonstrated in Figure 6.8, where coefficients are plotted against degrees of freedom for the simulation described in Section 6.2.3. It is seen that lasso fails to recognize the grouping structure in contrast to BlockBoost. For more details on the correlation-based approach in linear models, see Tutz and Ulbricht (2009). GLM-type models were considered by Ulbricht and Tutz (2008). An alternative way is to combine correlation-based penalties and the L_1-penalty into the form

$$J_c(\boldsymbol{\beta}) = \lambda_1 \sum_{j=1}^{p} |\beta_j| + \lambda_2 \boldsymbol{\beta}^T \boldsymbol{M} \boldsymbol{\beta}. \tag{6.9}$$

Anbari and Mkhadri (2008) demonstrated that the penalty shows good performance in many applications.

Fusion-Type Estimators

Daye and Jeng (2009) proposed the penalty

$$J(\boldsymbol{\beta}) = \lambda_1 \sum_{j=1}^{p} |\beta_j| + \lambda_2 \sum_{i<j} w_{ij} (\beta_i - \text{sign}(\varrho_{ij})\beta_j)^2,$$

where $\text{sign}(\varrho_{ij})$ denotes the sign of the correlation coefficient ϱ_{ij} taking values 1 or –1. The penalty combines the lasso with a term that enforces the fusion of variables. When λ_2 is large and weights are positive, the second term enforces $\hat{\beta}_i \approx \hat{\beta}_j$ for positive correlation between x_i and x_j and $\hat{\beta}_i \approx -\hat{\beta}_j$ for negative correlation. Because of the tendency to fuse predictors, Daye and Jeng (2009) call the resulting estimator a *weighted fusion estimator*. They use the correlation-driven weights $w_{ij} = |\varrho_{ij}|^{\gamma}/(1 - |\varrho_{ij}|^{\gamma})$, where γ is an additional tuning parameter, and derive conditions that make the estimator sign consistent for the linear model. Sign consistency is somewhat stronger than variable selection consistency and postulates that asymptotically the sign of the estimate is the same as the sign of the true parameter.

The penalty belongs to the general family of combination penalties

$$J(\boldsymbol{\beta}) = \lambda_1 \|\boldsymbol{\beta}\|_1 + \lambda_2 \|\boldsymbol{\beta}\|_{\boldsymbol{R}}^2, \tag{6.10}$$

where $\|\boldsymbol{\beta}\|_1 = (|\beta_1| + \cdots + |\beta_p|)$ is the L_1-norm and $\|\boldsymbol{\beta}\|_{\boldsymbol{R}}^2 = \boldsymbol{\beta}^T \boldsymbol{R} \boldsymbol{\beta}$ is the squared norm built with matrix $\boldsymbol{R}$. For $\boldsymbol{R} = \boldsymbol{I}$ one obtains the elastic net, for $\boldsymbol{R} = \boldsymbol{M}$ one obtains (6.9), and for specific $\boldsymbol{R}$ one obtains the weighted fusion estimator. It should be noted that all combination penalties (6.10) can be reformulated as lasso problems by simple data augmentation, and therefore algorithms that compute lasso solutions can be used (see, for example, Zou and Hastie, 2005).

In their derivation of the penalty Daye and Jeng (2009) refer to the fused lasso, which was proposed by Tibshirani et al. (2005). However, the latter enforces a fusion of predictors by using a lasso-type estimator instead of a ridge-type estimator for the differences. The corresponding *pairwise fused lasso* estimator has the form

$$J(\boldsymbol{\beta}) = \lambda_1 \sum_{j=1}^{p} |\beta_j| + \lambda_2 \sum_{i<j} |\beta_i - \beta_j|,$$

or with weights and correlation-based penalty,

$$J(\boldsymbol{\beta}) = \lambda_1 \sum_{j=1}^{p} |\beta_j| + \lambda_2 \sum_{i<j} w_{ij} |\beta_i - \text{sign}(\varrho_{ij})\beta_j|.$$

For applications see Petry et al. (2011). The fused lasso itself is considered in more detail in Section 10.4.4.

6.2.5 SCAD

An alternative penalty that yields simultaneous estimation and selection has been proposed by Fan and Li (2001). They identified three properties that a penalized estimator should have and derived an appropriate penalty. The resulting estimator should be nearly unbiased for large unknown coefficients (unbiasedness), it should automatically set small estimated coefficients to zero (sparsity), and it should be continuous in the data to avoid instability in model prediction (continuity). The penalty function considered by Fan and Li (2001) has the additive form

$$\lambda J(\boldsymbol{\beta}) = \lambda \sum_{j=1}^{p} p(|\beta_j|) = \sum_{j=1}^{p} p_\lambda(|\beta_j|),$$

where $p(.)$ are penalty functions that, in the most general form, can also depend on the variable. The functions $p_\lambda(|\beta|) = \lambda p(|\beta|)$ are introduced to allow that the penalty may depend on λ. Obviously, ridge and lasso are special cases with functions $p(|\beta_j|) = |\beta_j|^2$ and $p(|\beta_j|) = |\beta_j|$, respectively. The continuously differentiable penalty function proposed by Fan and Li is the *smoothly clipped absolute derivation (SCAD) penalty*, defined by its derivative:

$$p'_\lambda(\beta) = \lambda\{I(\beta \leq \lambda) + \frac{(a\lambda - \beta)_+}{(a-1)\lambda} I(\beta > \lambda)\},$$

for some $a > 2$ and $\beta > 0$, where $(x)_+ = x$ if $x > 0$ and 0 otherwise. The penalty corresponds to a quadratic spline function with knots at λ and $a\lambda$. Figure 6.11 shows the penalty and its derivative. It is seen that for small values the penalty is similar to the lasso penalty whereas for larger values the penalty levels off.

To gain some insight about the effect of penalty functions, it is common to study the linear regression model with orthonormal columns in the design matrix. Then it may be shown that the lasso penalty $p(|\beta|) = \lambda|\beta|$ yields the soft thresholding rule $\hat{\beta}_j = \text{sign}(\hat{\beta}_j^{ML})(|\hat{\beta}_j^{ML}| - \lambda)_+$, where β_j^{ML} denotes the ML estimate and SCAD yields

$$\hat{\beta}_j = \begin{cases} \text{sign}(\hat{\beta}_j^{ML})(|\hat{\beta}_j^{ML}| - \lambda)_+ & \text{if } \; |\hat{\beta}_j^{ML}| < 2\lambda \\ \{(a-1)\hat{\beta}_j^{ML} - \text{sign}(\beta_j^{ML})a\lambda\}/a - 2 & \text{if } \; 2\lambda < |\hat{\beta}_j^{ML}| \leq a\lambda \\ \hat{\beta}_j^{ML} & \text{if } \; |\hat{\beta}_j^{ML}| > a\lambda \end{cases}$$

(see Fan and Li, 2001). Usually these penalties are compared to the hard thresholding penalty function $p_\lambda(|\beta|) = \lambda^2 - (|\beta| - \lambda)^2 I(|\beta| < \lambda)$, which sets the estimate to zero if the ML estimate is below some threshold and retains the ML estimate if it is above the threshold, that is, $\hat{\beta}_j = \hat{\beta}_j^{ML} I(|\hat{\beta}_j^{ML}| > \lambda)$. It is seen from Figure 6.6 that SCAD and hard thresholding avoid bias for large coefficients, in contrast to lasso. SCAD shares with the lasso the continuity of the resulting function (for details see Fan and Li, 2001).

For generalized linear models, the penalized log-likelihood has to be minimized. Since the SCAD penalty functions are singular at the origin and do not have continuous second-order derivations for the computation, Fan and Li use that they can be locally approximated by a quadratic function. The proposed local quadratic approximations also applies to the lasso and hard thresholding penalties.

An advantage of the SCAD penalty is its oracle property. Let the parameter vector $\boldsymbol{\beta}$ be partitioned into $\boldsymbol{\beta}^T = (\boldsymbol{\beta}_1^T, \boldsymbol{\beta}_2^T)$ and assume $\boldsymbol{\beta}_2 = \mathbf{0}$. With $I(\boldsymbol{\beta})$ denoting the full Fisher matrix and $J_1(\boldsymbol{\beta}_1)$ the Fisher matrix and knowing $\boldsymbol{\beta}_2 = \mathbf{0}$, it may be shown that $\hat{\boldsymbol{\beta}}^T = (\hat{\boldsymbol{\beta}}_1^T, \hat{\boldsymbol{\beta}}_2^T)$ must asymptotically satisfy $\hat{\boldsymbol{\beta}}_2 = \mathbf{0}$ and $\hat{\boldsymbol{\beta}}_1$ is asymptotic normal with covariance matrix $J_1(\boldsymbol{\beta}_1)^{-1}$ if $n^{1/2}\lambda_n \to \infty$. This means that asymptotically the estimator performs as well as if $\boldsymbol{\beta}_2 = \mathbf{0}$ were known. The penalized estimate is root-n consistent if $\lambda_n \to 0$ and converges at the rate $O_p(n^{-1/2} + a_n)$, where $a_n = \max\{p'_{\lambda_n}(|\beta_j|), \beta_j \neq 0\}$.

The penalized score function has the form $s_p(\boldsymbol{\beta}) = s(\boldsymbol{\beta}) - \Sigma_j p'_\lambda(|\beta_j|)$, where $s(\boldsymbol{\beta})$ is the usual score function of a GLM. The derivative on the first term on the right-hand side yields the usual negative information matrix. The derivative of the second term has to be approximated

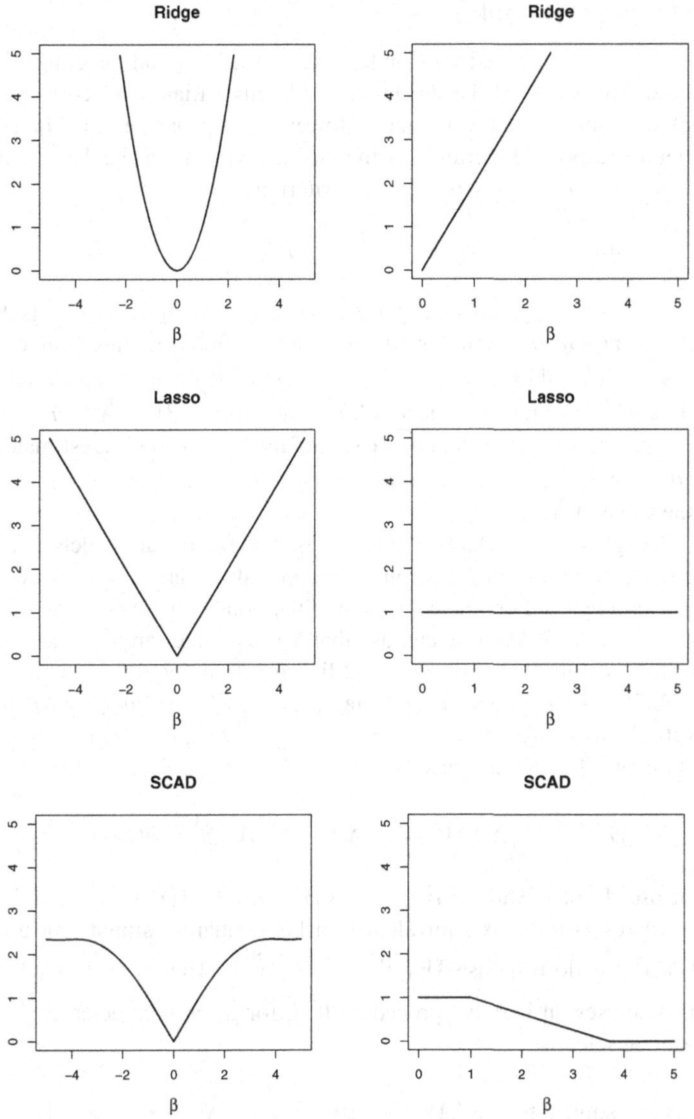

FIGURE 6.11: Components $p_\lambda(\beta)$ (left) and derivations $p'_\lambda(\beta)$ (right) of penalty functions for ridge, lasso, and SCAD.

because derivatives do not exist. Fan and Li (2001) used a quadratic approximation based on an initial estimate $\boldsymbol{\beta}_0$, which is given by $p_\lambda(|\beta_j|) = p'_\lambda(|\boldsymbol{\beta}_j|)\,\text{sign}(\beta_j) \approx \{p'_\lambda(|\beta_{0j}|)/|\beta_{0j}|\}\beta_j$ when $\beta_j \neq 0$. Let $\hat{\boldsymbol{\beta}}_1^T = (\hat{\beta}_{11}, \ldots \hat{\beta}_{1p})$ denote the non-vanishing components of $\hat{\boldsymbol{\beta}}$, obtained from $s_p(\hat{\boldsymbol{\beta}}) = 0$. Then the corresponding sandwich formula that approximates the covariance of $\hat{\boldsymbol{\beta}}_1$ is

$$\hat{\text{cov}}(\hat{\boldsymbol{\beta}}) \approx [I(\boldsymbol{\beta}_1) + P_\lambda(\hat{\boldsymbol{\beta}}_1)]^{-1}\hat{\text{cov}}(sp(\hat{\beta}_1))[I(\boldsymbol{\beta}_1) + P_\lambda(\hat{\boldsymbol{\beta}}_1)]^{-1},$$

where $I(\boldsymbol{\beta}_1) = -\partial l(\hat{\boldsymbol{\beta}}_1)/\partial\boldsymbol{\beta}\partial\boldsymbol{\beta}^T$ and $P_\lambda(\hat{\boldsymbol{\beta}}_1) = \text{diag}(p'_\lambda(|\beta_{01}|)/|\beta_{01}|, \ldots, p'_\lambda(|\beta_{0p}|)/|\beta_{0p}|)$. According to Fan and Li (2001), the formula has good accuracy for moderate sample sizes.

6.2.6 The Dantzig Selector

The Dantzig selector was proposed by Candes and Tao (2007) and generalized to GLMs by James and Radchenko (2008). Like the lasso, it obtains variable selection by using an L_1 penalty to shrink the coefficients toward zero. However, the penalty is used in a different way. For simplicity, let the response function h of the model $\mu = h(\boldsymbol{x}^T\boldsymbol{\beta})$ be the canonical response function. Then the generalized Dantzig selector criterion is

$$\min\|\tilde{\boldsymbol{\beta}}\|_1 \quad \text{subject to} \quad |\boldsymbol{x}_{.j}^T(\boldsymbol{y}-\tilde{\boldsymbol{\mu}})| \leq \lambda, j=1,\dots,p,$$

where $\|\tilde{\boldsymbol{\beta}}\|_1 = |\tilde{\beta}_1|+\dots+|\tilde{\beta}_p|$ represents, the L_1-norm, $\boldsymbol{x}_{.j}^T = (x_{1j},\dots,x_{nj})$ is the jth column of the design matrix, and $\boldsymbol{y}$, $\tilde{\boldsymbol{\mu}}$ denote the vector of observations and fitted values, respectively. The constraint region defined by $|\boldsymbol{x}_{.j}^T(\boldsymbol{y}-\tilde{\boldsymbol{\mu}})| \leq \lambda$, where λ is the tuning parameter, is based on the score function, which with a canonical link has the form $\boldsymbol{s}(\boldsymbol{\beta}) = \boldsymbol{X}^T(\boldsymbol{y}-\boldsymbol{\mu})$. Therefore, $\boldsymbol{x}_{.j}^T(\boldsymbol{y}-\boldsymbol{\mu})$ represents the jth component of the score function. While ML estimates are obtained when $\boldsymbol{x}_{.j}^T(\boldsymbol{y}-\hat{\boldsymbol{\mu}}) = 0$ for all j, the constraint $|\boldsymbol{x}_{.j}^T(\boldsymbol{y}-\tilde{\boldsymbol{\mu}})| \leq \lambda$ represents a weaker condition depending on the value of λ.

One of the strengths of the Dantzig selector is that for linear models it can be formulated as a linear programming problem and therefore also can be efficiently computed for high-dimensional problems. For linear problems, the constraint region has the simple form $|\boldsymbol{x}_{.j}^T(\boldsymbol{y}-\boldsymbol{X}\tilde{\boldsymbol{\beta}})| \leq \lambda$. For GLMs, one can use that ML estimates are obtained iteratively by a weighted least-squares algorithm. For a general link function, the score function has the form $\boldsymbol{s}(\boldsymbol{\beta}) = \boldsymbol{X}^T\boldsymbol{D}\,\boldsymbol{\Sigma}^{-1}(\boldsymbol{y}-\boldsymbol{\mu})$, where $\boldsymbol{D} =$ Diag $(\partial h(\eta_1)/\partial\eta,\dots\partial h(\eta_n)/\partial\eta)$ is the diagonal matrix of derivatives. With weight matrix $\boldsymbol{W} = \boldsymbol{D}\,\boldsymbol{\Sigma}^{-1}\boldsymbol{D}^T$, one step of the Fisher scoring iteration for obtaining ML estimates has the form

$$\hat{\boldsymbol{\beta}}^{(k+1)} = (\boldsymbol{X}^T\boldsymbol{W}(\hat{\boldsymbol{\beta}}^{(k)})\boldsymbol{X})^{-1}\boldsymbol{X}^T\boldsymbol{W}(\hat{\boldsymbol{\beta}}^{(k)})\tilde{\boldsymbol{\eta}}(\hat{\boldsymbol{\beta}}^{(k)})$$

with a vector of pseudo-observations $\tilde{\boldsymbol{\eta}}(\hat{\boldsymbol{\beta}}) = \boldsymbol{X}\hat{\boldsymbol{\beta}} + \boldsymbol{D}(\hat{\boldsymbol{\beta}})^{-1}(\boldsymbol{y}-\hat{\boldsymbol{\mu}})$. The implicitly used weighted least-squares estimate is equivalent to a least-squares estimate for a design matrix $\boldsymbol{W}(\hat{\boldsymbol{\beta}}^{(k)})^{1/2}\boldsymbol{X}$ and pseudo-response $\boldsymbol{W}(\hat{\boldsymbol{\beta}}^{(k)})^{1/2}\tilde{\boldsymbol{\eta}}(\hat{\boldsymbol{\beta}}^{(k)})$. Therefore, instead of solving the score equations, one uses an iterative procedure that, for given parameter $\hat{\boldsymbol{\beta}}^{(k)}$, computes the linear model Dantzig selector:

$$\min\|\tilde{\boldsymbol{\beta}}\|_1 \quad \text{subject to} \quad |\boldsymbol{x}_{.j}^T\boldsymbol{W}(\hat{\boldsymbol{\beta}}^{(k)})(\tilde{\boldsymbol{\eta}}(\hat{\boldsymbol{\beta}}^{(k)})-\boldsymbol{X}\tilde{\boldsymbol{\beta}})| \leq \lambda, j=1,\dots,p,$$

yielding the new estimate $\hat{\boldsymbol{\beta}}^{(k+1)}$. James and Radchenko (2008) also gave an algorithm for fitting the generalized Dantzig selector path. By computing the whole path efficiently, cross-validation on a fine grid can be performed. Efficient computing is needed since James and Radchenko (2008) use two shrinkage parameters. The reason is that the Dantzig selector tends to overshrink the coefficients. If strong shrinkage (large λ) is applied so that noisy variables are excluded, the estimates of coefficients are too small. If small λ is selected, noisy variables tend to be included.

Example 6.4: Heart Disease

Figure 6.12 shows the coefficient buildups for SCAD ($a = 3$; package *lqa*) and the Dantzig selector (standardized explanatory variables) plotted against $\|\boldsymbol{\beta}\|/\max\|\boldsymbol{\beta}\|$. Although the paths are differing, selection based on 10-fold cross-validation yields similar coefficients. □

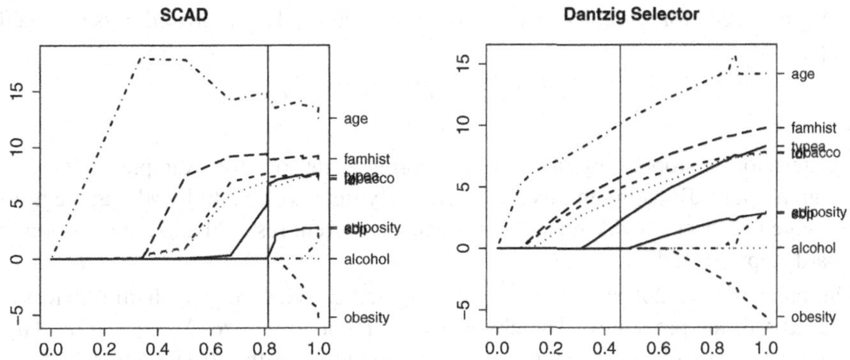

FIGURE 6.12: SCAD and Dantzig coefficient paths for heart disease data (package *lqa*; vertical line shows estimate selected by 10-fold cross-validation).

6.3 Boosting Methods

Boosting methods were originally developed in the machine learning community as a means to improve classification (e.g., Shapire, 1990). They have been proposed as ensemble methods, which rely on generating multiple predictions and averaging across the individual predictions. Later it was shown that boosting can be seen as the fitting of an additive structure by minimizing specific loss functions (see Friedman, 2001; Friedman et al., 2000). Bühlmann and Yu (2003) and Bühlmann (2006) proposed and investigated boosted estimators in the context of a linear regression with the focus on L_2 loss. In regressions, boosting may be seen as a regularization technique that also allows one to select predictors. For more background on boosting see also Section 15.5.3.

6.3.1 Boosting for Linear Models

Before considering the boosting of generalized linear models, we consider briefly the familiar case of normal regression models. Let the underlying regression structure be given by $\mathrm{E}(y|\boldsymbol{x}) = \eta(\boldsymbol{x})$ and data be given by $(y_i, \boldsymbol{x}_i), i = 1, \dots, n$.

Boosting is based on fitting a structured function that is supposed to approximate $\eta(\boldsymbol{x})$. The fitting of a structured function (a learner in machine learning terminology) is considered as a base procedure. Ensemble methods are based on averaging across several such procedures. Let $\hat{g}(\boldsymbol{x}, \{u_i, \boldsymbol{x}_i\})$ denote the base procedure at value $\boldsymbol{x}$ based on input data $\{u_i, \boldsymbol{x}_i\}$, which are not necessarily the original data $\{y_i, \boldsymbol{x}_i\}$. When fitting linear models, the base procedure uses a linear function $g(\boldsymbol{x}, \{u_i, \boldsymbol{x}_i\}) = \tilde{\boldsymbol{x}}^T\boldsymbol{\gamma}$, where $\tilde{\boldsymbol{x}}$ is usually a subvector of $\boldsymbol{x}$. In one step of the procedure one does not aim at estimating the whole vector $\boldsymbol{\beta}$ (from predictor $\eta = \boldsymbol{x}^T\boldsymbol{\beta}$) but at improving the estimate of the parameters $\boldsymbol{\gamma}$ that correspond to a subset of $\boldsymbol{\beta}$. A basic boosting algorithm then is given by:

Step 1 (Initialization)

Given data $\{y_i, \boldsymbol{x}_i\}$, fit the base procedure to yield the function estimate $\eta^{(0)}(.) = \hat{g}(., \{y_i, \boldsymbol{x}_i\})$.

Step 2 (Iteration)

For $l = 0, 1, 2, \dots,$ compute the residuals $u_i = y_i - \hat{\eta}^{(l)}(\boldsymbol{x}_i)$ and fit the base procedure to the current data $\{u_i, \boldsymbol{x}_i\}$. The fit $\hat{g}(., \{u_i, \boldsymbol{x}_i\})$ is an estimate based on the original predictor variables and the current residuals. The improved fit is obtained by the update

$$\hat{\eta}^{(l+1)}(.) = \hat{\eta}^{(l)}(.) + \hat{g}(., \{u_i, \boldsymbol{x}_i\}).$$

The iteration is stopped by applying a stopping criterion, for example, AIC or a cross-validation measure. Boosting in this form iteratively improves the fit by adding the fit of a base learner, which reduces the discrepancy between the current residual and the fit. It may be seen as forward stepwise additive modeling.

Bühlmann and Yu (2003) thoroughly investigated L_2 boosting, which utilizes least-squares estimates as a fitting procedure. Thereby, in one step one minimizes $\Sigma_i(u_i - \hat{g}(\boldsymbol{x}_i, \{u_i, \boldsymbol{x}_i\})^2$. L_2 boosting may also be derived as a gradient descent algorithm (see Section 15.5.3) and, as mentioned by Bühlmann and Yu, is nothing else more than repeated least-squares fitting of residuals, which for one boosting step has already been proposed by Tukey (1977) under the name "twicing." For linear models, which assume that the conditional mean $\eta(\boldsymbol{x}) = \mathrm{E}(y|\boldsymbol{x})$ has the form $\eta(\boldsymbol{x}) = \boldsymbol{x}^T\boldsymbol{\beta}$, a simple least squares-fitting of a linear predictor $g(\boldsymbol{x}, \{u_i, \boldsymbol{x}_i\}) = \boldsymbol{x}^T\boldsymbol{\beta}$ within one boosting step would simply yield the usual least-squares estimate after one step. Therefore, to obtain a regularized estimate, alternative base procedures have to be used.

An approach that implicitly selects variables is *componentwise boosting*. Componentwise boosting means that each part of the linear predictor is refitted separately, and among the fitted parts one is selected to be used in the update. In its simplest form, componentwise boosting refits only one coefficient, which is selected by some optimality criterion. Bühlmann (2006) proposed a componentwise linear least-squares algorithm for linear models by using the base learner $g(\boldsymbol{x}, \{u_i, x_i\}) = \gamma_{\hat{s}} x_{\hat{s}}$, where $\hat{\gamma}_j$ is the usual least-squares estimate resulting from using only the jth variable, $\hat{\gamma}_j = \Sigma_i u_i x_{ij} / \Sigma_i x_{ij}^2$ (centered predictors), and

$$\hat{s} = \arg\min_{1 \le j \le p} \sum_{i=1}^{n} (u_i - \hat{\gamma}_j x_{ij})^2$$

determines which variable is selected. Thus, the base procedure in componentwise linear least-squares boosting performs a linear least-squares regression against the one selected variable that reduces the residual sum of squares the most. The actual refit typically uses $\hat{g}(\boldsymbol{x}, \{u_i, x_i\}) = \nu\gamma_{\hat{s}} x_{\hat{s}}$, where the parameter ν is a fixed shrinkage parameter, in order to obtain a weak learner (see next section). The corresponding refit of parameters is given by $\hat{\beta}_j^{(l+1)} = \hat{\beta}_j^{(l)}$, $j \neq \hat{s}$, $\hat{\beta}_{\hat{s}}^{(l+1)} = \hat{\beta}_{\hat{s}}^{(l)} + \nu\gamma_{\hat{s}}$. Since in high-dimensional settings usually not all of the predictors are selected before the stopping criterion is reached, the procedure selects variables automatically. Bühlmann (2006) showed that the procedure is consistent for underlying regression functions, which are sparse in terms of the L_1-norm.

Boosting procedures are based on "weak" learners, a concept that has been derived in the machine learning community (Freund and Schapire, 1997). In classification, a weak learner may be considered as an estimator that is slightly better than guessing. In regression, a weak learner refers to small step sizes within the algorithm. Therefore, the update step in linear least-squares boosting uses

$$\hat{\eta}^{(r+1)}(.) = \eta^{(r)}(.) + \nu\hat{g}(., \{u_i, x_i\}),$$

where ν is a shrinkage parameter, for example, $\nu = 0.1$. Small step sizes (small ν) make the boosting algorithm slow and require a larger number of iterations, but improve the performance. Small values of ν have been shown to avoid early overfitting of the procedure.

6.3.2 Boosting for Generalized Linear Models

In generalized linear models least-squares estimates are not the best choice because they do not relate adequately to the underlying error structure. A better choice is likelihood-based boosting, which more generally aims at maximizing the log-likelihood rather than minimizing the squared residuals. For fixed link functions, likelihood-based approaches iteratively estimate the linear predictor, $\eta_i = \boldsymbol{x}_i^T \boldsymbol{\beta}$, which is linked to the mean $\mu_i = \mathrm{E}(y_i|\boldsymbol{x}_i)$ by $\mu_i = h(\eta_i)$.

One difference between the L_2 boost and a generalized linear model boosting is that in the iteration step one cannot fit a GLM to the residuals because, for example, with binary data, residuals are not from $\{0, 1\}$. The role of the residuals is taken by the offset. The basic likelihood-based boosting algorithm (GenBoost), which is also used in Chapter 15, has the following form:

Likelihood Boosting (GenBoost)

Step 1 (Initialization)

For given data $(y_i, \boldsymbol{x}_i), i = 1, \dots, n$, fit the intercept model $\mu^{(0)}(\boldsymbol{x}) = h(\beta_0)$ by maximizing the likelihood, yielding $\eta^{(0)} = \hat{\beta}_0, \hat{\mu}^{(0)} = h(\hat{\beta}_0), \hat{\boldsymbol{\beta}}^{(0)} = (\hat{\beta}_0, 0, \dots, 0)^T$.

Step 2 (Iteration)

For $l = 0, 1, 2, \dots$, fit the model

$$\mu_i = h(\hat{\eta}^{(l)}(\boldsymbol{x}_i) + \eta(\boldsymbol{x}_i, \boldsymbol{\gamma}))$$

to data $(y_i, \boldsymbol{x}_i), i = 1, \dots, n$, where $\hat{\eta}^{(l)}(\boldsymbol{x}_i)$ is treated as an offset and the predictor is estimated by fitting the parametrically structured term $\eta(\boldsymbol{x}_i, \boldsymbol{\gamma})$, obtaining $\hat{\boldsymbol{\gamma}}$. The improved fit is obtained by

$$\hat{\eta}^{(l+1)}(\boldsymbol{x}_i) = \hat{\eta}^{(l)}(\boldsymbol{x}_i) + \hat{\eta}(\boldsymbol{x}_i, \hat{\boldsymbol{\gamma}}), \quad \hat{\mu}_i^{(l+1)} = h(\hat{\eta}^{(l+1)}(\boldsymbol{x}_i)).$$

The improved parameter $\hat{\boldsymbol{\beta}}^{(l+1)}$ is obtained by adding $\hat{\boldsymbol{\gamma}}$ to the components of $\hat{\boldsymbol{\beta}}^{(l)}$.

One candidate for fitting is Fisher scoring, which is familiar from generalized linear model fitting. One first has to compute the pseudo-responses and weights:

$$\tilde{\eta}_i^{(l)} = \frac{y_i - \hat{\mu}_i^{(l)}}{\partial h(\hat{\eta}_i^{(l)})/\partial \eta}, \quad w_i^{(l)} = \frac{(\partial h(\hat{\eta}_i^{(l)})/\partial \eta)^2}{\sigma_i^2},$$

and then compute the weighted regression with weights $w_i^{(l)}$ and dependent variables $\tilde{\eta}_i^{(l)}$ to obtain $\hat{\boldsymbol{\gamma}}$. It is noteworthy that the pseudo-responses, in contrast to their usual definition, do not include a linear term, because the previous fit is contained in the offset. For the logit model one has $\partial h(\hat{\eta}_i)/\partial \eta = h(\eta_i)/(1 - h(\eta_i))$, and therefore the pseudo-responses and weights simplify to

$$\tilde{\eta}_i = \frac{y_i - \hat{\mu}_i^{(l)}}{\hat{\pi}^{(l)}(1 - \hat{\pi}^{(l)})}, \quad w_i = \hat{\pi}^{(l)}(1 - \hat{\pi}^{(l)}).$$

More concretely, let us consider the linear predictor $\eta(\boldsymbol{x}_i, \boldsymbol{\gamma}) = \tilde{\boldsymbol{x}}_i^T \boldsymbol{\gamma}$, where $\tilde{\boldsymbol{x}}_i$ is a specified subvector of $\boldsymbol{x}_i$. For example, when using $\eta(\boldsymbol{x}_i, \boldsymbol{\gamma}) = x_{ij}\gamma_j$, $\tilde{\boldsymbol{x}}_i$ contains only the jth covariate as a candidate for updating. One computes within the iteration steps one-step Fisher scoring estimates:

$$\hat{\gamma} = (\tilde{\boldsymbol{X}}^T \boldsymbol{W}^{(l)} \tilde{\boldsymbol{X}})^{-1} \tilde{\boldsymbol{X}}^T \boldsymbol{W}^{(l)} \tilde{\boldsymbol{\eta}}^{(l)},$$

where $\tilde{\boldsymbol{X}}$ is the design matrix built from the vectors $\tilde{\boldsymbol{x}}_i$, $\boldsymbol{W}^{(l)}$ is a diagonal matrix that contains the weights, $w_i^{(l)}$, and $\tilde{\boldsymbol{\eta}}^{(l)}$ contains the pseudo-responses $\tilde{\eta}_i^{(l)}$ (for Fisher scoring see Chapter 3, Section 3.9). Since the linear predictor $\eta(\boldsymbol{x}_i, \gamma) = \tilde{\boldsymbol{x}}_i^T \gamma$ is fitted, refitting refers only to the components contained in $\tilde{\boldsymbol{x}}_i$, that is, $\hat{\beta}_j^{(l+1)} = \hat{\beta}_j^{(l)} + \hat{\gamma}_j$ if x_{ij} is contained in $\tilde{\boldsymbol{x}}_i$ (with $\hat{\gamma}_j$ denoting the corresponding estimate) and $\hat{\beta}_j^{(l+1)} = \hat{\beta}_j^{(l)}$ if x_{ij} is not contained in $\tilde{\boldsymbol{x}}_i$.

To obtain a weak learner $\hat{\gamma}$ can be replaced by $\nu\hat{\gamma}$ with ν denoting a shrinkage parameter. One-step Fisher scoring, starting with the zero vector, can also be given in the form $\hat{\gamma} = (\boldsymbol{F}^{(l)})^{-1} \boldsymbol{s}^{(l)}$, where $\boldsymbol{F}^{(l)} = \tilde{\boldsymbol{X}}^T \boldsymbol{W}^{(l)} \tilde{\boldsymbol{X}}$ is the Fisher matrix and $\boldsymbol{s}^{(l)} = \tilde{\boldsymbol{X}}^T \boldsymbol{W}^{(l)} \tilde{\boldsymbol{\eta}}^{(l)}$. Weak learners may also be obtained in the spirit of ridge estimation by using a penalized Fisher matrix $\boldsymbol{F}^{(l)} + \lambda \boldsymbol{I}$, where λ is chosen large.

In componentwise boosting within the fitting step, a selection step is included that determines which of the parameters is refitted. In the simplest case one fits all one-covariate models $\eta(\boldsymbol{x}_i, \gamma) = \gamma_j x_{ij}$, $j = 0, \ldots, p$, as candidates, obtaining $\hat{\gamma}_j$, and then selects the variable that has the strongest impact on the improvement of the fit. A criterion is the improvement in deviance:

$$Dev(\hat{\eta}^{(l)}) - Dev(\hat{\eta}_{new(j)}),$$

where $\hat{\eta}_{new(j)}$ is based on the parameter vector in which only the jth component is updated to $\hat{\beta}_j^{(l)} + \hat{\gamma}_j$. When the sth variable is selected, the new parameter vector is $\hat{\boldsymbol{\beta}}^{(l+1)} = (\hat{\beta}_0^{(l)}, \ldots, \hat{\beta}_s^{(l)} + \hat{\gamma}_s, \ldots)^T$. Selection of the parameter that is actually updated within one step can also be based on information criteria like AIC or BIC.

In the case of the logit model, likelihood-based boosting is equivalent to the *LogitBoost* algorithm for two classes that were proposed by Friedman et al. (2000). The advantage of GenBoost is that it applies to all kinds of link functions and exponential family responses. The first extension to the exponential family setting was given by Ridgeway (1999). He gave two similar, though slightly different algorithms for boosting exponential family models (for details of Fisher scoring-based algorithms, see also Tutz and Binder, 2007).

Example 6.5: Heart Disease

Figure 6.13 shows the coefficient buildups for likelihood-based boosting with 500 iterations. The left plot was computed with the package *GAMBoost*, and the right plot uses the quadratic approximation used in the package *mboost*. The resulting paths are quite similar; however, *mboost* is much faster. □

Blockwise Boosting

The strategy to update just one variable is rather limited, and it is especially inadequate if categorical variables are in the predictor. A categorical predictor that takes k categories will be represented by $k-1$ dummy variables in the linear predictor. If one wants to avoid having the resulting selection depend on the coding scheme, the parameters for all the dummy variables representing one variable should be refitted simultaneously. Then the structured term to be fitted is $\eta(\boldsymbol{x}_i, \gamma) = \boldsymbol{x}_{ir}^T \gamma_r$, where $\boldsymbol{x}_{ir}$ is a vector of dummy variables corresponding to a categorical variable.

In general, to obtain a selection of relevant terms, the base procedures that are used typically contain only a small number of variables. In the extreme case only one coefficient is refitted; in

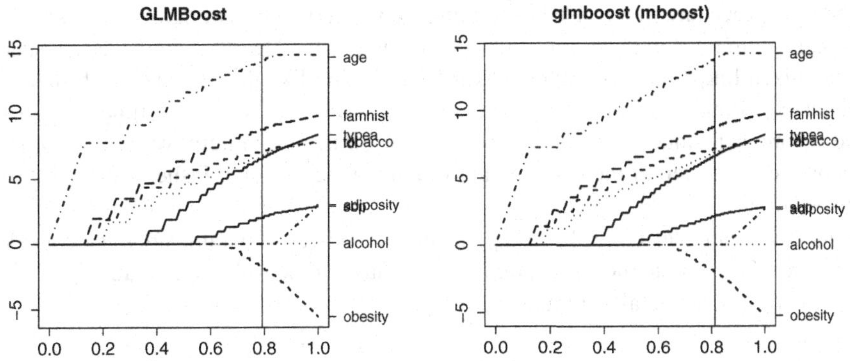

FIGURE 6.13: GlmBoost coefficient paths for heart disease data (package *mboost* (right), vertical line shows estimate selected by 10-fold cross-validation).

other cases, it can be a group of coefficients. A list of parametrically structured terms that may be fitted is

- $\eta(\boldsymbol{x}_i, \boldsymbol{\gamma}) = x_{ir}\gamma_r$, which specifies the linear effect of the rth covariate;
- $\eta(\boldsymbol{x}_i, \boldsymbol{\gamma}) = \gamma_0 + x_{ir}\gamma_r$, which specifies the intercept and the linear effect of the rth covariate;
- $\eta(\boldsymbol{x}_i, \boldsymbol{\gamma}) = \boldsymbol{x}_{ir}^T\boldsymbol{\gamma}_r$, where $\boldsymbol{x}_{ir}$ is a vector of dummy variables corresponding to a categorical variable;
- $\eta(\boldsymbol{x}_i, \boldsymbol{\gamma}) = x_{ir}x_{is}\gamma_{rs}$, representing an interaction between the rth and the sth covariates;
- $\eta(\boldsymbol{x}_i, \boldsymbol{\gamma}) = \boldsymbol{x}_{ir}\boldsymbol{x}_r^T\boldsymbol{\gamma}_{rs}$, representing an interaction between the rth variable and the sth categorical variable given by a vector of dummy variables.

These parametrically structured terms define the learner that is used. In general, the parameter $\boldsymbol{\gamma}$ is a vector that refers to a group or block of variables that define the design matrix $\tilde{\boldsymbol{X}}$. The update by adding $\hat{\boldsymbol{\gamma}}$ is done blockwise. When one fits all one-covariate models with intercepts $\eta(\boldsymbol{x}_i, \boldsymbol{\gamma}) = \gamma_0 + \gamma_r x_{ir}$, $r = 1, \dots, p$, as candidates, after selection of the best update, s, the update is given by $\hat{\boldsymbol{\beta}}^{(l+1)} = (\hat{\beta}_0^{(l)} + \hat{\gamma}_0, \dots, \hat{\beta}_j^{(l)} + \hat{\gamma}_s, \dots)^T$.

In addition, in some cases it is not sensible to let the procedure select among all the variables. For example, in treatment studies, the treatment should be considered as a mandatory variable that is always included in the predictor. Therefore, one should distinguish between *mandatory* and *optional* predictors. The more general concept of blockwise boosting allows one to distinguish between these types of variables and to refit groups of variables.

Blockwise (or partial) boosting means that in the lth iteration selected components of the parameter vector are re-estimated. The selection is determined by a specific structuring of the parameters (variables). Let the parameter indices $V = \{1, \dots, p\}$ be partitioned into disjoint sets by $V = V_c \cup V_{o1} \cup \dots \cup V_{oq}$, where V_c stands for the (mandatory) parameters (variables) that have to be included in the analysis, and $V_{o1}, \dots, V_{oq}$ represent blocks of parameters that are optional. A block V_{or} may refer to all the parameters that refer to a multicategorical variable, such that not only parameters but variables are evaluated. Candidates in the refitting process are all combinations $V_c \cup V_{or}, r = 1, \dots, q$, representing combinations of necessary and optional variables. Componentwise boosting that refers to single coefficients is the special case where $V_c = \emptyset, V_{oj} = \{j\}$.

The base procedure that is used in refitting steps refers to the fitting of a shrinked estimator (weak learner) to the candidate sets and the selection of the "least" candidate. As usual, fitting for generalized linear models means maximizing the log-likelihood. Since a shrinked version has better performance, a shrinkage estimator, for example, the ridge estimator with large λ, is used. Moreover, since the boosting procedure itself means an iterative refitting of residuals, within one refitting step of the boosting algorithm we use one-step Fisher scoring rather than a complete fit.

More technically, let $V_m = V_c \cup V_{om}$ denote the indices of parameters to be considered for refitting and $\boldsymbol{X}_{V_m}$ denote the corresponding submatrix of the full design matrix $(\boldsymbol{x}_{.1}, \dots, \boldsymbol{x}_{.p})$. Then the partial boosting algorithm is given by:

Blockwise/Partial Likelihood Boosting (GenPartBoostR)

Step 1: Initialization

Fit model $\mu_i = h(\beta_0)$ by iterative Fisher scoring to obtain $\hat{\boldsymbol{\beta}}^{(0)} = (\hat{\beta}_0, 0, \dots, 0)^T$, $\hat{\boldsymbol{\eta}}_{(0)} = \boldsymbol{X}\hat{\boldsymbol{\beta}}^{(0)}$.

Step 2: Iteration

For $l = 1, 2, \dots$

(a) Estimation: Estimation for candidate sets V_m, $m = 1, \dots, q$, corresponds to fitting of the model

$$\boldsymbol{\mu} = h(\hat{\boldsymbol{\eta}}^{(l-1)} + \boldsymbol{X}_{V_m}\boldsymbol{\gamma}_{V_m}),$$

where $\hat{\boldsymbol{\eta}}^{(l-1)} = \boldsymbol{X}\hat{\boldsymbol{\beta}}^{(l-1)}$ is treated as an offset (fixed constant). Fitting is performed by one step of Fisher scoring by use of a weak learner for $\boldsymbol{\gamma}_{V_m}$.

(b) Selection: For candidate sets V_m, $m = 1, \dots, q$, the set V_{m_0} is selected that improves the fit maximally.

(c) Update: One sets

$$\hat{\boldsymbol{\gamma}}^{(l)} = \begin{cases} \hat{\gamma}_{V_m,j} & j \in V_{m_0} \\ 0 & j \notin V_{m_0}, \end{cases}$$

$\hat{\boldsymbol{\beta}}^{(l)} = \hat{\boldsymbol{\beta}}^{(l-1)} + \hat{\boldsymbol{\gamma}}^{(m)}$, $\hat{\boldsymbol{\eta}}^{(l)} = \boldsymbol{X}\hat{\boldsymbol{\beta}}^{(l)}$, $\hat{\boldsymbol{\mu}}^{(l)} = h(\boldsymbol{X}\hat{\boldsymbol{\beta}}^{(l)})$, where h is applied componentwise.

Example 6.6: Abortion of Treatment

We illustrate the application of simple ridge boosting and partial boosting with real data from 344 admissions at a psychiatric hospital. The (binary) response variable to be investigated is whether treatment is aborted by the patient against physicians' advice (about 55% for this group of patients). From a total of 101 variables available, 8 variables likely to be relevant were identified: age, number of previous admissions ("noupto"), cumulative length of stay ("losupto") and the 0/1-variables indicating a previous ambulatory treatment ("prevamb"), no second diagnosis ("nosec"), second diagnosis "personality disorder" ("persdis"), somatic problems ("somprobl"), homelessness ("homeless"), and joblessness ("jobless"). Based on subject matter considerations, the two variables that relate to secondary diagnosis ("nosec" and "persdis") are mandatory members of the response set and no penalty is applied to their estimates. This illustrates

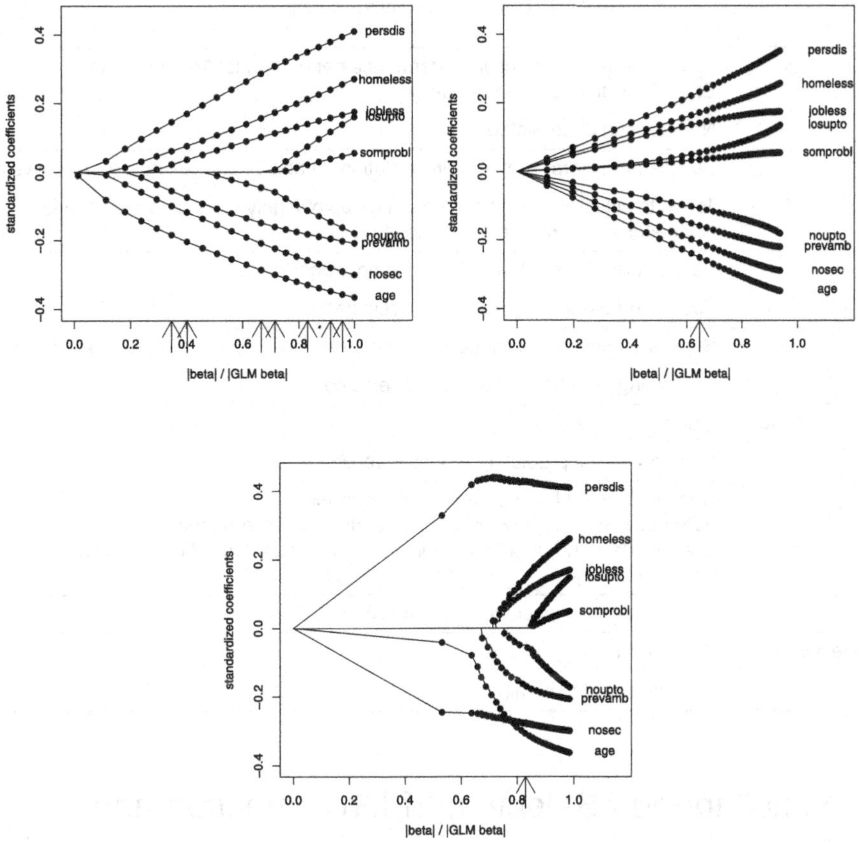

FIGURE 6.14: Coefficient buildups for abortion of treatment data when using lasso (upper left), ridge (upper right), and boosted ridge (lower panel) with mandatory variables.

the effect of augmenting an unpenalized model with a few mandatory variables with optional predictors. Figure 6.14 shows the coefficient buildup in the course of the boosting steps for partial boosting contrasted with simple ridge boosting (lower panel) and the lasso (upper left panel). The arrows indicate the number of steps chosen by AIC (for partial boosting and simple ridge boosting) and 10-fold cross-validation (for the lasso repeated 10 times). It can be seen that the mandatory components introduce a very different structure in coefficient buildups. One interesting feature is the slow decrease of the estimate for "persdis" beginning with boosting step 8. This indicates some overshooting of the initial estimate that is corrected when additional predictors are included. To identify relevant variables we used all 101 predictors and divided the data into a training set of size 270 and a test set of size 74. The lasso (with cross-validation) returned 13 predictors with a prediction error of 0.392. Partial boosting (using six mandatory response set elements relating to the secondary diagnosis) with penalty varying from 500 to 10000 returned 15 to 19 predictors and a prediction error between 0.378 and 0.392. □

Variable selection by regularization is a very active research area. Modifications and improvements are proposed and properties of existing estimates are investigated in many journals. Therefore, consideration of advantages and disadvantages tends to be preliminary. Nevertheless, in Table 6.1 some properties of currently available procedures are listed.

TABLE 6.1: Properties of regularized estimators.

Ridge	Estimates exist. Prediction performance better than for ML estimates. Explicit solution for linear model. No selection of predictors.
Lasso	Selects predictors, sparse representation. Oracle properties hold for adaptive lasso. Tends to select one predictor from a group of highly correlated predictors. Not necessarily consistent (but adaptive lasso is).
Elastic net	Selects predictors. Shows the grouping property. Two tuning parameters have to be selected.
Oscar	Selects predictors, exact grouping property. Clustering of predictors available. Two tuning parameters have to be selected.
Correlation-based	Grouping property. Does not select predictors (boosted version does)
SCAD	Nearly unbiased for large unknown coefficients. Automatically sets small estimated coefficients to zero (sparsity). Continuous in the data to avoid instability in model prediction (continuity). Oracle property.
Dantzig selector	Variable selection included. Can be computed efficiently.
Componentwise boosting	Selects variables. Inference hard to obtain.

6.4 Simultaneous Selection of Link Function and Predictors

When predictors are selected one typically assumes that the link function is known. However, if the assumed link function is wrong, the performance of the selection procedures can be strongly affected. For illustration, let us consider a small simulation study. Let the generating model be a Poisson model with the true response function having sigmoidal form $h_T(\eta) = 10/(1+\exp(-5\cdot\eta))$. Let the parameter vector of length $p = 20$ be given by $\boldsymbol{\beta}^T = (0.2, 0.4, -0.4, 0.8, 0, \dots, 0)$ and covariates be drawn from a normal distribution $\boldsymbol{x} \sim N(\boldsymbol{0}_p, \boldsymbol{\Sigma})$ with $\boldsymbol{\Sigma} = \{\sigma_{ij}\}_{i,j\in\{1,\dots,p\}}$, where $\sigma_{ij} = 0.5$, $i \neq j$, $\sigma_{ii} = 1$. We generate $N = 50$ datasets with $n = 200$ observations and fit the model by using the usual maximum likelihood (ML) procedure based on the canonical log-link (without variable selection). In addition, we apply three alternative fitting methods that include variable selection: a non-parametric flexible link procedure considered in the following, the lasso for generalized linear models, and a componentwise boosting procedure. While the flexible link procedure selects a link function, ML estimates as well as lasso and boosting use the canonical link. It is seen from Figure 6.15 that the best results are obtained if the link function is estimated non-parametrically. In particular, the parameters of the predictors that are not influential are estimated more stable and closer to zero. The dominance of the flexible procedure is also seen in Figure 6.16, which shows the mean squared error for the estimation of the parameter vector and the predictive deviance on an independently drawn test dataset with $n = 1000$.

The flexible procedure shown in Figure 6.15 is an extension of the non-parametric estimation procedure considered in Section 5.2. One fits the model $\mu_i = h_0(h(\eta_i))$, where $h_0(.)$ is a fixed transformation function, for example, the canonical link, and the inner function

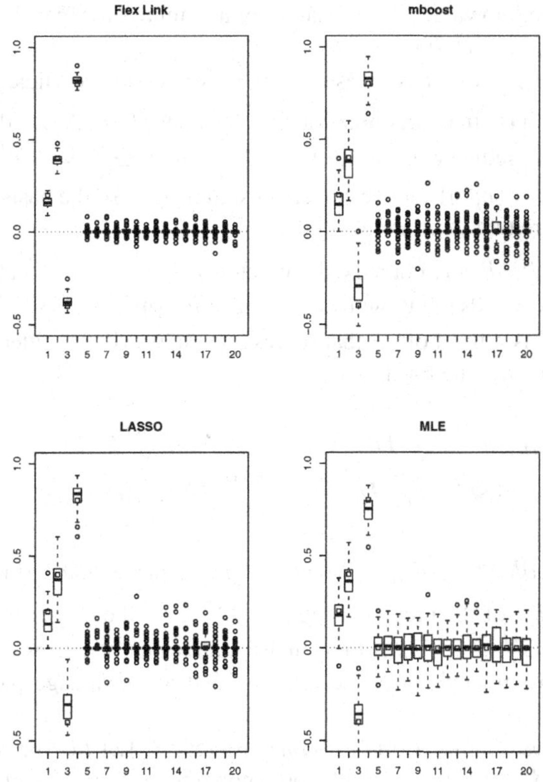

FIGURE 6.15: Resulting estimates of coefficient vector in simulation study for flexible link, boosting, lasso, and ML.

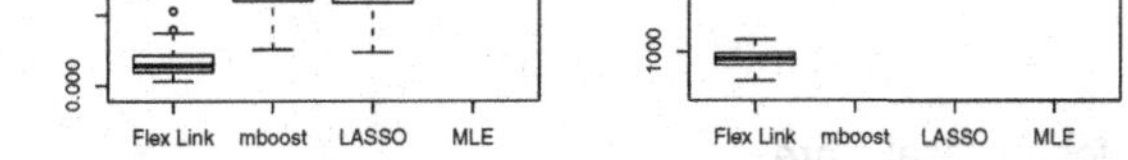

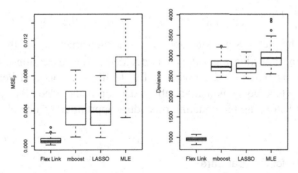

FIGURE 6.16: Mean squared error for parameter vector and predictive deviance for simulation setting.

$h(.)$ is considered as unknown and is estimated by assuming an expansion in basis functions $h(\eta_i) = \sum_{j=1}^m \alpha_j \phi_j(\eta_i) = \boldsymbol{\alpha}^T \boldsymbol{\Phi}_i$.

Estimates are obtained by iteratively estimating the regression coefficients $\boldsymbol{\beta}$ and the parameters of the link function $\boldsymbol{\alpha}$. In matrix notation, let $\hat{\boldsymbol{\beta}}^{(l)}$ and $\hat{\boldsymbol{\eta}}^{(l)} = \boldsymbol{X}\hat{\boldsymbol{\beta}}^{(l)}$ denote the parameter estimate and the fitted predictor in the lth step. Moreover, $\boldsymbol{\Phi}_i^{(l)} = (\boldsymbol{\Phi}_1^{(l)}, \dots, \boldsymbol{\Phi}_n^{(l)})^T$ with $\boldsymbol{\Phi}^{(l)} = (\phi_1(\hat{\eta}_i^{(l)}), \dots, \phi_m(\hat{\eta}_i^{(l)}))^T$ is the current design matrix for the basis functions. Within a boosting-type procedure two steps are iterated:

Boosting for Fixed Predictor. For a fixed predictor $\hat{\boldsymbol{\eta}}^{(l-1)} = \boldsymbol{X}\hat{\boldsymbol{\beta}}^{(l-1)}$, the estimation of the response function corresponds to fitting the model $\boldsymbol{\mu} = h_0((\boldsymbol{\Phi}^{(l-1)})^T \hat{\boldsymbol{\alpha}}^{(l-1)} + (\boldsymbol{\Phi}^{(l-1)})^T \hat{\boldsymbol{a}}^{(l)})$, where $\Phi^{(l-1)})^T \hat{\boldsymbol{\alpha}}^{(l-1)}$ is a fixed offset that represents the previously fitted value. One step of penalized Fisher scoring has the form

$$\begin{aligned}\hat{\boldsymbol{a}}^{(l)} &= \nu_h((\boldsymbol{\Phi}^{(l-1)})^T \hat{\boldsymbol{D}}^{(l-1)} (\hat{\boldsymbol{\Sigma}}^{(l-1)})^{-1} \hat{\boldsymbol{D}}^{(l-1)} \boldsymbol{\Phi}^{(l-1)} + \lambda_h \mathbf{P}_h)^{-1} \cdot \\ &\quad \cdot \left(\boldsymbol{\Phi}^{(l-1)}\right)^T \hat{\boldsymbol{D}}^{(l-1)} (\hat{\boldsymbol{\Sigma}}^{(l-1)})^{-1} (\boldsymbol{y} - \hat{\boldsymbol{\mu}}^{(l-1)}),\end{aligned}$$

where $\hat{\mathbf{D}}^{(l-1)} = \text{diag}(\partial h_0(\hat{h}^{(l-1)}(\hat{\eta}_i^{(l-1)})/\partial h^{(l-1)}(\eta))$ is the estimate of the derivative matrix evaluated at the estimate of the previous step and $\hat{\boldsymbol{\Sigma}}^{(l-1)}$ is the diagonal matrix of variances evaluated at $h_0(\hat{h}^{(l-1)}(\eta_i^{(l-1)}))$. $\mathbf{P}_h$ is the penalty matrix that penalizes the second derivation of the estimated (approximated) response function and the shrinkage parameter is fixed by $\nu_h = 0.1$.

Componentwise Boosting for Fixed Response Function. Let $h(.)$ be fixed and the design matrix have the form $\mathbf{X} = (\boldsymbol{x}_1|...|\boldsymbol{x}_p)$ with corresponding response vector $\boldsymbol{y} = (y_1, ..., y_n)^T$. Componentwise boosting means to update one parameter within one boosting step. Therefore, one fits the model $\mu = h_0(h(\boldsymbol{X}\hat{\boldsymbol{\beta}}^{(l-1)} + \boldsymbol{x}_j b_j))$, where $\boldsymbol{X}\hat{\boldsymbol{\beta}}^{(l-1)}$ is a fixed offset and only the variable $\boldsymbol{x}_j$ is included in the model. Then penalized Fisher scoring for the parameter b_j has the form

$$\hat{b}_j^{(l)} = \nu_p(\boldsymbol{x}_j^T \hat{\boldsymbol{D}}_\eta^{(l-1)} (\hat{\boldsymbol{\Sigma}}^{(l-1)})^{-1} \hat{\boldsymbol{D}}_\eta^{(l-1)} \boldsymbol{x}_j)^{-1} \boldsymbol{x}_j^T \hat{\boldsymbol{D}}_\eta^{(l-1)} (\hat{\boldsymbol{\Sigma}}^{(l-1)})^{-1} (\boldsymbol{y} - \hat{\boldsymbol{\mu}}^{(l-1)}),$$

where $\nu_p = 0.1$, $\hat{\boldsymbol{D}}_\eta^{(l-1)} = \text{diag}(\partial h_0(\hat{h}^{(l-1)}(\hat{\eta}_i^{(l-1)}))/\partial \eta)$ is the matrix of derivatives evaluated at the values of the previous iteration and $\hat{\boldsymbol{\Sigma}}^{(l-1)}$ is the variance from the previous step.

In each step of the boosting algorithm it is decided if the regression coefficients or the coefficients of the basis functions are updated; for details see Tutz and Petry (2011). The advantage of boosting techniques is that variable selection is included. By updating only one of the coefficients and stopping the updating procedure appropriately, one obtains the relevant predictors.

Example 6.7: Demand for Medical Care

In Example 7.6, count data with the number of physician office visits as the response variable were considered. Here we consider a Poisson model with flexible link and the same predictors as in Example 7.6. Figure 6.17 shows the estimated response functions plotted against the linear predictor. The canonical log-link is a strictly increasing function while the non-parametrically estimated response function becomes flat for large values of the predictor. The canonical link seems not to be appropriate. This is supported by an improved prediction error in subsamples when the flexible link function is used (see Tutz and Petry, 2011). □

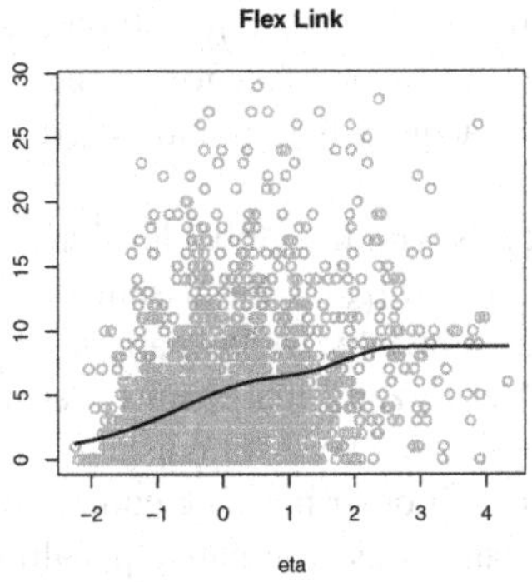

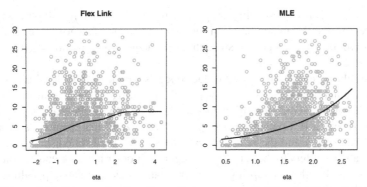

FIGURE 6.17: Response functions for medical care data against linear predictor: flexible link function (left) and canonical log link (right).

6.5 Categorical Predictors

The selection of categorical variables has already been briefly discussed for lasso-type penalties and boosting approaches. In the following we will consider further approaches. When predictors are categorical selection should distinguish between two cases: the selection of variables and the selection of effects within variables. If one wishes to select variables one has to select groups of variables because a categorical variable is represented by several dummy variables. If, however, one enforces selection among all terms in the linear predictor, including dummies, a selection strategy like the lasso will select single dummy variables with the effect that the coding scheme that has been used on the categorical predictors determines the result. To avoid such effects one should distinguish between the two problems:

- Which categorical predictors should be included in the model?
- Which categories within one categorical predictor should be distinguished?

The latter problem is concerned with one single variable and poses the question of which categories differ from one another with respect to the dependent variable. Or, to put it in a different way, which categories should be collapsed? The answer to that question depends on the scale level of the predictor; one should distinguish between nominal and ordered categories because of their differing information content. We will first consider selection strategies for effects within variables and then the selection of variables as groups of possibly regularized effects.

6.5.1 Selection within Categorical Predictors

Let us first consider just one categorical predictor $A \in \{1, \dots, k\}$, which is included in the predictor by the use of dummy variables in the form $\eta = \beta_0 + \sum_j x_{A(j)}\beta_j$. Then, when computing a penalized estimate, for example, by use of a lasso-type penalty,

$$J(\boldsymbol{\beta}) = \sum_j |\beta_j|,$$

the shrinkage effect depends on the coding scheme that is used. For simplicity, let the categorical predictor A have only three categories, $A \in \{1, 2, 3\}$, which are coded by two dummy variables $x_{A(2)}, x_{A(3)}$. If (0–1)-coding ($x_{A(j)} = 1$ if $A = j$ and $x_{A(j)} = 0$ otherwise) is used, shrinkage refers to the difference between the first and second categories and the difference between the third and first categories, since the first category is implicitly used as a reference ($\beta_{A(1)} = 0$). If effect coding is used ($x_{A(j)} = 1$ if $A = j$, $x_{A(j)} = -1$ if $A = k$, $x_{A(j)} = 0$ otherwise), shrinkage refers to the effect of factor levels with the global level across categories as the reference point because implicitly $\beta_{A(1)} + \beta_{A(2)} + \beta_{A(3)} = 0$ is assumed. Therefore, the selection of parameters, which is enforced by the lasso penalty, yields parameters that depend on the coding scheme. If a parameter, say $|\beta_j|$, is set to zero, that means that in the case of (0–1)-coding, category j and the reference category cannot be distinguished. But collapsing the categories always refers to the reference category; all other possible combinations of categories are ignored. To allow collapsing of any two categories, alternative penalties, which are considered in the following, have to be used.

Clustering of Categories for Nominal Predictor

For *nominal* predictor variables with many categories, a useful strategy is to search for clusters of categories with similar effects. The objective is to reduce the k categories to a smaller number of categories that form clusters; the effect of categories within one cluster is supposed to be the same, but responses will differ across clusters. With (0–1)-coding and reference category 1, that is, $\beta_1 = 0$, a fusion-type penalty that enforces clustering is

$$J(\boldsymbol{\beta}) = \sum_{i>j} w_{ij}|\beta_i - \beta_j| = w_{21}|\beta_2| + \cdots + w_{k1}|\beta_k| + w_{32}|\beta_3 - \beta_2| + \ldots, \tag{6.11}$$

where w_{ij} is an additional weight that may depend on the sample sizes within the categories. The penalty enforces the selection among effects $\theta_{ij} = \beta_i - \beta_j$, $i = 1, \ldots, k-1, i > j$. Since the ordering of the dummy varables $x_{A(1)}, \ldots, x_{A(k)}$ is arbitrary, all differences $\beta_i - \beta_j$ are used. For large λ, the penalty $\lambda J(\boldsymbol{\beta})$ tends to form clusters of categories; for $\lambda \to \infty$, all parameter estimates become zero and the categorical predictor is excluded.

The penalty is very useful when the predictor has many categories. For small sample size as compared to the number of categories, the ML estimate becomes unstable. In contrast, regularized estimates are much more stable, and with the selection effect of the L_1-penalty on differences they allow one to form clusters.

Ordered Categories

An interesting case is selection strategies for ordered predictors. Ordered categories contain more information than unordered categories, but the information has not been used in penalty (6.11). Since now the ordering of dummy coefficients is meaningful, a useful penalty for (0-1)-coding is

$$J(\boldsymbol{\beta}) = \sum_{i=2}^{k} w_i|\beta_i - \beta_{i-1}| = w_2|\beta_2| + w_3|\beta_3 - \beta_2| + \ldots, \tag{6.12}$$

with $\beta_1 = 0$. By putting the L_1-penalty on differences of adjacent categories, the procedure tends to fuse adjacent categories and select groups of categories that may actually be distinguished. Therefore, an ordered categorical predictor with many categories is reduced to a categorical predictor that is formed by the resulting clusters of categories. Typically the number of clusters is much smaller than the original number of categories. For $\lambda \to \infty$, all parameter estimates will become zero and the predictor is excluded since categories cannot be distinguished.

It should be noted that the penalty can also be given in the simpler form of a (weighted) lasso when split-coding of predictors is used. When using the split-coded predictors $\tilde{x}_{A(1)}, \ldots,$

$\tilde{x}_{A(k-1)}$ with $\tilde{x}_{A(i)} = 1$ if $A > i$ and $\tilde{x}_{A(i)} = 0$ otherwise, the corresponding penalty for parameters $\tilde{\beta}_1 = \beta_2$, $\tilde{\beta}_2 = \beta_3 - \beta_2$, $\tilde{\beta}_{k-1} = \beta_k - \beta_{k-1}$ used in the predictor $\eta = \beta_0 + \sum_{i=1}^{k-1} \tilde{x}_{A(i)}\tilde{\beta}_i$ is

$$\sum_{j=1}^{k-1} w_i |\tilde{\beta}_j|,$$

which is a weighted lasso-type penalty. For the transformation between (0-1)-coding and split-coding see also Section 4.4.3.

Both penalties, (6.11) for nominal predictors and (6.12) for ordinal predictors, show good clustering properties and allow one to reduce the number of categories. Fusion methodology goes back at least to Land and Friedman (1997). The penalty for ordered categories is a modification of the fused lasso penalty proposed by Tibshirani et al. (2005) (see also Section 10.4.4). The penalty for nominal predictors was considered by Bondell and Reich (2009) and Gertheiss and Tutz (2010). A further advantage of these penalties is that they have desirable asymptotic properties.

Let us consider nominal factors first. Let $\boldsymbol{\theta} = (\theta_{21}, \theta_{31}, \dots, \theta_{k,k-1})^T$ denote the vector of pairwise differences $\theta_{ij} = \beta_i - \beta_j$. Furthermore, let $\mathcal{C} = \{(i,j) : \beta_i^* \neq \beta_j^*, i > j\}$ denote the set of indices $i > j$ corresponding to differences of (true) dummy coefficients β_i^* that are truly non-zero, and let $\mathcal{C}_n$ denote the set corresponding to those difference that are estimated to be non-zero with sample size n. Let $\boldsymbol{\theta}_{\mathcal{C}}^*$ denote the true vector of pairwise differences included in $\mathcal{C}$, and $\hat{\boldsymbol{\theta}}_{\mathcal{C}}$ the corresponding estimate based on $\hat{\boldsymbol{\beta}}$. Moreover, let weights have the form

$$w_{ij} = \phi_{ij}(n)|\hat{\beta}_i^{(LS)} - \hat{\beta}_j^{(LS)}|^{-1},$$

where $\hat{\beta}_i^{(LS)}$ denotes the ordinary least-squares estimates, and for increasing n one has $\phi_{ij}(n) \to q_{ij}$ $(0 < q_{ij} < \infty)$ for all i, j. If $\lambda = \lambda_n$ with $\lambda_n/\sqrt{n} \to 0$ and $\lambda_n \to \infty$, and all class-wise sample sizes n_i satisfy $n_i/n \to c_i$, where $0 < c_i < 1$, then one obtains for the linear model $\sqrt{n}(\hat{\boldsymbol{\theta}}_{\mathcal{C}} - \boldsymbol{\theta}_{\mathcal{C}}^*) \to_d N(\mathbf{0}, \boldsymbol{\Sigma})$ (for specific matrix $\boldsymbol{\Sigma}$) and $\lim_{n\to\infty} P(\mathcal{C}_n = \mathcal{C}) = 1$. Therefore, asymptotically the right clusters are identified.

A similar property holds for ordered predictors. Now let $\mathcal{C} = \{i > 1 : \beta_i^* \neq \beta_{i-1}^*\}$ denote the set of indices corresponding to the differences of neighboring (true) dummy coefficients β_i^* that are truly non-zero, and again let $\mathcal{C}_n$ denote the set corresponding to those differences that are estimated to be non-zero. The vector of first differences $\delta_i = \beta_i - \beta_{i-1}$, $i = 2, \dots, k$, is now denoted as $\boldsymbol{\delta} = (\delta_2, \dots, \delta_k)^T$. In analogy to the unordered case, let $\boldsymbol{\delta}_{\mathcal{C}}^*$ denote the true vector of (first) differences included in $\mathcal{C}$, and $\hat{\boldsymbol{\delta}}_{\mathcal{C}}$ the corresponding estimate. With weights

$$w_i = \phi_i(n)|\hat{\beta}_i^{(LS)} - \hat{\beta}_{i-1}^{(LS)}|^{-1}$$

and the same conditions for $\phi_i(n)$ and n as for nominal factors, one obtains $\sqrt{n}(\hat{\boldsymbol{\delta}}_{\mathcal{C}} - \boldsymbol{\delta}_{\mathcal{C}}^*) \to_d N(\mathbf{0}, \boldsymbol{\Sigma})$ and $\lim_{n\to\infty} P(\mathcal{C}_n = \mathcal{C}) = 1$ (see Gertheiss and Tutz, 2010.).

6.5.2 Selection of Variables Combined with Clustering of Categories

When several categorical predictors are available, with the lth variable having categories $1, \dots, k_l$ and the corresponding parameter vector $\boldsymbol{\beta}_l^T = (\beta_{l1}, \dots, \beta_{lk_l})$, a combination of variable selection and clustering is obtained by the penalty

$$J(\boldsymbol{\beta}) = \sum_{l=1}^{p} J_l(\boldsymbol{\beta}_l), \tag{6.13}$$

with

$$J_l(\boldsymbol{\beta}_l) = \sum_{i>j} w_{ij}^{(l)} |\beta_{li} - \beta_{lj}|, \quad \text{or} \quad J_l(\boldsymbol{\beta}_l) = \sum_{i} w_{i}^{(l)} |\beta_{li} - \beta_{l,i-1}|,$$

depending on the scale level of the predictor x_l. The first expression refers to nominal covariates, the second to ordinal ones. At first sight the penalty seems to enforce clustering only. However, since $\beta_{l1} = 0$ is fixed for $l = 1, \dots, p$, a predictor is automatically excluded if all of its categories form one cluster. Due to the (additive) form of the penalty, theoretic results generalize to the case of multiple categorial inputs. For ordered categories the grouping into adjacent categories is enforced, while for unordered categories the clustering into not necessarily adjacent categories is enforced. With an appropriately chosen smoothing parameter, one automatically selects variables and the relevant information within variables.

In applications the weight function has to be specified. Bondell and Reich (2009) proposed weights determined by $w_{ij}^{(l)} = (k_l + 1)^{-1}\{(n_i^{(l)} + n_j^{(l)})/n\}^{1/2}$, where $n_i^{(l)}$ denotes the number of observations on level i of predictor x_l.

Example 6.8: Munich Rent

All larger German cities compose so-called rent standards to obtain a decision-making instrument available to tenants, landlords, renting advisory boards, and experts. These rent standards are used in particular for the determination of the local comparative rent. For the composition of the rent standards, a representative random sample is drawn from all relevant households. The data analyzed here come from 2053 households interviewed for the Munich rent standard 2003. The response is monthly rent per square meter in Euro. The predictors are ordered as well as unordered and also include binary factors. They include urban district (nominal, labeled by numbers $1, \dots, 25$), year of construction (ordered classes $[1910, 1919]$, $[1920, 1929], \dots$), number of rooms (taken as an ordinal factor with levels $1, 2, \dots, 6$), quality of residential area (ordinal, with levels "fair," "good," "excellent"), floor space (square meters, given in ordered classes $(0, 30)$, $[30, 40)$, $[40, 50)$, ..., $[140, \infty)$), hot water supply (binary, yes/no), central heating (binary, yes/no), tiled bathroom (binary, yes/no), supplementary equipment in bathroom (binary, yes/no), and well-equipped kitchen (binary, yes/no). The data can be downloaded from the data archive of the Department of Statistics at the University of Munich (http://www.stat.uni-muenchen.de/service/datenarchiv). In order to find relevant variables and identify clusters of categories that have the same effect on the predictor regularized estimates with penalty term (6.13) are computed. With weights chosen by cross-validation, the only predictor that is completely excluded from the model is the binary factor, which indicates if supplementary equipment in the bathroom is available. However, some categories of nominal and ordinal predictors are clustered, for example, houses constructed in the 1930s and 1940s, or urban districts 14, 16, 22, and 24. The original 25 districts of Munich have been reduced to merely 10 categories, which have differing rent levels (see Table 6.2). The regularization paths given in Figure 1.4 show how categories are combined. For the districts, which are treated as nominal, any combination of categories is allowed. For the year of construction ordering over decades is assumed. The regularization paths show how adjacent categories are fused to build clusters. A map of Munich with clusters (Figure 6.18) illustrates the 7 found clusters.

□

6.5.3 Selection of Variables

The selection of whole variables refers to the selection of groups of corresponding dummy variables. Within the framework of boosting, the selection of groups is easily obtained by the use of blockwise boosting, where blocks refer to one categorical predictor. An alternative strategy is the group lasso considered in Section 6.2.2. It tends to select the whole group of

TABLE 6.2: Estimated regression coefficients for Munich rent standard data using adaptive weights with refitting, and (cross-validation score minimizing) $s/s_{\max} = 0.61$.

predictor	label	coefficient
intercept		12.597
urban district	14, 16, 22, 24	−1.931
	11, 23	−1.719
	7	−1.622
	8, 10, 15, 17, 19, 20, 21, 25	−1.361
	6	−1.061
	9	−0.960
	13	−0.886
	2, 4, 5, 12, 18	−0.671
	3	−0.403
year of construction	1920s	−1.244
	1930s, 1940s	−0.953
	1950s	−0.322
	1960s	0.073
	1970s	0.325
	1980s	1.121
	1990s, 2000s	1.624
number of rooms	4, 5, 6	−0.502
	3	−0.180
	2	0.000
quality of residential area	good	0.373
	excellent	1.444
floor space (m^2)	$[140, \infty)$	−4.710
	$[90, 100)$, $[100, 110)$, $[110, 120)$, $[120, 130)$, $[130, 140)$	−3.688
	$[60, 70)$, $[70, 80)$, $[80, 90)$	−3.443
	$[50, 60)$	−3.177
	$[40, 50)$	−2.838
	$[30, 40)$	−1.733
hot water supply	no	−2.001
central heating	no	−1.319
tiled bathroom	no	−0.562
suppl. equipment in bathroom	yes	0.506
well equipped kitchen	yes	1.207

coefficients linked to one categorical predictor. The basic group lasso uses the penalty

$$J(\boldsymbol{\beta}) = \sum_{j=1}^{G} \sqrt{df_j}\|\boldsymbol{\beta}_j\|_2,$$

where $\|\boldsymbol{\beta}_j\|_2 = (\beta_{j1}^2 + \cdots + \beta_{j,df_j}^2)^{1/2}$ and $\boldsymbol{\beta}_j$ denotes the parameter of the jth group from the partitioned predictor $\boldsymbol{x}_i^T = (\boldsymbol{x}_{i1}^T, \ldots, \boldsymbol{x}_{i,G}^T)$. Thus the group of coefficients collected in $\boldsymbol{\beta}_j$ is shrunken (by use of a ridge-type penalty), but there is no selection effect within the group.

For ordered categories one can incorporate smoothing across categories by using the transformation from the preceding section. With (0-1)-coding and the first category as a reference category of a predictor with k_j categories, one replaces the term $\|\boldsymbol{\beta}_j\|_2$ in the penalty by

$$\|\boldsymbol{\beta}_j^T \boldsymbol{K}_j \boldsymbol{\beta}_j\|_2,$$

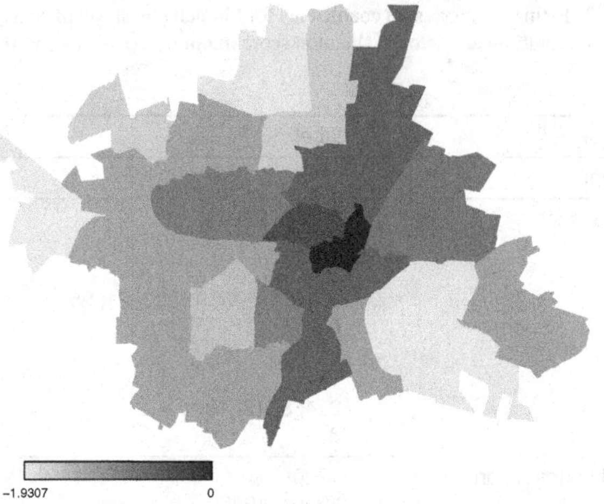

FIGURE 6.18: Map of Munich indicating clusters of urban districts; colors correspond to estimated dummy coefficients from Table 6.2.

where $\boldsymbol{K}_j = \boldsymbol{D}_j^T \boldsymbol{D}_j$ and $\boldsymbol{D}_j$ is the $((k_j - 1) \times (k_j - 1))$-matrix given in equation (4.9) without the first column and $\boldsymbol{\beta}_j^T = (\beta_{j2}, \ldots, \beta_{jk})$. As shown in Section 4.4.3, the penalty can be transformed into a penalty that uses split-coding of covariates. The transformation is helpful because then software designed for the group lasso can be used to fit the model (for more details and examples see Gertheiss et al., 2011).

6.6 Bayesian Approach

In Bayesian approaches to regularization, a prior distribution $p(\theta)$ is specified together with a sampling model $p(\boldsymbol{y}|\theta)$ for observations $\boldsymbol{y}$. Then the updated knowledge after the data have been seen is given by the posterior distribution

$$p(\theta|\boldsymbol{y}) = \frac{p(\boldsymbol{y}|\theta)p(\theta)}{\int p(\boldsymbol{y}|\theta)p(\theta)d\theta} \propto p(\boldsymbol{y}|\theta)p(\theta).$$

In regression modeling, typically the unknown parameter is the vector of coefficients $\boldsymbol{\beta}$ and one obtains

$$p(\boldsymbol{\beta}|\boldsymbol{y}) \propto p(\boldsymbol{y}|\boldsymbol{\beta})p(\boldsymbol{\beta}).$$

The posterior mode estimator, which maximizes the posterior distribution, may be obtained by maximizing the logarithm of the posterior, yielding

$$\text{argmax}_{\boldsymbol{\beta}} = \text{argmax}_{\boldsymbol{\beta}}(l(\boldsymbol{\beta}) + \log(p(\boldsymbol{\beta}))),$$

where $l(\boldsymbol{\beta}) = \log(p(\boldsymbol{y}|\boldsymbol{\beta}))$ is the log-likelihood. Therefore, the posterior mode estimator is equivalent to the penalized likelihood estimator for an appropriately chosen prior distribution.

For the Gaussian prior $\boldsymbol{\beta} \sim N(\boldsymbol{0}, \tau^2 \boldsymbol{I})$, the log-prior has the form of the ridge penalty, which, apart from additive constants, has the form

$$\log(p(\boldsymbol{\beta})) = -\frac{1}{2\tau^2}\boldsymbol{\beta}^T\boldsymbol{\beta}.$$

For the normal distribution linear model with fixed variance σ^2, the posterior is given by

$$\boldsymbol{\beta}|\boldsymbol{y} \sim \mathrm{N}((\boldsymbol{X}^T\boldsymbol{X} + \lambda \boldsymbol{I})^{-1}\boldsymbol{X}^T\boldsymbol{y}, \sigma^2(\boldsymbol{X}^T\boldsymbol{X} + \lambda \boldsymbol{I})^{-1}),$$

where $\lambda = \sigma^2/\tau^2$. The assumption of an iid Laplace prior (also called double-exponential prior), which has density

$$p(\boldsymbol{\beta}) = \prod_{j=1}^{p} \frac{\lambda}{2} e^{-\lambda|\beta_j|}$$

yields, apart from constants, the lasso penalty

$$\log(p(\boldsymbol{\beta})) = -\lambda \sum_{j=1}^{p} |\beta_j|.$$

Therefore, frequentist regularization corresponds to posterior mode estimation if all the other parameters, for example dispersion parameters, are fixed. However, from a Bayesian point of view, maximizing the posterior is not the best way to obtain estimates. A fully Bayesian approach will use the mean or median of the posterior to estimate the coefficient vector. The Bayesian lasso, proposed by Park and Casella (2008), uses posterior median estimates. It does not automatically perform variable selection but provides standard errors and Bayesian credible intervals that can be used to select variables. Park and Casella (2008) used the representation of the double-exponential distribution as a mixture of normals to generate a simple Gibbs sampler. A direct representation of the posterior distribution was used by Hans (2009). An overview of regularization techniques, including bridge regression penalties and elastic net penalties, is given by Fahrmeir and Kneib (2009).

6.7 Further Reading

Alternative Methods and Surveys. One of the first methods to obtain variable selection by shrinkage was the non-negative garotte, proposed by Breiman (1995). Although it is consistent (Zou, 2006), its performance is often poor in highly correlated settings. An overview on regularization in linear models is found in Hastie, Tibshirani, and Friedman (2009). More recently, Bühlmann and Van De Geer (2011) gave a thorough mathematical treatment of regularization methods.

R Packages. Lasso and elastic-net regularized generalized linear models can be fitted with the R package *glmnet*, which allows one to fit Gaussian, binomial, Poisson, and *multinomial* responses (Friedman et al., 2008). An alternative is *glmpath* (Park and Hastie, 2007), which fits models with Gaussian, binomial, and Poisson responses. The package *penalized* (Goeman, 2010) is designed for the Cox model but also fits logit and Poisson distribution models. The group lasso for metric response, the logit, and the log-linear Poisson model can be fitted by use of the R package *grplasso* (Meier et al., 2008). The package *ordPens* is able to handle ordinal predictors. Boosting is available in the packages *mboost* (Hothorn et al., 2009) and *GAMBoost*.

6.8 Exercises

6.1 Consider the simple linear regression model.

(a) Give the lasso and ridge estimators as functions of the ML estimate for a one-dimensional predictor.

(b) Give the lasso and ridge estimators as functions of the ML estimate for an orthogonal design.

6.2 Show by using a second-order Taylor approximation that the log-likelihood of a GLM at the maximum likelihood estimate $\hat{\boldsymbol{\beta}}_{ML}$ can be approximated by $l(\hat{\boldsymbol{\beta}}_{ML}) + \frac{1}{2}(\boldsymbol{\beta} - \hat{\boldsymbol{\beta}}_{ML})^T \boldsymbol{F}(\hat{\boldsymbol{\beta}}_{ML})(\boldsymbol{\beta} - \hat{\boldsymbol{\beta}}_{ML})$, where $\boldsymbol{F}(\hat{\boldsymbol{\beta}}_{ML})$ denotes the Fisher matrix.

6.3 The birth weight data, which have also been considered by Hosmer and Lemeshow (1989) and Venables and Ripley (2002), are available from the R package MASS (dataset birthwt). The data contain 189 observations that were collected at Baystate Medical Center, Springfield, Massachusetts, during 1986. The binary response is an indicator of birth weight less than 2.5 kg. The predictor variables are mother's age in years (age), mother's weight in pounds at last menstrual period (lwt), mother's race (1 = white, 2 = black, 3 = other), smoking status during pregnancy (smoke), number of previous premature labours (ptl), history of hypertension (ht), presence of uterine irritability (ui), and number of physician visits during the first trimester (ftv).

(a) Use regularized estimates to fit a binary regression model; use in particular ML, ridge, lasso, SCAD, elastic net, and boosting. R packages that might be useful are mentioned in the previous section.

(b) Compare the performance of the fitting methods in terms of prediction error by splitting the dataset several times into training data (for fitting) and test data (for evaluation of prediction).

Chapter 7

Regression Analysis of Count Data

In many applications the response variable is given in the form of event counts, where an event count refers to the number of times an event occurs. Simple examples are

- number of insolvent firms within a fixed time internal,
- number of insurance claims within a given period of time,
- number of epileptic seizures per day,
- number of cases with a specific disease in epidemiology.

In all of theses examples the response y may be viewed as a non-negative integer-valued random variable with $y \in \{0, 1, 2, \dots\}$. Although in many applications there is an upper bound for the response, because the number of firms or potential insurance claims is finite, the upper bound is often very large and considered as irrelevant in modeling. In other cases, for example, for the number of epileptic seizures, no upper bound is given. In the following some further examples are given.

Example 7.1: Number of Children

The German General Social Survey Allbus provides micro data, which allow one to model the dependence of the number of children on explanatory variables. We will consider women only and the predictors age in years (age), duration of school education (dur), nationality (nation, 0: German, 1: otherwise), religion (answer categories to "God is the most important in man", 1: strongly agree,..., 5: strongly disagree, 6: never thought about it), university degree (univ, 0: no, 1: yes). □

Example 7.2: Encephalitis

In a study on the occurrence of encephalitis in Central Europe (Karimi et al., 1998), the number of cases of herpes encephalitis in children was observed between 1980 and 1993 in Bavaria and Lower Saxony. Table 7.1 shows the resulting counts. □

The count data in Example 7.2 form a contingency table. Therefore, one might think of using tools for the analysis of contingency tables, a topic that is treated extensively in Chapter 12. Classical analysis of contingency tables treats the rows and columns as factors and thus does not not use the full information available in the potential predictors. It seems more appropriate to model time as a metric variable rather than a qualitative variable. Then the regression problem is determined by the qualitative explanatory variable country and the metric covariate time.

TABLE 7.1: Encephalitis infection in children.

	Bavaria	Lower Saxony
1980	1	2
1981	0	1
1982	1	2
1983	2	5
1984	2	4
1985	3	–
1986	8	–
1987	5	6
1988	13	7
1989	12	7
1990	6	7
1991	13	3
1992	10	4
1993	12	2

The benchmark model for count data is the Poisson distribution. Therefore, we will first consider the Poisson regression model, which can be treated within the framework of generalized linear models. One of the disadvantages of the Poisson distribution is that it is a one parameter distribution. Consequently, the Poisson regression is frequently not flexible enough to adapt to the given data. One step to more flexible models is the inclusion of an overdispersion parameter (Section 7.5). An alternative, more flexible model is the negative binomial model, which can be motivated as a mixture model (Section 7.6). Models for data that show overdispersion through excess zeros are considered in Section 7.7 and Section 7.8.

7.1 The Poisson Distribution

The Poisson distribution is a standard model for count data and was derived as a limiting case of the binomial by Poisson (1837). The discrete random variable Y is Poisson-distributed with intensity or rate parameter λ, $\lambda > 0$, if the density is given by

$$P(Y=y) = \begin{cases} \frac{\lambda^y}{y!} e^{-\lambda} & \text{for } y \in \{0,1,2,\dots\} \\ 0 & \text{otherwise.} \end{cases} \tag{7.1}$$

An abbreviation is $Y \sim \text{P}(\lambda)$. Figure 7.1 shows several examples of densities of Poisson-distributed variables. A closer look at the form of the densities is obtained by considering the proportion of probabilities for counts y and $y-1$ $(y \geq 1)$ given by

$$\frac{P(Y=y)}{P(Y=y-1)} = \frac{\lambda^y e^{-\lambda}/y!}{\lambda^{y-1} e^{-\lambda}/(y-1)!} = \frac{\lambda}{y}.$$

If $\lambda < 1$ one has $\lambda/y < 1$ and therefore the density is decreasing across integers, the largest probability occurs at $y = 0$. If $\lambda > 1$, the probabilities are increasing up to the integer value of λ, $[\lambda]$; for $y > [\lambda]$ the density is decreasing. Thus, for non-integer-valued λ the density is unimodal with the mode given by $[\lambda]$. If λ is integer-valued, the probabilities $P(Y=\lambda)$ and $P(Y=\lambda-1)$ are equal.

The first two central moments of the Poisson distribution are given by $\text{E}(Y) = \text{var}(Y) = \lambda$. Equality of the mean and variances is often referred to as the *equidispersion property* of the Poisson distribution. Thus, in contrast to the normal distribution, for which the mean and variance are unlinked, the Poisson distribution implicitly models stronger variability for larger

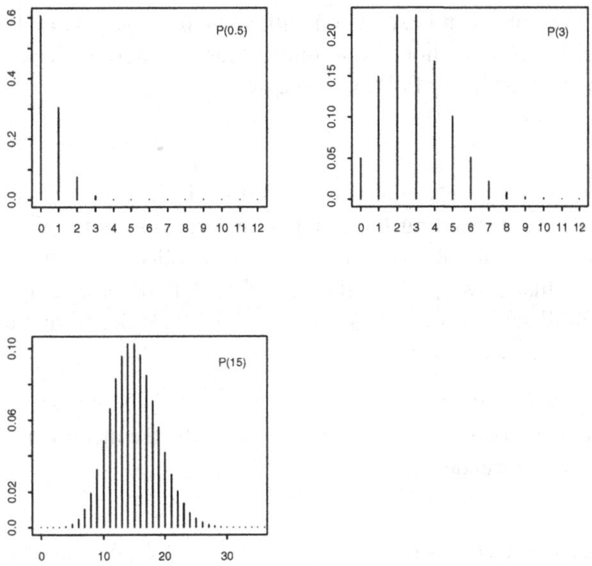

FIGURE 7.1: Probability mass functions of Poisson distributions with $\lambda = 0.5, 3, 15$.

means, a property that is often found in real-life data. On the other hand, in real-life data one frequently finds that the variance exceeds the mean, and the effect is overdispersion, which has to be modeled separately (see Section 7.5).

In the following some connections between distributions and some additional properties of the Poisson distribution are given. The connections between distributions help us to get a clearer picture of the range of applications of the Poisson distribution.

Poisson Distribution as the Law of Rare Events

The Poisson distribution may be obtained as a limiting case of the binomial distribution. Let Y denote the total number of successes in a large number n of independent Bernoulli trials with the successes probability π being small and linked to the number of trials by $\pi = \lambda/n$. As an example, one may consider the number of incoming telephone calls in a fixed time internal of unit length. Let λ denote the fixed mean number of calls. Now consider the division of the time interval into n subintervals with equal width. For small intervals, each interval may be considered as one trial with the success being defined as an incoming call. Then the number of total incoming calls can be modeled by a binomial distribution with the probability specified by λ/n. Then it may be shown that for increasing n the binomial distribution becomes the Poisson distribution.

More formally, let Y have a binomial distribution with parameter n and $\pi = \lambda/n, Y \sim$ $\mathrm{B}(n, \pi = \lambda/n)$. Then one has

$$\lim_{\substack{n\pi=\lambda \\ n\to\infty}} \binom{n}{y} \pi^y (1-\pi)^{n-y} = \frac{\lambda^y}{y!} e^{-\lambda}.$$

The law of rare events refers to this derivation from the binomial distribution, where the number of trials increases while the probability of success decreases correspondingly. However, the term is somewhat misleading because the mean λ may be arbitrarily large. The Poisson distribution is not restricted to small values of the mean.

The phone calls example is an idealization with some missing details. For example, in one time subinterval more than one phone call could occur. A more concise derivation for this example is obtained by considering Poisson processes.

Poisson Process

The Poisson distribution is closely linked to the Poisson process. Let $\{N(t), t \geq 0\}$ be a counting process with $N(t)$ denoting the event counts up to time t. $N(t)$ is a non-negative and integer-valued random variable, and the process is a collection of these random variables satisfying the property that $N(s) \leq N(t)$ if $s < t$. The Poisson process is a specific counting process that has to fulfill several properties. With $N(t, t + \Delta t)$ denoting the number of counts in interval $(t, t + \Delta t)$, one postulates

(a) *Independence of intervals*
For disjunct time intervals $(s, s + \Delta s)$ and $(t, t + \Delta t)$, the increments $N(s, s + \Delta s)$ and $N(t, t + \Delta t)$ are independent.

(b) *Stationarity*
The distribution of the counts in the interval $(t, t + \Delta t)$ depends only on the length of the interval Δt (it does not depend on t).

(c) *Intensity rate*
The probability of no or one event in interval $(t, t + \Delta t)$ is given by

$$P(N(t, t + \Delta t) = 1) = \lambda \Delta t + o(\Delta t),$$
$$P(N(t, t + \Delta t) = 0) = 1 - \lambda \Delta t + o(\Delta t),$$

where $o(h)$ denotes a remainder term with the property $o(h)/h \to 0$ as $h \to 0$.

For this process, the number of events occurring in the interval $(t, t + \Delta t)$ is Poisson-distributed with mean $\lambda \Delta t$:

$$N(t, t + \Delta t) \sim \mathrm{P}(\lambda \Delta t).$$

In particular, one has $N(t) = N(0, \Delta t) \sim \mathrm{P}(\lambda \Delta t)$. To obtain a Poisson distribution in the phone call example one has to postulate not only that the probability of occurrence in one time interval depends only on the length of the interval but also the independence of counts in intervals, stationarity, and that the probability that no or one event has a specific form in the limit.

The Poisson process is also strongly connected to the exponential distribution. If the Poisson process is a valid model, the waiting time between events follows an exponential distribution. It is immediately seen that for the waiting time for the first event, W_1, the outcome $W_1 > t$ occurs if no events occur in the interval $(0, t)$:

$$P(W_1 > t) = P(N(0, t) = 0) = \mathrm{e}^{-\lambda t}.$$

Therefore, W_1 follows the exponential distribution with parameter λ. The same distribution may be derived for the waiting time between any two events. Alternative characterizations of the Poisson distributions are considered, for example, in Cameron and Trivedi (1998).

Further Properties

(1) Sums of independent Poisson-distributed random variables are Poisson-distributed. More concrete, if $Y_i \sim \mathrm{P}(\lambda_i), i = 1, 2, \ldots$, are independent and $\sum_i \lambda_i < \infty$, then $\sum_i Y_i \sim \mathrm{P}(\sum_i \lambda_i)$.

(2) There is a strong connection between the Poisson and the multinomial distribution. If $Y_1, \ldots, Y_N$ are independent Poisson variables, $Y_i \sim \mathrm{P}(\lambda_i)$, and one conditions on the total sum $n_0 = Y_1, + \cdots + Y_N$, one obtains for $(Y_1, \ldots, Y_N)$ *given* n_0 the multinomial distribution $\mathrm{M}(n_0, (\pi_1, \ldots, \pi_N))$ with $\pi_i = \lambda_i/(\lambda_1 + \cdots + \lambda_N)$.

(3) For large values of λ the Poisson distribution $\mathrm{P}(\lambda)$ may be approximated by the normal distribution $\mathrm{N}(\lambda, \lambda)$. It is seen from Figure 7.1 that for small λ the distribution is strongly skewed. For larger values of λ the distribution becomes more symmetric. For details of approximations see, for example, McCullagh and Nelder (1989), Chapter 6.

7.2 Poisson Regression Model

The Poisson regression model is the standard model for count data. Let $(y_i, \boldsymbol{x}_i)$ denote n independent observations and $\mu_i = \mathrm{E}(y_i|\boldsymbol{x}_i)$. One assumes that $y_i|\boldsymbol{x}_i$ is Poisson-distributed with mean μ_i, and that the mean is determined by

$$\mu_i = h(\boldsymbol{x}_i^T \boldsymbol{\beta}) \quad or \quad g(\mu_i) = \boldsymbol{x}_i^T \boldsymbol{\beta}, \tag{7.2}$$

where g is a known link function and $h = g^{-1}$ denotes the response function. Since the Poisson distribution is from the simple exponential family, the model is a generalized linear model. The most widely used model uses the canonical link function by specifying

$$\mu_i = \exp(\boldsymbol{x}_i^T \boldsymbol{\beta}) \quad \text{or} \quad \log(\mu_i) = \boldsymbol{x}_i^T \boldsymbol{\beta}.$$

Since the logarithm of the conditional mean is linear in the parameters, the model is called a *log-linear* model.

The log-linear version of the model is particulary attractive because interpreting the parameters is very easy. The model implies that the conditional mean given $\boldsymbol{x}^T = (x_1, \ldots, x_p)$ has a multiplicative form given by

$$\mu(\boldsymbol{x}) = \exp(\boldsymbol{x}^T \boldsymbol{\beta}) = \mathrm{e}^{x_1 \beta_1} \ldots \mathrm{e}^{x_p \beta_p}.$$

Thus e^{β_j} represents the multiplicative effect on $\mu(\boldsymbol{x})$ if the variable x_j changes by one unit to $x_j + 1$ (given that the rest of the variables are fixed). One obtains

$$\frac{\mu(x_1, \ldots, x_j + 1, \ldots, x_p)}{\mu(x_1, \ldots, x_j, \ldots, x_p)} = \mathrm{e}^{\beta_j}$$

or, equivalently,

$$\log \mu(x_1, \ldots, x_j + 1, \ldots, x_p) - \log \mu(x_1, \ldots, x_j, \ldots, x_p) = \beta_j.$$

While β_j is the change in log-means if x_j increases by one unit, e^{β_j} is the multiplicative effect, which is easier to interpret because it directly effects upon the mean.

For illustration, let us consider Example 1.5, where the dependent variable is the number of insolvent firms and there is only one covariate, namely, time. The log-linear Poisson model

$$\log(\mu) = \beta_0 + \text{time}\beta$$

specifies the number of insolvent firms in dependence on time (1 to 36 for January 1994 to December 1996). One obtains the estimates $\hat{\beta}_0 = 4.25$ and $\hat{\beta} = 0.0097$, yielding $e^{\hat{\beta}} = 1.01$. Therefore, the log-mean increases additively by 0.0097 every month or, more intuitively, the mean increases by the factor 1.01 every month.

The canonical link has the additional advantage that the mean $\mu(\boldsymbol{x})$ is always positive whatever values the (estimated) parameters take. For alternative models like the linear model, $\mu_i = \boldsymbol{x}_i^T\boldsymbol{\beta}$ problems may occur when the mean is predicted for new observations $\boldsymbol{x}_i$, although μ_i may take admissible values in the original sample. Therefore, in most applications one uses the log-link. Nevertheless, the results can be misleading if the link is grossly misspecified. When nothing is known about the link one can also use non-parametric approaches to link specification (see Section 5.2 and Section 6.4).

7.3 Inference for the Poisson Regression Model

Maximum Likelihood Estimation

For inference, the whole machinery of generalized linear models may be used. For model (7.2) one obtains the log-likelihood

$$l(\boldsymbol{\beta}) = \sum_{i=1}^{n} y_i \log(\mu_i) - \mu_i - \log(y_i!) = \sum_{i=1}^{n} y_i \log(h(\boldsymbol{x}_i^T\boldsymbol{\beta})) - h(\boldsymbol{x}_i^T\boldsymbol{\beta}) - \log(y_i!).$$

The score function $\boldsymbol{s}(\boldsymbol{\beta}) = \partial l(\boldsymbol{\beta})/\partial\boldsymbol{\beta}$ is given by

$$\boldsymbol{s}(\boldsymbol{\beta}) = \sum_{i=1}^{n} \boldsymbol{x}_i \frac{h'(\boldsymbol{x}_i^T\boldsymbol{\beta})}{h(\boldsymbol{x}_i\boldsymbol{\beta})}(y_i - h(\boldsymbol{x}_i^T\boldsymbol{\beta})),$$

and the Fisher matrix is given by

$$\boldsymbol{F}(\boldsymbol{\beta}) = \mathrm{E}(-\partial^2 l(\boldsymbol{\beta})/\partial\boldsymbol{\beta}\partial\boldsymbol{\beta}^T) = \sum_{i=1}^{n} \boldsymbol{x}_i\boldsymbol{x}_i^T \frac{h'(\boldsymbol{x}_i^T\boldsymbol{\beta})^2}{h(\boldsymbol{x}_i^T\boldsymbol{\beta})}. \tag{7.3}$$

For the canonical link $h(\eta) = \exp(\eta)$, these terms simplify to

$$l(\boldsymbol{\beta}) = \sum_{i=1}^{n} y_i\boldsymbol{x}_i^T\boldsymbol{\beta} - \exp(\boldsymbol{x}_i^T\boldsymbol{\beta}) - \log(y_i!),$$

$$\boldsymbol{s}(\boldsymbol{\beta}) = \sum_{i=1}^{n} \boldsymbol{x}_i(y_i - \exp(\boldsymbol{x}_i^T\boldsymbol{\beta})),$$

$$\boldsymbol{F}(\boldsymbol{\beta}) = \sum_{i=1}^{n} \boldsymbol{x}_i\boldsymbol{x}_i^T \exp(\boldsymbol{x}_i^T\boldsymbol{\beta}).$$

Under regularity conditions, $\hat{\boldsymbol{\beta}}$ defined by $\boldsymbol{s}(\hat{\boldsymbol{\beta}}) = \boldsymbol{0}$ is consistent and asymptotically normal distributed:

$$\hat{\boldsymbol{\beta}} \overset{(a)}{\sim} \mathrm{N}(\boldsymbol{\beta}, \boldsymbol{F}(\boldsymbol{\beta})^{-1}),$$

where $\boldsymbol{F}(\boldsymbol{\beta})$ may be replaced by $\boldsymbol{F}(\hat{\boldsymbol{\beta}})$ to obtain standard errors.

Deviance and Goodness-of-Fit

The deviance as a measure of discrepancy between the fit and the data compares the log-likelihood of the fitted value for observation y_i, denoted by $l_i(\hat{\mu}_i) = y_i \log(\hat{\mu}_i) - \hat{\mu}_i - \log(y_i!)$, to the log-likelihood of the perfect fit $l_i(y_i) = y_i \log(y_i) - y_i - \log(y_i!)$, yielding

$$D = -2\sum_i l_i(\hat{\mu}_i) - l_i(y_i) = 2\sum_i \{y_i \log\left(\frac{y_i}{\hat{\mu}_i}\right) + [(\hat{\mu}_i - y_i)]\}.$$

If an intercept is included, the term in brackets, $\hat{\mu}_i - y_i$, may be omitted. Within the framework of GLMs the deviance is used as a goodness-of-fit statistic with a known asymptotic distribution when the observations are grouped. Let $y_{i1}, \dots, y_{in_i}$, $i = 1, \dots, N$, denote independent observations at a fixed measurement point $\boldsymbol{x}_i$ with $y_{it} \sim \mathrm{P}(\tilde{\mu}_i), \tilde{\mu}_i = h(\boldsymbol{x}_i^T \boldsymbol{\beta})$. Then $y_i = n_i \bar{y}_i = \sum_{t=1}^{n_i} y_{it} \sim \mathrm{P}(n_i \tilde{\mu}_i)$, which may be written as $y_i \sim \mathrm{P}(\mu_i)$, where $\mu_i = n_i \tilde{\mu}_i$. When using μ_i, the deviance for grouped observations has the same form as for single observations:

$$D = 2 \sum_{i=1}^{N} \{ y_i \log \left(\frac{y_i}{\hat{\mu}_i} \right) + [(\hat{\mu}_i - y_i)] \}. \tag{7.4}$$

D is asymptotically χ^2-distributed with $N - p$ degrees of freedom, where p is the dimension of the parameter vector. The underlying asymptotic concept is fixed cells asymptotic, where N is fixed and $n_i \to \infty$ for all i. In the form (7.4), where n_i is only implicitly contained in $\mu_i = n_i \tilde{\mu}_i$, one assumes $\mu_i \to \infty$. The assumption $\mu_i \to \infty$ is slightly more general because one does not have to assume that y_i is composed from repeated measurements. As an alternative goodness-of fit statistic one may also use the Pearson statistic considered in Chapter 3. The specific form is given in the box.

Goodness-of-Fit for Poisson Regression Model

$$D = 2 \sum_{i=1}^{N} y_i \log \left(\frac{y_i}{\hat{\mu}_i} \right)$$

$$\chi_P^2 = \sum_{i=1}^{N} \frac{(y_i - \hat{\mu}_i)^2}{\hat{\mu}_i}$$

For $\mu_i \to \infty$ one obtains the approximation

$$D, \chi_P^2 \sim \chi^2(N - p)$$

The use of the deviance and the Pearson statistic depends on whether asymptotic results apply. Usually one expects all of the means to be larger than three. Fienberg (1980) showed that the approximation might work even if for a small percentage the mean is only one; see also Read and Cressie (1988). If D and χ_P^2 are quite different, one might always suspect that the approximation is inadequate.

Testing of Hierarchical Models

The deviance may also be used to test hierarchical models $\tilde{M} \subset M$, where $\tilde{M}$ is determined by a linear hypothesis $\boldsymbol{C\beta} = \boldsymbol{\xi}$. The difference between deviances for the fit of model $\tilde{M}$ (yielding $\tilde{\boldsymbol{\beta}}$ and $\tilde{\mu}_i$) and model M (yielding $\hat{\boldsymbol{\beta}}$ and $\hat{\mu}_i$) is

$$D(\tilde{M}|M) = 2 \sum_i y_i \log(\frac{\hat{\mu}_i}{\tilde{\mu}_i}) + (\hat{\mu}_i - \tilde{\mu}_i),$$

which has asymptotically a χ^2-distribution with the degrees of freedom given by the rank of $\boldsymbol{C}$.

TABLE 7.2: Parameter estimates and log-linear Poisson model for number of children.

	Estimate	Std. Error	z-Value	Pr(>|z|)
Intercept	−12.280	1.484	−8.27	0.0000
age	0.935	0.124	7.55	0.0000
age^2	−0.025	0.004	−6.57	0.0000
age^3	0.000	0.000	5.78	0.0000
age^4	−0.000	0.000	−5.14	0.0000
dur	0.112	0.067	1.68	0.0929
dur^2	−0.008	0.003	−2.77	0.0054
nation	0.056	0.138	0.41	0.6816
god2	−0.010	0.059	−1.73	0.0826
god3	−0.144	0.068	−2.14	0.0327
god4	−0.128	0.071	−1.80	0.0711
god5	−0.036	0.067	−0.54	0.5886
god6	−0.092	0.075	−1.23	0.2182
univ	0.637	0.173	3.68	0.0002

TABLE 7.3: Deviances for Poisson model, number of children as response.

	DF	Difference	DF	Deviance
All effects			1747	1718.6
age	4	215.5	1751	1940.7
dur	2	40.3	1749	1758.9
nationality	1	0.2	1748	1718.8
religion	5	6.7	1752	1725.3
univ	1	13.5	1748	1732.1

Example 7.3: Number of Children

For the data described in Example 7.1 a log-linear Poisson model with the number of children as the dependent variable has been fitted. Table 7.2 shows the estimates. Since it is hardly to be expected that the metric predictors age and duration of school education have a linear effect, the polynomial terms are included in the predictor, which turn out to be highly significant. The only variable that seems to be not influential is nationality. However, the analysis of deviance, given in Table 7.3, shows that the effect of the variable religion is also not significant. The table gives the deviance of the model that contains all the predictors and the differences between deviances when one predictor is omitted. Since the effects of polynomial terms are hard to see from the estimates, the effects are plotted in Figure 7.2, with all other covariates considered as fixed. It is seen that especially for women between 20 and 40, age makes a difference as far as the expected number of children is concerned. The duration of school years shows that unfortunately the time spent in school decreases the number of children. For a more flexible modelling see Example 10.4, Chapter 10. □

Example 7.4: Encephalitis

For the encephalitis dataset, the number of infections is modeled in dependence on country (BAV=1: Bavaria, BAV=0: Lower Saxony) and TIME (1–14, corresponding to 1980–1993). The compared models are the log-linear Poisson model, the normal distribution model with log-link and the identity link. Table 7.4 shows the fits with an interaction effect between country and time. When using a normal distribution model, the log-linear model is to be preferred because its log-likelihood is larger. Comparison across distributions cannot be recommended because log-likelihoods are not comparable. The points in

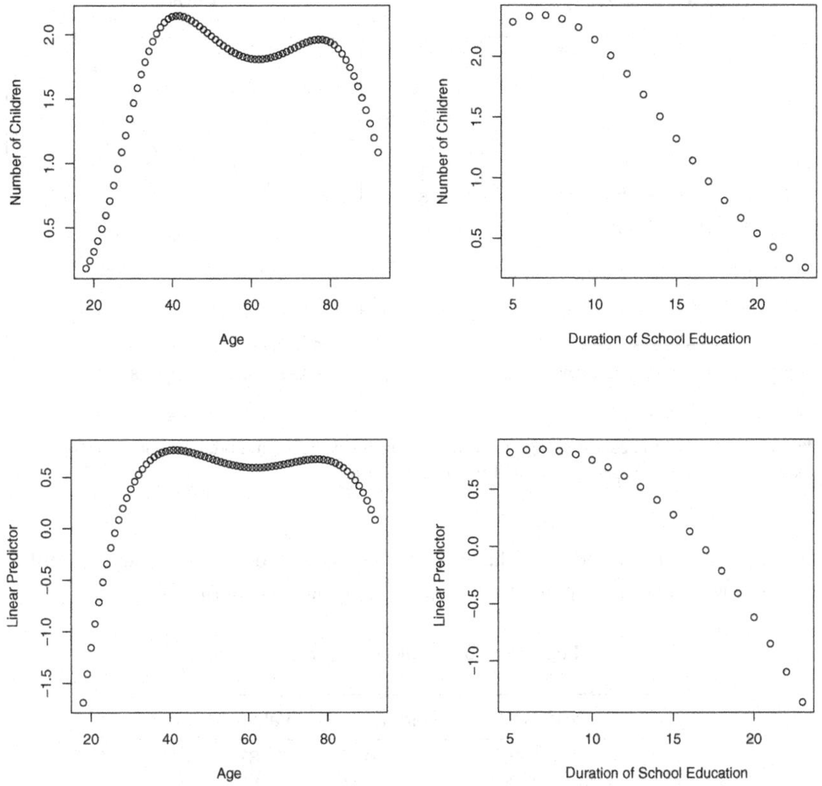

FIGURE 7.2: Number of children versus age and duration of education

TABLE 7.4: Models for encephalitis data.

	Log-Linear Poisson Model		Log-Linear Normal Model		Linear Normal Model	
	Estimate	p-Value	Estimate	p-value	Estimate	p-Value
Intercept	-0.255	0.622	-0.223	0.705	0.397	0.815
TIME	0.513	0.000	0.499	0.0002	1.154	0.014
TIME^2	-0.030	0.0001	-0.029	0.0002	-0.065	0.030
BAV	-1.587	0.006	-1.478	0.017	-4.414	0.014
BAV.TIME	0.211	0.003	0.198	0.001	0.853	0.000
Log-likelihood	-47.868		-51.398		-54.905	

favor of the Poisson model are that data are definitely discrete and the equidispersion property of the Poisson model. It is seen from Figure 7.3 that large means tend to have larger variability. □

Example 7.5: Insolvent Firms

For the number of insolvent firms between 1994 and 1996 (see Example 1.5) a log-linear Poisson model is fitted with time as the predictor. Time is considered as a number from 1 to 36, denoting months, starting with January 1994 and ending with December 1996. Since the counts are not too small, one might also fit a model that assumes normally distributed responses. The models that are compared are the log-linear model $\log(\mu) = \beta_0 + x\beta_1$ and the model $\log(\mu) = \beta_0 + x\beta_1 + x^2\beta_2$ with $x \in \{1, \ldots, 36\}$.

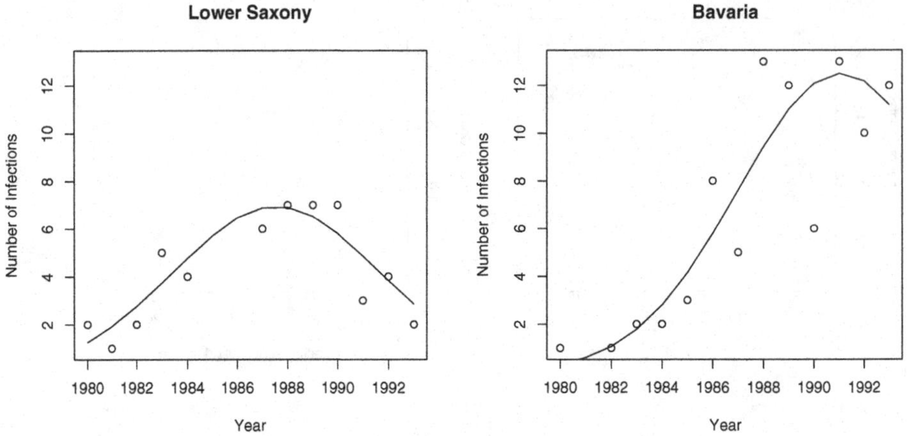

FIGURE 7.3: Estimated means against time for log-linear Poisson model for encephalitis data, Lower Saxony (upper panel) and Bavaria (lower panel).

The results given in Table 7.5 show that the quadratic terms seem to be unnecessary. Nevertheless, in Figure 7.4, which shows the mean response against months, the quadratic term is included. □

TABLE 7.5: Log-linear Poisson models for insolvency data.

	Estimate	Standard Error	z-Value
β_0	4.192	0.062	67.833
β_1	0.020	0.007	2.677
β_2	−0.00027	0.00019	−1.408

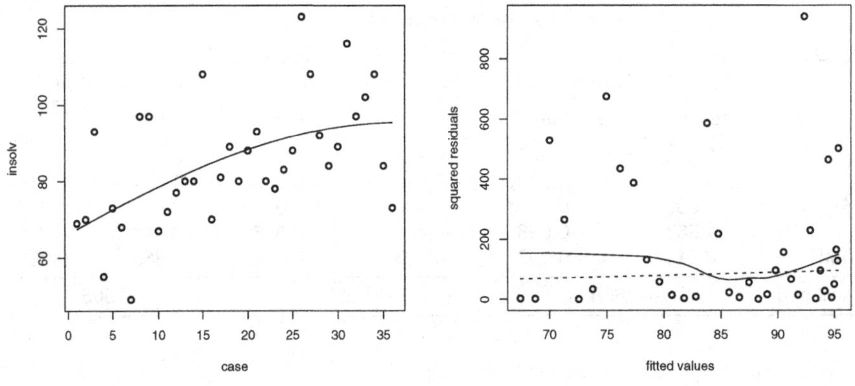

FIGURE 7.4: Log-linear model for insolvency data plotted against months (left) and squared residuals (right) against fitted values.

7.4 Poisson Regression with an Offset

In the standard log-linear Poisson model it is assumed that the log-mean of the response depends directly on the covariates in linear form, $\log(\mu_i) = \boldsymbol{x}_i^T\boldsymbol{\beta}$. In many applications, the response results from different levels of aggregation, and it is more appropriate to model the underlying driving force. For example, in epidemiology, if the incidence of infectious diseases is studied, count data may refer to geographical districts with varying population sizes. Thus the number of people at risk has to be taken into account. The same effect is found if the counts are observed in different time intervals. Let us consider the latter case. Then, one can use the strong connection between the Poisson distribution and the Poisson process as described in Section 7.1. By assuming a Poisson process with intensity rate λ, one obtains for the counts in intervals of length Δ the Poisson distribution $y \sim \mathrm{P}(\lambda\Delta)$. Consequently, the mean $\mu = \lambda\Delta$ depends on the length of the interval. Let data be given by $(y_i, \boldsymbol{x}_i)$, where y_i denotes the counts in intervals of length Δ_i. If counts arise from a Poisson process with intensity rate λ_i (depending on $\boldsymbol{x}_i$), one obtains $y_i \sim \mathrm{P}(\Delta_i\lambda_i)$. If the dependence of the intensity rate is modeled in log-linear form, $\log(\lambda_i) = \boldsymbol{x}_i^T\boldsymbol{\beta}$, one obtains for mean counts $\mu_i = \mathrm{E}(y_i|\boldsymbol{x}_i)$

$$\log(\mu_i) = \log(\Delta_i) + \boldsymbol{x}_i^T\boldsymbol{\beta} \tag{7.5}$$

or, equivalently,

$$\mu_i = \exp(\log(\Delta_i) + \boldsymbol{x}_i^T\boldsymbol{\beta}).$$

Therefore, the specification of the driving force, represented by λ_i, yields a model with a fixed term $\log(\Delta_i)$ in it. For an application, see Santner and Duffy (1989).

A similar form of the model results if one has different levels of aggregation. Let independent responses $y_{i1}, \dots, y_{in_i}$ be observed for fixed values of explanatory variables $\boldsymbol{x}_i$. If $y_{it} \sim \mathrm{P}(\lambda_i)$, one obtains for the sum of responses

$$y_i = \sum_{t=1}^{n_i} y_{it} = n_i\bar{y}_i \sim \mathrm{P}(n_i\lambda_i),$$

where $\bar{y}_i = y_i/n$. Since the local sample sizes n_i vary across measurements, one obtains with $\log(\lambda_i) = \boldsymbol{x}_i^T\boldsymbol{\beta}$ for the mean responses $\mu_i = \mathrm{E}(y_i|\boldsymbol{x}_i)$

$$\log(\mu_i) = \log(n_i) + \boldsymbol{x}_i^T\boldsymbol{\beta}. \tag{7.6}$$

The number n_i can be given implicitly, for example, as the population size in a given geographical district when one looks at the number of cases of a specific disease. From the form $\log(\mu_i/n_i) = \boldsymbol{x}_i^T\boldsymbol{\beta}$ it is seen that the appropriately standardized rate of occurrence is modeled rather than the number of cases itself.

Models (7.5) and (7.6) have the general form

$$\log(\mu_i) = \gamma_i + \boldsymbol{x}_i^T\boldsymbol{\beta},$$

where γ_i is a known parameter $\gamma_i = \log(\Delta_i)$ for the varying length of time intervals, and $\gamma_i = \log(n_i)$ for varying sample sizes. The parameter γ_i may be treated as an offset that remains fixed across interactions. The log-likelihood and the score function are given by $l(\boldsymbol{\beta}) = \sum_i y_i \log(\gamma_i + \boldsymbol{x}_i^T\boldsymbol{\beta}) - \exp(\gamma_i + \boldsymbol{x}_i^T\boldsymbol{\beta}) - \log(y_i!)$ and $\boldsymbol{s}(\boldsymbol{\beta}) = \sum_i \boldsymbol{x}_i(y_i - \exp(\gamma_i + \boldsymbol{x}_i^T\boldsymbol{\beta}))$. The Fisher matrix has the form

$$\boldsymbol{F}(\boldsymbol{\beta}) = \sum_{i=1}^{N} \boldsymbol{x}_i\boldsymbol{x}_i^T \exp(\gamma_i + \boldsymbol{x}_i^T\boldsymbol{\beta})) = \sum_{i=1}^{N} \boldsymbol{x}_i\boldsymbol{x}_i^T \exp(\boldsymbol{x}_i^T\boldsymbol{\beta})\exp(\gamma_i).$$

It is directly seen how γ_i determines the accuracy of the estimates. Since the ML estimate $\hat{\boldsymbol{\beta}}$ has approximate covariance $cov(\hat{\boldsymbol{\beta}}) \approx \boldsymbol{F}(\hat{\boldsymbol{\beta}})^{-1}$, standard errors decrease with increasing parameters γ_i. As is to be expected, larger time intervals or larger sample sizes yield better estimates.

7.5 Poisson Regression with Overdispersion

In many applications count data are overdispersed, with the conditional variance exceeding the conditional mean. One cause for this can be unmodeled heterogeneity among subjects. In the following several modeling approaches that account for overdispersion are considered. The first one is based on quasi-likelihood, and the second one models the heterogeneity among subjects explicitly. A specific model that models heterogeneity explicitly is the Gamma-Poisson or negative binomial model considered in Section 7.6. Also, models for excess zeros like the zero-inflated model (Section 7.7) and the hurdle model (Section 7.8) imply overdispersion.

7.5.1 Quasi-Likelihood Methods

Maximum likelihood estimates are based on the assumption $\mu_i = h(\boldsymbol{x}_i^T\boldsymbol{\beta})$ and $y_i|\boldsymbol{x}_i \sim \mathrm{P}(\mu_i)$. The estimates are obtained by setting the score function equal to zero. These assumptions may be weakened within the quasi-likelihood framework (see Section 3.11, Chapter 3). The link between the mean and the linear predictor has the usual GLM form $\mu_i = h(\boldsymbol{x}_i^T\boldsymbol{\beta})$, but instead of assuming a fixed distribution for y_i only a mean–variance relationship is assumed. The ML estimation equation $\boldsymbol{s}(\hat{\boldsymbol{\beta}}) = \mathbf{0}$ has the general form

$$\sum_{i=1}^{n} \boldsymbol{x}_i \frac{\partial \mu_i}{\partial \eta} \frac{y_i - \mu_i}{\sigma_i^2} = \mathbf{0}, \tag{7.7}$$

where $\mu_i = h(\eta_i)$ and σ_i^2 is the variance that has the form $\sigma_i^2 = \phi v(\mu_i)$ with variance function $v(\mu_i)$. Since ϕ cancels out, the estimation depends only on the specification of the mean and the variance function. For the Poisson distribution, the latter has the form $v(\mu_i) = \mu_i$. For alternative variance functions, which do not necessarily correspond to a Poisson distribution, (7.7) is considered as the estimation equation yielding quasi-likelihood estimates.

Model with Overdispersion Parameter

A simple quasi-likelihood approach uses the variance function $v(\mu_i) = \mu_i$, which yields the variance $\sigma_i^2 = \phi\mu_i$ for some unknown constant ϕ. The case $\phi > 1$ represents the *overdispersion* of the Poisson model, and the case $\phi < 1$, which is rarely found in applications, is called the *underdispersion*. If $\sigma_i^2 = \phi\mu_i$ is used in (7.7), ϕ drops out. Thus the estimation equation is identical to the likelihood equation for Poisson models. Consequently, parameter estimates are identical. However, the variance is inflated by overdispersion, since one obtains the asymptotic covariance

$$\mathrm{cov}(\hat{\boldsymbol{\beta}}) = \phi \boldsymbol{F}(\boldsymbol{\beta})^{-1},$$

with $\boldsymbol{F}(\boldsymbol{\beta})$ denoting the Fisher matrix from equation (7.3). Wedderburn (1974) proposed estimating the dispersion parameter by

$$\hat{\phi} = \frac{1}{n-p} \sum_{i=1}^{n} \frac{(y_i - \hat{\mu}_i)^2}{\hat{\mu}_i},$$

where p is the number of model parameters and $n - p$ is a degrees-of-freedom correction. The motivation for this estimator is that the variance function $v(\mu_i) = \mu_i$ implies $\mathrm{E}(y_i - \mu_i)^2 = \phi\mu_i$ and hence $\phi = \mathrm{E}((y_i - \mu_i)^2/\mu_i)$. $\hat{\phi}$ is motivated as a moment estimator with a degrees-of-freedom correction. There is a strong connection to the Pearson statistic, because $\hat{\phi} = \chi_P^2/(n-p)$. The approximation $\mathrm{E}(\chi_P^2) \approx n - p$ holds if $\sigma^2 = \mu$ is the underlying variance. For $\sigma^2 = \phi\mu$, one has $\mathrm{E}(\chi_P^2/\phi) \approx n - p$ and therefore $\mathrm{E}(\chi_P^2/(n-p)) \approx \phi$.

In summary, the variance $\sigma^2 = \phi\mu$ is easy to handle. One fits the usual Poisson model and uses the ML estimate. To obtain the correct covariance matrix of $\hat{\beta}$ one multiplies the maximum likelihood covariance by $\hat{\phi}$. Maximum likelihood standard errors are multiplied by $\sqrt{\hat{\phi}}$ and t-statistics are divided by $\sqrt{\hat{\phi}}$.

Alternative Variance Functions

Alternative variance functions usually continue to model the variance as a function of the mean. A general variance function that is in common use has the form

$$v(\mu_i) = \mu_i + \gamma\mu_i^m$$

for a fixed value of m. The choice $m = 2$ corresponds to the assumption of the negative binomial distribution (see Section 7.6) while the choice $m = 1$ yields $v(\mu_i) = (1+\gamma)\mu_i$. Hence, the case $m = 1$ is equivalent to assuming $v(\mu_i) = \phi\mu_i$. Breslow (1984) used the negative binomial type variance within a quasi-likelihood approach.

Example 7.6: Demand for Medical Care

Deb and Trivedi (1997) analyzed the demand for medical care for individuals, aged 66 and over, based on a dataset from the U.S. National Medical Expenditure survey in 1987/88. The data are available from the archive of the *Journal of Applied Econometrics* and the *Journal of Statistical Software*; see also Kleiber and Zeileis (2008), and Zeileis et al. (2008). Like Zeileis et al. (2008) we consider the number of physician/non-physician office and hospital outpatient visits (ofp) as dependent variable. The regressors used in the present analysis are the number of hospital stays (hosp), self-perceived health status (poor, medium, excellent), number of chronic conditions (numchron), age, marital status, and number of years of education (shool). Since the effects vary across gender, only male patients are used in the analysis. Table 7.6 shows the fits of a log-linear Poisson model without and with overdispersion (residual deviance is 9665.7 on 1770 degrees of freedom). With an estimated overdispersion parameter $\hat{\phi} = 7.393$ the data are highly overdispersed. The negative binomial model (Table 7.8) shows similar effects but slightly smaller standard errors ($\hat{\nu} = 1.079$, with standard error 0.048, and the residual deviance is -9607.73). The Poisson model yields the log-likelihood value -7296.398 ($df = 8$), and the negative binomial model, which uses just one more parameter, reduces the likelihood to -4803.867($df = 9$). □

TABLE 7.6: Log-linear Poisson and quasi-Poisson models for health care data (males).

		Poisson		Quasi-Poisson	
	Estimate	Std. Error	p-Value	Std. Error	p-Value
Intercept	0.746	0.136	0.000	0.370	0.044
hosp	0.188	0.009	0.000	0.025	0.000
healthpoor	0.221	0.030	0.000	0.081	0.006
healthexcellent	–0.229	0.045	0.000	0.122	0.060
numchron	0.153	0.007	0.000	0.020	0.000
age	0.004	0.017	0.833	0.047	0.938
married[yes]	0.132	0.027	0.000	0.073	0.072
school	0.043	0.003	0.000	0.008	0.000

In the preceding example overdispersion was found, which occurs quite frequently in applications. An exception is the log-linear model for the number of children in Example 7.3. When fitting a log-linear model with variance $\phi\mu_i$, one obtains the estimate $\hat{\phi} = 0.847$, which means weak underdispersion. Therefore, p-values for the corresponding parameter estimates are slightly smaller than the values given in Table 7.2.

7.5.2 Random Effects Model

One possible cause for overdispersion in the Poisson model is unobserved heterogeneity among subjects. A way of handling heterogeneity is to model it explicitly. It is assumed that the mean of observation y_i is given by

$$\lambda_i = b_i \mu_i = b_i \exp(\boldsymbol{x}_i^T \boldsymbol{\beta}), \tag{7.8}$$

where b_i is a subject-specific parameter, which is itself drawn from a mixing distribution. It represents the heterogeneity of the population that is not captured by the observed variables $\boldsymbol{x}_i$. The model assumption $y_i \sim \mathrm{P}(\lambda_i)$ is understood conditionally *for given* b_i (and $\boldsymbol{x}_i$).

Model (7.8) may also be written in the form

$$\lambda_i = \exp(a_i + \boldsymbol{x}_i^T \boldsymbol{\beta}),$$

where $a_i = \log(b_i)$ is a random intercept within the linear predictor. With $f(b_i)$ denoting the density of b_i, the marginal probability of the response value y_i is obtained in the usual way as

$$P(y_i) = \int f(y_i|b_i) f(b_i) db_i.$$

There are various ways of specifying the distribution of b_i and a_i, respectively. Hinde (1982) assumes a normal distribution for $a_i = \log(b_i)$. Then b_i follows the log-normal distribution. For the specific normal distribution $a_i \sim \mathrm{N}(-\sigma^2/2, \sigma^2)$ one obtains for b_i the mean $\mathrm{E}(b_i) = 1$ and the variance $\mathrm{var}(b_i) = \exp(\sigma^2) - 1$. In particular $\mathrm{E}(b_i) = 1$ is a sensible choice for the model (7.8), where λ_i is given by $\lambda_i = b_i \exp(\boldsymbol{x}_i^T \boldsymbol{\beta})$, since then the log-linear Poisson model is the limiting case when the variance of b_i tends to zero. Dean et al. (1989) consider a random effects model by using the inverse normal distribution. An alternative choice that yields an explicit marginal distribution is based an the Gamma mixing distribution. The corresponding Poisson-Gamma model is considered in the next section.

In general, maximization of the marginal likelihood is computationally intensive because the integrals have to be approximated, for example, by Gauss-Hermite integration (see Chapter 14).

7.6 Negative Binomial Model and Alternatives

Quasi-likelihood methods seem to provide a sufficiently flexible tool for the estimation of overdispersed count data. The assumptions are weak; by using only the first two moments one does not have to specify a distribution function. Nevertheless, parametric models have advantages. In particular, they are useful as building blocks of mixture models as considered in Sections 7.7 and 7.8. The type of overdispersion found in these mixture models cannot be modeled within the framework of quasi-likelihood methods. Therefore, in the following we will consider alternative parametric models.

There are several distribution models that are more flexible than the Poisson model but include it as a limiting case. A frequently used model is the negative binomial distribution. In contrast to the Poisson distribution, it is a two-parameter distribution and therefore more flexible than the Poisson model; in particular, it can model overdispersed counts. In the following we first consider the negative binomial model, which can be derived as a mixture of Poisson distributions. The second extension that will be considered is the generalized Poisson distribution.

Negative Binomial Model as Gamma-Poisson-Model

A specific choice for the mixing distribution in model (7.8), which allows a closed form of the marginal distribution, is the Gamma-distribution. The Gamma-distribution $b_i \sim \Gamma(\nu, \alpha)$ is given by the density

$$f(b_i) = \begin{cases} 0 & b_i \leq 0 \\ \frac{\alpha^\nu}{\Gamma(\nu)} b_i^{\nu-1} \mathrm{e}^{-\alpha b_i} & b_i > 0. \end{cases}$$

The mean and variance are $\mathrm{E}(b_i) = \nu/\alpha, \mathrm{var}(b_i) = \nu/\alpha^2$. If one assumes for the random parameter b_i the Gamma-distribution $\Gamma(\nu, \nu)$, the mean fulfills $\mathrm{E}(b_i) = 1$ and one obtains for the marginal probability

$$\begin{aligned} P(y_i) &= \int f(y_i|b_i) f(b_i) d b_i \\ &= \int \left(\mathrm{e}^{-b_i \mu_i} \frac{(b_i \mu_i)^{y_i}}{y_i!} \right) \left(\frac{\nu^\nu}{\Gamma(\nu)} b_i^{\nu-1} \mathrm{e}^{-\nu b_i} \right) d b_i \\ &= \frac{\Gamma(y_i + \nu)}{\Gamma(\nu)\Gamma(y_i + 1)} \left(\frac{\mu_i}{\mu_i + \nu} \right)^{y_i} \left(\frac{\nu}{\mu_i + \nu} \right)^{\nu}. \end{aligned} \tag{7.9}$$

The density (7.9) represents the *negative binomial distribution* $\mathrm{NB}(\nu, \mu_i)$, with mean and variance given by

$$\mathrm{E}(y_i) = \mu_i = \exp(\boldsymbol{x}_i^T \boldsymbol{\beta}), \quad \mathrm{var}(y_i) = \mu_i + \mu_i^2/\nu.$$

While the mean is the same as for the simple Poisson model, the variance exceeds the Poisson variances by μ_i^2/ν. The Poisson model may be seen as a limiting case ($\nu \to \infty$). The scaling of ν is such that small values signal strong overdispersion when compared to the Poisson model, while for large values of ν the model is similar to the Poisson model. Therefore, $1/\nu$ is considered the dispersion parameter. For illustration, Figure 7.5 shows three densities of the negative binomial distribution. It is seen that NB(100,3), which is is close to the Poisson distribution, is much more concentrated around the mean than NB(5,3). For *known* ν the *negative binomial model* can be estimated within the GLM framework.

In summary, the negative binomial model was motivated by the assumptions $y_i|\lambda_i \sim \mathrm{P}(\lambda_i)$, $b_i \sim \Gamma(\nu, \nu), \lambda_i = b_i \mu_i$ and is given by

$$y_i|\boldsymbol{x}_i \sim \mathrm{NB}(\nu, \mu_i), \quad \mu_i = \exp(\boldsymbol{x}_i^T \boldsymbol{\beta}). \tag{7.10}$$

The additional dispersion parameter makes the model more flexible than the simple Poisson model. If the parameter ν is fixed, for example, by $\nu = 1$, which corresponds to the geometric distribution, the additional flexibility is lost, but nevertheless variance functions that do not postulate equidispersion are used.

Alternatively, the Gamma-Poisson model may be derived from assuming that y_i is conditionally Poisson-distributed $P(\lambda_i)$ *for given* λ_i and specifying λ_i as a random variable that is Gamma-distributed $\Gamma(\nu_i, \frac{\nu_i}{\mu_i})$ with density function

$$f(\lambda_i) = \frac{1}{\Gamma(\nu_i)} (\frac{\nu_i}{\mu_i})^{\nu_i} \lambda_i^{\nu_i - 1} \exp(-\frac{\nu_i}{\mu_i} \lambda_i)$$

for $\lambda_i > 0$. Then one has mean $\mathrm{E}(\lambda_i) = \mu_i$ and variance $\mathrm{var}(\lambda_i) = \mu_i^2/\nu_i$. If the link between the mean and the linear predictor is specified by $\mu_i = \exp(\boldsymbol{x}_i^T \boldsymbol{\beta})$, one obtains for the conditional distribution $y_i|\lambda_i \sim \mathrm{P}(\exp(\boldsymbol{x}_i^T \boldsymbol{\beta}))$ and for the marginal distribution the discrete

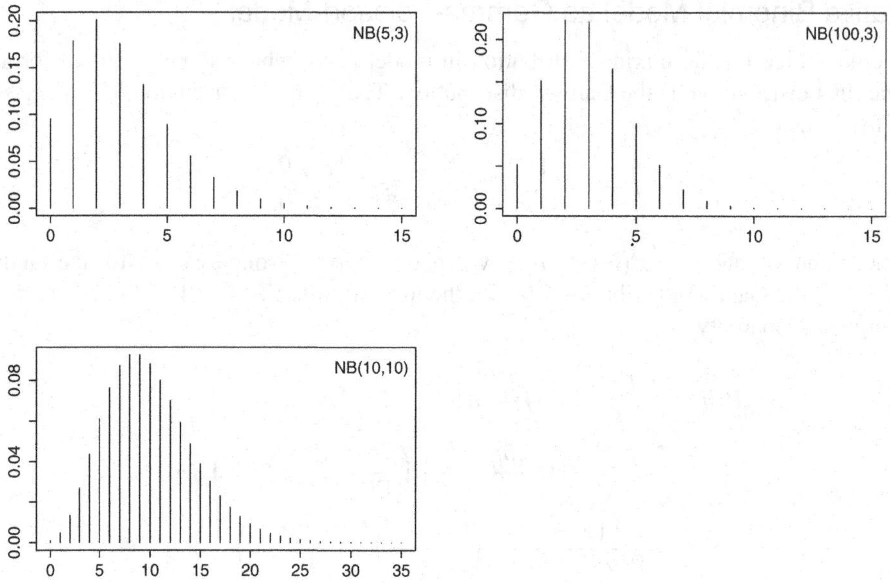

FIGURE 7.5: Probability mass functions of negative binomial distributions.

density

$$
\begin{aligned}
P(y_i; \nu_i, \mu_i) &= \int f(y_i|\lambda_i) f(\lambda_i) d\lambda_i \\
&= \frac{\Gamma(y_i + \nu_i)}{\Gamma(y_i + 1)\Gamma(\nu_i)} \left(\frac{\mu_i}{\mu_i + \nu_i}\right)^{y_i} \left(\frac{\nu_i}{\mu_i + \nu_i}\right)^{\nu_i}.
\end{aligned} \tag{7.11}
$$

Density (7.11) is the negative binomial distribution function with mean $\mathrm{E}(y_i) = \mu_i$ and variance $\mathrm{var}(y_i) = \mu_i + \frac{1}{\nu_i}\mu_i^2$. Therefore, the negative binomial model (7.10) may also be motivated by the assumptions $y_i|\lambda_i \sim \mathrm{P}(\lambda_i), \lambda_i \sim \Gamma(\nu, \frac{\nu}{\mu_i}), \mu_i = \exp(\boldsymbol{x}_i^T \boldsymbol{\beta})$.

The essential advantage of the negative binomial model over the Poisson model is that, by introducing a second parameter, more flexible variance structures are possible. However, the variance $\mathrm{var}(y_i) = \mu_i + \mu_i^2/\nu$ is also restrictive in a certain sense. Since $\nu > 0$, the variance can only be larger than assumed in the Poisson model. Therefore, underdispersion cannot be modeled adequately by the negative binomial model. When underdispersion occurs, as in Example 7.3, where $\hat{\phi} = 0.847$, the fitting of the negative binomial value typically yields the Poisson model with a very large value $\hat{\nu}$.

Example 7.7: Insolvent Firms

As in Example 7.5, a log-linear link is assumed, $\log(\mu) = \beta_0 + x\beta_1 + x^2\beta_2$, where x denotes the months ranging from 1 to 36. The fitted models are the log-linear Poisson model, the log-linear Poisson model with dispersion parameter ϕ, the log-linear negative binomial model, and a mixture model that assumes a normal distribution of individual effects a_i. Since the counts are rather large, in addition a normal distribution model with log-link has been fitted. It is seen from Table 7.7 that the quadratic effect seems negligible for all models. The Poisson model is certainly not the best choice since the data are overdispersed. The estimate $\hat{\phi} = 2.313$ signals that the variance is about twice what one would expect for the Poisson model. A check for overdispersion is shown in Figure 7.4 (right panel), where the quadratic residuals are plotted against the fitted values. The straight line shows what is to be expected when the Poisson model holds, namely, $\mathrm{E}(y_i - \mu_i)^2 = \mu_i$. The smooth fit shows that squared residual tends to be much larger than expected. Overdispersion is also seen from the fit of the negative binomial model. When compared to the Poisson model, the variance increases by μ_i^2/ν. For values $\mu_i \in [70, 100]$ and $\hat{\nu} = 77.93$, that means a substantial increase in variance between 63 and 128. The estimated effects for the models compare well with the exception of the Poisson model, for which the standard errors are definitely too small. AIC for the Poisson model was 306.82; for the normal distribution model, which is more flexible, one gets the smaller value 296.54; the smallest value, 296.27, is obtained by the negative binomial model. □

TABLE 7.7: Log-linear model for insolvencies (standard errors in brackets).

	Log-Linear Poisson Model $\phi = 1$	Log-Linear Dispersion Poisson Model	Negative Binomial Model	Log-Linear Normal Distribution Model	Mixture Gauss-Hermite
Variance	$\mathrm{var}(y_i) = \mu_i$	$\mathrm{var}(y_i) = \phi\mu_i$	$\mathrm{var}(y_i) = \mu_i + \frac{\mu_i^2}{\nu}$	σ^2	
Intercept	4.192 (0.062)	4.192 (0.094)	4.195 (0.086)	4.184 (0.101)	4.186 (0.086)
Month	0.020 (0.007)	0.020 (0.0112)	0.019 (0.011)	0.021 (0.01?)	0.0196 (0.0105)
Month^2	–0.00026 (0.00019)	–0.00026 (0.00028)	–0.00025 (0.00027)	–0.00029 (0.00029)	–0.00026 (0.00027)
Dispersion	—	$\hat{\phi} = 2.313$	$\hat{\nu} = 77.93$ (35.49)	$\hat{\sigma} = 13.90$	$\hat{\sigma} = 0.113$ (0.025)

Example 7.8: Demand for Medical Care

Table 7.8 shows the estimates of the negative binomial model for the medical care data (Example 7.6). It is seen that the effects are similar to the effects of a quasi-Poisson model, but the standard errors are slightly smaller ($\hat{\nu} = 1.079$, with standard error 0.048, and the residual deviance is -9607.73). The Poisson model yields the log-likelihood value -7296.398 ($df = 8$), and the negative binomial model, which uses just one more parameter, reduces the likelihood to -4803.867 ($df = 9$). □

TABLE 7.8: Negative binomial model for health care data (males).

	Estimate	Std. Error	p-Value
Intercept	0.556	0.333	0.094
hosp	0.245	0.033	0.000
healthpoor	0.255	0.083	0.002
healthexcellent	–0.206	0.096	0.032
numchron	0.182	0.020	0.000
age	0.021	0.042	0.622
married[yes]	0.148	0.063	0.020
school	0.040	0.007	0.000

Generalized Poisson Distribution

An alternative distribution that allows for overdispersion is the generalized Poisson distribution, which was investigated in detail in Consul (1998). A random variable Y follows a generalized Poisson distribution with parameters $\mu > 0$ and γ, $Y \sim GP(\mu, \gamma)$, if the density is given by

$$P(Y = y) = \begin{cases} \dfrac{\mu(\mu + y(\gamma - 1))^{y-1}\gamma^{-y}e^{-(\mu+y(\gamma-1))/\gamma}}{y!} & \text{for } y \in \{0, 1, 2, \dots\} \\ 0 & \text{for } y > m, \quad \text{if } \gamma < 1. \end{cases}$$

Additional constraints on the parameters are $\gamma \geq \max\{1/2, 1 - \mu/m\}$, where $m \geq 4$ is the largest natural number such that $\mu + m(\gamma - 1) > 0$ if $\gamma < 1$.

It is seen that the distribution becomes the Poisson distribution for $\gamma = 1$. For small values of γ the generalized Poisson distribution is very similar to the negative binomial distribution; for large values the negative binomial distribution puts more mass on small values of y. For the generalized Poisson distribution one obtains

$$\mathrm{E}(Y) = \mu \quad \mathrm{var}(Y) = \gamma^2\mu.$$

The parameter γ^2 can be seen as a dispersion parameter; for $\gamma^2 > 1$ one obtains greater dispersion than for the Poisson model, and for $\gamma^2 < 1$ one obtains underdispersion. An advantage of the generalized Poisson distribution is that the dispersion parameter also allows for underdispersion in contrast to the negative binomial model, for which the variance is $\mathrm{var}(Y) = \mu + \mu^2/\nu$. Like the negative binomial model, the generalized Poisson distribution can be derived as a mixture of Poisson distributions (Joe and Zhu, 2005). Gschoessl and Czado (2006) fitted a regression model based on the generalized Poisson distribution and compared several models for overdispersion from a Bayesian perspective.

7.7 Zero-Inflated Counts

In many applications one observes more zero counts than is consistent with the Poisson (or an alternative count data) model; the data display overdispersion through excess zeros. Often one may think of data as resulting from a mixture of distributions. If a person is asked, "How many times did you eat mussels in the past 3 months?" one records zero responses from people who never eat mussels and from those who do but happen not to have done so during the time interval in question.

In general, a zero-inflated count model may be motivated from a mixture of two subpopulations, the non-responders who are "never at risk" and the responders who are at risk. With C denoting the class indicator of subpopulations ($C_i = 1$ for responders and $C_i = 0$ for non-responders) one obtains the mixture distribution

$$P(y_i = y) = P(y_i = y|C_i = 1)\pi_i + P(y_i = y|C_i = 0)(1 - \pi_i),$$

where $\pi_i = P(C_i = 1)$ are the mixing probabilities. When one assumes that counts within the responder subpopulation are Poisson-distributed, one obtains with $P(y_i = 0|C_i = 0) = 1$

$$P(y_i = 0) = P(y_i = 0|C_i = 1)\pi_i + (1 - \pi_i) = \pi_i \,\mathrm{e}^{-\mu_i} + 1 - \pi_i,$$

and for $y > 0$

$$P(y_i = y) = P(y_i = y|C_i = 1)\pi_i = \pi_i \,\mathrm{e}^{-\mu_i} \mu_i^y/y!,$$

where μ_i is the mean of the Poisson distribution of population $C_i = 1$. One obtains

$$\mathrm{E}(y_i) = \pi_i\mu_i, \quad \mathrm{var}(y_i) = \pi_i\mu_i + \pi_i(1 - \pi_i)\mu_i^2 = \pi_i\mu_i(1 + \mu_i(1 - \pi_i))$$

(Exercise 7.4). Since $\text{var}(y_i) > \text{E}(y_i)$, excess zeros imply overdispersion if $\pi_i < 1$. Of course, the Poisson model is included as the special case where all observations refer to responders and $\pi_i = 1$.

When covariates are present one may specify a Poisson distribution model for $y|C_i = 1$ and a binary response model for $C_i \in \{0, 1\}$, for example,

$$\log(\mu_i) = \boldsymbol{x}_i^T \boldsymbol{\beta}, \quad \text{logit}(\pi_i) = \boldsymbol{z}_i^T \boldsymbol{\gamma},$$

where $\boldsymbol{x}_i, \boldsymbol{z}_i$ may be different sets of covariates. The simplest mixture model assumes only an intercept in the binary model, $\text{logit}(\pi_i) = \gamma_0$. For increasing γ_0 one obtains in the limit the Poisson model without zero inflation.

The joint log-likelihood function after omitting constants is given by

$$\begin{aligned} l &= \sum_{i=1}^{n} l_i(y_i) \\ &= \sum_{i=1}^{n} I(y_i = 0) \log\{1 + \frac{\exp(\boldsymbol{z}_i^T \boldsymbol{\gamma})}{1 + \exp(\boldsymbol{z}_i^T \boldsymbol{\gamma})} (\exp(-\exp(\boldsymbol{x}_i^T \boldsymbol{\beta}) - 1))\} \\ &+ (1 - I(y_i = 0))\{\boldsymbol{z}_i^T \boldsymbol{\gamma} - \log(1 + \exp(\boldsymbol{z}_i^T \boldsymbol{\gamma})) - \exp(\boldsymbol{x}_i^T \boldsymbol{\beta}) + y_i(\boldsymbol{x}_i^T \boldsymbol{\beta})\}, \end{aligned}$$

where $I(y_i = 0)$ denotes an indicator variable that takes value 1 if $y_i = 0$, and 0 otherwise. Lambert (1992) suggested the use of the EM algorithm to maximize the log-likelihood. Zeileis et al. (2008) obtain ML estimates by using optimization functions from R and allow to specify starting values estimated by the EM algorithm. They compute the covariance matrix as the numerically determined Hessian matrix.

The zero-inflated Poisson model has been extended to the zero-inflated generalized Poisson model in which the Poisson distribution is replaced by the generalized Poisson distribution (Famoye and Singh, 2003; Famoye and Singh, 2006; Gupta et al., 2004; Czado et al., 2007). The resulting family of models is rather large, comprising a zero-inflated Poisson regression and a generalized Poisson regression. Min and Czado (2010) discussed the use of the Wald and likelihood ratio tests for the investigation of zero inflation (or zero deflation).

Example 7.9: Demand for Medical Care

Table 7.9 shows the fit of a zero-inflated model for the health care data. The counts are modeled as a log-linear Poisson model, with all variables included. The binary response uses the logit link with intercept only. Therefore, it is assumed that all probabilities π_i are equal. The estimate -1.522 corresponds to a probability of 0.295, which means that the portion of responders is not very large and overdispersion has to be expected. That is in agreement with the fitted quasi-Poisson model (Table 7.6). Table 7.10 shows the fit of the zero-inflated model when all the variables can have an effect on the mixture component. With the exception of the health status, all variables seem to contribute to the inflation component. For the log-likelihood of the fitted models one obtains -6544 on 9 *df* (zero-inflated Poisson model with an intercept for the logit model) and -6455 on 16 *df* (zero-inflated Poisson model with all variables in both components). Compared to the Poisson model with log-likelihood -7296.398 ($df = 8$), already the zero-inflated model with just an intercept in the binary component is a distinct improvement. It is noteworthy that the results differ with respect to significance. The predictor married is not significant when a quasi-Poisson or a negative binomial model is fitted but seems not neglectable when a zero-inflated model is fitted. □

TABLE 7.9: Zero-inflated models, Poisson and logit, for health care data (males).

Count Model Coefficients (Poisson with Log Link)				
	Estimate	Std. Error	z-Value	p-Value
Intercept	1.639	0.140	11.678	0.0
hosp	0.165	0.009	17.477	0.0
healthpoor	0.241	0.030	8.078	0.0
healthexcellent	−0.176	0.047	−3.771	0.0
numchron	0.103	0.008	13.527	0.0
age	−0.048	0.018	−2.721	0.007
married[yes]	0.009	0.027	0.342	0.732
school	0.029	0.003	10.134	0.0
Zero-Inflation Model Coefficients (Binomial with Logit Link)				
	Estimate	Std. Error	z-Value	p-Value
Intercept	-1.522	0.063	-24.10	0.0

TABLE 7.10: Zero-inflated models, Poisson and logit, for health care data (males).

Count Model Coefficients (Poisson with Log Link)				
	Estimate	Std. Error	z-Value	p-Value
(Intercept)	1.672	0.140	11.980	0.0
hosp	0.165	0.009	17.451	0.0
healthpoor	0.240	0.030	8.057	0.0
healthexcellent	−0.163	0.045	−3.587	0.0003
numchron	0.101	0.008	13.351	0.0
age	−0.050	0.017	−2.839	0.0045
married[yes]	0.005	0.027	0.168	0.8663
school	0.028	0.003	9.919	0.0
Zero-Inflation Model Coefficients (Binomial with Logit Link)				
	Estimate	Std. Error	z-Value	p-Value
(Intercept)	3.152	0.890	3.541	0.0004
hosp	−0.604	0.156	−3.869	0.0001
healthpoor	0.214	0.245	0.874	0.3822
healthexcellent	0.260	0.213	1.221	0.2221
numchron	−0.477	0.065	−7.305	0.0
age	−0.348	0.115	−3.042	0.0024
married[yes]	−0.700	0.148	−4.745	0.0
school	−0.092	0.017	−5.505	0.0

7.8 Hurdle Models

An alternative model that is able to account for excess zeros is the hurdle models (Mullahy, 1986; Creel and Loomis, 1990). It allows one to model overdispersion through excess zeros for baseline models such as the Poisson model and the negative binomial model. The model specifies two processes that generate the zeros and the positives. The combination of both models, a binary model that determines whether the outcome is zero or positive and a truncated-at-zero count model, gives the model.

In general, one assumes that f_1, f_2 are the probability mass functions with support $\{0, 1, 2, \dots\}$. The hurdle model is given by

$$P(y=0) = f_1(0),$$
$$P(y=r) = f_2(r)\frac{1-f_1(0)}{1-f_2(0)}, \quad r = 1, 2, \dots.$$

The model may be seen as a stage-wise decision model. At the first stage a binary variable C determines whether a count variable has a zero or a positive outcome. $C = 1$ means that the "hurdle is crossed" and the outcome is positive, while $C = 0$ means that zero will be observed. The binary decision between zero and a positive outcome is determined by the f_1-distribution in the form

$$P(C=1) = 1 - f_1(0), \quad P(C=0) = f_1(0).$$

At the second stage the condition distribution given C is specified. If the hurdle is crossed, the response is determined by the truncated count model with probability mass function

$$P(y=r|C=1) = f_2(r)/(1-f_2(0)) \qquad r = 1, 2, \dots$$

If the hurdle is not crossed, the probability for zero outcome is 1, $P(y=0|C=0) = 1$. One obtains the hurdle model from $P(y=r) = P(y=r|C=0)P(C=0) + P(y=r|C=1)P(C=1)$, which yields

$$P(y=0) = P(C=0) = f_1(0)$$
$$P(y=r) = P(y=r|C=1)P(C=1)$$
$$= \{f_2(r)/(1-f_2(0))\}(1-f_1(0)), \; r = 1, 2, \dots.$$

The derivation shows that the hurdle model is a finite mixture of the truncated count model $P(y=r|C=1)$ and the degenerate distribution $P(y=r|C=0)$. In contrast to the zero inflated-counts models from Section 7.7, C is an observed variable and not an unobservable mixture. The truncated count model is determined by the probability mass function f_2, which has been called the *parent process* by Mullahy (1986). If $f_1 = f_2$, the model collapses to the parent model f_2.

The model is quite flexible and allows for both under- and overdispersion. This is seen by considering the mean and variance. With $\gamma = (1-f_1(0))/(1-f_2(0)) = P(y>0)/(1-f_2(0))$, the mean is given by

$$\mathrm{E}(y) = \sum_{r=1}^{\infty} r\; f_2(r)\gamma = P(y>0)\,\mathrm{E}(y|y>0)$$

and the variance has the form

$$\mathrm{var}(y) = P(y>0)\,\mathrm{var}(y|y>0) + P(y>0)(1-P(y>0))\,\mathrm{E}(y|y>0)^2.$$

Let us consider as a specific model, the *hurdle Poisson model*, which assumes that f_2 is the probability mass function of a Poisson distribution with mean μ_2. Let y_2 denote the corresponding random variable (Poisson distribution with mean μ_2). Then one has $\mathrm{E}(y_2) = \mu_2$ and $\mu_2 = \mathrm{var}(y_2) = \mathrm{E}(y_2^2) - \mathrm{E}(y_2)^2$, yielding $\mathrm{E}(y_2^2) = \mu_2 + \mu_2^2 = \mu_2(1+\mu_2)$. One obtains for the mean and variance of y

$$\mathrm{E}(y) = \gamma\mu_2,$$
$$\mathrm{var}(y) = \sum_{r=1}^{\infty} r^2 f_2(r)\gamma - (\sum_{r=1}^{\infty} r\, f_2(r)\gamma)^2 = \mu_2(1+\mu_2)\gamma - \mu_2^2\gamma^2,$$

and therefore

$$\frac{\text{var}(y)}{\text{E}(y)} = 1 + \mu_2(1-\gamma).$$

This means that for the non-trivial case $\mu_2 > 0$ one obtains *overdispersion* if $0 < \gamma < 1$ and *underdispersion* if $1 < \gamma < (1+\mu_2)/\mu_2$, where the upper threshold is determined by the restriction $\text{var}(y) > 0$. For $\gamma = 1$, the hurdle Poisson becomes the Poisson model.

The hurdle model is determined by the choices of f_1 and f_2. There is much flexibility because f_1 and f_2 may be Poisson, geometric, or negative binomial distributions. Moreover, the distributions do not have to be the same. One can also combine a binary logit model for the truncated (right-censored at $y = 1$) distribution of f_1 and a Poisson or negative binomial model for f_2.

Concrete parameterizations are obtained by linking the two distributions to explanatory variables. For illustration we consider the hurdle Poisson model where both f_1 and f_2 correspond to Poisson distributions with means μ_1 and μ_2, respectively. For observations $(y_i, \boldsymbol{x}_i)$ one may specify for

$$\mu_{i1} = \exp(\boldsymbol{x}_i^T \boldsymbol{\beta}_1), \quad \mu_{i2} = \exp(\boldsymbol{x}_i^T \boldsymbol{\beta}_2),$$

yielding the model

$$P(y_i = 0) = \exp(-\mu_{i1}),$$
$$P(y_i = r) = \frac{\mu_{i2}^r}{r!}\, \mathrm{e}^{-\mu_{i2}} \frac{1-\exp(-\mu_{i1})}{1-\exp(-\mu_{i2})}.$$

The log-likelihood is given by

$$l(\boldsymbol{\beta}_1, \boldsymbol{\beta}_2) = -\sum_{y_i=0} \mu_{i1} + \sum_{y_i>0} \log\left(\frac{1-\mathrm{e}^{-\mu_{i1}}}{1-\mathrm{e}^{-\mu_{i2}}} \frac{\mu_{i2}^{y_i}}{y_i!} \mathrm{e}^{-\mu_{i2}}\right),$$

which decomposes into $l(\boldsymbol{\beta}_1, \boldsymbol{\beta}_2) = l_1(\boldsymbol{\beta}_1) + l_2(\boldsymbol{\beta}_2)$ with

$$l_1(\boldsymbol{\beta}_1) = -\sum_{y_i=0} \mu_{i1} + \sum_{y_i>0} \log(1-\mathrm{e}^{-\mu_{i1}}),$$
$$l_2(\boldsymbol{\beta}_2) = \sum_{y_i>0} y_i \log(\mu_{i2}) - \mu_{i2} - \log(1-\mathrm{e}^{-\mu_{i2}}) - \log(y_i!).$$

Since the components depend on a one-parameter vector, only the two components can be maximized separately. In general, the regressors for the two model components do not have to be the same. But, if the same regressors as well as the same count models are used, as in the preceding Poisson example, a test of the hypothesis $\boldsymbol{\beta}_1 = \boldsymbol{\beta}_2$ tests whether the hurdle is needed. Although most hurdle models use the hurdle zero, the specification of more general models where the hurdle is some positive number is straightforward.

Example 7.10: Demand for Medical Care

Table 7.11 and Table 7.12 show the fits of hurdle models for the health care data. The counts are modeled as a log-linear Poisson model, with all variables included, and the binary response uses the logit link with intercept only (Table 7.11) or with all the covariates included (Table 7.12). For the log-likelihood of the fitted models one obtains -6549 on 9 *df* (hurdle model with an intercept for the logit model) and -6456 on 16 *df* (hurdle model with all variables in both components). Both models have a much better fit than the simple Poisson model (-7296.398 on $df = 8$). Moreover, the model with all covariates in the binary component is to be preferred over the pure intercept model. A comparison of the hurdle model and the

TABLE 7.11: Hurdle model, Poisson and logit, for health care data (males).

Count Model Coefficients (Truncated Poisson with Log Link)				
	Estimate	Std. Error	z-Value	p-Value
Intercept	1.673	0.139	12.001	0.0
hosp	0.165	0.009	17.450	0.0
healthpoor	0.240	0.030	8.062	0.0
healthexcellent	−0.164	0.046	−3.592	0.0
numchron	0.101	0.008	13.346	0.0
age	−0.050	0.017	−2.848	0.0
marriedyes	0.005	0.027	0.170	0.86
school	0.028	0.003	9.920	0.0
Zero Hurdle Model Coefficients (Binomial with Logit Link)				
	Estimate	Std. Error	z-Value	p-Value
Intercept	1.483	0.061	24.28	0.0

TABLE 7.12: Hurdle model, Poisson and logit, for health care data (males).

Count Model Coefficients (Truncated Poisson with Log Link)				
	Estimate	Std. Error	z-Value	p-Value
(Intercept)	1.673	0.139	12.001	0.0
hosp	0.165	0.009	17.450	0.0
healthpoor	0.240	0.030	8.062	0.0
healthexcellent	−0.164	0.046	−3.592	0.0003
numchron	0.101	0.008	13.346	0.0
age	−0.050	0.017	−2.848	0.0043
marriedyes	0.005	0.027	0.170	0.8652
school	0.028	0.003	9.920	0.0
Zero Hurdle Model Coefficients (Binomial with Logit Link):				
	Estimate	Std. Error	z-Value	p-Value
(Intercept)	−3.104	0.871	−3.564	0.0004
hosp	0.611	0.155	3.944	0.0
healthpoor	−0.199	0.244	−0.815	0.4151
healthexcellent	−0.274	0.208	−1.318	0.1876
numchron	0.482	0.064	7.476	0.0
age	0.336	0.118	3.010	0.0026
married[yes]	0.690	0.146	4.743	0.0
school	0.094	0.016	5.711	0.0

zero-inflated model shows that the parameter estimates and p-values are comparable; only the signs for the parameters of the mixture have changed since the response $y = 0$ is modeled in the hurdle models. □

7.9 Further Reading

Surveys and Books. A source book for the modeling of count data that includes many applications is Cameron and Trivedi (1998). An econometric view on count data is outlined in Winkelmann (1997) and Kleiber and Zeileis (2008). The negative binomial model is treated extensively in Hilbe (2011)

Tests on Zero Inflation. Tests that investigate the need for zero inflation have been suggested for the case of constant overdispersion. The most widely used test is the score test, because it requires only the fit under the null model; see van den Broek (1995), Deng and Paul (2005), and Gupta et al. (2004).

Hurdle Models. The Poisson hurdle and the geometric and hurdle have been examined by Mullahy (1986), and hurdle negative binomial models have been considered by Pohlmeier and Ulrich (1995). Zeileis et al. (2008) describe how regression models for count data, including zero-inflated and hurdle models, can be fitted in R.

R Packages. GLMs as the Poisson model can be fitted by using of the model fitting functions *glm* from the *MASS* package. Many tools for diagnostic and inference are available. *MASS* also allows one to fit negative binomial models with fixed dispersion parameters (function *negative.binomial*) and for estimating regression parameters and dispersion parameters (function *glm.nb*). Estimation procedures for zero-inflated and hurdle models are available in the *pscl* package (for details see Zeileis et al., 2008).

7.10 Exercises

7.1 In Example 1.5, the dependent variable is the number of insolvent firms depending on year and month (see Table 1.3).

(a) Consider time as the only covariate ranging from 1 to 36. Fit a log-linear Poisson model, an overdispersed model, a negative binomial model, and a log-linear normal distribution model with linear time (compare to Example 7.7).

(b) Fit the models from part (a) with the linear predictor determined by the factors year and month and an interaction effect if needed.

(c) Discuss the difference between the models fitted in parts (a) and (b).

TABLE 7.13: Cellular differentiation data from Piegorsch et al. (1988).

Number of Cells Differentiating	Dose of TNF(U/ml)	Dose of IFN(U/ml)
11	0	0
18	0	4
20	0	20
39	0	100
22	1	0
38	1	4
52	1	20
69	1	100
31	10	0
68	10	4
69	10	20
128	10	100
102	100	0
171	100	4
180	100	20
193	100	100

7.2 The R package *pscl* provides the dataset *bioChemists*.

(a) Use descriptive tools to learn about the data.

(b) Fit a zero-inflated Poisson model and a hurdle model by using the R package *pscl*.

7.3 Investigate the effect of explanatory variables on the number of children for men in analogy to Example 7.3 by using the dataset *children* from the the package *catdat*.

7.4 For the zero-inflated count model a mixture of two subnopulations is assumed, with C denoting the class indicator ($C_i = 1$, for responders and $C_i = 0$ for non-responders). When one assumes a Poisson model if $C_i = 1$, one has $P(y_i = 0) = \pi_i \, \mathrm{e}^{-\mu_i} + 1 - \pi_i$, $P(y_i = y) = \pi_i \, \mathrm{e}^{-\mu_i} \, \mu_i^y / y!$. Show that mean and variance have the form $\mathrm{E}(y_i) = \pi_i \mu_i$, $\mathrm{var}(y_i) = \pi_i \mu_i (1 + \mu_i (1 - \pi_i))$.

7.5 Table 7.13, which is reproduced from Piegorsch et al. (1988), shows data from a biomedical study of the immuno-activating ability of two agents, TNF (tumor necrosis factor) and IFN (interferon). Both agents induce cell differentiation. The number of cells that exhibited markers of differentiation after exposure to TNF and/or IFN was recorded. At each of the 16 dose combinations of TNF/INF, 200 cells were examined. It is of particular interest to investigate if the two agents stimulate cell differentiation synergistically or independently.

(a) Fit a log-linear Poisson model that includes an interaction term and investigate the effects.

(b) Use diagnostic tools to investigate the model fit.

(b) Fit alternative log-linear models that allow for overdispersion and compare the results to the Poisson model.

Chapter 8

Multinomial Response Models

In many regression problems the response is restricted to a fixed set of possible values, the so-called response categories. Response variables of this type are called *polytomous* or *multicategory* responses. In economical applications, the response categories may refer to the choice of different brands or to the choice of the transport mode (Example 1.3). In medical applications, the response categories may represent different side effects of medical treatment or several types of infection that may follow an operation. Most rating scales have fixed response categories that measure, for example, the medical condition after some treatment in categories like good, fair, and poor or the severeness of symptoms in categories like none, mild, moderate, marked. These examples show that there are at least two cases to be distinguished, namely, the case where response categories are mere labels that have no inherent ordering and the case where categories are ordered. In the first case, the response Y is measured on a *nominal scale*. Instead of using the numbers $1, \dots, k$ for the response categories, any set of k numbers would do. In the latter case, the response is measured on an *ordinal scale*, where the ordering of the categories and the corresponding numbers may be interpreted but not the distance or spacing between categories. Figures 8.1 and 8.2 illustrate different scalings of response categories. In the nominal case the response categories are given in an unsystematic way, while in the ordinal case the response categories are given on a straight line, thus illustrating the ordering of the categories.

Another type of response category that contains more structure than the nominal case but is not captured by simple ordering occurs in the form of nested or hierarchical response categories. Figure 8.3 shows an example where the basic response is in the categories "no infection," "infection type I", and "infection type II." However, for infection type I two cases have to be distinguished, namely, infection with and without additional complications. Thus, one has splits on two levels, first the split into basic categories and then the conditional split within outcome "infection type I."

In this chapter we will consider the modeling of responses with unordered categories. Modeling of ordered response categories is treated in Chapter 9. In the following some examples are given.

Example 8.1: Preference for Political Parties

Table 8.1 shows counts from a survey on the preference for political parties. The four German parties were the Christian Democratic Union (CDU), the Social Democratic Party (SPD), the Green Party, and the Liberal Party (FDP). The covariates are gender and age in categories. □

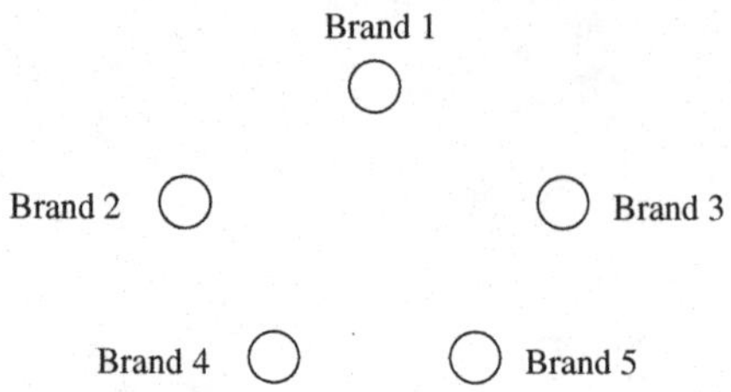

FIGURE 8.1: Choice of brand as nominal response categories.

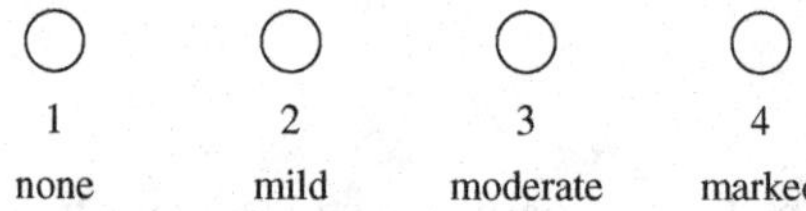

FIGURE 8.2: Severness of symptoms as ordered categories.

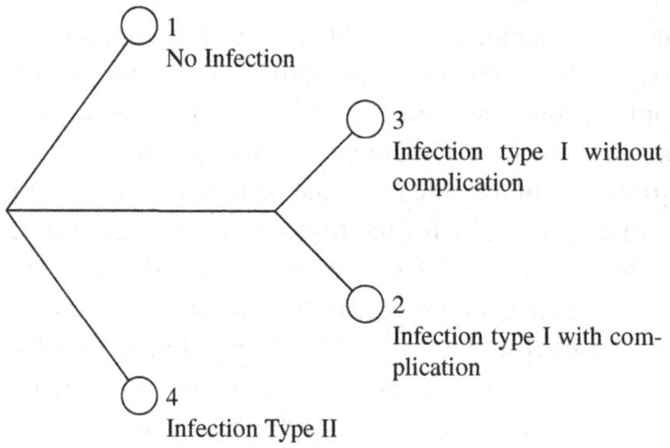

FIGURE 8.3: Type of infection as nested structure.

TABLE 8.1: Cross-classification of preference for political parties and gender.

		Preferred Party			
Gender	Age	CDU/CSU	SPD	Green Party	FDP
male	1	114	224	53	10
	2	134	226	42	9
	3	114	174	23	8
	4	339	414	13	30
female	1	42	161	44	5
	2	88	171	60	10
	3	90	168	31	8
	4	413	375	14	23

Example 8.2: Addiction

In a survey people were asked, "Is addiction a disease or are addicts weak-willed ?" The response was in three categories, "addicts are weak-willed," "addiction is a disease," or both alternatives hold. One wants to investigate how the response depends on predictors like gender, age, and education level. The dataset is available at http://www.stat.uni-muenchen.de/service/datenarchiv/sucht/sucht.html. □

8.1 The Multinomial Distribution

The multinomial distribution is a natural generalization of the binomial distribution. It allows for more than two possible outcomes. For example, in a sample survey respondents might be asked for their preference for political parties. Then the number of outcomes will depend on the number of competing parties.

Let the possible outcomes be denoted by $1,\dots,k$, which occur with probabilities $\pi_1,\dots,\pi_k$. For the random variable Y, which takes values $1,\dots,k$, one has the simple relationship $P(Y=r)=\pi_r$. However, the categories of the random variable Y hide that the response is genuinely multivariate, since each response category refers to a dimension of its own. A more appropriate representation is by a vector-valued random variable. In the general form of the multinomial distribution one usually considers a sample of, say, m responses. Then the components of the vector $\boldsymbol{y}^T=(y_1,\dots,y_k)$ give the cell counts in categories $1,\dots,k$. The vector $\boldsymbol{y}^T=(y_1,\dots,y_k)$ has probability mass function

$$f(y_1,\dots,y_k)=\begin{cases}\frac{m!}{y_1!\cdots y_k!}\pi_1^{y_1}\dots\pi_k^{y_k} & y_i\in\{0,\dots,m\},\ \Sigma_i y_i=m\\ 0 & \text{otherwise.}\end{cases}$$

A response (vector) with this probability mass function follows *a multinomial distribution* with parameters m and $\boldsymbol{\pi}^T=(\pi_1,\dots,\pi_k)$. Of course the probabilities are restricted by $\pi_i\in[0,1],\Sigma_i\pi_i=1$.

Since $\Sigma_i y_i=m$ there is some redundancy in the representation. Therefore, one often uses for the representation the shorter vector $\boldsymbol{y}^T=(y_1,\dots,y_q)$, $q=k-1$, obtaining for the relevant part of the mass function

$$f(y_1,\dots,y_q)=\frac{m!}{y_1!\dots y_q!(m-y_1\cdots-y_q)!}\pi_1^{y_1}\dots\pi_q^{y_q}\cdot(1-\pi_1-\dots-\pi_q)^{m-y_1-\dots-y_q}.$$

In the following the abbreviation $\boldsymbol{y}\sim \mathrm{M}(m,\boldsymbol{\pi})$ for the multinomial distribution will always refer to the latter version with $q=k-1$ components. In this representation it also becomes obvious that the binomial distribution is a special case of the multinomial distribution where $k=2$ $(q=1)$, since

$$f(y_1)=\frac{m!}{y_1!(m-y_1)!}\pi_1^{y_1}(1-\pi_1)^{m-y_1}=\binom{m}{y_1}\pi_1^{y_1}(1-\pi_1)^{m-y_1}.$$

For the components of the multinomial distribution $\boldsymbol{y}^T=(y_1,\dots y_q)$ one derives

$$\mathrm{E}(y_i)=m\pi_i,\quad \mathrm{var}(y_i)=m\pi_i(1-\pi_i),\quad \mathrm{cov}(y_i,y_j)=-m\pi_i\pi_j$$

(Exercise 8.1). In vector form, the covariance is given as $\mathrm{cov}(\boldsymbol{y})=m(\mathrm{diag}(\boldsymbol{\pi})-\boldsymbol{\pi}\boldsymbol{\pi}^T)$, where $\boldsymbol{\pi}^T=(\pi_1,\dots,\pi_q)$ and $\mathrm{diag}(\boldsymbol{\pi})$ is a diagonal matrix with entries $\pi_1,\dots,\pi_q$.

For *single* observations, where only one respondent ($m = 1$) is considered, one obtains $Y = r \Leftrightarrow y_r = 1$, with probabilities given by $\pi_r = P(Y = r) = P(y_r = 1)$ and possible outcome vectors of length $k - 1$ given by $(1, 0, \dots), (0, 1, 0, \dots) \dots (0, 0, \dots, 1)$.

The *scaled multinomial* distribution uses the vector of relative frequencies $\bar{\boldsymbol{y}} = (y_1/m, \dots, y_q/m) = \boldsymbol{y}^T/m$. It has mean $\mathrm{E}(\bar{\boldsymbol{y}}) = \boldsymbol{\pi}$ and covariance matrix $\mathrm{cov}(\bar{\boldsymbol{y}}) = (\mathrm{diag}(\boldsymbol{\pi}) - \boldsymbol{\pi}\boldsymbol{\pi}^T)/m$.

8.2 The Multinomial Logit Model

The multinomial logit model is the most widely used regression model that links a categorical response variable with unordered categories to explanatory variables. Again let $Y \in \{1, \dots, k\}$ denote the response in categories $1, \dots, k$ and $\boldsymbol{y}^T = (y_1, \dots, y_k)$ the corresponding multinomial distribution (for $m = 1$). Let $\boldsymbol{x}$ be a vector of explanatory variables. The binary logit model (Chapter 2) has the form

$$P(Y = 1|\boldsymbol{x}) = \frac{\exp(\boldsymbol{x}^T\boldsymbol{\beta})}{1 + \exp(\boldsymbol{x}^T\boldsymbol{\beta})}$$

or, equivalently,

$$\log\left(\frac{P(Y = 1|\boldsymbol{x})}{P(Y = 2|\boldsymbol{x})}\right) = \boldsymbol{x}^T\boldsymbol{\beta}.$$

The multinomial logit model uses the same linear form of logits. But instead of only one logit, one has to consider $k - 1$ logits. One may specify

$$\log\left(\frac{P(Y = r|\boldsymbol{x})}{P(Y = k|\boldsymbol{x})}\right) = \boldsymbol{x}^T\boldsymbol{\beta}_r, \quad r = 1, \dots, q, \tag{8.1}$$

where the log-odds compare $P(Y = r|\boldsymbol{x})$ to the probability $P(Y = k|\boldsymbol{x})$. In this presentation k serves as the reference category since all probabilities are compared to the last category. It should be noted that the vector $\boldsymbol{\beta}_r$ depends on r because comparison of $Y = r$ to $Y = k$ should be specific for r. The q logits $\log(P(Y = 1|\boldsymbol{x})/P(Y = k|\boldsymbol{x})), \dots, \log(P(Y = q|\boldsymbol{x})/P(Y = k|\boldsymbol{x}))$ specified in (8.1) determine the response probabilities $P(Y = 1|\boldsymbol{x}), \dots, P(Y = k|\boldsymbol{x})$ uniquely. From $P(Y = r|\boldsymbol{x}) = P(Y = k|\boldsymbol{x})\exp(\boldsymbol{x}^T\boldsymbol{\beta}_r)$ one obtains $\sum_{r=1}^{k-1} P(Y = r|\boldsymbol{x}) = P(Y = k|\boldsymbol{x})\sum_{r=1}^{k-1}\exp(\boldsymbol{x}^T\boldsymbol{\beta}_r)$. By adding $P(Y = k|\boldsymbol{x})$ on the left- and right-hand sides one obtains

$$P(Y = k|\boldsymbol{x}) = \frac{1}{1 + \sum_{r=1}^{k-1}\exp(\boldsymbol{x}^T\boldsymbol{\beta}_r)}.$$

When $P(Y = k|\boldsymbol{x})$ is inserted into (8.1) one obtains the probabilities of the multinomial model given in the following box.

Multinomial Logit Model with Reference Category k

$$\log\left(\frac{P(Y=r|\boldsymbol{x})}{P(Y=k|\boldsymbol{x})}\right)=\boldsymbol{x}^T\boldsymbol{\beta}_r,\qquad r=1,\dots,k-1, \tag{8.2}$$

or, equivalently,

$$P(Y=r|\boldsymbol{x})=\frac{\exp(\boldsymbol{x}^T\boldsymbol{\beta}_r)}{1+\sum_{s=1}^{k-1}\exp(\boldsymbol{x}^T\boldsymbol{\beta}_s)},\qquad r=1,\dots,k-1,$$
$$P(Y=k|\boldsymbol{x})=\frac{1}{1+\sum_{s=1}^{k-1}\exp(\boldsymbol{x}^T\boldsymbol{\beta}_s)}. \tag{8.3}$$

The representation of the nomnial logit model depends on the choice of the reference category. Instead of k, any category from $1,\dots,k$ could have been chosen as the reference category. The necessity to specify a reference category is due to the constraint $\Sigma_r P(Y=r|\boldsymbol{x})=1$. The consequence of this constraint is that only $q=k-1$ response categories may be specified; the remaining probability is implicitly determined. A generic form of the logit model is given by

$$P(Y=r|\boldsymbol{x})=\frac{\exp(\boldsymbol{x}^T\boldsymbol{\beta}_r)}{\sum_{s=1}^{k}\exp(\boldsymbol{x}^T\boldsymbol{\beta}_s)}, \tag{8.4}$$

where additional side constraints have to be specified to fulfill $\Sigma_r P(Y=r|\boldsymbol{x})=1$. It is obvious that without side constraints the parameters $\boldsymbol{\beta}_1,\dots\boldsymbol{\beta}_k$ are not identifiable. If $\boldsymbol{\beta}_r$ is replaced by $\boldsymbol{\beta}_r+\boldsymbol{c}$ with $\boldsymbol{c}$ denoting some fixed vector, the form (8.4) also holds with parameters $\tilde{\boldsymbol{\beta}}_r=\boldsymbol{\beta}_r+\boldsymbol{c}$.

Generic Multinomial Logit Model

$$P(Y=r|\boldsymbol{x})=\frac{\exp(\boldsymbol{x}^T\boldsymbol{\beta}_r)}{\Sigma_{s=1}^{k}\exp(\boldsymbol{x}^T\boldsymbol{\beta}_s)} \tag{8.5}$$

with optional side constraints

$\boldsymbol{\beta}_k^T=(0,\dots,0)$	reference category k
$\boldsymbol{\beta}_{r_0}^T=(0,\dots,0)$	reference category r_0
$\sum_{s=1}^{k}\boldsymbol{\beta}_s=(0,\dots,0)$	symmetric side constraint

Side Constraints

The side constraint $\boldsymbol{\beta}_k = \mathbf{0}$ immediately yields the logit model with reference category k. If one chooses the side constraints $\boldsymbol{\beta}_{r_0} = \mathbf{0}$, one obtains

$$\log\left(\frac{P(Y=r|\boldsymbol{x})}{P(Y=r_0|\boldsymbol{x})}\right) = \boldsymbol{x}^T\boldsymbol{\beta}_r,$$

which is equivalent to choosing r_0 as the reference category. It should be noted that the choice of the reference category is essential for interpreting the parameters. If r_0 is the reference category, $\boldsymbol{\beta}_r$ determines the logits $\log(P(Y=r|\boldsymbol{x})/P(Y=r_0|\boldsymbol{x}))$.

A symmetric form of the side constraint is given by

$$\sum_{s=1}^{k}\boldsymbol{\beta}_s = \mathbf{0}.$$

Then parameter interpretation is quite different; it refers to the "median" response. Let the median response be defined by the geometric mean

$$GM(\boldsymbol{x}) = \sqrt[k]{\prod_{s=1}^{k} P(Y=s|\boldsymbol{x})} = (\prod_{s=1}^{k} P(Y=s|\boldsymbol{x}))^{1/k}.$$

Then one can derive from (8.5)

$$\log\left(\frac{P(Y=r|\boldsymbol{x})}{GM(\boldsymbol{x})}\right) = \boldsymbol{x}^T\boldsymbol{\beta}_r$$

(Exercise 8.2). Therefore, $\boldsymbol{\beta}_r$ reflects the effects of $\boldsymbol{x}$ on the logits when $P(Y=r|\boldsymbol{x})$ is compared to the geometric mean response $GM(\boldsymbol{x})$.

It should be noted that whatever side constraint is used, the log-odds between two response probabilities and the corresponding weight are given by

$$\log\left(\frac{P(Y=r|\boldsymbol{x})}{P(Y=s|\boldsymbol{x})}\right) = \boldsymbol{x}^T(\boldsymbol{\beta}_r - \boldsymbol{\beta}_s),$$

which follows from (8.5) for any choice of response categories $r, s \in \{1,\dots,k\}$. The transformations between different side constraints are rather simple. Let $\boldsymbol{\beta}_1,\dots,\boldsymbol{\beta}_q$ denote the vectors with side constraint $\boldsymbol{\beta}_k = \mathbf{0}$ and $\boldsymbol{\beta}_1^*,\dots,\boldsymbol{\beta}_q^*$ denote the vectors with symmetric side constraints. Then one obtains $\boldsymbol{\beta}_r = 2\boldsymbol{\beta}_r^* + \sum_{s\neq r, s<k}\boldsymbol{\beta}_s^*$ (Exercise 8.3).

The following example illustrates the interpretation of effects in the simple case with just one categorical covariate.

Example 8.3: Preference for Political Parties

Let us model the data from Table 8.1 with gender as the single explanatory variable (1: female, 0: male). The effect of gender on the preference is investigated by use of the logit model

$$\log\left(\frac{P(Y=r|x)}{P(Y=1|x)}\right) = \beta_{r0} + x_G\beta_r,$$

where $x_G = 1$ for female respondents and $x_G = 0$ for male respondents. Implicitly category 1 (CDU) has been chosen as the reference category ($\beta_{10} = 0$). The interpretation of parameters follows from

$$\beta_{r0} = \log\left(\frac{P(Y=r|x_G=0)}{P(Y=1|x_G=0)}\right), \qquad e^{\beta_{l0}} = \frac{P(Y=r|x_G=0)}{P(Y=1|x_G=0)},$$

$$\beta_r = \log\left(\frac{P(Y=r|x_G=1)/P(Y=1|x_G=1)}{P(Y=r|x_G=0)/P(Y=1|x_G=0)}\right), \quad e^{\beta_r} = \frac{P(Y=r|x_G=1)/P(Y=1|x_G=1)}{P(Y=r|x_G=0)/P(Y=1|x_G=0)}.$$

Thus $e^{\beta_{r0}}$ represents the odds of preference for party r instead of reference party 1 for male respondents, and e^{β_r} represents the odds ratio that compares the odds for female respondents to the odds for male respondents. The parameters are given in Table 8.2. For example, for male respondents, the odds of preference for party 3 instead of reference party 1 are 0.187. A comparison of the odds from female and male respondents yields 1.259, signaling that female respondents prefer the Green Party stronger than male respondents. □

TABLE 8.2: Parameter estimates for party preference data with covariate gender and reference category CDU.

	β_{0r}	$e^{\beta_{0r}}$	β_r	e^{β_r}
CDU (1)	0	1	0	1
SPD (2)	0.392	1.480	−0.068	0.934
Greens (3)	−1.677	0.187	0.230	1.259
Liberals (4)	−2.509	0.081	−0.112	0.894

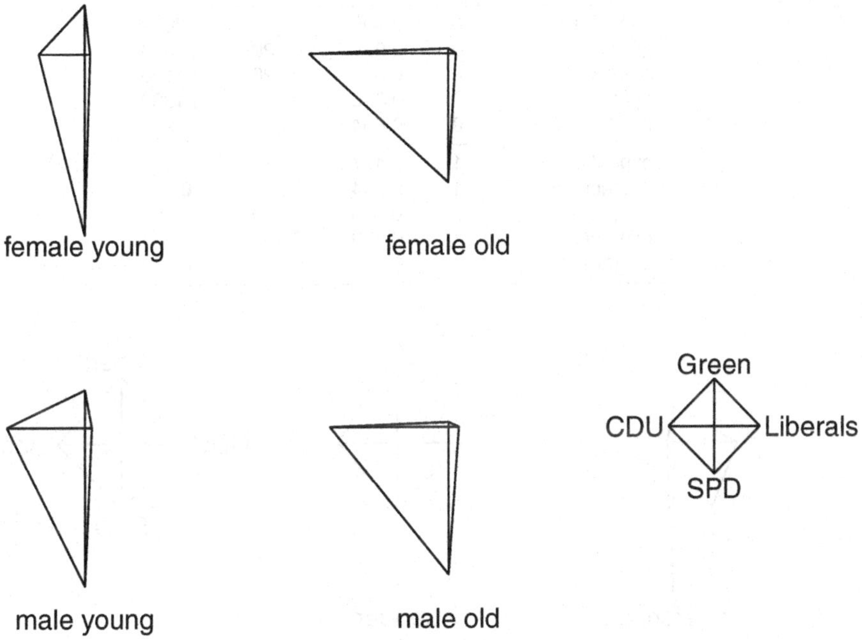

FIGURE 8.4: Star-plots for subpopulations of party preference data.

A simple way to visualize the response probabilities is by use of star-plots, which are in common use in multivariate statistics. The star-plot applied to response probabilities codes the probabilities or relative frequencies into the length of the rays emanating from the center of the plot. Figure 8.4 shows the resulting star-plots of relative frequencies for four subpopulations of the party preference data including one symmetric plot that serves to label the rays. It illustrates the strong effects of gender and age. In all the plots a strong preference for SPD is seen. But in the younger population there is a much stronger tendency toward the Green Party than in the older population; older voters prefer the CDU. The shifting of preference toward the Green Party is much stronger for females.

In more complex models, and when continuous predictors are included, it can be advantageous to represent the exponentials of parameters rather than subpopulations as star-plots. For illustration we will consider the main effect model for the party preference data (see also Exercise 8.5). Table 8.3 shows the fitted parameters and the exponentials. The latter represent the odds ratios and therefore the modification of the probabilities in comparison to the reference category. Figure 8.5 shows the corresponding star-plots. The first star-plot, which gives the exponentials of the intercept, represents the fitted odds in the reference population (male, age category 1). In all the plots the reference category among responses is CDU and the corresponding ray length is 1. The other plots of the exponentials of parameters show the modifications resulting from the covariates. It is in particular seen that females have a stronger tendency toward the Green Party when compared to the reference category of gender (male). For age, with reference category 1, it is seen that especially in age category 4 the tendency toward the Green Party is strongly reduced.

TABLE 8.3: Parameter estimates and exponentials for party preference data with covariates gender and age and reference category CDU.

	CDU	SPD	Greens	Liberals
intercept	0	0.905	−0.656	−2.3090
gender	0	−0.006	0.429	−0.0916
age2	0	−0.321	−0.328	−0.1100
age3	0	−0.386	−0.898	−0.1930
age4	0	−0.854	−2.910	−0.2970
exp(intercept)	1	2.471	0.518	0.099
exp(gender)	1	0.994	1.535	0.912
exp(age2)	1	0.725	0.720	0.895
exp(age3)	1	0.679	0.407	0.824
exp(age4)	1	0.425	0.054	0.743

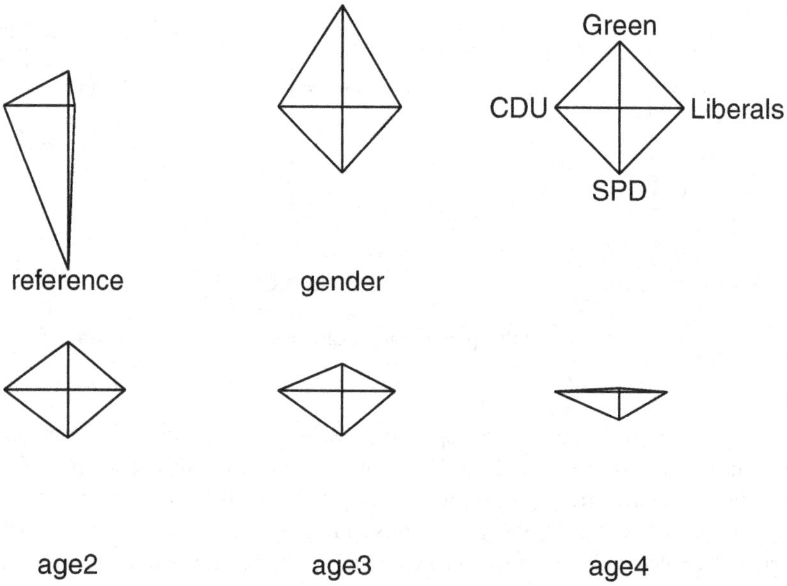

FIGURE 8.5: Star-plots of exponentials of fitted parameters for main effect model of party preference data.

8.3 Multinomial Model as Random Utility Model

In Section 8.2 the multinomial logit model was considered as a generalization of the binary logit model. There is an alternative motivation of the model that is not based on binary models but on random utilities. Although random utilities are considered more extensively within the framework of discrete choice models (Section 8.8), they are briefly sketched here because it helps to structure the linear predictor.

Let U_r be an unobservable random utility associated with the rth response category. For example, U_r is the (subjective) utility of a brand when the choice is among brands $1, \ldots, k$ or the "attractiveness" of the rth political party. Let U_r be given by

$$U_r = u_r + \varepsilon_r,$$

where u_r is a fixed value, representing the utility associated with the rth response category, and $\varepsilon_1, \ldots, \varepsilon_k$ are iid random variables with distribution function F. Now let the response Y be determined by the *principle of maximum random utility*, which specifies the link between the observable Y and the unobservable random utility by

$$Y = r \quad \Leftrightarrow \quad U_r = \max_{j=1,\ldots,k} U_j.$$

Therefore, the alternative r is chosen that maximizes the random utility. If one assumes that $\varepsilon_r, \ldots, \varepsilon_k$ are iid variables with distribution function $F(x) = \exp(-\exp(-x))$, which is the Gumbel or maximum extreme value distribution, one obtains

$$P(Y = r) = \frac{\exp(u_r)}{\sum_{j=r}^{k} \exp(u_j)}$$

(e.g., Yellott, 1977; McFadden, 1973). The resulting model corresponds to the generic form of the logit model (8.5). The fixed utilities are unique only up to additive constants. Therefore, one needs some additional side constraints; for example, one may consider the differences $u_r - u_k, r = 1, \ldots, k-1$, which is equivalent to considering k as the reference category. As shown in the next section, if one lets the fixed utilities $u_1, \ldots, u_k$ depend linearly on covariates one obtains the parametric multinomial logit model.

8.4 Structuring the Predictor

The covariates come into the multinomial logit model in the form of linear predictors:

$$\eta_r = \boldsymbol{x}^T \boldsymbol{\beta}_r.$$

In the same way as in univariate response models, the linear predictor may contain dummy variables for categorical covariates, polynomial terms for continuous variables, and interaction terms between both types of variables.

Apart from these transformations of the original observations for the structuring of the linear predictor, it is often useful to distinguish between two types of covariates, namely, *global* and *category-specific* variables. For example, when an individual chooses among alternatives $1, \ldots, k$, one may model the effect of characteristics of the individual like age and gender, which are global variables, but also account for measured attributes of the alternatives $1, \ldots, k$, which are category-specific variables. When the choice refers to transportation mode, the potential attributes are price and duration, which vary across the alternatives and therefore are category-specific.

Let $\boldsymbol{x}$ denote the individual characteristics and $\boldsymbol{w}_1, \ldots, \boldsymbol{w}_k$ denote the set of attributes of alternatives, where $\boldsymbol{w}_r$ are the attributes of category r. The first type of variable is called *global*, and the latter type *category-specific*. Then the set of linear predictors may be generalized to

$$\eta_r = \boldsymbol{x}^T \boldsymbol{\beta}_r + (\boldsymbol{w}_r - \boldsymbol{w}_k)^T \boldsymbol{\alpha}, \qquad r = 1, \ldots, k-1, \tag{8.6}$$

where k is chosen as the reference category. The first term specifies the effect of the global variables, and the second term specifies the effect of the difference $\boldsymbol{w}_r - \boldsymbol{w}_k$ on the choice between category r and the reference category. When $\boldsymbol{w}_r$ stands for the price of alternative r, it is quite natural to assume that the choice between alternatives r and k is determined by the difference.

The predictor (8.6) may be derived by maximizing the latent utilities. Let the latent utility of category r be specified by $u_r = \boldsymbol{x}^T \boldsymbol{\gamma}_r + \boldsymbol{w}_r^T \boldsymbol{\alpha}$. Then the difference is

$$u_r - u_k = \boldsymbol{x}^T(\boldsymbol{\gamma}_r - \boldsymbol{\gamma}_k) + (\boldsymbol{w}_r - \boldsymbol{w}_k)^T \boldsymbol{\alpha},$$

which has the form given in (8.6) with $\boldsymbol{\beta}_r = (\boldsymbol{\gamma}_r - \boldsymbol{\gamma}_k)$. Of course one may also specify interactions between the two types of variables. Let, for example, x_G be a dummy variable for gender and w_r denote the price of alternative r. Then the model

$$\eta_r = \beta_{0r} + x_G \beta_G + (w_r - w_k)\alpha_1 + x_G(w_r - w_k)\alpha_2$$

allows that the effect of prices depends on gender.

Formally, it is not necessary to distinguish between global and category-specific variables. One may always define one long vector of variables that contains all the specified variables. For example, the predictor with only global variables $\eta_r = \boldsymbol{x}^T \boldsymbol{\beta}_r$ may also be written as $\eta_r = (\boldsymbol{0}^T, \ldots, \boldsymbol{x}^T, \ldots, \boldsymbol{0}^T)\boldsymbol{\beta}$, where $\boldsymbol{\beta}^T = (\boldsymbol{\beta}_1^T, \ldots, \boldsymbol{\beta}_q^T)$. The model with category-specific variables $\eta_r = \boldsymbol{x}^T \boldsymbol{\beta}_r + (\boldsymbol{w}_r^T - \boldsymbol{w}_k^T)\boldsymbol{\alpha}$ has the form $\eta_r = (\boldsymbol{0}^T, \ldots, \boldsymbol{x}^T, \ldots, \boldsymbol{0}^T, \boldsymbol{w}_r^T - \boldsymbol{w}_k^T)\boldsymbol{\beta}$, where $\boldsymbol{\beta}$ is now given as $\boldsymbol{\beta}^T = (\boldsymbol{\beta}_1^T, \ldots, \boldsymbol{\beta}_q^T, \boldsymbol{\alpha}^T)$. Thus one always obtains the form $\eta_r = \boldsymbol{x}_r^T \boldsymbol{\beta}$, where $\boldsymbol{x}_r$ may or may not depend on r.

In econometrics, sometimes also for category-specific predictors category-specific effects are assumed. The latent utility $u_r = \boldsymbol{x}^T \boldsymbol{\gamma}_r + \boldsymbol{w}_r^T \boldsymbol{\alpha}_r$ with category-specific effect $\boldsymbol{\alpha}_r$ yields the difference of utilities that defines the linear predictor for reference category k

$$\eta_r = u_r - u_k = \boldsymbol{x}^T \boldsymbol{\beta}_r + \boldsymbol{w}_r^T \boldsymbol{\alpha}_r - \boldsymbol{w}_k^T \boldsymbol{\alpha}_k,$$

where $\boldsymbol{\beta}_r = (\boldsymbol{\gamma}_r - \boldsymbol{\gamma}_k)$. The total set of parameters that defines the total vector $\boldsymbol{\beta}$ now contains the $k-1$ $\boldsymbol{\beta}$-parameters $\boldsymbol{\beta}_1, \ldots, \boldsymbol{\beta}_{k-1}$, and the k $\boldsymbol{\alpha}$-vectors $\boldsymbol{\alpha}_1, \ldots, \boldsymbol{\alpha}_k$.

Example 8.4: Travel Mode

The choice of travel mode of $n = 840$ passengers in Australia was investigated by Greene (2003). The data are available from the R package *Ecdat*. The alternatives of travel mode were air, train, bus, and car, which have frequencies $0.276, 0.300, 0.142$, and 0.280. Air serves as the reference category. As category-specific variables we consider travel time in vehicle (timevc) and cost, and as the global variable we consider household income (income). The estimates given in Table 8.4 show that income seems to be influential for the preference of train and bus over airplane. Moreover, time in vehicle seems to matter for the preference of the travel mode. Cost turns out to be non-influential if income is in the predictor (see also Exercise 8.10). □

TABLE 8.4: Estimated coefficients for travel mode data.

	Estimate	Std. Error	z-Value	Pr(>|z|)
train	3.525	0.654	5.381	0.0
bus	2.278	0.717	3.174	0.001
car	1.533	0.706	2.170	0.029
train:income	−0.056	0.012	−4.588	0.0
bus:income	−0.035	0.013	−2.705	0.006
car:income	−0.002	0.010	−0.226	0.820
timevc	−0.003	0.001	−3.274	0.001
cost	−0.001	0.005	−0.293	0.769

8.5 Logit Model as Multivariate Generalized Linear Model

For simplicity, let $\pi_r = P(Y = r|\boldsymbol{x}, \{\boldsymbol{w}_j\})$ denote the response probability for one observation Y with covariates $\boldsymbol{x}, \{\boldsymbol{w}_j\}$, where $\boldsymbol{w}_j$ are category-specific attributes. Then, with $\tilde{\boldsymbol{w}}_r = \boldsymbol{w}_r - \boldsymbol{w}_k$ and the linear predictor $\eta_r = \boldsymbol{x}^T\boldsymbol{\beta}_r + \tilde{\boldsymbol{w}}_r^T\boldsymbol{\alpha}$, one may write the q equations that specify the nominal logit model with reference category k in matrix form by

$$\begin{pmatrix} \log(\pi_1/(1-\pi_1-\cdots-\pi_q)) \\ \vdots \\ \log(\pi_q/(1-\pi_1-\cdots-\pi_q)) \end{pmatrix} = \begin{pmatrix} \boldsymbol{x}^T & & 0 & \tilde{\boldsymbol{w}}_1^T \\ & \ddots & & \vdots \\ 0 & & \boldsymbol{x}^T & \tilde{\boldsymbol{w}}_q^T \end{pmatrix} \begin{pmatrix} \boldsymbol{\beta}_1 \\ \vdots \\ \boldsymbol{\beta}_q \\ \boldsymbol{\alpha} \end{pmatrix}. \tag{8.7}$$

The predictor for the rth logit $\log(\pi_r/(1-\pi_1-\ldots-\pi_q))$ has the form

$$\eta_r = \boldsymbol{x}^T\boldsymbol{\beta}_r + \tilde{\boldsymbol{w}}_r^T\boldsymbol{\alpha} = (0,\ldots,0,\boldsymbol{x}^T,0,\ldots,\tilde{\boldsymbol{w}}_r^T)\boldsymbol{\beta} = \boldsymbol{x}_r\boldsymbol{\beta},$$

where $\boldsymbol{\beta}^T = (\boldsymbol{\beta}_1^T,\ldots\boldsymbol{\beta}_q^T,\boldsymbol{\alpha}^T)$ and $\boldsymbol{x}_r = (0,\ldots,0,\boldsymbol{x}^T,0,\ldots,0,\tilde{\boldsymbol{w}}_r)$ is the corresponding design vector. Thus the general form of (8.7) is

$$g(\boldsymbol{\pi}) = \boldsymbol{X}\boldsymbol{\beta},$$

where $\boldsymbol{\pi}^T = (\pi_1,\ldots,\pi_q)$ is the vector the of response probabilities, $\boldsymbol{X}$ is a design matrix that corresponds to the total parameter vector $\boldsymbol{\beta}$, and g is the link function. For the logit model (8.7) the vector-valued *link function* $g = (g_1,\ldots,g_q) : \mathbb{R}^q \to \mathbb{R}^q$ is given by

$$g_r(\pi_1,\ldots,\pi_q) = \log\left(\frac{\pi_r}{1-\pi_1-\cdots-\pi_q}\right).$$

As usual in generalized linear models, an equivalent form is

$$\boldsymbol{\pi} = h(\boldsymbol{X}\boldsymbol{\beta}),$$

where $h = (h_1,\ldots,h_q) = g^{-1}$ is the response function, which in the present case has components

$$h_r(\eta_1,\ldots,\eta_q) = \frac{\exp(\eta_r)}{1+\sum_{s=1}^q \exp(\eta_s)}.$$

Thus, for *one* observation, the nominal logit model has the general form

$$g(\boldsymbol{\pi}) = \boldsymbol{X}\boldsymbol{\beta} \quad \text{or} \quad \boldsymbol{\pi} = h(\boldsymbol{X}\boldsymbol{\beta})$$

for an appropriately chosen vector-valued link function, design matrix, and parameter vector. For the given data $(\boldsymbol{y}_i, \boldsymbol{x}_i), i = 1, \dots, n$, one has

$$g(\boldsymbol{\pi}(\boldsymbol{x}_i)) = \boldsymbol{X}_i\boldsymbol{\beta} \quad \text{or} \quad \boldsymbol{\pi}(\boldsymbol{x}_i) = h(\boldsymbol{X}_i\boldsymbol{\beta}),$$

where $\boldsymbol{\pi}(\boldsymbol{x})^T = (\pi_1(\boldsymbol{x}), \dots, \pi_q(\boldsymbol{x}))$, $\pi_r(\boldsymbol{x}_i) = P(Y_i = r|\boldsymbol{x}_i)$, and $\boldsymbol{X}_i$ is composed from the covariates $\boldsymbol{x}_i$.

8.6 Inference for Multicategorical Response Models

Let the data be given by $(\boldsymbol{y}_i, \boldsymbol{x}_i), i = 1, \dots, N$, where $\boldsymbol{x}_i$ contains all the covariates that are observed, including category-specific variables. Given $\boldsymbol{x}_i$, one assumes a multinomial distribution, $\boldsymbol{y}_i \sim \mathrm{M}(n_i, \boldsymbol{\pi}_i)$. This means that at measurement point $\boldsymbol{x}_i$ one has n_i observations. In particular, if $\boldsymbol{x}_i$ contains categorical variables, usually more than one observation is collected at a fixed value of covariates. For example, in a sample survey respondents may be characterized by gender and educational level. Then, for fixed values of gender and educational level, one usually observes more than one response. This may be seen as a grouped data case, where the number of counts stored in $\boldsymbol{y}_i$ is the sum of the responses of single respondents, with each one having multinomial distribution $\mathrm{M}(1, \boldsymbol{\pi}_i)$. Instead of the multinomial distribution $\boldsymbol{y}_i \sim \mathrm{M}(n_i, \boldsymbol{\pi}_i)$ one may also consider the scaled multinomials or proportions $\bar{\boldsymbol{y}}_i = \boldsymbol{y}_i/n_i$, which are also denoted by $\boldsymbol{p}_i = \bar{\boldsymbol{y}}_i$. The components of $\boldsymbol{p}_i^T = (p_{i1}, \dots, p_{iq})$ contain the relative frequencies, and p_{ir} is the proportion of observations in category r. The proportions $\bar{\boldsymbol{y}}_i = \boldsymbol{p}_i$ have the advantage that $\mathrm{E}(\boldsymbol{p}_i) = \boldsymbol{\pi}_i$, whereas for $\boldsymbol{y}_i = n_i\boldsymbol{p}_i$ one has $\mathrm{E}(\boldsymbol{y}_i) = n_i\boldsymbol{p}_i$. The model that is assumed to hold has the form

$$g(\boldsymbol{\pi}_i) = \boldsymbol{X}_i\boldsymbol{\beta} \quad \text{or} \quad \boldsymbol{\pi}_i = h(\boldsymbol{X}_i\boldsymbol{\beta}).$$

8.6.1 Maximum Likelihood Estimation

The multinomial distribution has the form of a multivariate exponential family. Let $\boldsymbol{y}_i^T = (y_{i1}, \dots, y_{iq}) \sim \mathrm{M}(n_i, \boldsymbol{\pi}_i)$, $i = 1, \dots, N$, denote the multinomial distribution with $k = q+1$ categories. Then the probability mass function is

$$\begin{aligned} f(\boldsymbol{y}_i) &= \frac{n_i!}{y_{i1}! \cdots y_{iq}!(n_i - y_{i1} - \cdots - y_{iq})!} \pi_{i1}^{y_{i1}} \cdot \ldots \cdot \pi_{iq}^{y_{iq}} (1 - \pi_{i1} - \ldots - \pi_{iq})^{(n_i - y_{i1} - \ldots - y_{iq})} \\ &= \exp\left(\boldsymbol{y}_i^T\boldsymbol{\theta}_i + n_i \log(1 - \pi_{i1} - \ldots - \pi_{iq}) + \log(c_i)\right) \\ &= \exp\left([\boldsymbol{p}_i^T\boldsymbol{\theta}_i + \log(1 - \pi_{i1} - \ldots - \pi_{iq})]/(1/n_i) + \log(c_i)\right), \end{aligned}$$

where the canonical parameter vector is $\boldsymbol{\theta}_i^T = (\theta_{i1}, \dots, \theta_{iq})$, $\theta_{ir} = \log(\pi_{ir}/(1 - \pi_{i1} - \cdots - \pi_{iq}))$, the dispersion parameter is $1/n_i$, and $c_i = n_i!/(y_{i1}! \dots y_{iq}!(n_i - y_{i1} - \cdots - y_{iq})!)$. One obtains the likelihood

$$L(\boldsymbol{\beta}) = \prod_{i=1}^{N} c_i \, \pi_{i1}^{y_{i1}} \cdot \ldots \cdot \pi_{iq}^{y_{iq}} (1 - \pi_{i1} - \cdots - \pi_{iq})^{n_i - y_{i1} - \cdots - y_{iq}}$$

and the log-likelihood $l(\boldsymbol{\beta}) = \sum_{i=1}^{N} l_i(\boldsymbol{\pi}_i)$ with

$$\begin{aligned} l_i(\boldsymbol{\pi}_i) = n_i &\left\{ \sum_{r=1}^{q} p_{ir} \log\left(\frac{\pi_{ir}}{1 - \pi_{i1} - \cdots - \pi_{iq}}\right) + \log(1 - \pi_{i1} - \cdots - \pi_{iq}) \right\} \\ &+ \log(c_i). \end{aligned} \tag{8.8}$$

The score function $\boldsymbol{s}(\boldsymbol{\beta}) = \partial l(\boldsymbol{\beta})/\partial \boldsymbol{\beta}$ for the model $\boldsymbol{\pi}_i = h(\boldsymbol{X}_i\boldsymbol{\beta})$ has the form

$$\boldsymbol{s}(\boldsymbol{\beta}) = \sum_{i=1}^{N} \boldsymbol{X}_i^T \boldsymbol{D}_i(\boldsymbol{\beta})\boldsymbol{\Sigma}_i^{-1}(\boldsymbol{\beta})(\boldsymbol{p}_i - \boldsymbol{\pi}_i),$$

where $\boldsymbol{\eta}_i = \boldsymbol{X}_i\boldsymbol{\beta}$ and $\boldsymbol{D}_i(\boldsymbol{\beta}) = \partial h(\boldsymbol{\eta}_i)/\partial \boldsymbol{\eta} = (\partial g(\boldsymbol{\pi}_i)/\partial \boldsymbol{\pi})^{-1}$. It should be noted that the matrix $\boldsymbol{D}_i(\boldsymbol{\beta})$ with entries $\partial h_r(\boldsymbol{\eta}_i)/\partial \eta_s$ is not a symmetric matrix. The covariance $\boldsymbol{\Sigma}_i(\boldsymbol{\beta})$ is determined by the multinomial distribution and has the form

$$\begin{aligned}\boldsymbol{\Sigma}_i(\boldsymbol{\beta}) &= \frac{1}{n_i}\begin{pmatrix} \pi_{i1}(1-\pi_{i1}) & -\pi_{i1}\pi_{i2} & \dots & -\pi_{i1}\pi_{iq} \\ & \pi_{i2}(1-\pi_{i2}) & & \\ & & \ddots & \\ -\pi_{iq}\pi_{i1} & & & \pi_{iq}(1-\pi_{iq}) \end{pmatrix} \\ &= [\,\text{Diag}\,(\boldsymbol{\pi}_i) - \boldsymbol{\pi}_i\boldsymbol{\pi}_i^T]/n_i.\end{aligned}$$

In closed form one obtains

$$\boldsymbol{s}(\boldsymbol{\beta}) = \boldsymbol{X}^T\boldsymbol{D}(\boldsymbol{\beta})\boldsymbol{\Sigma}(\boldsymbol{\beta})^{-1}(\boldsymbol{p}-\boldsymbol{\pi}) = \boldsymbol{X}^T\boldsymbol{W}(\boldsymbol{\beta})\boldsymbol{D}(\boldsymbol{\beta})^{-T}(\boldsymbol{p}-\boldsymbol{\pi}),$$

where $\boldsymbol{X}^T = (\boldsymbol{X}_1^T, \dots, \boldsymbol{X}_N^T)$, $\boldsymbol{D}(\boldsymbol{\beta})$ and $\boldsymbol{\Sigma}(\boldsymbol{\beta})$ are block-diagonal matrices with blocks $\boldsymbol{D}_i(\boldsymbol{\beta})$, $\boldsymbol{\Sigma}_i(\boldsymbol{\beta})$, respectively, and $\boldsymbol{W}(\boldsymbol{\beta}) = \boldsymbol{D}(\boldsymbol{\beta})\boldsymbol{\Sigma}(\boldsymbol{\beta})^{-1}\boldsymbol{D}(\boldsymbol{\beta})^T$ is a block-diagonal matrix with blocks $\boldsymbol{W}_i(\boldsymbol{\beta}) = \boldsymbol{D}_i(\boldsymbol{\beta})\boldsymbol{\Sigma}_i(\boldsymbol{\beta})^{-1}\boldsymbol{D}_i(\boldsymbol{\beta})^T$. For $\boldsymbol{\Sigma}_i(\boldsymbol{\beta})^{-1}$ an explicit form is available (see Exercise 8.6).

The expected information or Fisher matrix $\boldsymbol{F}(\boldsymbol{\beta}) = \mathrm{E}\left(-\partial l/\partial\boldsymbol{\beta}\partial\boldsymbol{\beta}^T\right) = \mathrm{cov}(\boldsymbol{s}(\boldsymbol{\beta}))$ has the form

$$\boldsymbol{F}(\boldsymbol{\beta}) = \sum_{i=1}^{N} \boldsymbol{X}_i^T\boldsymbol{W}_i(\boldsymbol{\beta})\boldsymbol{X}_i = \boldsymbol{X}^T\boldsymbol{W}(\boldsymbol{\beta})\boldsymbol{X}.$$

The blocks $\boldsymbol{W}_i(\boldsymbol{\beta})$ in the weight matrix can also be given in the form $\boldsymbol{W}_i(\boldsymbol{\beta}) = (\frac{\partial g(\boldsymbol{\pi}_i)}{\partial \boldsymbol{\pi}^T}\boldsymbol{\Sigma}_i(\boldsymbol{\beta})\frac{\partial g(\boldsymbol{\pi}_i)}{\partial \boldsymbol{\pi}})^{-1}$, which is an approximation to the inverse of the covariance of $g(\boldsymbol{p}_i)$ when the model holds. For the logit model, which corresponds to the canonical link, simpler forms of the score function and the Fisher matrix can be found (see Exercise 8.4).

The estimate $\hat{\boldsymbol{\beta}}$ is under regularity conditions asymptotically $(n_1+\dots+n_N \to \infty)$ normally distributed with

$$\hat{\boldsymbol{\beta}} \overset{(a)}{\sim} \mathrm{N}(\boldsymbol{\beta}, \boldsymbol{F}(\hat{\boldsymbol{\beta}})^{-1});$$

for details see Fahrmeir and Kaufmann (1985). The score function and Fisher matrix have the same forms as in univariate GLMs, namely, $\boldsymbol{s}(\boldsymbol{\beta}) = \boldsymbol{X}^T\boldsymbol{D}(\boldsymbol{\beta})\boldsymbol{\Sigma}(\boldsymbol{\beta})^{-1}(\boldsymbol{p}-\boldsymbol{\pi})$ and $\boldsymbol{F}(\boldsymbol{\beta}) = \boldsymbol{X}^T\boldsymbol{W}(\boldsymbol{\beta})\boldsymbol{X}$. But for multicategorical responses the design matrix is composed from matrices for single observations, and the weight matrix $\boldsymbol{W}(\boldsymbol{\beta})$ as well as the matrix of derivatives $\boldsymbol{D}(\boldsymbol{\beta})$ are block-diagonal matrices in contrast to the univariate models, where $\boldsymbol{W}(\boldsymbol{\beta})$ and $\boldsymbol{D}(\boldsymbol{\beta})$ are diagonal matrices.

Separate Fitting of Binary Models

When one considers only two categories, say r and k, the multinomial model looks like a binary logit model. If k is chosen as reference category, one has

$$\log(\frac{P(Y=r|\boldsymbol{x})}{P(Y=k|\boldsymbol{x})}) = \boldsymbol{x}^T\boldsymbol{\beta}_r.$$

Therefore, the parameters $\boldsymbol{\beta}_r$ can also be estimated by fitting a binary logit model using only observations in categories r and k. The resulting estimates are conditional on classification in categories r and k. The estimates obtained by fitting $k-1$ separate binary models differ from the estimates obtained from the full likelihood of the multinomial model. In particular, they tend to have larger standard errors, although the effect is usually small if the reference category is chosen as the category with most of the observations (Begg and Gray, 1984). Another advantage of the multinomial likelihood is that the testing of hypotheses that refer to parameters that are linked to different categories is straightforward, for example, by using likelihood ratio tests.

8.6.2 Goodness-of-Fit

As in univariate GLMs, the goodness-of-fit may be checked by the Pearson statistic and the deviance. Again asymptotic results are obtained only for grouped data, where the number of repetitions n_i taken at observation vector $\boldsymbol{x}_i$ is not too small.

Pearson Statistic

When considering the discrepancy between observations and fit, one should have in mind that responses are vector-valued. One wants to compare the observation vectors $\boldsymbol{p}_i = \boldsymbol{y}_i/n$ and the fitted vector $\boldsymbol{\pi}_i$, where both vectors have dimension q, since one category (in our case the last one) is omitted from the vector. By using for the last category the observation $p_{ik} = 1 - p_{i1} - \cdots - p_{iq}$ and the fit $\hat{\pi}_{ik} = 1 - \hat{\pi}_{i1} - \cdots - \hat{\pi}_{iq}$, one defines the quadratic *Pearson residual* of the ith observation by

$$\chi_P^2(\boldsymbol{p}_i, \hat{\boldsymbol{\pi}}_i) = \sum_{r=1}^{k} n_i \frac{(p_{ir} - \hat{\pi}_{ir})^2}{\hat{\pi}_{ir}}.$$

The corresponding Pearson statistic is given by

$$\chi_P^2 = \sum_{i=1}^{N} \chi_P^2(\boldsymbol{p}_i, \hat{\boldsymbol{\pi}}_i) = \sum_{i=1}^{N} (\boldsymbol{p}_i - \hat{\boldsymbol{\pi}}_i)^T \boldsymbol{\Sigma}_i^{-1}(\hat{\boldsymbol{\beta}})(\boldsymbol{p}_i - \hat{\boldsymbol{\pi}}_i),$$

where the last form is derived by explicitly deriving the inverse of the $q \times q$-matrix $\boldsymbol{\Sigma}_i(\hat{\boldsymbol{\beta}}) = [\,\text{Diag}\,(\hat{\boldsymbol{\pi}}_i) - \boldsymbol{\pi}_i \boldsymbol{\pi}_i^T]/n_i$.

Deviance

The deviance for multinomial (grouped) observations is given by $D = 2\sum_{i=1}^{N} l_i(\boldsymbol{p}_i) - l_i(\hat{\boldsymbol{\pi}}_i)$, yielding

$$D = 2\sum_{i=1}^{N} n_i \sum_{r=1}^{k} p_{ir} \log\left(\frac{p_{ir}}{\hat{\pi}_{ir}}\right).$$

The corresponding quadratic *deviance residuals* are given by

$$\chi_D^2(\boldsymbol{p}_i, \hat{\boldsymbol{\pi}}_i) = 2n_i \sum_{r=1}^{k} p_{ir} \log\left(\frac{p_{ir}}{\hat{\pi}_{ir}}\right).$$

Under the assumptions of the fixed cells asymptotic ($n_i/N \to \lambda_i \in (0,1)$) and regularity conditions, χ_P^2 and D are asymptotically χ^2-distributed with $N(k-1)-p$ degrees of freedom, where N is the number of (grouped) observations, k is the number of response categories and p is the number of estimated parameters.

Goodness-of-Fit Tests

Pearson statistic

$$\chi_P^2 = \sum_{i=1}^{N} \chi_P^2(\boldsymbol{p}_i, \hat{\boldsymbol{\pi}}_i)$$

with $$\chi_P^2(\boldsymbol{p}_i, \hat{\boldsymbol{\pi}}_i) = n_i \sum_{r=1}^{k} (p_{ir} - \hat{\pi}_{ir})^2 / \hat{\pi}_{ir}$$

Deviance

$$D = \sum_{i=1}^{N} \chi_D^2(\boldsymbol{p}_i, \hat{\boldsymbol{\pi}}_i)$$

with $$\chi_D^2(\boldsymbol{p}_i, \hat{\boldsymbol{\pi}}_i) = 2n_i \sum_{r=1}^{k} p_{ir} \log\left(\frac{p_{ir}}{\hat{\pi}_{ir}}\right)$$

D, χ_P^2 are compared to the asymptotic χ^2-distribution with $N(k-1) - p$ degrees of freedom $(n_i/N \to \lambda_i \in [0, 1])$.

As in the case of the binomial distribution, one should be cautious when using these goodness-of-fit statistics if n_i is small. For the ungrouped case, for example, if covariates are continuous, one has $n_i = 1$ for all observations and the deviance becomes

$$D = -2 \sum_{r=1}^{k} I(Y_i = r) \log(\hat{\pi}_{ir}) = -2 \sum_{i=1}^{N} \log(\hat{\pi}_{iY_i}),$$

where $Y_i \in \{1, \ldots, k\}$ denotes the ith observation, and I is the indicator function with $I(a) = 1$ if a holds and $I(a) = 0$ otherwise. The value of D should certainly not be compared to quantiles of the χ^2-distribution since one has N observations and $N(k-1)$ degrees of freedom. For small values of n_i alternative asymptotic concepts should be used (see also next section).

Power-Divergence Family

A general family of goodness-of-fit statistics that comprises the deviance and the Pearson statistic is the power-divergence family, which for $\lambda \in (-\infty, \infty)$ has the form

$$S_\lambda = \sum_{i=1}^{N} SD_\lambda(\boldsymbol{p}_i, \hat{\boldsymbol{\pi}}_i),$$

where

$$SD_\lambda(\boldsymbol{p}_i, \hat{\boldsymbol{\pi}}_i) = \frac{2n_i}{\lambda(\lambda+1)} \sum_{r=1}^{k} p_{ir} \left[\left(\frac{p_{ir}}{\hat{\pi}_{ir}}\right)^\lambda - 1\right]. \tag{8.9}$$

As special cases, one obtains for $\lambda = 1$ the Pearson statistic and for the limit $\lambda \to 0$ the deviance. However, the family includes further statistics that have been proposed in the literature; in particular one obtains for the limit $\lambda \to -1$ Kullback's minimum discrimination information statistic with $SD_{-1}(\boldsymbol{p}_i, \hat{\boldsymbol{\pi}}_i) = n_i \sum_r \hat{\pi}_{ir} \log(\hat{\pi}_{ir}/p_{ir})$ and for $\lambda = -2$ Neyman's minimum modified χ^2-statistic with $SD_{-2}(\boldsymbol{p}_i, \hat{\boldsymbol{\pi}}_i) = n_i \sum_r (p_{ir} - \hat{\pi}_{ir})^2 / p_{ir}$.

Under the assumptions of fixed cells asymptotics one obtains the same asymptotic χ^2-distribution for any λ, if estimates $\hat{\pi}_{ir}$ are best asymptotically normal (BAN-) distributed estimates, as for example the ML estimates or estimates obtained from minimizing S_λ. Fixed cells asymptotics postulates in particular fixed numbers of groups and large values of n_i. If the number of observations n_i at a fixed value $\boldsymbol{x}_i$ is small, the usual asymptotic fails. An alternative is *increasing cells asymptotics*, which allows that with increasing sample size $n = n_1 + \cdots + n_N \to \infty$ the number of groups also increases $N \to \infty$. However, under increasing cells asymptotics the asymptotic distribution is normal and depends on λ. Therefore, if goodness-of-fit statistics like the deviance and Pearson statistic differ strongly, one might suspect that fixed asymptotics does not apply. For more details on increasing cells asymptotics see Read and Cressie (1988) and Osius and Rojek (1992).

Further test statistics are variations of the tests for binary responses considered in Section 4.2.3. One is the Hosmer-Lemeshow statistic, which has been extended to the multinomial model by Pigeon and Heyse (1999), and the other is based on smoothing of residuals. It has been adapted to the multinomial model by Goeman and le Cessie (2006). Of course problems found in the binary case, namely, low power of the Hosmer-Lemeshow statistic and the restriction to low dimensions for smoothed residuals, carry over to the multinomial case.

8.6.3 Diagnostic Tools

Hat Matrix

For multicategory response models, the iteratively reweighted least-squares fitting procedure (see Chapter 3) has the form

$$\hat{\boldsymbol{\beta}}^{(l+1)} = (\boldsymbol{X}^T \boldsymbol{W}(\boldsymbol{\beta}^{(l)})\boldsymbol{X})^{-1}\boldsymbol{X}^T \boldsymbol{W}(\boldsymbol{\beta}^{(l)})\tilde{\boldsymbol{\eta}}(\hat{\boldsymbol{\beta}}^{(l)}),$$

with $\tilde{\boldsymbol{\eta}}(\hat{\boldsymbol{\beta}}^{(l)}) = \boldsymbol{X}\hat{\boldsymbol{\beta}} + \boldsymbol{D}^{-1}(\hat{\boldsymbol{\beta}})^T(\boldsymbol{p}_i - \boldsymbol{\pi}_i(\boldsymbol{\beta}))$, and one obtains at convergence

$$\hat{\boldsymbol{\beta}} = (\boldsymbol{X}^T \boldsymbol{W}(\boldsymbol{\beta})\boldsymbol{X})^{-1}\boldsymbol{X}^T \boldsymbol{W}(\boldsymbol{\beta})\tilde{\boldsymbol{\eta}}(\hat{\boldsymbol{\beta}}).$$

The corresponding hat matrix is

$$\boldsymbol{H} = \boldsymbol{W}^{T/2}(\hat{\boldsymbol{\beta}})\boldsymbol{X}\boldsymbol{F}^{-1}(\hat{\boldsymbol{\beta}})\boldsymbol{X}^T\boldsymbol{W}^{1/2}(\hat{\boldsymbol{\beta}}).$$

$\boldsymbol{H}$ is an $(Nq \times Nq)$-matrix with blocks $\boldsymbol{H}_{ij}$. The $(q \times q)$-matrix $\boldsymbol{H}_{ii}$ corresponds to the ith observation. As indicators for leverage one can use $\det(\boldsymbol{H}_{ii})$ or $\mathrm{tr}(\boldsymbol{H}_{ii})$.

Residuals

The vector of raw residuals is given by $\boldsymbol{p}_i - \hat{\boldsymbol{\pi}}_i$. Correcting for the variances of $\boldsymbol{p}_i$ yields the *Pearson residual*:

$$\boldsymbol{r}_P(\boldsymbol{p}_i, \hat{\boldsymbol{\pi}}_i) = \boldsymbol{\Sigma}_i^{-1/2}(\hat{\boldsymbol{\beta}})(\boldsymbol{p}_i - \hat{\boldsymbol{\pi}}_i),$$

which forms the Pearson statistic $\chi^2_P = \sum_i \boldsymbol{r}_P(\boldsymbol{p}_i, \hat{\boldsymbol{\pi}}_i)^T \boldsymbol{r}_P(\boldsymbol{p}_i, \hat{\boldsymbol{\pi}}_i)$. A plot of the squared Pearson residuals shows which observations have a strong impact on the goodness-of-fit. The Pearson residuals themselves are vector-valued and show how well categories are fitted.

Standardized Pearson residuals, which correct for the variance of the Pearson residual, have the form

$$r_P(\boldsymbol{p}_i, \hat{\boldsymbol{\pi}}_i) = I - \boldsymbol{H}_{ii}^{-1/2}\boldsymbol{\Sigma}_i^{-1/2}(\hat{\boldsymbol{\beta}})(\boldsymbol{p}_i - \hat{\boldsymbol{\pi}}_i),$$

which uses the approximation $\text{cov}(\boldsymbol{p}_i - \hat{\boldsymbol{\pi}}_i) \simeq \boldsymbol{\Sigma}_i^{1/2}(\hat{\boldsymbol{\beta}})(\boldsymbol{I} - \boldsymbol{H}_{ii})\boldsymbol{\Sigma}_i^{T/2}(\hat{\boldsymbol{\beta}})$. The approximation may be derived in the same way as in the univariate case (see Section 3.10). More details on regression diagnostics are found in Lesaffre and Albert (1989).

Testing Components of the Linear Predictor

Linear hypotheses concerning the linear predictor have the form

$$H_0 : \boldsymbol{C\beta} = \boldsymbol{\xi} \text{ against } H_1 : \boldsymbol{C\beta} \neq \boldsymbol{\xi},$$

where $\boldsymbol{C}$ is a fixed matrix of full rank $s \leq p$ and $\boldsymbol{\xi}$ is a fixed vector. Let, for example, the linear predictor contain only two variables, such that $\eta_r = \beta_{ro} + x_1\beta_{r1} + x_2\beta_{r2}$. Then the hypothesis that variable x_1 has no influence has the form

$$H_0 : \beta_{11} = \cdots = \beta_{q1} = 0 \text{ against } H_1 : \beta_{r1} \neq 0 \text{ for one } r.$$

It is easy to find a matrix $\boldsymbol{C}$ and a vector $\boldsymbol{\xi}$ that form the corresponding linear hypothesis. Hypotheses like that make it necessary to treat the multicategorical model as a multivariate model. Since the hypothesis involves parameters that correspond to more than one response category, the fitting of q separate binary models could not be used directly to test if H_0 holds. The test statistics in common use are the same as in univariate GLMs, namely, the likelihood ratio statistic, the Wald test, and the score test. The form is the same as given in Section 4.4; one just has to replace the score function and the Fisher matrix by their multivariate analogs.

Test procedures serve to determine if the variables have significant weights. Another aspect is the explanatory value of the predictors, which can be evaluated for example by R-squared measures. For measures of this type see Amemiya (1981).

Example 8.5: Addiction

We refer to Example 8.2. In the survey people were asked, "Is addiction a disease or are addicts weak-willed?" The response was in three categories, 0: addicts are weak-willed, 1: addiction is a disease, 2: both. Table 8.5 shows the coefficients of the multinomial logit model with the covariates gender (0: male; 1: female), age in years, and university degree (0: no; 1: yes). Category 0 was chosen as the reference category. It is seen that women show a stronger tendency to accept addiction as a disease than men. The same effect is found for respondents with a university degree. Age also shows a significant effect, and at least a quadratic effect should be included since the inclusion of a quadratic effect reduces the deviance by 32.66. Figure 8.6 shows the estimated probabilities against age for males and females with a university degree (compare also the smooth modeling in Example 10.5). □

8.7 Multinomial Models with Hierarchically Structured Response

In some applications the response categories have some inherent grouping. For example, when the response categories in a clinical study are given by {no infection, infection type I, infection type II}, it is natural to consider the two latter response categories as one group {infection}. When investigating the effect of the predictors on these responses, one might want to take

TABLE 8.5: Estimated coefficients for addiction survey with quadratic effect of age.

	Intercept	Gender	Univ	Age	Age^2
	Estimates				
1	−3.720	0.526	1.454	0.184	−0.002
2	−3.502	0.356	0.936	0.135	−0.001
	Standard Errors				
1	0.546	0.201	0.257	0.029	0.0003
2	0.596	0.224	0.290	0.030	0.0003

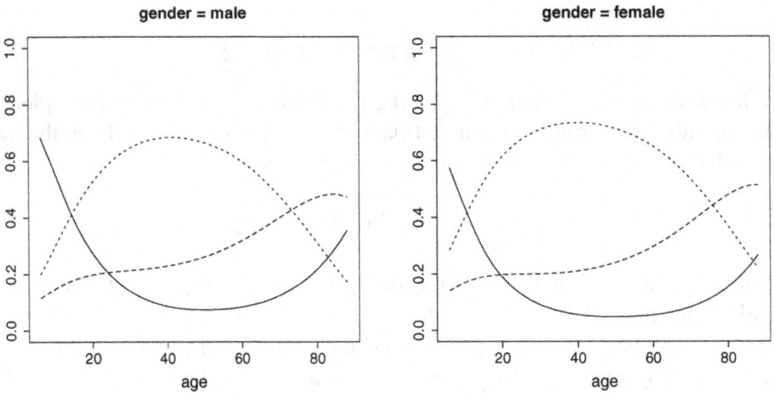

FIGURE 8.6: Estimated probabilities for addiction data with quadratic effect of age (category 0: solid line; category 1: dotted line; category 2: dashed line).

the similarities between the response categories into account to obtain effects with a simple interpretation.

A hierarchical model is obtained by first modeling the response in groups of homogeneous response categories, and in a second step the response within the groups is modeled. In general, let the response categories $K = \{1, \ldots, k\}$ be subdivided into basic sets $S_1, \ldots, S_m$, where $K = S_1 \cup \cdots \cup S_m$. In the first step, let the logit model be

$$P(Y \in S_t|\boldsymbol{x}) = \frac{\exp(\boldsymbol{x}^T\boldsymbol{\beta}_t)}{\sum_{s\in K}\exp(\boldsymbol{x}^T\boldsymbol{\beta}_s)}.$$

In the second step, the conditional response given S_t is modeled as

$$P(Y = r|Y \in S_t, \boldsymbol{x}) = \frac{\exp(\boldsymbol{x}^T\boldsymbol{\beta}_r^{(S_t)})}{\sum_{s\in S_t}\exp(\boldsymbol{x}^T\boldsymbol{\beta}_s^{(S_t)})}.$$

The parameters have to be restricted by side constraints on two levels, for the responses in $S_1, \ldots, S_m$ and for the conditional response. For example, one might use S_m as a reference on the first level by setting $\boldsymbol{\beta}_m$ to zero and choose one category from each set S_t as a reference on the second level by setting $\boldsymbol{\beta}_{r_t}$ to zero for one category $r_t \in S_t$.

For the derivation of the maximum likelihood estimates one needs the marginal probability of response on category r, which for $r \in S_t$ has the simple form

$$P(Y = r|\boldsymbol{x}) = P(Y = r|Y \in S_t)P(Y \in S_t).$$

Therefore, assuming multinomially distributed responses $\boldsymbol{y}_i \sim \mathrm{M}(n_i, \boldsymbol{\pi}_i)$, $i = 1, \dots, N$, $\boldsymbol{\pi}_i = \boldsymbol{\pi}(\boldsymbol{x}_i)$, the log-likelihood is

$$l = \sum_{i=1}^{N} \sum_{r=1}^{k} y_{ir} \log(\pi_{ir}) = \sum_{i=1}^{N} \sum_{t=1}^{m} \sum_{r \in S_t} y_{ir} \log(\pi_{ir}).$$

It decomposes into $l = l_g(\{\boldsymbol{\beta}_t\}) + \sum_{t=1}^{m} l_t(\{\boldsymbol{\beta}_r^{(S_t)}\})$, where

$$l_g(\{\boldsymbol{\beta}_t\}) = \sum_{i=1}^{N} \sum_{t=1}^{m} y_{iS_t} \log(P(Y_i \in S_t)),$$

with $y_{iS_t} = \sum_{r \in S_t} y_{ir}$, is the log-likelihood for the grouped observations on the first level, depending only on $\boldsymbol{\beta}_1, \dots, \boldsymbol{\beta}_m$, and

$$l_t(\{\boldsymbol{\beta}_r^{(S_t)}\}) = \sum_{i=1}^{N} \sum_{r \in S_t} y_{ir} \log(P(Y_i \in r | Y_i \in S_t))$$

is the log-likelihood for responses within S_t depending on $\{\boldsymbol{\beta}_r^{(S_t)}, r \in S_t\}$. Since the log-likelihood decomposes into additive terms, each part of the log-likelihood may be fitted separately by fitting the corresponding logit model. Likelihood ratio tests for individual parameters apply on the level of each model. However, Wald tests, which are typically used to examine single parameters, and the corresponding p-values are not trustworthy because they are based only on the components of the total model (Exercise 8.8).

It should be noted that the assumption of a logit model on both levels yields a model that is not equivalent to a one-step logit model. If a logit model holds for all categories, one easily derives that the conditional model $P(Y = r | Y \in S_t, \boldsymbol{x})$ is again a logit model, but that does not hold for the probabilities $P(Y \in S_t | \boldsymbol{x})$.

Example 8.6: Addiction

We refer again to the addiction data (Example 8.5). The response was in three categories, 0: addicts are weak-willed; 1: addiction is a disease; 2: both, which are grouped into $S_1 = \{0, 1\}, S_2 = \{2\}$. Therefore, the binary logit model that distinguishes between S_1 and S_2 models if the respondents think that a single cause is responsible for addiction. In the second step, the logit model compares category 1 to category 0, given that the response is in categories $\{0, 1\}$. From the estimates in Tables 8.6 and 8.7 it is seen that the covariates have no effect on the distinction between categories $\{0, 1\}$ and $\{2\}$ but have strong effects on the distinction between causes given that they think that one cause is behind addiction. □

TABLE 8.6: Estimated probabilities for addiction data with response in category 2 compared to categories $\{0, 1\}$.

	Estimate	Std. Error	z-Value	Pr(>|z|)
Intercept	2.1789	0.5145	4.23	0.000
gender	−0.0172	0.1828	−0.09	0.925
university	0.0895	0.2067	0.43	0.665
age	−0.0342	0.0255	−1.34	0.179
age2	0.0001	0.0003	0.45	0.650

TABLE 8.7: Estimated probabilities for addiction data comparing category 1 to category 0, given the response is in categories $\{0, 1\}$.

	Estimate	Std. Error	z-Value	Pr(>|z|)
Intercept	−3.5468	0.5443	−6.52	0.000
gender	0.5433	0.2055	2.64	0.008
university	1.4656	0.2601	5.64	0.000
age	0.1720	0.0284	6.06	0.000
age2	−0.0017	0.0003	−5.19	0.000

8.8 Discrete Choice Models

Since many economic decisions involve choice among discrete alternatives, probabilistic choice models have become an important research area in contemporary econometrics. In contrast to earlier approaches to model demand on an aggregate level, probabilistic choice models focus on the modeling of individual behavior. In the following some basic concepts are given. They provide a motivation for the multinomial logit model but also give rise to alternative models. More extensive treatments of probabilistic choice have been given, for example, by McFadden (1981).

Let $K = \{1, \dots, k\}$ denote the total set of alternatives. For a subset $B \subset K$ of available alternatives let $P_B(r)$ denote the probability of choosing $r \in B$, given that a selection must be made from set B. A *probabilistic choice system* may be described as a tupel:

$$(K, \mathcal{B}, \{P_B, B \in \mathcal{B}\}),$$

where $\mathcal{B}$ is a family of subsets from K. A simple example is the one-member family $\mathcal{B} = \{K\}$. Then one considers only selections from the full set of alternatives K. In a pair comparison system with $\mathcal{B} = \{\{i, j\}, i, j \in K\}\}$, the selection is among two alternatives, where all combinations of alternatives are presented. In a complete choice experiment $\mathcal{B}$ contains all subsets $K \subset \mathcal{B}$ with $|K| \geq 2$.

A probabilistic choice system is called a *random utility model* if there exist random variables $U_r, r \in K$, such that

$$P_B(r) = P(U_r = \max_{s \in B}\{U_s\}). \tag{8.10}$$

The random utility U_r represents the utility attached to alternative r. Usually it is a latent variable that cannot be measured directly. Its interpretation depends on the application; it may be the subjective utility of a travel mode or of a brand in commodity purchases. If the choice probabilities have a representation (8.10), one says that the *random utility maximization hypothesis* holds.

Let the random utility U_r have the form $U_r = u_r + \varepsilon_r$, where u_r is the structural part or non-random fixed utility and ε_r is a noise variable. The fixed utility u_r is determined by the characteristics of the decision maker and the attributes of the alternatives, while the random variable ε_r represents the residual variation.

Random Utility Models

Equation (8.10) may also be seen as a way of constructing choice probabilities. Let us consider a set $B = \{i_1, \dots, i_m\}, i_1 < \dots < i_m$. Then the choice probabilities $P(r) = P_B(r)$ are obtained as

$$
\begin{aligned}
P_B(i_r) &= P(U_{i_r} \geq U_j \quad \text{for} \quad j \in B) \\
&= P(U_{i_r} \geq U_{i_1}, \dots, U_{i_r} \geq U_{i_m}) \\
&= P(u_{i_r} - u_{i_1} \geq \varepsilon_{i_1} - \varepsilon_{i_r}, \dots, u_{i_r} - u_{i_m} \geq \varepsilon_{i_m} - \varepsilon_{i_r}) \\
&= \int_{-\infty}^{u_{i_r}-u_{i_1}} \dots \int_{-\infty}^{u_{ir}-u_{i_m}} f_{B,r}(\varepsilon_{i_1 i_r}, \dots, \varepsilon_{i_m i_r}) d\varepsilon_{i_1 i_r} \dots d\varepsilon_{i_m i_r} \\
&= F_{B,r}(u_{i_r} - u_{i_1}, \dots, u_{i_r} - u_{i_m}),
\end{aligned} \tag{8.11}
$$

where $\boldsymbol{\varepsilon}_{B,r}^T = (\varepsilon_{i_1 i_r}, \dots, \varepsilon_{i_m i_r})$, with $\varepsilon_{i_s i_r} = \varepsilon_{i_s} - \varepsilon_{i_r}$, is the $(m-1)$-dimensional vector of differences; $f_{B,r}$ is the density of $\boldsymbol{\varepsilon}_{B,r}$; and $F_{B,r}$ is the cumulative distribution function of $\boldsymbol{\varepsilon}_{B,r}$.

Let us consider the simple case where the full set of alternatives K forms the choice set. Then any distribution of $\boldsymbol{\varepsilon}^T = (\varepsilon_1, \dots, \varepsilon_k)$ will generate a discrete choice model. A familiar model results when one assumes that $\varepsilon_1, \dots, \varepsilon_k$ are iid variables with marginal distribution function $F(x) = \exp(-\exp(-x))$, which is the Gumbel or maximum extreme value distribution. Then one obtains for the cumulative distribution function of the differences $(\varepsilon_1 - \varepsilon_r, \dots, , \varepsilon_k - \varepsilon_r)$

$$
F_{m-1}(x_1, \dots, x_{m-1}) = \frac{1}{1 + \sum_{i=1}^{m-1} \exp(-x_i)}
$$

and therefore the logit model

$$
P_K(r) = \frac{1}{1 + \sum_{j \neq r} \exp(-(u_r - u_j))} = \frac{\exp(u_r)}{\sum_{j=1}^{k} \exp(u_j)}
$$

(e.g. Yellott, 1977). One obtains the familiar multinomial logit model with predictors by assuming that one has a vector $\boldsymbol{x}$ that characterizes the decision maker and the attributes $\boldsymbol{w}_r$ connected to alternative r that form the latent fixed utility:

$$
u_r = \boldsymbol{x}^T \boldsymbol{\gamma}_r + \boldsymbol{w}_r^T \boldsymbol{\alpha}.
$$

Then the model has the familiar form

$$
\begin{aligned}
\log\left(\frac{P_K(r)}{P_K(k)}\right) &= u_r - u_k = \boldsymbol{x}^T(\boldsymbol{\gamma}_r - \boldsymbol{\gamma}_k) + (\boldsymbol{w}_r - \boldsymbol{w}_k)^T \boldsymbol{\alpha} \\
&= \boldsymbol{x}^T \boldsymbol{\beta}_r + (\boldsymbol{w}_r - \boldsymbol{w}_k)^T \boldsymbol{\alpha},
\end{aligned}
$$

which is equivalent to the logit model considered in Section 8.4.

Random utility models that assume iid distributions for the noise variables have a long history. In psychophysics, Thurstone (1927) proposed the law of comparative judgement, which is based on normally distributed variables. Generally, in the iid case, all differences have the same distribution and one obtains with $\sigma_0^2 = \text{var}(\varepsilon_i)$ for the covariance of differences:

$$
\begin{aligned}
\text{cov}(\varepsilon_i - \varepsilon_r, \varepsilon_j - \varepsilon_r) &= \text{E}(\varepsilon_i - \varepsilon_r)(\varepsilon_j - \varepsilon_r) \\
&= \text{E}(\varepsilon_i \varepsilon_j - \varepsilon_i \varepsilon_r - \varepsilon_r \varepsilon_j + \varepsilon_r^2) = \text{E}(\varepsilon_r^2) = \sigma_0^2.
\end{aligned}
$$

Then the covariance matrix of the differences is given by $\boldsymbol{\Sigma}_0 = \sigma_0^2(\mathbf{I} + \mathbf{1}\mathbf{1}^T)$. If one assumes that ε_i is normally distributed with $\varepsilon_i \sim \text{N}(0, \sigma_0^2)$, one obtains from maximizing the random utility the *multinomial probit model*

$$
P_K(r) = \phi_{\mathbf{0}, \boldsymbol{\Sigma}_0}(u_r - u_1, \dots, u_r - u_k),
$$

where $\phi_{\mathbf{0},\boldsymbol{\Sigma}_0}$ is the $(k-1)$-dimensional cumulative distribution function of the normal distribution $\mathrm{N}(\mathbf{0}, \boldsymbol{\Sigma}_0)$. In psychology, the model is also known as *Thurstone's case V*. The multinomial probit model is harder to use for higher numbers, of alternatives because the integral has to be computed numerically.

Independence from Irrelevant Alternatives

Simple models like the multinomial logit model imply a property that has caused some discussion in economics. The problem occurs if decisions for more than just one fixed set of alternatives are investigated. For example, in a complete choice experiment, the choice probabilities are determined for every subset of alternatives. Then the logit model for subset B, derived from the maximization of random utilities U_r, has the form

$$P_B(r) = \frac{\exp(u_r)}{\sum_{j\in B}\exp(u_j)},$$

and one obtains for any subset $B, C, r, s \in B \cap C$

$$\frac{P_B(r)}{P_B(s)} = \frac{P_C(r)}{P_C(s)} = \frac{\exp(u_r)}{\exp(u_s)}. \tag{8.12}$$

This means that the proportion of probabilities is identical when different sets of alternatives are considered. The system of choice probabilities satisfies Luce's Choice Axiom (Luce, 1959), which implies that the choice probabilities are *independent from irrelevant alternatives*.

McFadden (1986) calls the independence from irrelevant alternatives a blessing and a curse. An advantage is that if it holds, it makes it possible to infer choice behavior with multiple alternatives using data from simple experiments like paired comparisons. A disadvantage is that it is a rather strict assumption that may not hold for heterogeneous patterns of similarities encountered in economics. A famous problem that illustrates the case is the "red bus–blue bus" problem that has been used by McFadden (see, e.g., Hausman and Wise, 1978). Suppose a commuter has the initial alternatives of driving or taking a red bus with the odds given by

$$\frac{P_{\{\text{driving,red bus}\}}(\text{driving})}{P_{\{\text{driving,red bus}\}}(\text{red bus})} = 1.$$

Then an additional alternative becomes available, namely, a blue bus that is identical in all respects to the red bus, except color. If the logit model holds, it is seen from (8.12) that the odds of choosing the driving alternative over the red bus remain the same. Since the odds for the choice between the red and the blue bus are

$$\frac{P_{\{\text{red bus, blue bus}\}}(\text{red bus})}{P_{\{\text{red bus, blue bus}\}}(\text{blue bus})} = 1,$$

one obtains for any B for all paired comparisons

$$\frac{P_B(\text{driving})}{P_B(\text{red bus})} = \frac{P_B(\text{driving})}{P_B(\text{blue bus})} = \frac{P_B(\text{red bus})}{P_B(\text{blue bus})} = 1$$

and therefore

$$P_{\{1,2,3\}}(\text{driving}) = P_{\{1,2,3\}}(\text{red bus}) = P_{\{1,2,3\}}(\text{blue bus}) = 1/3.$$

This is a counterintuitive result because the additional "irrelevant" alternative blue bus has decreased the choice probability of driving substantially. Problems of this type occur not only for the logit model. The probit model based on iid distributions has a similar property as demonstrated by Hausman and Wise (1978). In fact, the same sort of counterintuitive results are found for all choice systems that share a property called *simple scalability* (Krantz, 1964). Simple scalability means that there exist scales $v_1, \ldots, u_k$ and functions $F_2, \ldots, F_k$ that determine the choice probability for $B = \{i_1, \ldots, i_m\}$ by

$$P_B(r) = F_m(v_{i_r}, \ldots, v_{i_m}),$$

where F_m is strictly increasing in the first argument and strictly decreasing in the remaining $m - 1$ arguments. It is easily shown that simple scalability holds for the multinomial logit model.

Tversky (1972) shows that simple scalability is equivalent to order independence that holds whenever for all $r, s \in B \setminus C$ and $t \in C$

$$P_B(r) \geq P_B(s) \qquad \Leftrightarrow \qquad P_{C \cup \{r\}}(t) \leq P_{C \cup \{s\}}(t).$$

This is a weaker version of the independence of irrelevant alternatives, which implies that only the order of $P_B(r)$ and $P_B(s)$, and not necessarily their ratio, is independent of B.

The independence of irrelevant alternatives raises problems when one wants to combine results from different choice sets, which is frequently wanted in econometric applications. However, if the choice set is fixed, and each person faces the full set of alternatives, the counterintuitive results have no relevance. For the simultaneous treatment of different choice sets, the most widely used model in econometrics is McFadden's nested multinomial model, which is sketched in the following section.

Pair Comparison Models

In pair comparison systems only two alternatives are compared at a time. Therefore, the family of subsets that is considered is given by $\mathcal{B} = \{\{i, j\}, i, j \in K\}\}$. The resulting pair comparison models are useful in psychometrics to scale stimuli or in marketing to scale the attractiveness of product brands. Moreover, it is often used in sport competitions when one wants to measure the ability of a team or a player. The most frequently used model is the logistic model

$$P_{\{r,s\}}(r) = \frac{\exp(u_r - u_s)}{1 + \exp(u_r - u_s)}, \tag{8.13}$$

which results from the assumption of a Gumbel distribution for the residual variation. It is also called the *BTL (Bradley-Terry-Luce) model*, with reference to Bradley and Terry (1952) and Luce (1959).

If $u_r > u_s$, the probability that stimuli r is preferred over s (or that player r will win against player s) increases with the difference of attractiveness (or ability) $u_r - u_s$. Since ties are not allowed, in the simple model the probability is 0.5 if $u_r = u_s$. The advantage of the model is that one obtains the attractiveness of stimuli on a one-dimensional scale, and estimates can be used to predict the future outcome. When one assumes that the comparisons are independent, simple logit models apply for estimation. The set of stimuli can be treated as a factor and the design matrix specifies the differences between the alternatives. Of course a reference alternative has to be chosen, for example, by setting $u_1 = 0$.

The model is easily extended to incorporate an order effect. By specifying

$$P_{\{r,s\}}(r) = \frac{\exp(\alpha + u_r - u_s)}{1 + \exp(\alpha + u_r - u_s)}, \tag{8.14}$$

one obtains $P_{\{r,s\}}(r) = \exp(\alpha)/1 + \exp(\alpha)$ for the preference of r over s. In sports competitions, α represents the home advantage; when stimuli are rated it refers to the order in which the stimuli are presented. Moreover, category-specific variables can be included in the model; for literature see Section 8.11.

Example 8.7: Paired Comparison

Rumelhart and Greeno (1971) asked 234 college students for their preferences concerning nine famous persons. For each of the 36 pairs the subjects were instructed to choose the person with whom they would rather spend an hour discussing a topic of their choosing. Table 8.8 shows the data, and Table 8.9 gives the estimates of the fitted BTL model. One might want to distinguish between the effect of the profession and the effect of the person by including profession in the predictor. Let the scale value be structured as $u_r = w_{r1}\gamma_r + w_{r2}\gamma_2 + \delta_r$, where $w_{r1} = 1$ if person r is a politician (0 otherwise), and $w_{r2} = 1$ if person r is a sportsman (0 otherwise). Actor is the reference category and δ_r is the additional effect of the person with reference $\delta_2 = 0$ for politicians, $\delta_5 = 0$ for sportsmen, and $\delta_9 = 0$ for actors. Table 8.10 shows that sportsmen are distinctly preferred over actors, but not politicians. □

TABLE 8.8: Pair comparison referring to politicians Harold Wilson (1), Charles de Gaulle (2), and Lyndon B. Johnson (3); sporstmen Johnny Unitas (4), Carl Yastrazemski (5), and A. Foyt (6); and actors Brigitte Bardot (7), Elizabeth Taylor (8), and Sophia Loren (9).

	1	2	3	4	5	6	7	8	9	Σ
1	–	159	163	175	183	179	173	160	142	1334
2	75	–	138	164	172	160	156	122	122	1109
3	71	96	–	145	157	140	138	122	120	989
4	59	70	89	–	176	115	124	86	61	780
5	51	62	77	58	–	77	95	72	61	553
6	55	74	94	119	157	–	134	92	71	796
7	61	78	96	110	139	100	–	67	48	699
8	74	112	112	148	162	142	167	–	87	1004
9	92	112	114	173	173	163	186	147	–	1160

TABLE 8.9: Estimated coefficients for pair comparison.

	Persons	u_i	Standard Deviation
	1 WI	-0.382	0.066
politicians	2 GA	0.106	0.064
	3 JO	0.350	0.064
	4 UN	0.772	0.065
sportsmen	5 YA	1.260	0.067
	6 FO	0.739	0.065
	7 BB	0.940	0.065
actors	8 ET	0.319	0.066
	9 SL	0.000	0.064

TABLE 8.10: Pair comparison data with effects of profession separated.

	Effects	Standard Deviation
γ_1 (politicians)	0.087	0.093
γ_2 (sportsmen)	1.250	0.078
δ_1 (WI)	−0.491	0.067
δ_3 (JO)	0.246	0.064
δ_4 (UN)	−0.491	0.066
δ_6 (FO)	−0.518	0.067
δ_7 (BB)	0.935	0.065
δ_8 (ET)	0.317	0.065

8.9 Nested Logit Model

The assumption of iid noise variables ε_i yields models with certain weaknesses when some of the alternatives are similar. These weaknesses can be ameliorated when using the nested logit model that has been proposed in several papers by McFadden (1981, 1978). For illustration we adopt Amemiya's (1985) approach to start with the red bus–blue bus problem. Let $U_r = u_r + \varepsilon_r, r = 1, 2, 3$, be the utilities associated with driving, red bus, and blue bus. McFadden proposed taking into account the similarity between alternatives 2 and 3 by assuming the bivariate extreme-value distribution function

$$F(\varepsilon_2, \varepsilon_3) = \exp\{-[\exp(-\varepsilon_2/\delta) + \exp(-\varepsilon_3/\delta)]^\delta\}$$

with $\delta \in (0, 1]$. For the distribution of ε_1 the usual extreme value distribution $F(\varepsilon_1) = \exp(-\exp(-\varepsilon_1))$ is assumed. The total distribution function for all residuals is

$$F(\varepsilon_1, \varepsilon_2, \varepsilon_3) = \exp\{-\mathrm{e}^{\varepsilon_1} - [\mathrm{e}^{-\varepsilon_2/\delta} + \mathrm{e}^{-\varepsilon_3/\delta}]^\delta\}.$$

It may be shown that the correlation between ε_1 and ε_2 is $\varrho^2 = 1 - \delta^2$. In the independence case $\varrho = 0$ $(\delta = 1)$, $F(\varepsilon_1, \varepsilon_2, \varepsilon_3)$ becomes the product of the three extreme value (Gumbel-) distributions and the multinomial logit model follows. Otherwise, one obtains the trichotomous nested logit model

$$P(Y = 1) = \frac{\exp(u_1)}{\exp(u_1) + [\exp(u_2/\delta) + \exp(u_3/\delta)]^\delta}, \tag{8.15}$$

and

$$P(Y = 2|Y \in \{2, 3\}) = \frac{\exp(u_2/\delta)}{\exp(u_2/\delta) + \exp(u_3/\delta)}. \tag{8.16}$$

The conditional model (8.16) that specifies the choice between the two similar alternative is again a logit model. The choice between driving and taking buses, given by (8.15), is not a logit model in the strict sense, although it is similar to a logit model, but with a sort of weighted average of $\exp(u_2)$ and $\exp(u_3)$ in the denominator. For $\delta = 1$ the model given by (8.15) and (8.16) simplifies to a simple logit model. Therefore, the hypothesis of independence from irrelevant alternatives may be tested by testing $\delta = 1$ against the alternative of the nested logit model (for an example see Hausman and McFadden, 1984).

Example 8.8: Red Bus–Blue Bus

When using the nested logit model (8.17) and (8.18), the model constructed from the maximization of the random utilities does not suffer from the independence of the irrelevant alternatives. By defining

$$G(y_1, y_2, y_3) = y_1 + (y_2^{1/\delta} + y_3^{1/\delta})^\delta,$$

the underlying distribution function is

$$F(\varepsilon_1, \varepsilon_2, \varepsilon_3) = \exp(-G(\mathrm{e}^{-\varepsilon_1}, \mathrm{e}^{-\varepsilon_2}, \mathrm{e}^{-\varepsilon_3})),$$

and one obtains for the choice among alternatives $\{1, 2, 3\}$

$$P_{\{1,2,3\}}(Y = 1) = \frac{e^{u_1}}{G(e^{u_1}, e^{u_2}, e^{u_3})},$$

$$P_{\{1,2,3\}}(Y = 2) = \frac{e^{u_2/\delta}}{(e^{u_2/(\delta)} + e^{u_3/\delta})^{\delta} G(e^{u_1}, e^{u_2}, e^{u_3})},$$

$$P_{\{1,2,3\}}(Y = 3) = \frac{e^{u_3/\delta}}{(e^{u_2/(\delta)} + e^{u_3/\delta})^{\delta} G(e^{u_1}, e^{u_2}, e^{u_3})}.$$

For the choice between alternatives 1 and 2 one obtains

$$P_{\{1,2\}}(Y = 1) = \frac{e^{u_1}}{e^{u_1} + e^{u_2}}, \qquad P_{\{1,2\}}(Y = 2) = 1 - P_{\{1,2\}}(Y = 1),$$

and therefore

$$\frac{P_{\{1,2,3\}}(Y = 1)}{P_{\{1,2,3\}}(Y = 2)} \neq \frac{P_{\{1,2\}}(Y = 1)}{P_{\{1,2\}}(Y = 2)}.$$

□

In the general case let the set of alternatives $K = \{1, \ldots, k\}$ be partitioned into distinct sets $S_1, \ldots, S_m$ of similar alternatives, such that $K = S_1 \cup \cdots \cup S_m$. The joint distribution proposed by McFadden has the form

$$F(\varepsilon_1, \ldots, \delta_k) = \exp\left\{-\sum_{t=1}^{m} \alpha_t [\sum_{r \in S_t} \exp(-\varepsilon_r/\delta_t)]^{\delta_t}\right\}.$$

One obtains the nested logit model

$$P(Y \in S_t) = \frac{\alpha_t \left(\sum_{r \in S_t} \exp(u_r/\delta_t)\right)^{\delta_t}}{\sum_{i=1}^{m} \alpha_i [\sum_{r \in S_i} \exp(u_r/\delta_i)]^{\delta_i}}, \tag{8.17}$$

$$P(Y = r | Y \in S_t) = \frac{\exp(u_r/\delta_t)}{\sum_{i \in S_t} \exp(u_i/\delta_t)}. \tag{8.18}$$

The predictors come into the model by specifying $u_r = \boldsymbol{x}_r^T \boldsymbol{\beta}$. The resulting model may be estimated by maximum likelihood methods. However, for large-scale models the computation becomes troublesome. Therefore, often a two-step method is used that first estimates $\boldsymbol{\beta}/\delta_1$ by using (8.18). Then the estimates are plugged into (8.17) to obtain estimates of δ_1 and α_t (for details see McFadden, 1981).

McFadden (1978) introduced a more general model based on the generalized extreme-value distribution. The distributions considered in the three alternatives case and the more general case of partioning into m subsets may be generalized to

$$F(\varepsilon_1, \ldots, \varepsilon_k) = \exp(-G(\mathrm{e}^{-\varepsilon_1}, \ldots, \mathrm{e}^{-\varepsilon_k}));$$

where G satisfies the conditions

(1) $G(y_1, \ldots, y_k) \geq 0$ for $y_i \geq 0$;

(2) $\lim_{y_i \to \infty} G(y_1, \dots, y_k) = \infty, \qquad i = 1, \dots, k;$

(3) $G(\alpha y_1, \dots, \alpha y_k) = \alpha^{\psi} G(y_1, \dots, y_k);$

(4) $\partial G^s(y_1, \dots, y_k)/(\partial y_{i_1}, \dots, \partial y_{i_s}) \begin{cases} \geq 0 & \text{if } s \text{ is odd} \\ \leq 0 & \text{if } k \text{ is even.} \end{cases}$

When $U_r = u_r + \varepsilon_r$ the probabilities following from maximization of the random utility are given by

$$P(Y = r) = e^{u_r} \frac{\partial G}{\partial y_r}(e^{u_1}, \dots, e^{u_k})/G(e^{u_1}, \dots, e^{u_k}). \tag{8.19}$$

The multinomial logit model follows as the special case where $G(y_1, \dots, y_m) = y_1 + \dots + y_k$. The nested logit model with the similar alternatives given by the partition $S_1, \dots, S_m$ is based on $G(y_1, \dots, y_m) = \sum_{t=1}^{m} \alpha_t [\sum_{r \in S_t} y_r^{1/\delta}]^{\delta}$.

The models considered so far may be seen as implying two levels of nesting with the original set being subdivided into the sets $S_1, \dots, S_m$. In the general concept, three and higher level nested models also may be constructed (see McFadden, 1981). Although models based on generalized extreme-value distributions are more general, in applications the nested logit model is the dominating model from these class of models.

8.10 Regularization for the Multinomial Model

In Chapter 6 various regularization methods, including variable selection, were discussed. In the following the methods are extended to the case of multicategorical responses. Let us consider the multinomial logit model in symmetrical form:

$$P(Y = r|\boldsymbol{x}) = \frac{\exp(\beta_{r0} + \boldsymbol{x}^T \boldsymbol{\beta}_r)}{\Sigma_{s=1}^{k} \exp(\beta_{s0} + \boldsymbol{x}^T \boldsymbol{\beta}_s)}, \tag{8.20}$$

where the intercept has been separated from the predictors. Regularization methods based on penalization are again based on the *penalized log-likelihood*:

$$l_p(\boldsymbol{\beta}) = \sum_{i=1}^{n} l_i(\boldsymbol{\beta}) - \frac{\lambda}{2} J(\boldsymbol{\beta}),$$

where $l_i(\boldsymbol{\beta})$ is the usual log-likelihood contribution of the ith observation, λ is a tuning parameter, and $J(\boldsymbol{\beta})$ is a functional that penalizes the size of the parameters. Maximizing the penalized log-likelihood $l_p(\boldsymbol{\beta})$ rather than the unpenalized log-likelihood is advantageous in particular when many predictors are available and ML estimates do not exist.

Ridge-Type Penalties

For a one-dimensional response, one of the simplest penalties is the ridge penalty, which uses the functional $J(\boldsymbol{\beta}) = \sum_{i=1}^{p} \beta_i^2$. It penalizes the length of the parameter $\boldsymbol{\beta}$ and yields estimates that are shrinked towards zero. In the multicategorical case one has not only one parameter vector but a collection of parameter vectors $\boldsymbol{\beta}_1, \dots, \boldsymbol{\beta}_k$, which, for reasons of identifiability, have to be constrained. When parameters are penalized it is natural to use the symmetric side constraint $\sum_{s=1}^{k} \boldsymbol{\beta}_s^T = (0, \dots, 0)$ with a ridge penalty that has the form

$$J(\boldsymbol{\beta}) = \sum_{r=1}^{k} \sum_{j=1}^{p} \beta_{rj}^2 = \sum_{j=1}^{p} \boldsymbol{\beta}_{.j}^T \boldsymbol{\beta}_{.j}, \tag{8.21}$$

where $\boldsymbol{\beta}_{.j}^T = (\boldsymbol{\beta}_{1j}, \dots, \boldsymbol{\beta}_{kj})$ collects the parameters associated with the jth variable and the side constraints $\mathbf{1}^T\boldsymbol{\beta}_{.j} = 0$ have to hold for $j = 0, \dots, p$.

It is noteworthy that then the penalty for the side constraint $\boldsymbol{\beta}_k = \mathbf{0}$, which specifies k as reference category, does not have the usual form of the ridge penalty. This may be seen from looking at the transformations of parameters when changing constraints. Let $\tilde{\boldsymbol{\beta}}_{.j}^c = (\tilde{\beta}_{1j}, \dots, \tilde{\beta}_{k-1,j})^T$ denote the shortened vector of parameters with side constraints $\tilde{\beta}_{kj} = 0$, $j = 0, \dots, p$, and $\boldsymbol{\beta}_{.j}^c = (\beta_{1j}, \dots, \beta_{k-1,j})^T$ the shortened vector with symmetric side constraints $\beta_{k,j} = -\beta_{1j} - \dots - \beta_{qj}$, $q = k-1$. Then the transformation $\boldsymbol{\beta}_{.j}^c = \boldsymbol{T}\tilde{\boldsymbol{\beta}}_{.j}^c$ is given by

$$\begin{pmatrix} \beta_{1j} \\ \beta_{2j} \\ \vdots \\ \beta_{k-1,j} \end{pmatrix} = \begin{pmatrix} \frac{k-1}{k} & -\frac{1}{k} & \cdots & -\frac{1}{k} \\ -\frac{1}{k} & \ddots & & \\ \vdots & & \ddots & \\ -\frac{1}{k} & & & \frac{k-1}{k} \end{pmatrix} \begin{pmatrix} \tilde{\beta}_{1j} \\ \tilde{\beta}_{2j} \\ \vdots \\ \tilde{\beta}_{k-1,j} \end{pmatrix}. \tag{8.22}$$

The inverse transformation $\tilde{\boldsymbol{\beta}}_{.j}^c = \boldsymbol{T}^{-1}\boldsymbol{\beta}_{.j}^c$ uses $\boldsymbol{T}^{-1} = \boldsymbol{I} + \mathbf{1}\mathbf{1}^T$, where $\mathbf{1}^T = (1, \dots, 1)$. For single parameters the transformations are

$$\tilde{\beta}_{rj} = 2\beta_{rj} + \sum_{s \neq q} \beta_{sj}, \quad \beta_{rj} = ((k-1)\tilde{\beta}_{rj} - \sum_{s \neq q} \tilde{\beta}_{sj})/k,$$

$r = 1, \dots, k-1$. Some derivation shows that penalty (8.21) is equivalent to

$$J(\boldsymbol{\beta}) = \sum_{j=1}^{p} (\boldsymbol{\beta}_{.j}^c)^T \boldsymbol{T}^{-1} \boldsymbol{\beta}_{.j}^c \quad \text{or} \quad J(\tilde{\boldsymbol{\beta}}) = \sum_{j=1}^{p} (\tilde{\boldsymbol{\beta}}_{.j}^c)^T \boldsymbol{T} \tilde{\boldsymbol{\beta}}_{.j}^c, \tag{8.23}$$

which uses only the $k-1$ parameters that can be identified.

When computing estimates it is helpful to have the penalty in a closed form with the parameters given in the usual ordering. Let $\boldsymbol{\beta}^T = (\boldsymbol{\beta}_1^T, \dots, \boldsymbol{\beta}_q^T)$, $\boldsymbol{\beta}_r^T = (\beta_{r0}, \dots, \beta_{rp})$ denote the whole vector with symmetric side constraints, where it is assumed that the covariate vector $\boldsymbol{x}$ contains an intercept and p predictors. The corresponding parameter vector with reference category k is $\tilde{\boldsymbol{\beta}}^T = (\tilde{\boldsymbol{\beta}}_1^T, \dots, \tilde{\boldsymbol{\beta}}_q^T)$, $\tilde{\boldsymbol{\beta}}_r^T = (\tilde{\beta}_{r0}, \dots, \tilde{\beta}_{rp})$. Then a closed form of the penalty is

$$J(\boldsymbol{\beta}) = \boldsymbol{\beta}^T \boldsymbol{T}_0 \boldsymbol{\beta} \quad \text{or} \quad J(\tilde{\boldsymbol{\beta}}) = \tilde{\boldsymbol{\beta}}^T \boldsymbol{T}_1 \tilde{\boldsymbol{\beta}}, \tag{8.24}$$

where $\boldsymbol{T}_0 = \boldsymbol{T}^{-1} \otimes \boldsymbol{I}_0$, $\boldsymbol{T}_1 = \boldsymbol{T} \otimes \boldsymbol{I}_0$, with $\boldsymbol{I}_0$ denoting a $(p+1) \times (p+1)$-matrix that differs from the usual identity matrix by having values zero in the first column.

Then penalized estimates for the constraint $\tilde{\beta}_{kj} = 0$ can be computed by using the design matrix and link function given in Section 8.5, where the logit model is shown to be a multivariate GLM. The penalized estimates are obtained by solving $\boldsymbol{s}_p(\tilde{\boldsymbol{\beta}}) = \mathbf{0}$, where $\boldsymbol{s}_p(\tilde{\boldsymbol{\beta}})$ is the penalized score function $\boldsymbol{s}_p(\tilde{\boldsymbol{\beta}}) = \boldsymbol{s}(\tilde{\boldsymbol{\beta}}) - \lambda \sum_{j=1}^{p} \boldsymbol{T}_1 \tilde{\boldsymbol{\beta}}$. Iterative Fisher scoring uses the corresponding penalized Fisher matrix $\boldsymbol{F}_p(\tilde{\boldsymbol{\beta}}) = \boldsymbol{F}(\tilde{\boldsymbol{\beta}}) + \lambda \boldsymbol{T}_1$. For the definition of $\boldsymbol{s}(\tilde{\boldsymbol{\beta}})$ and $\boldsymbol{F}(\tilde{\boldsymbol{\beta}})$, see Section 8.6.1.

Alternatively, one can work with symmetric side constraints. However, then the design matrix has to be adapted to symmetric side constraints. With the design matrix $\boldsymbol{X}$ for reference category k (see Section 8.5) one has $\boldsymbol{X}\tilde{\boldsymbol{\beta}}$, which transforms into $\boldsymbol{X}_0\boldsymbol{\beta}$, where $\boldsymbol{X}_0 = \boldsymbol{X}(\boldsymbol{T}^{-1} \otimes \boldsymbol{I})$ since $\tilde{\boldsymbol{\beta}} = (\boldsymbol{T}^{-1} \otimes \boldsymbol{I})\boldsymbol{\beta}$. Zahid and Tutz (2009) demonstrated in simulation studies that ridge estimates outperform ML estimates in terms of mean squared errors that refer to the estimation of the parameter vector and the underlying probability vector. A further advantage of ridge estimation is that estimates can be computed even when ML estimates fail to exist. A penalized regression for the multinomial logit model with ridge-type penalties was also investigated by Zhu and Hastie (2004).

Penalties Including Selection of Parameters and Predictors

A general penalty for multicategorical responses has the additive form

$$\lambda J(\boldsymbol{\beta}) = \sum_{r=1}^{k} \sum_{j=1}^{p} p_\lambda(|\beta_{rj}|),$$

where $p_\lambda(|\beta|)$ are penalty functions that may depend on λ. The multicategorical ridge and lasso are special cases with function $p_\lambda(|\beta_{rj}|) = \lambda|\beta_{rj}|^2$ and $p_\lambda(|\beta_{rj}|) = \lambda|\beta_{rj}|$, respectively. Since shrinkage should not depend on the reference category, the penalties should use the symmetric constraints that transform to different functions when reference categories are used.

Multinomial logistic regressions with lasso-type estimates that use a reference category were considered by Krishnapuram et al. (2005) with the focus on classification. Friedman et al. (2010) used the elastic net penalty $p_\lambda(|\beta_{rj}|) = \lambda\{(1/2)(1-\alpha)|\beta_{rj}|^2 + \alpha|\beta_{rj}|\}$ and proposed using coordinate descent in the form of partial Newton steps. However, in that approach selection refers to parameters and not to predictors. Therefore, it might occur that a predictor is kept in the model because only one of the parameters linked to that predictor is needed. With the focus on variable selection it seems more appropriate to penalize the group of parameters that is associated with one variable. Penalties of that type are a modification of (8.21). A lasso-type penalty that is similar to the grouped lasso but where grouping refers to response categories rather than groups of dummy variables associated to a categorical predictor is

$$J(\boldsymbol{\beta}) = \sum_{j=1}^{p} ||\boldsymbol{\beta}_{.j}||_2 = \sum_{j=1}^{p} (\beta_{1j}^2 + \cdots + \beta_{kj}^2)^{1/2}.$$

It enforces the selection of the predictors in the sense that whole predictors are deleted. The penalty is a special case of the composite absolute penalty family proposed by Zhao et al. (2009). In multicategorical modeling the predictor can also include category-specific covariates. The corresponding predictor of the multinomial model has the form $\eta_r = \boldsymbol{x}^T\boldsymbol{\beta}_r + (\boldsymbol{w}_{rk})^T\boldsymbol{\alpha}$, where $\boldsymbol{\alpha}^T = (\alpha_1, \ldots, \alpha_m)$ weights the category-specific variables $\boldsymbol{w}_{rk}$ (see Section 8.4). Selection of both types of variables, global and category-specific ones, is enforced by using

$$J(\boldsymbol{\beta}) = \gamma \sum_{j=1}^{p} ||\boldsymbol{\beta}_{.j}||_2 + (1-\gamma) \sum_{j=1}^{m} ||\alpha_j||,$$

where γ steers the amount of penalization exerted on the two types of variables.

Example 8.9: Party Choice

In spatial election theory it is assumed that each voter has a utility function in a finite-dimensional space that characterizes the profiles of parties and voters (see, for example, Thurner and Eymann, 2000). Let the latent utility for voter i and party r be given by $u_{ir} = \boldsymbol{x}_i^T\boldsymbol{\gamma}_r + \boldsymbol{w}_{ir}^T\boldsymbol{\alpha}$, where the $\boldsymbol{x}_i$'s are global variables that characterize the voter and the $\boldsymbol{w}_{ir}$'s represent category-specific variables that are specific for the party (compare Section 8.4). In spatial election theory the category-specific variables are determined as the distance between the position of the voter on a specific scale (the "ideal point") and the perceived position of the party on this scale. The scales represent policy dimensions, for example, the attitude toward the use of nuclear energy. We consider data from the German Longitudinal Election Study with the

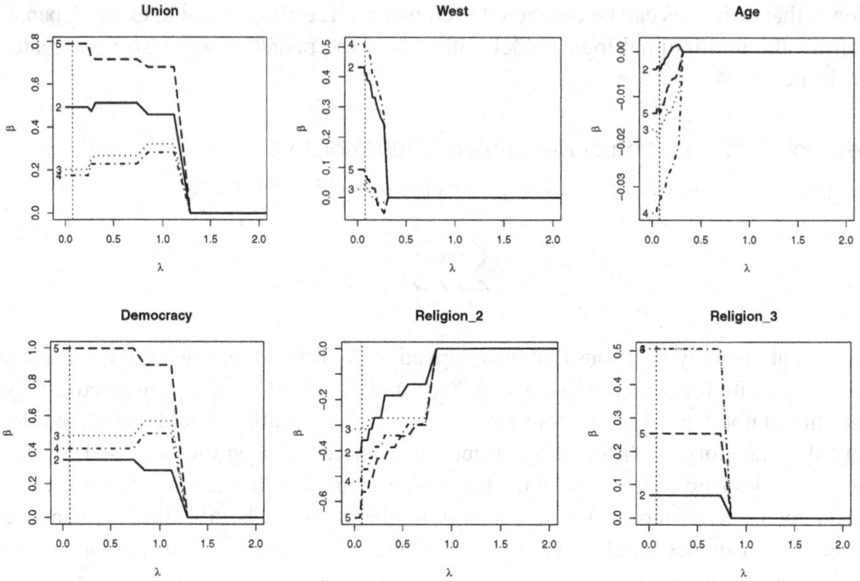

FIGURE 8.7: Coefficient buildups for selected global variables of party choice data.

alternatives Christian Democratic Union (CDU: 1), the Social Democratic Party (SPD: 2), the Green Party (3), the Liberal Party (FDP: 4), and the Left Party (Die Linke: 5). The global predictors are age, political interest (1: less interested; 0: very interested), religion (1: evangelical; 2: catholic; 3: otherwise), regional provenance (west; 1: former West Germany; 0: otherwise), gender (1: male; 0: female), union (1: member of a union; 0: otherwise), satisfaction with the functioning of democracy (democracy; 1: not content; 0: content), unemployment (1: currently unemployed; 0: otherwise), and high school degree (1: yes; 0: no). The included policy dimensions are attitude toward the immigration of foreigners, attidude toward the use of nuclear energy, and the positioning on a political left-right scale (left). Figure 8.7 shows the buildups of global variables resulting from lasso-type regularization. Only variables that turned out to be influential are shown. The vertical line shows the selected smoothing parameter based on cross-validation. Figure 8.8 shows the buildups for the category-specific variables. It is seen that only the left-right scale is influential; the other two dimensions can be neglected. For the interpretation it is useful to consider again the construction of the predictor. The linear predictor has the form $\eta_{ir} = \beta_{r0} + \boldsymbol{x}_i^T \boldsymbol{\beta}_r + (\boldsymbol{w}_{ir} - \boldsymbol{w}_{ik})^T \boldsymbol{\alpha}$ with the predictor $\boldsymbol{w}_{ir}$ denoting the distance between the position of the voter on a specific scale and the perceived position of the party on this scale. Therefore, if for the rth party this distance is larger than for

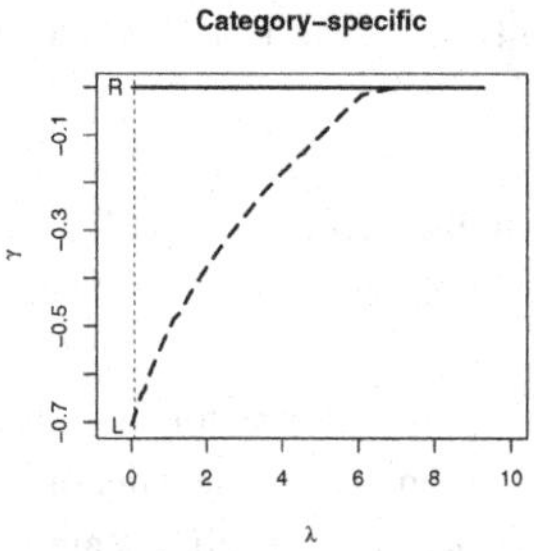

FIGURE 8.8: Coefficient buildups for category-specific variables of party choice data (L denotes left right scale, R denotes the rest).

the kth party, the probability for the rth party is reduced because the weight α on the left-right scale is negative. □

Alternatively, one can use likelihood-based boosting techniques as given in Section 6.3.2. The likelihood procedure is essentially the same as GenBoost, but the underlying GLM now is multivariate. To obtain predictor selection rather than parameter selection, one includes within one step just one variable and selects the variable that improves the fit maximally. By including one variable at a time, all the parameters that refer to that variable are updated. With k response categories that means $k-1$ parameters for one metric variable. If predictors are categorical, the number increases because all the dummy variables for all the response categories are included. For details see Zahid and Tutz (2010).

TABLE 8.11: Parameter estimates for the contraceptive methods data (response categories: no-use, long-term, and short-term) with lasso approach and boosting. For the boosting estimates, deviance was used as the stopping criterion with 10-fold cross-validation, and deviance was also used for predictor selection.

	Lasso			Boosting		
Predictor	No-Use	Long-Term	Short-Term	No-Use	Long-Term	Short-Term
wife.age	0.0468	0	−0.0593	0.3683	0.0715	−0.4397
wife.edu2	−0.0371	0.6058	0	−0.1072	0.1303	−0.0231
wife.edu3	−0.3581	1.0610	0	−0.2872	0.3390	−0.0518
wife.edu4	−0.9607	1.3694	0	−0.5445	0.5760	−0.0315
husband.edu2	0	−0.7669	1.0697	−0.0397	−0.2146	0.2543
husband.edu3	0	−0.5997	1.2333	−0.0872	−0.2770	0.3642
husband.edu4	0	−0.5686	1.0176	−0.0764	−0.2628	0.3392
children	−0.3449	0	0	−0.4979	0.2407	0.2572
wife.religion	0.3522	−0.1795	0	0	0	0
wife.working	−0.0130	0	0.1527	0	0	0
husband.job2	0	−0.4072	0.0162	0	0	0
husband.job3	0	−0.2178	0.2897	0	0	0
husband.job4	−0.4590	0	0.0312	0	0	0
sol.index2	−0.3402	0.0219	0	0	0	0
sol.index3	−0.4528	0.2433	0	0	0	0
sol.index4	−0.6841	0.2440	0	0	0	0
media	0.5316	0	0	0	0	0

Example 8.10: Contraceptive Method

A subset of the 1987 National Indonesia Contraceptive Prevalence Survey is available from the UCI machine learning repository. The sample comprises 1473 married women who were either not pregnant or did not know if they were at the time of interview. The problem is to analyze the current contraceptive method choice (no use, long-term methods, or short-term methods) of a woman based on 10 demographic and socio-economic characteristics as: wife's age, wife's education (wife.edu; $1 =$ low, 2, 3, $4 =$ high), husband's education (husband.edu; $1 =$ low, 2, 3, $4 =$ high), number of children ever born, wife's religion ($0 =$ non-Islam, $1 =$ Islam), wife's current working ($0 =$ yes, $1 =$ no), husband's occupation (categorical: $1, 2, 3, 4$), standard of living (sol.index; $1 =$ low, 2, 3, $4 =$ high), and media exposure ($0 =$ good, $1 =$ not good).

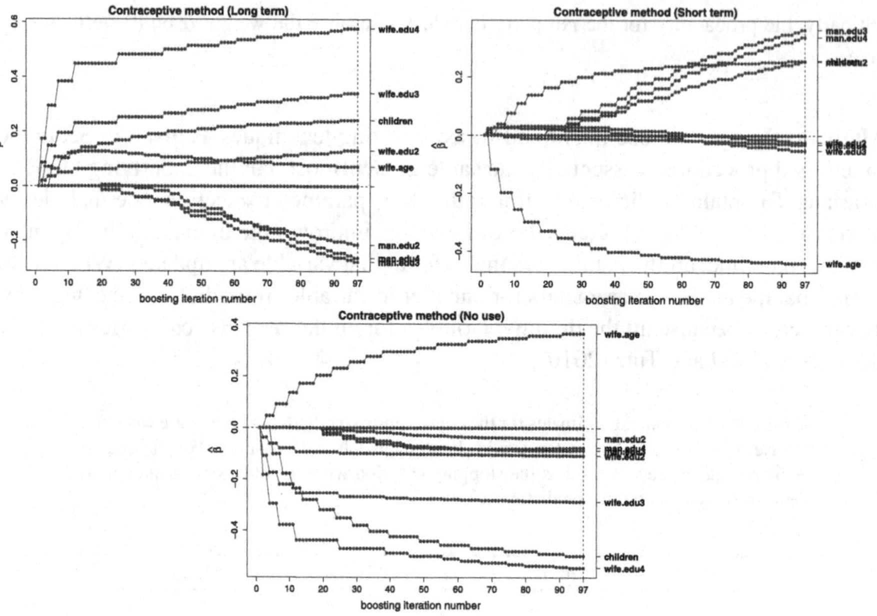

FIGURE 8.9: Coefficient buildups obtained by boosting for contraceptive method choice data. The vertical dotted line represents the optimal boosting iteration number on the basis of 10-fold cross-validation when deviance is used as a stopping criterion.

A multinomial logit model with selection of variables was fitted by using the blockwise boosting algorithm with the deviance being used for variable selection among the candidate predictors. The optimal number of boosting iterations was decided by deviance with 10-fold cross-validation. For metric and nominal predictors, a ridge-type estimator was used for candidate predictors; for ordered predictors, adjacent categories were penalized to obtain a smooth estimate of effects. Alternatively, lasso estimates, which select parameters instead of variables, were obtained by use of the *glmnet* package (Friedman et al., 2010). The optimal value of the penalty term was also based on 10-fold cross-validation.

Estimates are given in Table 8.11. Since in the lasso approach all variables are found relevant when at least one predictor (or dummy associated with the categorical predictor) has non-zero estimate(s) for at least one response category, one obtains no reduction of predictors. In contrast, boosting performs variable selection by grouping all the parameters associated with one predictor. Consequently, it recommends just four informative predictors. The informative predictors include two continuous predictors, that is, wife's age and number of children ever born, and two categorical predictors (with all of their categories), that is, wife's education and husband's education. The coefficient buildup for boosting is shown in Figure 8.9. For more details and further examples see Zahid and Tutz (2010).

□

8.11 Further Reading

Discrete Choice. An overview on probabilistic choice models was given by McFadden (1981). Models for discrete choice are mostly treated in the econometric literature, for example, by Greene (2003) and Maddala (1983). Estimation of nested logit models was investigated by McFadden (1981), Amemiya (1978), Ben-Akiva and Lerman (1985), Hausman and McFadden (1984), Börsch-Supan (1987), and Brownstone and Small (1989).

Paired Comparison Models. An overview on pair comparison models was given by Bradley (1976, 1984). Basic modeling concepts were considered by Yellott (1977) and Colonius (1980).

Marketing applications are found in Dillon et al. (1993). Extensions to incorporate ties were given by Davidson (1970) and Rao and Kupper (1967); more general model with ordered response categories were proposed by Agresti (1992a), Tutz (1986), and Böckenholt and Dillon (1997). The inclusion of category-specific variables was discussed in Dittrich et al. (1998).

R Packages. Multinomial response models with global predictors can be fitted using the function *multinom* from the package *nnet*. Models that contain global and category-specific predictors can be fitted by using the function *mlogit* from the package *mlogit*. Paired comparison models can be fitted by using *prefmod* or *BradleyTerry2*. The package *glmnet* (Friedman et al., 2010) fits the lasso and elastic net for multionmial responses.

8.12 Exercises

8.1 Consider the multinomial distribution $\boldsymbol{y}^T = (y_1, \dots, y_k) \sim \mathrm{M}(m, \boldsymbol{\pi})$, where $\boldsymbol{\pi}^T = (\pi_1, \dots, \pi_k)$ is a vector of probabilities restricted by $\pi_i \in [0,1], \sum_i \pi_i = 1$.

(a) Give an experiment, for example an urn model, for which the outcomes follow a multinomial distribution.

(a) How are single components y_i distributed?

(b) Derive the mean $\mathrm{E}(\boldsymbol{y})$, the variance of components $\mathrm{var}(y_i)$, and the covariance $\mathrm{cov}(y_i, y_j)$.

8.2 Assume the symmetric side constraint $\sum_{s=1}^k \boldsymbol{\beta}_s = \mathbf{0}$ for the multinomial logit model. Show that then $\log(P(Y=r|\boldsymbol{x})/GM(\boldsymbol{x})) = \boldsymbol{x}^T\boldsymbol{\beta}_r$ holds, where $GM(\boldsymbol{x}) = (\prod_{s=1}^k P(Y=s|\boldsymbol{x}))^{1/k}$ is the geometric mean response.

8.3 Consider the multinomial logit model for k categories with parameters $\boldsymbol{\beta}_1, \dots, \boldsymbol{\beta}_q, q = k-1$, and side constraint $\boldsymbol{\beta}_k = \mathbf{0}$. Alternatively, one can use parameters $\boldsymbol{\beta}_1^*, \dots, \boldsymbol{\beta}_q^*$ and side constraint $\sum_j \boldsymbol{\beta}_j^* = \mathbf{0}$.

(a) Derive the transformations between the parameters with reference category k and the symmetric side constraint.

(b) Interpret the parameters in Table 8.2.

(c) Give the parameters of the logit model that was used to obtain Table 8.2 when the symmetric side constraint is used.

8.4 Equation (8.8) shows the likelihood contribution of the ith observation for multinomial data in terms of relative frequencies for the categories (p_{ir} denotes the relative frequency of category r, and n_i the number of observations for fixed covariate predictor $\boldsymbol{x}_i$).

(a) Show that the score function of the multinomial logit model has the form $\boldsymbol{s}(\boldsymbol{\beta}) = \sum_{i=1}^N n_i \boldsymbol{X}_i^T(\boldsymbol{p}_i - \boldsymbol{\pi}_i)$.

(b) Show that the Fisher matrix of the multinomial logit model has the form $\boldsymbol{F}(\boldsymbol{\beta}) = \sum_{i=1}^N n_i^2 \boldsymbol{X}_i^T \boldsymbol{\Sigma}_i(\boldsymbol{\beta})\boldsymbol{X}_i$.

8.5 Investigate if an interaction between gender and age is needed for the party preference data in Table 8.1.

8.6 The covariance of a multinomial distribution with k categories, n observations, and vector of probabilities $\boldsymbol{\pi}^T = (\pi_1, \dots, \pi_q)$, $q = k-1$, has covariance matrix $\boldsymbol{\Sigma} = [\,\mathrm{Diag}\,(\hat{\boldsymbol{\pi}}) - \boldsymbol{\pi}\boldsymbol{\pi}^T]/n$.

(a) Show that the inverse covariance matrix $\boldsymbol{\Sigma}^{-1} = (\sigma^{ij})$ has entries $\sigma^{ii} = n/\pi_i + n/(1 - \sum_{r=1}^q \pi_r)$, $\sigma^{ij} = n/(1 - \sum_{r=1}^q \pi_r), i \neq j$.

(b) Show that the squared Pearson residual $\chi^2_{p_i} = n_i(p_{ir} - \hat{\pi}_{ir})^2/\hat{\pi}_{ir}$ is equivalent to $(\boldsymbol{p}_i - \hat{\boldsymbol{\pi}}_i)^T \boldsymbol{\Sigma}_i(\hat{\boldsymbol{\beta}})^{-1}(\boldsymbol{p}_i - \hat{\boldsymbol{\pi}}_i)$, where $\hat{\boldsymbol{\pi}}_i^T = (\hat{\pi}_{i1}, \dots, \hat{\pi}_{iq}), \boldsymbol{p}_i^T = (p_{i1}, \dots, p_{iq})$, and $\boldsymbol{\Sigma}_i(\hat{\beta}) = [\,\text{Diag}\,(\hat{\boldsymbol{\pi}}_i) - \hat{\boldsymbol{\pi}}_i \hat{\boldsymbol{\pi}}_i^T]/n_i$.

8.7 Let $\tilde{\boldsymbol{\beta}}^c_{.j} = (\tilde{\boldsymbol{\beta}}_{1j}, \dots, \tilde{\boldsymbol{\beta}}_{k-1,j})^T$ denote the vector of parameters corresponding to variable j with side constraint $\tilde{\beta}_{kj} = 0$, $j = 0, \dots, p$ and $\boldsymbol{\beta}^c_{.j} = (\boldsymbol{\beta}_{1j}, \dots, \boldsymbol{\beta}_{k-1,j})^T$ the vector with symmetric side constraints. Derive the matrix $\boldsymbol{T}_0$ that transforms $\tilde{\boldsymbol{\beta}}^c_{.j}$ into $\boldsymbol{\beta}^c_{.j}$, $\boldsymbol{\beta}^c_{.j} = \boldsymbol{T}\tilde{\boldsymbol{\beta}}^c_{.j}$ and its inverse $\boldsymbol{T}^{-1}$. In addition, show that (8.21) is equivalent to (8.23).

8.8 The hierarchically structured multinomial model for subsets of response categories $S_1, \dots, S_m$ can be given as $\log(P(Y \in S_t|\boldsymbol{x})/P(Y \in S_m|\boldsymbol{x})) = \boldsymbol{x}^T\boldsymbol{\beta}_t$, $\log(P(Y = r|Y \in S_t, \boldsymbol{x})/P(Y = r_t|Y \in S_t, \boldsymbol{x})) = \boldsymbol{x}^T\boldsymbol{\beta}_r^{(S_t)}$, where $r_t \in S_t$.

(a) Consider the case of four response categories with subsets given by $S_1 = \{1, 2\}$ and $S_2 = \{3, 4\}$. Show that the model has the form of a multivariate GLM $g(\boldsymbol{\pi}) = \boldsymbol{X}\boldsymbol{\beta}$ and derive the link function.

(b) Derive the score function and the Fisher matrix for the general case. Do the standard errors of coefficients differ from the standard errors obtained by the fitting of model components (for example fitting of $\log(P(Y \in S_t|\boldsymbol{x})/P(Y \in S_m|\boldsymbol{x})) = \boldsymbol{x}^T\boldsymbol{\beta}_t$)?

(c) Use the dataset *addiction* from the package *catdata* with subsets $S_1 = \{0, 1\}$ and $S_2 = \{2\}$ (compare Example 8.5). Fit the model with the covariates gender, university, and age and test if covariates can be omitted by use of the likelihood ratio test.

8.9 The R package *MASS* contains data on the cross-classification of people in Caithness, Scotland, by eye and hair color. The dataset, named *caith*, is given as a 4 by 5 table with rows the eye colors (blue, light, medium, dark) and columns the hair colors (fair, red, medium, dark, black); see also Venables and Ripley (2002). Consider eye color as the predictor and hair colour as the response. Fit a multinomial logit model and investigate the significance of the effects.

8.10 The R package *mlogit* can be used to fit multinomial models with global and category-specific predictors. The dataset *ModeChoice* (package *Ecdat*) contains various predictors for the choice of travel mode in Australia. Fit the appropriate models and interpret the results.

8.11 The R library *faraway* contains the dataset nes96, which is a subset of the 1996 American National Election Study (see Faraway, 2006; Rosenstone et al., 1997). Consider the response party identification with the categories democratic, liberal, and republican.

(a) Fit a multinomial logit model containing age education level and income group of the respondents and visualize the dependence of the response categories on the explanatory variable

(b) Include further explanatory variables and decide which are relevant.

Chapter 9

Ordinal Response Models

When the response categories in a regression problem are ordered one can find simpler models than the multinomial logit model. The multinomial logit wastes information because the ordering of categories is not explicitly used. Therefore, often more parameters than are really needed are in the model. In particular with categorical data, parsimonious models are to be preferred because the information content in the response is always low.

Data analysts who are not familiar with ordinal models usually seek solutions by inadequate modeling. If the number of response categories is high, for example in rating scales, they ignore that the response is ordinal and use classical regression models that assume that the response is at least on an interval scale. Thereby they also ignore that the response is categorical. The result is often spurious effects. Analysts who are aware of the ordinal scale but are not familiar with ordinal models frequently use binary regression models by collapsing outcomes into two groups of response categories. The effect is a loss of information. Armstrong and Sloan (1989) demonstrated that the binary model may attain only between 50 and 75% efficiency relative to an ordinal model for a five-level ordered response; see also Steadman and Weissfeld (1998), who in addition consider polytomous models as alternatives.

One may distinguish between two types of ordinal categorical variables, *grouped continuous variables* and *assessed ordinal categorical variables* (Anderson, 1984). The first type is a mere categorized version of a continuous variable, which in principle can be observed itself. For example, when the duration of unemployment (Example 9.1) is categorized into short-term, medium-term, and long-term unemployment, the underlying response variable is the duration of unemployment in days. The second type of ordered variable arises when an assessor processes an unknown amount of information, leading to the judgement of the grade of the ordered categorical scale. This sort of variable is found in the knee injury study (Example 1.4), where pain after treatment is assumed on a five-point scale representing an assessed categorical variable. Also, the retinopathy status in Example 9.2 was assessed by the physician without reference to an underlying continuous measurement.

TABLE 9.1: Cross-classification of pain and treatment for knee data.

	No Pain 1	2	3	4	Severe Pain 5	
Placebo	17	8	14	20	4	63
Treatment	19	26	11	6	2	64

Example 9.1: Unemployment

In unemployment studies one often distinguishes between short-term, medium-term, and long-term unemployment, obtaining a categorical response variable with three categories $Y \in \{1, 2, 3\}$. Then one may investigate the effect of gender, age, education level, and other differentiating covariates on the three levels. □

Example 9.2: Retinopathy

In a 6-year followup study on diabetes and retinopathy status reported by Bender and Grouven (1998) the interesting question is how the retinopathy status is associated with risk factors. The considered risk factor is smoking (SM = 1: smoker, SM = 0: non-smoker) adjusted for the known risk factors diabetes duration (DIAB) measured in years, glycosylated hemoglobin (GH), which is measured in percent, and diastolic blood pressure (BP) measured in mmHg. The response variable retinopathy status has three categories (1: no retinopathy; 2: nonproliferative retinopathy; 3: advanced retinopathy or blind). □

TABLE 9.2: Types of ordinal models and dichotomous variables y_r that are used.

Cumulative-Type Model, Dichotomization into Groups

$$[1, \dots, r | r+1, \dots, k] \qquad y_r = \begin{cases} 1 & Y \in \{1, \dots, r\} \\ 0 & Y \in \{r+1, \dots, k\} \end{cases}$$

Sequential-Type Model, Dichotomization Given $Y \geq r$

$$1, \dots, [r | r+1, \dots, k] \qquad y_r = \begin{cases} 1 & Y = r \\ 0 & Y > r \end{cases} \quad \text{given } Y \geq r$$

Adjacent-Type Model, Dichotomization Given $Y \in \{r, r+1\}$

$$1, \dots, [r | r+1], \dots, k \qquad y_r = \begin{cases} 1 & Y = r \\ 0 & Y \geq r+1 \end{cases} \quad \text{given } Y \in \{r, r+1\}$$

One way of modeling ordinal responses is to start from binary models. There are several approaches to construct ordinal response models from binary response models. These approaches differ in how the ordered categories $1, \dots, k$ are transformed into a binary response. The simplest approach is to consider a split between categories r and $r+1$, yielding the grouped response categories $\{1, \dots, r\}$ and $\{r+1, \dots, k\}$. The *cumulative approach* uses these groupings by considering the binary variable $y_r = 1$ if $Y \leq r$ and $y_r = 0$ if $Y > r$. For the response Y, the binary response model $P(y_r = 1|\boldsymbol{x}) = F(\boldsymbol{x}^T\boldsymbol{\beta}_r)$ turns into

$$P(Y \leq r|\boldsymbol{x}) = F(\boldsymbol{x}^T\boldsymbol{\beta}_r), r = 1, \dots, q,$$

where F is the response function of the binary model and β_r may depend on the splitting that is modeled. Alternatively, one may consider it as a sequential decision where the transition from category r to category $r+1$ given category r or higher follows a binary model. The binary model distinguishes between $Y = r$ and $Y > r$ given $Y \geq r$. One obtains the *sequential-type model*

$$P(Y = r|Y \geq r, \boldsymbol{x}) = F(\boldsymbol{x}^T\boldsymbol{\beta}_r), r = 1, \dots, q.$$

The third approach is based on the consideration of *adjacent categories*. Given two adjacent categories, the binary model distinguishes between these two categories by using

$$P(Y = r|Y \in \{r, r+1\}) = F(\boldsymbol{x}^T\boldsymbol{\beta}_r), r = 1, \ldots, q.$$

Table 9.2 illustrates the three types of dichotomizations that determine the models. Brackets denote which response categories are used when dichotomizing and the "|" determines the (conditional) split. In the following the use of these models is motivated and applications are given.

9.1 Cumulative Models

The most frequently used model is the cumulative model, which was propagated by McCullagh (1980).

9.1.1 Simple Cumulative Model

A simple form of the cumulative type model may be derived from the assumption that the observed categories represent a coarser (categorical) version of an underlying (continuous) regression model. Let $\tilde{Y}$ be an underlying latent variable that follows a regression model:

$$\tilde{Y} = -\boldsymbol{x}^T\boldsymbol{\gamma} + \epsilon,$$

where ϵ is a noise variable with continuous distribution function F. Furthermore, let the link between the observable categories and the latent variable be given by

$$Y = r \quad \Leftrightarrow \quad \gamma_{0,r-1} < \tilde{Y} \leq \gamma_{0r},$$

where $-\infty = \gamma_{00} < \gamma_{01} < \cdots < \gamma_{0k} = \infty$ are thresholds on the latent scale. One obtains immediately

$$P(Y \leq r|\boldsymbol{x}) = P(-\boldsymbol{x}^T\boldsymbol{\gamma} + \epsilon \leq \gamma_{0r}) = P(\epsilon \leq \gamma_{0r} + \boldsymbol{x}^T\boldsymbol{\gamma}) = F(\gamma_{0r} + \boldsymbol{x}^T\boldsymbol{\gamma}).$$

The model is essentially a univariate response model since it is assumed that a univariate response $\tilde{Y}$ is acting in the background. The response Y is just a coarser version of $\tilde{Y}$ where the thresholds γ_{0r} determine the preference for categories and the covariates produce a shifting on the latent scale. Figure 9.1 demonstrates the shifting for the simple example where the response is given in three categories (short-term, medium-term, long-term unemployment) and the only covariate is age. The slope γ of the latent variable is negative, yielding an increase of $E(\tilde{Y}) = -x\gamma$ with increasing age. It is seen how the probability of category 1 (short-term unemployment) decreases with increasing age. Often one might give some interpretation for the latent variable. In the case of unemployment, $\tilde{Y}$ might represent the lack of opportunities in the labour market, resulting in high categories if $\tilde{Y}$ is high. In medical examples, where differing degrees of illness are observed, $\tilde{Y}$ may stand for the latent damage arising from the patient history that is contained in the covariates. Although often there is an interpretation of the latent variable, it is essential to note that the resulting model may be used without reference to latent variables, which may be considered only as a motivation for the model. The advantage of the model is its simplicity. Due to the derivation from the latent regression model with slope $\boldsymbol{\gamma}$ the effect of $\boldsymbol{x}$ does not depend on the category; it is a so-called *global* effect. Since the cumulative probabilities $P(Y \leq r|\boldsymbol{x})$ are parameterized, the model is called the (simple) *cumulative* or *threshold model*.

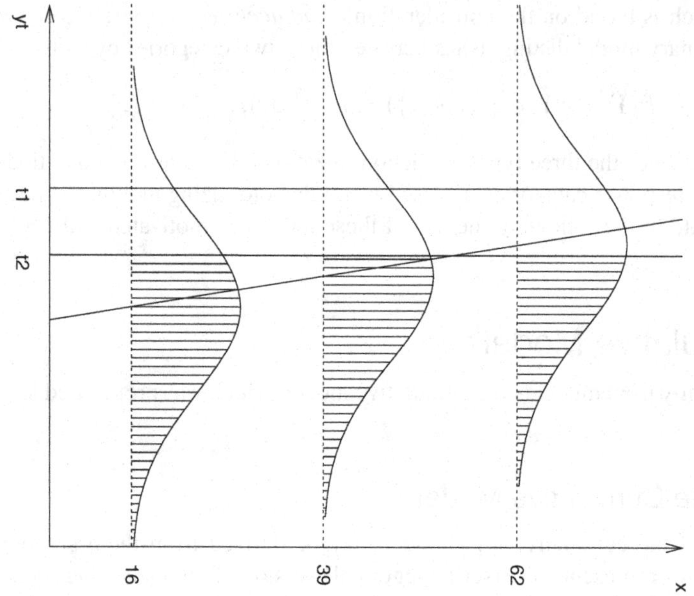

FIGURE 9.1: Cumulative model for predictor age and three response categories of duration of unemployment.

Threshold (Simple Cumulative) Model

$$P(Y \leq r|\boldsymbol{x}) = F(\gamma_{0r} + \boldsymbol{x}^T\boldsymbol{\gamma}) \tag{9.1}$$

$$P(Y = r|\boldsymbol{x}) = F(\gamma_{0r} + \boldsymbol{x}^T\boldsymbol{\gamma}) - F(\gamma_{0,r-1} + \boldsymbol{x}^T\boldsymbol{\gamma}),$$

$$\text{where } -\infty = \gamma_{00} \leq \gamma_{01} \leq \cdots \leq \gamma_{0k} = \infty$$

The model may be seen as a series of binary regression models with common regression parameters. By considering the split of the response categories into $\{1, \dots, r\}$ and $\{r+1, \dots, k\}$ and defining $y_r = 1$ if $Y \in \{1, \dots, r\}$, $y_r = 0$ if $Y = \{r+1, \dots, k\}$, one has $P(Y \leq r|\boldsymbol{x}) = P(y_r = 1|\boldsymbol{x})$ and (9.1) is a binary regression model for $y_r \in \{0, 1\}$. Of course it is implied that the regression parameter $\boldsymbol{\gamma}$ is the same for all splits. That makes the model simple and allows easy interpretation of the regression parameter. One consequence of having one regression parameter is that the intercepts have to be ordered, $\gamma_{01} \leq ... \leq \gamma_{0k}$. Otherwise, the probabilities could be negative. The ordering results naturally from the derivation as a coarser version of the underlying continuous response $\tilde{Y}$.

The most widely used model is the *cumulative logit model* or the *proportional odds model*, where F is the logistic distribution function.

Cumulative Logit Model (Proportional Odds Model)

$$\log\left(\frac{P(Y \leq r|\boldsymbol{x})}{P(Y > r|\boldsymbol{x})}\right) = \gamma_{0r} + \boldsymbol{x}^T\boldsymbol{\gamma}$$

The name "proportional odds model" is due to a specific property called *strict stochastic ordering*, which holds for all simple cumulative models. Consider two populations that are characterized by the covariate values $\boldsymbol{x}$ and $\tilde{\boldsymbol{x}}$. Then one obtains

$$F^{-1}(P(Y \leq r|\boldsymbol{x})) - F^{-1}(P(Y \leq r|\tilde{\boldsymbol{x}}) = (\boldsymbol{x} - \tilde{\boldsymbol{x}})^T\boldsymbol{\gamma}, \tag{9.2}$$

meaning that the comparison of (transformed) cumulative probabilities $P(Y \leq r|\boldsymbol{x})$ and $P(Y \leq r|\tilde{\boldsymbol{x}})$ does not depend on the category. For the logit model, (9.2) takes the form

$$\frac{P(Y \leq r|\boldsymbol{x})/P(Y > r|\boldsymbol{x})}{P(Y \leq r|\tilde{\boldsymbol{x}})/P(Y > r|\tilde{\boldsymbol{x}})} = \exp((\boldsymbol{x} - \tilde{\boldsymbol{x}})^T\boldsymbol{\gamma}).$$

Thus the comparison of populations in terms of *cumulative odds* $P(Y \leq r|\boldsymbol{x})/P(Y > r|\boldsymbol{x})$ does not depend on the category. That means that if, for example, the cumulative odds in population $\boldsymbol{x}$ are twice the cumulative odds in population $\tilde{\boldsymbol{x}}$, and this holds for all the categories. This makes interpreting the effects simple because interpretation only refers to the effects of $\boldsymbol{x}$ without refering to specific categories. The same holds for all models of the form (9.1); however, a comparison for alternative models does not refer to cumulative odds but to different scalings.

TABLE 9.3: Proportional odds model with linear age for knee data.

Variable	Estimate	Standard Error	Wald	p-Value	Odds Ratio
Therapy	0.943	0.335	7.91	0.004	2.568
Gender	−0.049	0.373	0.02	0.893	0.985
Age	0.015	0.016	0.87	0.349	1.015

TABLE 9.4: Proportional odds model with quadratic age effect for knee data.

Variable	Estimate	Standard Error	Wald	p-Value	Odds Ratio
Therapy	0.944	0.338	7.78	0.005	2.570
Gender	0.082	0.378	0.04	0.826	1.085
Age	−0.001	0.018	0.01	0.924	0.999
Age^2	0.006	0.002	8.86	0.002	1.006

Example 9.3: Knee Injuries

As a simple application let us first investigate the treatment effect from Table 9.1, thereby ignoring all the covariates. Without using ordinal models one can test if there is an association between treatment and pain level. The simple χ^2-test of independence yields 18.20 on 4 degrees of freedom. With a p-value of 0.001 the effect of treatment is highly significant. However, association tests do not show how the treatment effect is linked to the pain level.

The quantification of effect strength may be obtained by fitting an ordinal model. The fitting of the proportional odds model yields for the treatment effect 0.893 with standard error 0.328, showing a distinct treatment effect. The corresponding odds ratio is given by $e^{0.893} = 2.442$, which means that, for any dichotomization of response categories, the (estimated) cumulative odds ratio comparing treatment to a placebo is

$$\frac{P(Y \leq r|\text{Treatment})/P(Y > r|\text{Treatment})}{P(Y \leq r|\text{Placebo})/P(Y \geq r|\text{Placebo})} = 2.442.$$

Thus the odds for the low response category, meaning lesser pain, as compared to higher categories are distinctly higher in the treatment group.

More generally, the ordinal regression model allows one to model the treatment effect (1: treatment; 0: placebo) as well as the effect of the covariates on pain during movement. The covariates are gender (1: male; 0: female) and age in years, where age has been centered around 30 years. We consider the main effect proportional odds model with predictor

$$\eta_r = \text{Intercept}_r + \text{Treatment}\gamma_T + \text{Gender}\gamma_G + \text{Age}\gamma_A$$

and a model with a quadratic effect of age having predictor

$$\eta_r = \text{Intercept}_r + \text{Treatment}\gamma_T + \text{Gender}\gamma_G + \text{Age}\gamma_A + \text{Age}^2\gamma_{A^2}.$$

Tables 9.3 and 9.4 show the parameter estimates. In the main effect model only therapy seems to be influential; therapy and age have non-significant effects. However, the assumption of a linear effect of age is not supported by the data. Deviances are given by 372.24 for the linear age effect model and 362.88 for the quadratic effect of age. With a difference in deviances of 9.36, one concludes that the quadratic effect should not be omitted, since $\chi^2_{.95}(1) = 3.84$. The significance of the quadratic effect is also supported by the Wald test, which shows p-value 0.002. When the covariate effects of gender and age are taken into account the treatment effect is slightly stronger. One obtains 0.944, corresponding to odds ratio 2.57, as compared to 0.893 and odds ratio 2.44 for the simple model where covariates are ignored. □

The cumulative model was propagated by McCullagh (1980) and has been widely used since this influential paper has been published. Earlier versions of the cumulative logistic model have been suggested by Snell (1964), Walker and Duncan (1967), and Williams and Grizzle (1972). Since interpretation is very simple, it is tempting to use the simple cumulative model. However, the results could be misleading if the proportional odds assumption, implied by the model, does not hold (see Example 9.5). Therefore, it is advisable to check if the assumption is appropriate. This may be done by embedding the model into a more general model (see Section 9.1.3). However, first we will consider alternative cumulative models, which have different link functions.

9.1.2 Alternative Link Functions

The link function of the cumulative model results from the distribution of the noise variable ε when the model is derived from the latent regression model $\tilde{Y} = -\boldsymbol{x}^T\boldsymbol{\gamma} + \varepsilon$. Alternative models are obtained by assuming different distributions for the noise variable ε. Figure 9.2 shows the densities of the underlying regression model $\tilde{Y} = -\boldsymbol{x}^T\boldsymbol{\gamma} + \varepsilon$ for two subpopulations characterized by different values of the covariates $\boldsymbol{x}_1$ and $\boldsymbol{x}_2$ and four different distributions. The dashed lines represent the cutoff points when the continuous response $\tilde{Y}$ is cut into slices that represent the coarser categorical responses.

Cumulative Extreme Value Models

If one assumes for ε the minimum extreme value (or Gompertz) distribution $F(\eta) = 1 - \exp(-\exp(\eta))$, one obtains the *cumulative minimum extreme value model,* which has several representations.

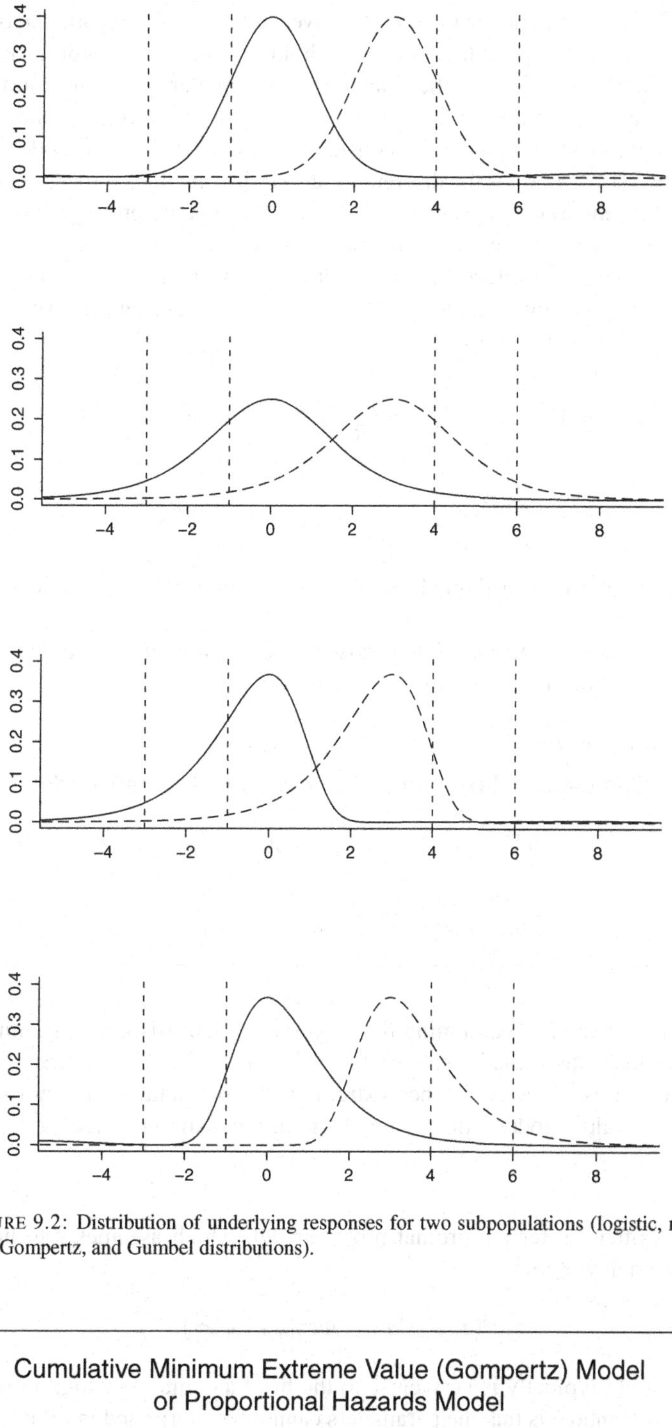

FIGURE 9.2: Distribution of underlying responses for two subpopulations (logistic, normal, Gompertz, and Gumbel distributions).

Cumulative Minimum Extreme Value (Gompertz) Model or Proportional Hazards Model

$$P(Y \leq r|\boldsymbol{x}) = 1 - \exp(-\exp(\gamma_{0r} + \boldsymbol{x}^T\boldsymbol{\gamma}))$$

or

$$P(Y = r|Y \geq r, \boldsymbol{x}) = 1 - \exp(-\exp(\tilde{\gamma}_{0r} + \boldsymbol{x}^T\boldsymbol{\gamma}))$$

or

$$\log(-\log(P(Y > r|\boldsymbol{x})) = \gamma_{0r} + \boldsymbol{x}^T\boldsymbol{\gamma}$$

The second form of the model is simply derived by using the reparameterization $\tilde{\gamma}_{0r} = \log\{\exp(\gamma_{0r}) - \exp(\gamma_{0,r-1})\}$. Although the thresholds γ_{0r} have to be ordered by $\gamma_{01} \leq \cdots < \gamma_{0q}$, there is no such restriction on the transformed parameter $\tilde{\gamma}_{0r}$. The formulation of the model in conditional probabilities $P(Y = r|Y \geq r, \boldsymbol{x})$ shows the strong relation to sequential-type models (Section 9.2), which can be considered as discrete hazard models. By defining $\lambda(r, \boldsymbol{x}) = P(Y = r|Y \geq r, \boldsymbol{x})$ as the discrete hazard, reflecting that a process stops in category r, given it is reached, the model represents a discrete version of the proportional hazards or Cox model, which is widely used in continuous time survival analysis (see Section 9.2). Therefore the model is also called the *grouped Cox model* or the *proportional hazards model.*

For the proportional hazards model, the strict stochastic ordering property for two populations $\boldsymbol{x}, \tilde{\boldsymbol{x}}$ may be given in the form

$$\log\left(\frac{P(Y > r|Y \geq r, \boldsymbol{x})}{P(Y > r|Y \geq r, \tilde{\boldsymbol{x}})}\right) = -\exp((\boldsymbol{x} - \tilde{\boldsymbol{x}})^T\boldsymbol{\gamma})$$

or

$$\frac{\log(P(Y > r|\boldsymbol{x}))}{\log(P(Y > r|\tilde{\boldsymbol{x}}))} = \exp((\boldsymbol{x} - \tilde{\boldsymbol{x}})^T\boldsymbol{\gamma}).$$

Thus the proportion of the logarithmic hazards $P(Y > r|\boldsymbol{x})/P(Y \geq r|\boldsymbol{x})$ does not depend on the categories.

An alternative model is the *cumulative maximum extreme-value model*, which is based on the (Gumbel) distribution function $F(\eta) = \exp(-\exp(\eta))$.

Cumulative Maximum Extreme-Value (Gumbel) Model

$$P(Y \leq r|\boldsymbol{x}) = \exp(-\exp(\gamma_{0r} + \boldsymbol{x}^T\boldsymbol{\gamma}))$$

or

$$\log(-\log(P(Y \leq r|\boldsymbol{x})) = \gamma_{0r} + \boldsymbol{x}^T\boldsymbol{\gamma}$$

The model is equivalent to assuming the proportional hazards model (minimum extreme-value model) for the transformed response $Y_r = k + 1 - Y$, which reverses the ordering of the response categories. However, since extreme-value distributions are not symmetric, the maximum extreme-value model differs from the minimum extreme-value model.

Probit Model

Econometricians often prefer the ordinal probit model, which assumes ε to have a standard normal distribution Φ yielding

$$P(Y \leq r|\boldsymbol{x}) = \Phi(\gamma_{0r} + \boldsymbol{x}^T\boldsymbol{\gamma}).$$

In applications the fit typically is very close to the fit of a cumulative logit model. What may be seen as an disadvantage is that the parameters cannot be interpreted in terms of (cumulative) odds.

Example 9.4: Knee Injuries

A comparison of link functions for the knee injuries data is given in Table 9.5. The underlying model includes a quadratic term of age since for all the models the quadratic effect cannot be omitted. It is seen that the logit model and the Gompertz model have the best fit. Although the Gompertz model has a slightly better fit, the logit model might be a better choice because interpretation of the parameters is in terms of the (cumulative) odds ratios. □

TABLE 9.5: Deviances for ordinal models with alternative link functions (knee injuries data).

	Cumulative	Sequential
Logit	362.88	360.45
Probit	364.45	361.12
Gumbel	368.89	361.77
Gompertz	361.31	361.31

9.1.3 General Cumulative Models

The cumulative models considered in the previous sections assume that the effect of covariates does not depend on the category. More general, however, the effect the of covariates may vary across categories. The general cumulative model may be seen as a combination of the binary response models for the splitting of categories into groupings $\{1,\dots,r\},\{r+1,\dots,k\}$.

Cumulative Model

$$P(Y \le r|\boldsymbol{x}) = F(\gamma_{0r} + \boldsymbol{x}^T\boldsymbol{\gamma}_r), \quad r = 1,\dots,q$$

Since for each split a binary regression model *with separate parameters* is assumed, the model may be seen as a series of binary regression models for *dependent* binary response variables. In its general form it has as many parameters as the multinomial response model. Moreover, it is implied that $\gamma_{0,r-1} + \boldsymbol{x}^T\boldsymbol{\gamma}_{r-1} \le \gamma_{0r} + \boldsymbol{x}^T\boldsymbol{\gamma}_r$ for all $\boldsymbol{x}$ and all categories r, since $P(Y \le r-1|\boldsymbol{x}) \le P(Y \le r|\boldsymbol{x})$ has to hold for all categories. The model serves as a general model into which simple cumulative models may be embedded and which may be simplified by investigating if at least some of the predictors have regression parameters that do not depend on the category. The most used model is the cumulative logit model.

Cumulative Logit Model

$$P(Y \le r|\boldsymbol{x}) = \frac{\exp(\gamma_{0r} + \boldsymbol{x}^T\boldsymbol{\gamma}_r)}{1+\exp(\gamma_{0r} + \boldsymbol{x}^T\boldsymbol{\gamma}_r)}$$

or

$$\log\frac{P(Y \le r|\boldsymbol{x})}{P(Y > r|\boldsymbol{x})} = \gamma_{0r} + \boldsymbol{x}^T\boldsymbol{\gamma}_r$$

The parameter $\boldsymbol{\gamma}_r$ may be interpreted as in the binary model, where "binary" refers to the grouping $\{1,\dots,r\},\{r+1,\dots,k\}$. However, the simple interpretation from the simple cumulative model is lost. Since in general $\boldsymbol{\gamma}_r \ne \boldsymbol{\gamma}_{r+1}$, the effect of $\boldsymbol{x}$ has to be interpreted with respect to the dichotomization it refers to.

A mixture of types of effects results, if part of the covariates has an effect that does not depend on the category. Let $\boldsymbol{x}$ be partitioned into $\boldsymbol{x}_1$ and $\boldsymbol{x}_2$, $\boldsymbol{x}^T = (\boldsymbol{x}_1, \boldsymbol{x}_2)$. Then the logit-type model or *partial proportional odds model* is given by

$$\log\left(\frac{P(Y \le r|\boldsymbol{x})}{P(Y > r|\boldsymbol{x})}\right) = \gamma_{0r} + \boldsymbol{x}_1^T\boldsymbol{\gamma} + \boldsymbol{x}_2^T\boldsymbol{\gamma}_r,$$

where $\boldsymbol{x}_1$ has a *global effect* and $\boldsymbol{x}_2$ has a *category-specific effect*. Then the cumulative odds are only proportional (independent of the category) for the variables collected in $\boldsymbol{x}_1$. Consider two populations that differ only in the first set of variables, with the populations specified by $(\boldsymbol{x}_1, \boldsymbol{x}_2)$ and $(\tilde{\boldsymbol{x}}_1, \boldsymbol{x}_2)$. Then one obtains that

$$\frac{P(Y \le r|\boldsymbol{x}_1, \boldsymbol{x}_2)/P(Y > r|\boldsymbol{x}_1, \boldsymbol{x}_2)}{P(Y \le r|\tilde{\boldsymbol{x}}_1, \boldsymbol{x}_2)/P(Y > r|\tilde{\boldsymbol{x}}_1, \boldsymbol{x}_2)} = \exp((\boldsymbol{x}_1 - \tilde{\boldsymbol{x}}_1)^T\boldsymbol{\gamma})$$

does not depend on r. Therefore, the (cumulative) odds for the variables in $\boldsymbol{x}_1$ are still proportional.

The simple- and the general cumulative-type models differ in the linear predictor. While the simple model uses the predictor

$$\eta_r = \gamma_{0r} + \boldsymbol{x}^T\boldsymbol{\gamma}, \quad r = 1, \dots, q, \tag{9.3}$$

the general model is based on

$$\eta_r = \gamma_{0r} + \boldsymbol{x}^T\boldsymbol{\gamma}_r, \quad r = 1, \dots, q. \tag{9.4}$$

Sometimes it is useful to reparameterize the effects such that the general predictor has the form

$$\eta_r = \gamma_{0r} + \boldsymbol{x}^T\boldsymbol{\gamma} + \boldsymbol{x}^T\tilde{\boldsymbol{\gamma}}_r,$$

where $\tilde{\boldsymbol{\gamma}}_r$ has to fulfill some side constraints. By using category 1 as the reference category and setting $\tilde{\boldsymbol{\gamma}}_1 = \boldsymbol{0}$, the parameters are given by $\boldsymbol{\gamma} = \boldsymbol{\gamma}_1$, and $\tilde{\boldsymbol{\gamma}}_r = \boldsymbol{\gamma}_r - \boldsymbol{\gamma}_1$, $r = 1, \dots, q$, represent the deviation of $\boldsymbol{\gamma}_r$ from $\boldsymbol{\gamma}_1$. An alternative side constraint is $\sum_j \tilde{\boldsymbol{\gamma}}_j = \boldsymbol{0}$.

In an extended Wilkinson Rogers notation the predictor (9.3), where only the intercept depends on the category, may be abbreviated by

$$y + x_1 + \dots + x_p$$

and the predictor (9.4) by

$$y * x_1 + \dots + y * x_p.$$

The use of y instead of Y is motivated by the fact that $\boldsymbol{y} = (y_1, \dots, y_q)$ is multinomially distributed with $y_r = 1$ if $Y = r$ and $y_r = 0$ if $Y \ne r$. Thus $y + x_1 + \dots + x_p$ may be read as $y_1 + \dots + y_q + x_1 + \dots + x_p$ and $y * x_1 + \dots + y * x_p$ as $y_1 + \dots + y_q + y_1 x_1 + \dots + y_q x_1 + \dots + y_q x_p$, corresponding to the terms actually needed in the predictor.

Abbreviation	Linear Predictor $Y = r$
$y + x_1 + \dots + x_p$	$\gamma_{0r} + x^T\boldsymbol{\gamma}$
$y * x_1 + \dots + y * x_p$	$\gamma_{0r} + \boldsymbol{x}'\boldsymbol{\gamma}_r + x^T\tilde{\boldsymbol{\gamma}}_r.$

9.1.4 Testing the Proportional Odds Assumption

A crucial assumption of the proportional odds models is that cumulative odds are proportional, meaning that for two values of predictors $\boldsymbol{x}$ and $\tilde{\boldsymbol{x}}$ the proportion

$$\frac{P(Y \le r|\boldsymbol{x})/P(Y > r|\boldsymbol{x})}{P(Y \le r|\tilde{\boldsymbol{x}})/P(Y > r|\tilde{\boldsymbol{x}})} = \exp((\boldsymbol{x} - \tilde{\boldsymbol{x}})^T \gamma)$$

does not depend on the category. When the cumulative model is seen as a submodel of the corresponding general cumulative logit model with category specific effects γ_r, the assumption is equivalent to testing the null hypothesis

$$H_0 : \gamma_1 = \cdots = \gamma_q.$$

More specifically, one may investigate if the proportional odds assumption holds for single variables by testing

$$H_0 : \gamma_{1j} = \cdots = \gamma_{qj},$$

which means that the jth predictor has the global effect and therefore fulfills that for two values $x_j, \tilde{x}_j$

$$\frac{P(Y \le r|x_1, \dots, x_j, \dots x_p)/P(Y > r|x_1, \dots, x_j, \dots, x_p)}{P(Y \le r|x_1, \dots, \tilde{x}_j, \dots, x_p)/P(Y > r|x_1, \dots, \tilde{x}_j, \dots, x_p)} \tag{9.5}$$

does not depend on the category. The odds ratio (9.5) compares the odds for two proportions that differ only in the jth covariate. Since the hypotheses are linear, one can use the likelihood ratio test, the Wald test, or the score test, applied to the parameters of the category-specific model.

Example 9.5: Retinopathy Data

The risk factors in the retinopathy study from Bender and Grouven (1998) are smoking (SM), duration of diabetes (DIAB), glycosylated hemoglobin (GH), and blood pressure (BP). In a first step the proportional odds model is fitted. From the estimates in Table 9.6 one might infer that smoking has no effect (−0.254 with standard deviation 0.192). However, the inference is based on the assumption that the proportional odds model holds. If it does not hold, the results are not very trustworthy. Therefore, one should first investigate if the proportional odds assumption is realistic. It turns out that the deviance for the proportional odds model is 904.14. For the cumulative model with all parameters specified as category-specific one obtains 892.45. A comparison yields 11.69 on 4 degrees of freedom, which means that the proportional odds assumption should be rejected. Table 9.7 shows the model fits for a hierarchy of models, where [ ] denotes that the effect is category-specific; otherwise it is assumed to be global. It is seen that the effects of DIAB and SM should be specified as category-specific. Table 9.6 shows the estimates for the corresponding model and the proportional odds model. While the effect of smoking seems neglectable when fitting the proportional odds model, it seems to be influential when fitting the partial proportional odds model. For the split between the first category and the others, the effect of smoking is −0.399 with a standard error of 0.205 yielding a z-value of 1.946, which makes smoking at least suspicious. Since the effect is very weak for the split between the first two categories and the last one (0.062 with standard error 0.804), it appears to be neglectable when the category-specific effect of smoking is ignored by using the ill-fitting proportional odds model. □

As is seen from the retinopathy example, inference from the proportional odds model might be misleading if the model has a poor fit. Bender and Grouven (1998) already used the retinopathy example to demonstrate this effect. They proposed considering the fitting of separate binary

TABLE 9.6: Parameter estimates for proportional odds model and partial proportional odds model with category-specific effects of SM and DIAB for retinopathy data.

	Proportional Odds Model		Partial Proportional Odds Model	
	Estimates	Standard Error	Estimates	Standard Error
γ_{01}	12.302	1.290	12.188	1.293
γ_{02}	13.673	1.317	13.985	1.356
GH	−0.459	0.074	−0.468	0.074
BP	−0.072	0.013	−0.071	0.014
SM[1]	−0.254	0.192	−0.399	0.205
SM[2]			0.062	0.804
DIAB[1]	−0.140	0.013	−0.129	0.014
DIAB[2]			−0.163	0.017

TABLE 9.7: Model fits for partial proportional odds model (retinopathy data).

Partial Proportional Odds Models	Deviance	Proport. Odds Effect	Diff.	Df
$[SM], [DIAB], [GH], [BP]$	892.45			
$[SM], [DIAB], [GH], BP$	892.67	BP	0.22	1
$[SM], [DIAB], GH, BP$	893.77	GH	1.10	1
$[SM], DIAB, GH, BP$	897.97	DIAB	4.20	1
$SM, DIAB, GH, BP$	904.14	SM	6.17	1

models to groupings of response categories. A problem with fitting separate models is that likelihood ratio fits cannot be based directly on the separate fits. However, Brant (1990) showed a way to base test procedures for the proportional odds assumption on separate fits.

Of course, even if the test on proportional odds shows no significant result, it is no guarantee that the proportional odds model holds, at least when continuous predictors are involved and no goodness-of-fit tests are available.

One problem that occurs when testing the proportional odds assumption by likelihood ratio statistics is that estimates for the full model with category-specific regression parameters have to exist. Since strong restrictions on the parameter space are involved, that is not always the case. An alternative is the score statistic:

$$u = \boldsymbol{s}(\tilde{\boldsymbol{\beta}})^T \boldsymbol{F}^{-1}(\tilde{\boldsymbol{\beta}})\boldsymbol{s}(\tilde{\boldsymbol{\beta}}),$$

where $\boldsymbol{s}(\tilde{\boldsymbol{\beta}})$ is the score function and $\boldsymbol{F}(\tilde{\boldsymbol{\beta}})$ denotes the Fisher matrix for the larger model, evaluated at $\tilde{\boldsymbol{\beta}}$, which is the maximum likelihood estimates of the submodel, that is, the proportional or partial proportional odds model. The advantage is that only the fit of the proportional odds model is needed.

9.2 Sequential Models

In many applications one can imagine that the categories $1, \dots, k$ are reached successively. Often this is the only way to obtain a response in a higher category. For example, in duration models where categories refer to the duration of unemployment, it is obvious that long-term unemployment can only be observed if previously short-term unemployment was a step of the process. Analogously, if the response categories refer to the number of cars in a household, one can assume that the cars are bought successively. The observed number of cars represents the (preliminary) end of a process. Consequently, the modeling might reflect the successive transition to higher categories in a stepwise model.

9.2.1 Basic Model

Let the process start in category 1. The decision between category $\{1\}$ and categories $\{2, \ldots, k\}$ is determined in the first step by a dichotomous response model

$$P(Y = 1|\boldsymbol{x}) = F(\gamma_{01} + \boldsymbol{x}^T \boldsymbol{\gamma}_1).$$

If $Y = 1$, the process stops. If $Y \geq 2$, the second step is a decision between category $\{2\}$ and categories $\{3, \ldots, k\}$ and is determined by

$$P(Y = 2|Y \geq 2, \boldsymbol{x}) = F(\gamma_{02} + \boldsymbol{x}^T \boldsymbol{\gamma}_2).$$

In general in the rth step the decision between category $\{r\}$ and categories $\{r+1, \ldots, k\}$ is modeled by the binary model

$$P(Y = r|Y \geq r, \boldsymbol{x}) = F(\gamma_{0r} + \boldsymbol{x}^T \boldsymbol{\gamma}_r).$$

In addition to the stepwise modeling it is only assumed that the decision between the category reached and higher categories is determined by the same binary model that has response function F. The collection of steps represents the sequential model.

> **Sequential Model**
>
> $$P(Y = r|Y \geq r, \boldsymbol{x}) = F(\gamma_{0r} + \boldsymbol{x}^T \boldsymbol{\gamma}_r), r = 1, \ldots, q$$

In this general form the transition to higher categories in the rth step is determined by category-specific effects $\boldsymbol{\gamma}_r$. Thus the effects of the covariates may depend on the category that makes it a model with (possibly too) many parameters. A simplifying assumption is that the effects are the same in each step, that is, $\boldsymbol{\gamma}_1 = \ldots \boldsymbol{\gamma}_q = \boldsymbol{\gamma}$ and only intercepts $\gamma_{0r}, \ldots, \gamma_{0q}$ depend on the step under consideration. This simplification corresponds to the simple cumulative model where the effects of covariates $\boldsymbol{x}$ are also global, meaning that $\boldsymbol{\gamma}$ does not depend on the category.

The response probabilities in the sequential model are given by

$$\begin{aligned} P(Y = r|\boldsymbol{x}) &= P(Y = r|Y \geq r, \boldsymbol{x}) \prod_{s=1}^{r-1} P(Y > s|Y \geq s, \boldsymbol{x}) \\ &= F(\gamma_{0r} + \boldsymbol{x}^T \boldsymbol{\gamma}_r) \prod_{s=1}^{r-1} (1 - F(\gamma_{0s} + \boldsymbol{x}^T \boldsymbol{\gamma}_s)). \end{aligned}$$

(see also Section 9.5.2). Sequential models are closely related to discrete time survival models. Let the categories refer to discrete time, measured in days, months, or years. Then the conditional probabilities $\lambda(r, \boldsymbol{x}) = P(Y = r|Y \geq r, \boldsymbol{x})$ represent the discrete hazard function for given predictor $\boldsymbol{x}$. The discrete hazard $\lambda(r, \boldsymbol{x})$ is the probability that a duration ("survival") ends in the time interval that corresponds to category r given this time interval is reached. For example, it represents the probability of finding a job in month r given the person was unemployed during the previous months. The response probability, given by $P(Y = r|\boldsymbol{x}) = \lambda(r, \boldsymbol{x}) \prod_{i=1}^{r-1} \lambda(i, \boldsymbol{x})$, has an easy interpretation. It represents that the first $r - 1$ intervals were "survived," but the duration ends in category r; the transition to the next category was not successful. For discrete survival models many extensions have been considered that also can be used in ordinal modeling; see, for example, Fahrmeir (1994), and Tutz and Binder (2004); an overview is given in Fahrmeir and Tutz (2001).

For the modeling of ordinal data, the most common sequential model is the sequential logit model, where $F(\eta) = \exp(\eta)/(1+\exp(\eta))$ is the logistic distribution function.

Sequential Logit Model (Continuation Ratio Logits Model)

$$P(Y = r|Y \geq r, \boldsymbol{x}) = \frac{\exp(\gamma_{0r} + \boldsymbol{x}^T\boldsymbol{\gamma}_r)}{1+\exp(\gamma_{0r} + \boldsymbol{x}^T\boldsymbol{\gamma}_r)}$$

or

$$\log\left(\frac{P(Y = r|\boldsymbol{x})}{P(Y > r|\boldsymbol{x})}\right) = \gamma_{0r} + \boldsymbol{x}^T\boldsymbol{\gamma}_r$$

The logits $\log(P(Y = r|\boldsymbol{x})/P(Y > r|\boldsymbol{x}))$ compare the response category r to the response categories $\{r+1,\dots,k\}$ and may be considered as conditional logits, given $Y \geq r$, since the dichotomization into $\{r\}$ and $\{r+1,\dots,k\}$ refers only to categories $r, r+1\dots,k$. They are also referred to as *continuation ratio logits*.

In contrast to the cumulative model, there is no restriction on the parameters, which makes estimation easier. However, the general model contains many parameters because each transition has its own parameter $\boldsymbol{\gamma}_r$. A simpler version is the sequential model with global effects:

$$\log\left(\frac{P(Y = r|\boldsymbol{x})}{P(Y > r|\boldsymbol{x})}\right) = \gamma_{0r} + \boldsymbol{x}^T\boldsymbol{\gamma},$$

where $\boldsymbol{\gamma}$ does not depend on the category. As for the simple cumulative model, the model implies strict stochastic ordering. For two populations $\boldsymbol{x}$ and $\tilde{\boldsymbol{x}}$ one obtains that the proportion of odds

$$\frac{P(Y = r|\boldsymbol{x})/P(Y > r|\boldsymbol{x})}{P(Y = r|\tilde{\boldsymbol{x}})/P(Y > r|\tilde{\boldsymbol{x}})} = \exp((\boldsymbol{x}-\tilde{\boldsymbol{x}})^T\boldsymbol{\gamma})$$

does not depend on the category.

TABLE 9.8: Deviances for several models of retinopathy data.

	Deviance	Effect	Diff	Df
y+y∗SM+y∗DIAB+y∗GH+y∗BP	891.419			
y+y∗SM+DIAB+y∗GH+y∗BP	891.443	y∗DIAB	0.024	1
y+y∗SM+DIAB+GH+y∗BP	891.469	y∗GH	0.026	1
y+y∗SM+DIAB+GH+BP	891.977	y∗BP	0.508	1
y+SM+DIAB+GH+BP	897.710	y∗SM	5.679	1

Example 9.6: Retinopathy Data

Table 9.8 shows deviances for several sequential logit models for the retinopathy data from Examples 9.2 and 9.5. If the variable has a category-specific parameter γ_r, it is denoted as an interaction with y. For example, $y\star$ SM means that $x_{\text{SM}}\gamma_r$ is in the linear predictor. It is seen that one has to specify a category-specific effect of smoking (SM). For the rest of the variables, – diabetes (DIAB), glycosylated hemoglobin (GH), and blood pressure (BP) – there is no need for category-specific effects. For the model

$$\log(\frac{P(Y = r|\boldsymbol{x})}{P(Y > r|\boldsymbol{x})}) = \gamma_{0r} + x_{\text{SM}}\beta_{\text{SM},r} + \text{DIAB}\beta_D + \text{GH}\beta_{\text{GH}} + \text{BP}\beta_{\text{BP}}$$

one obtains deviance 891.977. The model fits slightly better than the corresponding cumulative model. The estimates are given in Table 9.9. It is seen that smoking seems to have an effect on the transition between category 1 and category 2 such that the probability of a transition is reduced when smoking. □

TABLE 9.9: Parameter estimates for the sequential logit model for retinopathy data.

	Estimates	StdError	Wald-Chi-Square	P-Value	Odds-Ratio
γ_{01}	11.127	1.168	90.67	0.00	
γ_{02}	10.915	1.213	80.92	0.00	
DIAB	−0.128	0.012	108.78	0.00	0.87
GH	−0.424	0.067	39.83	0.00	0.65
BP	−0.062	0.012	26.04	0.00	0.93
SM[1]	−0.377	0.202	3.47	0.06	0.68
SM[2]	0.490	0.312	2.46	0.11	1.63

9.2.2 Alternative Links

Of course any binary response model may be used to model the transition between categories r and $r+1$, yielding different sequential models. An especially interesting model follows from assuming a Gompertz link. The link function $F(\eta) = 1 - \exp(-\exp(\eta))$, which corresponds to the Gompertz distribution, yields the sequential minimum extreme-value model

$$P(Y = r|Y \geq r, \boldsymbol{x}) = 1 - \exp(-\exp(\gamma_{0r} + \boldsymbol{x}^T\boldsymbol{\gamma}_r)).$$

If $\boldsymbol{\gamma}_1 = \cdots = \boldsymbol{\gamma}_q = \boldsymbol{\gamma}$, the model is equivalent to the proportional hazards model from Section 9.1.3. Thus, for this special distribution and assumed global effects, the cumulative model is equivalent to the sequential model.

9.3 Further Properties and Comparison of Models

9.3.1 Collapsibility

Since the categories of an ordinal response variable are sometimes slightly arbitrary, it is preferable that grouping the response categories does not change the conclusions drawn from the fitted models. Let the response categories $\{1, \ldots, k\}$ be partitioned into subsets $S_i = \{m_{i-1}+1, \ldots, m_i\}, i = 1, \ldots, t$, where $m_0 = 0, m_t = k$, and consider the simpler response variable $\bar{Y} = i$ if $Y \in S_i$. A model is called *collapsible* with reference to a parameter if the parameter is unchanged by the transition from Y to the coarser version $\bar{Y}$. For the simple cumulative model one gets immediately

$$P(\bar{Y} \leq i|\boldsymbol{x}) = F(\gamma_{0m_i} + \boldsymbol{x}^T\boldsymbol{\gamma}).$$

Thus the cumulative model is collapsible with reference to the parameters $\boldsymbol{\gamma}$. Moreover, it is collapsible for the threshold parameters γ_{0r}, where of course some rearranging of categories has to be kept in mind. In general, the sequential model is not collapsible. Of course, exceptions are the models where cumulative and sequential approaches coincide (see next section). However, the importance of collapsibility should not be overrated. It is merely a secondary criterion for ordinal variables. If a response pattern is changed by collapsing categories, the influence of explanatory variables may also change.

9.3.2 Equivalence of Cumulative and Sequential Models

Cumulative and sequential models are based on quite different approaches to the modeling of the response variable. The response in cumulative models may be seen as a coarser version of the latent variable $\tilde{Y}$, whereas the sequential model is constructed by assuming that the response is determined step-by-step. Thus one might conclude that the resulting models are distinct.

Indeed, in general, the cumulative and the sequential approaches yield different models. However, there are two exceptions. The first is the proportional hazards model (sequential minimum extreme-value model). For the simple versions with only global parameters γ, the cumulative model is equivalent to the sequential model, which was first noted by Laara and Matthews (1985), for the general case see Tutz (1991).

The second exception is the exponential response model, which uses the exponential distribution $F(\eta) = 1 - \exp(-\eta)$. A simple derivation shows that the general sequential model with global and category-specific effects for covariates $\boldsymbol{x}$ and $\boldsymbol{z}$,

$$P(Y = r|Y \geq r) = 1 - \exp(-(\gamma_{0r} + \boldsymbol{x}^T\boldsymbol{\gamma} + \boldsymbol{z}^T\boldsymbol{\gamma}_r)),$$

is equivalent to the cumulative model

$$P(Y \leq r|\boldsymbol{x}) = 1 - \exp(-(\tilde{\gamma}_{0r} + \boldsymbol{x}^T\tilde{\boldsymbol{\gamma}}_{r,z} + \boldsymbol{z}^T\tilde{\boldsymbol{\gamma}}_r)),$$

where $\tilde{\gamma}_{0r} = \sum_{j=1}^{r} \gamma_{0j}, \tilde{\boldsymbol{\gamma}}_{r,z} = r\boldsymbol{\gamma}, \tilde{\boldsymbol{\gamma}}_r = \sum_{j=1}^{r} \boldsymbol{\gamma}_r$. It should be noted that global parameters in the sequential model turn into category-specific parameters in the cumulative model. If all the parameters are category-specific, the two models are equivalent. However, the model is not often used because the restriction $\tilde{\eta}_r = \tilde{\gamma}_{0r} + \boldsymbol{z}^T\boldsymbol{\gamma}_{r,z} + \boldsymbol{x}^T\boldsymbol{\gamma}_r \geq 0$ may often be violated and special software is needed to fit the model with that restriction.

9.3.3 Cumulative Models versus Sequential Models

Although the cumulative model is often considered as *the* ordinal model, both types of models, the cumulative model and the sequential, are ordinal models. Both model types have their advantages and drawbacks. In the following we give some points concerning the differences. For simplicity, logit models are considered.

Interpretation of parameters. The interpretation of parameters clearly depends on the model type. For the cumulative model the parameters refer to the cumulative odds $P(Y \leq r|\boldsymbol{x})/P(Y > r|\boldsymbol{x})$, while for the sequential model they refer to continuation ratios $P(Y = r|\boldsymbol{x})/P(Y > r|\boldsymbol{x})$. It depends on the application and the viewpoint of the user which parameterization is more appropriate. If the underlying process is a sequential process, it is tempting to use the sequential model.

Flexible modeling and existence of estimates. A drawback of the cumulative model is that for complex models (with category-specific parameters) iterative estimation procedures frequently fail to converge. This is due to the restriction on the parameter space that makes more carefully designed iterative procedures necessary, in particular when higher dimensional predictors are modeled. A consequence is that often less flexible models with global rather than category-specific effects are fitted for cumulative models. Since parameters do not have to be ordered, the sequential model can be more easily extended to more complex modeling problems, for example, to model monotonicity (Tutz, 2005).

Collapsibility. An advantage of the cumulative model is that the parameters remain the same when the categories are taken together. Due to the model structure

$$\log(P(Y \leq r|\boldsymbol{x})/P(Y > r|\boldsymbol{x})) = \gamma_{0r} + \boldsymbol{x}^T\boldsymbol{\gamma}_r,$$

which is based on the dichotomization $\{1, 2, \ldots, r\}, \{r + 1, \ldots, k\}$, the parameter values remain the same if one combines two or more categories using, for example, responses $\{1, 2\}$, $3, 4, \ldots, k$ rather than $1, 2, \ldots, k$.

9.4 Alternative Models

9.4.1 Cumulative Model with Scale Parameters

McCullagh (1980) introduced the cumulative-type model

$$P(Y \leq r|\boldsymbol{x}) = F\left(\frac{\gamma_{0r} + \boldsymbol{x}^T\boldsymbol{\gamma}}{\tau_x}\right), \tag{9.6}$$

which contains an additional scale parameter τ_x. In cases where the concentration in response categories varies across populations, the model is more appropriate than the simple cumulative model. The simple cumulative model was motivated by an underlying continuous regression model $\tilde{Y} = -\boldsymbol{x}^T\boldsymbol{\gamma} + \varepsilon$, where the distribution of ε does not depend on $\boldsymbol{x}$. Thus the model assumes that with varying $\boldsymbol{x}$ the probability mass is merely shifted on the latent scale. If the probability mass is more concentrated in one population and spread out in other populations, the simple cumulative model will be unable to model the varying dispersion. The following example has been used by McCullagh (1980) to motivate the model.

Example 9.7: Eye Vision

Table 9.10 gives Stuart's (1953) quality of vision data for men and women as treated by McCullagh (1980). From the data it is obvious that women are more concentrated in the middle categories while men have relatively high proportions in the extreme categories. Since the cumulative model is based on a shifting of distributions on the underlying continuum, the model is not appropriate. □

TABLE 9.10: Quality of right eye vision in men and women.

	Highest (1)	Vision 2	Quality 3	Lowest (4)
Men	1053	782	893	514
Women	1976	2256	2456	789

9.4.2 Hierarchically Structured Models

In many examples the ordered response categories may naturally be divided into subgroups of categories such that the categories within groups are homogenous but the groups themselves differ in interpretation. For example, the response categories of ordered responses might be in distinct groups indicating improvement, no change, and change for the worse.

Example 9.8: Arthritis

Table 9.11 shows data that have been analyzed previously by Mehta et al. (1984). For patients with acute rheumatoid arthritis a new agent was compared with an active control. Each patient was evaluated on a five-point assessment scale ranging from "much improved" to "much worse." There are three groups of categories that are strictly different: the improvement group, the no change group, and the group of change for the worse. □

In examples like the arthritis data it might be appropriate to fit a model with more structure in the response than is used in simple models for ordinal data. A hierarchical model is obtained by first modeling the response groups of homogenous response categories and then modeling the response within groups. Generally, let the categories $1, \dots, k$ be subdivided into basic sets $S_1, \dots, S_t$´, where $S_i = \{m_{i-1}+1, \dots, m_i\}$, $m_0 = 0, m_t = k$. In the *first step* the response in

TABLE 9.11: Clinical trial of a new agent and an active control (Mehta et al., 1984).

Drug	Global Assessment				
	Much Improvement	Improvement	No Change	Worse	Much Worse
New agent	24	37	21	19	6
Active control	11	51	22	21	7

one of the sets is determined by a cumulative model with γ_0. In the *second step* the conditional response given S_i is determined by a cumulative model with parameters that are linked to S_i. From these two steps one obtains the model

$$
\begin{aligned}
P(Y \in T_i|\boldsymbol{x}) =& F(\theta_i + \boldsymbol{x}^T\gamma_0), \\
P(Y \leq r|Y \in S_i, \boldsymbol{x}) =& F(\theta_{ir} + \boldsymbol{x}^T\gamma_i),
\end{aligned}
\tag{9.7}
$$

where $T_i = S_1 \cup \dots \cup S_i, \theta_1 < \ldots < \theta_{t-1} < \theta_t = \infty, \theta_{i,m_{i-1}+1} < \ldots < \theta_{i,m_i-1} < \theta_{i,m_i} = \infty, i = 1, \ldots\ldots, t.$

In model (9.7) the effect of the explanatory variables on the dependent variable is modeled in two ways. First the effect on the chosen subsets is parameterized, and then the conditional effect within the subsets is investigated. An advantage of the model is that different parameters are involved at different stages. In the arthritis example, and the choice between the basic sets is determined by the parameter γ_0, and the choice on the finer level is determined by the parameters γ_i. The choice between the basic sets, for example, the alternative improvement or no improvement, may be influenced by different strengths or even by different variables than the choice within the "improvement" set and the "no improvement" set. The first step that models the global effect may refer to a different biological mechanism than the second step. In attitude questionnaires the individual might first choose the global response level, that is, agreement or no agreement, and in a subsequent process decide for the strength of agreement or no agreement. Different covariates may be responsible for decisions on different levels. After rearranging the data, estimation can be based on common software for the fitting of simple cumulative models (Exercise 9.7).

The hierarchically structured model may also be used as an alternative to the scale model (9.6). In the eye vision example, it allows one to model the varying dispersion and in addition to test the symmetry of the underlying process.

TABLE 9.12: Analysis of eye data; *z*-values in brackets.

	Hierarchically Structured Model	Model with $\gamma_0 = 0$	Model with $\gamma_1 = -\gamma_2$
θ_1	0.27	0.27	0.27
θ_{11}	0.08	0.08	0.09
θ_{23}	0.84	0.84	0.86
γ_0	0.0 (0.0)	-	0.0
γ_1	0.22(7.63)	0.22	0.25
γ_2	−0.29 (−8.47)	−0.29	−0.25
Deviance	0.0	$3.5 \cdot 10^{-7}$	2.96

Example 9.9: Eye Vision

The basic sets that determine the analysis here are chosen by a grouping in categories of eye vision above average, that is, $S_1 = \{1, 2\}$, and below average, that is, $S_2 = \{3, 4\}$. With those basic sets the hierarchically structured cumulative logit model has been fitted. Let the explaining variable sex be given by 1 (male) and -1 (female). For this example, the two-step model is equivalent to the saturated model. However, it is a saturated model with a specific parameterization that allows for reduction within this parameterization.

Let $\lambda(x) = \log(P(Y \in S_1|x)/P(Y \in S_2|x))$ denote the log-odds that the quality of eye vision is above average. For the estimated log-odds one obtains $\hat{\lambda}(1) - \hat{\lambda}(-1) = 0.000025$. Thus there is almost no difference between men and women with respect to quality of eye vision when quality is considered as a dichotomous variable with the categories "above average" and "below average." For the interpretation of γ_1 and γ_2 one has to distinguish between two populations: the population with quality of eye vision above average and the population with quality of eye vision below average. For $S \subset \{1, 2, 3, 4\}$ let $\lambda_S(r|x) = \log(P(Y = r|Y \in S, x)/P(Y \neq r|Y \in S, x))$ denote the conditional log-odds and let $\lambda_S(r) = \lambda_S(r|x = 1) - (\lambda_S(r|x = -1))$ denote the difference of conditional log-odds between men and women. From $\lambda_{(1,2)}(1) = -\lambda_{(1,2)}(2) = 2\gamma_1$ and $\lambda_{(3,4)}(4) = -\lambda_{(3,4)}(3) = -2\gamma_2$ we get the estimates $\hat{\lambda}_{(1,2)}(2) = -2\hat{\gamma}_1 = -0.430$ and $\lambda_{(3,4)}(3) = 2\hat{\gamma}_2 = -0.583$. The parameter γ_1 represents the difference between men and women with respect to category 1 in the population "above average," $-\gamma_2$ represents the same difference with respect to category 4 in the population "below average." From the values of γ_1 and $-\gamma_2$ it is seen that the effect strength of gender is about the same in both populations. In the discussion of McCullagh's (1980) paper, Atkinson refers to the finding of Heim (1970) that, for many characteristics, men tend to be more extreme than women, even if the means of the two sexes are the same. This effect is captured by the hierachically structured response with the parameters γ_1 and $-\gamma_2$ representing measures for the strength of the tendency versus the extremes. The hypothesis of no differences between the two sexes with respect to performance below and above average has the form $H_0 : \gamma_0 = 0$. Though the number of observations is very high and therefore the χ^2-statistic generally will tend to rejection of non-saturated models, the hierarchically structured model with $\gamma_0 = 0$ fits the data so well that the deviance is almost vanishing (see Table 9.12). Another interesting hypothesis formalizes whether the tendency to extreme categories is the same in both populations. The corresponding hypothesis takes the form $H_0 : \gamma_1 = -\gamma_2$. With deviance 2.96, the model is not rejected. Quite different from the fit of the non-saturated versions of the hierarchically structured cumulative model, the cumulative model yields deviance 128.39 and, as expected, has a very poor fit, whereas the model with $\gamma_1 = -\gamma_2$ fits the data very well. The hierarchically structured model allows one to test for the tendency versus the middle. In contrast, the scale model (9.6) does not allow for different strengths toward the middle because it does not distinguish between the upper and lower categories (see also Exercise 9.7). □

TABLE 9.13: Analysis of arthritis data.

	$\hat{\gamma}_0$	$\hat{\gamma}_1$	$\hat{\gamma}_2$	Log-Likelihood	Deviance	Df
Hierarchically structured model	0.073	1.101	0.054	−315.502	0.008	1
Model with $\gamma_0 = 0$	-	0.55	−0.02	−315.541	0.087	2
Model with $\gamma_0 = \gamma_2 = 0$	-	0.55	-	−315.545	0.094	3
Model with $\gamma_0 = \gamma_1 = \gamma_2 = 0$				−319.132	7.269	4

Example 9.10: Arthritis

Table 9.13 shows the fit for the hierarchically structured response model where the grouping is in $S_1 = \{1, 2\}, S_2 = \{3\}, S_3 = \{4, 5\}$ and the predictor is 1 for agent and 0 for control. It is seen that the model with $\gamma_0 = \gamma_2 = 0$ fits well. The only relevant effect seems to be γ_1, which distinguishes between response categories 1 and 2. □

Hierarchically structured models are conditional models, in which the conditioning is on subgroups of categories. Very similar models apply in cases where the response categories are partially ordered. Sampson and Singh (2002) give an example from psychiatry where anxiety is measured in the categories "no anxiety," "mild anxiety," "anxiety with depression," and "severe anxiety." Only parts of these responses can be considered ordered. For a general framework for the modeling of partially ordered data see Zhang and Ip (2011).

9.4.3 Stereotype Model

Anderson (1984) introduced the so-called stereotype regression model, which in the simple one-dimensional form is given by

$$P(Y = r|x) = \frac{\exp(\gamma_{0r} - \phi_r \boldsymbol{x}^T \boldsymbol{\gamma})}{1 + \sum_{i=1}^{q} \exp(\gamma_{0r} - \phi_i \boldsymbol{x}^T \boldsymbol{\gamma})},$$

$r = 1, \dots, q$. To get an ordered regression model, the parameters $\phi_1, \dots, \phi_k$ must fulfill the constraints

$$1 = \phi_1 > \cdots > \phi_k = 0.$$

Most often the model is estimated without imposing the constraints a priori. However, if the estimated values $\hat{\phi}_i$ are allowed to determine the ordering of categories, the order is a result of the model and not a trait of the variable considered. Then the model is not an ordinal regression model because it makes no use of the information provided by the ordering. It is a model that generates an ordering rather than a model for ordered response categories. Anderson (1984) also considered the concept of indistinguishability, meaning that response categories are indistinguishable if $\boldsymbol{x}$ is not predictive between these categories. A comparison of the proportional odds model and the stereotype model was given by Holtbrügge and Schuhmacher (1991); see also Greenland (1994).

9.4.4 Models with Scores

Another type of model assumes given scores for the categories of the response Y. Williams and Grizzle (1972) consider the model

$$\sum_{r=1}^{k} s_r P(Y = r|x) = \boldsymbol{x}^T \boldsymbol{\gamma},$$

where $s_1, \dots, s_k$ are given scores. Instead of using the support $\{1, \dots, k\}$, one may consider $Y \in \{s_1, \dots, s_k\}$ and write the model as

$$\sum_{r=1}^{k} s_r P(Y = s_r|x) = \boldsymbol{x}^T \boldsymbol{\gamma}.$$

Obviously, models of this type are not suited for responses that are measured on the ordinal scale level. By introducing scores, a higher scale level is assumed for the discrete response. For further scores in ordinal modeling see Agresti (2009).

9.4.5 Adjacent Categories Logits

An alternative type of model is based on adjacent categories logits (e.g., Agresti, 2009). The model

$$\log\left[P(Y = r|\boldsymbol{x})/P(Y = r-1|\boldsymbol{x})\right] = \boldsymbol{x}^T \boldsymbol{\gamma}_r$$

is based on the consideration of the adjacent categories $\{r-1, r\}$. Logits are built locally for these adjacent categories. Another form of the model is

$$P(Y = r | Y \in \{r, r+1\}, \boldsymbol{x}) = F(\boldsymbol{x}^T \boldsymbol{\gamma}_r),$$

where F is the logistic distribution function. The latter form shows that the logistic distribution function may be substituted for any strictly monotone increasing distribution function. Moreover, it shows that it may be considered as a dichotomous response model given $Y \in \{r, r+1\}$. Very similar models are used in item response theory (e.g., Masters, 1982,) and are often misinterpreted as sequential process models. The model may also be considered as the corresponding regression model that is obtained from the row-column (RC) association model considered by Goodman (1979, 1981a, b).

9.5 Inference for Ordinal Models

Ordinal regression models are specific cases of multivariate generalized linear models. Hence, the machinery of multivariate GLMs as given in Section 8.6 can be used. In the following, first the embedding into the framework of GLMs is outlined. For sequential models maximum likelihood estimates may be obtained in a simpler way. Since transitions between categories correspond to binary decisions, estimation may be performed by using binary models. This approach is given in Section 9.5.2.

The general form of the multivariate GLM for categorical responses is given by

$$g(\boldsymbol{\pi}_i) = \boldsymbol{X}_i \boldsymbol{\beta} \quad \text{or} \quad \boldsymbol{\pi}_i = h(\boldsymbol{X}_i \boldsymbol{\beta}),$$

where $\boldsymbol{\pi}_i^T = (\pi_{i1}, \dots, \pi_{iq})$ is the vector of response probabilities, g is the (multivariate) link function, and $h = g^{-1}$ is the inverse link function. The representation of ordinal models as multivariate GLMs is given separately for cumulative-type models (Section 9.5.1) and sequential models (Section 9.5.2).

9.5.1 Cumulative Models

The simple cumulative model has the form

$$P(Y_i \le r | \boldsymbol{x}_i) = F(\gamma_{0r} + \boldsymbol{x}_i^T \boldsymbol{\gamma}),$$

where F is a fixed distribution function that, in a multivariate GLM, should not be confused with the response function. One obtains immediately for the probabilities $\pi_{ir} = P(Y_i = r | \boldsymbol{x}_i)$

$$\pi_{ir} = F(\gamma_{0r} + \boldsymbol{x}_i^T \boldsymbol{\gamma}) - F(\gamma_{0,r-1} + \boldsymbol{x}_i^T \boldsymbol{\gamma}),$$

where $-\infty = \gamma_{00} < \gamma_{01} < \dots < \gamma_{0k} = \infty$. Let $\eta_{ir} = \gamma_{0r} + \boldsymbol{x}_i^T \boldsymbol{\gamma}$, $r = 1, \dots, q$, denote the rth linear predictor. Then

$$\pi_{ir} = F(\eta_{ir}) - F(\eta_{i,r-1}).$$

One obtains the matrix form

$$\begin{pmatrix} \pi_{i1} \\ \vdots \\ \pi_{iq} \end{pmatrix} = h\left\{ \begin{pmatrix} 1 & & & \boldsymbol{x}_i^T \\ & 1 & & \vdots \\ & & \ddots & \vdots \\ & & & \vdots \\ & & 1 & \boldsymbol{x}_i^T \end{pmatrix} \begin{pmatrix} \gamma_{01} \\ \gamma_{02} \\ \vdots \\ \gamma_{0q} \\ \boldsymbol{\gamma} \end{pmatrix} \right\},$$

where all parameters are collected in $\boldsymbol{\beta}^T = (\gamma_{01}, \dots, \gamma_{0q}, \boldsymbol{\gamma}^T)$. The components of the q-dimensional response function $h = (h_1, \dots, h_q) : \mathbb{R}^q \to \mathbb{R}^q$ are given by

$$h_r(\eta_{i1}, \dots, \eta_{iq}) = F(\eta_{ir}) - F(\eta_{i,r-1}).$$

The link function (inverse response function) is easily derived from the form

$$F^{-1}(P(Y_i \le r|\boldsymbol{x}_i)) = \gamma_{0r} + \boldsymbol{x}_i^T \boldsymbol{\gamma}$$

or, equivalently,

$$F^{-1}(\pi_{i1} + \dots + \pi_{ir}) = \eta_{ir}. \tag{9.8}$$

One immediately obtains for $g = (g_1, \dots, g_q) : \mathbb{R}^q \to \mathbb{R}^q$

$$g_r(\pi_{i1}, \dots, \pi_{iq}) = F^{-1}(\pi_{i1} + \dots + \pi_{ir}).$$

For example, for the logistic distribution function $F(\eta) = \exp(\eta)/(1 + \exp(\eta))$, the inverse is given by $F^{-1}(\pi) = \log(\pi/(1 - \pi))$. Therefore, (9.8) is equivalent to

$$\log\left(\frac{\pi_{i1} + \dots + \pi_{ir}}{1 - \pi_{i1} - \dots - \pi_{ir}}\right) = \eta_{ir},$$

and hence

$$g_r(\pi_{i1}, \dots, \pi_{iq}) = \log\left(\frac{\pi_{i1} + \dots + \pi_{ir}}{1 - \pi_{i1} - \dots - \pi_{ir}}\right).$$

One obtains

$$g\left\{\begin{pmatrix} \pi_{i1} \\ \vdots \\ \pi_{iq} \end{pmatrix}\right\} = \begin{pmatrix} 1 & & & & \boldsymbol{x}_i^T \\ & 1 & & & \vdots \\ & & \ddots & & \vdots \\ & & & \ddots & \vdots \\ & & & 1 & \boldsymbol{x}_i^T \end{pmatrix} \begin{pmatrix} \gamma_{01} \\ \gamma_{02} \\ \vdots \\ \gamma_{0q} \\ \boldsymbol{\gamma} \end{pmatrix}.$$

The more general model with category-specific parameters has the form

$$P(Y_i \le r|\boldsymbol{x}_i) = F(\gamma_{0r} + \boldsymbol{x}_i^T \boldsymbol{\gamma}_r).$$

The derivation of the response and link function is the same as for the simple model. The only difference is in the linear predictor, which now has components

$$\eta_{ir} = \gamma_{0r} + \boldsymbol{x}_i^T \boldsymbol{\gamma}_r = (1, \boldsymbol{x}_i^T)\begin{pmatrix} \gamma_{0r} \\ \boldsymbol{\gamma}_r \end{pmatrix}.$$

Hence, in the general model, the design matrix is different. One obtains with an identical response function as for the simple cumulative model

$$\begin{pmatrix} \pi_{i1} \\ \vdots \\ \pi_{iq} \end{pmatrix} = h\left\{\begin{pmatrix} 1 & & & \boldsymbol{x}_i^T & & & \\ & 1 & & & \boldsymbol{x}_i^T & & \\ & & \ddots & & & \ddots & \\ & & & 1 & & & \boldsymbol{x}_i^T \end{pmatrix}\begin{pmatrix} \gamma_{01} \\ \vdots \\ \gamma_{0q} \\ \boldsymbol{\gamma}_1 \\ \vdots \\ \boldsymbol{\gamma}_q \end{pmatrix}\right\},$$

where the collection of all parameters is given by $\boldsymbol{\beta}^T = (\gamma_{01}, \dots, \gamma_{0q}, \boldsymbol{\gamma}_1^T, \dots, \boldsymbol{\gamma}_q^T)$.

9.5.2 Sequential Models

Sequential Model as Multivariate GLM

The general sequential model has the form

$$P(Y_i = r|Y_i \geq r, \boldsymbol{x}_i) = F(\gamma_{0r} + \boldsymbol{x}_i^T \boldsymbol{\gamma}_r)$$

or, equivalently,

$$\pi_{ir} = P(Y_i = r|\boldsymbol{x}_i) = F(\gamma_{0r} + \boldsymbol{x}_i^T \boldsymbol{\gamma}_r) \prod_{j=1}^{r-1} (1 - F(\gamma_{0j} + \boldsymbol{x}_i^T \boldsymbol{\gamma}_j)).$$

Let $\eta_{ir} = \gamma_{0r} + \boldsymbol{x}_i^T \boldsymbol{\gamma}_r$ denote the rth linear predictor. Then one obtains directly for the multivariate response function $h = (h_1, \dots, h_q) : \mathbb{R}^q \to \mathbb{R}^q$

$$h_r(\eta_{i1}, \dots, \eta_{iq}) = F(\eta_{ir}) \prod_{j=1}^{r-1} (1 - F(\eta_{ij})) \tag{9.9}$$

and for $g = h^{-1}$

$$g_r(\pi_{i1}, \dots, \pi_{iq}) = F^{-1}(\pi_{ir}/(1 - \pi_{i1} - \dots - \pi_{i,r-1})).$$

The design matrix is the same as for the category-specific cumulative model. The simpler form of the model with $\boldsymbol{\gamma}_1 = \dots = \boldsymbol{\gamma}_q$ also has response function (9.9), but the design matrix simplifies to the form of the design matrix of the simple cumulative model.

Maximum Likelihood Estimation by Binary Response Models

Maximum likelihood estimates for sequential models may be obtained by fitting binary regression models. The binary responses that are used correspond to the transition between categories. In the following the equivalence of likelihoods is derived by representing the probability of a categorical response as a product of discrete hazards. Let $\lambda(r)$ denote the conditional transition probability

$$\lambda(r) = P(Y = r|Y \geq r),$$

which is parameterized in the sequential model. λ is also called a discrete hazard, in particular when categories refer to discrete time. For the responses one generally obtains $P(Y \geq r) = \prod_{s=1}^{r-1}(1 - \lambda(s))$ (Exercise 9.1). Therefore one has

$$P(Y = r) = P(Y = r|Y \geq r)P(Y \geq r) = \lambda(r) \prod_{s=1}^{r-1} (1 - \lambda(s)).$$

Let the ith observation Y_i take value r_i and $\lambda(r|\boldsymbol{x}_i)$ denote the conditional hazard, given the covariates $\boldsymbol{x}_i$. Then the likelihood contribution of observation Y_i is given by

$$L_i = P(Y_i = r_i|\boldsymbol{x}_i) = \lambda(r_i|\boldsymbol{x}_i) \prod_{s=1}^{r_i-1} (1 - \lambda(s|\boldsymbol{x}_i)).$$

By defining the response vector $(y_{i1}, \dots, y_{ir_i}) = (0, \dots, 0, 1)$, the contribution to the likelihood may be written as

$$L_i = \prod_{s=1}^{r_i} \lambda(s|\boldsymbol{x}_i)^{y_{is}} (1 - \lambda(s|\boldsymbol{x}_i))^{1-y_{is}}.$$

L_i is equivalent to the likelihood of a binary response model with observations $y_{i1}, \dots, y_{ir_i}$. Since for a sequential model the hazard $\lambda(s|\boldsymbol{x}_i)$ has the form of a binary regression model with response function $F(.)$, one obtains with $\lambda(s|\boldsymbol{x}_i) = F(\eta_{is})$ the likelihood

$$L_i = \prod_{j=1}^{r_i} F(\eta_{ij})^{y_{ij}} (1 - F(\eta_{ij}))^{1-y_{ij}}.$$

The total likelihood, given by $L = L_1 \cdot \ldots \cdot L_n$, yields the log-likelihood

$$l = \sum_{i=1}^{n} \sum_{j=1}^{r_i} y_{ij} \log(F(\eta_{ij})) + (1 - y_{ij}) \log(1 - F(\eta_{ij})). \tag{9.10}$$

Therefore, the likelihood of the sequential model

$$P(Y_i = r | Y_i \geq r, \boldsymbol{x}_{ir}) = F(\eta_{ir})$$

is equivalent to the likelihood of the binary model

$$P(y_{ir} = 1 | \boldsymbol{x}_{ir}) = F(\eta_{ir})$$

for observations $y_{i1} = 0, \dots, y_{i,r_i-1} = 0, y_{ir} = 1, i = 1, \dots, n$.

It should be noted that the log-likelihood (9.10) has $r_1 + \cdots + r_n$ binary observations even though the original number of observations is n. A simple recipe for fitting the sequential model is to generate for each observation the binary observations $y_{i1}, \dots y_{ir_i}$ and the corresponding design variables and then fit the binary model as if the observations were independent.

Let us consider the example of a simple sequential model for which the linear predictor has the form $\eta_{ir} = \gamma_{or} + \boldsymbol{x}_i^T \boldsymbol{\gamma}$. To have the usual form $\boldsymbol{x}^T \boldsymbol{\beta}$ for the linear predictor of a binary model one defines the parameter vector $\boldsymbol{\beta}^T = (\gamma_{01}, \dots \gamma_{0q}, \boldsymbol{\gamma}^T)$ and adapts the covariate vector to the observations under consideration. If $Y = r_i$, the r_i binary observations and design variables are given by

Binary Observations	Design Variables					
0	1	0	0	...	0	$\boldsymbol{x}_i^T$
0	0	1	0	...	0	$\boldsymbol{x}_i^T$
0	0	0	1	...	0	
⋮						
1	0	0	0	1	0	$\boldsymbol{x}_i^T$

As a simple example, let us consider a response $Y \in \{1, \dots, 5\}$ and the covariates age (in years) and height (in cm). For a given covariate vector (35,180) let the response be in category 3. Then one obtains three binary observations of the binary model given by

Binary Observations	Design Variables					
0	1	0	0	0	35	180
0	0	1	0	0	35	180
1	0	0	1	0	35	180

The equivalence of the likelihood of the multinomial response model and the binary model (with derived observations) may be used to obtain maximum likelihood estimates. The binary observations that are used are strongly related to the underlying multinomial response variables. Let $\boldsymbol{y}_i \sim M(1, (\pi_{i1}, \dots, \pi_{iq}))$ denote the original observation. Then an observation in category r_i yields $\boldsymbol{y}_i^T = (0, \dots, 1, \dots, 0)$. The binary observations used in (9.10) are the first r_i

observations of the vector. Thus the binary observations are a truncated version of the underlying multinomial response vector (Exercise 9.2). Although these observations yield a likelihood where $y_{i1}, \dots, y_{ir}$ look like independent observations, they are not. This means that one cannot use the usual methods of inference for binary data. When computing standard errors and investigating the goodness-of-fit one should use the instruments for multinomial distributions given in Chapter 8, Section 8.6.

9.6 Further Reading

Surveys and Basic Papers. A careful survey that covers the methods for the analysis of ordinal categorical data is Agresti (2009). A shorter overview on ordered categorical data was given by Liu and Agresti (2005). An important paper that had a strong influence on the development of regression models for ordered response categories is McCullagh (1980).

Efficiency. Comparisons between the cumulative model and the binary model are found in Armstrong and Sloan (1989) and Steadman and Weissfeld (1998). Agresti (1986) gave an extension of R^2-type association measures to ordinal data.

Prediction. The use of ordinal models for the prediction of ordinal responses was investigated by Rudolfer et al. (1995), Campbell and Donner (1989), Campbell et al. (1991), Anderson and Phillips (1981), and Tutz and Hechenbichler (2005). Prediction for ordinal data is further investigated in Section 15.9.

Cumulative Link Models. Genter and Farewell (1985) consider a generalized link function that includes the probit model, complementary log-log models, and other links. They show that the links may be discriminated for moderate sample sizes. Models with a dispersion effect of the type (9.6) have been considered by Nair (1987) and Hamada and Wu (1996).

Models with Assigned Scores. Williams and Grizzle (1972) proposed a model with assigned scores for the ordered response categories. Lipsitz, Fitzmaurice, and Molenberghs (1996) consider goodness-of-fit tests for models of this type.

Bayesian Models. Albert and Chib (2001) present a Bayesian analysis of the sequential probit model.

Partial Proportional Odds Models. Partial proportional odds models have been investigated by Cox (1995), Brant (1990), and Peterson and Harrell (1990).

Goodness-of-Fit Tests. Lipsitz et al. (1996) extended the Hosmer-Lemeshow statistic, which assesses goodness-of-fit when continuous covariates are present, to the ordinal setting. However, the tests are based on assigned scores, which is somewhat beyond ordinal response modeling. Pulkstenis and Robinson (2004) proposed a modification for cases when continuous and categorical variables are present as an extension of their test in the binary setting (Pulkstenis and Robinson, 2002).

R Packages. The proportional odds model can be fitted with function *vglm* from the package *VGAM* and with function *lrm* from the package *Design*. It can be also be fitted with the function *polr* from the *MASS* library. Attention has to be paid to the algebraic signs of the coefficients. These are inverse to the definition of the porportional odds models used here. With the function *vglm* from the library *VGAM*, both cumulative and sequential models can be fitted.

9.7 Exercises

9.1 Let $Y \in \{1, \dots, k\}$ denote an ordered categorical response variable and $\lambda(r) = P(Y = r|Y \geq r)$ denote conditional transition probabilities. Show that $P(Y \geq r) = \prod_{s=1}^{r-1}(1 - \lambda(s))$ holds and the probability for category r is given by $P(Y = r) = \lambda(r) \prod_{s=1}^{r-1}(1 - \lambda(s))$.

9.2 Let $(Y_1 \dots Y_k)$ follow a multinomial distribution $M(m, (\pi_1, \dots, \pi_k))$, where $\sum_r \pi_r = 1$. Let $\lambda(r) = P(Y = r | Y \geq r)$ denote the discrete hazard, that is, the probability that category r is observed given $Y \geq r$.

(a) Show that the probability mass function of the multinomial distribution can be represented as the product of binomial distributions,

$$P(Y_1 = y_1, \dots, Y_q = y_q) = \prod_{i=1}^{q} b_i \lambda(i)^{y_i} (1 - \lambda(i))^{m - y_1 - \dots - y_{i-1}},$$

where $b_i = (m - y_1 - \dots - y_{i-1})!/(y_i!(m - y_1 - \dots - y_i)!)$.

(b) Let $m = 1$ and $k = 5$. Show that the probability for the multinomial response vector $(0, 0, 1, 0, 0)$ depends only on the first three hazards.

9.3 Table 9.1 gives the cross-classification of pain and treatment for the knee data.

(a) Fit a cumulative model with general predictor $\eta_r = \gamma_{0r} + \text{pain}\gamma_r$ and investigate if parameters have to be category-specific.

(b) Fit a sequential model with general predictor $\eta_r = \gamma_{0r} + \text{pain}\gamma_r$ and investigate if parameters have to be category-specific.

(c) Interpret the parameters for both modeling approaches and compare the interpretations.

9.4 Ananth and Kleinbaum (1997) consider data from a clinical trial of a single-dose post-operative analgesic clinical trial. Four drugs were randomized to patients and the responses were recorded on a five-level ordinal scale that has been reduced to four categories because of sparse cell counts. The data are given in Table 9.14.

(a) Fit the proportional odds model and test if the proportional odds assumption holds.

(b) Fit the sequential model and compare with the cumulative model fits.

TABLE 9.14: Rating of drugs from clinical trial.

Drug	Rating of the Drugs				
	Poor	Fair	Good	Very Good	Total
$C15\&C60$	17	18	20	5	61
$Z100\&EC4$	10	4	13	34	60

9.5 In a questionnaire it was evaluated what effect the expectation of students for finding adequate employment has on their motivation. The data are given in Table 9.15.

(a) Fit the proportional odds model and test if the proportional odds assumption holds.

(b) Fit the sequential model and compare with the cumulative model fits.

9.6 The data set *children* from the package *catdata* contains the response variable number of children and covariates age, duration of school education, and nationality (see Example 7.1). Consider the categorized response number of children 1, 2, 3, or above.

(a) Fit the proportional odds model and test if the proportional odds assumption holds.

(b) Fit the sequential model and compare with the cumulative model fits.

(c) Discuss the differences between treating the response as Poisson-distributed and as an ordinal response.

TABLE 9.15: Motivation of students.

Faculty	Age	Often Negative (1)	Sometimes Negative (2)	None or Mixed (3)	Sometimes Positive (4)	Often Positive (5)
		Effect on Motivation				
Psychology	$\leq 21(1)$	5	18	29	3	0
(1)	22–23 (2)	1	5	13	2	1
	> 24 (3)	3	3	11	3	5
Physics	≤ 21	4	9	39	6	2
(2)	22–23	3	6	33	7	2
	> 24	1	7	28	7	2
Teaching	≤ 21	4	10	11	0	3
(3)	22–23	8	6	10	0	0
	> 24	14	4	14	0	1

9.7 The hierarchically structured cumulative model has the form $P(Y \in S_1 \cup \cdots \cup S_i|\boldsymbol{x}) = F(\theta_i + \boldsymbol{x}^T\boldsymbol{\gamma}_0)$, $P(Y \leq r|Y \in S_i, \boldsymbol{x}) = F(\theta_{ir} + \boldsymbol{x}^T\boldsymbol{\gamma}_i)$.

(a) Show that the likelihood decomposes into a sum of two components, the first containing the parameters $\{\theta_i, \boldsymbol{\gamma}_0\}$ and the second containing the parameters $\{\theta_{ir}, \boldsymbol{\gamma}_i\}$.

(b) Explain how common software for the fitting of cumulative models can be used to fit the hierarchically structured cumulative model.

(c) Fit the hierarchically structured cumulative model to the eye data given in Table 9.10 by specifying $S_1 = \{1, 2\}, S_2 = \{3, 4\}$. Also fit the simplified version with $\boldsymbol{\gamma}_0 = 0$. Compare the fit of the models to the fit of the simple cumulative model.

(d) Can the hypothesis $H_0 : \boldsymbol{\gamma}_i = \mathbf{0}$ be tested by using the separate fits of the components?

Chapter 10

Semi- and Non-Parametric Generalized Regression

Most of the models considered in the previous chapters are members of the generalized linear models family and have the form $g(\mu) = \boldsymbol{x}^T\boldsymbol{\beta}$, with link function g. The models are non-linear because of the link function, but nonetheless they are parametric, because the effect of covariates is based on the linear predictor $\eta = \boldsymbol{x}^T\boldsymbol{\beta}$. In many applications, parametric models are too restrictive. For example, in a linear logit model with a unidimensional predictor it is assumed that the response probability is either strictly increasing or decreasing over the whole range of the predictor given that the covariate has an effect at all.

Example 10.1: Duration of Unemployment

When duration of unemployment is measured by two categories, short-term unemployment (1: below 6 months) and long-term employment (0: above 6 months), an interesting covariate is age of the unemployed person. Figure 10.1 shows the fits of a linear logistic model, a model with additional quadratic terms, and a model with cubic terms. The most restrictive model is the linear logistic model, which implies strict monotonicity of the probability depending on age. It is seen that the fit is rather crude and unable to fit the observations at the boundary. The quadratic and the cubic logistic models show better fit to the data but still lack flexibility. Non-parametric fits, which will be considered in this chapter, are also given in Figure 10.1. They show that the probability of short-term unemployment seems to be rather constant up to about 45 years of age but then strongly decreases. The methods behind these fitted curves will be considered in Section 10.1.3 □

In this chapter less restrictive modeling approaches, which allow for a more flexible functional form of the predictor, are considered. We start with the simple case of a generalized non-parametric regression with a univariate predictor variable. The extension to multiple predictors is treated in the following section. Then structured additive approaches that also work in higher dimensions are considered.

10.1 Univariate Generalized Non-Parametric Regression

In this section we consider non-parametric regression models for the simple case of a univariate predictor variable. The functional form of $\mu(x) = \mathrm{E}(y|x)$ is not restricted by a linear term, but often it is implicitly assumed that it is smoothly varying when x is a continuous variable. Rather than considering $\mu(x)$ itself, we consider more flexible forms of the predictor $\eta(x)$, which is

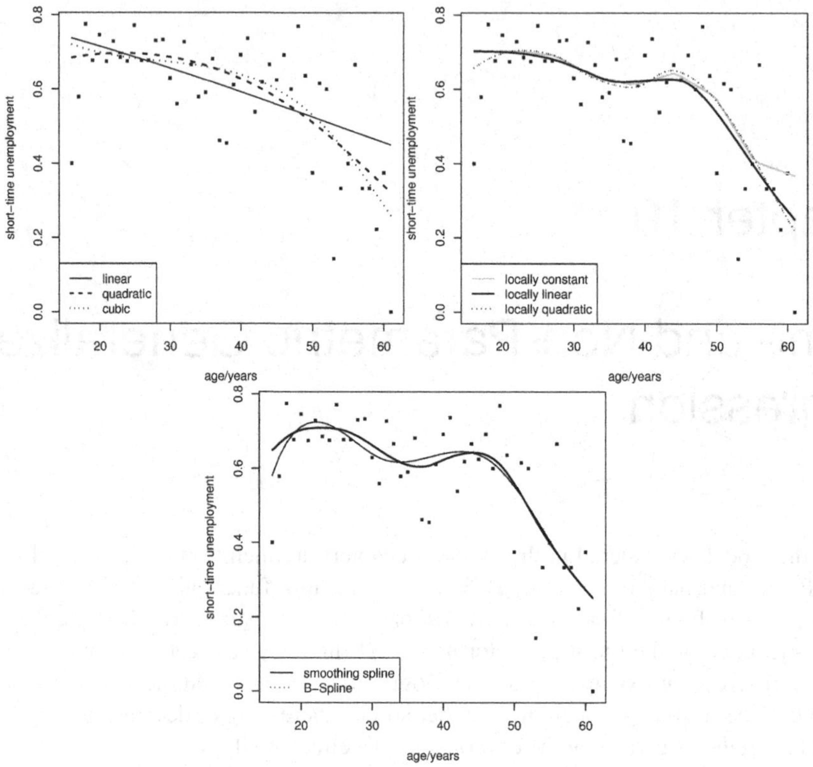

FIGURE 10.1: Probability of short-term unemployment plotted against age. Left upper panel shows the fit of a linear, quadratic, and cubic logistic regression model; right upper panel shows local fits, and the lower panel shows fits based on the expansion in basis functions.

linked to $\mu(x)$ by the usual link or response function:

$$\mu(x) = h(\eta(x)) \quad \text{or} \quad g(\mu(x)) = \eta(x)$$

for fixed (strictly increasing) h and $g = h^{-1}$. The choice of h is rather arbitrary, because its main role is to ensure that μ is in an admissible range. For example, the logit link ensures that μ is in the interval $[0, 1]$ whatever value $\eta(x)$ takes. Therefore, when modeling $\eta(x)$ one does not have to worry about admissible ranges. Most often, the canonical link is used because it allows for a somewhat simpler estimation.

10.1.1 Regression Splines and Basis Expansions

A simple way to obtain non-linear predictors $\eta(x)$ is to use non-linear transformations of x. A convenient transformation is provided by polynomials. That means one uses predictors of the form $\eta(x) = \beta_0 + \beta_1 x + \ldots + \beta_m x^m$ rather than the simple linear predictor $\eta(x) = \beta_0 + \beta_1 x$. By including a finite number of power functions one obtains the familiar form of a generalized polynomial regression. However, this strategy has some severe drawbacks. For example, single observations may have a strong influence on remote areas of the fitted function, and increasing the degree of the polynomial, although adding flexibility, may yield highly fluctuating curves.

An alternative is polynomial splines, which are considered in the following. Rather than fitting a polynomial to the whole domain of x, polynomial splines fit polynomials only within small ranges of the predictor domain. They have the nice property that they still may be represented in the form

$$\eta(x) = \sum_{j=1}^{m} \beta_j \phi_j(x), \tag{10.1}$$

where ϕ_j is the jth fixed transformation of x, also called the jth basis function. The general form (10.1) is a linear basis expansion in x that makes estimation rather easy, because, once the basis functions ϕ_j have been determined, the predictor is linear in these new variables. Fitting procedures may use the whole framework of generalized linear models.

Regression Splines: Truncated Power Series Basis

Polynomial regression splines are obtained by dividing the domain of x into contiguous intervals and representing the unknown function $\eta(x)$ by a separate polynomial in each interval. In addition, the polynomials are supposed to join smoothly at the knots.

The boundaries of the intervals are determined by a chosen sequence of breakpoints or knots $\tau_1 < \ldots < \tau_{m_s}$ from the domain of the predictor $[a, b]$. With the additional knots $\tau_0 = a, \tau_{m_s+1} = b$, a function s defined on $[a, b]$ is called a *spline function of degree k (order $k+1$)* if

> s is a polynomial of degree k on each interval $[\tau_j, \tau_{j+1}], j = 0, \ldots, m_s$, and
>
> the derivatives of $s(x)$ up to order $k-1$ exist and are continuous on $[a, b]$.

Thus polynomial regression splines of degree k form polynomials on the intervals and are continuous on the whole interval $[a, b]$; more technically, $s \in C_{k-1}[a, b]$, where $C_{k-1}[a, b]$ denotes the set of functions for which the $(k-1)$th derivative $s(x)^{(k-1)}$ is continuous on the interval $[a, b]$. An important special case is *cubic splines*, which are polynomials of degree 3 in each interval and have first and second derivatives at the knots and therefore on the whole interval.

For given knots, splines of degree k may be represented by the so-called *truncated power series basis* in the form

$$s(x) = \beta_0 + \beta_1 x + \ldots + \beta_k x^k + \sum_{i=1}^{m_s} \beta_{k+i}(x - \tau_i)_+^k, \tag{10.2}$$

where $(x - \tau_i)_+^k$ are truncated power functions defined by

$$(x-\tau)_+^k = \begin{cases} (x-\tau)^k & \text{if } \; x \geq \tau \\ 0 & \text{if } \; x < \tau. \end{cases}$$

It is easily seen that $s(x)$ from (10.2) has derivations up to order $k-1$ that are continuous. The representation (10.2) is a linear combination of the basis functions

$$\phi_1(x) = 1, \phi_2(x) = x, \ldots, \phi_{k+1}(x) = x^k,$$
$$\phi_{k+2}(x) = (x - \tau_1)_+^k, \ldots, \phi_{k+m_s+1}(x) = (x - \tau_k)_+^k,$$

which form the truncated power series basis of degree k (order $k+1$). In general, for given knots, splines of fixed degree form a vector space of dimension $k+m_s+1$. The truncated power series functions form a basis of that vector space. Therefore, in the space of spline functions, the representation is unique, in the sense that only one combination of parameters is able to represent $s(x)$.

Now let (10.2) model the unknown predictor $\eta(x)$ and the data be given by (y_i, x_i), $i = 1, \dots, n$. Then the parameters $\boldsymbol{\beta}^T = (\beta_0, \dots, \beta_{k+m_s})$ may be estimated by maximizing the log-likelihood for the model $g(\mu_i) = \eta(x_i)$ with the linear predictor

$$\eta(x_i) = \boldsymbol{\phi}_i^T \boldsymbol{\beta},$$

where $\boldsymbol{\phi}_i^T = (\phi_1(x_i), \dots, \phi_{k+m_s+1}(x_i))$. The log-likelihood function, the score function, the Fisher matrix, and the iterative fitting procedure from a generalized linear model are directly applicable by using for the ith observation the multiple design vector $\boldsymbol{x}_i = \boldsymbol{\phi}_i$.

Of course, first knots have to be selected. Typically the knots are chosen to be evenly spaced through the range of observed x-values or placed at quantiles of the distribution of x-values. More elaborate strategies to select knots have also been proposed (for references see Section 10.5).

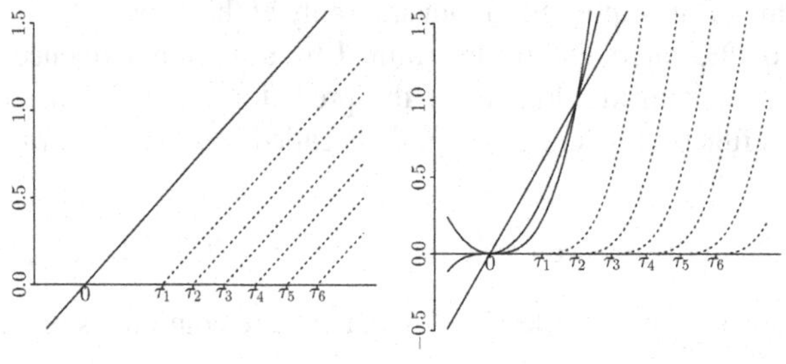

FIGURE 10.2: Truncated power series basis of degrees 1 and 3.

Representation by B-Splines

The truncated power series representation of splines has the disadvantage that they are numerically unstable for a large number of knots. A basis that is numerically more stable and is in common use is the *B-spline basis*.

A *B-spline function of degree* k with knots $\tau_i, \dots \tau_{i+k+1}$ is defined recursively by

$$B_{i,1}(x) = \begin{cases} 1 & \text{if } x \in [\tau_i, \tau_{i+1}] \\ 0 & \text{otherwise,} \end{cases}$$

$$B_{i,r+1}(x) = \frac{x - \tau_i}{\tau_{i+r} - \tau_i} B_{i,r}(x) + \frac{\tau_{i+r+1} - x}{\tau_{i+r+1} - \tau_{i+1}} B_{i+1,r}(x),$$

$r = 1, \dots, k$ (Dierckx, 1993, p. 8). The last function of this iteration, $B_{i,k+1}(x)$, represents the B-spline function of degree k (order $k+1$). By introducing additional knots $\tau_{-k} \le \dots \le \tau_0 = a, b = \tau_{m_s+1} \le \dots \le \tau_{m_s+k+1}$, every spline of degree k has a unique representation in $k + m_s + 1$ basis functions:

$$s(x) = \sum_{i=-k}^{m_s} \beta_i B_i(x),$$

where in $B_i(x) = B_{i,k+1}(x)$ the index $k+1$ is omitted. B-splines are very appealing because the basis functions are strictly local. Therefore, single fitted basis functions have no effect on remote areas. Figure 10.3 shows two B-spline bases, for degree 1 and degree 3 for equally and unequally spaced knots.

Let us consider B-splines of degree k (order $k+1$). Then the essential properties of B-splines may be summarized as follows:

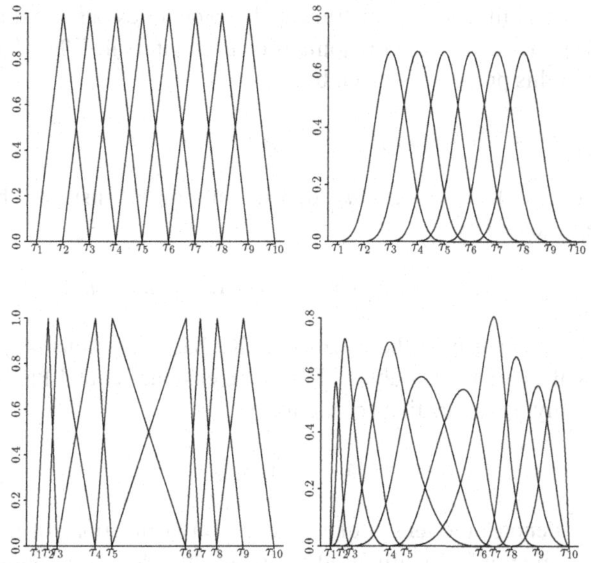

FIGURE 10.3: B-splines of degrees 1 (left panels) and 3 (right panels). In the upper panels the knots are equally spaced; in the lower panels they are not equally spaced.

- A B-spline consists of $k+1$ polynomial pieces, each of degree k.

- The polynomial pieces join at k inner knots; at the joining knots, derivates up to order $k-1$ are continuous.

- The spline function is positive on a domain spanned by $k+2$ knots; elsewhere it is zero.

- At a given x, $k+1$ B-splines are non-zero.

- Except at boundaries, the spline function overlaps with $2k$ polynomial pieces of its neighbors.

- B-splines sum up to 1 at each point, $\sum_i B_i(x) = 1$.

It should be noted that B-splines as defined above form a basis of the vector space of splines with fixed degrees and knots. Any spline may be represented by the truncated power series basis *or* B-splines. The transformation between bases has the form $\boldsymbol{\phi}_T(x) = \boldsymbol{L}\boldsymbol{\phi}_B(x)$, where $\boldsymbol{\phi}_T(x)^T = (\phi_{T,1}(x), \phi_{T,2}(x), \dots)$ is a vector containing the truncated power series basis, $\boldsymbol{\phi}_B(x)^T = (\phi_{B,1}(x), \phi_{B,2}(x), \dots)$ contains the B-spline basis, and $\boldsymbol{L}$ is a square invertible matrix. For the design matrices built from the data, $\boldsymbol{\phi}_T^T = (\boldsymbol{\phi}_T(x_1) \dots \boldsymbol{\phi}_T(x_n))$, $\boldsymbol{\phi}_B^T = (\boldsymbol{\phi}_B(x_1) \dots \boldsymbol{\phi}_B(x_n))$, one obtains $\boldsymbol{\phi}_T = \boldsymbol{\phi}_B \boldsymbol{L}$.

Alternative Basis Functions

The linear basis functions approach assumes that $\eta(x)$ may be approximated by

$$\eta(x) = \sum_{i=1}^{m} \beta_i \phi_i(x), \tag{10.3}$$

where $\phi_i(.)$ are fixed basis functions. The truncated power series basis and B-splines span the space of polynomial splines and yield polynomials on each interval. When k is odd, another set of basis functions with this property is given by

$$\phi_1(x) = 1, \phi_2(x) = x, \ldots, \phi_{k+1}(x) = x^k, \phi_{k+i}(x) = |x - \tau_i|^k, i = 1, \ldots, m_s.$$

The basis functions $\phi_{k+i}(x), i = 1, \ldots, m_s$, depend only on the distances between x and the knots and may be represented by

$$|x - \tau_i|^k = r(|x - \tau_i|), \quad \text{where} \quad r(u) = u^k.$$

An advantage of the dependence on the distance is that basis functions may be easily defined for higher dimensional predictor variables. For p-dimensional $\boldsymbol{x}$ and knots $\boldsymbol{\tau}_1, \ldots, \boldsymbol{\tau}_m$ one considers more generally basis functions of the form

$$r(\|\boldsymbol{x} - \boldsymbol{\tau}_i\|/\lambda_i),$$

where $\|\boldsymbol{z}\| = \sqrt{\boldsymbol{z}^T\boldsymbol{z}}$ denotes the length of vector $\boldsymbol{z}$. Basis functions of this form are called *radial basis functions* and are frequently used in the machine learning community. Often they have the form of a localized density. For example, the Gaussian radial basis function has the form

$$\phi_i(\boldsymbol{x}) = \exp(-\frac{\|\boldsymbol{x} - \boldsymbol{\tau}_i\|^2}{2\sigma_i^2}),$$

where $\boldsymbol{\tau_i}$ is the center of the basis function and σ_i^2 is an additional parameter that determines the spread of the basis function. For simplicity one may use $\sigma_i^2 = \sigma^2$.

When using radial basis functions on a non-equally spaced grid it is often useful to use renormalized basis functions to obtain more support in areas where all the basis functions are small. Renormalized basis functions of the form

$$\phi_i(\boldsymbol{x}) = \frac{r(\|\boldsymbol{x} - \boldsymbol{\tau}_i\|/\lambda)}{\sum_{i=1}^m r(\|\boldsymbol{x} - \boldsymbol{\tau}_i\|/\lambda)}$$

share with B-splines the property $\sum_{j=1}^m \phi_j(x) = 1$.

An alternative choice of basis function that is particularly useful when modeling seasonal effects is the Fourier basis $\phi(x) = 1$, $\phi_{2k}(x) = \sqrt{2}cos(2\pi kx)$, $\phi_{2k+1}(x) = \sqrt{2}sin(2\pi kx)$. It is an orthonormal basis that fulfills $\int \phi_i(x)\phi_j(x) = \delta_{ij}$, where δ_{ij} denotes the Kronecker delta function.

Further alternatives are wavelet bases, which are very efficient for metric responses (e.g., Ogden, 1997; Vidakovic, 1999) or thin plate splines. The latter provide an elegant way to estimate in multiple predictor cases (see, e.g., Wood, 2006b, 2006c).

As has been shown, the fitting of basis functions is easily done within the framework of generalized linear models when the basis function (including knots) has been chosen. However, since one has to estimate as many parameters as basis functions, one has to restrict oneself to rather few basis functions to obtain estimates. Thus the form of the unknown predictor is severely restricted. More flexible approaches are obtained when many basis functions are used but estimates are regularized (see Section 10.1.3).

10.1.2 Smoothing Splines

When using regression splines it is necessary to choose knot locations and basis functions. This adds some degree of subjectivity into the model fitting process. An alternative approach that is not based on pre-specified knots starts with a criterion that explicitly contains the two conflicting aims of fitting, faith with the data and low roughness of the fitted function. One aims at finding the function $\eta(x)$ that maximizes

$$\sum_{i=1}^{n} l_i(y_i, \eta(x_i)) - \frac{1}{2}\lambda \int (\eta''(u))^2 du, \tag{10.4}$$

where $l_i(y_i, \eta(x_i))$ is the log-likelihood contribution of observation y_i, fitted by $\eta(x_i)$, and $\eta(.)$ is assumed to have continuous first and second derivatives, η' and η'', with η'' being square integrable. The generalized spline smoothing criterion (10.4) is a compromise between faith with the data, represented by a large log-likelihood, and smoothness of the estimated function, with smoothness defined by the penalty term $\lambda/2 \int (\eta''(u)du)$. The influence of the penalty term is determined by the tuning parameter λ, with $\lambda = 0$ meaning no restriction and $\lambda \to \infty$ postulating that the second derivative η'' becomes zero. By using the deviance $D(\{y_i\}, \{\eta(x_i)\}) = -\phi \Sigma_i l_i(y_i, \eta(x_i))$ as a measure for the discrepancy between the fit and the data, maximization of (10.4) turns into the minimization of

$$D(\{y_i\}, \{\eta(x_i)\}) + \frac{1}{2}\lambda \int (\eta''(u))^2 du,$$

where the dispersion ϕ is included in the tuning parameter λ. For normally distributed responses one obtains the penalized least-squares criterion:

$$\sum_{i=1}^{n} (y_i - \eta(x_i))^2 + \lambda \int (\eta''(u))^2 du.$$

The solution $\hat{\eta}_\lambda(x)$ is a so-called *natural cubic smoothing spline*. This means that the function $\hat{\eta}_\lambda(x)$ is a cubic polynomial on intervals $[x_i, x_{i+1}]$, first and second derivatives are continuous at the observation points, and the second derivative is zero at the boundary points x_1, x_n. Smoothing splines arising from the penalized least-squares problem have been considered by Reinsch (1967); see also Green and Silverman (1994) for the more general case. For the resulting splines, the observations themselves correspond to the knots of the polynomial splines. Since the solution has the form of cubic splines with n knots, optimization with respect to a set of functions reduces to a finite-dimensional optimization problem. It may be shown that the evaluations $\hat{\eta}_i = \hat{\eta}(x_i)$ at the observed points $x_1 < \cdots < x_n$ may be computed by maximization of a penalized likelihood:

$$l_p(\boldsymbol{\eta}) = \sum_{i=1}^{n} l_i(y_i, \eta_i) - \frac{1}{2}\lambda \boldsymbol{\eta}^T \boldsymbol{K} \boldsymbol{\eta}, \tag{10.5}$$

where $\boldsymbol{\eta}^T = (\eta_1, \ldots, \eta_n)$ and $\boldsymbol{K}$ is a matrix that depends only on the differences $x_{i+1} - x_i$ (see Green and Silverman, 1994). Thus, in principle, estimates $\hat{\eta}$ may be computed by using the Fisher scoring algorithm with a simple penalty term. However, if the number of observations and therefore knots is large, computation by Fisher scoring is not very efficient. In particular, if one has to handle more than one covariate, the computational costs become very severe. For the penalized least-squares problem, an algorithm that explicitly uses that the matrix $\boldsymbol{K}$ implies banded matrices has been given by Reinsch (1967).

For normally distributed responses, maximization of $l_p(\boldsymbol{\eta})$ corresponds to the penalized least-squares problem that has the solution $\hat{\boldsymbol{\eta}} = (\boldsymbol{I} + \lambda \boldsymbol{K})^{-1} \boldsymbol{y}$, where $\boldsymbol{I}$ is the identity matrix and $\boldsymbol{y}^T = (y_1, \ldots, y_n)$. Therefore one has a simple linear smoother $\hat{\boldsymbol{\eta}} = \boldsymbol{S}_\lambda \boldsymbol{y}$ with a smoother matrix $S_\lambda = (I + \lambda K)^{-1}$. However, computation should not be based on this explicit form; Reinsch (1967) proposed an efficient algorithm that makes efficient use of the band matrices that constitute the matrix K.

Smoothing splines provide some background for the use of spline functions. The maximization problem is clearly defined on functional spaces and does not imply that the function one seeks is a spline. Nevertheless, the solution is cubic splines with as many basis functions as observations. In practice, regression splines, in particular when estimated by discrete penalization techniques (see next section), yield similar estimates with fewer knots than smoothing splines. In addition, when using regression splines, the number of knots may be determined by the user.

10.1.3 Penalized Estimation

The big advantage of the linear basis function approach $\eta(x) = \sum_j \beta_j \phi_j(x)$ is that for fixed basis functions one has a linear predictor and estimation may be based on procedures that are familiar from generalized linear models. However, to be sufficiently flexible, one has to use a fine grid of intervals for regression splines or many basis functions when using radial basis functions.

One strategy is to select basis functions in a stepwise way; alternatively, one uses many basis functions right from the start. Using as many basis functions as possible should ameliorate the dependence on the chosen knots. The difficulty with many basis functions is the instability or even non-existence of estimates. One approach to overcome this problem is to regularize estimation by including a penalty term as in the derivation of smoothing splines. Rather than using simple maximum likelihood estimation one maximizes the penalized log-likelihood

$$l_p(\beta) = l(\beta) - \frac{\lambda}{2} J(\eta),$$

where $l(\beta)$ is the familiar log-likelihood function, λ is a tuning parameter, and $J(\eta)$ is a penalty functional measuring the roughness of η.

A roughness penalty that has been used in the derivation of smoothing splines is $J(\eta) = \int (\eta''(u))^2 \, du$ as a global measure for the curvature of the function $\eta(x)$. When maximizing l_p, one looks for a compromise between data fit and smoothness of the estimated predictor $\eta(x)$. The parameter λ controls the trade-off between the smoothness of $\eta(x)$ and the faith with the data. Large values of λ enforce smooth functions with small variance but possibly big bias, whereas small values of λ allow wiggly function estimates with high variance.

A somewhat more general penalty is

$$J(\eta) = \int_a^b (\eta^{(d)}(u))^2 du,$$

where $\eta^{(d)}$ denotes the dth derivative and $[a, b]$ is the range of observations. When using linear basis functions $\eta(x) = \Sigma_j \beta_j \phi_j(x)$ one obtains with $\phi^{(d)}(u)^T = (\phi_1^{(d)}(u), \dots \phi_m^{(d)}(u))$ the parameterized penalty

$$J(\eta) = \int \|\beta^T \phi^{(d)}(u)\|^2 du = \beta^T K \beta, \tag{10.6}$$

where $K = (k_{ij})$ has entries

$$k_{ij} = \int_a^b \phi_i^{(d)}(u) \phi_j^{(d)}(u) du.$$

The penalty (10.6) has the advantage that it may be used for any spacing of knots. The penalized likelihood takes the simple form

$$l_p(\boldsymbol{\beta}) = l(\boldsymbol{\beta}) - \frac{\lambda}{2}\boldsymbol{\beta}^T \boldsymbol{K}\boldsymbol{\beta},$$

which is familiar from shrinkage estimators (see Chapter 6).

The effect of a penalty usually becomes clear by considering the limit. When polynomial splines of degree k are used, k should be chosen such that $d < k$, since otherwise derivatives would not exist. For $d < k$, the penalty (10.6) shrinks toward a polynomial of degree $d-1$. Thus, for $\lambda \to \infty$, a polynomial of degree $d-1$ is fitted. For example, when using cubic splines and derivatives of the second order, $d = 2$, $\lambda \to \infty$ produces a linear fit. This means that by fitting splines, part of the function is not penalized. By reparameterization the unpenalized part can be separated from the rest, yielding a simpler penalty (for details see Appendix C).

For specific patterns of knots alternative penalties may be used. However, the choice of a sensible penalty should be linked to the choice of the basis functions. If one uses the truncated power function basis (10.2) and equally spaced knots, an adequate penalty is also

$$J = \sum_{i=k+1}^{m_s} \beta_i^2,$$

which penalizes only coefficients of the truncated power functions. Since only coefficients of the truncated power functions are penalized, the fit is shrunk toward the log-likelihood fit of a polynomial of degree k. For $\lambda \to \infty$ one fits the polynomial model $\eta(x) = \beta_0 + \beta_1 x + ... + \beta_k x^k$.

Penalized Splines

An alternative and widely used penalty is based on differences. Marx and Eilers (1998) proposed using B-splines with equally spaced knots and penalties of the form

$$J_d = \sum_{j=d+1}^{m} (\Delta^d \beta_j)^2 = \boldsymbol{\beta}^T \boldsymbol{K}_d \boldsymbol{\beta}, \tag{10.7}$$

where Δ is the difference operator, operating on adjacent B-spline coefficients, that is, $\Delta\beta_j = \beta_j - \beta_{j-1}, \Delta^2\beta_i = \Delta(\beta_j - \beta_{j-1}) = \beta_j - 2\beta_{j-1} + \beta_{j-2}$. The method is referred to as *P-splines* (for penalized splines). The corresponding matrix $\boldsymbol{K}_d$ has a banded structure and simply represents the differences in matrix form.

Let the $(m-1) \times m$-matrix $\boldsymbol{D}_1$ be given by

$$\boldsymbol{D}_1 = \begin{pmatrix} -1 & 1 & 0 & 0 & \dots & 0 \\ 0 & -1 & 1 & 0 & \dots & 0 \\ 0 & 0 & -1 & 1 & \dots & 0 \\ \vdots & & & \ddots & -1 & 1 \end{pmatrix}.$$

Then the vector of first differences is $\boldsymbol{D}_1\boldsymbol{\beta}$ and the corresponding penalty term is $\boldsymbol{\beta}^T \boldsymbol{K}_1 \boldsymbol{\beta} = \boldsymbol{\beta}^T \boldsymbol{D}_1^T \boldsymbol{D}_1 \boldsymbol{\beta}$. The *differences of order* d as given in (10.7) are obtained from $\boldsymbol{D}_d\boldsymbol{\beta} = \boldsymbol{D}_1 \boldsymbol{D}_{d-1}\boldsymbol{\beta}$ and $\boldsymbol{K}_d = \boldsymbol{D}_d^T \boldsymbol{D}_d$. By building differences of differences, the dimension reduces such that $\boldsymbol{K}_d$ is a $(m-d) \times (m-d)$-matrix. The choice of the order of differences determines the smoothness of the intended estimate. Gijbels and Verhaselt (2010) proposed a data-driven procedure for the choice of the order based on the AIC criterion.

The penalty (10.7) is useful for B-splines but also for bases like the radial basis functions. The only restriction is that the knots should be equally spaced. B-splines have the advantage that in the limit, with strong smoothing, a polynomial is fitted. If a penalty of order d is used and the degree of the B-spline is higher than, d, for large values of λ the fit will approach a polynomial of degree $d-1$. Marx and Eilers (1998) also use equally spaced knots when the data are far from being uniformly distributed. However, when the spacings of the data are very irregular, knots at quantiles seem more appropriate.

If one changes the basis, of course, the corresponding penalty changes. In general, when using basis functions, the linear predictor has the form $\boldsymbol{\eta} = \boldsymbol{\phi\beta}$, where $\boldsymbol{\phi}$ is the design matrix composed from evaluations of basis functions, more precisely, $\boldsymbol{\phi} = (\phi_{ij})$ is a $(n \times m)$-matrix with entries $\phi_{ij} = \phi_j(x_i)$. The transformation between the truncated power series and the B-splines is given by $\boldsymbol{\phi}_T = \boldsymbol{\phi}_B \boldsymbol{L}$, where $\boldsymbol{L}$ is a square invertible matrix, and $\boldsymbol{\phi}_T$ and $\boldsymbol{\phi}_B$ denote the matrices for the truncated power series and B-splines, respectively. Let the predictor for B-splines be $\boldsymbol{\eta} = \boldsymbol{\phi}_B\boldsymbol{\beta}_B$ with penalty $\boldsymbol{\beta}_B^T \boldsymbol{K}_B \boldsymbol{\beta}_B$. Then one obtains for the truncated power series $\boldsymbol{\eta} = \boldsymbol{\phi}_T\boldsymbol{\beta}_T = \boldsymbol{\phi}_B\boldsymbol{L}\boldsymbol{\beta}_T = \boldsymbol{\phi}_B\boldsymbol{\beta}_B$ with $\boldsymbol{\beta}_B = \boldsymbol{L}\boldsymbol{\beta}_T$. The corresponding penalty in $\boldsymbol{\beta}_T$-parameters is $\boldsymbol{\beta}_B^T \boldsymbol{K}_B \boldsymbol{\beta}_B = \boldsymbol{\beta}_T^T \boldsymbol{L}^T \boldsymbol{K}_B \boldsymbol{L} \boldsymbol{\beta}_T = \boldsymbol{\beta}_T^T \boldsymbol{K}_T \boldsymbol{\beta}_T$, where $\boldsymbol{K}_T = \boldsymbol{L}^T \boldsymbol{K}_B \boldsymbol{L}$.

Bayesian View of Difference Penalties

Within a Bayesian framework parameters are restricted by assuming a prior distribution on the space of parameters. The Bayesian analog to differences of order d are random walks of order d. First-and second-order random walks for equidistant knots may be specified by

$$\beta_j = \beta_{j-1} + u_j, \qquad j = 2, \dots$$

and

$$\beta_j = 2\beta_{j-1} - \beta_{j-2} + u_j, \qquad j = 3, \dots,$$

where $u_j \sim N(0, \tau^2)$. For initial values one assumes diffuse priors $p(\beta_1)$ and $p(\beta_1, \beta_2)$, respectively. Alternatively, one can assume $\beta_j | \beta_{j-1} \sim N(\beta_{j-1}, \tau^2)$ for the first-order random walk and $\beta_j | \beta_{j-1}, \beta_{j-2} \sim N(2\beta_{j-1} - \beta_{j-2}, \tau^2)$ for the second-order random walk. The order of the random walk determines the structures within the fitted parameters. The first-order walk tends to fit a constant for the coefficients, whereas a second-order random walk implies a linear trend, which becomes obvious from $\beta_j - \beta_{j-1} = \beta_{j-1} - \beta_{j-2} + \varepsilon_i$. The variance parameter acts as a smoothing parameter. If τ is very small, the coefficients have to be close to a constant or a linear trend. If τ is large, the coefficients can deviate from the specified trend.

The corresponding joint distributions of the parameter vectors have the form of an (improper) Gaussian distribution:

$$p(\boldsymbol{\beta}|\tau^2) \propto \exp(-\frac{1}{2\tau^2} \boldsymbol{\beta}^T \boldsymbol{K}_d \boldsymbol{\beta}),$$

where the precision matrix $\boldsymbol{K}_d$ is given by $\boldsymbol{K}_d = \boldsymbol{D}_d^T \boldsymbol{D}_d$. For more details, see Lang and Brezger (2004a), Brezger and Lang (2006), and Rue and Held (2005).

Maximizing the Penalized Likelihood

For fixed knots and a given λ, the penalties considered in this section yield the penalized likelihood

$$l_p(\boldsymbol{\beta}) = l(\boldsymbol{\beta}) - \frac{\lambda}{2} \boldsymbol{\beta}^T \boldsymbol{K} \boldsymbol{\beta}, \tag{10.8}$$

where $\boldsymbol{K}$ is a symmetric matrix and $l(\boldsymbol{\beta})$ is the log-likelihood of the linear basis predictors $\eta_i = \eta(x_i) = \boldsymbol{\phi}_i^T \boldsymbol{\beta}, \boldsymbol{\phi}_i^T = (\phi_1(x_i), \ldots, \phi_m(x_i))$. In matrix notation, the corresponding penalized score function is

$$\boldsymbol{s}_p(\boldsymbol{\beta}) = \boldsymbol{\phi}^T \boldsymbol{D}(\boldsymbol{\beta})\ \boldsymbol{\Sigma}(\boldsymbol{\beta})^{-1}(\boldsymbol{y} - \boldsymbol{\mu}) - \lambda \boldsymbol{K}\boldsymbol{\beta},$$

where $\boldsymbol{\phi}^T = (\boldsymbol{\phi}_1, \ldots, \boldsymbol{\phi}_n)$ is the design matrix, $\boldsymbol{D}(\boldsymbol{\beta}) = \text{Diag}\ (\partial h(\eta_1)/\partial \eta, \ldots, \partial h(\eta_n)/\partial \eta)$ is the diagonal matrix of derivatives, $\boldsymbol{\Sigma}(\boldsymbol{\beta}) = \text{Diag}\ (\sigma_1^2, \ldots, \sigma_n^2)$ is the covariance matrix, and $\boldsymbol{y}^T = (y_1, \ldots, y_n), \boldsymbol{\mu}^T = (\mu_1, \ldots, \mu_n)$ are the vectors of observations and means.

The estimation equation $\boldsymbol{s}_p(\hat{\boldsymbol{\beta}}) = \boldsymbol{0}$ may be solved by iterative pseudo-Fisher scoring:

$$\hat{\boldsymbol{\beta}}^{(k+1)} = \hat{\boldsymbol{\beta}}^{(k)} + \boldsymbol{F}_p(\hat{\boldsymbol{\beta}}^{(k)})^{-1} \boldsymbol{s}_p(\hat{\boldsymbol{\beta}}^{(k)}),$$

where $\boldsymbol{F}_p(\boldsymbol{\beta}) = \boldsymbol{F}(\boldsymbol{\beta}) + \lambda \boldsymbol{K}$ and $\boldsymbol{F}(\boldsymbol{\beta}) = E(-\partial l / \partial \boldsymbol{\beta} \partial \boldsymbol{\beta}^T) = \boldsymbol{\phi}^T \boldsymbol{W}(\boldsymbol{\beta}) \boldsymbol{\phi}$ with $\boldsymbol{W}(\boldsymbol{\beta}) = \boldsymbol{D}(\boldsymbol{\beta})\ \boldsymbol{\Sigma}(\boldsymbol{\beta})^{-1} \boldsymbol{D}(\boldsymbol{\beta})^T$. In the same way as for generalized linear models, pseudo-Fisher scoring can be written as iteratively reweighted least-squares fitting:

$$\hat{\boldsymbol{\beta}}^{(k+1)} = (\boldsymbol{\phi}^T \boldsymbol{W}(\hat{\boldsymbol{\beta}}^{(k)}) \boldsymbol{\phi} + \lambda \boldsymbol{K})^{-1} \boldsymbol{\phi}^T \boldsymbol{W}(\hat{\boldsymbol{\beta}}^{(k)}) \tilde{\boldsymbol{\eta}}(\hat{\boldsymbol{\beta}}^{(k)})$$

with the adjusted variable $\tilde{\boldsymbol{\eta}}(\hat{\boldsymbol{\beta}}) = \hat{\boldsymbol{\eta}} + \boldsymbol{D}(\hat{\boldsymbol{\beta}})^{-1}(\boldsymbol{y} - \boldsymbol{\mu}(\hat{\boldsymbol{\beta}}))$, where $\hat{\boldsymbol{\eta}} = \boldsymbol{\phi}\hat{\boldsymbol{\beta}}$. At convergence one obtains

$$\hat{\boldsymbol{\beta}} = (\boldsymbol{\phi}^T \boldsymbol{W}(\hat{\boldsymbol{\beta}}) \boldsymbol{\phi})^{-1} \boldsymbol{\phi}^T \boldsymbol{W}(\hat{\boldsymbol{\beta}}) \tilde{\boldsymbol{\eta}}(\hat{\boldsymbol{\beta}}).$$

The hat matrix may be approximated by

$$\boldsymbol{H} = \boldsymbol{W}(\hat{\boldsymbol{\beta}})^{T/2} \boldsymbol{\phi} (\boldsymbol{\phi}^T \boldsymbol{W}(\hat{\boldsymbol{\beta}}) \boldsymbol{\phi} + \lambda \boldsymbol{K})^{-1} \boldsymbol{\phi}^T \boldsymbol{W}(\hat{\boldsymbol{\beta}})^{1/2}.$$

Approximate covariances are obtained by the sandwich matrix:

$$\text{cov}(\hat{\boldsymbol{\beta}}) \approx (\boldsymbol{F}(\hat{\boldsymbol{\beta}}) + \lambda \boldsymbol{K})^{-1} \boldsymbol{F}(\hat{\boldsymbol{\beta}}) (\boldsymbol{F}(\hat{\boldsymbol{\beta}}) + \lambda \boldsymbol{K})^{-1}.$$

In the case of normally distributed responses and an identity link, maximization of the penalized likelihood is equivalent to solving the penalized least-squares problem $\sum_i (y_i - \boldsymbol{\phi}_i^T \boldsymbol{\beta})^2 + \tilde{\lambda} \boldsymbol{\beta}^T \boldsymbol{K} \boldsymbol{\beta}$, where $\tilde{\lambda} = \lambda \sigma^2$. The solution has the simple form

$$\hat{\boldsymbol{\beta}}_\lambda = (\boldsymbol{\phi}^T \boldsymbol{\phi} + \tilde{\lambda} \boldsymbol{K})^{-1} \boldsymbol{\phi}^T \boldsymbol{y}$$

with design matrix $\boldsymbol{\phi}^T = (\boldsymbol{\phi}_1, \ldots \boldsymbol{\phi}_n)$ and $\boldsymbol{y}^T = (y_1, \ldots, y_n)$. Then the fitted values are given by $\hat{\boldsymbol{y}} = \boldsymbol{\phi} \hat{\boldsymbol{\beta}}_\lambda = \boldsymbol{S}_\lambda \boldsymbol{y}$ with $\boldsymbol{S}_\lambda = \boldsymbol{\phi}(\boldsymbol{\phi}^T \boldsymbol{\phi} + \tilde{\lambda} \boldsymbol{K})^{-1} \boldsymbol{\phi}^T$.

Effective Degrees of Freedom of a Smoother

In linear regression the trace of the hat matrix, which is defined by $\hat{\boldsymbol{y}} = \boldsymbol{H}\boldsymbol{y}$, may be used to compute the *degrees of freedom* of the fit, since

$$tr(\boldsymbol{H}) = \text{number of fitted parameters}.$$

With an $n \times p$ full rank design matrix $\boldsymbol{X}$, the hat matrix in linear regression is $\boldsymbol{X}(\boldsymbol{X}^T \boldsymbol{X})^{-1} \boldsymbol{X}^T$. For linear smoothers, given by $\hat{\boldsymbol{\mu}} = \boldsymbol{S}_\lambda \boldsymbol{y}$, the trace of $\boldsymbol{S}_\lambda$ plays the same role and $tr(\boldsymbol{S}_\lambda)$ may be interpreted as the number of parameters used in the fit or the *effective degrees of freedom of the fit*. In particular, when responses are normally distributed and h is the identity, one obtains linear smoothers. For penalized regression splines one obtains $\boldsymbol{S}_\lambda = \boldsymbol{\phi}(\boldsymbol{\phi}^T \boldsymbol{\phi} + \tilde{\lambda} \boldsymbol{K})^{-1} \boldsymbol{\phi}^T$.

When $\tilde{\lambda} = 0$ $\boldsymbol{S}_\lambda$ is the usual hat matrix, $\boldsymbol{\phi}(\boldsymbol{\phi}^T\boldsymbol{\phi})^{-1}\boldsymbol{\phi}^T$ with design matrix $\boldsymbol{\phi}$ and the number of columns of the full rank matrix $\boldsymbol{\phi}$ are the degrees of freedom. If $\tilde{\lambda}$ is very large, the model is less flexible and the effective degrees of freedom will be much smaller.

An alternative way to determine the degrees of freedom is associated with the unpenalized estimate $\hat{\boldsymbol{\beta}}_u$ resulting from $\tilde{\lambda} = 0$. Wood (2006a) considers the penalized estimate of $\boldsymbol{\beta}$ in the form

$$\begin{aligned}\hat{\boldsymbol{\beta}} &= (\boldsymbol{\phi}^T\boldsymbol{\phi} + \tilde{\lambda}\boldsymbol{K})^{-1}\boldsymbol{\phi}^T\boldsymbol{y} = (\boldsymbol{\phi}^T\boldsymbol{\phi} + \tilde{\lambda}\boldsymbol{K})^{-1}\boldsymbol{\phi}^T\boldsymbol{\phi}(\boldsymbol{\phi}^T\boldsymbol{\phi})^{-1}\boldsymbol{\phi}^T\boldsymbol{y}\\ &= (\boldsymbol{\phi}^T\boldsymbol{\phi} + \tilde{\lambda}\boldsymbol{K})^{-1}\boldsymbol{\phi}^T\boldsymbol{\phi}\hat{\boldsymbol{\beta}}_u\end{aligned}$$

since the unpenalized estimate is $\hat{\boldsymbol{\beta}}_u = (\boldsymbol{\phi}^T\boldsymbol{\phi})^{-1}\boldsymbol{\phi}^T\boldsymbol{y}$. Then the transformation from $\hat{\boldsymbol{\beta}}_u$ to $\hat{\boldsymbol{\beta}}$ has the form $\hat{\boldsymbol{\beta}} = \boldsymbol{M}\hat{\boldsymbol{\beta}}_u$ with

$$\boldsymbol{M} = (\boldsymbol{\phi}^T\boldsymbol{\phi} + \tilde{\lambda}\boldsymbol{K})^{-1}\boldsymbol{\phi}^T\boldsymbol{\phi}.$$

The diagonal elements m_{ii} of $\boldsymbol{M}$ may be seen as an approximate measure of how much the penalized estimate will change when $\hat{\boldsymbol{\beta}}$ changes by one unit. In the unpenalized setting the ith parameter β_i has one degree of freedom, which changes approximately by a factor m_{ii}. However, as Wood (2006a) notes, there is no general guarantee that $m_{ii} > 0$. As an approximation m_{ii} is useful to investigate the effective degrees of freedom associated with the ith parameter. Moreover, the effective degrees of freedom, determined by the trace, $tr(\boldsymbol{M})$, is equivalent to the degrees of freedom obtained by $tr(\boldsymbol{S}_\lambda)$ since $tr(\boldsymbol{\phi}^T\boldsymbol{\phi}+\tilde{\lambda}\boldsymbol{K})^{-1}\boldsymbol{\phi}^T\boldsymbol{\phi} = tr(\boldsymbol{\phi}(\boldsymbol{\phi}^T+\lambda\boldsymbol{K})^{-1}\boldsymbol{\phi}^T)$.

In the general case, when $\mu = h(\eta)$ is fitted, smoothers are not linear and the effective degrees of freedom may be defined as the trace of the corresponding generalized hat matrix:

$$\boldsymbol{H} = \boldsymbol{\phi}^T(\boldsymbol{\phi}^T\boldsymbol{W}\boldsymbol{\phi} + \lambda\boldsymbol{K})^{-1}\boldsymbol{\phi}.$$

Individual degrees of freedom for separate components may again be derived by considering the fitting at convergence. With a vector of pseudo-observations $\tilde{\boldsymbol{\eta}}(\hat{\boldsymbol{\beta}}) = \boldsymbol{\phi}\hat{\boldsymbol{\beta}} + \boldsymbol{D}^{-1}(\boldsymbol{y} - \hat{\boldsymbol{\mu}})$ one has

$$\begin{aligned}\hat{\boldsymbol{\beta}} &= (\boldsymbol{\phi}^T\boldsymbol{W}\boldsymbol{\phi} + \lambda\boldsymbol{K})^{-1}\boldsymbol{\phi}^T\boldsymbol{W}\tilde{\boldsymbol{\eta}}(\hat{\boldsymbol{\beta}})\\ &= (\boldsymbol{\phi}^T\boldsymbol{W}\boldsymbol{\phi} + \lambda\boldsymbol{K})^{-1}\boldsymbol{\phi}^T\boldsymbol{W}\boldsymbol{\phi}(\boldsymbol{\phi}^T\boldsymbol{W}\boldsymbol{\phi})^{-1}\boldsymbol{\phi}^T\boldsymbol{W}\tilde{\boldsymbol{\eta}}(\hat{\boldsymbol{\beta}})\\ &= \boldsymbol{M}(\boldsymbol{\phi}^T\boldsymbol{W}\boldsymbol{\phi})^{-1}\boldsymbol{\phi}^T\boldsymbol{W}\tilde{\boldsymbol{\eta}}(\hat{\boldsymbol{\beta}}_u),\end{aligned}$$

where $\boldsymbol{M} = (\boldsymbol{\phi}^T\boldsymbol{W}\boldsymbol{\phi} + \lambda\boldsymbol{K})^{-1}\boldsymbol{\phi}^T\boldsymbol{W}\boldsymbol{\phi}$ and $\tilde{\boldsymbol{\eta}}(\hat{\boldsymbol{\beta}}_u) = (\boldsymbol{\phi}^T\boldsymbol{W}\boldsymbol{\phi})^{-1}\boldsymbol{\phi}^T\boldsymbol{W}\tilde{\boldsymbol{\eta}}(\hat{\boldsymbol{\beta}})$ is an approximation to the unpenalized estimate $(\boldsymbol{\phi}^T\boldsymbol{W}\boldsymbol{\phi})^{-1}\boldsymbol{\phi}^T\boldsymbol{W}\boldsymbol{\eta}(\hat{\boldsymbol{\beta}}_u)$. Again, one has $tr(\boldsymbol{H}) = tr(\boldsymbol{M})$.

Example 10.2: Duration of Unemployment

For the duration of unemployment data one obtains the fitted functions given in Figure 10.4. The fitted basis functions were cubic B-splines on an equally spaced grid. The left panel shows the unpenalized fit, and the right panel shows the fit based on a first-order difference penalty (smoothing parameter chosen by generalized cross-validation). In addition, the weighted basis functions that produce the fit are shown. It is seen that the unpenalized fit is close to the data but very wiggly. As an estimate of the underlying function it is hardly convincing. The penalized estimate is much smoother and also aesthetically more satisfying. □

The penalized likelihood for classical smoothing splines (10.5) has the same form as the likelihood for regression splines (10.8). The main difference is that for smoothing splines the vector of evaluations $\boldsymbol{\eta}$ is n-dimensional, whereas for regression splines it is much smaller, depending on the number of basis functions. The latter approach uses a smaller number of basis functions. Smoothers that use considerably less than n basis functions are usually called *low-rank* smoothers, while smoothers that use n basis functions are called *full-rank* smoothers. Small-rank smoothers may reduce the computation time without loosing much in accuracy. For linear smoothers, the effect of the reduction to fewer dimensions was treated extensively by Hastie (1996).

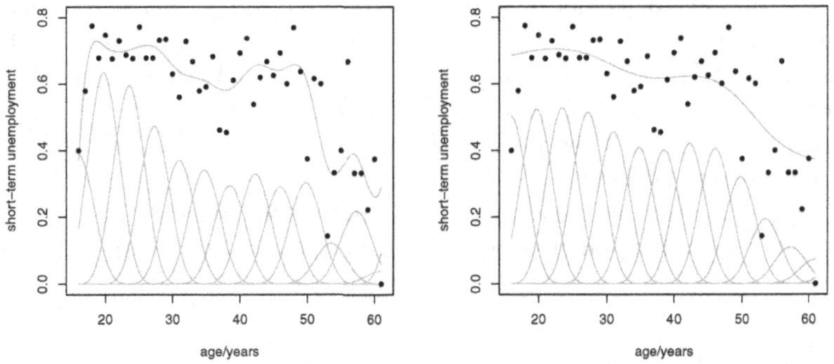

FIGURE 10.4: Probability of short-term unemployment plotted against age; cubic splines on equally spaced grid, not penalized (left panel) and penalized (right panel) by first-order differences.

10.1.4 Local Regression

Localizing techniques provide a quite different approach to the estimation of the unknown predictor function $\eta(x)$. In a local regression estimation of $\eta(x)$ at a target value, x is obtained by fitting a parametric model locally in a neighborhood around x. The parametric model to be fitted is a model with polynomial terms, giving the procedure its name, *local polynomial regression*. Taylor expansions provide some motivation for the approach: If one assumes that the function $\eta(x)$ is sufficiently smooth, it may be approximated by a Taylor expansion (see Appendix B):

$$\eta(x_i) = \eta(x) + \eta'(x)(x_i - x) + \eta''(x)(x_i - x)^2/2 + \dots,$$

which, for fixed x, has the form

$$\eta(x_i) \approx \beta_0 + \beta_1(x_i - x) + \beta_2(x_i - x)^2 + \dots,$$

where the dependence of the parameters β_j on the target value x is suppressed. Since $\eta(x_i)$ may be approximated by a polynomial, one fits (for fixed x) locally the model with polynomial predictors. When an approximation by second-order terms is used, one uses the predictors

$$\eta_i = \eta(x_i) = \tilde{\boldsymbol{x}}_i^T \boldsymbol{\beta},$$
$$\tilde{\boldsymbol{x}}_i^T = (1, (x_i - x), (x_i - x)^2), \ \boldsymbol{\beta}^T = (\beta_0, \beta_1, \beta_2).$$

Local fitting means maximization of the local log-likelihood:

$$l_x(\boldsymbol{\beta}) = \sum_{i=1}^{n} l_i(\boldsymbol{\beta}) w_\lambda(x, x_i),$$

where $l_i(\boldsymbol{\beta})$ is the log-likelihood for ith observation of the model with predictors $\eta(x_i)$, and $w_\lambda(x, x_i)$ is a weighting function depending on the target value x and the observation x_i, and, in addition, contains a smoothing parameter λ. Common weight functions are based on kernels. When a kernel or density function, which is a continuous symmetric function that fulfills $\int K(u)du = 1$, the corresponding kernel weight is

$$w_\lambda(x, x_i) \propto K(\frac{x - x_i}{\lambda}).$$

Candidates are the Gaussian kernel, where K is the standard Gaussian density, or more refined kernels like the Epanechnikov kernel $K(u) = \frac{3}{4}(1 - u^2)$ for $|u| \leq 1$ and zero otherwise. The smoothing parameter (or window width) λ determines how local the estimate is. For small λ the weights decrease fast with increasing distance $|x - x_i|$ and estimates are based solely on observations from the close neighborhood of x. For large λ the decrease is very slow; in the extreme case ($\lambda \to \infty$) each observation has the same weight and one fits the polynomial regression model to the full dataset.

Let $\hat{\boldsymbol{\beta}} = (\hat{\beta}_0, \hat{\beta}_1, \hat{\beta}_2)^T$ denote the estimate resulting from the maximization of $l_x(\boldsymbol{\beta})$ based on an approximation by second-order terms. Then the estimates $\hat{\eta}(x)$ and $\hat{\mu}(x)$ at target value x are given by

$$\hat{\eta}(x) = \hat{\beta}_0, \;\; \hat{\mu}(x) = h(\hat{\beta}_0).$$

Of course the estimates have to be computed on a grid of target values to obtain the functions $\hat{\mu}(.)$ and $\hat{\eta}(.)$.

Since local likelihood is a likelihood with multiplicative weights, computing $\hat{\boldsymbol{\beta}}$ (for target value x) is straightforward by using weighted scoring techniques. The local score function $\boldsymbol{s}_x(\boldsymbol{\beta}) = \partial l_x(\boldsymbol{\beta})/\partial\boldsymbol{\beta}$ has the form

$$\boldsymbol{s}_x(\boldsymbol{\beta}) = \sum_{i=1}^{n} \boldsymbol{s}_i(\boldsymbol{\beta}) w_\lambda(x, x_i) = \sum_{i=1}^{n} \tilde{\boldsymbol{x}}_i (\partial h(\eta_i)/\partial\eta)\sigma_i^{-2}(y_i - \mu_i) w_\lambda(x, x_i).$$

The solution of $\boldsymbol{s}_x(\hat{\boldsymbol{\beta}}) = \boldsymbol{0}$ can be obtained by weighted iterative Fisher scoring:

$$\hat{\boldsymbol{\beta}}^{(k+1)} = \hat{\boldsymbol{\beta}}^{(k)} + \boldsymbol{F}_\lambda(\hat{\boldsymbol{\beta}}^{(k)})^{-1} \boldsymbol{s}_x(\hat{\boldsymbol{\beta}}^{(k)}), \tag{10.9}$$

where $\boldsymbol{F}_\lambda(\beta) = \sum_{i=1}^{n} \tilde{\boldsymbol{x}}_i \tilde{\boldsymbol{x}}_i^T (\partial h(\eta_i)/\partial\eta)\sigma_i^{-2}$. The simplest polynomial that can be fitted in a local regression is a polynomial of degree zero. Then one fits the model $\mu_i = h(\beta_0)$ for all observations. When using the canonical link, the estimation equation $\boldsymbol{s}_x(\hat{\beta}_0) = \boldsymbol{0}$ for target value x is $\sum_i (y_i - \hat{\mu}(x)) w_\lambda(x, x_i) = 0$ and the resulting estimate is a weighted sum of observations:

$$\hat{\mu}(x) = \sum_{i=1}^{n} s_\lambda(x, x_i) y_i, \tag{10.10}$$

where the weights are given by $s_\lambda(x, x_i) = w_\lambda(x, x_i)/(\sum_j w_\lambda(x, x_j))$, for which $\sum_i s_\lambda(x, x_i) = 1$ holds. The estimate (10.10) illustrates nicely the effect of localizing. Rather than fitting a polynomial parametric regression model, one estimates $\mu(x)$ as a weighted mean over observations from a neighborhood of x. The resulting estimate is not based on the assumption of an underlying parametric model (which might not be identical to the data generating model). When a polynomial of degree larger than zero is fitted, the effect is also an averaging over neighborhood observations, but with a much more complicated weighting scheme. The simple estimate (10.10) with kernel weights has been proposed for metric responses by Nadaraya (1964) and Watson (1964) and is called the *Nadaraya-Watson estimator*.

Fitting of higher polynomials often results in smoother estimates and has less problems with the bias in boundary regions. In particular, polynomials with odd degree show so-called boundary carpentry, which means that local fitting automatically adapts to the special data situation at the boundary of the design (see Hastie and Loader, 1993). Local polynomial regression for metric responses has a long tradition, which is nicely reviewed in Cleveland and Loader (1996). The extension to a wider class of distributions started with the concept of local likelihood estimation, which was formulated by Tibshirani and Hastie (1987), who gave an extensive treatment of asymptotic properties of local regression. Loader (1994) gave an overview of localizing techniques including classification problems.

One problem with local estimates is that they work well only in low to moderate dimensions because they suffer from the so-called *curse of dimensionality* (Bellman, 1961). It has many manifestations, which are nicely examined in Hastie et al. (2009). One demonstration they used considers data that are uniformly distributed in a p-dimensional unit hypercube. If one wants to capture a fraction r of observations about a target point, the needed edge length (for each dimension) is $r^{1/p}$. Therefore, if $r = 0.1$, that is, 10 percent of the data are to be captured, with $p = 10$ one needs a cube with edge length 0.8, which is hardly local. If one wants to estimate with the same accuracy as in low dimensions, the sample size typically has to grow exponentially with the dimension (for examples and details see Hastie et al., 2009, pp. 22ff).

10.1.5 Selection of Smoothing Parameters

When smoothers are applied several quantities have to be specified. With regression splines one has to specify the locations of knots and the degree of the polynomial; for localized estimates one has to choose the polynomial and the type of localization. But, most important, one has to select a smoothing parameter that determines the smoothness of the estimate and therefore crucially determines the performance of the smoothing procedure. In particular, for linear smoothers there is a wide body of literature investigating asymptotically efficient smoothing procedures. In the following we consider some simple data-driven procedures that may be also used for non-linear smoothers.

Cross-Validation

The naive approach to smoothing parameter selection is to select the parameter that optimizes the goodness-of-fit. The back side of this approach is that the data at hand are well fitted but the fit is rather wiggly, hardly approximates the underlying curve, and generalizes badly. A simple cure for this problem is cross-validation, where the data are split into disjoint sets. Part of the data is used to fit a curve and the performance is evaluated on the data that have not been used in fitting. Thus the fit has not seen the new data. In *K-fold cross-validation* the data are split into K roughly equal-sized parts; then, in turn, $K - 1$ of these parts are used to fit the curve and the performance is evaluated on the part of the data that has been left out. The extreme case is *leaving-one-out cross-validation*, where only one observation is used to evaluate future performance.

In the general setting a measure for the performance is predictive deviance. Let $S = \{(y_i, \boldsymbol{x}_i), i = 1, \dots, n\}$ denote the full dataset, let $\hat{\eta}_{\setminus i}(\boldsymbol{x})$ be the fitted value at $\boldsymbol{x}$ based on observations $S \setminus (y_i, \boldsymbol{x}_i)$, and let $\hat{\mu}_{\setminus i}(\boldsymbol{x}) = h(\hat{\eta}_{\setminus i}(\boldsymbol{x}))$, where the dependence on λ is suppressed in the notation. Then performance on future observations may be measured by the *leaving-one-out predictive deviance:*

$$D_{CV}(\lambda) = \sum_{(y_i, \boldsymbol{x}_i) \in S} d(y_i, \hat{\mu}_{\setminus i}(\boldsymbol{x}_i)),$$

where $d(y_i, \hat{\mu}_{\setminus i}(\boldsymbol{x}_i))$ is the contribution to the deviance at value y_i when $\hat{\mu}_{\setminus i}(\boldsymbol{x}_i)$ is the fitted value. For a normally distributed y one obtains the usual squared error, $d(y_i, \hat{y}_{i\setminus i}) = (y_i - \hat{\mu}_{\setminus i}(\boldsymbol{x}_i))^2$, and for binary distributions one obtains the likelihood value

$$d(y_i, \hat{\pi}_{\setminus i}(\boldsymbol{x}_i)) = -\log(1 - |y_i - \hat{\pi}_{\setminus i}(\boldsymbol{x}_i)|) = \begin{cases} -\log(\hat{\pi}_{\setminus i}(\boldsymbol{x}_i)) & y_i = 1 \\ -\log(1 - \hat{\pi}_{\setminus i}(\boldsymbol{x}_i)) & y_i = 0. \end{cases}$$

For non-normal distributions the deviance is more appropriate than the simple squared error because it measures the discrepancy between the data and the fit by taking the underlying distribution into account.

Simple cross-validation is often replaced by *generalized cross-validation.* With $D(\lambda) = \sum_{(y_i, \boldsymbol{x}_i) \in S}(y_i, \hat{\mu}(\boldsymbol{x}_i))$ denoting the deviance for the smoothing parameter λ, the criterion to be minimized is

$$\text{GCV}(\lambda) = \frac{D(\lambda)/n}{(1 - tr(\boldsymbol{H})/n)^2},$$

where $tr(\boldsymbol{H})$ is the effective degrees of freedom, determined by the hat matrix $\boldsymbol{H}$ (see next section). An information-based criterion is the extension of AIC from Section 4.5, which has the form

$$\text{AIC}(\lambda) = D(\lambda) + 2tr(\boldsymbol{H})\phi,$$

where ϕ is the dispersion parameter, which is equal to 1 for binary data. For more details see Hastie and Tibshirani (1990) and Wood (2006a).

Likelihood-Based Approaches

Likelihood-based approaches to the selection of the smoothing parameter use the connection to mixed models. An excellent introduction to spline-based smoothing with mixed model representations was given by Wand (2003). An extensive treatment is found in the book of Ruppert et al. (2003).

Let us consider the truncated power series as a basis function. Then the predictor may be spilt into components by $\eta(\boldsymbol{x}_i) = \boldsymbol{x}_i^T\boldsymbol{\gamma} + \boldsymbol{z}_i^T\boldsymbol{b}$, where $\boldsymbol{x}_i^T = (1, x_i, \ldots, x_i^k)$, $\boldsymbol{\gamma}^T = (\beta_0, \ldots, \beta_k)$, $\boldsymbol{z}_i^T = ((x_i - \tau_1)_+^k, \ldots, (x_i - \tau_{m_s})_+^k)$, $\boldsymbol{b}^T = (\beta_{k+1}, \ldots, \beta_{k+m_s})$. In matrix form one has $\boldsymbol{\eta} = \boldsymbol{X}\boldsymbol{\gamma} + \boldsymbol{Z}\boldsymbol{b}$, where $\boldsymbol{X}$ has rows $\boldsymbol{x}_i^T$ and $\boldsymbol{Z}$ has rows $\boldsymbol{z}_i^T$. Only the parameters collected in $\boldsymbol{b}$ are penalized by use of the penalty term $(\lambda/2)J = (\lambda/2)\|\boldsymbol{b}\| = (\lambda/2)\boldsymbol{\beta}^T\boldsymbol{K}_T\boldsymbol{\beta}$, where $\boldsymbol{\beta}^T = (\boldsymbol{\gamma}^T, \boldsymbol{b}^T)$ and $\boldsymbol{K}_T$ is a block-diagonal matrix:

$$\boldsymbol{K}_T = \begin{pmatrix} \boldsymbol{0} & \boldsymbol{0} \\ \boldsymbol{0} & \boldsymbol{I}_{k \times k} \end{pmatrix}. \tag{10.11}$$

The penalized log-likelihood $l_p(\boldsymbol{\beta}) = l(\boldsymbol{\beta}) - (\lambda/2)\boldsymbol{\beta}^T\boldsymbol{K}_T\boldsymbol{\beta}$ has the same form as the (approximative) log-likelihood obtained in Chapter 14 for the mixed model $l_p = \log(f(\boldsymbol{y}|\boldsymbol{b}, \boldsymbol{\gamma})) - (1/2)\boldsymbol{b}^T\boldsymbol{Q}_b^{-1}\boldsymbol{b}$ (see equation (14.31) in Chapter 14).

For the mixed model considered in Chapter 14, the parameters in the decomposition $\boldsymbol{\eta} = \boldsymbol{X}\boldsymbol{\gamma} + \boldsymbol{Z}\boldsymbol{b}$ are considered as fixed effects and random effects, respectively. While $\boldsymbol{\gamma}$ is a fixed parameter, $\boldsymbol{b}$ is a random effect, for which normal distribution, $\boldsymbol{b} \sim N(\boldsymbol{0}, \boldsymbol{Q}_b)$, is assumed. Therefore, maximization of the penalized likelihood corresponds to fitting a mixed model with covariance matrix $\text{cov}(\boldsymbol{b}) = \boldsymbol{I}_{k\times k}/\lambda$. Thus the smoothing parameter λ corresponds to the inverse variance $1/\sigma_b^2$ of iid random effects $\beta_{k+i}, i = 1, \ldots, m_s$. Large variances correspond to small values of λ and therefore weak smoothing, whereas small variances correspond to strong smoothing. The mixed model approach may be used to select the smoothing parameter by computing restricted ML (REML) estimates estimates of σ_b^2 to obtain $\hat{\lambda}_{\text{REML}} = 1/\hat{\sigma}_{b,\text{REML}}^2$ (see

Chapter 14) if $\phi = 1$. When ϕ is not given it has also to be estimated, and one obtains $\hat{\lambda}_{\text{REML}} = \hat{\phi}/\hat{\sigma}^2_{b,\text{REML}}$. In the case of GLMs, the mixed model approach to the selection of smoothing parameters relies on the assumption that the approximation to the log-likelihood, which uses the Laplace approximation, holds. Although the approximation can be bad for typical mixed models with few observations in one cluster, for the selection of smoothing parameters the scenario is different. Clusters refer to spline functions, and with increasing sample size the number of observations for each spline function (cluster) increases. In that scenario the Laplace approximation works (for details see Kauermann et al., 2009). A method that allows for locally adaptive smoothing was proposed by Krivobokova, Crainiceanu, and Kauermann (2008).

10.2 Non-Parametric Regression with Multiple Covariates

In generalized linear models it is assumed that the mean $\mu(x) = E(y|\boldsymbol{x})$ is determined by $\mu(\boldsymbol{x}) = h(\eta(\boldsymbol{x}))$ or $g(\mu(\boldsymbol{x})) = \eta(\boldsymbol{x})$ with the predictor having the linear form $\eta(\boldsymbol{x}) = \boldsymbol{x}^T\boldsymbol{\beta}$. In a non-parametric regression the link between the mean and the predictor is still determined by the link function g, but the functional form of $\eta(\boldsymbol{x})$ is much more flexible. In the most flexible case $\eta(\boldsymbol{x})$ is estimated non-parametrically without imposing further constraints. The methods for one-dimensional predictors, which were considered in the proceeding sections, can be extended to that case. We will start with the expansion in basis functions and then consider localizing techniques. More structured modeling of predictors will be considered in the following sections.

10.2.1 Expansion in Basis Functions

The linear basis functions approach now assumes that $\eta(\boldsymbol{x})$ can be approximated by

$$\eta(\boldsymbol{x}) = \sum_{j=1}^{m} \beta_j \phi_j(\boldsymbol{x}),$$

where $\phi_j(.)$ are fixed basis functions with p-dimensional argument $\boldsymbol{x}$. Basis functions in higher dimensions may be generated from one-dimensional basis functions. For simplicity, let us consider the two-dimensional case. If $\phi_{1j}, j = 1, \dots, m_1$, is a basis for representing functions of x_1, and $\phi_{2l}, l = 1, \dots, m_2$, is a basis for x_2, the *tensor-product basis* is defined by

$$\phi_{jl}(x_1,\ x_2) = \phi_{1j}(x_1)\phi_{2l}(x_2). \tag{10.12}$$

Figure 10.5 illustrates two-dimensional basis functions that are built as products of cubic B-splines. The design matrix for the tensor-product basis easily derives from the matrices for the marginal smooths. Let $\boldsymbol{\phi}_1 = (\phi_{1j}(x_{i1}))_{i,j}$ denote the $n \times m_1$ design matrix for the first variable and $\boldsymbol{\phi}_2 = (\phi_{2l}(x_{i2}))_{i,l}$ denote the $n \times m_2$ design matrix for the second variable. Then, by ordering the β_{jl}'s in $\boldsymbol{\beta}^T = (\boldsymbol{\beta}^T_{1.}, \dots, \boldsymbol{\beta}^T_{m_1.})$, $\boldsymbol{\beta}^T_{j.} = (\beta_{j1}, \dots, \beta_{jm_2})$, one obtains the design matrix $\boldsymbol{\phi} = \boldsymbol{\phi}_1 \otimes \boldsymbol{\phi}_2$, where $\otimes$ is the usual Kronecker product and $\eta(x_i)$ corresponds to the ith entry in $\boldsymbol{\phi\beta}$. The function $\eta(\boldsymbol{x})$ is then represented by

$$\eta(\boldsymbol{x}) = \sum_{j,l} \beta_{jl}\phi_{jl}(\boldsymbol{x}).$$

Alternatively, basis functions may be derived by selecting a set of knots $\boldsymbol{\tau}_1, \dots, \boldsymbol{\tau}_m \in \mathbb{R}^2$ and defining

$$\phi_j(\boldsymbol{x}) = \phi(\|\boldsymbol{x} - \boldsymbol{\tau}_j\|/\gamma),$$

where ϕ is a one-dimensional basis function, for example, a radial basis function, and γ determines the spread of the function.

Since the representation is linear in known basis functions, it is straightforward to apply maximum likelihood techniques for obtaining estimates of coefficients. It is somewhat harder to formulate appropriate penalty terms, however. Before specifying the form of the penalty let us consider a general functional penalty approach.

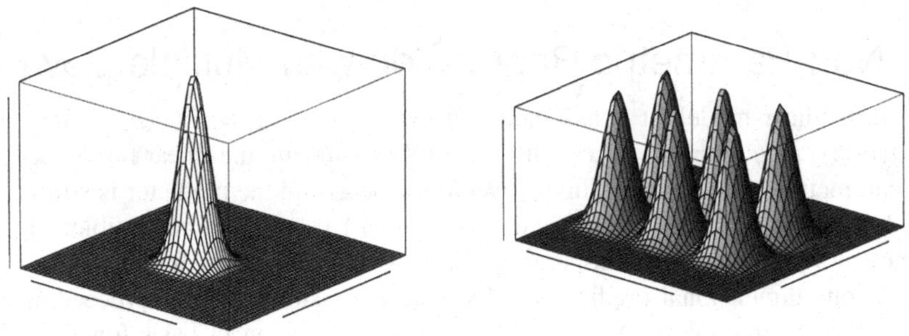

FIGURE 10.5: Two-dimensional B-splines of degree 3. One single basis function (left), functions built from two B-splines in the first dimension, and three B-splines in the second dimension (right).

10.2.2 Smoothing Splines

One-dimensional smoothing splines seek to find a regularized function that in some sense is close to the data. The concept generalizes to higher dimensions. For data $(y_i, \boldsymbol{x}_i)$, $\boldsymbol{x}_i \in \mathbb{R}^2$, and $\eta_i = \eta(\boldsymbol{x}_i)$ one considers the criterion

$$\sum_{i=1}^{n}(y_i - \eta_i)^2 + \lambda J(\eta), \tag{10.13}$$

where

$$J(\eta) = \int\int (\frac{\partial^2 \eta}{\partial x_1^2})^2 + 2(\frac{\partial^2 \eta}{\partial x_1 \partial x_2})^2 + (\frac{\partial^2 \eta}{\partial x_2^2})^2 dx_1 dx_2$$

is a penalty functional for stabilizing a function η in $\mathbb{R}^2$. The smooth two-dimensional surface resulting from minimizing (10.13) is known as a thin-plate spline. The penalty $J(\eta)$ is a roughness penalty that penalizes the wiggliness of the function by using squared derivatives. The general form for p dimensions is found, for example, in Duchon (1977).

The effect of λ is to find a compromise between faith to the data and smoothness. For $\lambda \rightarrow \infty$ the solution approaches a plane, and for $\lambda \rightarrow \infty$ one obtains an interpolating function of the data. It is remarkable that while the criterion is defined over an infinitesimal space, the solution has a finite-dimensional representation:

$$\eta(\boldsymbol{x}) = \beta_0 + \beta_1 x_1 + \beta_2 x_2 + \sum_{j=1}^{n} \tilde{\beta}_j \phi_j(\boldsymbol{x}),$$

where $\phi_j(\boldsymbol{x}) = \phi(\|\boldsymbol{x}-\boldsymbol{x}_j\|), \phi(z) = z^2\log(z^2)$ are specific radial basis functions. The vectors of coefficients, $\boldsymbol{\beta}$ and $\tilde{\boldsymbol{\beta}}$, have to be estimated, subject to the linear constraints that $\boldsymbol{T}^T\tilde{\boldsymbol{\beta}} = \boldsymbol{0}$, where $T_{ij} = x_i^{j-1}$ corresponds to the design of the polynomial term. The polynomial term represents the space of functions for which the penalty is zero; penalized estimates are shrunk toward the corresponding plane. The criterion to be minimized becomes

$$\|\boldsymbol{y} - \boldsymbol{P}\tilde{\boldsymbol{\beta}} - \boldsymbol{T}\boldsymbol{\beta}\|^2 + \lambda\tilde{\boldsymbol{\beta}}\boldsymbol{P}\tilde{\boldsymbol{\beta}}$$

subject to $\boldsymbol{T}^T\tilde{\boldsymbol{\beta}} = \boldsymbol{0}$, where the penalty matrix is $\boldsymbol{P} = (p_{ij}),\ p_{ij} = \phi_j(\boldsymbol{x}_i)$.

The problem with thin splines is the high computational burden. Therefore, one truncates the space of basis functions ϕ_j while leaving the polynomial term. The properties of the resulting thin plate regression splines are nicely summarized in Wood (2006a), pp. 157ff. For more details on thin plate splines see Wahba (1990) and Green and Silverman (1994). A general treatment of regularization methods connected to reproducing Hilbert spaces is found in Wahba (1990), and a brief introduction was given by Hastie et al. (2001).

10.2.3 Penalized Regression Splines

The finite-dimensional representation of smoothing splines has as many parameters as observations. Therefore, the use of regression splines together with appropriate smoothing offers a low-dimensional alternative with a reduced computational burden. The choices of the basis functions and the basis dimension may be seen as part of the model specification. In particular, the basis dimension should be chosen not too small in order to provide flexibility of the fit. The number of basis functions itself is not so crucial because the smoothing parameters constrain the actual degrees of freedom. A knot-based approximation to smoothing splines in two dimensions is obtained by specifying knots $\boldsymbol{\tau}_1, \dots, \boldsymbol{\tau}_m \in \mathbb{R}^2$ and using the approximation

$$\eta(\boldsymbol{x}) = \beta_0 + \beta_1 x_1 + \beta_2 x_2 + \sum_{j=1}^{m} \tilde{\beta}_j \phi(\|\boldsymbol{x} - \boldsymbol{\tau}_j\|),$$

where the ϕ's are the radial basis functions, $\phi(z) = z^2\log(z^2)$. The fitting uses the penalty term $\lambda\tilde{\boldsymbol{\beta}}^T\boldsymbol{P}\tilde{\boldsymbol{\beta}}$, where $\boldsymbol{P} = (\phi(\|\boldsymbol{\tau}_i - \boldsymbol{\tau}_j\|))_{ij}$. The radial basis functions approach treats smoothness in all directions equally. The resulting fit is invariant to rotation. What comes as an advantage in bivariate geographical applications, where the variables correspond to longitude and latitude, might be problematic if the two variables have quite different scalings. An alternative is to use tensor-product basis functions ϕ, which are given by (10.12) for two dimensions. The corresponding predictor is

$$\eta(\boldsymbol{x}) = \sum_{j,l} \beta_{jl}\phi_{1j}(x_1)\phi_{2l}(x_2).$$

If the basis functions ϕ_{1j} and ϕ_{2l} are linked to equally spaced knots (m_1 for x_1 and m_2 for x_2), one can use the difference penalty in two dimensions:

$$\lambda_1\boldsymbol{\beta}\boldsymbol{P}_1\boldsymbol{\beta} + \lambda_2\boldsymbol{\beta}\boldsymbol{P}_2\boldsymbol{\beta} = \lambda_1\sum_{l=1}^{m_2}\sum_{j=2}^{m_1}(\beta_{jl} - \beta_{j-1,l})^2 + \lambda_2\sum_{j=1}^{m_1}\sum_{l=2}^{m_2}(\beta_{jl} - \beta_{j,l-1})^2.$$

The penalty controls the variation of parameters β_{jl} in both directions; the first term controls the smoothness in the x_1-direction and the second in the x_2-direction. More generally, one might use differences of order d. Let Δ_1 denote the age difference operator for the first index. Then the first-order difference $\Delta_1\beta_{jl} = (\beta_{jl} - \beta_{j-1,l})$ generalizes to $\Delta_1^2\beta_{j2} = \Delta_1(\Delta_1\beta_{jl})$ and

$\Delta_1^d \beta_{jl} = \Delta_1^{d-1}\Delta_1\beta_{jl}$. By defining the corresponding difference matrices one obtains a penalty of the form

$$\lambda\boldsymbol{\beta}^T\boldsymbol{P}_1\boldsymbol{\beta} + \lambda_2\boldsymbol{\beta}^T\boldsymbol{P}_2\boldsymbol{\beta} = \lambda_1\boldsymbol{\beta}^T(\boldsymbol{D}^T\boldsymbol{D}\otimes\boldsymbol{I}_{m_2})\boldsymbol{\beta},$$

where $\boldsymbol{\beta} = (\boldsymbol{\beta}_{1.}^T,\dots,\boldsymbol{\beta}_{m_1.}^T)$, $\boldsymbol{\beta}_{j.}^T = (\beta_{j1},\dots,\beta_{jm_2})$, and $\boldsymbol{D}$ represents the difference matrices. Penalties of this type with product B-splines were used, for example, by Currie et al. (2004) when smoothing two-dimensional mortality ranks; see also Eilers and Marx (2003) for an application to signal regression.

10.2.4 Local Estimation in Two Dimensions

Localizing techniques generalize easily to two and higher dimensions. In the two-dimensional case let $\boldsymbol{x}^T = (x_1, x_2)$ denote the target value and $\boldsymbol{x}_i^T = (x_{i1}, x_{i2})$ denote the ith observation. One can fit locally the linear model

$$\eta_i = \eta(\boldsymbol{x}_i) = \tilde{\boldsymbol{x}}_i^T\boldsymbol{\beta},$$

where $\tilde{\boldsymbol{x}}_i^T = (1, (x_{i1}-x_1), (x_{i2}-x_2), (x_{i1}-x_1)(x_{i2}-x_2)$ and $\boldsymbol{\beta}^T = (\beta_0, \beta_1, \beta_2, \beta_{12})$. When fitting the model by maximizing the local likelihood one uses vector-valued weights determined by kernels. The candidates are product kernels,

$$w_\lambda(\boldsymbol{x}, \boldsymbol{x}_i) \propto \prod_{j=1}^{p} K(\frac{x_{ij}-x_j}{\lambda_j}),$$

and distance-based kernels,

$$w_\lambda(\boldsymbol{x}, \boldsymbol{x}_i) \propto K(\frac{\|x_{ij}-x_j\|}{\lambda}).$$

Estimates may be obtained by local Fisher scoring as given in (10.9).

Although smoothing that is free of any structured assumptions may be extended to more than two dimensions, the computational burden increases, and, more severe, it is hard to visualize the resulting fit. Bivariate smoothing yields a surface estimate that is easily visualized and corresponds to the important case of a two-variable interaction, that is, more specifically, the interaction between two continuous variables, effecting on a response variable.

Example 10.3: Exposure to Dust (Smokers)

In Example 4.6 the effects of dust on bronchitis were investigated. The observed covariates were mean dust concentration at working place in mg/m^3 (dust), duration of exposure in years (years), and smoking (1: yes; 0: no). The binary response refers to the presence of bronchitis (1: present; 0: not present). Figure 10.6 shows the fit of a logit model including the effects of dust concentration and years without further structuring for smokers. The method used is penalized splines by using default values of the R package *mgcv*. It is seen that both covariates seem to effect on the probability of bronchitis. When using the more structured generalized additive model that assumes the linear predictor

$$\eta = s(\text{dust}) + s(\text{years})$$

(see Section 10.3.1), the effects of the two variables are separated. Figure 10.7 shows the fitted smooth effects. □

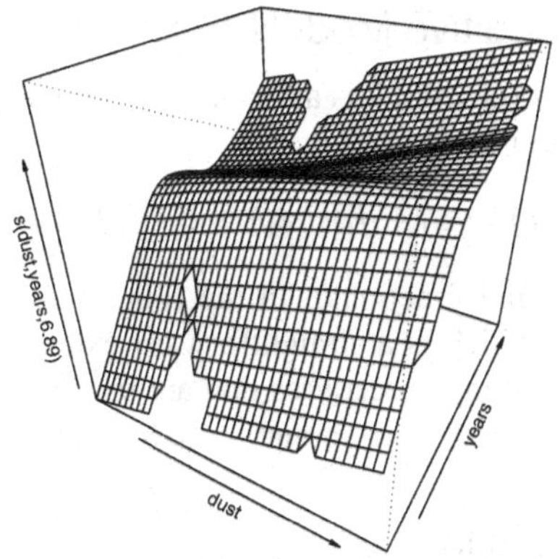

FIGURE 10.6: Effect of concentration of dust and years of employment on probability of bronchitis (smokers).

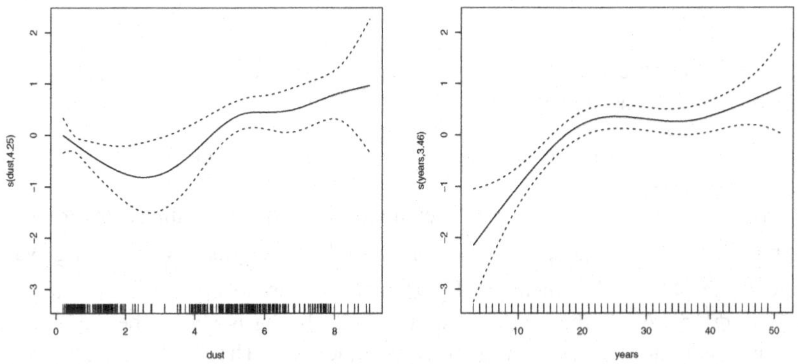

FIGURE 10.7: Generalized additive model including the effects of concentration of dust and years of employment on probability of bronchitis (smokers).

10.3 Structured Additive Regression

With a moderate to large number of explanatory variables it is often useful to assume some weak structure in the predictor. In a structured additive regression (STAR) one keeps the additive form but lets the components be determined in a much more flexible way than in a linear regression. Therefore one assumes that the predictor has the form

$$\eta(\boldsymbol{x}) = f_{(1)}(v_1) + \dots f_{(m)}(v_m),$$

with unspecified unknown functions $f_{(1)}(.), \dots, f_{(m)}(.)$ and the v_j denoting generic covariates of different types and dimensions. In the simplest form v_j represents one covariate x_j, yielding generalized additive models. In the following, we will start with that model and consider the general model at the end of the section.

10.3.1 Generalized Additive Models

In generalized additive models (GAMs) it is assumed that the underlying predictor $\eta(\boldsymbol{x})$ can be represented in the additive form

$$\eta(\boldsymbol{x}) = \beta_0 + f_{(1)}(x_1) + \cdots + f_{(p)}(x_p), \tag{10.14}$$

where the $f_{(j)}(.)$ are unspecified functions. In the following we consider fitting GAMs by penalized regression splines, which is a straightforward extension of the penalized fitting of one-dimensional models, and fitting procedures that are based on backfitting. Various alternative procedures have been proposed; for references see Section 10.5.

Penalized Regression Splines

One way to estimate a GAM is to set it up as a penalized generalized linear model. To this end one specifies a basis for each smooth function and defines what is meant by the smoothness of a function by specifying a corresponding penalty. Let the functions be represented by

$$f_{(j)}(x_j) = \sum_{s=1}^{m_j} \beta_{js}\phi_{js}(x_j),$$

where $\phi_{j1},\dots\phi_{jm_j}$ denote the basis functions for the jth variable. For data $(y_i, \boldsymbol{x}_i), i = 1,\dots,n$, one obtains the linear predictor

$$\eta_i = \eta(\boldsymbol{x}_i) = \beta_0 + \boldsymbol{\phi}_{i1}^T\boldsymbol{\beta}_1 + \cdots + \boldsymbol{\phi}_{ip}^T\boldsymbol{\beta}_p = \boldsymbol{\phi}_i^T\boldsymbol{\beta},$$

where $\boldsymbol{\phi}_{ij}^T = (\phi_{j1}(x_{ij})\dots\phi_{jm_i}(x_{ij}))$ are the evaluations of the basis functions for the jth variable at x_{ij} and $\boldsymbol{\beta}_j^T = (\beta_{j1},\dots\beta_{jm_j})$ are the corresponding weights. By collecting evaluations and parameters one obtains the linear predictor $\boldsymbol{\phi}_i^T\boldsymbol{\beta}$, which specifies the generalized linear model $g(\mu_i) = \boldsymbol{\phi}_i^T\boldsymbol{\beta}$. As in the unidimensional case, the basis should be sufficiently large to be able to approximate the unknown underlying functions. The resulting high-dimensional fitting problem typically needs penalization. The penalty term specifies what is meant by the smoothness of the underlying function and has to be linked to the chosen basis. For example, if one chooses B-splines with equally spaced knots, an appropriate penalty is given by penalizing the squared differences between the adjacent basis functions:

$$J(\boldsymbol{\beta}) = \sum_{j=1}^{p} \lambda_j \sum_{s} (\beta_{j,s+1} - \beta_{j,s})^2,$$

which has the general form $J(\boldsymbol{\beta}) = \sum_j \lambda_j \boldsymbol{\beta}_j^T \boldsymbol{P}_j \boldsymbol{\beta}_j$. If $\boldsymbol{P}_j = \boldsymbol{K}_d$ from (10.7), one penalizes the differences of dth order. It is most convenient to work with penalties that are given as a quadratic form in the coefficients, since penalized likelihood estimation becomes rather simple. Maximization of the penalized likelihood

$$l_p(\boldsymbol{\beta}) = l(\boldsymbol{\beta}) - \frac{1}{2}\sum_{j=1}^{p} \lambda_j \boldsymbol{\beta}_j^T \boldsymbol{P}_j \boldsymbol{\beta}_j$$

uses the penalized score function $\boldsymbol{s}_p(\boldsymbol{\beta}) = \boldsymbol{s}(\boldsymbol{\beta}) - \sum_{j=1}^{p} \lambda_j \boldsymbol{P}_j \boldsymbol{\beta}_j$ and estimates may be obtained by iterative Fisher scoring by using the penalized Fisher matrix $\boldsymbol{F}_p(\boldsymbol{\beta}) = \boldsymbol{F}(\boldsymbol{\beta}) + \sum_j \lambda_j \boldsymbol{P}_j$. In matrix notation, the linear predictor $\boldsymbol{\eta}^T = (\eta_1,\dots,\eta_n)$ has the form

$$\boldsymbol{\eta} = \beta_0 \mathbf{1} + \boldsymbol{\phi}_{(1)}\boldsymbol{\beta}_1 + \cdots + \boldsymbol{\phi}_{(p)}\boldsymbol{\beta}_p = \boldsymbol{\phi}\boldsymbol{\beta}$$

and the penalty matrix is $\tilde{\boldsymbol{K}} = BlockDiag(\lambda_1 \boldsymbol{P}_1, \ldots, \lambda_p \boldsymbol{P}_p)$. Then approximate covariances are obtained by the sandwich matrix:

$$\text{cov}(\hat{\boldsymbol{\beta}}) \approx (\boldsymbol{F}(\hat{\boldsymbol{\beta}}) + \tilde{\boldsymbol{K}})^{-1} \boldsymbol{F}(\hat{\boldsymbol{\beta}})(\boldsymbol{F}(\hat{\boldsymbol{\beta}}) + \tilde{\boldsymbol{K}})^{-1}. \tag{10.15}$$

(comare Section 10.1.3).

In the preceding presentation it has been ignored that the functions in (10.14) are not identifiable because the shifting of functions can be compensated by the shifting of other functions or the intercept. For example, the linear predictor is unchanged if one uses the intercept $\beta_0 + c$ and the function $f_{(1)}(x_1) - c$ for x_1. To obtain an identifiable model one has to center the functions. A suitable constraint is that the evaluations at observations sum up to zero:

$$\sum_{i=1}^{n} f_{(j)}(x_{ij}) = 0, \; j = 1, \ldots, p,$$

or, equivalently, $\boldsymbol{1}^T \boldsymbol{\phi} \boldsymbol{\beta} = \boldsymbol{0}$. Since this is a linear constraint on the parameters, it is easily handled by reparameterization (see Appendix C). One can find a matrix $\boldsymbol{Z}$ that satisfies $\boldsymbol{1}^T \boldsymbol{\phi} \boldsymbol{Z} = \boldsymbol{0}$ and $\boldsymbol{\beta} = \boldsymbol{Z}\boldsymbol{\beta}_u$, where the $\boldsymbol{\beta}_u$'s are unconstrained parameters. The corresponding penalty term takes the form $\boldsymbol{\beta}^T \boldsymbol{Z}\boldsymbol{P}\boldsymbol{Z}\boldsymbol{\beta}$, where $\boldsymbol{P}$ is the block-diagonal matrix with entries $\boldsymbol{P}_1, \ldots \boldsymbol{P}_p$.

An advantage of the penalized regression splines approach is that it also works in more general additive models. In the *partially linear model*,

$$\eta(\boldsymbol{x}, \boldsymbol{\gamma}) = \beta_0 + f_{(1)}(x_1) + \cdots + f_{(p)}(x_p) + \boldsymbol{z}^T \boldsymbol{\gamma}, \tag{10.16}$$

in addition to the additive term, a linear term $\boldsymbol{z}^T \boldsymbol{\gamma}$ is included that specifies the linear effect of additional covariates $\boldsymbol{z}$. In particular, when some of the covariates are categorical, the simple additive model has to be extended, since the effect of a binary covariate is determined by one parameter, not by a function (when no interaction effects are assumed). Therefore, within an additive approach, categorical variables are included by dummy variables in the form of a linear term. For the partially linear model the linear predictor has the form

$$\eta_i = \eta(\boldsymbol{x}_i, \boldsymbol{\gamma}) = \boldsymbol{\phi}_i^T \boldsymbol{\beta} + \boldsymbol{z}_i^T \boldsymbol{\gamma}.$$

The penalized log-likelihood is the same as in the additive model, but the parameter to be estimated is $(\boldsymbol{\beta}, \boldsymbol{\gamma})$. In particular, the penalty term is the same since only the weights $\boldsymbol{\beta}$ have to be penalized.

It should be noted that for GAMs the choice of the link function might be relevant. When one fits the model $\mu(x) = h(\eta(x))$ with a one-dimensional predictor, there is not much to choose when $\eta(x)$ is unspecified. But the link function helps to confine the mean $\mu(x)$ within the admissible but otherwise determines only the scaling of the predictor. This is different for an additively structured predictor. When additivity holds for a specific link function, for example, the logit transformation, it does not hold for other link functions. Therefore, it might be useful to compare the fits of GAMs with alternative link functions.

Selection of Smoothing Parameters

When fitting a generalized additive model one has to select the smoothing parameters $\lambda_1, \ldots, \lambda_p$, where λ_j determines the amount of smoothing for the jth component. Therefore, one has to select smoothing parameters from a p-dimensional space. Although cross-validation is straightforward, it is limited to the case of small p, since the selection of smoothing parameters on a grid implies heavy computational effort. An alternative is to rely on likelihood-based approaches that make use of the mixed model methodology. When using truncated power series, the penalized likelihood becomes $l_p(\boldsymbol{\beta}) = l(\boldsymbol{\beta}) - \sum_j (\lambda_j/2)\boldsymbol{\beta}_j^T \boldsymbol{K}_T \boldsymbol{\beta}_j$ with $\boldsymbol{K}_T$ from (10.11). In the same way as for univariate smoothing problems, the form of the penalized likelihood is the same as the (approximative) likelihood obtained for the mixed model $l_p = log f(\boldsymbol{y}|\boldsymbol{b}, \boldsymbol{\gamma}) - (1/2)\boldsymbol{b}^T \boldsymbol{Q}_b^{-1} \boldsymbol{b}$, but now $\boldsymbol{Q}$ is a block-diagonal matrix with the blocks corresponding to the components to be smoothed. REML estimates may be used to obtain the smoothing parameters.

Distributional Results

A general reliable theory for the smooth components of a GAM does not seem to be available, although some results were given by Hastie and Tibshirani (1990), and Wood (2006b, 2006a). In the following we sketch an approximate test procedure following Wood (2006a). Let $\boldsymbol{\beta}_j$ denote the coefficients for a single smooth term. Then, if $\boldsymbol{\beta}_j = \boldsymbol{0}$, the expectation $\mathrm{E}(\hat{\boldsymbol{\beta}}_j)$ will be approximately zero. As an approximate covariance matrix of the estimate one uses the corresponding submatrix of (10.15), denoted by $\boldsymbol{V}_{\boldsymbol{\beta}_j}$. If $\boldsymbol{V}_{\boldsymbol{\beta}_j}$ is of full rank, p-values for testing the null hypothesis $\boldsymbol{\beta}_j = \boldsymbol{0}$ are obtained by using the fact that $\boldsymbol{\beta}_j^T \boldsymbol{V}_{\boldsymbol{\beta}_j}^{-1} \boldsymbol{\beta}_j$ is approximately $\chi^2(d)$-distributed, where d is the dimension of $\boldsymbol{\beta}_j$. If $\boldsymbol{V}_{\boldsymbol{\beta}_j}$ has rank $r < d$, one uses the test statistic $\boldsymbol{\beta}_j^T \boldsymbol{V}_{\boldsymbol{\beta}_j}^{-} \boldsymbol{\beta}_j$, where $\boldsymbol{V}_{\boldsymbol{\beta}_j}^{-}$ is the rank r pseudo-inverse of $\boldsymbol{V}_{\boldsymbol{\beta}_j}$. Testing relies on the approximate $\chi^2(r)$-distribution. If an unknown scale parameter ϕ is involved, one can use an approximate F-distribution $F(r, edf)$ for the test statistic $(\boldsymbol{\beta}_j^T \boldsymbol{V}_{\boldsymbol{\beta}_j}^{-} \boldsymbol{\beta}_j / r)/(\hat{\phi}/(n - edf))$, where edf is the estimated degrees of freedom of the model.

The resulting p-values rely on several approximations. In particular, the estimation of the smoothing parameters is not sufficiently accounted for. Therefore, when smoothing parameters are estimated they tend to be smaller than they should be, and thus smooth terms are found relevant that are not (for more details see Wood, 2006a). Alternative testing procedures can be based on the mixed model approach to the fitting of GAMs; see, for example, Ruppert et al. (2003) or the Bayesian methods outlined in Wood (2006a).

Simultaneous Selection of Variables and Amount of Smoothing

An alternative approach to the fitting of generalized additive models that implicitly selects the relevant variables and the amount of smoothing utilizes boosting techniques. Boosting concepts were already considered in Section 6.3 and are treated extensively in Section 15.5.3; here we only briefly consider the extension of the variable selection method from Section 6.3. The main idea of componentwise boosting is greedy forward stagewise additive modeling. For the fitting of additive models, this means that, in an iterative process within one stage, just one component of the additive term is selected and refitted without adjusting the other components. Thus, within one step the model

$$\mu_i = h(\hat{\eta}_{(l)}(\boldsymbol{x}_i) + \boldsymbol{x}_{ij}^T \boldsymbol{\gamma})$$

is fitted, where $\hat{\eta}_{(l)}(\boldsymbol{x}_i)$ is the fit from the previous step (treated as an offset) and $\boldsymbol{x}_{ij}$ is the vector of basis functions for the jth component. Therefore, only the jth component is refitted. When using one-step Fisher scoring, the fit has the form

$$\hat{\boldsymbol{\eta}}^{j,new} = \boldsymbol{\phi}_j(\boldsymbol{\phi}_j^T \hat{\boldsymbol{W}} \boldsymbol{\phi}_j + \lambda \boldsymbol{K})^{-1} \boldsymbol{\phi}_j^T \hat{\boldsymbol{W}} \hat{\boldsymbol{D}}^{-1}(\boldsymbol{y} - \hat{\boldsymbol{\mu}}),$$

where $\boldsymbol{\phi}_j$ is the design matrix of the jth component and $\hat{\boldsymbol{W}}, \hat{\boldsymbol{D}}, \hat{\boldsymbol{\mu}}$ are evaluated at the values from the previous step, $\hat{\boldsymbol{\eta}}^T = (\hat{\eta}_{(l)}(\boldsymbol{x}_1), \ldots, \hat{\eta}_{(l)}(\boldsymbol{x}_n))$. It should be noted that the Fisher step starts with $\gamma = 0$, since the previous fits are contained in the off-set. Boosting becomes efficient by using so-called weak learners, which means that in each step one tries to improve the fit only slightly by using large λ. Therefore, λ is not an optimally chosen smoothing parameter but has to be chosen simply large. The different amounts of smoothing for the components are obtained be updating only that component that gains the most within one step of the algorithm; complex functions are updated more often than simple functions. In addition, the selection of components to update implies variable selection for the whole procedure. The actual tuning of the fitting procedure lies in the stopping criterion. When the iterative procedure is stopped based on cross-validation or an information-based criterion only variables are included that have been selected previously.

In vectorized form, the fitting of a generalized model means determining of the additive linear predictor $\hat{\boldsymbol{\eta}} = \boldsymbol{f}_1 + \cdots + \boldsymbol{f}_p$, where $\hat{\boldsymbol{\eta}}^T = (\hat{\eta}_1, \ldots, \hat{\eta}_n)$ is the fitted predictor and $\boldsymbol{f}_j^T = (f_{1j}, \ldots, f_{nj}), f_{ij} = f_{(j)}(x_{ij})$ is the fitted vector of the jth variable at observation $x_{1j}, \ldots, x_{nj}$. With $Dev(\hat{\boldsymbol{\eta}})$ denoting the deviance when $\hat{\boldsymbol{\eta}}$ is the fitted predictor, the fitting step including variable selection may given in the following form (for details see Tutz and Binder, 2006).

GAMBoost with Penalized Regression Splines

For $l = 0, 1, \ldots$:

1. *Estimation step:* For $s = 1, \ldots, p$ compute

$$\hat{\boldsymbol{f}}_{s,new} = \boldsymbol{\phi}_j(\boldsymbol{\phi}_j^T \hat{\boldsymbol{W}} \boldsymbol{\phi} + \lambda \boldsymbol{K})^{-1} \boldsymbol{\phi}_j^T \hat{\boldsymbol{W}} \hat{\boldsymbol{D}}^{-1} (\boldsymbol{y} - \hat{\boldsymbol{\mu}}), \tag{10.17}$$

where $\hat{\boldsymbol{W}}, \hat{\boldsymbol{D}}, \hat{\boldsymbol{\mu}}$ are evaluated at $\hat{\boldsymbol{\eta}}^T = (\hat{\eta}_{(l)}(x_1), \ldots, \hat{\eta}_{(l)}(x_n))$.

2. *Selection step:* Set $\boldsymbol{f}_s = \boldsymbol{f}_s^{(l)} + \boldsymbol{f}_{s,new}$ yielding $\hat{\boldsymbol{\eta}}_{s,new}$.
Compute $j = \arg\max_s \{Dev(\hat{\boldsymbol{\eta}}_{(l)})\} - Dev(\hat{\boldsymbol{\eta}}_{s,new}))$.

3. *Update:* Set $\boldsymbol{f}_j^{(l+1)} = \boldsymbol{f}_j^{(l)} + \boldsymbol{f}_{j,new}$.

Variable selection for GAMs by removing a smooth term from the predictor can also be based on the use of properly chosen penalty terms. The smoothing penalty in the penalized likelihood has the general form $\lambda_j \boldsymbol{\beta}_j^T \boldsymbol{P}_j \boldsymbol{\beta}_j$, with $\boldsymbol{P}_j$ denoting the smoothing matrix for the jth variable. The penalty generates a smooth function $f_j(.)$ but does not select variables. Even in the limiting case $\lambda_j \to \infty$, the function component in the null space of the penalty is not penalized and therefore not selected. For example, if one uses natural cubic splines, the linear part of the function is not penalized and will remain in the predictor. Therefore, an appropriate penalty has to separate the function component in the null space, which is not penalized from the shrinking component of the function. One can consider the eigenvalue decomposition $\boldsymbol{P}_j = \boldsymbol{U}_j \boldsymbol{\Lambda}_j \boldsymbol{U}_j^T$ with eigenvector matrix $\boldsymbol{U}_j$ and diagonal eigenvalue matrix $\boldsymbol{\Lambda}_j$. Since part of the function is not penalized, $\boldsymbol{\Lambda}_j$ contains zero eigenvalues. Marra and Wood (2011) included an extra term into the penalty and obtained $\lambda_j \boldsymbol{\beta}_j^T \boldsymbol{P}_j \boldsymbol{\beta}_j + \tilde{\lambda}_j \boldsymbol{\beta}_j^T \tilde{\boldsymbol{P}}_j \boldsymbol{\beta}_j$, where $\tilde{\boldsymbol{P}}_j = \tilde{\boldsymbol{U}}_j \tilde{\boldsymbol{U}}_j^T$, with $\tilde{\boldsymbol{U}}_j$ denoting the matrix of eigenvectors corresponding to zero eigenvalues. The smoothing parameter $\tilde{\lambda}_j$ can shrink the function components in the null space of the penalty to zero. Marra and Wood also gave an alternative version that avoids the double penalty. Alternatively, Avalos et al. (2007) included a lasso term for the component in the null space. An approach that modifies the backfitting procedure considered in the next section to include variable selection was given by Belitz and Lang (2008).

Univariate Smoothers and the Backfitting Algorithm

When Hastie and Tibshirani (1990) propagated additive modeling, the backfitting algorithm was the most widely used algorithm for the fitting of additive models. The basic idea behind the algorithm is to fit one component at a time based on the present partial residuals. Therefore, only univariate smoothers are needed and all kinds of smoothers can be used.

For a description of the algorithm let us start with the additive model (where the link function is the identity function) and then consider the more general case. For data $(y_i, \boldsymbol{x}_i), i = 1, \dots, n$, with $\boldsymbol{x}_i^T = (x_{i1}, \dots, x_{ip})$, the additive model has the form

$$\mu_i = \eta_i = \beta_0 + f_{(1)}(x_{i1}) + \cdots + f_{(p)}(x_{ip}).$$

In matrix notation one obtains

$$\boldsymbol{y} = \beta_0 \boldsymbol{1} + \boldsymbol{f}_1 + \cdots + \boldsymbol{f}_p + \boldsymbol{\varepsilon},$$

where $\boldsymbol{y}^T = (y_1, \dots, y_n)$, $\boldsymbol{f}_j^T = (\boldsymbol{f}_{(j)}(x_{1j}), \dots, \boldsymbol{f}_{(j)}(x_{nj}))$ is the vector of evaluations, and $\boldsymbol{\varepsilon}^T = (\varepsilon_1, \dots, \varepsilon_n)$ is a noise vector. By omitting the noise vector one has approximatively

$$\boldsymbol{f}_j \approx \boldsymbol{y} - \sum_{i \neq j} \boldsymbol{f}_i.$$

Therefore, when $\boldsymbol{f}_i, i \neq j$, have been estimated, $\boldsymbol{y} - \sum_{i \neq j} \hat{\boldsymbol{f}}_i$ may be seen as a vector of partial residuals. Now one may use a univariate smoother treating $\boldsymbol{y} - \sum_{i \neq j} \hat{\boldsymbol{f}}_i$ as the vector of responses that depends smoothly on the jth variable. For a simple linear smoother with smoother matrix $\boldsymbol{S}_j$ one obtains in the sth cycle the update

$$\hat{\boldsymbol{f}}_j^{(s)} = \boldsymbol{S}_j \left(\boldsymbol{y} - \sum_{i \neq j} \hat{\boldsymbol{f}}_i^{(s-1)} \right). \tag{10.18}$$

Within one cycle of the iterative procedure all of the estimates of $\boldsymbol{f}_1, \dots, \boldsymbol{f}_p$ are updated. There are two modifications to this basic procedure. The first concerns the residuals within one iteration cycle. When estimating $\boldsymbol{f}_j$, $j > 1$, within cycle s the estimates $\hat{\boldsymbol{f}}_1^{(s)}, \dots, \hat{\boldsymbol{f}}_{j-1}^{(s)}$ are already computed. Therefore, instead of (10.18) one uses the update

$$\boldsymbol{f}_j^{(s)} = \boldsymbol{S}_j \left(y - \sum_{i<j} \boldsymbol{f}_i^{(s)} - \sum_{i>j} \boldsymbol{f}_i^{(s-1)} \right).$$

The second modification concerns the constant term, β_0, which could be incorporated within any of the functions. Therefore, one first fits a vector of constants $\hat{\boldsymbol{f}}_0' = (\bar{y}, \dots, \bar{y})$, where $\bar{y} = \sum_i y_i / n$.

Backfitting Algorithm

(1) Initialize for $s = 0$ by $\boldsymbol{f}_j^{(0)} = (0, \dots, 0)$, $j = 1, \dots, p$.

(2) a. Increase s by one.

b. Compute for $j = 1, \ldots, p$ the update

$$\hat{\boldsymbol{f}}_j^{(s)} = \boldsymbol{S}_j \left(\boldsymbol{y} - \hat{\boldsymbol{f}}_0 - \sum_{i<j} \hat{\boldsymbol{f}}_i^{(s)} - \sum_{i>j} \hat{\boldsymbol{f}}_i^{(s-1)} \right).$$

c. Stop when $\hat{\boldsymbol{f}}_1^{(s)}, \ldots, \hat{\boldsymbol{f}}_p^{(s)}$ do not change when compared to $\hat{\boldsymbol{f}}_1^{(s-1)}, \ldots, \hat{\boldsymbol{f}}_p^{(s-1)}$.

The backfitting or Gauss-Seidel algorithm may be shown to solve the system of equations

$$\begin{pmatrix} \boldsymbol{I} & \boldsymbol{S}_1 & \ldots & \boldsymbol{S}_1 \\ \boldsymbol{S}_2 & \boldsymbol{I} & \ldots & \boldsymbol{S}_2 \\ \vdots & \vdots & & \vdots \\ \boldsymbol{S}_p & \boldsymbol{S}_p & \ldots & \boldsymbol{I} \end{pmatrix} \begin{pmatrix} \boldsymbol{f}_1 \\ \boldsymbol{f}_2 \\ \vdots \\ \boldsymbol{f}_p \end{pmatrix} = \begin{pmatrix} \boldsymbol{S}_1\boldsymbol{y} \\ \boldsymbol{S}_2\boldsymbol{y} \\ \vdots \\ \boldsymbol{S}_p\boldsymbol{y} \end{pmatrix}, \tag{10.19}$$

which has in the jth row

$$\sum_{i \neq j} \boldsymbol{S}_j \boldsymbol{f}_i + \boldsymbol{f}_j = \boldsymbol{S}_j \boldsymbol{y} \quad \text{or} \quad \boldsymbol{f}_j = \boldsymbol{S}_j \left(\boldsymbol{y} - \sum_{i \neq j} \boldsymbol{S}_j \boldsymbol{f}_i \right).$$

When fitting generalized additive models one has to take the link function and the form of the response distribution into account. Let the penalized likelihood that is to be maximized be given in matrix notation as

$$l_p(\boldsymbol{f}_1, \ldots, \boldsymbol{f}_p) = l(\boldsymbol{f}_1, \ldots, \boldsymbol{f}_p) - \frac{1}{2} \sum_j \lambda_j \boldsymbol{f}_j^T \boldsymbol{K}_j \boldsymbol{f}_j, \tag{10.20}$$

where $\boldsymbol{K}_j$ is a penalty matrix for the jth component. Maximization of (10.20) may be based on Fisher scoring, which generates a new estimate $\hat{\boldsymbol{f}}^{(k+1)}$ by computing $\hat{\boldsymbol{f}}^{(k+1)} = \hat{\boldsymbol{f}}^{(k)} + \{E(\partial l_p/\partial \boldsymbol{f})\}^{-1} \partial l_p / \partial \boldsymbol{f}$, $\hat{\boldsymbol{f}}^T = (\boldsymbol{f}_1^T, \ldots, \boldsymbol{f}_p^T)$. In each iteration one has a system of equations of the form (10.19), which may be solved by an inner loop of Fisher scoring steps. A simple form of the algorithm that uses linear smoothers to fit pseudo-observations is given in the following.

Backfitting Algorithm with Fisher Scoring

(1) Initialize by setting $\hat{\boldsymbol{f}}_0^{(0)} = (g(\bar{y}), \ldots, g(\bar{y}))^T$ $\hat{\boldsymbol{f}}_j^{(0)} = (0, \ldots, 0)^T$, $j = 1, \ldots, p$, $s = 0$.

(2) For $s = 0, 1, \ldots$:

(i) Compute the current pseudo-observations

$$z_i = \eta_i^{(s)} + \frac{y_i - h(\eta_i^{(s)})}{d_i^{(s)}},$$

where $d_i^{(s)} = \partial h(\eta_i^{(s)}) / \partial \eta$ with $\boldsymbol{\eta}_i^{(s)} = \hat{\boldsymbol{f}}_0^{(s)} + \sum_{j=1}^p \hat{\boldsymbol{f}}_j^{(s)}$ and the weights $w_i^{(s)} = (d_i^{(s)} / \sigma^{(s)})^2$.

(ii) *Inner backfitting loop for observations* $z_1, \ldots, z_n$:

(a) Initialize $\tilde{\boldsymbol{f}}_j^{(0)} = \hat{\boldsymbol{f}}_j^{(s)}$, $j = 1, \ldots, p$, $\tilde{\boldsymbol{f}}_0^{(0)} = (\bar{z}, \ldots, \bar{z})^T$, $\bar{z} = \sum_i z_i/n$.

(b) For $j = 1, \ldots, p$ compute

$$\tilde{\boldsymbol{f}}_j^{(1)} = \boldsymbol{S}_j \left(z_i - \sum_{i<j} \tilde{\boldsymbol{f}}_i^{(1)} - \sum_{i>j} \tilde{\boldsymbol{f}}_i^{(0)} \right),$$

with the corresponding smoother matrix $\boldsymbol{S}_j$ including the weights.

(iii) Set $\hat{\boldsymbol{f}}_j^{(s+1)} = \tilde{\boldsymbol{f}}_j^{(1)}$, $j = 1, \ldots, p$.

(3) Comparison of $\hat{\boldsymbol{f}}_j^{(s+1)}$ and $\hat{\boldsymbol{f}}_j^{(s)}$, $j = 1, \ldots, p$ determines upon termination.

A careful derivation and discussion of properties of the backfitting algorithm was given by Buja et al. (1989). Opsomer and Ruppert (1997), Opsomer (2000), and Kauermann and Opsomer (2004) derived further results and investigated asymptotic properties.

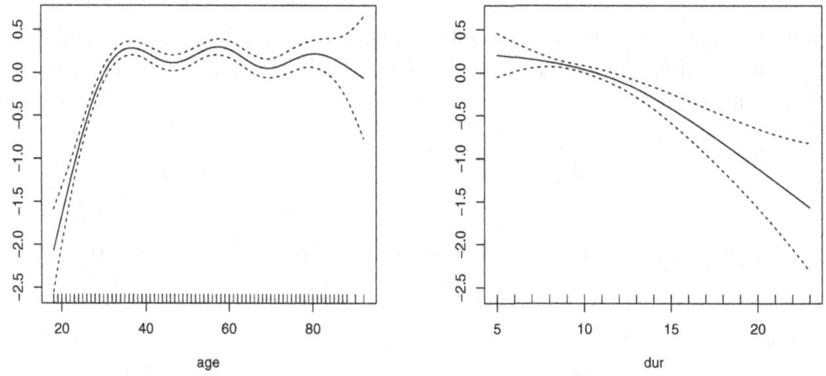

FIGURE 10.8: Estimated effects of age and duration of education on number of children.

Example 10.4: Number of Children

In Example 7.3 the number of children was modeled in dependence on the predictors age in years (age), duration of school education (dur), nationality (nation; 0: German; 1: otherwise), religion (answer categories to "God is the most important in man"; 1: strongly agree,...; 5: strongly disagree; 6: never thought about it), university degree (univ; 0: no; 1: yes). The fitted model was a log-linear Poisson model with polynomial terms. The partially additive model considered here has the predictor

$$\eta = s(\text{age}) + s(\text{dur}) + \text{nation} * \beta_n + \text{god}_2 * \beta_{g2} + \cdots + \text{god}_6 * \beta_{g6} + \text{univ} * \beta_u.$$

The curves in Figure 10.8 differ from the curves with polynomial terms (Figure 7.2). While the curves with polynomial terms decrease for large values of age, the estimates assuming the smooth effects of age without being restricted to polynomials show a stable level above 40 years of age. Parameter estimates for the partially additive model are given in Table 10.1.

□

TABLE 10.1: Parameter estimates for Poisson model with number of children as response.

	Estimate	Std. Error	z-value	Pr(>|z|)
(Intercept)	0.42292	0.04970	8.510	0.0
nation	0.08040	0.13876	0.579	0.56232
god2	−0.10819	0.05914	−1.829	0.06736
god3	−0.14317	0.06784	−2.110	0.03482
god4	−0.13138	0.07092	−1.852	0.06396
god5	−0.04899	0.06703	−0.731	0.46481
god6	−0.10644	0.07517	−1.416	0.15675
univ1	0.55647	0.17130	3.249	0.00116
	edf	Ref.df	Chi.sq	p-Value
s(age)	7.369	8.289	172.71	0.0
s(dur)	2.322	2.997	31.77	0.0

Example 10.5: Addiction

Here we consider again the addiction data (Example 8.2). The response was in three categories (0 : addicts are weak-willed; 1: addiction is a disease; 2: both). Two additive logit models were fitted, one comparing category $\{1\}$ and category $\{0\}$ and the other comparing category $\{2\}$ and category $\{0\}$. In both models the predictor has the form

$$\eta = s(\text{age}) + \text{gender} * \beta_g + \text{universitydegree} * \beta_u.$$

The resulting probabilities for the subpopulation with university degree are shown in Figure 10.9. In contrast to Figure 8.6, which shows the fit of a parametric model with the quadratic effect of age, it is seen from Figure 10.9 that the probability of category $\{0\}$ remains almost constant above 30 years of age. The increase seen in Figure 8.6 seems to be caused by the restrictive quadratic modeling of age.

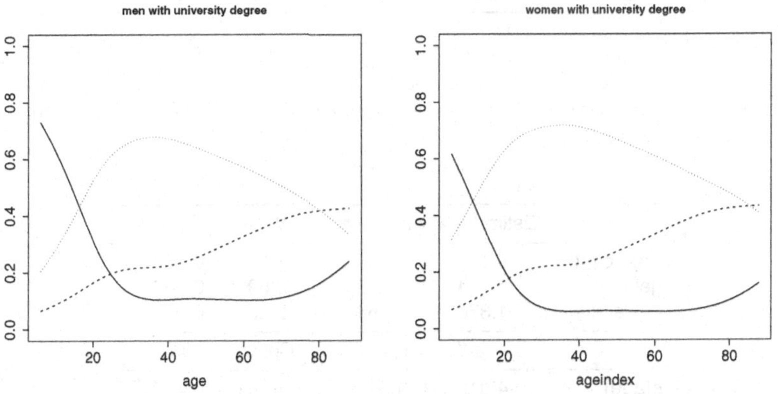

FIGURE 10.9: Estimated probabilities for addiction data with smooth effect of age (category 0: solid line; category 1: dotted line; category 2: dashed line).

In addition, a hierarchical model was fitted (compare Example 8.6). The model comprises two binary models. In the first, category $\{2\}$ was compared with $\{0, 1\}$; in the second, category $\{1\}$ was compared with category $\{0\}$, given that the response was in $\{0, 1\}$. In both models a predictor with a smooth effect of age was fitted. Figure 10.10 shows the corresponding effects of age for both binary models. Age seems to have an almost linear effect on the grouped binary response $\{2\}$ against $\{0, 1\}$, but given a single cause it is assumed (either weak-willed or a disease) that the effect is non-linear with a strong increase in the beginning and a rather stable effect above 25 years of age. Tables 10.2 and 10.3 show the estimated effects. □

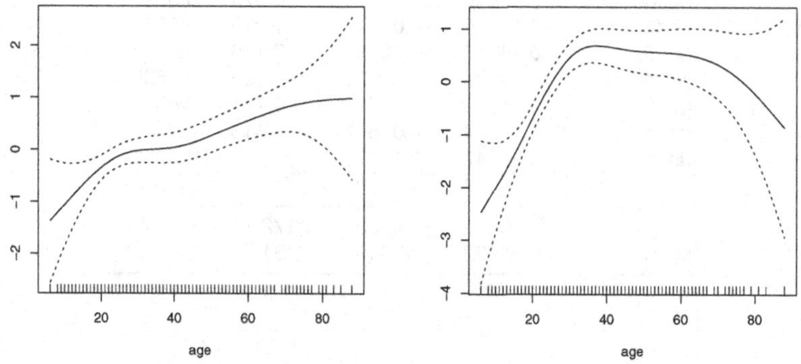

FIGURE 10.10: Estimated effects of age for addiction data, category $\{2\}$ against category $\{0,1\}$ (left) and $\{1\}$ against $\{0\}$ given response in $\{0,1\}$ (right).

TABLE 10.2: Parameter estimates for addiction data, hierarchical model, category $\{2\}$ against categories $\{0,1\}$.

	Estimate	Std. Error	z-Value	Pr(> $\|z\|$)
(Intercept)	−1.154	0.148	−7.768	0.0
gender	0.025	0.183	0.139	0.889
university	−0.130	0.209	−0.622	0.534
	edf	Ref.df	Chi.sq	p-Value
s(age)	3.526	4.411	21.34	0.0004

TABLE 10.3: Parameter estimates for addiction data, hierarchical model, $\{1\}$ against $\{0\}$ given response in $\{0,1\}$.

	Estimate	Std. Error	z-Value	Pr(> $\|z\|$)
(Intercept)	−0.175	0.158	−1.103	0.270
gender	0.604	0.208	2.899	0.003
university	1.376	0.264	5.203	0.0
	edf	Ref.df	Chi.sq	p-Value
s(age)	4.018	4.518	50.41	0.0

10.3.2 Extension to Multicategorical Response

When the response is in categories with an underlying multinomial distribution it is useful to distinguish between ordered response categories (Chapter 9) and unordered response categories (Chapter 8). The extension to the additive structuring of covariates is very easy for simple ordinal models. For example, the simple cumulative model has the form $P(Y \leq r|\boldsymbol{x}) = F(\eta_r)$, where the rth predictor is $\eta_r = \gamma_{0r} + \boldsymbol{x}^T\boldsymbol{\gamma}$. Additive structuring uses

$$\eta_r = \gamma_{0r} + f_{(1)}(x_1) + \cdots + f_{(p)}(x_p),$$

where $\gamma_{01} \leq \cdots \leq \gamma_{0k}$ has to hold for the cumulative model. By expanding the unknown functions in basis functions, estimation procedures including a penalty term are developed straightforwardly. If one assumes the form $f_{(j)}(x_j) = \sum_{s=1}^{m_j} \beta_{js}\phi_{js}(x_j)$, one obtains for data $(y_i, \boldsymbol{x}_i), i = 1, \ldots, n$, the linear predictor $\eta_{ir} = \gamma_{0r} + \boldsymbol{\phi}_{i1}^T\boldsymbol{\beta}_1 + \cdots + \boldsymbol{\phi}_{ip}^T\boldsymbol{\beta}_p$, where $\boldsymbol{\phi}_{ij}^T = (\phi_{j1}(x_{ij}) \ldots \phi_{jm_j}(x_{ij}))$ are the evaluations of basis functions for the jth variable at x_{ij}. By collecting all the evaluations into one matrix and the parameters into one vector one obtains for the ith observation the vector $\boldsymbol{\eta}_i = (\eta_{i1}, \ldots, \eta_{i,k-1})^T = \boldsymbol{\phi}_i\boldsymbol{\beta}$. By using the estimation procedures from Chapter 9 and including a penalty term that penalizes differences of adjacent basis functions one obtains a smoothed estimate. In this simple case all of the effects of covariates are assumed to be global; therefore, the effects do not vary over response categories. The more general model with category-specific effects uses the $k-1$ unknown functions

$$\eta_r = \gamma_{0r} + f_{(1),r}(x_1) + \cdots + f_{(p),r}(x_p), \quad , r = 1, \ldots, k-1.$$

Now the effects of the covariates are category-specific, which is denoted by the subscript r. For this model and all combinations of global and category-specific covariates, expansion in basis functions again yields a linear predictor of the form $\boldsymbol{\eta}_i = \boldsymbol{\phi}_i\boldsymbol{\beta}$ and appropriate penalty terms will stabilize estimation.

In the case of k unordered categories, a basic model is the multinomial logit model $P(Y = r|\boldsymbol{x}) = \exp(\eta_r)/\Sigma_s \exp(\eta_s)$ with predictors $\eta_r = \boldsymbol{x}^T\boldsymbol{\beta}_r$. The extension to additive predictors also has to use category-specific functions:

$$\eta_r = \gamma_{0r} + f_{(1),r}(x_1) + \cdots + f_{(p),r}(x_p),$$

where, for identifiability reasons, one predictor is set to zero (for example, $\eta_k = 0$). While the simple ordinal model with global effects only is basically one-dimensional, the multinomial model is always $(k-1)$-dimensional, and the functions compare the effect of covariates to a reference category. But, again, the estimation methods for additive models can be extended by using the multivariate model presentation of the logit model given in Chapter 8. Specific software is needed for the fitting. In simple cases where one can structure the response categories hierarchically, one can also use software for binary models (see Example 10.5).

Hastie and Tibshirani (1990) considered backfitting procedures for the fitting of the ordinal proportional-odds model, Yee and Wild (1996) put the estimation of multinomial models within the framework of vector generalized additive models and used the backfitting algorithm, Yee and Hastie (2003) considered the reduced rank version. Tutz and Scholz (2004) distinguished between global and category-specific effects; flexible modeling for discrete choice data was considered by Abe (1999). General semiparametrically structured regression models have also been considered by Tutz (2003), and Kauermann and Tutz (2003).

Example 10.6: Preference for Political Parties

In Section 8.2 the preference for political parties dependent on gender and age was investigated with age given in four categories. Now we include age as a continuous variable and allow for smooth effects. The considered German parties are the Christian Democratic Union (CDU), which serves as the reference; the Social Democratic Party (SPD, category 1); the Green Party (category 2); and the Liberal Party (FDP, category 3). Figure 10.11 shows the fitted effects of age by use of the package *VGAM*. The preference for the SPD over the CDU seems to decrease between 30 and 70 years of age, whereas the preference for the Green Party over the CDU decreases over the whole range, showing that in particular younger people prefer the Greens. The comparison between the Liberal Party and the CDU is less distinct (see also Figure 8.5, which was built with categorized age). □

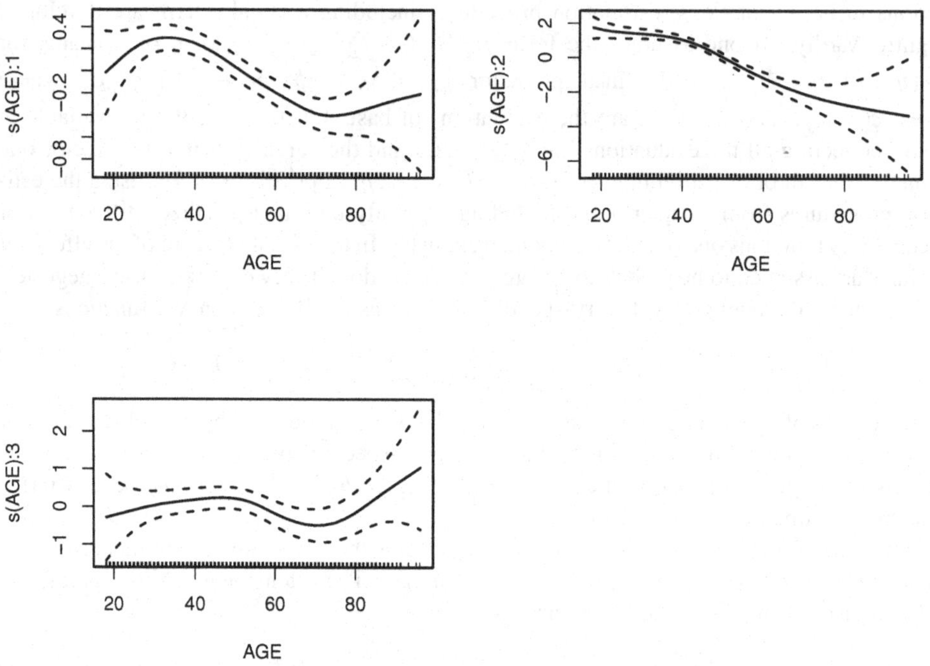

FIGURE 10.11: Effect of age resulting from the fitting of an additive model for party preference with reference category CDU.

10.3.3 Structured Interactions

Additive models have the attractive feature that effects of single predictors can be plotted separately. Thus, the effect of predictors is easily visualized. However, the strong underlying assumption is that the effect of one predictor does not depend on the values at which other variables are fixed. For arbitrary values of the other variables it is assumed that the effect of one predictor is the same. Although the assumption is much weaker than the linear predictor assumption, it is still strong enough to sometimes make the fit to the data unsatisfactory.

Interaction effects occur if the effect of one predictor on the response depends on the values of other predictors. Then the effect of the predictor cannot be separated from the fixed values of the other predictors. In the following we consider some approaches to structured interaction terms. An alternative approach to model interactions based on tree-based methods is given in Chapter 11.

Varying-Coefficients Models

A specific way of modeling interactions is by so-called varying-coefficients models, which were proposed by Hastie and Tibshirani (1993). In varying-coefficient models a two-way interaction is specified by modeling the effect of one predictor linearly but the slope may vary non-parametrically as a function of the second predictor. The general form of the predictor for binary or continuous variables $z_1, \dots, z_p$ and $x_0, x_1, \dots, x_p$ is

$$\eta = \beta_0(x_0) + z_1\beta_1(x_1) + \cdots + z_p\beta_p(x_p), \tag{10.21}$$

where the $\beta_j(x_j)$ are unspecified functions of predictions. Thus the z-variables have a linear effect, but the effects are modified by the x-variables. The x-variables are called *effect modifiers*. It should be noted that (10.21) may contain simple additive terms. For example, if z_1 and z_2 are constant, $z_1 \equiv z_2 \equiv 1$, the linear predictor contains the additive terms $\beta_1(x_1) + \beta_2(x_2)$. For simplicity in the following we consider only single interaction terms.

Discrete-by-Continuous Interaction

A simple case of the varying coefficient model specifies the interaction of a binary variable $z \in \{0, 1\}$ and a continuous variable x by

$$\eta(z, x) = \beta_0(x) + z\beta_1(x),$$

where the intercept term $\beta_0(x)$ and the slope are functions of x. Formally, one obtains the model from (10.21) by setting $z_1 = z, x_0 = x_1 = x$. The model assumes that the effect of z may vary with x. It also implies that the effect of x is determined by different functions for the two values of z:

$$\eta(1, x) = \beta_0(x) + \beta_1(x), \qquad \eta(0, x) = \beta_0(x).$$

Thus $\beta_0(x)$ represents the curve for the reference population $z = 0$, and $\beta_1(x)$ represents the additive modification of the "baseline" function $\beta_0(x)$. When the effects $\beta_j(x_i)$, $j = 0, 1$, are centered, for example by $\sum_i \beta_j(x_i) = 0$, the model can also be given in a form that separates main effects and interaction terms. One obtains $\eta(z, x) = \alpha_0 + z\alpha + \beta_0(x) + z\beta_1(x)$, yielding $\eta(1, x) = \alpha_0 + \alpha + \beta_0(x) + \beta_1(x)$, $\eta(0, x) = \alpha_0 + \beta_0(x)$.

In the more general case of a discrete variable $A \in \{1, \dots, k\}$, the varying-coefficients model may be written with (0-1)-dummy variables $x_{A(1)}, \dots, x_{A(k)}$ as

$$\eta(A, x) = \beta_0(x) + x_{A(2)}\beta_2(x) + \cdots + x_{A(k)}\beta_k(x),$$

or with centered effects as $\eta(A, x) = \alpha_1 + \sum_{j=2}^{k} x_{A(j)}\alpha_j + \beta_0(x) + \sum_{j=2}^{k} x_{A(j)}\beta_j(x)$.

Continuous-by-Discrete Interaction

For the binary effect-modifier $x \in \{0, 1\}$ and the continuous variable z, the predictor $\eta(z, x) = \beta_0(x) + z\beta(x)$ implies

$$\eta(z, 1) = \beta_0(1) + z\beta(1), \qquad \eta(z, 0) = \beta_0(0) + z\beta(0),$$

which means that the effect of z varies across values of x. Simple reparameterization shows that the predictor is the same as in simple interaction modeling, which uses the predictor $\beta_0 + x\beta_x + z\beta_z + xz\beta_{xz}$. In the more general case of a discrete variable $A \in \{1, \dots, k\}$ one obtains $\eta(z, A = i) = \beta_0(i) + z\beta(i)$.

Continuous-by-Continuous Interactions

When both interacting variables are continuous one can distinguish between the fully non-parametric approach and interaction terms where one variable is modeled parametrically. With variables x_i and x_j the fully non-parametric approach assumes a term $f(x_i, x_j)$ in the additive predictor (10.14), yielding

$$\eta(\boldsymbol{x}) = \beta_0 + f_{(1)}(x_1) + \cdots + f_{(p)}(x_p) + f_{(ij)}(x_i, x_j).$$

These interaction terms, which treat both variables *symmetrically*, can be modeled by tensor products or radial basis functions as considered in Section 10.2.

In the *asymmetrical* case, one has to decide which of the two variables is modeled parametrically. Let x and z denote the two variables, and let x be modeled non-parametrically. Then one specifies a varying coefficient model with predictor

$$\eta(z, x) = \beta_0(x) + z\beta(x).$$

Of course the model assumes more structure than a model where the combined effect of x, z is unspecified.

Estimation

Estimation in varying-coefficients models with continuous effect modifiers is very similar to estimation in additive models. The main difference is that the unknown functions are multiplied by some known predictor. When using an expansion in basis functions, the unknown functions have the form

$$\beta_j(x) = \sum_l \delta_l \phi_l^{(j)}(x).$$

Then the interaction term $z_j\beta_j(x_j)$ in predictor (10.21) becomes, for data $x_{ij}, z_{ij}, i = 1, \ldots, n$,

$$z_{ij}\beta_j(x_{ij}) = \sum_l \delta_l (z_{ij}\phi_l^{(j)}(x_{ij})),$$

which is again linear since $z_{ij}\phi_l^{(j)}(x_{ij})$ is known for fixed basis functions.

Marx and Eilers (1998) used B-splines with penalization, and Ruppert et al. (2003) gave several examples based on the truncated power series and an extensive discussion of interaction modeling. Alternatively, in particular, if only one effect modifier is included, local polynomial regression methods may be used; see, for example, Kauermann and Tutz (2001) and Tutz and Kauermann (1997). For examples that use alternative smoothers with the backfitting algorithm see Hastie and Tibshirani (1993). Many alternative methods have been proposed, in particular for the linear model; see, for example, Fan and Zhang (1999), Wang and Xia (2009), and Leng (2009).

In the case of categorical effect modifiers, regularization techniques for categorical predictors are useful because the number of parameters to be estimated tends to be large. If p predictors and an effect modifying variable $A \in \{1, \ldots, k\}$ are available, $(p+1)k$ parameters have to be estimated. To restrict estimation to the relevant effects, regularization techniques as in Section 6.5 can be used. In particular one wants to know which categories of A have to be distinguished and which variables are influential. With the predictor $\eta(\boldsymbol{z}, A = i) = \sum_j z_j\beta_j(i)$ and nominal effect modifier A one includes into the log-likelihood the penalty term

$$J(\boldsymbol{\beta}) = \sum_{j=0}^{p}\sum_{r>s} |\beta_j(r) - \beta_j(s)| + \sum_{j=1}^{p}\sum_{r=1}^{k} |\beta_j(r)|$$

together with a smoothing parameter that steers the strength of regularization. The first term enforces the collapsing of categories, while the second term enforces the selection of covariates.

Collapsing of categories yields clusters of categories for which the covariate z_j has the same effect on the dependent variable. For ordered effect modifier a modified penalty term that enforces clustering of adjacent categories should be used; for details see Gertheiss and Tutz (2011). In the following we give an example for the case of a categorical effect modifier.

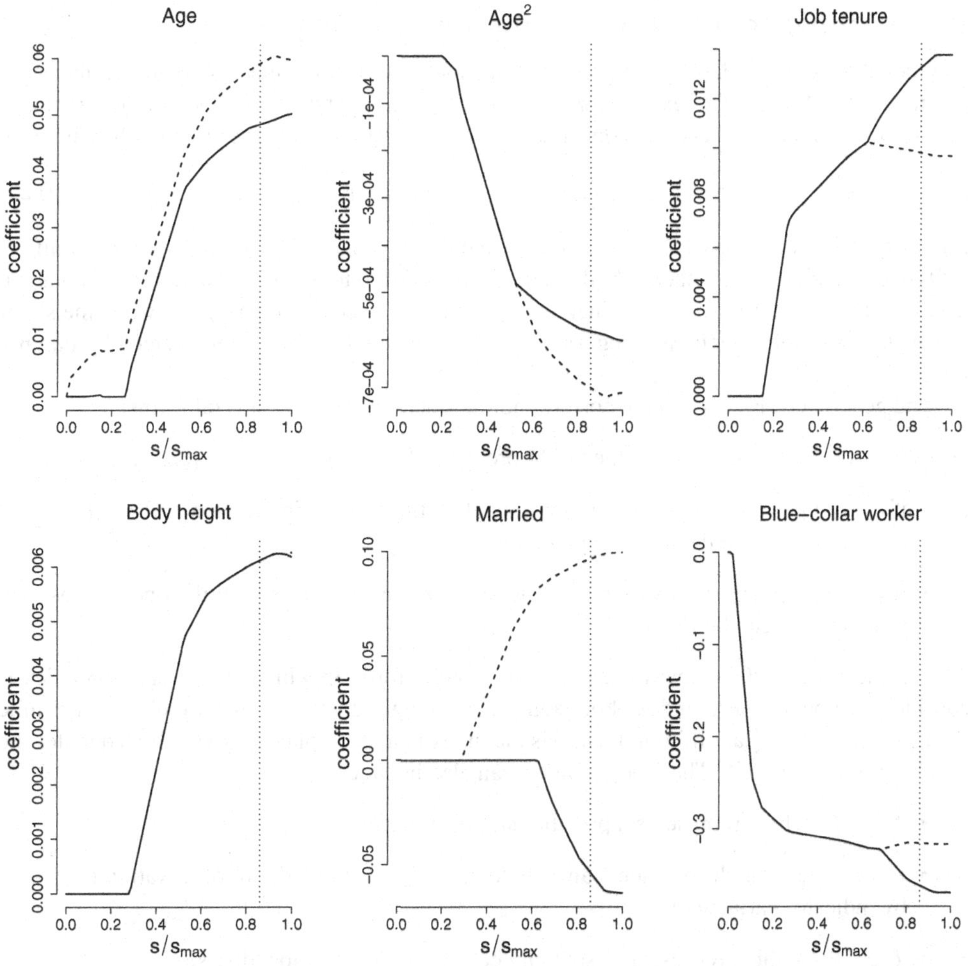

FIGURE 10.12: Coefficient buildups for income data with effect modifier gender, dashed lines denote coefficients for men, solid lines for women.

Example 10.7: Income Data
As response we consider monthly log-income depending on age (in years), squared age, job tenure (in months), body height (cm), married (yes/no), Abitur (allows access to university, yes/no), and blue-collar worker (yes/no). As the effect modifier we consider gender. The data stem from the Socio-Economic Panel Study (SOEP), which is a longitudinal study of private households in Germany. In Figure 10.12 six of the eight gender-dependent coefficients are plotted against varying degrees of smoothing. The dashed lines refer to males and the solid lines to females. The vertical lines indicate the chosen degree of regularization based on cross-validation. It is seen that the effects of most predictors are modified by gender; only for

body height there is no difference between sexes. A strong effect is found in particular for the variable married. For males the effect of being married is positive and for females negative. Therefore, as far as income is concerned, being married has an positive effect for males but not for females. For more details see Gertheiss (2011) and Gertheiss and Tutz (2011).

□

10.3.4 Structured Additive Regression Modeling

The model terms considered in the previous sections can be combined to obtain a rather general model for the effect of explanatory variables on the dependent variable. In the structured additive regression (STAR) model one assumes for observation i that the predictor has the form

$$\eta(\boldsymbol{x}_i) = f_{(1)}(\boldsymbol{v}_{i1}) + \dots f_{(m)}(\boldsymbol{v}_{im}) + \boldsymbol{v}_{i,m+1}^T \gamma, \tag{10.22}$$

with unspecified functions $f_{(1)}(.), \dots, f_{(p)}(.)$ and the $\boldsymbol{v}_{ij}$ denoting generic covariates of different types and dimensions that are built from $\boldsymbol{x}_i$. In addition to the term that contains unknown functions a linear term $\boldsymbol{v}_{i,m+1}^T \gamma$ is included. That term is needed in particular when the set of explanatory variables contains categorical variables, which have to be represented by dummy variables.

The generic covariates and the corresponding functions can take several forms:

- with $\boldsymbol{v}_{ij} = x_{ij}$ one obtains the GAM predictor $f_{(1)}(x_{i1}) + \dots + f_{(p)}(x_{ip})$;
- if $\boldsymbol{v}_{ij} = (x_{ir}, x_{is})^T$ with two-dimensional function $f_{(j)}$, an interaction term $f_{(j)}(x_{ir}, x_{is})$ can be included in the additive predictor;
- if $\boldsymbol{v}_{ij} = (x_{ir}, x_{is})^T$, a varying coefficients term $x_{ir} f_{(j)}(x_{is})$ with the one-dimensional function $f_{(j)}$ can be included.

Fahrmeir et al. (2004) investigated a more general form, in which the observation index is considered to be a generic index. For example, for longitudinal data that are observed at time points $t \in \{1, \dots, T\}$ at different locations the index i can be replaced by it, which denotes the ith observation at time t. Then the predictor can also include

- $f_{time}(t)$, which represents a possibly nonlinear trend;
- $\boldsymbol{v}_i^T \gamma = \gamma_{i0}$, which represents an individual parameter for the ith observation (specified by indicator variables $\boldsymbol{v}_i$);
- $f_{spat}(s_{it})$, which represents a spatially correlated effect of location s_{it}.

These extended versions represent repeated measurements, which will be considered in Chapters 13 and 14.

Regularized estimation of a STAR model typically uses an additive penalty that includes a separate penalty for each included component that contains an unspecified function. If the linear term contains many variables or categorical variables with many categories, the parameter in the linear term also should be penalized. Bayesian estimates for very general generalized additive regression approaches were considered by Fahrmeir et al. (2004) and Fahrmeir and Kneib (2009).

In particular, if many variables are available, selection of the relevant terms is important. An approach that combines estimation and selection is boosting, which we considered previously for simple additive models. A basic boosting algorithm for generally structured models is the following.

GAMBoost for Structured Additive Regression

For $l = 0, 1, \ldots$:

1. *Estimation step:* For $s = 1, \ldots, m$, fit the model

$$\mu_i = h(\hat{\eta}^{(l)}(\boldsymbol{x}_i) + f_{(s)}(\boldsymbol{v}_{is})),$$

where $\hat{\eta}^{(l)}(\boldsymbol{x}_i)$ is the fit from the previous step (treated as an off-set) to obtain estimates $\hat{f}_{(s),new}(\boldsymbol{v}_{is})$.

2. *Selection step:* Compare the fit of the models $\mu_i = h(\hat{\eta}^{(l)}(\boldsymbol{x}_i) + \hat{f}_{(s),new}(\boldsymbol{v}_{is}))$, $s = 1, \ldots, m$, by using the deviance or some information criterion. Let j denote the index of that model for which the criterion is minimized.

3. *Update:* Set for j only $\hat{f}_{(j)}^{(l+1)}(.) = \hat{f}_{(j)}^{(l)}(.) + \hat{f}_{(j),new}(.)$, and for $s \neq j$ set $\hat{f}_{(s)}^{(l+1)}(.) = \hat{f}_{(s)}^{(l)}(.)$.

How the model is fitted depends on the type of function $f_{(s)}(.)$. If the function is one-dimensional, other methods are to be used than for two-dimensional functions. Some care is needed to obtain learners (estimators) that show the same degree of weakness. Kneib et al. (2009) showed how the complexity of the fit can be adapted by using simple quadratic fitting procedures. The next example shows how components are selected starting from a complex candidate model.

Example 10.8: Forest Health

The health status of beeches at 83 observation plots located in a northern Bavarian forest district has been assessed in visual forest health inventories carried out between 1983 and 2004 (see also Kneib and Fahrmeir, 2006; Kneib and Fahrmeir, 2008). The health status is classified on an ordinal scale, where the nine possible categories denote different degrees of defoliation. The domain is divided in 12.5% steps, ranging from healthy trees (0% defoliation) to trees with 100% defoliation. Since the data become relatively sparse already for a medium amount of defoliation, we will model the dichotomized response variable defoliation with categories 1 (defoliation above 25%) and 0 (defoliation less or equal to 25%). The collected data have a temporal and a spatial component. It has to be assumed that trees measured at the same plot are correlated. The covariates in the dataset are age (age of the tree in years, continuous), time (calendar time, continuous, $1983 \leq \text{time} \leq 2004$), elevation (elevation above sea level in meters, continuous), inclination (inclination of slope in percent, continuous), soil (depth of soil layer in centimeters, continuous), ph (ph value in 0–2 cm depth, continuous), canopy (density of forest canopy in percent, continuous), stand (type of stand; categorical; 1: deciduous forest; -1: mixed forest), fertilization (categorical, 1: yes; -1: no), humus (thickness of humus layer in five categories, ordinal; higher categories represent higher proportions), moisture (level of soil moisture, categorical; 1: moderately dry; 2: moderately moist; 3: moist or temporary wet), and saturation (base saturation, ordinal; higher categories indicate higher base saturation).

The data include a temporal and a spatial component. In particular, the longitudinal structure calls for the incorporation of plot-specific effects. Previous studies described in Kneib and Fahrmeir (2006) suggest the presence of interaction effects and the non-linear influences of some continuous predictors. Based on these results we consider a logit model with the candidate predictor

$$\begin{aligned}\eta(\boldsymbol{x}) &= \boldsymbol{v}^T\boldsymbol{\beta} + f_1(\text{ph}) + f_2(\text{canopy}) + f_3(\text{soil}) + f_4(\text{inclination}) + f_5(\text{elevation}) \\ &\quad + f_6(\text{time}) + f_7(\text{age}) + f_8(\text{time, age}) + f_9(s_x, s_y) + b_{\text{plot}},\end{aligned}$$

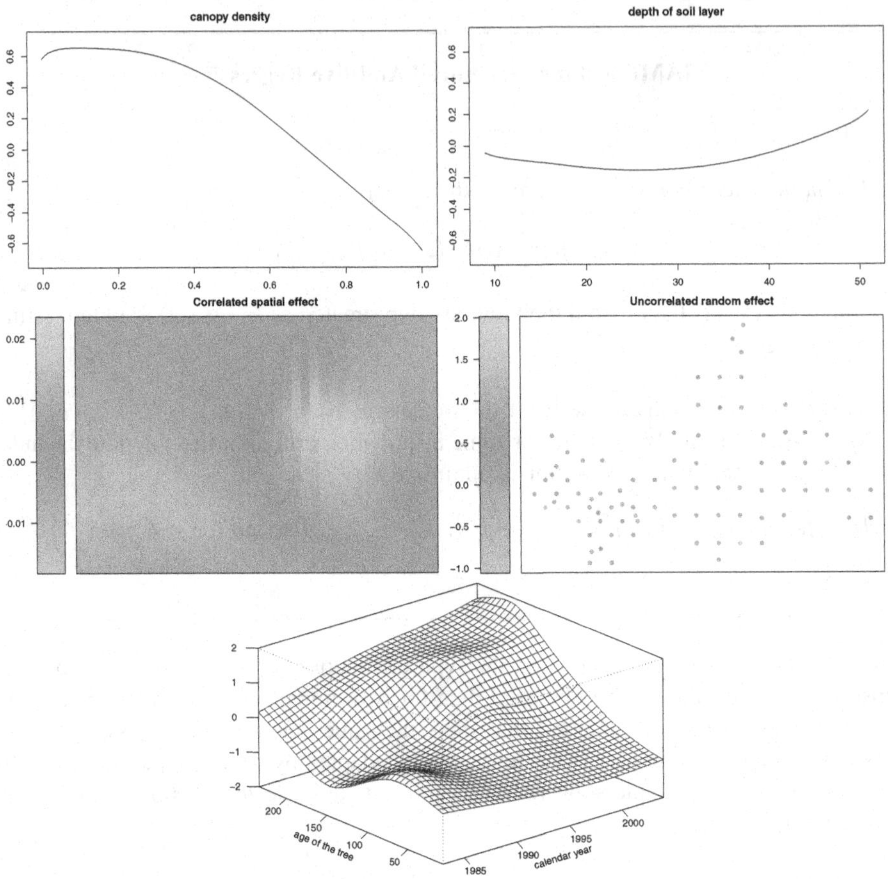

FIGURE 10.13: Identified effects for forest health data: non-linear effects of canopy density and depth of soil, spatial effect, plot-specific effects, and interaction between age and year.

where $\boldsymbol{v}$ contains the parametric effects of the categorical covariates and the base learners for the smooth effects $f_1, \ldots, f_7$ are specified as univariate cubic penalized splines with 20 inner knots and a second-order difference penalty. For both the interaction effect f_8 and the spatial effect f_9 we assume bivariate cubic penalized splines with first-order difference penalties and 12 inner knots for each of the directions. The plot-specific random effect b_{plot} is assumed to be Gaussian with the random effects variance fixed such that the base learner has one degree of freedom. Similarly, all univariate and bivariate non-parametric effects are decomposed into parametric parts and non-parametric parts with one degree of freedom each. The stopping criterion was determined by a booststrapping procedure.

After applying the stopping rule, no effect was found for the ph value, inclination of slope, and elevation above sea level. The univariate effects for age and calendar time where strictly parametric linear but the interaction effect turned out to be very influential. The sum of both the linear main effects and the non-parametric interaction is shown in Figure 10.13. The spatial effect was selected only in a relatively small number of iterations, whereas the random effect was the component selected most frequently. We can therefore conclude that the spatial variation in the data set seems to be present mostly very locally. For canopy density and soil depth, non-linear effects where identified as visualized in Figure 10.13 (for more details see Kneib et al., 2009). □

10.4 Functional Data and Signal Regression

In recent years functional data analysis (FDA) has become an important tool to model data where the unit of observation is a curve or in general a function. Functional responses are related to repeated measurements, which are treated in Chapters 13 and 14. Here we focus on functional predictors, which can be considered a strongly structured and high-dimensional form of predictor. We will start with some examples.

Example 10.9: Canadian Weather Data

The data taken from Ramsey and Silverman (2005) give the average daily precipitation and temperature at 35 Canadian weather stations. It can be downloaded from the related website http://www.functionaldata.org. Like Ramsay and Silverman, we try to predict the logarithm of the total annual precipitation from the pattern of temperature variation through the year. Figure 10.14 shows the temperature profiles (in degrees Celsius) of the weather stations across the year, averaged over the years 1960 to 1994. Like Ramsey and Silverman (2005) we will consider the base 10 logarithm of the total annual precipitation as response variable. □

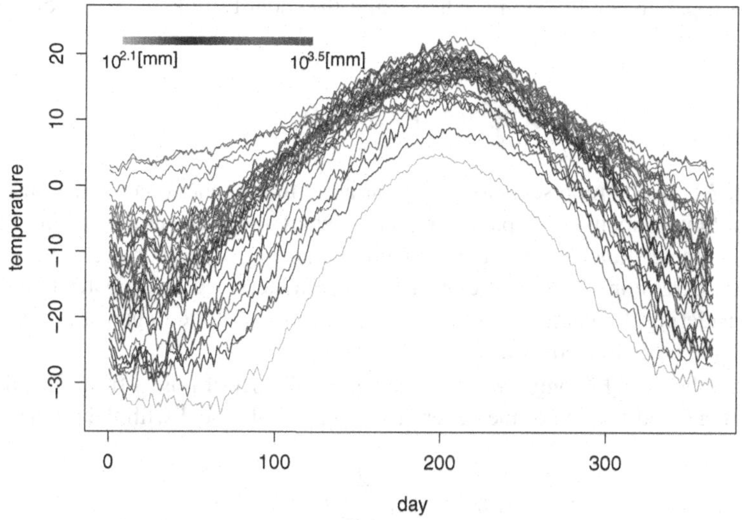

FIGURE 10.14: Temperature profiles of 35 Canadian weather stations.

Example 10.10: Mass Spectrometry

In mass spectrometry–based predictive proteomics one often wants to distinguish between healthy patients and severely ill patients by using mass spectrometry data. We use data from Petricoin et al. (2002), which is available from the National Cancer Institute via http://home.ccr.cancer.gov/ncifdaproteomics. The goal is to discriminate prostate cancer from benign prostate conditions by proteomic pattern diagnosis. All in all, 69 SELDI-TOF blood serum spectra from cancer patients and 253 from patients with benign conditions are given. Each spectrum is composed of measurements at approximately 15,200 points defined by mass over charge ratio m/z values; see Petricoin et al. (2002) for more details. In some applications the full spectrum is used as input to a classifier that itself selects variables. Then features are selected in terms of single m/z values. Due to horizontal variability, however, the results are difficult to interpret, because the same m/z values do not necessarily correspond to the same feature. Therefore, Hoefsloot et al. (2008)

manually clustered m/z values at the end of the analysis. Alternatively, feature selection in terms of peak detection and peak alignment Tibshirani et al. (2004) is often performed before employing a classifier, as, for example, described in Barla et al. (2008). □

10.4.1 Functional Model for Univariate Response

Let the data be given by $(y_i, x_i(t))$, $i = 1, \ldots, n$, where y_i is the response variable and $x_i(t), t \in I$, denotes a function defined on an interval $I \in R$, also called the signal. A *generalized functional linear model* for scalar responses assumes that the mean μ_i depends on the functional predictor $x_i(t)$ in the form $g(\mu_i) = \eta_i$ or $\mu_i = h(\eta_i)$, with the linear predictor determined by

$$\eta_i = \beta_0 + \int x_i(t)\beta(t)dt, \tag{10.23}$$

where $\beta(.)$ is a parameter function. When using the identity link, $h = id$, one obtains the *functional linear model*

$$y_i = \beta_0 + \int x_i(t)\beta(t)dt + \varepsilon_i, \tag{10.24}$$

where ε_i with $E(\varepsilon_i) = 0$ represents the noise variable. In the functional model, the usual summation over a finite-dimensional space is replaced by an integral over the infinite-dimensional space. The parameter function $\beta(.)$ is the functional analog of the parameter vector in a generalized linear regression. Instead of considering parameters one investigates the effect of the signal by considering the parameter function that is usually plotted against t. Since $\beta(.)$ is a function, it has to be estimated as such.

In practice, the signal is only ever observed at a finite set of points from I, and one might consider the dicretized version of the generalized functional model with the predictor

$$\eta_i = \beta_0 + \sum_{j=1}^{p} x_{ij}\beta_j, \tag{10.25}$$

where $x_{ij} = x_i(t_j), \beta_j = \beta(t_j)$ for values $t_1 < \cdots < t_p, t_j \in I$. Most often the values $t_1, \ldots, t_p$ are equidistant, $t_{j+1} - t_j = \Delta$.

However, estimation within the framework of generalized linear models is hardly an option, since the number of predictors typically is too large when the predictor is functional. Of course, variable selection and regularization methods like ridge regression, the lasso, and the elastic net (see Chapter 6) could be used. But they are not up to the challenge of functional predictors because they do not make use of the ordering of the predictor. In most applications the space I on which the signals and the coefficient functions are defined is a metric space, and a distance measure is available that measures the distance between two measurement points t_i and t_j. Thus, the appropriate methods should use the metric information that is available. Variable selection strategies like the lasso may be seen as *equivariant* methods, referring to the fact that they are equivariant to permutations of the predictor indices and therefore less appropriate for functional predictors. In contrast, *spatial* methods regularize by utilizing the spatial nature of the predictor index (compare Land and Friedman, 1997).

10.4.2 The FDA Approach

Since the signal $x_i(t)$ and the unknown parameter function $\beta(t)$ are functions in t, one might expand both functions in basis functions:

$$x_i(t) = \sum_{j=1}^{m_1} \xi_{ij}\phi_j(t), \quad \beta(t) = \sum_{l=1}^{m} \beta_l \phi_l^\beta(t),$$

where $\phi_j(.)$ and $\phi_i^\beta(.)$ are fixed known basis functions defined on I. Then one obtains for the predictor

$$\eta_i = \beta_0 + \int x_i(t)\beta(t)dt = \beta_0 + \int \sum_j \sum_l \xi_{ij}\beta_l\phi_j(t)\phi_l^\beta(t)dt$$
$$= \beta_0 + \sum_l \beta_l \sum_j \xi_{ij} \int \phi_j(t)\phi_l^\beta(t)dt.$$

With $\phi_{jl} = \int \phi_j(t)\phi_l^\beta(t)dt$ and $z_{it} = \sum_{j=1}^{m_1} \xi_{ij}\phi_{jl}$ one has the linear predictor

$$\eta_i = \beta_0 + \sum_{l=1}^{m} \beta_l z_{il},$$

which is from m-dimensional space. Therefore, the problem is reduced to a linear predictor problem in m dimensions. The reduction works only when the parameters ξ_{ij} are known. Therefore, in the *first step* of FDA, the corresponding parameters ξ_{ij} for each signal are estimated (yielding ξ_{ij}). In the *second step*, one estimates the parameters of the linear predictor

$$\eta_i = \beta_0 + \sum_{l=1}^{m} \beta_l \tilde{z}_{il} = \tilde{\boldsymbol{z}}_i^T\boldsymbol{\beta},$$

where $\tilde{\boldsymbol{z}}_i^T = (1, \tilde{z}_{i1}, \ldots, \tilde{z}_{im_1})$, $\tilde{z}_{il} = \sum_j \hat{\xi}_{ij}\phi_{jl}$, $\boldsymbol{\beta}^T = (\beta_0, \beta_1, \ldots \beta_m)$. The first step is an unsupervised learning step in the sense that the response is not used to fit the expansion of the signal in basis functions. The second step explicitly connects the parameters to the observed responses (compare Ramsey and Silverman, 2005).

An alternative approach has been proposed by James (2002). He uses the expansion of $x_i(t)$ to obtain

$$\eta_i = \beta_0 + \sum_{j=1}^{m_1} \xi_{ij} \int \phi_j(t)\beta(t)dt = \boldsymbol{z}_i^T\boldsymbol{\xi}_i,$$

with $\boldsymbol{z}_i^T = (1, \int \phi_1(t)\beta(t)dt, \int \phi_2(t)\beta(t)dt, \ldots)$, $\boldsymbol{\xi}_i^T = (\beta_0, \xi_{i1}, \xi_{i2}, \ldots)$. He treats the parameters $\boldsymbol{\xi}_i$ as missing data and uses the EM algorithm (see Appendix B) to maximize the likelihood.

10.4.3 Penalized Signal Regression

In penalized signal regression only the parameter function is expanded in known basis functions, $\beta(t) = \sum_l \beta_l\phi_l^\beta(t)$, yielding the linear predictor

$$\eta_i = \int x_i(t)\beta(t)dt = \sum_{l=1}^{m} \beta_l \int x_i(t)\phi_l^\beta(t)dt = \sum_{l=1}^{m} \beta_l z_{il} = \boldsymbol{z}_i^T\boldsymbol{\beta},$$

where $z_{il} = \int x_i(t)\phi_l^{\beta}(t)dt$ can be computed from the observed signal $x_i(i)$ and the known basis functions $\phi_l^{\beta}(i)$. In contrast to the FDA approach, the signal is not approximated by an expansion in basis functions. The dimension of the predictor is again determined by the number of basis functions that are used to describe the parameter function. For fixed basis functions estimation can be carried out by using the methods from the preceding sections. One maximizes the penalized log-likelihood,

$$l_p(\boldsymbol{\beta}) = l(\boldsymbol{\beta}) - \frac{\lambda}{2}\boldsymbol{\beta}^T \boldsymbol{K}\boldsymbol{\beta},$$

where $\boldsymbol{K}$ is a penalty matrix that penalizes the differences of parameters linked to adjacent basis functions (see Section 10.1.3).

The signal regression approach with B-splines was propagated by Marx and Eilers (1999). Marx and Eilers (2005) extended the method to a multidimensional signal regression.

10.4.4 Fused Lasso

Tibshirani et al. (2005) proposed a regularization method called the fused lasso that deals with metrically structured data and applies to functional data. In general, they consider the model $\eta_i = \beta_0 + \sum_j x_{ij}\beta_j$, with an ordering in the predictors $x_{i1}, x_{i2}, \ldots$. A special case is the functional model when the signal has been discretized (equation (10.25)). The fused lasso uses the penalized log-likelihood

$$l_p(\boldsymbol{\beta}) = \sum_{i=1}^{n} l_i(\boldsymbol{\beta}) - \frac{\lambda_1}{2}\sum_{j=1}^{p} |\beta_j| + \frac{\lambda_2}{2}\sum_{j=2}^{p} |\beta_j - \beta_{j-1}|,$$

where $l_i(\boldsymbol{\beta})$ is the usual log-likelihood contribution of the ith observation and λ_1, λ_2 are tuning parameters. In the original form Tibshirani et al. (2005) used the constraint representation and restricted themselves to the linear model. The effect of the penalty is twofold: The first penalty encourages sparsity in the coefficients while the second penalty encourages sparsity in the differences between adjacent parameters. Therefore, the second term lets few values of parameters be different, and the resulting parameter curve tends to be a step function. It is seen from Figure 10.15, which shows among other procedures the fused lasso for the Canadian weather data, that solutions can be very sparse; only three parameter values are distinguished.

Tibshirani et al. (2005) gave an algorithm for the linear model and discussed the asymptotic results. Land and Friedman (1997) coined the term "variable fusion" in an earlier paper, in which only the second penalty term was used. The method is related to the methods considered in Chapter 6.

10.4.5 Feature Extraction in Signal Regression

Most signal regression methods fit a smooth parameter function. Areas of the signal that are not relevant are supposed to be seen from the fitted curve. When the fitted parameter function is close to zero one might infer that these parts of the signal do not exert influence on the response. If, however, one suspects that the whole signal is not relevant, this assumption can be incorporated in the fitting procedure by trying to select relevant features from the signal. In some applications, as, for example, in proteomics, where one wants to find influential proteins, feature selection is the central issue; identification of the relevant parts in the signal is more important than prediction.

Common selection procedures like the lasso and the elastic net will not perform best because the order information of signals is not used. The fused lasso can be seen as a better feature selection method but is restricted to step functions. An alternative method is blockwise boosting, which can be seen as a generalization of componentwise boosting. In common componentwise boosting (compare Chapter 15) the selection step usually refers to the single components of the vector $x_i = (x_{i1}, \dots, x_{ip})^T$. Predictors that are never selected get value zero. In blockwise boosting, the selection and refitting refer to groups or blocks of variables.

With the focus on signal regression one has to define what the relevant parts of the signal might be. Since a signal $x_i(.)$ may be seen as a mapping $x_i : I \to R$, one can utilize the metric that is available on I. A potentially relevant part of the signal $x_i(.)$ may be characterized by $\{x_{i,U}(t) | t \in U(t_0)\}$, where $U(t_0)$ is defined as a neighborhood of $t_0 \in I$, that is, $U(t_0) = \{t| \|t - t_0\| \le \delta\}$ for the metric $\|.\|$ on I. For the digitized signal, the potentially relevant signal part turns into groups of variables $\{x_{ij} | t_j \in U(t_0)\}$. Simple subsets that may be used in the refitting procedure are $U_s = U_k(t_s) = [t_s, t_s + (k-1)\Delta], k \in \{1, 2, \dots\}$. For $k = 1$ one obtains the limiting case of single variables $\{x_i(t_s)\}$, for $k = 2$ one gets pairs of variables, and so on.

Let $\boldsymbol{X}^{(s)}$ denote the design matrix of variables from U_s, that is, $\boldsymbol{X}^{(s)}$ has rows $(x_i(t_s), \dots, x_i(t_{s+k-1})) = (x_{is}, \dots, x_{i,s+k-1})$. An update step of blockwise boosting will be based on estimating the vector $\boldsymbol{b}^{(s)} = (b(t_s), \dots, b(t_{s+k-1}))^T$ from the data $(\boldsymbol{u}, X^{(s)})$, where $\boldsymbol{u} = (u_1, \dots, u_n)^T$ denotes the current residual. Straightforward maximum likelihood estimation cannot be recommended, since variables in $X^{(s)}$ tend to be highly correlated. Smoother coefficients can be obtained by penalizing differences between the adjacent coefficients. As a parameter estimate within one boosting step one may use the generalized ridge estimator with penalty term $(\lambda/2)\boldsymbol{\beta}^T \boldsymbol{\Omega} \boldsymbol{\beta}$, where the penalty matrix is $\boldsymbol{\Omega} = \boldsymbol{D}^T \boldsymbol{D}$, with $D_{11} = 1$, $D_{r,r-1} = -1$, $D_{k+1,k} = 1$ and zero otherwise, $r = 2, \dots, k$. By upper left and lower right 1's in D differences to zero coefficients of neighboring but not selected values are penalized.

Update steps based on residuals use in the mth iteration estimates

$$\boldsymbol{b}^{(m)} = (\boldsymbol{X}^{(s_m)T} \boldsymbol{X}^{(s_m)} + \lambda \boldsymbol{\Omega})^{-1} \boldsymbol{X}^{(s_m)T} \boldsymbol{u}^{(m)}.$$

Additional weights are needed for binary predictors and one uses

$$\boldsymbol{b}^{(m)} = (\boldsymbol{X}^{(s_m)T} \boldsymbol{W}^{(m)} \boldsymbol{X}^{(s_m)} + \lambda \boldsymbol{\Omega})^{-1} \boldsymbol{X}^{(s_m)T} \boldsymbol{W}^{(m)} \boldsymbol{z}^{(m)},$$

where the working response is $\boldsymbol{z}^{(m)} = (z_1^{(m)}, \dots, z_n^{(m)})^T$, $z_i^{(m)} = (y_i - \pi_i^{(m-1)})/w_i^{(m)})$ with weight matrix $\boldsymbol{W}^{(m)} = \text{diag}(w_1^{(m)}, \dots, w_n^{(m)})$, $w_i^{(m)} = \pi_i^{(m-1)} \cdot (1 - \pi_i^{(m-1)})$. In iteration m block s_m producing minimum error is selected.

For details of the blockwise boosting procedure in signal regression see Tutz and Gertheiss (2010) for continuous responses and Gertheiss and Tutz (2009c) for binary responses. An alternative approach that performs feature selection within a quantile regression framework by use of a structured elastic net was given by Slawski (2010).

Example 10.11: Canadian Weather Data

An illustration of the differences between methods is given in Figure 10.15, where the coefficient functions resulting from the lasso, fused lasso, ridge regression, generalized ridge regression with first-difference penalty, functional data approaches (Ramsey and Silverman, 2005), and BlockBoost are shown for the Canadian weather data, which has become some sort of benchmark dataset. It is seen that the lasso selects only few variables, that is, measurement points, theoretically at most n variables. Here, selecting only a few variables means selecting only few days, whose mean temperature is assumed to be relevant for the

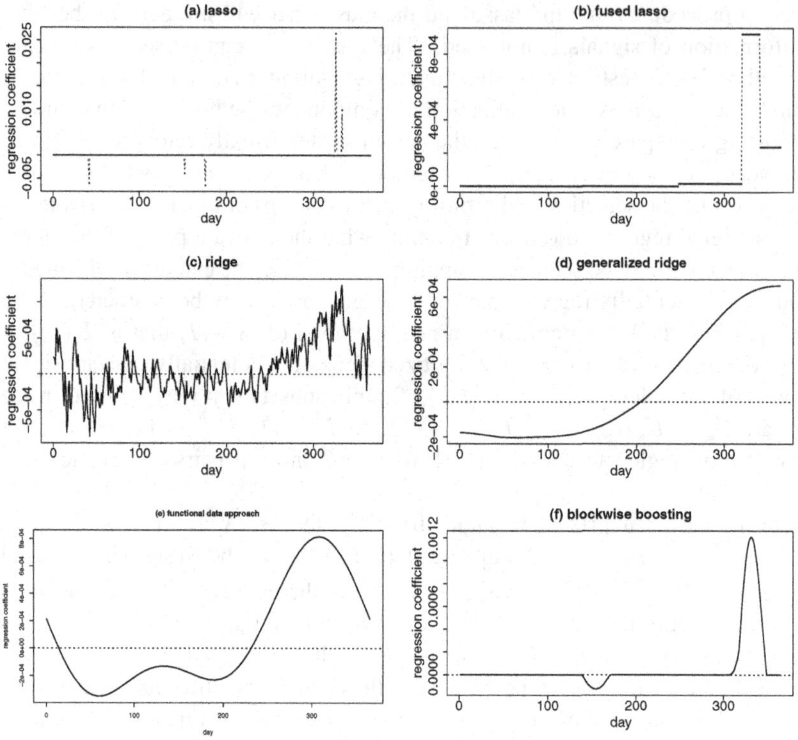

FIGURE 10.15: Regression coefficients estimated by various methods: lasso, fused lasso, ridge, generalized ridge with first-difference penalty, functional data approach, and blockwise boosting. Temperature profiles of 35 Canadian weather stations as predictors; log total annual precipitation as response.

total annual precipitation. By construction, a ridge regression takes into account all variables. Smoothing the coefficient function is possible by penalizing differences between adjacent coefficients, but still almost every day's temperature is considered to be important. The fused lasso and BlockBoost select only some periods instead, that is, some weeks in late autumn / early winter. The smooth BlockBoost estimates result from penalizing (first) differences between adjacent coefficients. Details of the procedures are given in Tutz and Gertheiss (2010). The functional data approach here uses Fourier basis functions for smoothing both the functional regressors and the coefficient function, which has the effect that the end of December shows the same effect as the beginning of January. □

Example 10.12: Prostate Cancer

Figure 10.16 shows the the mass spectrometry data (Example 10.10) together with the parameters that were extracted as relevant. The data show the mean curves for healthy patients (solid line) and patients suffering from prostate cancer (dashed line). The curves themselves hardly show how to discriminate the two groups of patients. However, in particular, the area containing small values of mass charge ratio seem to be relevant for discrimination. Therefore, the area between 0 and 800 is given separately, together with the corresponding estimated values. The major discriminatory power is contained in rather few values of the signal. For a comparison with alternative methods in terms of discrimination power, see Gertheiss and Tutz (2009c). □

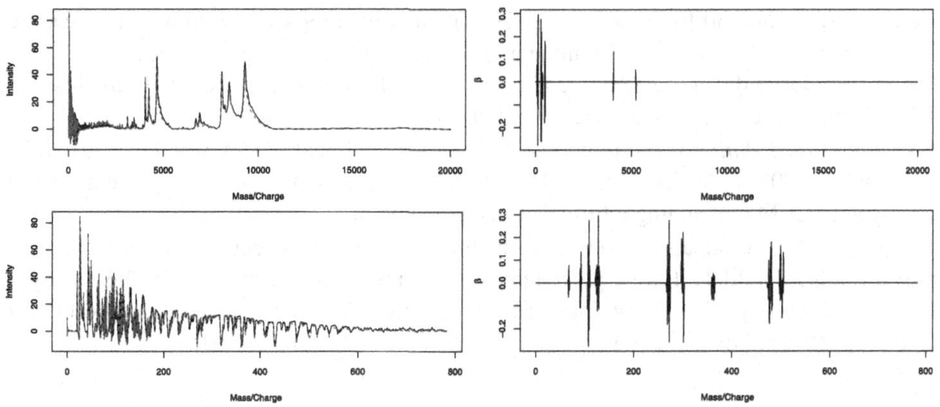

FIGURE 10.16: Mass spectrometry data together with selected parameter values by boosting procedure for the whole spectrum (above) and selected area (below).

10.5 Further Reading

Surveys and Books. Generalized additive models are treated extensively in Hastie and Tibshirani (1990). Green and Silverman (1994) give a thorough account of the roughness penalty approach to smoothing, and localizing techniques are extensively discussed by Fan and Gijbels (1996) and Loader (1999). Semiparametric modeling with the focus on low-rank smoothers and the mixed model representation of penalized splines is outlined by Ruppert et al. (2003), and Ruppert et al. (2009) review the developments in the field during 2003–2007. A source book for generalized additive models and the accompanying R package *mgcv* is Wood (2006a).

Knot Selection Approaches. Stepwise selection of knots in regression splines goes back at least to Smith (1982). Wand (2000) gives an overview and proposed a procedure that generalizes the method of Stone et al. (1997). An alternative procedure, proposed by Osborne et al. (1998), is based on the least absolute shrinkage and regression operator (Tibshirani, 1996). A strategy based on boosting techniques was given by Leitenstorfer and Tutz (2007). He and Ng (1999) proposed a method that is based on quantile regression techniques for smoothing problems (see, e.g., Koenker et al., 1994). Knot selection from a Bayesian perspective was treated by Smith and Kohn (1996), Denison et al. (1998), and Lang and Brezger (2004b).

Fitting of GAMs. Various estimation procedures have been proposed for the estimation of functions $f_{(j)}$. We considered penalized regression splines that were used by Marx and Eilers (1998), and Wand (2000), and backfitting procedures in which the component functions are estimated iteratively by unidimensional smoothers (Hastie and Tibshirani, 1990). Alternative estimators were proposed by Linton and Härdle (1996), who use the marginal integration method. Gu and Wahba (1993) and Gu (2002) treat estimation problems within the framework of smoothing splines, and Bayesian approaches for flexible semiparametric models have been considered, for example, by Fahrmeir and Lang (2001). For methods of smoothing parameter selection see Gu and Wahba (1991, 1993), and Wood (2000). Wood (2004) gave a nice overview of available methods.

High-Dimensional Additive Models and Variable Selection. Avalos et al. (2007) proposed a selection procedure for additive components based on lasso-type penalties, and an alternative approach with lasso penalties was given by Zheng (2008). Marra and Wood (2011) proposed a garrote-type estimator, and Belitz and Lang (2008) a modification of backfitting. A procedure that uses boosting techniques to select influential components was proposed by Tutz and Binder (2006), and Binder and Tutz (2008). Meier et al. (2009) proposed an efficient method for variable selection based on the group lasso and gave asymptotic results.

Asymptotics for P-Splines. Asymptotic properties for the linear model were investigated by Li and Ruppert (2008) and Claeskens et al. (2009). Kauermann et al. (2009) also considered the extension to GLMs, including a fully Bayesian viewpoint.

R Packages. A very versatile package for the fitting of GAMs is *mgcv*; the function *gam* fits generalized additive models. Fitting procedures based on boosting are available in the packages *GAMBoost* and *mboost*. Functional data may be fitted by use of the package *fda*. The *VGAM* package (Yee, 2010) allows one to fit multinomial and ordinal additive models.

10.6 Exercises

10.1 For data (y_i, x_i), $i = 1, \dots, n$, a local estimate of $\mu(x)$ at target value x can be obtained by minimizing the weighted sum $\sum_i (y_i - \mu(x))^2 K_\lambda(x_i - x)$, where $K_\lambda(.) = K(.)/\lambda$, with K denoting a continuous symmetric kernel function fulfilling $\int K(u)du = 1$.

(a) Show that the mininimization problem corresponds to a local regression problem and derive the estimate $\hat{\mu}(x)$.

(b) Derive the criterion to be minimized in a local regression with a polynomial of degree zero and binary responses y_i. Find the estimate $\hat{\mu}(x)$.

(c) Interpret the estimates from (a) and (b) when the uniform kernel $K(u) = 1$ for $0 \leq u \leq 1$ is used.

(d) Derive the bias for the estimates from (a) and (b). Is the estimate unbiased in the limiting case $\lambda \to \infty$?

10.2 Let a penalized likelihood have the form $l_p(\boldsymbol{\eta}) = \sum_{i=1}^n l_i(y_i, \eta_i) - \frac{\lambda}{2}\boldsymbol{\eta}^T \boldsymbol{K} \boldsymbol{\eta}$, where $\boldsymbol{\eta}^T = (\eta_1, \dots, \eta_n)$ and $\boldsymbol{K}$ is a fixed matrix. Derive the maximizer of $l_p(\boldsymbol{\eta})$ if y_i is normally distributed, and $y_i = \eta_i + \epsilon_i$, with noise variable ϵ_i.

10.3 Table 7.1 shows the number of cases of encephalitis in children observed between 1980 and 1993 in Bavaria and Lower Saxony.

(a) Use local regression methods to investigate the variation of counts over years for each country separately.

(b) Compare the results for different degrees of the polynomial fit. Can the fit of a non-localized polynomial regression be obtained as a special case? What favors the local fit?

(c) Compare with the fitted curves for regression splines and smoothing splines.

(d) Discuss the dependence of the fit on the distributional assumption and the link.

10.4 Table 7.13 shows the number of differentiating cells together with the dose of TNF and IFN.

(a) Investigate the effect of TNF and IFN on the number of differentiating cells by fitting an additive model.

(b) Compare the fit of an additive model to the fit of a response surface without further constraints.

(c) Discuss the dependence of the fit on the distributional assumption and the link.

10.5 The dataset *dust* from the package *catdata* contains the binary response variable bronchitis and explanatory variables dust concentration and duration of exposure.

(a) Fit a smooth model without further restrictions and an additive model to the dust data for non-smokers and compare.

(b) Compare to the results of the fitting of a GLM.

10.6 The dataset *addiction* from the package *catdata* contains the response in the three categories 0: addicts are weak-willed; 1: addiction is a disease; 2: both. Fit two additive logit models, one comparing category $\{1\}$ and category $\{0\}$ and the other comparing category $\{2\}$ and category $\{0\}$.

(a) Fit smooth models with only predictor age allowing for an unspecified functional form. Visualize the resulting probabilities.

(b) Fit smooth models with predictor age allowing for an unspecified functional form and additional covariates. Visualize the resulting probabilities and compare with (a).

Chapter 11

Tree-Based Methods

Tree-based models provide an alternative to additive and smooth models for regression problems. The method has its roots in automatic interaction detection (AID), proposed by Morgan and Sonquist (1963). The most popular modern version is due to Breiman et al. (1984) and is known by the name *classification and regression trees*, often abbreviated as CART, which is also the name of a program package. The method is conceptually very simple. By binary recursive partitioning the feature space is partitioned into a set of rectangles, and on each rectangle a simple model (for example, a constant) is fitted.

The approach is different from that given by fitting parametric models like the logit model, where linear combinations of predictors are at the core of the method. Rather than getting parameters, one obtains a binary tree that visualizes the partitioning of the feature space. If only one predictor is used, the one-dimensional feature space is partitioned, and the resulting estimate is a step function that may be considered a non-parametric (but rough) estimate of the regression function.

11.1 Regression and Classification Trees

In the following the basic concepts of classification and regression trees (CARTs) are given. The term "regression" in CARTs refers to metrically scaled outcomes, while "classification" refers to the prediction of underlying classes. By considering the underlying classes as the outcomes of a categorical response variable, classification can be treated within the general framework of regression.

Let us first illustrate the underlying principle for a one-dimensional response variable y and a two-dimensional predictor. In the first step one chooses a predictor and a split-point to split the predictor space into two regions. In each region the response is modeled by the mean in that region, and choice of the split is guided by the best fit to the data. For example, in Figure 11.1 the first split yields the partition $\{x_1 \leq c_1\}$, $\{x_1 > c_1\}$. In the next step one or both of the obtained regions are split into new regions by using the same criterion for the selection of the predictor and the split-point. In Figure 11.1, the region $\{x_1 \leq c_1\}$ is split by variable x_2 at split-point c_2, yielding the partition $\{x_2 \leq c_2\} \cap \{x_1 \leq c_1\}$, $\{x_2 > c_2\} \cap \{x_1 \leq c_1\}$. The region $\{x_1 > c_1\}$ is split at c_3, yielding $\{x_2 \leq c_3\} \cap \{x_1 > c_1\}$, $\{x_2 > c_3\} \cap \{x_1 > c_1\}$. In our example, in a last step, the region $\{x_2 > c_2\} \cap \{x_1 \leq c_1\}$ is split into $\{x_1 \leq c_4\} \cap \{x_2 > c_2\} \cap \{x_1 \leq c_1\}$ and $\{x_1 > c_4\} \cap \{x_2 > c_2\} \cap \{x_1 \leq c_1\}$. In the end one obtains five regions, $R_1, \ldots, R_5$. Since in each region the response is modeled by the mean, one obtains a regression

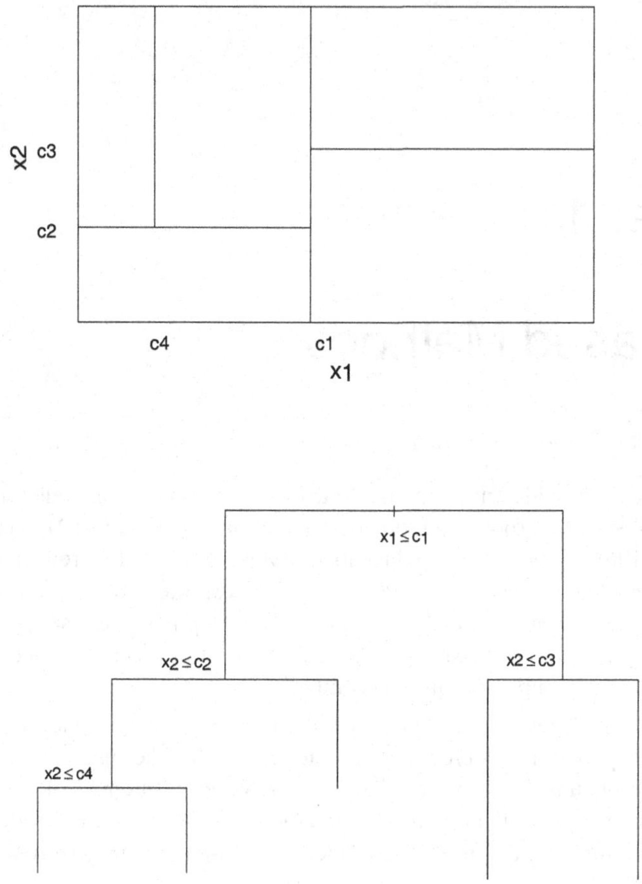

FIGURE 11.1: Partition of a two-dimensional space and the corresponding tree.

model with a piecewise constant fit:

$$\hat{\mu}(\boldsymbol{x}) = \sum_{i=1}^{5} \gamma_i I((x_1, x_2) \in R_i).$$

A node of the tree corresponds to a subset of the predictor space. The *root* is the top node consisting of $\mathbb{R}^2$, and the *terminal nodes* or *leaves* of the tree correspond to the regions $R_1, \dots, R_5$. The partition may be seen from the upper panel of Figure 11.1 or, simpler, from the tree below it.

In general, trees may be seen as a hierarchical way to describe a partition of the predictor space. For more than two predictor variables it is hard to visualize the partition of the space. But the tree represents the partition in a unique way. Moreover, the tree yields an interpretable structure of how the predictors are linked to the response since response is modeled constant within the nodes. A further advantage of the hierarchical structuring is that the growing of a tree may be described locally by specifying how a given node A, corresponding to a subset of predictor space, is partitioned into a left and right daughter nodes, which correspond to subsets A_1, A_2 of A.

To obtain an algorithm for the growing of a tree one has to decide on several issues. In particular, the type of partitioning and the split criterion have to be chosen, and one has to decide when to stop splitting.

11.1.1 Standard Splits

For the trees considered here, the splits are restricted to "standard splits," which means that each partition of node A into subsets A_1, A_2 is determined by only one variable. The splits to be considered depend on the scale of the variable:

- For *metrically scaled* and *ordinal* variables, the partition into two subsets has the form

  $$A \cap \{x_i \le c\}, \quad A \cap \{x_i > c\},$$

  based on the threshold c on variable x_i.

- For *categorical* variables without ordering $x_i \in \{1, \dots, k_i\}$, the partition has the form

  $$A \cap S, \quad A \cap \bar{S},$$

  where S is a non-empty subset $S \subset \{1, \dots, k_i\}$ and $\bar{S} = \{1, \dots, k_i\} \setminus S$ is the complement.

The number of possible splits is determined by the scaling of the variable and the data that are given by $(y_i, \boldsymbol{x}_i)$, $1 = 1, \dots, n$. For continuous variables one typically has $n - 1$ thresholds to consider; when the data are discrete the number of thresholds usually reduces. For categorical variables on a nominal scale (without ordering) one obtains $2^{n-1} - 1$ non-empty pairs $S, \bar{S}$, $S \subset \{1, \dots, k_i\}$, if all the categories have been observed. An alternative approach that was proposed by Quinlan (1986) uses multiple splits instead of binary splits when the predictors are categorical. Then one obtains as many nodes as there are categories.

11.1.2 Split Criteria

When considering all possible splits one has to decide on the value of a split, in order to select one. Several criteria for the values of a split have been proposed. One approach is based on test statistics, while others use impurity measures like the Gini index. We will start with test-based procedures.

Test-Based Splits

Let A be a given node that is to be split into two subsets A_1, A_2 and y be a one-dimensional response variable with mean $\mu(\boldsymbol{x}) = E(y|\boldsymbol{x})$. It should be noted that only the subset of observations $(y_i, \boldsymbol{x}_i)$ with $\boldsymbol{x}_i \in A$ is used when the splitting of A is investigated. Typically the model to be fitted specifies that the mean is constant over regions of the predictor space. One compares the fit of the model

$$M_A : \quad \mu(\boldsymbol{x}) = \mu \text{ for } \boldsymbol{x} \in A$$

to the fit of the model

$$M_{A_1,A_2} : \quad \mu(\boldsymbol{x}) = \mu_1 \text{ for } \boldsymbol{x} \in A_1, \quad \mu(\boldsymbol{x}) = \mu_2 \text{ for } \boldsymbol{x} \in A_2,$$

where A_1 and A_2 are subsets of A, defined by $A_1 = A \cap \{x_i \leq c\}$, $A_2 = A \cap \{x_i > c\}$ if x_i is a metrically scaled predictor. The model M_A specifies that the response is homogeneous across the region A, whereas M_{A_1,A_2} specifies that the response is homogeneous across the subsets A_1 and A_2. To evaluate of the improvement in fit when model M_{A_1,A_2} is fitted instead of model M_A, one needs some measure for the goodness-of-fit of these models. A measure that applies if the responses are from a simple exponential family is the difference of deviances:

$$D(M_A|M_{A_1,A_2}) = D(M_A) - D(M_{A_1,A_2})$$

where $D(M_A)$, $D(M_{A_1,A_2})$ denote the deviances of the models M_A and M_{A_1,A_2} respectively. Since the deviance of a model measures the discrepancy between the data and the model fit, the difference $D(M_A|M_{A_1,A_2})$ represents the increase in discrepancy when model M_A is fitted instead of model M_{A_1,A_2}. Therefore, one selects the partition A_1, A_2, for which the difference in discrepancies is maximal. The procedure is equivalent to the selection of the partition A_1, A_2, which shows the minimal discrepancy between the data and the fit. This is easily seen by rewriting the criterion in the form

$$D(M_A|M_{A_1,A_2}) = D(M_A) - \{D(M_{A_1}) + D(M_{A_2})\}, \qquad (11.1)$$

where $D(M_{A_i})$ denotes the deviance of the model $M_{A_i} : \mu(\boldsymbol{x}) = \mu_i$ for $\boldsymbol{x} \in A_i$, fitted for observation $(y_i, \boldsymbol{x}_i)$ with $\boldsymbol{x}_i \in A_i$. It should be noted that the representation of $D(M_{A_1,A_2})$ as $\{D(M_{A_1}) + D(M_{A_2})\}$ is possible because disjunct subsets of observations are used. Since $D(M_A)$ is fixed, one selects the partition A_1, A_2 that has the smallest deviance $\{D(M_{A_1}) + D(M_{A_2})\}$. This also means that one selects the partition that shows the best fit to the data.

Since M_A is a submodel of M_{A_1,A_2}, the difference of deviances is also a test that investigates if model M_A holds, given M_{A_1,A_2} holds. Therefore, the difference is abbreviated by $D(M_A|M_{A_1,A_2})$. Of course alternative test statistics may be used to this end. In the following we consider some examples for specific responses.

Dichotomous Responses

For dichotomous data, encoded by $Y \in \{1, 2\}$ or $y \in \{1, 0\}$, the data may be arranged in a (2×2)-table. With $n_i(A_j)$ denoting the number of observations with response $Y = i$ within region A_j one obtains the table in Table 11.1. The marginals are given by $n(A_j) = n_1(A_j) + n_2(A_j)$ and $n_i(A) = n_i(A_1) + n_i(A_2)$. The total sum $n(A) = n(A_1) + n(A_2)$ is the number of observations with $\boldsymbol{x}_i \in A$.

TABLE 11.1: Contingency table for node A, split into A_1, A_2.

	Y		
	1	2	Marginals
A_1	$n_1(A_1)$	$n_2(A_1)$	$n(A_1)$
A_2	$n_1(A_2)$	$n_2(A_2)$	$n(A_2)$
	$n_1(A)$	$n_2(A)$	$n(A)$

With $\pi(\boldsymbol{x}) = P(Y = 1|\boldsymbol{x})$ denoting the mean, the models to be compared are $M_A : \pi(\boldsymbol{x}) = \pi$ for $\boldsymbol{x} \in A$ and $M_{A_1,A_2} : \pi(\boldsymbol{x}) = \pi_i$ for $\boldsymbol{x} \in A_i$. The corresponding deviances are

$$D(M_A) = -2\{n_1(A)\log(p(A)) + n_2(A)\log(1 - p(A))\},$$
$$D(M_{A_s}) = -2\{n_1(A_s)\log(p(A_s)) + n_2(A_s)\log(1 - p(A_s))\},$$

yielding

$$D(M_A|M_{A_1,A_2}) = 2\sum_{i=1}^{2} n(A_i)\left\{p(A_i)\log(\frac{p(A_i)}{p(A)}) + \{1-p(A_i)\}\log(\frac{1-p(A_i)}{1-p(A)})\right\}, \qquad (11.2)$$

where $p(A) = n_1(A)/n(A)$ and $p(A_i) = n_1(A_i)/n(A_i)$ are the ML estimates under M_A and M_{A_1,A_2}, respectively.

For given nodes A and partition A_1, A_2, the *conditional* deviance has asymptotically a χ^2-distribution with 1 df. Thus the "best" partition may also be found by looking for the smallest p-value amongst all partitions. Alternative test statistics may be used to test the homogeneity of the response within A. One may use Pearson's χ^2-statistic (see Section 4.4.2) or an exact test, like Fisher's test. By using Fishers's test one obtains exact p-values, which seems advantageous.

The following example demonstrates the fitting of a tree when the predictor space is one-dimensional.

Example 11.1: Duration of Unemployment

In Example 10.1, the effects of age on the probability of short-term unemployment were investigated by non-parametric modeling. In contrast to these non-parametric smooth fits, the fitting of a tree yields a step function because the predictor space is partitioned into regions in which the response probability is held constant. The split-points show where the probability changes. Figure 11.2 shows the resulting fit obtained by the R package *party* and the corresponding tree. It is seen that the fitted tree distinguishes only three regions with the first split-point at 52 and the second split point at 29 years of age. □

Multicategorical Response

For multicategorical response $Y \in \{1, \dots, k\}$ and node A, partitioned into A_1, A_2, cross-classification is given by the contingency table in Table 11.2.

TABLE 11.2: Contingency table for node A, partitioned into A_1, A_2.

	Y				
	1	2	...	k	Marginals
A_1	$n_1(A_1)$	$n_2(A_1)$	...	$n_k(A_1)$	$n(A_1)$
A_2	$n_1(A_2)$	$n_2(A_2)$	...	$n_k(A_2)$	$n(A_2)$
	$n_1(A)$	$n_2(A)$	...	$n_k(A)$	$n(A)$

As before, let $p_r(A) = n_r(A)/n(A)$, $p_r(A_j) = n_r(A_j)/n(A_j)$ denote the relative frequencies. Then, the deviances have the form

$$D(M_A) = -2\sum_{r=1}^{k} n_r(A_i)\log(p_r(A))$$

and the difference of deviances is

$$D(M_A|M_{A_1,A_2}) = 2\sum_{i=1}^{2} n(A_i)\sum_{r=1}^{k} n_r(A_i)\log\left(\frac{p_r(A_i)}{p_r(A)}\right).$$

The test statistic is asymptotically χ^2-distributed with $k-1$ df, as is the corresponding Pearson statistic. For exact tests see, for example, Agresti (2002).

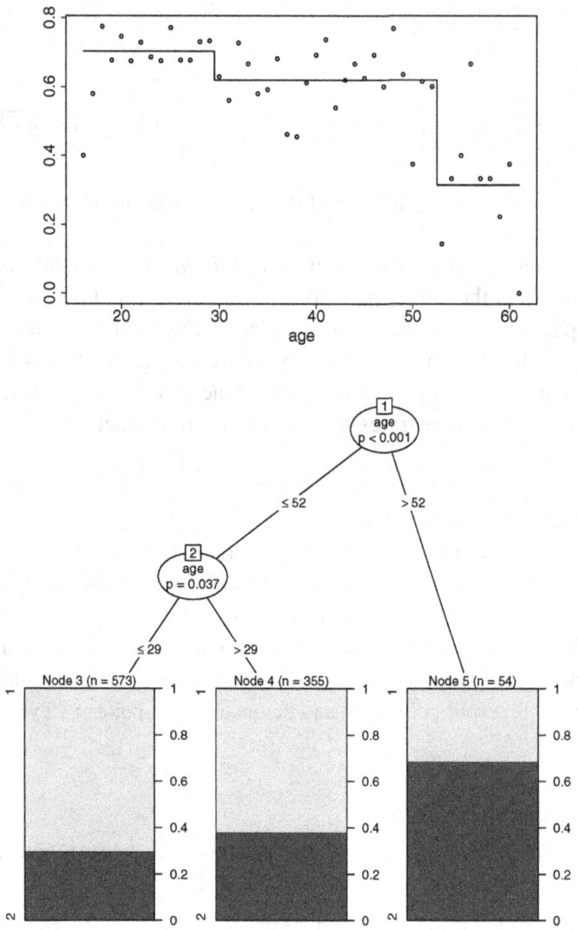

FIGURE 11.2: Fitted function and tree for the probability of short-term unemployment depending on age.

Splitting by Impurity Measures

An alternative way to obtain splitting rules is by considering the "impurities" of nodes. Let us consider a multicategorical response $Y \in \{1, \dots, k\}$ with response probabilities $\pi_1(A), \dots, \pi_k(A)$ within region A. Then the *impurity of node* A may be measured, according to Breiman et al. (1984), by measures of the form

$$I(A) = \phi(\pi_1(A), \dots, \pi_k(A)),$$

where ϕ is an *impurity function*, which is symmetrical in its arguments and takes its minimal value if the distribution is concentrated on one response category, that is, if one of the probabilities $\pi_1(A), \dots, \pi_k(A)$ takes value 1. Moreover, one postulates that $\phi(1/k, \dots, 1/k) \geq \phi(\pi_1, \dots, \pi_k)$ holds for all probabilities $\pi_1, \dots, \pi_k$. A commonly used measure of this type is the *Gini index*, which uses the impurity function $\phi(\boldsymbol{\pi}) = -\sum_{i \neq j} \pi_i \pi_j$, yielding the impurity

$$I_G(A) = 1 - \sum_{r=1}^{k} (\pi_r(A))^2.$$

Another is based on the *entropy* $\phi(\boldsymbol{\pi}) = -\sum_r \pi_r \log(\pi_r)$, yielding the impurity

$$I_E(A) = -\sum_{r=1}^{k} \pi_r(A)\log(\pi_r(A)).$$

It is easily seen that both measures take value zero if one of the probabilities $\pi_1(A), \ldots, \pi_k(A)$ has value 1 (and the rest 0). The node is considered as "purest" because the probability mass is concentrated within one response category. Consequently, the impurity takes its smallest value, namely, zero. For a uniform distribution across response categories, $\pi_1(A) = \ldots = \pi_k(A) = 1/k$, the impurity takes its maximal value.

Having defined what impurity means, we now can consider criteria that are apt to reduce impurities by splitting. The decrease in impurities may be measured by

$$\Delta_I(A|A_1, A_2) = I(A) - \{w_1 I(A_1) + w_2 I(A_2)\}, \tag{11.3}$$

where w_1, w_2 are additional weights that add up to one. By selecting the split that has minimal impurity $w_1 I(A_1) + w_2 I(A_2)$, one maximizes the decrease in impurity.

The measures considered so far are population versions of impurity, defined for probabilities $\pi_1(A), \ldots, \pi_k(A)$. In the empirical versions one replaces the probabilities $\pi_1(A), \ldots, \pi_k(A)$ by relative frequencies $\hat{\pi}_1(A) = n_1(A)/n(A), \ldots, \hat{\pi}_k(A) = n_k(A)/n(A)$ with $n_r(A), n(A)$ taken from Table 11.2. For example, the empirical version of the Gini-based impurity has the form

$$I_{G,emp}(A) = 1 - \sum_{r=1}^{k}(n_r(A)/n(A))^2.$$

The weights in (11.3) are typically chosen by $w_i = n(A_i)/n(A)$ and measure the proportion of observations in A_i and observations in A. If impurity is measured by entropy, one obtains for the empirical difference

$$\Delta_{I_E,emp} = \frac{1}{2n(A)} D(M_A).$$

Since $n(A)$ is fixed, the maximization of the decrease in impurity measured by entropy is equivalent to the deviance criterion (11.1). Therefore, for a given A, the deviance criterion may be seen as a special case of impurity minimization.

The criterion (11.3) is a local criterion that steers the selection in one step of the algorithm. It may also be linked to the concept of impurity of the whole tree. Let the empirical version of *impurity of a tree* be defined by

$$I(\text{Tree}) = \sum_{\text{terminal nodes } A} \hat{p}(A) I_{emp}(A),$$

where $\hat{p}(A) = n(A)/n$ is an estimate of the probability that an observation reaches node A and $I_{emp}(A)$ is the empirical version of $I(A)$, where the probabilities $\pi_r(A)$ have been replaced by $\hat{\pi}_r(A) = n_r(A)/n(A)$. If node A is split into A_1 and A_2, the difference in impurity is

$$I(\text{Tree before splitting of } A) - I(\text{Tree with splitting into } A_1, A_2) = \hat{p}(A)\Delta_I(A|A_1, A_2).$$

Therefore, for a fixed A, a maximal decrease of impurity of A by splitting maximally decreases the impurity of the tree. When using the entropy one obtains for the impurity of the tree

$$I(\text{Tree}) = -\frac{1}{n} \sum_{\text{terminal nodes}} \sum_{r=1}^{k} n_r(A) \log\left(\frac{n_r(A)}{n(A)}\right) = -\frac{1}{2n} D(\text{Tree}),$$

where $D(\text{Tree})$ is the deviance of the partition generated by the tree. The deviance $D(\text{Tree})$ compares the fit of the "model" partition into final nodes to the (ungrouped) saturated model for the training set; see Ciampi et al. (1987) and Clark and Pregibon (1992).

11.1.3 Size of a Tree

When growing a tree the impurity of the tree decreases and the partition gets finer. Therefore, it might be tempting to grow a rather large tree. However, there is a severe danger of overfitting. A tree with terminal nodes that contain only few observations often generalizes badly. Prediction in future datasets is worse than for moderate-sized trees. Therefore, the growing of a tree has to be stopped at an appropriate size. Simple step criteria are

- Stop if a node contains fewer than n_{STOP} observations.

- Stop if the splitting criterion is above or below a fixed threshold (for example, if the p-value is above p_{STOP} or the improvement in decrease is below a fixed value).

An alternative strategy to obtain a tree of an appropriate size is to grow a very large tree and then prune the tree. Growing a large tree typically means stopping when some small minimal node size (say 3 observations in the node) is reached. A procedure for tree pruning proposed by Breiman et al. (1984) is based on a cost complexity criterion of the form

$$C_\alpha(T) = C(T) + \alpha|T|,$$

where $|T|$ is the number of terminal nodes, also called the *size* of tree T, $\alpha \geq 0$ is a tuning parameter, and $C(T)$ is a measure of the goodness-of-fit or the performance of the partition given by the terminal nodes. Candidates for $C(T)$ are the deviance or the entropy of the partition or as a measure of performance of the prediction on the training set or a validation set. The tuning parameter α may be chosen by m-fold cross-validation or by the prediction in a validation set. It governs the trade-off between tree size and closeness to the data. For large α one obtains a small tree, whereas $\alpha = 0$ yields very large trees.

Pruning a tree T means that the nodes are collapsed, thereby obtaining a subtree of T or, more precisely, a rooted subtree, which means that the subtree has the same root as T. It may be shown that for each α there is a unique smallest subtree that minimizes $C_\alpha(T)$. The trees may be found by successively collapsing the nodes, yielding a finite sequence of trees that are optimal with respect to $C_\alpha(T)$ for specific intervals of α-values. For details see Breiman et al. (1984) or Ripley (1996), Chapter 7.

An alternative, quite attractive way to determine the size of a tree was proposed by Hothorn et al. (2006). They proposed a unified framework for recursive partitioning that embeds tree-structured regression models into a well-defined theory of conditional inference procedures. The splitting is stopped when the global null hypothesis of independence between the response and any of the predictors cannot be rejected at a pre-specified nominal significance level α. The method explicitly accounts for the involved multiple test problem. By separating variable selection and splitting procedure one arrives at an unbiased recursive partitioning scheme that also avoids the selection bias toward predictors with many possible splits or missing values (for selection bias see also Section 11.1.5). Since the method employs p-values for variable selection, it does not rely on pruning.

Example 11.2: Exposure to Dust

In Example 11.2 only the effects of duration on bronchitis were examined. Now we include the concentration of dust and smoking. We fit a conditional independence tree by using the *ctree* function of the R package *party* (Hothorn et al., 2006). In each node a significance test on independence between any of the predictors and the response is performed and a split is established when the p-value is smaller than a pre-specified significance level. Figure 11.3 shows the resulting tree. The first split distinguishes between lesser and more than 15 years of exposure. When exposure was for more than 15 years it is distinguished between smokers and non-smokers. Subpopulations are split further with respect to years and concentration of exposure. □

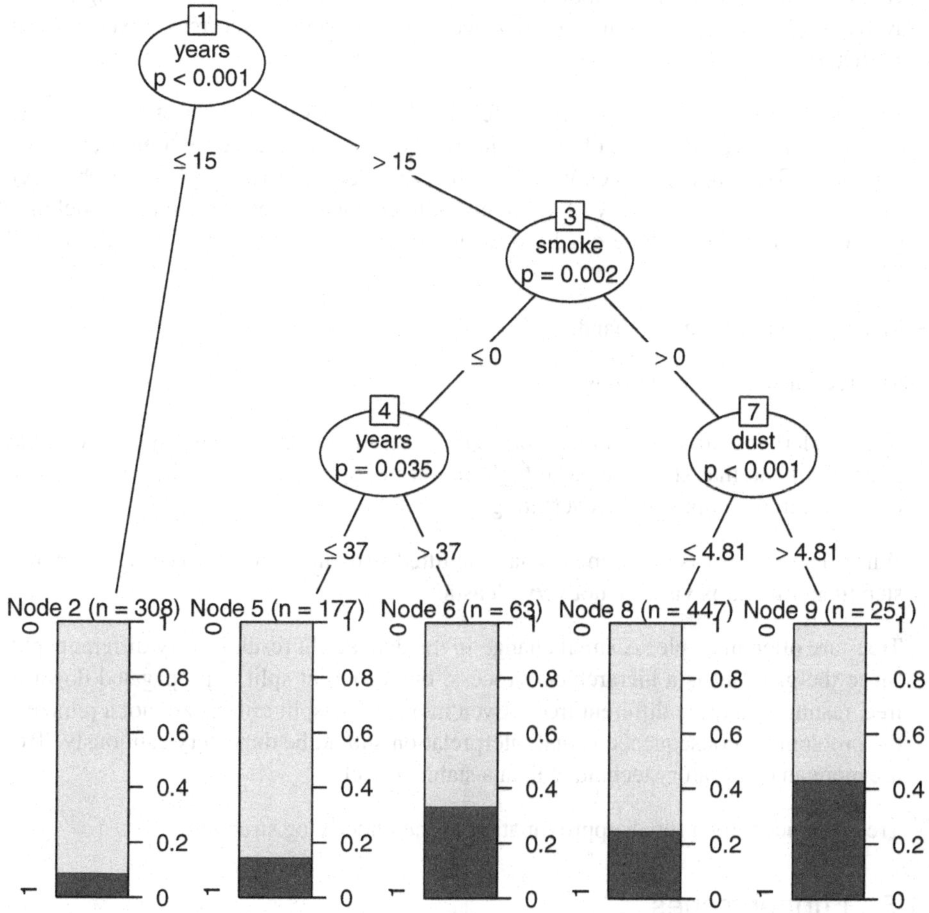

FIGURE 11.3: Trees for exposure data depending on years of exposure.

11.1.4 Advantages and Disadvantages of Trees

The major advantages of trees, which made them popular, are the following:

- Trees are easy to interpret and the visualization makes it easy to communicate the underlying structure to practitioners. Practitioners can understand what model is fitted and, depending on their statistical experience, are often happier with trees than with (generalized) linear models.

- Trees provide a non-parametric modeling approach with high flexibility.

- Trees have a built-in interaction detector. By splitting successively in different variables, interaction effects between two and more variables are captured.

- Trees are invariant to monotone transformations of predictor variables. This is a big advantage over most parametric models, where it seriously matters if predictors are, for example, log-transformed or not.

- Trees, in particular in the form of random trees (see Section 15.5.2), are among the best available classifiers with high prediction accuracy. Averaging over trees makes powerful predictors.

- Trees are not restricted to the case of few predictors. Even in the $p \gg n$ case, when one has more predictors than observations, trees apply, because they automatically select variables. The implicit selection of variables may be seen as an advantage by itself. Thus, trees turn out to be very useful in high-dimensional problems like the modeling of gene expression data, where one has few observations but sometimes several thousands of predictors.

- Missing values are easy to handle.

Of course trees also have disadvantages:

- If the underlying structure is linear or, more generally, additive, many splits are needed to approximate that structure. The high number of splits typically will not be reached because it might appear to be overfitting.

- When the predictor is one-dimensional, the fitted structure is not smooth. One obtains a step function that is visually not very pleasing.

- Trees are often unstable. A small change in the data might result in very different splits. Since the growing is a hierarchical process, one different split is propagated down the tree, resulting in quite different trees. Even more stable split criteria are not a remedy to the problem. A consequence is that interpretation should be done very cautiously. Trees are more an exploratory technique than a stable model.

- Trees provide only a rough approximation of the underlying structure.

11.1.5 Further Issues

Prediction

When a tree has been grown and one has decided on its size, a response is predicted in each terminal node. One uses all the observations in the terminal node A to obtain the average response value $y_A = ave(y_i|\boldsymbol{x}_i \in A)$ in regression and the majority vote $y_A = \text{argmax}_j(\sum_i I(y_i = j|\boldsymbol{x}_i \in A))$ in classification trees.

Random Forests

Random forests are a combination of many trees that show much better performance in prediction than single trees. They are composed as so-called ensemble methods, which means that various predictors are aggregated for prediction. Random methods are considered in the context of prediction (Section 15.5.2).

Missing Values

When considering the problem of missing values, one should distinguish between the fitting procedure and the application on future samples. As far as the fitting procedure is concerned, trees can easily handle missing values since splitting uses only one variable at a time. Therefore, in an available case strategy one uses all the observations that are available for the variable to be split. When predictor values are missing in future data one uses surrogate variables that best mimic the split in the training data achieved by the selected variable.

Selection Bias

When the predictor variables vary in the number of distinct observations, for example, due to missing values, variables are systematically preferred or penalized. Loh and Shih (1997) demonstrated that predictor variables with more distinct values are systematically preferred over variables with less distinct values when Pearson's χ^2-statistic is used for continuous predictor and binary outcomes. In contrast, Loh and Shih (1997) showed that in classification trees selection is biased toward selection of variables with a lower number of distinct values, caused by missing values, if the Gini index is used as a split criterion. Strobl et al. (2007) showed that partially counteracting effects are at work when the variables vary in the number of distinct values. One effect is due to multiple comparisons. When a variable has more categories, more binary partitions are evaluated and consequently variables with many categories are preferred over variables with a smaller number of possible values. A quite different effect is the bias toward variables with many missing values that has been observed for metric variables. The effect is somewhat counterintuitive because many missing values means that the number of possible splits is reduced. Of course splitting is based on the common available case strategy, which means that for each variable one uses the data that have been observed for that variable. Since the number of missing values may vary strongly across variables, the resulting selection bias may be severe. Strobl et al. (2007) investigated the effects for trees obtained by Gini-based splits by investigating bias and variance effects. In the binary response case the impurity based on the Gini index is given by $I_G(A) = 2\pi_1(A)(1 - \pi_1(A))$. The empirical version has the form $I_{G,emp} = 2\hat{\pi}_1(A)(1-\hat{\pi}_1(A))$, where $\hat{\pi}_1(A) = n_1(A)/n(A)$. It turns out that the mean is given by $E(I_{G,emp}) = \{(n(A)-1)/n(A)\}I_G(A)$, which means that the underlying impurity is underestimated. The effect carries over to the empirical decrease of impurity. It may be shown that $E(\Delta_{I_G,emp}(A|A_1, A_2)) = I_G(A)/n(A)$. This means that for uninformative variables, where the expectation should be zero, but with an underlying distinctly positive Gini index I_G, the decrease in impurity is a biased estimate. The bias is small for a large sample size but becomes influential if $n(A)$ is small, which occurs when many observations are missing. Therefore, selection by the difference in impurity tends to select uninformative variables when the number of available observations for that variable is small. Selection bias may be avoided by using unbiased classification trees like the ones proposed by Hothorn et al. (2006).

Maximally Selected Statistics

To avoid selection bias one should separate variable selection from cut-point selection. In particular, the comparison of test statistics across variables and cut-points yields biased results since different numbers of splits are invoked for different variables. One approach to separate variable selection from cut-point selection is to compute p-values for each variable by using the maximally selected statistic. The basic idea behind maximally selected statistics is to consider the distribution of the selection process. When a split-point is selected based on a test statistic or association measure T_i for all the possible split-points $i = 1, \dots, m$, one investigates the distribution of $T_{max} = max_{i=1,\dots,m}T_i$. The p-value of the distribution of T_{max} provides a measure for the relevance of a predictor that does not depend on the number of split-points since the number has been taken into account.

The maximally selected statistic approach was proposed by Miller and Siegmund (1982) and has been extended by Hothorn and Lausen (2003), Shih (2004), and Shih and Tsai (2004). Strobl et al. (2007) extended the approach to Gini-based splitting rules.

11.2 Multivariate Adaptive Regression Splines

Multivariate adaptive regression splines (MARS), introduced by Friedman (1991), may be seen as a modification of tree methodology. The first split in CARTs can be seen as fitting a model with predictor

$$\eta(\boldsymbol{x}) = \gamma_1 I((x_j \leq c_j) + \gamma_2 I((x_j > c_j),$$

where the variable x_j is split at split-point c_j. In the next step, one of the regions, $\{x_j \leq c_j\}$ or $\{x_j > c_j\}$, is split further. If $\{x_j \leq c_j\}$ is split by use of variable x_l, the fitted model has the form

$$\eta(\boldsymbol{x}) = \gamma_1 I((x_j \leq c_j) I((x_l \leq c_l) + \gamma_2 I((x_j \leq c_j) I((x_l > c_l) + \gamma_3 I((x_j > c),$$

where, of course, the denotation of the γ-parameters has changed. Therefore, CART methodology fits a model of the form

$$\eta(\boldsymbol{x}) = \sum_{i}^{m} \gamma_i h_i(\boldsymbol{x}),$$

where the functions $h_i(\boldsymbol{x})$ are built as products of simple (indicator) functions that depend on only one component of the vector $\boldsymbol{x}$. The functions themselves are found adaptively by recursive partitioning.

In MARS the products of indicator functions are replaced by piecewise linear basis functions given by

$$(x - t)_+ \quad \text{and} \quad (t - x)_+,$$

where the subscript denotes the positive part, that is, $(a)_+ = a$ if $a > 0$ and 0 otherwise. Both functions, $(x - t)_+$ and $(t - x)_+$, are piecewise linear with a knot at the value t and are called a *reflected pair*. The observation values of the corresponding variable are used as knots. As in CARTs, the given "model" is enlarged by adding products with basis functions, but in MARS the used basis functions are reflected pairs. At each stage one adds to the given model a term that is constructed from former basis functions $h_i(\boldsymbol{x})$ in the form

$$\gamma_{1,new} h_i(\boldsymbol{x})(x_j - t)_+ \gamma_{2,new} h_i(\boldsymbol{x})(t - x_j)_+.$$

Thus a pair of new basis functions is added. The new model is fit by common ML methods for linearly structured predictors. Selection of the basis function $h_i(\boldsymbol{x})$ that is enlarged to products, the variable that is used, and the knot are all based on goodness-of-fit.

The procedure typically starts with the basis function $h_1(\boldsymbol{x}) = 1$. Then, in the first step, it is enlarged to

$$\gamma_1 h_1(\boldsymbol{x}) + \gamma_2 h_2(\boldsymbol{x}) + \gamma_3 h_3(\boldsymbol{x}),$$

where $h_2(\boldsymbol{x}) = (x_j - t_j)_+$, $h_3(\boldsymbol{x}) = (t_j - x_j)_+$. In the next step one adds a term of the form

$$\gamma_4 h_i(\boldsymbol{x})(x_l - t_l)_+ \gamma_5 h_i(\boldsymbol{x})(t_l - x_l)_+,$$

where $h_i(\boldsymbol{x})$ is any previously used function $h_1(\boldsymbol{x})$, $h_2(\boldsymbol{x})$ or $h_3(\boldsymbol{x})$. In general, one adds a pair of basis functions that are built as products from basis functions that were already in the model. Thus multiway products are always built from previous basis functions.

Typically not all possible products are allowed. The inclusion of a pair of basis functions is restricted such that a predictor can appear at most once in a product. Otherwise higher order powers of a predictor, which would make the procedure more unstable, could occur. An additional optional restriction is often useful. By setting an upper limit on the order of interaction one obtains a fit that has an easier interpretation. For example, if the upper limit is two, only basis functions that contain products of two piecewise linear functions are allowed in the predictor term. The resulting fit includes the interaction between only two predictors. It may be seen as a less structured form of interaction modeling than the approaches considered in Section 10.3.3.

Aside from the use of piecewise linear functions instead of indicator functions the main difference from CARTs is that, in CARTs, given basis functions are always modified in one step of the fitting procedure, whereas in MARS new basis functions are added to the given basis functions. Therefore MARS does not yield a simple tree like CART does.

Similar as in trees, MARS may be based on increasing the predictor and subsequent pruning (Friedman, 1991). Stone et al. (1997) considered a modification of MARS called POLYMARS, which allows for a multidimensional response variable and defines an allowable space by listing its basis functions. Stone et al. (1997) gave a more general treatment of polynomial splines and tensor products within an approach called extended linear modeling.

11.3 Further Reading

Basic Literature and Overviews. The roots of classification and regression trees are in automatic interaction detection, as proposed by Morgan and Sonquist (1963). The most popular methods are CART, outlined in Breiman et al. (1984), and the C4.5 algorithm (and its predecessor ID3), which were proposed by Quinlan (1986, 1993). The main difference between the approaches is that C4.5 produces for categorical variables as many nodes as there are categories (multiple or *k*-ary splitting) whereas CART uses binary splitting rules. Zhang and Singer (1999) gave an overview on recursive partitioning in the health sciences, and Strobl, Malley, and Tutz (2009) gave an introduction including random forests with applications in psychology.

Multiple and Ordinal Regression. An extension of classification trees to multiple binary outcomes was given by Zhang (1998). Piccarreta (2008) considered classification trees for ordinal response variables.

High-Dimensional Predictors. Trees and random forests have been successfully used in genetics; see, for example, Diaz-Uriarte and de Andres (2006a), Lunetta et al. (2004), and Huang et al. (2005).

R Packages. Trees can be fitted by using the function *rpart* from the package *rpart*. Conditional unbiased recursive partitioning is available in the function *ctree* of the package party (Hothorn et al., 2006). The function *cforest* from the same package fits random forests, which are also available in *randomForest*. MARS can be fitted by use of *mars* from *mda*.

11.4 Exercises

11.1 For which values of $\pi_1, \dots, \pi_k$, $\sum_i \pi_i = 1$, do the Gini index $\phi(\boldsymbol{\pi}) = -\sum_{i \neq j} \pi_i \pi_j$ and the entropy $\phi(\boldsymbol{\pi}) = -\sum_r \pi_r \log(\pi_r)$, $\boldsymbol{\pi}^T = (\pi_1, \dots, \pi_k)$ take their maximal and minimal values?

11.2 Fit a tree for the dust data (package *catdata*) that investigates the effect of duration of exposure on bronchitis and compare it with the fit of a logit model with the linear and quadratic effects of duration of exposure.

11.3 Fit a tree for the heart disease data considered in Example 6.3 (dataset *heart* from package *catdata*) and discuss the results in comparison to the fit of a linear logistic model.

Chapter 12

The Analysis of Contingency Tables: Log-Linear and Graphical Models

Contingency tables, or cross-classified data, come in various forms, differing in dimensions, distributional assumptions, and margins. In general, they may be seen as a structured way of representing count data. They were already used to represent data in binary and multinomial regression problems when explanatory variables were categorical (Chapters 2 and 8). Also, count data with categorical explanatory variables (Chapter 7) may be given in the form of contingency tables.

In this chapter log-linear models are presented that may be seen as regression models or association models, depending on the underlying distribution. Three types of distributions are considered: the Poisson distribution, the multinomial, and the product-multinomial distribution. When the underlying distribution is a Poisson distribution, one considers regression problems as in Chapter 7. When the underlying distribution is multinomial, or product-multinomial, one has more structure in the multinomial response than in the regression problems considered in Chapters 2 and 8. In those chapters the response is assumed to be multinomial without further structuring, whereas in the present chapter the multinomial response arises from the consideration of several response variables that together form a contingency table. Then one wants to analyze the association between these variables. Log-linear models provide a common tool to investigate the association structure in terms of independence or conditional independence between variables. Several examples of contingency tables have already been given in previous chapters. Two more examples are the following.

Example 12.1: Birth Data

In a survey study several variables have been collected that are linked to the birth process (see also Boulesteix, 2006). Table 12.1 shows the data for the variables gender of the child (G, 1: male; 2: female), if membranes did rupture before the beginning of labor (M, 1: yes; 0: no), if a Cesarean section has been applied (C, 1: yes, 0: no) and if birth has been induced (I, 1: yes; 0: no). The association between the four variables is unknown and shall be investigated. □

Example 12.2: Leukoplakia

Table 12.2 shows data from a study on leukoplakia, which is a clinical term used to describe patches of keratosis visible as adherent white patches on the membranes of the oral cavity. It shows the alcohol intake in grams of alcohol, smoking habits, and presence of leukoplakia. The objective is the analysis of the association between disease and risk factors (data are taken from Hamerle and Tutz, 1980). □

TABLE 12.1: Contingency table for birth data with variables gender (G), membranes (M), Cesarean section (C), and induced birth (I).

			Induced	
Gender	Membranes	Cesarean	0	1
1	0	0	177	45
		1	37	18
	1	0	104	16
		1	9	7
2	0	0	137	53
		1	24	12
	1	0	74	15
		1	8	2

TABLE 12.2: Contingency table for oral leukoplakia.

		Leukoplakia (A)	
Alcohol	Smoker	Yes	No
no	yes	26	10
	no	8	8
0 g, 40 g	yes	38	8
	no	43	24
40 g, 80 g	yes	4	1
	no	14	17
> 80 g	yes	1	0
	no	3	7

12.1 Types of Contingency Tables

In particular, three types of contingency tables and their corresponding scientific questions will be studied. The *first type* of contingency table occurs if cell counts are Poisson-distributed given the configuration of the cells. Then the counts themselves represent the response, and the categorical variables that determine the cells are the explanatory variables. For example, in Table 1.3 in Chapter 1, the number of firms with insolvency problems may be considered as the response while the year and month represent the explanatory variables. The total number of insolvent firms is not fixed beforehand and is itself a realization of a random variable.

In the *second type* of contingency table a fixed number of subjects is observed and the cell counts represent the multivariate response given the total number of observations. The common assumption is that cell counts have a multinomial distribution. Contingency tables of this type occur if a fixed number of individuals is cross-classified with respect to variables like gender and preference for distinct political parties (see Table 8.1 in Chapter 8). The analysis for this type of contingency table may focus on the association between gender and preference for parties. Alternatively, one might be interested in modeling the preference as the response given gender as the explanatory variable.

The *third type* of contingency table is found, for example, in clinical trials. Table 9.1 shows cross-classified data that have been collected by randomly allocating patients to one of two groups, a treatment group and a group that receives a placebo. After 10 days of treatment the pain occuring during movement of the knee is assessed on a five-point scale. The natural response in this example is the level of pain given the treatment group. The number of people in the two groups is fixed, while the counts themselves are random variables. The level of pain, given the treatment group, is a multivariate response, usually modeled by a multinomial distribution.

In general, two-way $(I \times J)$-contingency tables with I rows and J columns may be described by

$$\begin{array}{ll} X_{ij} & = \text{counts in cell} \quad (i, j), \\ X_A \in \{1, \ldots, I\} & \text{representing the rows,} \\ X_B \in \{1, \ldots, J\} & \text{representing the columns.} \end{array}$$

The observed contingency table has the form

$$\begin{array}{cc|cccc|c} & & & X_B & & & \\ & & 1 & 2 & \ldots & J & \\ \hline & 1 & X_{11} & X_{12} & \ldots & X_{1J} & X_{1+} \\ & 2 & X_{21} & \ddots & & \vdots & \vdots \\ X_A & \vdots & \vdots & & \ddots & \vdots & \\ & I & X_{I1} & \ldots & & X_{IJ} & X_{I+} \\ \hline & & X_{+1} & \ldots & & X_{+J} & \end{array}$$

where $X_{i+} = \sum_{j=1}^{J} X_{ij}, X_{+j} = \sum_{i=1} \hat{I} X_{ij}$ denote the marginal counts. The subscript "+" denotes the sum over that index.

The three types of contingency tables may be distinguished by the distribution that is being assumed.

Type 1: Poisson Distribution (Number of Insolvent Firms)

It is assumed that $X_{11}, \ldots, X_{IJ}$ are independent Poisson-distributed random variables, $X_{ij} \sim P(\lambda_{ij})$. The total number of counts $X_{11} + \cdots + X_{IJ}$ as well as the marginal counts are random variables. The natural model considers the counts as the response and X_A and X_B as explanatory variables.

Type 2: Multinomial Distribution (Gender and Preference for Political Parties)

For a fixed number of subjects one observes independently the response tupel (X_A, X_B) with possible outcomes $\{(1, 1), \ldots, (I, J)\}$. The cells of the table represent the IJ possible outcomes. The resulting cell counts follow a multinomial distribution $(X_{11}, \ldots, X_{IJ}) \sim M(n, (\pi_{11}, \ldots, \pi_{IJ}))$, where $\pi_{ij} = P(X_A = i, X_B = j)$ denotes the probability of a response in cell (i, j). The probabilities $\{\pi_{ij}\}$, or, in vector form, $\boldsymbol{\pi}^T = (\pi_{11}, \ldots \pi_{IJ})$, represent the joint distribution of X_A and X_B.

Type 3: Product-Multinomial Distribution (Treatment and Pain)

In contrast to type 2, now one set of marginal counts is fixed. In the treatment and pain example the row tables are fixed by $n_1 = X_{1+}, n_2 = X_{2+}$. Given the treatment group, one observes for each individual the response $X_B \in \{1, \ldots, J\}$. The cell counts for the given treatment group i follow a multinomial distribution:

$$(X_{i1}, \ldots, X_{iJ}) \sim M(n_i, (\pi_{i1}, \ldots \pi_{iJ})),$$

where π_{ij} now denotes the *conditional probability* $\pi_{ij} = P(X_B = j | X_A = i)$. The cell counts of the total contingency table follow a *product-multinomial* distribution. The natural modeling is to consider X_B as the response variable and X_A as the explanatory variable. This modeling approach follows directly from the design of the study. Of course, if the column totals are fixed, the natural response is X_A with X_B as the explanatory variable.

Models and Types of Contingency Tables

The three types of contingency tables differ by the way the data are collected. When considering typical scientific questions, to be investigated by the analysis of contingency tables, one finds a hierarchy within the types of tables. While the Poisson distribution is the most general, allowing for various types of analyses, the product-multinomial contingency table is the most restrictive. The hierarchy is due to the possible transformations of distributions by conditioning.

Poisson and Multinomial Distributions

Let $X_{ij}, i = 1, \ldots, I, j = 1, \ldots, J$ follow independent Poisson distributions, $X_{ij} \sim P(\lambda_{ij})$. Then the *conditional distribution* of $(X_{11}, \ldots, X_{IJ})$ given $n = \sum_{ij} X_{ij}$ is multinomial. More concrete, one has

$$(X_{11}, \ldots, X_{IJ}) | \sum_{i,j} X_{ij} = n \sim M(n, (\frac{\lambda_{11}}{\lambda}, \ldots, \frac{\lambda_{IJ}}{\lambda})),$$

where $\lambda = \sum_{ij} \lambda_{ij}$. Therefore, given that one has Poisson-distributed cell counts, by conditioning one obtains a structured multinomial distribution that is connected to the response tupel (X_A, X_B). Hence, by conditioning on n one may study the response (X_A, X_B), its marginal distribution, and the association between (X_A, X_B).

Multinomial and Product-Multinomial Distributions

Let $(X_{11}, \ldots, X_{IJ})$ have multinomial distribution, $(X_{11}, \ldots, X_{IJ}) \sim M(n, (\pi_{11}, \ldots, \pi_{IJ}))$. By conditioning on the row margins $n_{i+} = \sum_j X_{ij}$ one obtains the product-multinomial distribution with probability mass function

$$f(x_{11}, \ldots, x_{IJ}) = \prod_{i=1}^{I} \frac{n_{i+}!}{x_{i1}! \ldots x_{iJ}!} \pi_{1|i}^{x_{i1}} \ldots \pi_{J|i}^{x_{IJ}},$$

where $\pi_{j|i} = \pi_{ij} / \sum_j \pi_{ij} = \pi_{ij} / \pi_{i+}$. Thus the cell counts of one row given n_{i+} have a multinomial distribution:

$$(X_{i1}, \ldots, X_{iJ}) \sim M(n_{i+}, (\pi_{1|i}, \ldots, \pi_{J|i}))$$

and the I multinomials corresponding to the rows are independent.

If the Poisson distribution generates the counts in the table, one may consider the counts given X_A and X_B within a regression framework, or one may condition on the total sample size n and model the marginal distribution and the association of X_A and X_B based on the multinomial distribution. One may also go one step further and condition on the marginal counts of the rows (columns) and consider the regression model where X_B (X_A) is the response and X_A (X_B) is the explanatory variable. If the multinomial distribution generates the contingency table, one may consider (X_A, X_B) as the response or choose one of the two variables by conditioning on the other one. In this sense the Poisson distribution contingency table is the most versatile. Since the Poisson, multinomial, and product-multinomial distributions may be treated within a general framework, in the following $\mu_{ij} = E(X_{ij})$ is used rather than $n\pi_{ij}$ (or $n_i\pi_{ij}$), which would be more appropriate for the multinomial (or product-multinomial) distribution. In Table 12.3 we summarize the types of distributions and the modeling approaches.

TABLE 12.3: Types of two-way contingency tables and modeling approaches.

Poisson distribution	Regression $(X_A, X_B) \to$ Counts
	Association between X_A and X_B (conditional on n)
	Regression $X_A \to X_B$ (conditional on X_{i+})
	Regression $X_B \to X_A$ (conditional on X_{+j})
Multinomial distribution	Association between X_A and X_B (conditional on n)
	Regression $X_A \to X_B$ (conditional on X_{i+})
	Regression $X_B \to X_A$ (conditional on X_{+j})
Product-multinomial distribution	
X_{i+} fixed	Regression $X_A \to X_B$
X_{+j} fixed	Regression $X_B \to A_A$

12.2 Log-Linear Models for Two-Way Tables

Consider an $(I \times J)$-contingency table $\{X_{ij}\}$. Let $\mu_{ij} = E(X_{ij})$ denote the mean, where $\mu_{ij} = n\pi_{ij} = nP(X_A = i, X_B = j)$ for the multinomial distribution, $\mu_{ij} = n_{i+}P(X_B = j|X_A = i)$ if one conditions on $n_{i+} = \sum_j X_{ij}$, and $\mu_{ij} = n_{+j}P(X_A = i|X_B = j)$ if one conditions on $n_{+j} = \sum_i X_{ij}$. The general log-linear model for two-way tables has the form

$$\log(\mu_{ij}) = \lambda_0 + \lambda_{A(i)} + \lambda_{B(j)} + \lambda_{AB(ij)} \tag{12.1}$$

or, equivalently,

$$\mu_{ij} = e^{\lambda_0} e^{\lambda_{A(i)}} e^{\lambda_{B(j)}} e^{\lambda_{AB(ij)}}.$$

Since model (12.1) contains too many parameters, identifiability requires constraints on the parameters. Two sets of constraints are in common use, the symmetrical constraints and constraints that use a baseline parameter:

Symmetrical constraints

$$\sum_{i=1}^{I} \lambda_{A(i)} = \sum_{j=1}^{J} \lambda_{B(j)} = \sum_{i=1}^{I} \lambda_{AB(ij)} = \sum_{j=1}^{J} \lambda_{AB(ij)} = 0 \quad \text{for all } i, j.$$

Baseline parameters set to zero

$$\lambda_{A(I)} = \lambda_{B(J)} = \lambda_{AB(iJ)} = \lambda_{AB(Ij)} = 0 \quad \text{for all } i, j.$$

The symmetrical constraints are identical to the constraints used in analysis-of-variance (ANOVA). In ANOVA the dependence of a response variable on categorical variables, called factors, is studied. In particular, one is often interested in interaction effects. There is a strong similarity between ANOVA and Poisson contingency tables, where the counts represent the response and the categorical variables X_A and X_B form the design. The main difference is that ANOVA models assume a normal distribution for the response whereas in log-linear models for count data the response is integer-valued.

These sets of constraints are closely related to the coding of dummy variables. Symmetrical constraints refer to effect coding whereas the choice of baseline parameters is equivalent to choosing a reference category in dummy coding (see Section 1.4.1). Model (12.1) may be also written with dummy variables, yielding

$$\begin{aligned}\log(\mu_{ij}) = \lambda_0 &+ \lambda_{A(1)} x_{A(1)} + \cdots + \lambda_{A(I-1)} x_{A(I-1)} \\ &+ \lambda_{B(1)} x_{B(1)} + \cdots + \lambda_{B(J-1)} x_{B(J-1)} \\ &+ \lambda_{AB(1,1)} x_{A(1)} x_{B(1)} + \cdots + \lambda_{AB(I-1,J-1)} x_{A(I-1)} x_{B(J-1)},\end{aligned}$$

where $x_{A(1)}, \ldots$ are dummy variables coding $A = i$ and $x_{B(1)}, \ldots$ are dummy variables coding $B = j$. This form is usually too clumsy and will be avoided. However, it is easily seen that the effect coding of dummy variables is equivalent to the symmetric constraints, and choosing $(X_A = I, X_B = J)$ as reference categories in dummy coding is equivalent to using baseline parameters. One should keep in mind that baseline parameters that refer to reference categories may be chosen arbitrarily; different software use different constraints.

The sets of constraints given above apply for Poisson distribution tables. For multinomial and product-multinomial tables, additional constraints are needed to ascertain that $\sum_{ij} X_{ij} = n$ (multinomial) and $\sum_j X_{ij} = n_{i+}$ (product-multinomial, fixed row sums) hold:

Additional constraint for multinomial tables

$$\sum_{i,j} e^{\lambda_0} e^{\lambda_{A(i)}} e^{\lambda_{B(j)}} e^{\lambda_{AB(ij)}} = n.$$

Additional constraints for product-multinomial tables

$$\sum_{j=1}^{J} e^{\lambda_0} e^{\lambda_{A(i)}} e^{\lambda_{B(j)}} e^{\lambda_{AB(ij)}} = n_{i+}, \quad i = 1, \ldots, I \quad (\text{for } n_{i+} \text{ fixed}),$$

$$\sum_{i=1}^{I} e^{\lambda_0} e^{\lambda_{A(i)}} e^{\lambda_{B(j)}} e^{\lambda_{AB(ij)}} = n_{+j}, \quad j = 1, \ldots, J \quad (\text{for } n_{+j} \text{ fixed}).$$

TABLE 12.4: Log-linear model for two-way tables.

Log-Linear Model for Two-Way Tables

$$\log(\mu_{ij}) = \lambda_0 + \lambda_{A(i)} + \lambda_{B(j)} + \lambda_{AB(i,j)}$$

Constraints:

$$\sum_{i=1}^{I} \lambda_{A(i)} = \sum_{j=1}^{J} \lambda_{B(j)} = \sum_{i=1}^{I} \lambda_{AB(ij)} = \sum_{j=1}^{J} \lambda_{AB(ij)} = 0$$

or

$$\lambda_{A(I)} = \lambda_{B(J)} = \lambda_{AB(iJ)} = \lambda_{AB(Ij)} = 0$$

Model (12.1) is the most general model for two-way contingency tables, the so-called *saturated model*. It is saturated since it represents only a reparameterization of the means $\{\mu_{ij}\}$, and any set of means $\{\mu_{ij}\}$ $(\mu_{ij} > 0)$ may be represented by the parameters $\lambda_B, \lambda_{A(i)}, \lambda_{B(j)}, \lambda_{AB(ij)}$, $i = 1, \ldots, I, j = 1, \ldots, J$.

Consequently, not much insight is gained by considering the saturated log-linear model. The most important submodel is the *log-linear model of independence*,

$$\log(\mu_{ij}) = \lambda_0 + \lambda_{A(i)} + \lambda_{B(j)}, \tag{12.2}$$

where it is assumed that $\lambda_{AB(ij)} = 0$. This is no longer a saturated model because it implies severe restrictions on the underlying regression or association structure. The restriction has different meanings, depending on the distribution of the cell counts X_{ij}. For the Poisson distribution it simply means that there is no interaction effect of the variables X_A and X_B when effecting on the cell counts. For the multinomial model it is helpful to consider the multiplicative form of (12.2):

$$\mu_{ij} = nP(X_A = i, X_B = j) = e^{\lambda_0} e^{\lambda_{A(i)}} e^{\lambda_{B(j)}}. \tag{12.3}$$

This means that the probability $P(X_A = i, X_B = j)$ may be written in a multiplicative form with factors depending only on X_A or X_B. Taking constraints into account, it is easily derived that (12.3) is equivalent to assuming that X_A and X_B are independent random variables, or, equivalently, that $P(X_A = i, X_B = j) = P(X_A = i)P(X_B = j)$ holds. That property gives the model its name.

For the product-multinomial table (row marginals n_{i+} fixed) one has

$$\mu_{ij} = n_{i+}P(X_B = j|X_A = i) = e^{\lambda_0} e^{\lambda_{A(i)}} e^{\lambda_{B(j)}}.$$

With the constraint $\sum_j e^{\lambda_0} e^{\lambda_{A(i)}} e^{\lambda_{B(j)}} = n_{i+}$ one obtains

$$P(X_B = j|X_A = i) = e^{\lambda_{B(j)}} / \sum_r e^{\lambda_{B(r)}},$$

which means that the response X_B does not depend on the variable X_A. Thus the model postulates that the response probabilities are identical across rows:

$$P(X_B = j|X_A = 1) = \ldots = P(X_B = j|X_A = I),$$

which means *homogeneity* across rows. Considering it is a regression model with X_B as the response and X_A as the explanatory variables, it means that X_A has no effect on X_B. The interpretation of the models is summarized in the following.

- *Poisson distribution:* No interaction effect of X_A and X_B on counts.
- *Multinomial distribution:* X_A and X_B are independent.
- *Product-multinomial distribution:* Response X_B does not depend on X_A (fixed row marginals), and response X_A does not depend on X_B (fixed column marginals).

Tests for the null hypothesis $H_0 : \mu_{AB(ij)} = 0$ for all i, j have different interpretations. If H_0 is not rejected, for Poisson distribution tables it means that the interaction term is not significant. For multinomial distribution tables, the test is equivalent to testing the independence between X_A and X_B. If X_A and X_B are random variables and data have been collected as Poisson counts, by conditioning on X_A and X_B, the interpretation as a test for independence also holds for Poisson tables (by conditioning on n). Of course, in applications where X_A and X_B refer to experimental conditions, that interpretation is useless. Consider Example 7.2 in Chapter 7 where the counts of cases of encephalitis are modeled depending on country and time. These explanatory variables are experimental conditions rather than random variables, and it is futile to try to investigate the independence of these conditions.

Parameters and Odds Ratio

The parameters of the saturated model (12.1) with symmetric constraints are easily computed as

$$\lambda = \frac{1}{IJ}\sum_{i,j}\log(\mu_{ij}), \quad \lambda_{A(i)} = \frac{1}{J}\sum_{j}\log(\mu_{ij}) - \lambda,$$

$$\lambda_{B(j)} = \frac{1}{I}\sum_{i}\log(\mu_{ij}) - \lambda, \quad \lambda_{AB(ij)} = \log(\mu_{ij}) - \lambda - \lambda_{A(i)} - \lambda_{B(j)}.$$

The parameters $\lambda_{A(i)}, \lambda_{B(j)}$ are the *main effects*, and $\lambda_{AB(ij)}$ is a *two-factor interaction.*

For multinomial and product-multinomial distributions, an independent measure of association that is strongly linked to two-factor interactions is the odds ratio. For the simple (2×2)-contingency table the odds ratio has the form

$$\gamma = \frac{\pi_{11}/\pi_{12}}{\pi_{21}/\pi_{22}} = \frac{P(X_A = 1, X_B = 1)/P(X_A = 1, X_B = 2)}{P(X_A = 2, X_B = 1)/P(X_A = 2, X_B = 2)} = \frac{P(X_B = 1|X_A = 1)/P(X_B = 2|X_A = 1)}{P(X_B = 1|X_A = 2)/P(X_B = 2|X_A = 2)}.$$

By using $\mu_{ij} = n\pi_{ij}$ (multinomial distribution) or $\mu_{ij} = n_{i+}\pi_{ij}$ (product-multinomial, fixed rows) one obtains for the log-linear model with symmetrical constraints

$$\log(\gamma) = 4\lambda_{AB(11)},$$

and for the model with the last category set to zero $\log(\gamma) = \lambda_{AB(11)}$. Thus γ is a direct function of the two-factor interaction. The connection to independence is immediately seen: $\lambda_{AB(11)} = 0$ is equivalent to $\gamma = 1$, which means independence of the variables X_A and X_B (multinomial distribution) or homogeneity (product-multinomial distribution).

In the general case of $(I \times J)$-contingency tables one considers the odds ratio formed by the (2×2)-subtable built from rows $\{i_1, i_2\}$ and columns $\{j_1, j_2\}$ with cells $\{(i_1, j_1), (i_1, j_2), (i_2, j_1), (i_2, j_2)\}$. The corresponding odds ratio

$$\gamma_{(i_1 i_2)(j_1 j_2)} = \frac{\pi_{i_1 j_1}/\pi_{i_1 j_2}}{\pi_{i_2 j_1}/\pi_{i_2 j_2}}$$

may be expressed in two-factor interactions by

$$\log(\gamma_{(i_1 i_2)(j_1 j_2)}) = \lambda_{AB(i_1 j_1)} + \lambda_{AB(i_2, j_2)} - \lambda_{AB(i_2 j_1)} - \lambda_{AB(i_1 j_2)}.$$

12.3 Log-Linear Models for Three-Way Tables

Three-way tables are characterized by three categorical variables, $X_A \in \{1, \ldots, I\}$, $X_B \in \{1, \ldots, J\}$, and $X_C \in \{1, \ldots, K\}$, which refer to rows, columns, and layers of the table. Let $\{X_{ijk}\}$ denote the collection of cell counts, where

X_{ijk} denotes the counts in cell (i, j, k),
that is, the number of observations with $X_A = i, X_B = j, X_C = k$.

The general form of three-way tables is given in Table 12.5. Throughout this section the convention is used that the subscript " $+$ " denotes the sum over that index, for example, $X_{ij+} = \sum_k X_{ijk}$. The types of contingency tables are in principle the same as for two-way tables. However, now there are more variants of conditioning.

TABLE 12.5: General form of three-way tables.

		X_C				
X_A	X_B	1	2	$\cdots$	K	
1	1	X_{111}	X_{112}	$\cdots$	X_{11K}	X_{11+}
	2	X_{121}	X_{122}			$\vdots$
	$\vdots$	$\vdots$				
	J	X_{1J1}	$\cdots$		X_{1JK}	X_{1J+}
2	1	X_{211}	X_{212}	$\cdots$	X_{21K}	X_{21+}
	2	X_{221}	X_{222}			$\vdots$
	$\vdots$	$\vdots$				
	J	X_{2J1}	$\cdots$		X_{2JK}	X_{2J+}
$\vdots$	$\vdots$	$\vdots$	$\vdots$	$\vdots$	$\vdots$	
I	1	X_{I11}	X_{I12}	$\cdots$	X_{I1K}	X_{I1+}
	2	X_{I21}	X_{I22}			$\vdots$
	$\vdots$	$\vdots$				
	J	X_{IJ1}	$\cdots$		X_{IJK}	X_{IJ+}

Type 1: Poisson Distribution

It is assumed that the X_{ijk} are independent Poisson-distributed random variables, $X_{ijk} \sim P(\lambda_{ijk})$. The total number of counts $n = \sum_{ijk} X_{ijk}$ as well as the marginal counts are random variables. The natural model considers the counts as the response and X_A, X_B, and X_C as explanatory variables, which might refer to experimental conditions or random variables.

Type 2: Multinomial Distribution

For a fixed number of subjects one observes the tupel (X_A, X_B, X_C) with possible outcomes $\{(1,1,1), \dots, (I,J,K)\}$. The count in cell (i,j,k) is the number of observations with $X_A = i, X_B = j, X_C = k$. The counts $\{X_{ijk}\}$ follow a multinomial distribution $M(n, \{\pi_{ijk}\})$, where $\pi_{ijk} = P(X_A = i, X_B = j, X_C = k)$. For three- and higher dimensional tables the notation $\{X_{ijk}\}$ and $\{\pi_{ijk}\}$ is preferred over the representation as vectors.

Type 3: Product-Multinomial Distribution

There are several variants of the product-multinomial distribution. Either one of the variables (X_A or X_B or X_C) is a design variable, meaning that the corresponding marginals are fixed, or two of them are design variables, meaning that two-dimensional margins are fixed. Let us consider as an example of the first variant the table that results from the design variable X_A. This means $n_{i++} = X_{i++}$ is fixed and

$$(X_{i11}, \dots, X_{iJK}) \sim M(n_{i++}, (\pi_{i11}, \dots, \pi_{iJK})), \tag{12.4}$$

where $\pi_{ijk} = P(X_B = j, X_C = k | X_A = i)$. An example of the second variant (two design variables) is obtained by letting X_A and X_B be design variables, that is, $n_{ij+} = X_{ij+}$ is fixed and

$$(X_{ij1}, \dots, X_{ijK}) \sim M(n_{ij+}, (\pi_{ij1}, \dots, \pi_{ijK})), \tag{12.5}$$

where $\pi_{ijk} = P(X_C = k|X_A = i, X_B = j)$. Hence, only X_C is a response variable, and the number of observations for $(X_A, X_B) = (i, j)$ is given.

It should be noted that there is again a hierarchy among distributions. If $\{X_{ijk}\}$ have a Poisson distribution, conditioning on $n = \Sigma_{ijk} X_{ijk}$ yields the multinomial distribution $\{X_{ijk}\} \sim M(n, \{\pi_{ijk}\})$, where $\pi_{ijk} = \lambda_{ijk}/\Sigma_{ijk}\lambda_{ijk}$. Further conditioning on $n_{i++} = \Sigma_{jk} X_{ijk}$ yields the product of the multinomial distributions (12.4). If, in addition, one conditions on $n_{ij+} = \Sigma_k X_{ijk}$, one obtains the product of distributions (12.5).

TABLE 12.6: Log-linear model for three-way tables with constraints.

Log-linear Model for Three-Way Tables

$$\log(\mu_{ijk}) = \lambda_0 + \lambda_{A(i)} + \lambda_{B(j)} + \lambda_{C(k)} + \lambda_{AB(ij)} + \lambda_{AC(ik)} + \lambda_{BC(jk)} + \lambda_{ABC(ijk)}.$$

Constraints:

$$\sum_i \lambda_{A(i)} = \sum_j \lambda_{B(j)} = \sum_k \lambda_{C(k)} = 0,$$

$$\sum_i \lambda_{AB(ij)} = \sum_j \lambda_{AB(ij)} = \sum_i \lambda_{AC(ik)} = \sum_k \lambda_{AC(ik)} = \sum_j \lambda_{BC(jk)} = \sum_k \lambda_{BC(jk)} = 0,$$

$$\sum_i \lambda_{ABC(ijk)} = \sum_j \lambda_{ABC(ijk)} = \sum_k \lambda_{ABC(ijk)} = 0,$$

or

$$\lambda_{A(I)} = \lambda_{B(J)} = \lambda_{C(K)} = 0,$$

$$\lambda_{AB(i,J)} = \lambda_{AB(I,j)} = \lambda_{AC(iK)} = \lambda_{AC(Ik)} = \lambda_{BC(jK)} = \lambda_{BC(Jk)} = 0,$$

$$\lambda_{ABC(Ijk)} = \lambda_{ABC(iJk)} = \lambda_{ABC(ijK)} = 0.$$

Let in general $\mu_{ijk} = \mathrm{E}(X_{ijk})$ denote the mean of the cell counts. Then the general form of the three-dimensional log-linear model is

$$\log(\mu_{ijk}) = \lambda_0 + \lambda_{A(i)} + \lambda_{B(j)} + \lambda_{C(k)} + \lambda_{AB(ij)} + \lambda_{AC(ik)} + \lambda_{BC(jk)} + \lambda_{ABC(ijk)}.$$

For the necessary constraints see Table 12.6, where two sets of constraints are given: the set of symmetric constraints corresponding to ANOVA models and the set of constraints based on reference categories. For the multinomial and the product-multinomial models, additional constraints are needed, which are easily derived from the restrictions $n = \sum_{ijk} X_{ijk}$, and so on. The model has three types of parameters: the three-factor interactions $\lambda_{ABC(ijk)}$, the two-factor interactions $\lambda_{AB(ij)}$, $\lambda_{AC(ik)}$ and $\lambda_{BC(jk)}$, and the main effects $\lambda_{A(i)}$, $\lambda_{B(j)}$, $\lambda_{C(k)}$. The general model is saturated, which means that it has as many parameters as means μ_{ijk} and consequently every dataset (without empty cells) yields a perfect fit by setting $\hat{\mu}_{ijk} = X_{ijk}$.

More interesting models are derived from the general model by omitting groups of parameters corresponding to interaction terms. The attractive feature of log-linear models is that most of the resulting models have an interpretation in terms of independence or conditional independence. In general, categorical variables X_A, X_B, X_C are *independent* if

$$P(X_A = i, X_B = j, X_C = k) = P(X_A = i)P(X_B = j)P(X_C = k)$$

holds for all i, j, k. *Conditional independence* of X_A and X_B given X_C (in short $X_A \perp X_B | X_C$) holds if, for all i, j, k,

$$P(X_A = i, X_B = j | X_C = k) = P(X_A = i | X_C = k)P(X_B = j | X_C = k).$$

Hierarchical Models

The interesting class of models that may be interpreted in terms of (conditional) independence are the *hierarchical models*. A model is called hierarchical if it includes all lower order terms composed from variables contained in a higher order term. For example, if a model contains $\lambda_{BC(jk)}$, it also contains the marginals $\lambda_{B(j)}$ and $\lambda_{C(k)}$. Hierarchical models may be abbreviated by giving the terms of highest order. For example, the symbol AB/AC denotes the model containing $\lambda_{AB}, \lambda_{AC}, \lambda_A, \lambda_B, \lambda_C$ (and a constant term). Further examples are given in Table 12.7. The notation is very similar to the Wilkinson-Rogers notation (see Section 4.4), which, for the model AB/AC is, $A * B + A * C$. The latter form is itself shorthand for the extended Wilkinson-Rogers notation $A.B + A.C + A + B + C$.

Graphical Models

Most of the hierarchical log-linear models for three-way tables are also *graphical models*, which are considered in more detail in Section 12.5. The basic concept is only sketched here. If a graph is drawn by linking variables for which the two-factor interaction is contained in the model, one obtains a simple graph. If in the resulting graph there is no connection between the groups of variables, these groups of variables are independent. If two variables are connected only by edges through the third variable, the two variables are conditionally independent given a third variable. For examples, see Table 12.7, and for a more concise definition of graphical models see Section 12.5.

12.4 Specific Log-Linear Models

In the following, the types of hierarchical models for three-way tables are considered under the assumption that X_A, X_B, X_C represent random variables (i.e., multinomial contingency tables).

Type 0: Saturated Model

The saturated model is given by

$$\log(\mu_{ijk}) = \lambda_0 + \lambda_{A(i)} + \lambda_{B(j)} + \lambda_{C(k)} + \lambda_{AB(i,j)} + \lambda_{AC(ik)} + \lambda_{BC(jk)} + \lambda_{ABC(ijk)}.$$

It represents a reparameterization of the means $\{\mu_{ijk}\}$ without implying any additional structure (except $\mu_{ijk} > 0$).

Type 1: No Three-Factor Interaction

The model

$$\log(\mu_{ijk}) = \lambda_0 + \lambda_{A(i)} + \lambda_{B(j)} + \lambda_{C(k)} + \lambda_{AB(ij)} + \lambda_{AC(ik)} + \lambda_{BC(jk)}$$

contains only two-factor interactions and is denoted by $AB/AC/BC$. Since the three-factor interaction is omitted, the model has to imply restrictions on the underlying probabilities. To see what the model implies, it is useful to look at the conditional association of two variables given a specific level of the third variable.

Let us consider the odds ratios of X_A and X_B given $X_C = k$ and for simplicity assume that all variables are binary. Then the conditional association measured by the odds ratio has the form

$$\gamma(X_A, X_B|X_C = k) = \frac{P(X_A = 1, X_B = 1|X_C = k)/P(X_A = 2, X_B = 1|X_C = k)}{P(X_A = 1, X_B = 2|X_C = k)/P(X_A = 2, X_B = 2|X_C = k)}$$

and is built from the (2×2)-table formed by X_A and X_B for a fixed level $X_C = k$. By using $\mu_{ijk} = n\pi_{ijk}$ and

$$\gamma(X_A, X_B|X_C = k) = \frac{\pi_{11k}/\pi_{21k}}{\pi_{12k}/\pi_{22k}} = \frac{\mu_{11k}/\mu_{21k}}{\mu_{12k}/\mu_{22k}}$$

one obtains for the model without three-factor interactions that all terms depending on k cancel out and therefore $\gamma(X_A, X_B|X_C = k)$ *does not depend on* k. This means that the conditional association between X_A and X_B given $X_C = k$ does not depend on the level k. Whatever the conditional association between these two variables, strong or weak or not present, it is the same for all levels of X_C, and thus X_C does not modify the association between X_A and X_B. The same holds if X_A and X_B have more than two categories, where conditional association is measured by odds ratios of (2×2)- subtables built from the total table. Moreover, since the model is symmetric in the variables, it is also implied that the conditional association between X_A and X_C given $X_B = j$ does not depend on j and the conditional association between X_B and X_C given $X_A = i$ does not depend on i.

It should be noted that the model without the three-factor interaction does *not* imply that two variables are independent of the third variable. There might be a strong dependence between $\{X_A, X_B\}$ and X_C, although the conditional association of X_A and X_B given $X_C = k$ does not depend on the level of X_C. The model is somewhat special because it is the only log-linear model for three-way tables that is not a graphical model and therefore cannot be represented by a simple graph. It is also the only model that cannot be interpreted in terms of independence or conditional independence of variables.

Type 2: Only Two Two-Factor Interactions Contained

A model of this type is the model AC/BC, given by

$$\log(\mu_{ijk}) = \lambda_0 + \lambda_{A(i)} + \lambda_{B(j)} + \lambda_{C(k)} + \lambda_{AC(ik)} + \lambda_{BC(jk)}.$$

If the model holds, the variables X_A and X_B are *conditionally independent*, given X_C, or , more formally,

$$P(X_A = i, X_B = j|X_C = k) = P(X_A = i|X_C = k)P(X_B = j|X_C = k).$$

This may be easily derived by using that the model is equivalent to postulating that $\mu_{ijk} = \mu_{i+k}\mu_{+jk}/\mu_{++k}$. It means that conditionally the variables X_A and X_B are not associated.

TABLE 12.7: Graphical models for three-way tables.

	Log-Linear Model	Regressors of Logit-Model (with response X_C)
AB/AC X_B, X_C conditionally independent, given X_A	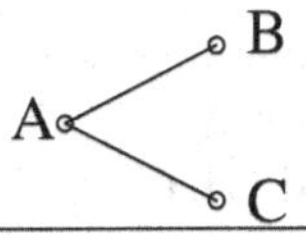	$1, x_A$
AB/BC X_A, X_C conditionally independent, given X_B	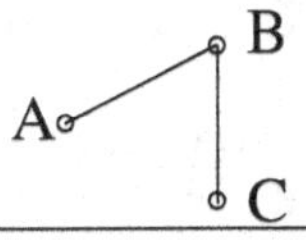	$1, x_B$
AC/BC X_A, X_B conditionally independent, given X_C	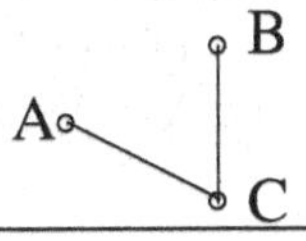	$1, x_A, x_B$
A/BC X_A independent of (X_B, X_C)	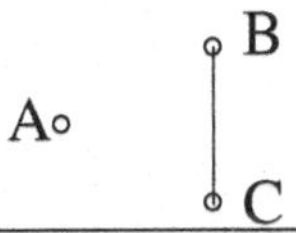	$1, x_B$
AC/B (X_A, X_C) independent of X_B	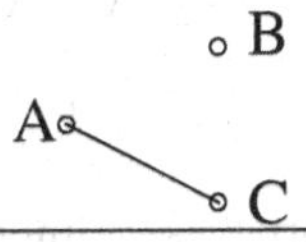	$1, x_A$
AB/C (X_A, X_B) independent of X_C	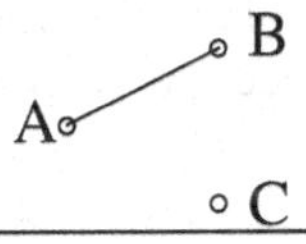	1
$A/B/C$ X_A, X_B, X_C are dependent	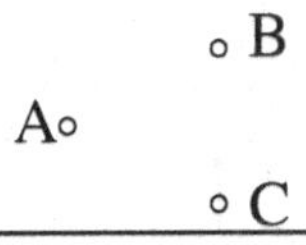	1

However, that does not mean that there is no marginal association between X_A and X_B. X_A and X_B may be strongly associated when X_C is ignored. The model is a graphical model, with the graph given in Table 12.7. The graph contains edges between X_A and X_C as well as between X_B and X_C but not between X_A and X_B. It illustrates that X_A and X_B have some connection through the common variable X_C. And that is exactly the meaning of the graph: Given X_C, the variables X_A and X_B are independent because the connection between X_A and X_B is only through X_C.

The other two models of this type are AB/AC and AB/BC (shown in Table 12.7). The first postulates that X_B and X_C are conditionally independent given X_A, and the latter postulates that X_A and X_C are conditionally independent given X_B.

Type 3: Only One Two-Factor Interaction Contained

A model of this type is the model A/BC, given by

$$\log(\mu_{ijk}) = \lambda_0 + \lambda_{A(i)} + \lambda_{B(j)} + \lambda_{C(k)} + \lambda_{BC(jk)},$$

which contains only main effects and one two-factor interaction. By simple derivation one obtains that the model postulates that X_A is *jointly independent* of X_B and X_C. This means that the groups of variables $\{A\}$ and $\{B, C\}$ are independent, or, more formally,

$$P(X_A = i, X_B = j, X_C = k) = P(X_A = i)P(X_B = j, X_C = k).$$

The model implies stronger restrictions on the underlying probability structure than the model AC/BC because now, in addition, the two-factor interaction λ_{AC} is omitted. The corresponding graph in Table 12.7 is very suggestive. There is no edge between the variable X_A and the two variables X_B, X_C; the two groups of variables are well separated, corresponding to the interpretation of the model that X_A and X_B, X_C are independent.

Type 4: Main Effects Model

The model has the form

$$\log(\mu_{ijk}) = \lambda_0 + \lambda_{A(i)} + \lambda_{B(j)} + \lambda_{C(k)}.$$

It represents the independence of variables X_A, X_B, X_C:

$$P(X_A = i, X_B = j, X_C = k) = P(X_A = i)P(X_B = j)P(X_C = k),$$

meaning in particular that all the variables are mutually independent.

Figure 12.1 shows the hierarchy of log-linear models. It is obvious that the model AB/BC is a submodel of $AB/BC/AC$ since the latter is less restrictive than the former. But not

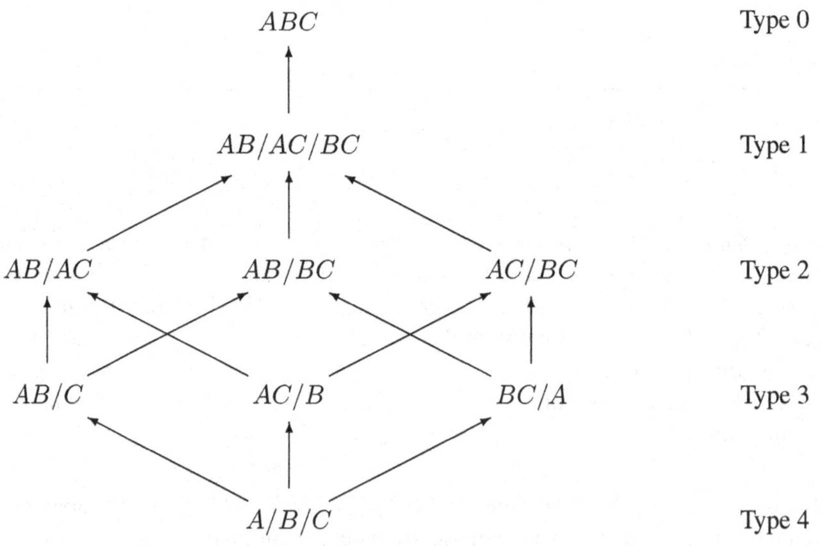

FIGURE 12.1: Hierarchy of three-dimensional log-linear models.

all models are nested. For example, there is no hierarchy between the models AB/AC and AB/BC. The possible models form a lattice with semi-ordering.

Table 12.8 shows what restrictions are implied by omitting interaction terms. For example, the model AB/AC implies that $\mu_{ijk} = \mu_{ij+}\mu_{i+k}/\mu_{i++}$ holds. When the sampling is multinomial, it is easily derived what that means for conditional probabilities and therefore for the interpretation of the assumed association structure. The corresponding graphs are given in Table 12.7.

TABLE 12.8: Graphical models and interpretation for three-way-tables.

AB/AC	$\lambda_{ABC} = \lambda_{BC} = 0$	$\mu_{ijk} = \frac{\mu_{ij+}\mu_{i+k}}{\mu_{i++}}$ $P(X_B, X_C \mid X_A) = P(X_B \mid X_A)P(X_C \mid X_A)$
AB/BC	$\lambda_{ABC} = \lambda_{AC} = 0$	$\mu_{ijk} = \frac{\mu_{+jk}\mu_{ij+}}{\mu_{+j+}}$ $P(X_A, X_C \mid X_B) = P(X_A \mid X_B)P(X_C \mid X_B)$
AC/BC	$\lambda_{ABC} = \lambda_{AB} = 0$	$\mu_{ijk} = \frac{\mu_{i+k}\mu_{+jk}}{\mu_{++k}}$ $P(X_A, X_B \mid X_C) = P(X_A \mid X_C)P(X_B \mid X_C)$
A/BC	$\lambda_{ABC} = \lambda_{AB} = \lambda_{AC} = 0$	$\mu_{ijk} = \frac{\mu_{i++}\mu_{+jk}}{\mu_{+++}}$ $P(X_A, X_B, X_C) = P(X_A)P(X_B, X_C)$
AC/B	$\lambda_{ABC} = \lambda_{AB} = \lambda_{BC} = 0$	$\mu_{ijk} = \frac{\mu_{i+k}\mu_{+j+}}{\mu_{+++}}$ $P(X_A, X_B, X_C) = P(X_A, X_C)P(X_B)$
AB/C	$\lambda_{ABC} = \lambda_{AC} = \lambda_{BC} = 0$	$\mu_{ijk} = \frac{\mu_{ij+}\mu_{++k}}{\mu_{+++}}$ $P(X_A, X_B, X_C) = P(X_A, X_B)P(X_C)$
A/B/C	$\lambda_{ABC} = \lambda_{AB} = \lambda_{AC} = 0$ $\lambda_{BC} = 0$	$\mu_{ijk} = \frac{\mu_{i++}\mu_{+j+}\mu_{++k}}{\mu_{+++}^2}$ $P(X_A, X_B, X_C) = P(X_A)P(X_B)P(X_C)$

Models for Product-Multinomial Contingency Tables

While all models in Figure 12.1 apply for multinomial contingency tables, not all models may be built for product-multinomial contingency tables, because marginal sums that are fixed by design have to be fitted by the model. In general, if margins are fixed by design, the corresponding interaction term has to be contained in the model. For example, if the two-dimensional margins $X_{ij+} = \sum_k X_{ijk}$ are fixed by design, the model has to contain the interaction λ_{AB}. The model AC/BC is not a valid model because it fits the margins X_{i+k} and X_{+jk} but does not contain λ_{AB} (see also Lang, 1996a ; Bishop et al., 1975; Agresti, 2002).

12.5 Log-Linear and Graphical Models for Higher Dimensions

Log-linear models for dimensions higher than three have basically the same structure, but the number of possible interaction terms and the number of possible models increase. For example, in four-way tables a four-factor interaction term can be contained. It is helpful that for hierarchical models the same notation applies as in lower dimensional models. An example of a four-way model is ABC/AD, which is given by

$$\log(\mu_{ijkl}) = \lambda_0 + \lambda_{A(i)} + \lambda_{B(j)} + \lambda_{C(k)} + \lambda_{D(l)} + \lambda_{AB(ij)} + \lambda_{AC(ik)} + \lambda_{BC(jk)} + \lambda_{AD(il)} + \lambda_{ABC(ijk)}.$$

The model contains only one three-factor interaction and only four two-factor interactions but all main effects. For interpreting higher dimensional tables, representing them as graphical models is a helpful tool.

Graphical Models

To obtain models that have simple interpretations in terms of conditional interpretations it is useful to restrict consideration to subclasses of log-linear models. We already made the restriction to hierarchical models. A log-linear model is hierarchical if the model includes all lower order terms composed from variables contained in a higher order term. A further restriction is pertaining to graphical models:

> A log-linear model is *graphical* if, whenever the model contains all two-factor interactions generated by a higher order interaction, the model also contains the higher order interaction.

In three-way tables there is only one log-linear model that is not graphical, namely, the model *AB*/*AC*/*BC*. That model contains all two-factor interactions λ_{AB}, λ_{AC}, λ_{BC}, which are generated as marginal parameters of the three-factor interaction λ_{ABC}, but λ_{ABC} itself is not contained in the model.

A graphical model has a graphical representation that makes it easy to see what types of conditional independence structures are implied. The representation is based on mathematical graph theory, outlined, for example, in Whittaker (1990) and Lauritzen (1996). In general, a *graph* consists of two sets: the sets of *vertices*, K, and the set of *edges*, E. The set of edges consists of pairs of elements from $K, E \subset K \times K$. In graphical log-linear models, the vertices correspond to variables and the edges correspond to pairs of variables. Therefore, we will set K to $K = \{A, B, C, \ldots\}$ and an element from E has the form (A, C). In undirected graphs, the type of graph that is considered here, if (A, B) is in E, (B, A) is also in E and the edge or line between A and B is undirected. A *chain* between vertices A and C is determined by a sequence of distinct vertices $V_1, \ldots, V_m, V_i \in K$. The chain is given by the sequence of edges $[AV_1/V_1V_2/\ldots/V_mC]$ for which (V_i, V_{i+1}) as well as $(AV_1), (V_mC)$ are in E. This means that a chain represents a sequence of variables leading from one variable to another within the graph. Although A and C may be identical, a vertex between A and C may not be included more than once. Therefore circles are avoided. The left graph in Figure 12.2 contains, for example, the chains $[BA/AC]$, $[CA/AB]$, $[AB]$, and the right graph contains the chain $[AB/BC/CA]$.

FIGURE 12.2: Graphs for log-linear models AB/AC (left) and ABC (right).

Chains are important for interpreting the model. The left graph in Figure 12.2 corresponds to the model AB/AC, which implies that X_B and X_C are conditionally independent given A. If one looks at the paths that connect B and C, one can see that any paths connecting B and C involve A. This property of the graph may be read as the conditional independence of X_B and X_C given X_A.

For the correspondence of graphical log-linear models and graphs it is helpful to consider the largest sets of vertices that include all possible edges between them. A set of vertices for which all the vertices are connected by edges is called *complete*. The corresponding vertices form a complete *subgraph*. A complete set that is not contained in any other complete set is called a *maximal complete set* or a *clique*. The cliques determine the graphical linear model and correspond directly to the notation defining the model. For example, the model AB/AC has the cliques $\{AB\}$ and $\{AC\}$. The saturated model ABC that contains all possible edges has the maximal complete set or clique $\{ABC\}$. An example of a higher dimensional model is the model ABC/AD, which is a graphical model. The model has the cliques $\{ABC\}$, $\{AD\}$ (see Figure 12.3 for the graph). Figure 12.3 also shows the graphs for the saturated model $ABCD$ and the model $ABC/ABD/DE$.

The strength of graphing log-linear models becomes obvious in higher dimensional tables. The basic tool for interpreting graphical log-linear models is a result by Darroch et al. (1980) :

> Let the sets F_0, F_1, F_2 denote disjoint subsets of the variables in a graphical log-linear model. The factors in F_1 are conditionally independent of the factors in F_2 given F_0 if and only if every chain between a factor in F_1 and a factor in F_2 involves at least one factor in F_0. Then F_0 is said to separate the subgraphs formed by F_1 and F_2.

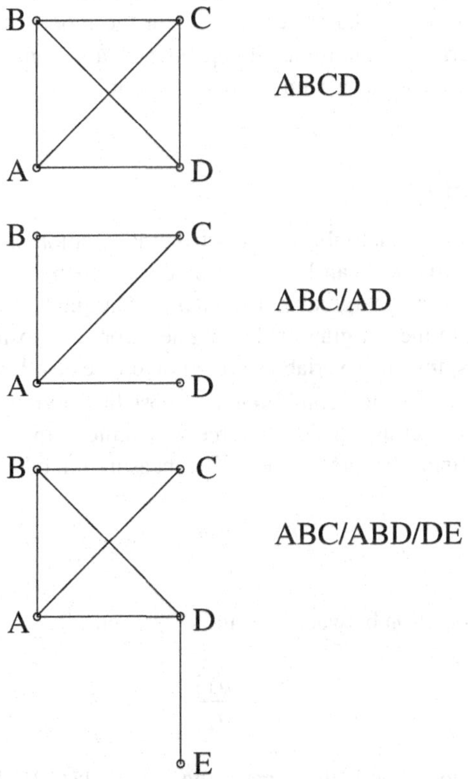

FIGURE 12.3: Graphs for log-linear models in multiway tables.

For the model AB/AC (see graph in Figure 12.2) one may consider $F_1 = \{B\}, F_2 = \{C\}$, and $F_0 = \{A\}$. The conditional independence of X_B and X_C given X_A, in short, $X_B \perp X_C | X_A$, follows directly from the result of Darroch et al. (1980). For the model $ABC/ABD/DE$ (see graph in Figure 12.3) one may build several subsets of variables. By considering $F_1 = \{A, B, C\}, F_2 = \{E\}$, and $F_0 = \{D\}$, one obtains that $\{X_A, X_B, X_C\}$ are conditionally independent of X_E given X_D, $\{X_A, X_B, X_C\} \perp X_E | X_D$. It is said that X_D separates the subgraphs formed by $\{X_A, X_B, X_C\}$ and $\{X_E\}$. By considering $F_1 = \{B, C\}, F_2 = \{E\}, F_0 = \{A, D\}$ one obtains that $\{X_B, X_C\}$ are conditionally independent of X_E given $\{X_A, X_D\}$. It is seen that for higher dimensional models several independence structures usually are involved when considering a graphical log-linear model.

Marginal independence occurs if there are no chains in the graph that connect two groups of variables. The graph corresponding to model AB/C (see graph in Table 12.7) contains no chain between $\{A, B\}$ and $\{C\}$. The implication is that the variables $\{X_A, X_B\}$ are independent of X_C, in short, $\{X_A, X_B\} \perp X_C$. A model may also imply certain marginal independence structures. For example, the model $AB/BC/CD$ implies conditional independence relations $X_A \perp X_D | \{X_B, X_C\}$ and $X_A \perp \{X_C, X_D\} | X_B$, which include all four variables, but also $X_A \perp X_C | X_B$, which concerns the marginal distribution of X_A, X_B, X_C.

As was already seen in three-way tables, not all log-linear models are graphical. That raises the question of how to interpret a log-linear model that is not graphical. Fortunately, any log-linear model can be embedded in a graphical model. For the interpretation one uses the smallest graphical model that contains the specific model. Since the specific model is a submodel of that graphical model, all the (conditional) independence structures of the larger model also have to hold for the specific model. That strategy does not always work satisfactorily. The smallest graphical model that contains the three-way table model $AB/AC/BC$ is the saturated model, which implies no independence structure. But the model $AB/AC/BC$ also has no simple interpretation in terms of conditional independence, although excluding the three-factor interaction restricts the association structure between variables.

12.6 Collapsibility

In general, association in marginal tables differs from association structures found in the full table. For example, X_A and X_B can be conditionally independent given $X_C = k$, even if the variables X_A and X_B are marginally dependent. Marginal dependence means that the association is considered in the marginal table obtained from collapsing over the categories of the other variables; that is, the other variables are ignored (see also Exercise 12.6).

The question arises under which conditions is it possible to infer on the association structure from marginal tables. Let us consider a three-way table with means μ_{ijk}. The marginal association between the binary factors X_A and X_B, measured in odds ratios, is determined by

$$\frac{\mu_{11+}/\mu_{12+}}{\mu_{21+}/\mu_{22+}},$$

while the conditional association between X_A and X_B given $X_C = k$ is determined by

$$\frac{\mu_{11k}/\mu_{12k}}{\mu_{21k}/\mu_{22k}}.$$

One can show that the association is the same if the model AB/AC holds (compare Exercise 12.13).

In general, the association is unchanged if groups of variables are separated:

> Let the sets F_1, F_2, F_0 denote the disjoint subsets of the variables in a graphical log-linear model. If every chain between a factor in F_1 and a factor in F_2 involves at least one factor in F_0, the association among the factors in F_1 and F_0 can be examined in the marginal table obtained from collapsing over the factors in F_2. In the same way, the association among the factors in F_2 and F_0 can be examined in the marginal table obtained from collapsing over the factors in F_1

(compare Darroch et al., 1980, and Bishop et al., 2007). Therefore, if F_0 separates the subgraphs formed by F_1 and F_2, one can collapse over F_2 (or F_1, respectively). In the model AB/AC considered previously, the association between X_A and X_B as well as the association between X_A and X_C can be examined from the corresponding marginal tables.

12.7 Log-Linear Models and the Logit Model

The log-linear models for contingency tables may be represented as logit models. Let us consider a three-way table with categorical variables X_A, X_B, X_C. The most general log-linear model is the saturated model:

$$\begin{aligned}\log(\mu_{ijk}) = \lambda_0 &+ \lambda_{A(i)} + \lambda_{B(j)} + \lambda_{C(k)} \\ &+ \lambda_{AB(ij)} + \lambda_{AC(ik)} + \lambda_{BC(jk)} + \lambda_{ABC(ijk)}.\end{aligned}$$

For multinomial tables, for which $\mu_{ijk} = n\pi_{ijk}$, one obtains the logit model with reference category $(X_A = I, X_B = J, X_C = K)$:

$$\log\left(\frac{\pi_{ijk}}{\mu_{IJK}}\right) = \gamma_{A(i)} + \gamma_{B(j)} + \gamma_{C(k)} + \gamma_{AB(ij)} + \gamma_{AC(ik)} + \gamma_{BC(jk)} + \gamma_{ABC(ijk)},$$

where the γ-parameters are obtained as differences, for example, $\gamma_{A(i)} = \lambda_{A(i)} - \lambda_{A(I)}$, $\gamma_{ABC(ijk)} = \lambda_{ABC(ijk)} - \lambda_{ABC(IJK)}$. The parametrization of the model, which uses reference categories for the variables X_A, X_B, X_C, reflects that a structured multinomial distribution is given. In contrast to the simple multinomial distributions considered in Chapter 8, the distribution of the response is determined by three separate variables that structure the multinomial distribution.

Logit Models with Selected Response Variables

Consider now that X_C is chosen as a response variable. Then one obtains for multinomial tables

$$\log\left(\frac{\mu_{ijr}}{\mu_{ijK}}\right) = \log\left(\frac{P(X_A = i, X_B = j, X_C = k)}{P(X_A = i, X_B = j, X_C = K)}\right) = \log\left(\frac{\pi_{r|ij}}{\pi_{K|ij}}\right),$$

where $\pi_{r|ij} = P(X_C = r|X_A = i, X_B = j)$. By using the saturated model, which always holds, one obtains from easy derivation the multinomial logit model

$$\log\left(\frac{P(X_C = r|X_A = i, X_B = j)}{P(X_C = K|X_A = i, X_B = j)}\right) = \gamma_{0r} + \gamma_{A(i),r} + \gamma_{B(j),r} + \gamma_{AB(ij),r},$$

where

$$\begin{aligned}\gamma_{0r} &= \lambda_{C(r)} - \lambda_{C(K)}, \quad \gamma_{A(i),r} = \lambda_{AC(ir)} - \lambda_{AC(iK)}, \\ \gamma_{B(j),r} &= \lambda_{BC(jr)} - \lambda_{BC(jK)}, \quad \gamma_{AB(ij),r} = \lambda_{ABC(ijr)} - \lambda_{ABC(ijK)}.\end{aligned}$$

An alternative form of the model, which uses dummy variables, is

$$\log\left(\frac{\pi_{r|ij)}}{\pi_{K|ij)}}\right) = \gamma_{0r} + \gamma_{A(1),r}x_{A(1)} + \cdots + \gamma_{B(1),r}x_{B(1)} + \quad \cdots +$$

$$\gamma_{AB(11),r}x_{A(1)}x_{B(1)} + \cdots + \gamma_{AB(I-1,J-1),r}x_{A(I-1)}x_{B(J-1)}.$$

The constraints on the λ-parameters and therefore the type of coding of dummy variables carry over to the γ-parameters. For example, the constraint $\sum_i \lambda_{AC(ik)} = 0$ transforms into $\sum_i \gamma_{A(i),r} = 0$.

In summary, by choosing one variable as the response variable, the log-linear model of association between X_A, X_B, X_C turns into a regression model. If the log-linear model is a submodel of the saturated model, some γ terms are not contained in the corresponding logit model. For example, by assuming the log-linear model AB/AC (meaning that X_B and X_C are conditionally independent) one obtains the logit model

$$\log\left(\frac{P(X_C = r|X_{A=i}, X_B = j)}{P(X_C = K|X_A = i, X_B = j)}\right) = \gamma_{0r} + \gamma_{A(i),r},$$

which contains only the explanatory variable X_A. Since X_B and X_C are conditionally independent given X_A, it is quite natural that X_B does not effect on X_C since it is associated with X_C only through X_A. In Table 12.7 the explanatory variables of logit models with response X_C are given together with the underlying log-linear model. It is seen that model AB/BC as well as model A/BC yield a logit model with X_B as the only explanatory variable. Model AB/BC is weaker than A/BC. Since the effect of the variable X_A on X_C is already omitted if model AB/BC holds, it is naturally omitted if an even stronger model holds.

12.8 Inference for Log-Linear Models

Log-linear models may be embedded into the framework of generalized linear models. For all three sampling schemes – Poisson distribution, multinomial distribution, and product-multinomial distribution – the response distribution is in the exponential family. The log-linear model has the form assumed in GLMs, where the mean is linked to the linear predictor by a transformation function. Thus maximum likelihood estimation and testing are based on the methods developed in Chapters 3 and 8. Advantages of log-linear models are that maximum likelihood estimates are sometimes easier to compute and sufficient statistics have a simple form. In the following the results are briefly stretched.

12.8.1 Maximum Likelihood Estimates and Minimal Sufficient Statistics

For simplicity, the Poisson distribution is considered for three-way models. Let all of the parameters be collected in one parameter vector $\boldsymbol{\lambda}$. From the likelihood function

$$L(\boldsymbol{\lambda}) = \prod_{i=1}^{I}\prod_{j=1}^{J}\prod_{k=1}^{K} \frac{\mu_{ijk}^{x_{ijk}}}{x_{ijk}!} e^{-\mu_{ijk}}$$

one obtains the log-likelihood

$$l(\boldsymbol{\lambda}) = \sum_{i,j,k} x_{ijk}\log(\mu_{ijk}) - \sum_{i,j,k}\mu_{ijk} - \sum_{i,j,k}\log(x_{ijk}!).$$

With μ_{ijk} parameterized as the saturated log-linear model one obtains by rearranging terms (and omitting constants)

$$\begin{aligned} l(\boldsymbol{\lambda}) &= n\lambda_0 + \sum_i x_{i++}\lambda_{A(i)} + \sum_j x_{+j+}\lambda_{B(j)} + \sum_k x_{++k}\lambda_{C(k)} \\ &+ \sum_{i,j} x_{ij+}\lambda_{AB(ij)} + \sum_{i,k} x_{i+k}\lambda_{AC(ik)} + \sum_{j,k} x_{+jk}\lambda_{BC(jk)} \\ &+ \sum_{i,j,k} x_{ijk}\lambda_{ABC(ijk)} - \sum_{i,j,k} \exp(\lambda_0 + \lambda_{A(i)} + \ldots + \lambda_{ABC(ijk)}). \end{aligned}$$

The form of the log-likelihood remains the same when non-saturated models are considered. For example, if $\lambda_{ABC} = 0$, the term $\sum_{i,j,k} x_{ijk}\lambda_{ABC(ijk)}$ is omitted. Since the Poisson is an exponential family distribution, the factors on parameters represent sufficient statistics that contain all the information about parameters. This means that for non-saturated models the parameter estimates are determined by marginal sums. For example, the likelihood of the independence model $A/B/C$ contains only the marginal sums $x_{i++}, x_{+j+}, x_{++k}$. It is noteworthy that the sufficient statistics, which are even minimal statistics, correspond directly to the symbol for the model. Table 12.9 gives the sufficient statistics for the various types of log-linear models for three-way tables.

As usual, maximum likelihood estimates are obtained by setting the derivations of the log-likelihood equal to zero. The derivative for one of the parameters, say $\lambda_{AB(ij)}$, is given by

$$\frac{\partial l(\boldsymbol{\lambda})}{\partial \lambda_{AB(ij)}} = x_{ij+} - \sum_k \exp(\lambda_0 + \lambda_{A(i)} + \ldots) = x_{ij+} - \mu_{ij+}.$$

From $\partial l(\boldsymbol{\lambda})/\partial \lambda_{AB(ij)} = 0$ one obtains immediately $x_{ij+} = \hat{\mu}_{ij+}$. Hence, computing the maximum likelihood estimates reduces to solving the equations that equal the sufficient statistics to their expected values. For example, for the independence model one has to solve the system of equations

$$x_{i++} = \hat{\mu}_{i++}, \quad x_{+j+} = \hat{\mu}_{+j+}, \quad x_{++k} = \hat{\mu}_{++k}, \tag{12.6}$$

$i = 1, \ldots, I, j = 1, \ldots J, k = 1, \ldots, K$.

If the log-linear model is represented in the general vector form

$$\log(\boldsymbol{\mu}) = \boldsymbol{X\lambda}$$

with $\boldsymbol{\mu}$ containing all the expected cell counts and $\boldsymbol{X}$ denoting the corresponding design matrix, the likelihood equations equating sufficient statistics to expected values have the form

$$\boldsymbol{X}^T\boldsymbol{x} = \boldsymbol{X}^T\boldsymbol{\mu},$$

where $\boldsymbol{x}$ is the vector of cell counts (for three-way tables $\boldsymbol{x}^T = (x_{111}, x_{112}, \ldots, x_{IJK})$).

TABLE 12.9: Log-linear models and sufficient statistics for three-way tables.

ABC	$\{x_{ijk}\}$
AB/AC/BC	$\{x_{ij+}\}, \{x_{i+k}\}\{x_{+jk}\}$
AB/AC	$\{x_{ij+}\}, \{x_{i+k}\}$
A/BC	$\{x_{i++}\}, \{x_{+jk}\}$
A/B/C	$\{x_{i++}\}, \{x_{+j+}\}, \{x_{++k}\}$

Solving these equations can be very easy. For example, the solution of (12.6) is directly given by

$$\hat{\mu}_{ijk} = n\frac{x_{i++}}{n}\frac{x_{+j+}}{n}\frac{x_{++k}}{n}.$$

The form mimics $\mu_{ijk} = n\pi_{i++}\pi_{+j+}\pi_{++k}$. Direct estimates are available for all log-linear models for three-way tables except $AB/AC/BC$. A general class of models for which direct estimates exist is decomposable models. A model is called decomposable if it is graphical and *chordal*, where chordal means that every closed chain $[AV_1/V_1V_2/\dots/V_mA]$ (for which the starting point and end point are identical) that involves at least four distinct edges has a shortcut. A simple example is the model $AB/BC/CD/AD$, which is represented by a rectangle. It is not decomposable because the chain $[AB/BC/CD/DA]$ has no shortcut. It becomes decomposable by adding one more edge, AC or BD, yielding the model ABC/ACD or ABD/BCD, respectively. A more extensive treatment of direct estimates was given, for example by Bishop et al. (1975).

If no direct estimates are available, iterative procedures as in GLMs can be used. An alternative, rather stable procedure that is still used for log-linear models is *iterative proportional fitting*, also called the *Deming-Stephan algorithm* (Deming and Stephan, 1940). It iteratively fits the marginals, which for the example of the independence model are given in 12.6. It works for direct estimates as well as for models, for which no direct estimates exist.

For graphical models the density can always be represented in the form

$$f(\{x_{ijk}\}) = \frac{1}{z_0}\prod_{C_l}\phi_{C_l}(x_{C_l}),$$

where the sum is over the cliques, z_0 is a normalizing constant, and the $\phi_{C_l}(x_{C_l})$ are so-called clique potentials depending on observations x_{C_l} within the subgraph formed by C_l. The clique potentials do not have to be density functions but contain the dependencies in C_l. Therefore, the estimation is based on marginals that are determined by the cliques (for general algorithms based on the decomposition see, for example, Lauritzen, 1996).

For Poisson sampling, the Fisher matrix $\boldsymbol{F}(\hat{\boldsymbol{\lambda}})$ has the simple form $\boldsymbol{X}^T diag(\boldsymbol{\mu})\boldsymbol{X}$, yielding the approximation

$$\text{cov}(\hat{\boldsymbol{\lambda}}) \approx (\boldsymbol{X}^T \text{diag}(\hat{\boldsymbol{\mu}})\boldsymbol{X})^{-1}.$$

For multionomial sampling one has to separate the intercept, which is fixed by the sample size. For the corresponding model $\log(\boldsymbol{\mu}) = \lambda_0\boldsymbol{1}+\boldsymbol{X}\boldsymbol{\gamma}$ the Fisher matrix is $\boldsymbol{X}^T(diag(\boldsymbol{\mu})-\boldsymbol{\mu}\boldsymbol{\mu}^T)\boldsymbol{X}$, yielding the approximation

$$\text{cov}(\hat{\boldsymbol{\gamma}}) \approx (\boldsymbol{X}^T(diag(\hat{\boldsymbol{\mu}}) - \hat{\boldsymbol{\mu}}\hat{\boldsymbol{\mu}}^T)\boldsymbol{X})^{-1}.$$

ML estimates for both sampling distributions can be computed within a closed framework. Let $\mu = \sum_i \mu_i = \sum_i \exp(\lambda_0 + \boldsymbol{x}_i^T\boldsymbol{\gamma})$ denote the total expected cell counts and

$$\pi_i = \frac{\mu_i}{\sum_j \mu_j} = \frac{\exp(\lambda_0 + \boldsymbol{x}_i^T\boldsymbol{\gamma})}{\sum_j \exp(\lambda_0 + \boldsymbol{x}_j^T\boldsymbol{\gamma})} = \frac{\exp(\boldsymbol{x}_i^T\boldsymbol{\gamma})}{\sum_j \exp(\boldsymbol{x}_j^T\boldsymbol{\gamma})}$$

the relative expected cell counts, which do not depend on λ_0. With $x = \sum_i x_i$ representing the total count one obtains the log-likelihood for the Poisson distribution:

$$l(\lambda_0, \boldsymbol{\gamma}) = \sum_i x_i \log(\mu_i) - \sum_i \mu_i = \sum_i x_i(\lambda_0 + \boldsymbol{x}_i^T\boldsymbol{\gamma}) - \mu = x\lambda_0 + \sum_i x_i(\boldsymbol{x}_i^T\boldsymbol{\gamma}) - \mu.$$

By including $x\log(\mu) - x\log(\mu)$ one obtains the additive decomposition

$$l(\lambda_0, \gamma) = \{\sum_i x_i(\boldsymbol{x}_i^T\gamma) - x\log(\sum_i \exp(\boldsymbol{x}_i^T\gamma))\} + \{x\log(\mu) - \mu\}.$$

The first term is the log-likelihood of a multinomial distribution $(x_1, x_2, \ldots) \sim M(x, \boldsymbol{\pi})$), and the term $x\log(\mu) - \mu$ is the log-likelihood of a Poison distribution $x \sim P(\mu)$. Therefore, maximization of the first term, which does not include the intercept, yields estimates $\hat{\gamma}$ for multinomial sampling, conditional on the number of cell counts x. Maximization of the Poisson log-likelihood yields $\hat{\mu} = x$, which determines the estimate of λ_0, since $\mu = c\exp(\lambda_0)$, where $c = \sum_j \exp(\boldsymbol{x}_j^T\gamma)$ is just a scaling constant determined by the maximization of the first term.

A similar decomposition of the Poisson log-likelihood holds for the product-multinomial distribution. Computation as well as inference can be based on the same likelihood with conditioning arguments. For details see Palmgren (1981) and Lang (1996a).

12.8.2 Testing and Goodness-of-Fit

Let the cell counts be given by the vector $\boldsymbol{x}^T = (x_1, \ldots, x_N)$, where N is the number of cells and only a single index is used for denoting the cell. The vector $\hat{\boldsymbol{\mu}} = (\hat{\mu}_1, \ldots, \hat{\mu}_N)$ denotes the corresponding fitted means. For models with an intercept the deviance has the form

$$D = 2\sum_{i=1}^{N} x_i \log(\frac{x_i}{\hat{\mu}_i}). \tag{12.7}$$

When considering goodness-of-fit an alternative is Pearson's χ^2:

$$\chi_P^2 = \sum_{i=1}^{N} \frac{(x_i - \hat{\mu}_i)^2}{\hat{\mu}_i}.$$

For fixed N, both statistics have an approximate χ^2-distribution if the assumed model holds and the means μ_i are large. The degrees of freedom are $N - p$, where p is the number of estimated parameters. The degrees of freedom are computed from the general rule

> Number of parameters in the saturated model – Number of parameters in the assumed model.

More concisely, the number of parameters is the number of linearly independent parameters. For example, the restriction $\sum_i \lambda_{A(i)} = 0$ implies that the effective number of parameters $\lambda_{A(i)}, i = 1, \ldots, I$, is $I - 1$ since $\lambda_{A(I)} = -\lambda_{A(1)} - \cdots - \lambda_{A(I-1)}$.

Let us consider an example for three-way tables. The saturated model has IJK parameters (corresponding to the cells) for Poisson data, but $IJK - 1$ parameters for multinomial data, since the restriction $\sum_{ijk} \mu_{ijk} = 1$ applies. For the independence model the number of parameters is determined by the $I - 1$ parameters $\lambda_{A(i)}$, the $J - 1$ parameters $\lambda_{B(j)}$, and the $K - 1$ parameters $\lambda_{C(k)}$. For Poisson data one has an additional intercept that yields the difference:

$$df = IJK - (1 + I - 1 + J - 1 + K - 1) = IJK - I - J - K + 2.$$

For multinomial data, the restriction $\sum_{ijk} \mu_{ijk} = 1$ applies (reducing the number of parameters by 1) and one obtains

$$df = \{IJK - 1\} - \{I - 1 + J - 1 + K - 1\} = IJK - I - J - K + 2,$$

which is the same as for Poisson data. In general, the degrees of freedom of the approximate χ^2-distribution are the same as for the sampling schemes. For obtaining asymptotically a χ^2-distribution one has to assume $\sum_i \mu_i \to \infty$ with μ_i/μ_j being constant for Poisson data and $n \to \infty$ for multinomial data and product-multinomial data, where in the latter case a constant ratio between n and the sampled subpopulation is assumed (for a derivation of the asymptotic distribution see, for example, Christensen, 1997, Section 2.3).

The analysis of deviance as given in Section 3.7.2 provides test statistics for the comparison of models. Models are compared by the difference in deviances. If $\tilde{M}$ is a submodel of M, one considers

$$D(\tilde{M}|M) = D(\tilde{M}) - D(M). \tag{12.8}$$

The deviance (12.7) may be seen as the difference between the fitted model and the saturated model since the deviance of the saturated model, which has a perfect fit, is zero.

A hierarchical submodel is always determined by assuming that part of the parameters equals zero. For example, the model AB/C assumes that

$$H_0 : \lambda_{AC(ik)} = \lambda_{BC(jk)} = \lambda_{ABC(ijk)} = 0 \quad \text{for all } i, j, k.$$

The deviance for the model AB/C may also be seen as a test statistic of the null hypothesis H_0. When using the difference (12.8) one implicitly tests that the parameters that are contained in M but not in $\tilde{M}$ are zero, given that model M holds.

Example 12.3: Birth Data

In the birth data example (Example 12.1) the variables are gender of the child (G; 1: male; 2: female), if membranes did rupture before the beginning of labour (M; 1: yes; 0: no), if Cesarean section has been applied (C; 1: yes; 0: no) and if the birth has been induced (I; 1: yes; 0: no). The search for an adequate model is started by fitting models that contain all interaction terms of a specific order. Let $M([m])$ denote the model that contains all m-factor interactions. For example, $M([1])$ denotes the main effect model $G/M/C/I$. From Table 12.10 it is seen that $M([3]), M([2])$ fit well but $M([1])$ should be rejected. Thus one considers models between $M([2])$ and $M([1])$. Starting from $M([2])$, reduced models are obtained by omitting one of the six two-factor interactions at a time. For example, $M([2])\backslash GM$ denotes the model that contains all two-factor interactions except GM. The difference of deviances, for example, for model $M([2])$ and model $M([2])\backslash GM$, is an indicator of the relevance of the interaction GM. It is seen that the interactions MC, CI, and MI should not be omitted. The model $G/MC/MI/CI$ shows a satisfying fit while further reduction by omitting G is inappropriate.

The model $G/MC/MI/CI$ is not a graphical model. The smallest graphical model that contains $G/MC/MI/CI$ is the model G/CMI, which is shown in Figure 12.4. It means that I, C, M are interacting but are independent of gender. The gender of the child seems not to be connected to the variables membranes, Cesarean section, and induced birth. □

12.9 Model Selection and Regularization

Model selection is usually guided by the objective of the underlying study. If a specific association structure is to be investigated, the analysis can be reduced to testing if certain interaction terms can be omitted, which is equivalent to testing the fit of the correspondingly reduced model or, more general, a sequence of models. When no specific hypotheses are to be investigated, model selection aims at a compromise between two competing goals: sparsity and goodness-of-fit. One wants to find models that are close to the data but have an economic representation that allows a simple interpretation.

TABLE 12.10: Deviances and differences for log-linear models for birth data.

Model	Dev.	Df	Differences	Diff-Df	Diff-Dev	p-Value
M([4])	0	0				
M([3])	0.834	1	M([3])−M([4])	1	0.834	0.361
M([2])	4.765	5	M([2])−M([3])	4	3.931	0.415
M([1])	28.915	11	M([1])−M([2])	6	24.150	0.000
M([2]\GM)	5.244	6	M([2]\GM) −M([2])	1	0.478	0.489
M([2]\MC)	9.965	6	M([2]\MC) − M([2])	1	5.200	0.023
M([2]\CI)	12.167	6	M([2]\CI) − M([2])	1	7.402	0.007
M([2]\GI)	6.971	6	M([2]\GI) − M([2])	1	2.206	0.137
M([2]\GC)	6.566	6	M([2]\GC) − M([2])	1	1.801	0.180
M([2]\MI)	10.100	6	M([2]\MI) - M([2])	1	5.334	0.021
M(G/MC/MI/CI)	8.910	8	M(G/CI/MI/CI)-M([2])	3	4.145	0.246
M(MC/MI/CI)	19.428	9	M(CI/MI/CI)-M([2])	4	14.663	0.005

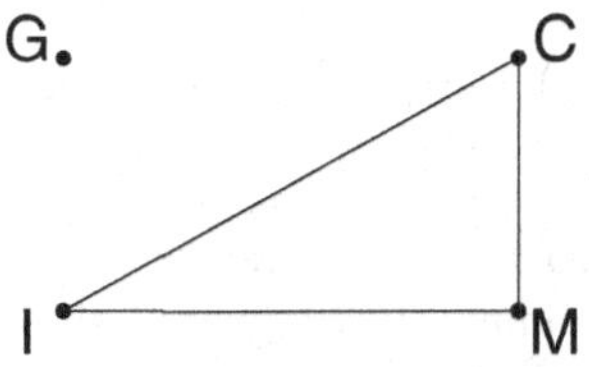

FIGURE 12.4: Graphical model for birth data.

Several model selection procedures for log-linear models were proposed. Some try to account for the selection error by using multiple testing strategies, while others rely on screening procedures (for references see Section 12.11). More recently, regularization methods for the selection of log-linear and grahical models have been developed. The methods are particularly attractive for finding sparse solutions that fit the data well. In particular, in bioinformatics the goal to identify relevant structure is very ambitious. With thousands of variables in genomics, it is to be seen if the selection strategies are sufficiently reliable. However, the strategies are also useful when the number of variables is much smaller but too large for the fitting of all possible models.

A strategy that is strongly related to the regularization methods in Chapter 6 has been given by Dahinden et al. (2007). Let $X_1, \dots, X_p$ denote the factors, where $X_j \in \{1, \dots, k_j\}$ and $I = \{1, \dots, p\}$ denote the index set of factors. By using subsets $A \subset I$ to define the main and interaction terms, the design matrix of the log-linear $\log(\boldsymbol{\mu}) = \boldsymbol{X}\boldsymbol{\lambda}$ can be decomposed into

$$\boldsymbol{X} = [\boldsymbol{X}_{A_1}|\dots|\boldsymbol{X}_{A_m}],$$

where $\boldsymbol{X}_{A_j}$ refers to a specific main or interaction term. For example, $\boldsymbol{X}_{\{1,2\}}$ refers to the interaction terms of variables X_1, X_2. Correspondingly, let $\boldsymbol{\lambda}_{A_j}$ denote the vector of main or interaction parameters. The penalized log-likelihood, considered in Chapter 6, has the form $l_p(\boldsymbol{\beta}) = l(\boldsymbol{\beta}) - \frac{\lambda}{2}J(\boldsymbol{\beta})$, where $l(\boldsymbol{\beta})$ is the usual log-likelihood, $J(\boldsymbol{\beta})$ represents a penalty term and λ is a tuning parameter. Then the grouped lasso (Section 6.2.2) can be applied by using the penalty

$$J(\boldsymbol{\lambda}) = \sum_{j=1}^{G} \sqrt{df_j} \|\boldsymbol{\lambda}_{A_j}\|_2, \tag{12.9}$$

where $\|\boldsymbol{\lambda}_{A_j}\|_2 = (\lambda^2_{A_j,1} + \cdots + \lambda^2_{A_j,df_j})^{1/2}$ is the L_2-norm of the parameters of the jth group of parameters, which comprises df_j parameters. The penalty encourages sparsity in the sense that either $\hat{\boldsymbol{\lambda}}_{A_j} = \mathbf{0}$ or $\lambda_{A_j,s} \neq 0$ for $s = 1, \dots, df_j$. If one has one binary variable X_1 and a variable X_2 with three categories, for example, the interaction term comprises two parameters $\lambda_{12(11)}, \lambda_{12(12)}$ and the L_2-norm of the parameters is $(\lambda^2_{12(11)} + \lambda^2_{12(12)})^{1/2}$.

When using the grouped lasso the resulting model will in general be non-hierarchical. Of course, it is easy to fit the corresponding hierarchical model with all the necessary marginal effects included. However, if one single high-order interaction term is selected, the resulting model can be quite complex. Therefore, Dahinden et al. (2007) proposed starting the selection procedure not only from the full model but from all models $M([m])$, which contains all m-factor interactions. Then the best model is selected.

A strategy that is quite common is to start from $M([2])$, which contains all two-factor interactions. For binary variables $X_i \in \{0, 1\}$, the approach is usually based on the *Ising model*, which assumes that the joint probabilities are given by

$$P(X_1, \dots, X_p) = \exp\Big(\sum_{(j,k) \in E} \theta_{jk} X_j X_k - \phi(\boldsymbol{\theta}) \Big),$$

where the normalizing function $\boldsymbol{\theta}$ contains the parameters θ_{jk}, and the sum is over the edges E of a graphical model (see, for example, Ravikumar et al., 2009). For technical reasons an artificial variable $X_0 = 1$ and edges between X_0 and all the other variables are included. For log-linear models, the Ising model with all possible edges is equivalent to the multinomial model that contains all two-factor interactions. For the conditional model, conditioned on the other variables, one obtains a main effect model, which in the parametrization of the Ising model is given by

$$P(X_j = 1 | X_1 = x_1, \dots, X_p = x_p) = \frac{\exp(\sum_{(j,k) \in E} \theta_{jk} x_k)}{1 + \exp(\sum_{(j,k) \in E} \theta_{jk} x_k)}.$$

(Exercise 12.15). The model is equivalent to a main effect logit model with response variable X_j and explanatory variables X_k that are linked to X_j within the graph. If the relevant two-factor interactions are identified, it is straightforward to identify the corresponding graphical model. However, starting from a two-factor interaction model has the disadvantage that all higher interaction terms are neglected during the selection procedure. It might be more appropriate to enforce sparse modeling by administering stronger penalties on higher interaction terms or by strictly fitting hierarchical models within a boosting procedure.

Example 12.4: Birth Data

Let us consider again the birth data (Example 12.3). Figure 12.5 shows the coefficient build ups for the fitting of a log-linear model with two-factor interactions, where the two-factor interactions are penalized by (12.9), while main effects are not penalized. The coefficient buildups show the parameter estimates for varying degrees of smoothing λ; here they are plotted against $\|\boldsymbol{\beta}\|$. At the right end no penalty is exerted and the model that contains all two-factor interactions is fitted. The solid lines show the two-factor interactions, and the dashed lines represent the main effects. Since the main effects are not penalized, they remain rather stable. The vertical lines in Figure 12.5 show the models that are selected by use of AIC and BIC. The stronger criterion, BIC, yields a model that contains only the strong interactions MC, MI, CI (Figure 12.4). The graphical model that contains these interactions is the same as found

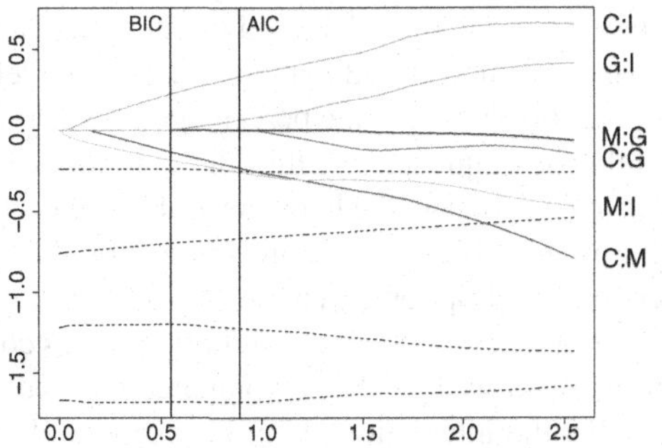

FIGURE 12.5: Coefficient buildups for a log-linear model with two-factor interactions (birth data).

before, $G/MC/MI/CI$. If one uses AIC, the rather weak interaction GI also has to be included. As was to be expected, BIC yields a sparser model.

□

Example 12.5: Leukoplakia

In Example 12.2 one wants to examine the association between the occurrence of leukoplakia (L), alcohol intake in grams of alcohol (A), and smoking habits (S). Figure 12.6 shows the coefficient buildups for the penalized fitting of a log-linear model with two-factor interactions. AIC as well as BIC (vertical line) suggest that the interaction between leucoplakia and alcohol intake is not needed. Leukoplakia and alcohol seem to be conditional independent given smoking habits, yielding the model SL/SA (see also Exercise 12.11). □

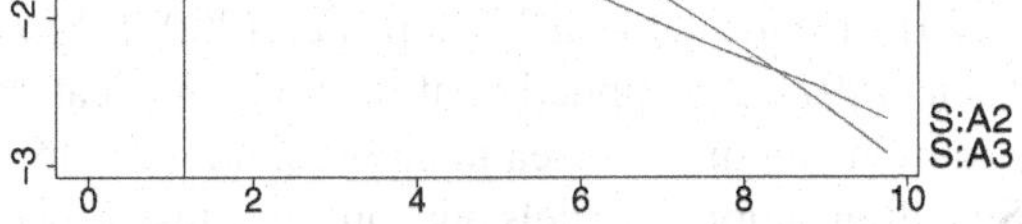

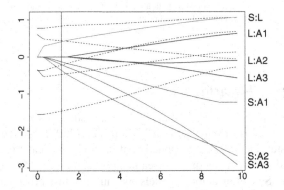

FIGURE 12.6: Coefficient buildups for a log-linear model with two-factor interactions, leucoplakia data; solid lines show the two-factor interactions, and the dashed lines show the main effects.

12.10 Mosaic Plots

For low-dimensional contingency tables as in the preceding example a helpful graphical representation is the *mosaic plot*. It starts as a square with length one. The square is divided first into bars whose widths are proportional to the relative frequencies associated with the first categorical variable. Then each bar is split vertically into bars that are proportional to the conditional probabilities of the second categorical variable. Additional splits can follow for further variables. In Figure 12.7, left panel, the first split distinguishes between leukoplakia, yes or no, and the widths of the columns are proportional to the percentage of observations. The vertical splits represent the conditional relative frequencies within the columns and are proportional to the heights of the boxes. Under independence, the heights would be the same, which is not the case in this example. In the right panel of Figure 12.7, one more split that represents the conditional relative frequencies of alcohol consumption given the status of leukoplakia and smoking is included. It is seen that the distribution of alcohol consumption varies considerably with smoking behavior, but the variation is much weaker if one compares the two groups with and without leukoplakia. This supports that the interaction effect between smoking and alcohol is needed whereas the association between leukoplakia and alcohol can be neglected.

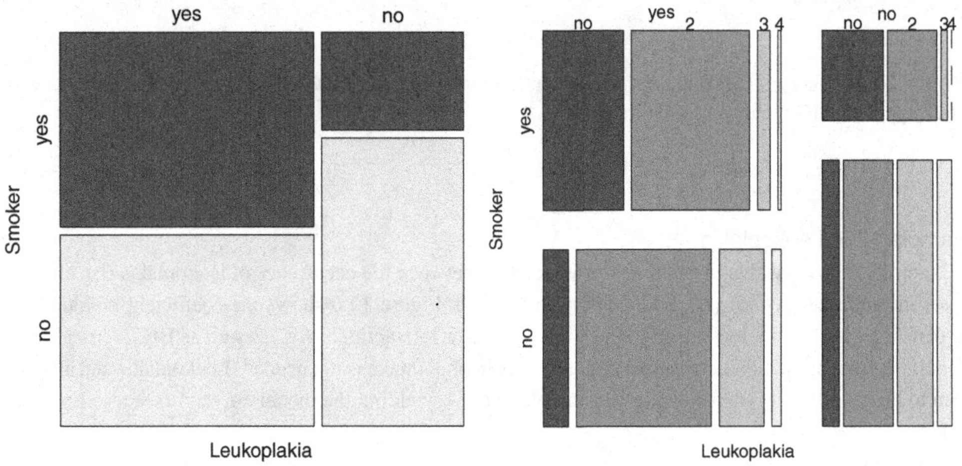

FIGURE 12.7: Mosaic plots for leukoplakia data. Left: leukoplakia and smoking; right: leukoplakia, smoking and alcohol consumption (alcohol in categories no, 1, 2, 4).

12.11 Further Reading

Surveys on Log-Linear and Graphical Models. An early summary of log-linear models was given by Bishop et al. (1975), also available as Bishop et al. (2007). More on log-linear models is also found in Christensen (1997). An applied treatment of graphical models is in the book of Whittaker (2008), and a more mathematical treatment is found in Lauritzen (1996).

Model Selection. Selection among models by multiple test procedures was considered by Aitkin (1979, 1980). Alternative strategies including screening procedures were given by Brown (1976), Benedetti and Brown (1978), and Edwards and Havranek (1987).

Ordinal Association and Smoothing. Ordinal association models, which use assigned scores, were considered by Goodman (1979, 1981b, 1983, 1985), Haberman (1974), and Agresti and Kezouh (1983). An overview was given by Agresti (2009). Smoothing for sparse tables was investigated by Simonoff (1983, 1995), overviews are found in Simonoff (1996), Simonoff and Tutz (2000).

Exact inference. A survey of exact inference for contingency tables was given by Agresti (1992b); see also Agresti (2001).

R Packages. Log-linear models can be fitted by use of the R-function *loglin* from the package stats, which applies an iterative-proportional-fitting algorithm. The function *loglm* from the package *MASS* provides a front-end to *loglin*, to allow log-linear models to be specified and fitted in a manner similar to that of other fitting functions, such as GLM. The function *mosaicplot* from the package *vcd* can be used to generate mosaic plots.

12.12 Exercises

12.1 Let $Y_1, \ldots, Y_N$ denote independent Poisson variables, $Y_i \sim \mathrm{P}(\lambda_i)$. Show that one obtains for $(Y_1, \ldots, Y_N)$ conditional on the total sum $n_0 = Y_1, + \cdots + Y_N$, the multinomial distribution $\mathrm{M}(n_0, (\pi_1, \ldots, \pi_N))$ with $\pi_i = \lambda_i/(\lambda_1 + \cdots + \lambda_N)$.

12.2 Consider the log-linear model for a (2×2)-contingency table:

$$\log(\mu_{ij}) = \lambda_0 + \lambda_{A(i)} + \lambda_{B(j)} + \lambda_{AB(ij)}$$

with multinomial distribution and appropriate constraints.

(a) Derive the parameters $\lambda_{A(i)}, \lambda_{B(j)}, \lambda_{AB(ij)}$ as functions of odds and odds ratios for symmetrical constraints and when the last category parameter is set to zero.

(b) Show that $\lambda_{AB(ij)} = 0$ is equivalent to the independence of the variables X_A, X_B, which generate the rows and columns.

12.3 Consider the log-linear model for a (2×2)-contingency table:

$$\log(\mu_{ij}) = \lambda_0 + \lambda_{A(i)} + \lambda_{B(j)} + \lambda_{AB(ij)},$$

which describes the distribution of product-multinomial sampling with fixed marginals $x_{i.}$.

(a) Specify the appropriate constraints for the parameters.

(b) Show that $\lambda_{AB(ij)} = 0$ is equivalent to the homogeneity of the response X_A across levels of X_B.

12.4 Show that the log-linear model AB/AC in a multinomial distribution $(I \times J \times K)$-contingency table is equivalent to assuming that the variables X_B and X_C are conditionally independent given X_A.

12.5 Show that for the log-linear model AB/AC, ML estimates of means are given by $\hat{\mu}_{ijk} = x_{ij+}x_{i+k}/x_{i++}$. Use the estimation equations that have to hold.

12.6 Find a set of probabilities $\{\pi_{ijk}\}$ for three-way tables, where $\pi_{ijk} = P(X_A = i, X_B = j, X_C = k)$, such that X_A and X_B are conditionally independent given $X_C = k$ but the variables X_A and X_B are (marginally) dependent.

12.7 Compute the parameters of a three-way contingency table as functions of the underlying means $\{\mu_{ijk}\}$ for symmetric side constraints.

12.8 Consider the log-linear models

$AB/AC/AD/DE, ABC/BCD/BDE/CDE, AB/BCE/CDE/AE.$

Are these models graphical? If they are, draw the graph; if not, give the smallest graphical model that includes the corresponding model and draw the graph.

12.9

(a) Interpret the model $AE/BC/CD/BD$.

(b) Give all the independence relations that are implied by the model $AB/BC/CD$.

TABLE 12.11: Regular reader of women's journal with employment, age, and education.

			Regular Reader	
			Yes	No
Working(W)	Age (A)	Education (E)		
Yes	18 – 29	L1	1	14
		L2	32	49
		L3	20	34
		L4	8	3
	30 – 39	L1	9	23
		L2	31	57
		L3	11	26
		L4	5	7
	40 – 49	L1	1	33
		L2	12	50
		L3	5	11
		L4	1	7
No	18 – 29	L1	3	24
		L2	12	41
		L3	19	20
		L4	14	13
	30 – 39	L1	1	37
		L2	12	68
		L3	14	43
		L4	4	7
	40 – 49	L1	11	54
		L2	14	53
		L3	8	15
		L4	1	3

12.10 The contingency table 12.11 shows data from a survey on the reading behavior of women (Hamerle and Tutz, 1980). The cells are determined by working (yes/no), age in categories, education level (L1 to L4), and if the women is a regular reader of a specific journal. Find an appropriate log-linear model for the data.

12.11 Fit log-linear models for the leucoplakia data (Table 12.2) and select an appropriate model (compare to Example 12.5).

12.12 In contingency table 12.12 defendants in cases of multiple murders in Florida between 1976 and 1987 are classified with respect to death penalty, race of defendent, and race of victim (see Agresti, 2002; Radelet and Pierce, 1991).

(a) Investigate the association between defendant's race and death penalty when the victim's race is ignored (from the marginal table).

(b) Investigate the association between defendant's race and death penalty when the victim's race is taken into account.

(c) Use mosaic plots to visualize the considered models.

12.13 Consider the marginal association between the binary factors X_A and X_B, measured in odds ratios:

$$\frac{\mu_{11+}/\mu_{12+}}{\mu_{21+}/\mu_{22+}}.$$

Show that the value is the same as for the conditional association between X_A and X_B given $X_C = k$:

$$\frac{\mu_{11k}/\mu_{12k}}{\mu_{21k}/\mu_{22k}}.$$

TABLE 12.12: Death penalty verdict by defendant's race and victim's race.

Victims's Race	Defendant's Race	Death Penalty Yes	Death Penalty No
White	White	53	414
	Black	11	37
Black	White	0	16
	Black	4	139

12.14 Consider the saturated log-linear model for three variables X, X_B, X_C and a multinomial distribution. Derive the parameters of the logit model with reference category $(X_A = I, X_B = J, X_C = K)$. Which constraints hold for the parameters?

12.15 For binary variables $X_i \in \{0, 1\}$, the Ising model specifies that the joint probabilities are given by $P(X_1, \ldots, X_p) = \exp(\sum_{(j,k)\in E} \theta_{jk} X_j X_k - \phi(\boldsymbol{\theta}))$, where the normalizing function $\boldsymbol{\theta}$ contains the parameters θ_{jk}, and the sum is over the edges E of a graphical model. Let $X_0 = 1$ denote an additional variable and let all edges $(0j)$, corresponding to the term $\theta_{0j} X_0 X_j$, be included.

(a) Show that the conditional probability $P(X_j = 1 | X_1 = x_1, \ldots, X_p = x_p)$ follows a main effect logit model that contains the parameters $\{\theta_{jk} | (j,k) \in E, k \neq j\}$.

(b) Show that the log-linear model with all two-factor interactions is an Ising model.

Chapter 13

Multivariate Response Models

In many studies the objective is to model more than one response variable. For each unit in the sample a vector of correlated response variables, together with explanatory variables, is observed. Two cases are most important:

- repeated measurements, when the same variable is measured repeatedly at different times or/and under different conditions;
- different response variables, observed on one subject or unit in the sample.

Repeated measurements occur in most longitudinal studies. For example, in a longitudinal study measurements on an individual may be observed at several times under possibly varying conditions. In Example 1.4 (Chapter 1) an active ingredient is compared to a placebo by observing the healing after 3, 7, and 10 days of treatment. In Example 13.1, the number of epileptic seizures is considered at each of four two-week periods. Although they often do, repeated responses need not refer to different times. Response variables may also refer to different questions in an interview or to the presence of different commodities in a household. In Example 13.2 the two, possibly correlated responses are the type of birth (Ceaserian or not) and the stay of the child in intensive care (yes or no). Responses may also refer to a *cluster* of subjects; for example, when the health status of the members of a family is investigated, the observed responses form a cluster linked to one family. In Example 10.8, where the health status of trees is investigated, clusters are formed by the trees measured at the same spot. It has to be expected that observations within a cluster are less different than observations from different clusters. Therefore, one has to assume that the observations are correlated.

Example 13.1: Epilepsy

Thall and Vail (1990) used data from a randomized clinical trial in which 59 patients with epilepsy were randomized into groups receiving an antiepileptic drug (Progabide) or a placebo. The number of seizures suffered in each of four two-week periods were recorded along with a baseline seizure count for the weeks prior to treatment. In addition, the age of the patient is known (see also Everitt and Hothorn, 2006; Fitzmaurice et al., 2004). □

Example 13.2: Birth Data

In the birth study (see also Example 12.1) there are several responses that can be linked to explanatory variables. We will focus on the bivariate response Cesarean section (C; 1: yes; 0: no) and intensive care (IC; 1: yes; 0: no). Explanatory variables are gender of child (G; 1: male; 2: female), weight of child, age of mother, and number of previous pregnancies. □

In regression modeling for correlated categorical responses, one can distinguish several main approaches:

- *Conditional* models specify the conditional distribution of each component of the response vector given other components and covariates. In the case of repeated measurements, conditional models often model the transition between categories and are called *transition models*.
- *Marginal models* focus on the specification of the marginal distribution, that is, the distribution of the single components of the response vector. The marginal response is modeled conditionally on covariates but not on other responses.
- *Cluster-specific approaches* like random effects models allow for cluster- or subject-specific effects. Thus each cluster has its own effect, and the responses are conditional on the covariates and cluster-specific effects.

In the following we give simple examples of these approaches. For simplicity, let the response variables $y_{i1},\dots,y_{iT}$, observed together with covariate vector $\boldsymbol{x}_i$, be binary with $y_{it} \in \{0,1\}$. Given $\boldsymbol{x}_i$, the binary variables $y_{i1},\dots,y_{iT}$ form a contingency table with 2^T cells, with the response being multinomially distributed. When T is large, it is hard to specify the full multinomial distribution with its $2^T - 1$ parameters, especially when the modeling should account for the effects of potentially continuous covariates.

Conditional Model

When measurements $y_{i1},\dots,y_{iT}$ are repeated measurements taken for unit i at times $t = 1,\dots,T$, the response at time t may depend on previous responses. The simple transition model

$$P(y_{it} = 1|\boldsymbol{x}_i, y_{i,t-1}) = h(\beta_0 + y_{i,t-1} + \boldsymbol{x}_i^T\boldsymbol{\beta})$$

assumes that the response at time t depends on the covariates $\boldsymbol{x}_i$ as well as on the previous response. The model contains an autoregressive term of order one, and the response is conditional on $y_{i,t-1}$ and $\boldsymbol{x}_i$. Interpretation of $\boldsymbol{\beta}$ should take the presence of the autoregressive term into account. Typically one assumes independence of the responses given the covariates and previous responses.

Marginal Model

A typical marginal model specifies the marginal responses in the form

$$P(y_{it} = 1|\boldsymbol{x}_{it}) = h(\beta_0 + \boldsymbol{x}_{it}^T\boldsymbol{\beta}),$$

where the simple explanatory variable $\boldsymbol{x}_i$ is replaced by a sequence of explanatory variables $\boldsymbol{x}_{i1},\dots,\boldsymbol{x}_{iT}$, which can vary across measurements. It should be noted that only the marginal probabilities are modeled and not the whole multinomial distribution. ML estimation becomes hard when T is large. Therefore, one frequently considers the association between $y_{i1},\dots,y_{iT}$ as nuisance parameters and estimates by solving generalized estimation equations.

Subject-Specific Models

A simple model that contains subject-specific intercepts is

$$P(y_{it} = 1|\boldsymbol{x}_{it}) = h(\beta_i + \boldsymbol{x}_{it}^T\boldsymbol{\beta}),$$

where $\boldsymbol{\beta}$ is a fixed effect that is common to all clusters but the β_i's are subject- or cluster-specific individual effects. Thus each cluster has its own response level, a flexibility that seems appropriate in many applications. By assuming individual effects, the number of parameters is considerably increased, in particular when the β_i's are assumed to be fixed effects. Therefore, in subject-specific approaches it is often assumed that the individual effects, denoted by b_i, are random following a specified distribution function, for example, the normal distribution $N(0, \sigma_b^2)$. The corresponding random effects model is

$$P(y_{it} = 1|\boldsymbol{x}_{it}, b_i) = h(b_i + \boldsymbol{x}_{it}^T\boldsymbol{\beta}).$$

The distribution of the total response vector $(y_{i1}, \ldots, y_{iT})$ is obtained by assuming the conditional independence of $y_{i1}, \ldots, y_{iT}$ given b_i.

Parameters in marginal and subject-specific approaches have different *interpretations*. If one of the explanatory variables increases by one unit, the linear predictor of the marginal model, $\beta_0 + \boldsymbol{x}_{it}^T\boldsymbol{\beta}$, increases or decreases by the corresponding parameter, which in turn changes the response probability. The same happens for the linear predictor of the random effects model $b_i + \boldsymbol{x}_{it}^T\boldsymbol{\beta}$. But since all other explanatory variables are fixed, the change in the linear predictor starts from different levels for different individuals. Because h typically is a non-linear function, for example, the logistic distribution function, the effect on the response probability depends on the individual parameter b_i. Therefore, the effect is *subject-specific* or what can be called a *within-subject effect*. It is conditional on the random effect value. In contrast, the effect of parameters in marginal models is the same across all clusters; the effects are also called *population-averaged effects*. The difference vanishes when h is the identity, as is often assumed in normal distribution models. For non-linear models like the logit model, however, the population-averaged effects of marginal models tend to be smaller than subject-specific effects (for details see Chapter 14). For more discussion on the distinction in interpretation see also Neuhaus et al. (1991) and Agresti (1993c).

In the remainder of the chapter we will consider conditional and marginal models. First, in Section 13.1, we give conditional approaches including transition models and symmetric conditional models. Then likelihood-based approaches to marginal modeling are briefly sketched. The general marginal model that treats association as a nuisance and uses generalized estimation equations for estimation is outlined in Section 13.2. In Section 13.4 we consider a specific problem in marginal modeling termed marginal homogeneity. We investigate if the distribution of responses changes over measurements. The modeling of random effects is considered in a separate chapter (Chapter 14).

13.1 Conditional Modeling

13.1.1 Transition Models and Response Variables with a Natural Order

In many applications the responses have some natural ordering. In particular, when measurements are taken at discrete time points the sequence $Y_t, t = 1, 2, \ldots$, denoting measurements at time t, forms a stochastic process. It is natural to assume that the responses at time t depend on previously observed responses, which serve as explanatory variables.

Models that make use of that dependence structure are based on the decomposition

$$P(Y_1, \ldots, Y_T|\boldsymbol{x}) = P(Y_1|\boldsymbol{x}) \cdot P(Y_2|Y_1, \boldsymbol{x}) \cdots P(Y_T|Y_1, \ldots, Y_{T-1}, \boldsymbol{x}), \tag{13.1}$$

where Y_t has discrete values that form the state space of the process.

A general type of model is obtained by modeling the transitions within the decomposition (13.1) through a generalized linear model:

$$P(Y_t = r|Y_1, \ldots, Y_{t-1}, \boldsymbol{x}) = h(\boldsymbol{z}_t^T\boldsymbol{\beta}), \tag{13.2}$$

where $\boldsymbol{z}_t = \boldsymbol{z}(Y_1, \dots, Y_{t-1}, \boldsymbol{x})$ is a function of previous outcomes $Y_1, \dots, Y_{t-1}$ and the vector of explanatory variables $\boldsymbol{x}$. Since the response is determined by previous outcomes, conditional models of this type are sometimes called *data-driven*.

Markov Chains

A *Markov chain* is a special transition model of the type (13.2). Without further explanatory variables one considers the discrete stochastic process $Y_t, t = 1, 2, \dots$, where $Y_t \in \{1, \dots, m\}$. The process is called a *kth-order Markov chain* if the conditional distribution of Y_t, given $Y_{t-1}, \dots, Y_1$, is identical to the distribution of Y_t, given $Y_{t-1}, \dots, Y_{t-k}$. Thus the future of the process depends on the past only through the states at the previous k measurements. The simplest model is the first-order Markov model, which assumes $P(Y_t|Y_{t-1}, \dots, Y_1) = P(Y_t|Y_{t-1})$ for all t. It can be specified as a multicategorical logit model:

$$\log(P(Y_t = r|Y_{t-1})/P(Y_t = m|Y_{t-1})) = x_{A(1)}\beta_{r1} + \dots + x_{A(m)}\beta_{rm},$$

where $x_{A(r)} = I(Y_{t-1} = r)$, with $I(.)$ denoting the indicator function, represents (0-1)-coding of the response at time $t - 1$. Since the parameters do not depend on time, the Markov chain is homogeneous.

When additional explanatory variables are available *Markov-type transition models* of the first order assume $P(Y_t = r|Y_1, \dots, Y_{t-1}, \boldsymbol{x}) = P(Y_t|Y_{t-1}, \boldsymbol{x})$. For binary outcomes a simple model is

$$\begin{aligned} \log\left(P(y_1 = 1|\boldsymbol{x})/P(y_1 = 0|\boldsymbol{x})\right) &= \beta_{01} + \boldsymbol{x}^T\boldsymbol{\beta}_1, \\ \log\left(\frac{P(y_t = 1|y_1, \dots, y_{t-1}, \boldsymbol{x})}{P(y_t = 0|y_1, \dots, y_{t-1}, \boldsymbol{x})}\right) &= \beta_{0t} + \boldsymbol{x}^T\boldsymbol{\beta} + y_{t-1}\gamma_t. \end{aligned}$$

Regressive Models

Markov chain models aim to model the transitions for a larger number of repeated measurements. However, conditional models also apply when the sequence of responses is small and fixed. In the birth study (Example 13.2) one considers various explanatory variables like age of mother and gender of child on more than one response. In particular, one considers Cesarian section (yes/no) and duration in intensive care (in categories). It is natural to model first the dependence of Cesarean section (Y_1) on the covariates and then the dependence of duration in intensive care (Y_2) on the covariates *and* whether it was a Cesarean section. Therefore, one models in the second step the conditional response $Y_1|Y_2$.

For binary responses $y_1, \dots, y_t$ Bonney (1987) called such models *regressive logistic models*. They have have the form

$$\log\left(\frac{P(y_t = 1|y_1, \dots, y_{t-1}, \boldsymbol{x}_t)}{P(y_t = 0|y_1, \dots, y_{t-1}, \boldsymbol{x}_t)}\right) = \beta_0 + \boldsymbol{x}_t\boldsymbol{\beta} + \gamma_1 y_1 + \dots + \gamma_{t-1}y_{t-1}.$$

Thus the number of included previous outcomes depends on the component y_t, but no Markov-type assumption is implied. If the variables are multicategorical (with possibly varying numbers of categories), regressive models may be based on the modeling approaches for categorical responses as building blocks.

Estimation

Let the data be given by $(y_{it}, \boldsymbol{x}_{it})$, $i = 1, \dots, n$, $t = 1, \dots, T_i$. Let $\mu_{it} = \mathrm{E}(y_{it}|H_{it})$ denote the conditional mean specified by a simple exponential family and $H_{it} = \{\boldsymbol{x}_{i1}, \dots, \boldsymbol{x}_{it},$

$y_{i1}, \dots, y_{i,t-1}\}$ denote the history of observation i. For a univariate response the model has the form

$$\mu_{it} = h(\boldsymbol{z}_{it}^T \boldsymbol{\beta}),$$

where $\boldsymbol{z}_{it} = \boldsymbol{z}_{it}(H_{it})$ is a function of the history. If $\boldsymbol{x}_i$ collects the fixed covariates linked to the ith observation, the density of responses $y_{i1}, \dots, y_{it}$ has the decomposition

$$f(y_{i1}, \dots, y_{iT_i}|\boldsymbol{x}_i) = \prod_{t=1}^{T_i} f(y_{it}|y_{i1}, \dots, y_{i,t-1}, \boldsymbol{x}_i).$$

For stochastic covariates one obtains

$$f(y_{i1}, \dots, y_{iT_i}|\boldsymbol{x}_i) = \prod_{t=1}^{T_i} f(y_{it}|y_{i1}, \dots, y_{i,t-1}, \boldsymbol{x}_i) \prod_{t=1}^{T_i} f(\boldsymbol{x}_{it}|\boldsymbol{x}_{i1}, \dots, \boldsymbol{x}_{i,t-1}),$$

where it is assumed that the covariate process $\boldsymbol{x}_{it}|\boldsymbol{x}_{i1}, \dots, \boldsymbol{x}_{i,t-1}$ does not depend on previous outcomes $y_{i1}, \dots, y_{i,t-1}$. If the last factor is not informative for the parameters, it may be omitted in inference, yielding a partial likelihood. Under the assumption $f(y_{it}|y_{i1}, \dots, y_{i,t-1}, \boldsymbol{x}_i) = f(y_{it}|H_{it})$, the (partial) log-likelihood has the form

$$l(\boldsymbol{\beta}) = \sum_{i=1}^{n} \sum_{t} \log(f(y_{it}|H_{it})).$$

Summation is over $t = 1, \dots, T_i$ if all the responses are specified but reduces if in a Markov-type model only transitions given lagged responses of order k are specified. Then the estimates are conditional given the starting values $y_{i1}, \dots, y_{i,k}$. The conditional score function is

$$\boldsymbol{s}(\boldsymbol{\beta}) = \sum_{i=1}^{n} \sum_{t} \boldsymbol{z}_{it} (\partial h(\eta_{it})/\partial \eta)(y_{it} - \mu_{it})/\sigma_{it}^2 = \boldsymbol{X}^T \boldsymbol{D} \boldsymbol{\Sigma}^{-1}(\boldsymbol{y} - \boldsymbol{\mu}),$$

where μ_{it} is the conditional expectation and $\sigma_{it}^2 = \text{cov}(y_{it}|H_{it})$ is the conditional variance. The matrices $\boldsymbol{X}, \boldsymbol{D}, \boldsymbol{\Sigma}$ collect design vectors, derivatives, and variances, and the vectors $\boldsymbol{y}, \boldsymbol{\mu}$ collect the responses and means. The conditional information matrix has the form

$$\boldsymbol{F}(\boldsymbol{\beta}) = \sum_{i=1}^{n} \sum_{t} \boldsymbol{z}_{it} \boldsymbol{z}_{it}^T \left(\frac{\partial h(\eta_{it})}{\partial \eta} \right)^2 /\sigma_{it}^2 = \boldsymbol{X}^T \boldsymbol{D} \boldsymbol{\Sigma}^{-1} \boldsymbol{D} \boldsymbol{X}.$$

For categorical responses the data are given in the form $(Y_{it}, \boldsymbol{x}_{it})$, where $Y_{it} \in \{1, \dots, m\}$. The corresponding mean is a vector of probabilities and the model has the form $\boldsymbol{\pi}_{it} = h(\boldsymbol{Z}_{it}^T \boldsymbol{\beta})$, where the covariate vector is replaced by the design matrix $\boldsymbol{Z}_{it}$. The score function and the conditional information matrix again have the form $\boldsymbol{s}(\boldsymbol{\beta}) = \boldsymbol{X}^T \boldsymbol{D} \boldsymbol{\Sigma}^{-1}(\boldsymbol{y} - \boldsymbol{\mu})$, $\boldsymbol{F}(\boldsymbol{\beta}) = \boldsymbol{X}^T \boldsymbol{D} \boldsymbol{\Sigma}^{-1} \boldsymbol{D} \boldsymbol{X}$, but the matrices are composed as in Section 8.6.1.

Models may be fitted by using standard software, where $y_{i1}, y_{i1}|y_{i2}, \dots, y_{iT_i}|y_{i1}, \dots, y_{i,T_i-1}$ are treated as separate observations. In the case of small T_i and large n, inference may be embedded into the framework of GLMs. For $n = 1$ the model with lagged variables represents an autoregressive time series model and asymptotically ($T \to \infty$) the covariance $\text{cov}(\hat{\boldsymbol{\beta}})$ can be approximated by the inverse conditional information matrix (for details see Kaufmann, 1987; Fahrmeir and Kaufmann, 1987). For an extensive treatment of categorical time series see also Kedem and Fokianos (2002). In cases with more than one unit observed across time, the inverse information matrix may serve as an approximation of the covariance of the estimate but rigorous proofs seem not to be available.

13.1.2 Symmetric Conditional Models

If there is no natural ordering of the response variables, one can model them in a symmetric way by conditioning on the outcome of the other variables. For simplicity let the response variables $y_1, \dots, y_T$, observed together with covariate vector $\boldsymbol{x}$, be binary with $y_i \in \{0,1\}$. Given $\boldsymbol{x}$, the binary variables $y_1, \dots, y_q$ form a contingency table with 2^T cells. The response is multinomially distributed, $M(n(\boldsymbol{x}), \{\pi_{r_1 \dots r_T}\})$, where $n(\boldsymbol{x})$ is the number of observations at covariate $\boldsymbol{x}$, and

$$\pi_{r_1 \dots r_T} = P(y_1 = r_1, \dots, y_T = r_T | \boldsymbol{x})$$

is the probability of observing the response $(r_1, \dots, r_T)$. Multinomial responses have already been treated in Chapter 8, but for less structured multinomials. The multinomial distribution considered here is determined by T binary distributions, therefore forming a strongly structured multinomial distribution. Also, models for multinomial distributions that are formed by several categorical responses have been considered before (Chapter 12), but without further covariates. In the following it is illustrated that the underlying parametrization is conditional, so that marginal models that include covariates have to be based on more general log-linear models.

Without covariates the multinomial distribution can be reparameterized as a saturated log-linear model (see Chapter 12). With (0-1)-coding one obtains

$$\log(\pi_{y_1 \dots y_T}) = \lambda_0 + y_1\lambda_1 + \dots + y_T\lambda_T + y_1 y_2 \lambda_{12} + \dots + y_1 \dots y_T \lambda_{1 \dots T}$$

or, in the form of the logit model,

$$\log(\frac{\pi_{y_1 \dots y_T}}{\pi_{0 \dots 0}}) = y_1\lambda_1 + \dots + y_T\lambda_T + y_1 y_2 \lambda_{12} + \dots + y_1 \dots y_T \lambda_{1 \dots T}, \tag{13.3}$$

where $\pi_{0 \dots 0} = P(y_0 = 0, \dots, y_T = 0)$ is the reference category.

		$y_2 = 1$	$y_2 = 0$
y_1	1	π_{11}	π_{10}
	0	π_{01}	π_{00}

In the simplest case, $T = 2$, one has a (2×2)-contingency table and one easily computes

$$\begin{aligned}
\lambda_1 &= \log(\pi_{10}/\pi_{00}) = \log(\frac{P(y_1 = 1|y_2 = 0)}{P(y_1 = 0|y_2 = 0)}) = \operatorname{logit}(P(y_1 = 1|y_2 = 0)),\\
\lambda_2 &= \log(\pi_{01}/\pi_{00}) = \log(\frac{P(y_2 = 1|y_1 = 0)}{P(y_2 = 0|y_1 = 0)}) = \operatorname{logit}(P(y_2 = 1|y_1 = 0)),\\
\lambda_{12} &= \log(\frac{\pi_{11}/\pi_{01}}{\pi_{10}/\pi_{00}}) = \frac{P(y_1 = 1|y_2 = 1)/P(y_1 = 0|y_2 = 1)}{P(y_1 = 1|y_2 = 0)/P(y_1 = 0|y_2 = 0)}.
\end{aligned}$$

Thus, λ_1 represents the log-odds that compare $y_1 = 1$ to $y_1 = 0$, *given* $y_2 = 0$, and λ_2 represents the log-odds that compare $y_2 = 1$ to $y_2 = 0$, *given* $y_1 = 0$. Both parameters contain the effect of one variable given that the other variable takes the value zero. In this sense the parametrization is *conditional*. The interaction parameter λ_{12} is a log-odds ratio where in the numerator $y_1 = 1$ is compared to $y_1 = 0$ for $y_2 = 1$ and in the denominator $y_1 = 1$ is compared to $y_1 = 0$ for $y_2 = 0$.

In the general case with T response variables one obtains for the main and two-factor interaction parameters

$$\lambda_j = \log(\frac{P(y_j = 1|y_l = 0, l \neq j)}{P(y_j = 0|y_l = 0, l \neq j)}) = \text{logit}(P(y_j = 1|y_l = 0, l \neq j)),$$
$$\lambda_{jr} = \log(\frac{P(y_j = 1|y_r = 1, y_l = 0, l \neq j, r)/P(y_j = 0|y_r = 1, y_l = 0, l \neq j, r)}{P(y_j = 1|y_r = 0, y_l = 0, l \neq j, r)/P(y_j = 0|y_r = 0, y_l = 0, l \neq j, r)}).$$

The crucial point is that the parameters characterize the conditional probabilities of the subsets of responses given particular values for all the other variables.

As shown by Fitzmaurice and Laird (1993) for binary vector $\boldsymbol{y}^T = (y_1, \dots, y_T)$, the density can also be given in the form

$$f(\boldsymbol{y}) = \exp(\boldsymbol{y}^T\boldsymbol{\lambda} + \boldsymbol{w}^T\gamma - c(\boldsymbol{\lambda}, \gamma)),$$

where $\boldsymbol{w}^T = (y_1y_2, y_1y_3, \dots, y_{T-1}y_T, \dots, y_1y_2\dots y_T)$, $\boldsymbol{\lambda}^T = (\lambda_1, \dots, \lambda_T)$, $\boldsymbol{\gamma}^T = (\lambda_{12}, \dots, \lambda_{12\dots T})$. $c(\boldsymbol{\lambda}, \boldsymbol{\gamma})$ is a normalizing constant given by

$$\exp(c(\boldsymbol{\lambda}, \boldsymbol{\gamma})) = \sum_{\boldsymbol{r}} \exp(\boldsymbol{r}^T\boldsymbol{\lambda} + \boldsymbol{w}^T\gamma),$$

where the sum is over all vectors $\boldsymbol{r} \in \{0, 1\}^T$. The density is a special case of the partly exponential family used by Zhao et al. (1992). In the case of two binary variables one obtains the simple form $f(\boldsymbol{y}) = \exp(y_1\lambda_1 + y_1\lambda_2 + y_1y_2\lambda_{12} - c(\boldsymbol{\lambda}, \lambda_{12}))$, with $c(\boldsymbol{\lambda}, \lambda_{12}) = \log(1 + \exp(\lambda_1) + \exp(\lambda_2) + \exp(\lambda_1 + \lambda_2 + \lambda_{12})$.

A parsimonious class of conditional logistic models, in which the individual binary variables are treated symmetrically, was considered by Qu et al. (1987). They considered the conditional response probability

$$P(y_t = 1|y_k,\ k \neq t;\ \boldsymbol{x}_t) = h(\alpha(w_t; \boldsymbol{\theta}) + \boldsymbol{x}_t^T\boldsymbol{\beta}_t), \tag{13.4}$$

where h is the logistic function and α is an arbitrary function of a parameter θ and the sum $w_t = \sum_{k\neq t} y_k$. In model (13.4), α depends on the conditioning y's, whereas the covariate effects are kept constant. For the case of two components (y_1, y_2), the sums w_1, w_2 reduce to y_2, y_1, respectively. Then the simplest choice is a logistic model that includes the conditioning response as a further covariate:

$$P(y_t = 1|y_k,\ k \neq t;\ \boldsymbol{x}_t) = h(\theta_0 + \theta_1 y_k + \boldsymbol{x}_t^T\boldsymbol{\beta}_t), \qquad t, k = 1, 2. \tag{13.5}$$

Other choices for $\alpha(w; \boldsymbol{\theta})$ were discussed by Qu et al. (1987) and Conolly and Liang (1988). The joint density $P(y_1, \dots, y_T;\ \boldsymbol{x}_1^T, \dots, \boldsymbol{x}_T^T)$ can be derived from (13.4); however, it involves a normalizing constant, which is a complicated function of the unknown parameters $\boldsymbol{\theta}$ and $\boldsymbol{\beta}$; see Prentice (1988) and Rosner (1984). Full likelihood estimation may be avoided by using a quasi-likelihood approach (Conolly and Liang, 1988).

When using symmetric conditional models one should be aware that one conditions on covariates *and* the other responses; one is modeling the effect of covariates given other responses. Therefore, the effect of covariates is measured having already accounted for the effect of the other responses. Then the covariate effects could be conditioned away because the other responses might also be related to the covariates. Moreover, the effects will change if one considers a subset of responses, because the conditioning changes. Therefore, the dimension of the response vector should be the same for all observations.

13.2 Marginal Parametrization and Generalized Log-Linear Models

Frequently the primary scientific objective is the modeling of covariates on marginal responses, for example, by parameterizing the marginal logits $\text{logit}(P(y_t = 1|y_t = 0))$. Then one has to use representations of the response patterns that differ from the ones considered in the previous section. Let us consider again the simple case of two binary responses y_1, y_2. The response is determined by the probabilities $\pi_{11}, \pi_{10}, \pi_{01}, \pi_{00}$, or, more precisely, by three of these probabilities. The representation as an exponential family uses the three parameters $\lambda_1, \lambda_2, \gamma$, where $\gamma = \lambda_{12}$ is the log-odds ratio for the variables y_1, y_2. Alternatively, one can also use the parameters

$$\pi_{1+} = \pi_{11} + \pi_{10}, \quad \pi_{+1} = \pi_{11} + \pi_{01}, \quad \gamma,$$

where $P(y_1 = 1) = \pi_{1+}$, $P(y_2 = 1) = \pi_{+1}$ are the marginal probabilities and γ is the log-odds ratio. The sets $\{\pi_{11}, \pi_{10}, \pi_{01}, \pi_{00}\}$, $\{\lambda_1, \lambda_2, \gamma\}$, and $\{\pi_{1+}, \pi_{+1}, \gamma\}$ are equivalent representations of the (2×2)-probability table. The set of parameters $\{\pi_{1+}, \pi_{+1}, \gamma\}$ is especially attractive when marginal probabilities are of interest. It is also straightforward to include covariates. A simple model that links covariates to these parameters is the model considered by Palmgren (1989):

$$\text{logit}(\pi_{1+}) = \boldsymbol{x}^T\boldsymbol{\beta}_1, \quad \text{logit}(\pi_{+1}) = \boldsymbol{x}^T\boldsymbol{\beta}_2, \quad \log(\gamma) = \boldsymbol{x}^T\boldsymbol{\beta}_3. \tag{13.6}$$

It should be noted that transforming between different representations is not always easy. Of course, marginal probabilities and log-odds ratios are defined as functions of the response probabilities $\pi_{11}, \pi_{10}, \pi_{01}, \pi_{00}$, but the representation of the response probabilities as functions of $\{\pi_{1+}, \pi_{+1}, \gamma\}$ is tedious (see equation (13.11)). The parameters from the set $\{\pi_{1+}, \pi_{+1}, \gamma\}$ are independent in the sense that any choice $\pi_{1+}, \pi_{+1} \in (0, 1)$, $\gamma \in \mathbb{R}$, yields valid probabilities. Moreover, the parameter vector (π_{1+}, π_{+1}) is orthogonal to γ, meaning that the information matrix is a block-diagonal matrix (Cox and Reid, 1987).

In the case of two binary response variables, the association can be modeled by just one parameter. In the general case, however, in addition to marginals, association terms of higher order are needed, so it is advisable to restrict parametric models that include explanatory variables to the specification of only a few of them. A class of models that can be used is generalized log-linear models, which are considered briefly in the following.

Log-linear models, as considered in Chapter 12, can be represented in the general form $\log(\boldsymbol{\pi}) = \boldsymbol{X}\boldsymbol{\lambda}$, with $\boldsymbol{\pi}$ containing all the probabilities for individual cells and $\boldsymbol{X}$ denoting a design matrix. For three binary response variables the vector $\boldsymbol{\pi}$ is given by $(\pi_{111}, \pi_{112}, \dots, \pi_{222})$. Marginals can be obtained by using a linear transformation $\boldsymbol{A}\boldsymbol{\pi}$, where $\boldsymbol{A}$ is a matrix that contains zeros and ones only. With an appropriately chosen $\boldsymbol{A}$ one can build the general vector of marginals:

$$\boldsymbol{A}\boldsymbol{\pi} = (\pi_{1++}, \pi_{+1+}, \dots, \pi_{11+}, \dots, \pi_{222})^T$$

or the shorter vector that contains only univariate and bivariate marginals:

$$\boldsymbol{A}\boldsymbol{\pi} = (\pi_{1++}, \pi_{+1+}, \dots, \pi_{11+}, \dots, \pi_{+22})^T.$$

In a second step one can specify the logarithmic contrasts of interest,

$$\boldsymbol{\eta} = \boldsymbol{C}\log(\boldsymbol{A}\boldsymbol{\pi}),$$

based on a chosen matrix $\boldsymbol{C}$. For example, the univariate contrasts are

$$\begin{aligned}
\eta_1 &= \log(\pi_{1++}) - \log(\pi_{2++}) = \text{logit}(P(y_1 = 1)),\\
\eta_2 &= \log(\pi_{+1+}) - \log(\pi_{+2+}) = \text{logit}(P(y_2 = 1)),\\
\eta_3 &= \log(\pi_{++1}) - \log(\pi_{++2}) = \text{logit}(P(y_3 = 1)),
\end{aligned}$$

and the bivariate contrasts are

$$\begin{aligned}
\eta_{12} &= \log(\pi_{11+}) - \log(\pi_{12+}) - \log(\pi_{21+}) + \log(\pi_{22+}),\\
\eta_{13} &= \log(\pi_{1+1}) - \log(\pi_{1+2}) - \log(\pi_{2+1}) + \log(\pi_{2+2}),\\
\eta_{23} &= \log(\pi_{+11}) - \log(\pi_{+12}) - \log(\pi_{+21}) + \log(\pi_{+22}).
\end{aligned}$$

The link to the explanatory variable is obtained by specifying a design matrix for the chosen contrasts. With the univariate marginals specified by

$$\eta_t = \text{logit}(P(y_t = 1)) = \boldsymbol{x}^T\boldsymbol{\beta}_t$$

one obtains a model that has the form of a *generalized log-linear model*:

$$\boldsymbol{C}\log(\boldsymbol{A\pi}) = \boldsymbol{X\lambda},$$

which is obviously a generalization of the log-linear model $\log(\boldsymbol{\pi}) = \boldsymbol{X\lambda}$. Of course, bivariate and trivariate marginals also can be linked to the explanatory variables, yielding more complex models. For multicategorical responses different contrasts have to be used.

Maximum likelihood estimation for the generalized log-linear model is not easy and specialized software is needed. For details see McCullagh and Nelder (1989), Chapter 6, Fitzmaurice and Laird (1993), Lang and Agresti (1994), Lang (1996b), Glonek and McCullagh (1995), and Glonek (1996). An extended outline of the full ML approach to marginal models was given by Bergsma et al. (2009). For the alternative marginalized random effects model see Section 14.5.

Example 13.3: Birth Data

Table 13.1 shows the fitted values of model (13.6) for the birth study (Example 13.2). The two response variables are Cesarean section (C, 1: yes; 0: no) and intensive care (IC; 1: yes; 0 :no). Explanatory variables are gender of child (G; 1: male; 2: female), weight, age of mother, and number of previous pregnancies (1: no previous pregnancy; 2: one pregnancy; 3: more than 2 pregnancies). It is seen that the age of the mother has an effect on both outcomes, and also the odds ratio depends on age. The number of previous pregnancies has an effect on treatment in intensive care in the case of more than one previous pregnancies. □

13.3 General Marginal Models: Association as Nuisance and GEEs

In the following we will consider an alternative methodology for the fitting of marginal models. The approach is more general and applies as well to alternative marginal distributions. The basic concept is to consider the association as a nuisance parameter and to find generalized estimation equations that use a working covariance matrix. Let $y_{i1}, \dots, y_{iT_i}$ denote the univariate measurements on the ith unit. The focus is on modeling how the marginal measurement y_{it}

TABLE 13.1: Estimated regression coefficients for birth data with response Ceaserian section (1), intensive care (2), and odds ratio between the two responses (3).

	Estimate	Standard Error	z-Value
(Intercept):1	3.651	1.036	3.521
(Intercept):2	−1.058	0.805	−1.314
(Intercept):3	6.101	2.848	2.142
Weight:1	−0.002	0.0002	−8.863
Weight:2	−0.0007	0.0002	−4.458
Weight:3	−0.0005	0.0005	−0.906
AgeMother:1	0.0118	0.0289	0.407
AgeMother:2	0.079	0.0231	3.442
AgeMother:3	−0.171	0.076	−2.258
as.factor(Gender)2:1	−0.165	0.247	−0.665
as.factor(Gender)2:2	−0.260	0.190	−1.372
as.factor(Gender)2:3	0.286	0.599	0.479
as.factor(Previous)1:1	−0.611	0.376	−1.621
as.factor(Previous)1:2	−0.592	0.255	−2.316
as.factor(Previous)1:3	1.398	0.905	1.543
as.factor(Previous)2:1	0.513	0.493	1.039
as.factor(Previous)2:2	−2.226	0.780	−2.852
as.factor(Previous)2:3	4.127	2.150	1.918

depends on a covariate vector $\boldsymbol{x}_{it}$. For ease of presentation let the variables be collected in the response vector $\boldsymbol{y}_i^T = (y_{i1}, \dots, y_{iT_i})$ and the covariate vector $\boldsymbol{x}_i^T = (\boldsymbol{x}_{i1}^T, \dots, \boldsymbol{x}_{iT_i}^T)$.

The marginal model assumes that the marginal means are specified correctly, whereas the variance structure is not necessarily the variance that generates the data. In detail, a marginal model is structured in the following way:

(1) The *marginal means* are given by

$$\mu_{it} = E(y_{it}|\boldsymbol{x}_{it}) = h(\boldsymbol{x}_{it}^T\boldsymbol{\beta}), \tag{13.7}$$

where h is a fixed response function.

(2) The *marginal variance* is specified by

$$\sigma_{it}^2 = \text{var}(y_{it}|\boldsymbol{x}_{it}) = \phi v(\mu_{it}), \tag{13.8}$$

where v is a known variance function and ϕ a dispersion parameter.

(3) In addition, a *working covariance* structure is specified by

$$\tilde{\text{cov}}(y_{is}, y_{it}) = c(\mu_{is}, \mu_{it}; \boldsymbol{\alpha}), s \neq t, \tag{13.9}$$

where c is a known function depending on an unknown covariate vector $\boldsymbol{\alpha}$.

In matrix notation the model may be given in the form $\boldsymbol{\mu}_i = h(\boldsymbol{X}_i\boldsymbol{\beta})$, where the response function h transforms the single components of $\boldsymbol{X}_i\boldsymbol{\beta}$ and the covariates are collected in the matrix $\boldsymbol{X}_i^T = (\boldsymbol{x}_{i1} \dots \boldsymbol{x}_{iT_i})$.

The essential point in marginal models is that the dependence of the response $\boldsymbol{y}_i$ on the covariates $\boldsymbol{x}_i$ can be investigated without assuming that the working covariance is correctly specified. While it is assumed that the relationship between the measurements $\boldsymbol{y}_i$ and the covariates is specified correctly by the marginal means given in (13.7), in general, the true variance

Marginal Models

Data: $\boldsymbol{y}_i^T = (y_{i1}, \ldots, y_{iT_i}),$

$\boldsymbol{x}_i^T = (\boldsymbol{x}_{i1}^T, \ldots, \boldsymbol{x}_{iT_i}^T), \quad i = 1, \ldots, n$

$\boldsymbol{y}_1, \ldots, \boldsymbol{y}_n | \boldsymbol{x}_1, \ldots, \boldsymbol{x}_n$ independent

(1) Marginal mean

$$\boldsymbol{\mu}_i = h(\boldsymbol{X}_i \boldsymbol{\beta})$$

(2) Working Covariance

$$\boldsymbol{W}_i = \boldsymbol{W}_i(\boldsymbol{\beta}, \boldsymbol{\alpha}, \phi),$$

with diagonal elements $\sigma_{it}^2 = \phi v(\mu_{it})$
and off-diagonal entries $c(\mu_{is}, \mu_{it}; \boldsymbol{\alpha})$, $s \neq t$

is not equal to the working variance. Therefore, it is distinguished between the underlying (but unknown) covariance of $\boldsymbol{y}_i$,

$$\boldsymbol{\Sigma}_i = \text{cov}(\boldsymbol{y}_i) = (\text{cov}(y_{is}, y_{it}))_{st},$$

and the working covariance of $\boldsymbol{y}_i$,

$$\boldsymbol{W}_i = \boldsymbol{W}_i(\boldsymbol{\beta}, \boldsymbol{\alpha}, \phi),$$

which has $\phi v(\mu_{i1}), \ldots, \phi v(\mu_{iT_i})$ in the diagonal and $c(\mu_{is}, \mu_{it}; \boldsymbol{\alpha})$, $s \neq t$, as off-diagonal entries. Since the true covariance is unknown, the matrix $\boldsymbol{W}_i$ is used to establish the estimation equations. However, in specific cases, for example, binary responses, it can happen that it is not even a valid covariance matrix. Therefore, in general, it should be considered as a weight matrix within the estimation equations. Although its main use is that of a weight matrix, we will use the traditional name "working covariance matrix," since it is structured as a covariance matrix. In the following the dependence on ϕ is often suppressed and the simpler notation $\boldsymbol{W}_i(\boldsymbol{\beta}, \boldsymbol{\alpha})$ is used.

There are two main approaches for specifying working covariances. Liang and Zeger (1986) and Prentice (1988) use correlations, whereas Lipsitz et al. (1991) and Liang et al. (1992) use the odds ratio, which is more appropriate for binary responses.

Working Correlation Matrices

The working covariance matrix may be determined by assuming a simple correlation structure between the components of $\boldsymbol{y}_i$. Since the relation between the covariance and correlation has the simple form $\text{cov}(y_{is}, y_{it}) = \varrho_{st}\sigma_{is}\sigma_{it}$, with the correlation between y_{is} and y_{it} given by $\varrho_{st} = 1$ for $s = t$, the working covariance matrix has the form

$$\boldsymbol{W}_i = \boldsymbol{W}_i(\boldsymbol{\beta}, \boldsymbol{\alpha}) = \boldsymbol{C}_i^{1/2}(\boldsymbol{\beta}) \boldsymbol{R}_i(\boldsymbol{\alpha}) \boldsymbol{C}_i^{1/2}(\boldsymbol{\beta}),$$

where $\boldsymbol{C}_i(\boldsymbol{\beta}) = diag(\sigma_{i1}^2, \ldots, \sigma_{iT_i}^2)$ is the diagonal matrix of variances, specified by (13.8), and $\boldsymbol{R}_i(\boldsymbol{\alpha})$ is the assumed correlation structure. The simplest choice is the *working independence model*, where $\boldsymbol{R}_i(\boldsymbol{\alpha}) = \boldsymbol{I}$, with $\boldsymbol{I}$ denoting the identity matrix. In that case the parameter $\boldsymbol{\alpha}$ may be dropped. In the *equicorrelation model* a more flexible covariance structure is obtained by using $\mathrm{corr}(y_{is}, y_{it}) = \alpha$, yielding a correlation matrix $\boldsymbol{R}_i(\boldsymbol{\alpha})$ that has ones in the diagonal and α in the off-diagonals. Correlation structures of that type occur if measurements share a common intercept (see Section 14.1.1). If the observations $y_{i1}, y_{i2}, \ldots$ represent repeated measurements over time, it is often appropriate to assume that the correlations will decrease with the distance between measurements. A correlation structure of this type is the *exponential correlation model* $\boldsymbol{R}_i(\boldsymbol{\alpha}) = (\alpha^{|s-t|})_{s,t},\ \alpha \geq 0$.

Working Correlation Matrices

Uncorrelated	$\boldsymbol{R}_i(\boldsymbol{\alpha}) = \boldsymbol{I}$
Equicorrelation	$\boldsymbol{R}_i(\boldsymbol{\alpha}) = (\alpha^{I(s \neq t)})_{s,t}, \alpha \geq 0$
Exponential correlation	$\boldsymbol{R}_i(\boldsymbol{\alpha}) = (\alpha^{|s-t|})_{s,t}, \alpha \geq 0$

When working correlation matrices are specified in the form of equicorrelation or exponential correlation models it is implicitly assumed that the correlation can be the same over the possible values of the covariates. What works for normally distributed responses, where the mean structure is separated from the covariance structure, fails when the responses are categorical. Especially for binary variables $y_{is}, y_{it} \in \{0, 1\}$, the specification of a correlation matrix implies rather strong constraints. By definition, the correlation $\varrho_{ist} = \mathrm{corr}(y_{is}, y_{it})$ is given by

$$\varrho_{ist} = \frac{P(y_{is} = 1, y_{it} = 1) - \pi_{is}\pi_{it}}{\{\pi_{is}(1 - \pi_{is})\pi_{it}(1 - \pi_{it})\}^{1/2}},$$

where $\pi_{is} = P(y_{is} = 1|\boldsymbol{x}_{is})$. Since $P(y_{is} = y_{it} = 1)$ is constrained by $max\{0, \pi_{is} + \pi_{it} - 1\} \leq P(y_{is} = y_{it} = 1) \leq min(\pi_{is}, \pi_{it})$, the range for admissible correlations may be narrowed down considerably. It may be shown that the range of ϱ_{ist} depends on the odds ratio of *marginal* probabilities $\gamma_{ist}^m = \{\pi_{is}/(1 - \pi_{is})\}/\{\pi_{it}/(1 - \pi_{it})\}$. One obtains the inequalities

$$max\left\{-\sqrt{\gamma_{ist}^m}, -\sqrt{1/\gamma_{ist}^m}\right\} \leq \varrho_{ist} \leq min\left\{\sqrt{\gamma_{ist}^m}, \sqrt{1/\gamma_{ist}^m}\right\}; \tag{13.10}$$

see, for example, McDonald (1993). This means that the constraint is very strong if the marginal probabilities differ strongly; for example, if $\pi_{is} = 0.1, \pi_{it} = 0.5$, the maximal correlation is $1/3$. If the association is weak, the constraint will not be all that important and the specification of correlation matrices will work fine. However, for a strong association the constraint is severe.

In particular, Crowder (1995) and Chaganty and Joe (2004) pointed out that the restriction (13.10) has the consequence that, for binary observations, working matrices that are specified by correlations can be far from proper covariance matrices. Since the probabilities are functions of the covariates, $\pi_{it} = \pi(\boldsymbol{x}_{it})$, the marginal odds ratios also depend on the covariates, $\gamma_{ist} = \gamma(\boldsymbol{x}_{is}, \boldsymbol{x}_{it})$. When the covariates vary across a wide range, for example, when the covariates are normally distributed, then inequality (13.10) may shrink the range of correlations to a single point 0. When the working covariance matrix is not a proper covariance matrix it serves merely as a weight matrix.

Specification by Odds Ratios

In addition to the problems with the range for admissible correlations, simple correlation structures like the equicorrelation model ignore that for binary variables the correlation depends on the marginals. An alternative and more appropriate way of specifying the dependence between binary observations is based on odds ratios. Odds ratios are easy to interpret and have desirable properties. In particular, odds ratios do not depend on the marginal probabilities. Let the odds ratio for the variables $y_{is}, y_{it}, s \neq t$ be given by

$$\gamma_{ist} = \frac{P(y_{is}=1, y_{it}=1)/P(y_{is}=0, y_{it}=1)}{p(y_{is}=1, y_{it}=0)/P(y_{is}=0, y_{it}=0)}.$$

The covariance between y_{is} and y_{it} is then given by $\text{cov}(y_{is}, y_{it}) = E(y_{is}y_{it}) - \pi_{is}\pi_{it}$, which is a function of γ_{ist}, since

$$\begin{aligned} E(y_{is}y_{it}) &= P(y_{it} = y_{it} = 1) \\ &= \begin{cases} \dfrac{1-(\pi_{is}+\pi_{it})(1-\gamma_{ist}) - s(\pi_{is}, \pi_{it}, \gamma_{ist})}{2(\gamma_{st}-1)}), & \gamma_{ist} \neq 1 \\ \pi_{is}\pi_{it} & \gamma_{ist} = 1, \end{cases} \end{aligned} \tag{13.11}$$

where $s(\pi_{is}, \pi_{it}, \gamma_{ist}) = \left[\{1-(\pi_{is}+\pi_{it})(1-\gamma_{ist})\}^2 - 4(\gamma_{ist}-1)\gamma_{ist}\pi_{is}\pi_{it}\right]^{1/2}$ (see Lipsitz et al., 1991; Liang et al., 1992). Thus $\text{cov}(y_{is}, y_{it})$ is a function of $\pi_{is}, \pi_{it}, \gamma_{ist}$ and the working covariance may be specified by

$$\tilde{\text{cov}}(y_{is}, y_{it}) = c(\pi_{is}, \pi_{it}, \boldsymbol{\alpha}), s \neq t,$$

where $\boldsymbol{\alpha}$ contains the odds ratios $\{\gamma_{ist}, s \neq t\}$. With the diagonals $\sigma_{it}^2 = \pi_{it}(1-\pi_{it})$ one obtains the working covariance $\boldsymbol{W}_i$. When γ_{ist} is fixed the variation of the marginal probabilities over the covariate vectors implies a variation of the correlation between y_{is} and y_{it}. Therefore, a constant correlation between variables, as is assumed, for example, by eqicorrelation is avoided.

To reduce the number of parameters odds ratios are often themselves parameterized. The simplest model is $\gamma_{ist} = 1$, which corresponds to uncorrelated components y_{is}, y_{it}. In most cases the model

$$\gamma_{ist} = \gamma \quad \text{for all} \quad s \neq t,$$

which corresponds to the assumption of equicorrelation, will be closer if not identical to the underlying true structure. Alternatively, odds ratios may be a function of the covariates following the log-linear model $\log(\gamma_{ist}) = \boldsymbol{\alpha}^T \boldsymbol{w}_{ist}$, where $\boldsymbol{w}_{ist}$ is a vector of covariates.

Odds Ratio Modeling

Uncorrelated	$\gamma_{ist} = 1$
Equicorrelation	$\gamma_{ist} = \gamma$
Flexible	$\gamma_{ist} = \exp(\boldsymbol{\alpha}^T \boldsymbol{w}_{ist})$

Some examples of marginal models, specified by a marginal dependence on the covariates, marginal variance, and working covariance, are the following, where $\text{corr}(y_{is}, y_{it})$ denotes the specified correlation between y_{is}, y_{it}:

Binary responses

(1) $\text{logit}(\mu_{it}) = \boldsymbol{x}_{it}^T\boldsymbol{\beta}$, where $\mu_{it} = \pi_{it} = P(y_{it} = 1|x_{it})$,

(2) $\sigma_{it}^2 = \pi_{it}(1 - \pi_{it})$,

(3) $\tilde{\text{cov}}(y_{is}, y_{it}) = 0, s \neq t$ (independence working matrix) or $\gamma_{ist} = \gamma(= \alpha)$ (equal odds ratio).

Count data

(1) $\mu_{it} = \exp(\boldsymbol{x}_{it}^T\boldsymbol{\beta})$,

(2) $\sigma_{it}^2 = \phi\mu_{it}$,

(3) $\text{corr}(y_{is}, y_{it}) = \alpha, s \neq t$ (equicorrelation).

Of course normal distribution models also can be specified as marginal models. For example, one can specify the mean by $\mu_{it} = \boldsymbol{x}_{it}^T\boldsymbol{\beta}$ or $\mu_{it} = \log(\boldsymbol{x}_{it}^T\boldsymbol{\beta})$ and the variances by $\sigma_{it}^2 = \phi = \sigma^2$, $\tilde{\text{cov}}(y_{is}, y_{it}) = \alpha_{st}$.

13.3.1 Generalized Estimation Approach

In linear models for Gaussian responses, the specification of means and covariances determines the likelihood. However, in the case of discrete dependent variables, the likelihood also depends on higher order moments. ML estimation usually becomes computationally cumbersome, even if additional assumptions are made. The *generalized estimation approach* circumvents these problems by using only the mean structure,

$$\boldsymbol{\mu}_i(\boldsymbol{\beta}) = h(\boldsymbol{X}_i\boldsymbol{\beta}),$$

where $\boldsymbol{\mu}_i = E(\boldsymbol{y}_i|\boldsymbol{x}_{i1}, \dots, \boldsymbol{x}_{iT_i}), \boldsymbol{X}_i^T = (\boldsymbol{x}_{i1}, \dots, \boldsymbol{x}_{iT_i})$, and the working covariance $\boldsymbol{W}_i(\boldsymbol{\beta}, \boldsymbol{\alpha}, \phi)$.

For fixed $\boldsymbol{\alpha}, \phi$, the *generalized estimation equation* (GEE) for parameter $\boldsymbol{\beta}$ is given by

$$\sum_{i=1}^{n} \boldsymbol{X}_i^T \boldsymbol{D}_i(\boldsymbol{\beta}) \boldsymbol{W}_i^{-1}(\boldsymbol{\beta}, \boldsymbol{\alpha})(\boldsymbol{y}_i - \boldsymbol{\mu}_i(\boldsymbol{\beta})) = \boldsymbol{0}, \tag{13.12}$$

where $\boldsymbol{D}_i(\boldsymbol{\beta}) = diag(\partial h(\eta_{i1})/\partial\eta, \dots, \partial h(\eta_{iT_i})/\partial\eta)$ contains the derivatives (and in $\boldsymbol{W}_i$ the dependence on ϕ is suppressed). Equation (13.12) may be seen as $\boldsymbol{s}_{\boldsymbol{\beta}}(\boldsymbol{\beta}, \boldsymbol{\alpha}) = \boldsymbol{0}$, where

$$\begin{aligned} \boldsymbol{s}_{\boldsymbol{\beta}}(\boldsymbol{\beta}, \boldsymbol{\alpha}) &= \sum_{i=1}^{n} \frac{\partial \boldsymbol{\mu}_i(\boldsymbol{\beta})}{\partial \boldsymbol{\beta}} \boldsymbol{W}_i^{-1}(\boldsymbol{\beta}, \boldsymbol{\alpha})(\boldsymbol{y}_i - \boldsymbol{\mu}_i(\boldsymbol{\beta})) \\ &= \sum_{i=1}^{n} \boldsymbol{X}_i^T \boldsymbol{D}_i(\boldsymbol{\beta}) \boldsymbol{W}_i^{-1}(\boldsymbol{\beta}, \boldsymbol{\alpha})(\boldsymbol{y}_i - \boldsymbol{\mu}_i(\boldsymbol{\beta})) \end{aligned}$$

is a quasi-score function. For a correctly specified covariance $\boldsymbol{W}_i = \boldsymbol{\Sigma}_i$ and the linear Gaussian model, $\boldsymbol{s}_{\boldsymbol{\beta}}$ represents the usual score function that is the derivative $\partial l/\partial\boldsymbol{\beta}$ of the log-likelihood l. Also, in the case of only two binary variables, $\boldsymbol{y}_i^T = (y_{i1}, y_{i2})$, and the correctly specified $\boldsymbol{W}_i$, solving (13.12) is equivalent to maximizing the likelihood. The reason is that for two binary variables the distribution of y_i is determined by π_{i1}, π_{i2} and γ_{i12}, whereas pairwise correlations do not determine the distribution completely for more than two components. The solution of (13.12) for fixed $\boldsymbol{\alpha}$ and ϕ may be obtained iteratively by

$$\hat{\boldsymbol{\beta}}^{(k+1)} = \hat{\boldsymbol{\beta}}^{(k)} + \tilde{\boldsymbol{F}}(\hat{\boldsymbol{\beta}}^{(k)}, \boldsymbol{\alpha})^{-1} s_{\boldsymbol{\beta}}(\hat{\boldsymbol{\beta}}^{(k)}, \boldsymbol{\alpha}),$$

where $\tilde{\boldsymbol{F}}(\hat{\boldsymbol{\beta}}, \boldsymbol{\alpha}) = \sum_{i=1}^{n} \boldsymbol{X}_i^T \boldsymbol{D}_i(\hat{\boldsymbol{\beta}}) \boldsymbol{W}_i(\hat{\boldsymbol{\beta}})^{-1} \boldsymbol{D}_i(\hat{\boldsymbol{\beta}}) \boldsymbol{X}_i$ represents the quasi-information matrix. The estimation of $\boldsymbol{\alpha}$ depends on the assumed structure. In the following we give some examples.

Working Independence Model

If the working correlations are specified by $\boldsymbol{R}_i(\boldsymbol{\alpha}) = \boldsymbol{I}$, no parameter $\boldsymbol{\alpha}$ has to be estimated. The working covariance has the form

$$\boldsymbol{W}_i(\boldsymbol{\beta}) = diag(\sigma_{i1}^2, \ldots \sigma_{iT_i}^2)$$

with the variances $\sigma_{it}^2 = var(y_{it})$. Then, solving equation (13.12) is equivalent to a usual ML estimation under the assumption that all observations $y_{i1}, \ldots, y_{nT_n}$ are independent and the variances are correctly specified. Therefore, the usual software for ML estimation may be used. If an (over-)dispersion parameter is present, it may be estimated by the method of moments.

Structured Correlation Model

If $\boldsymbol{R}_i(\boldsymbol{\alpha})$ depends on a parameter $\boldsymbol{\alpha}$, several approaches have been proposed for the estimation. Liang and Zeger (1986) use a method of moments based on Pearson residuals:

$$\hat{r}_{it} = \frac{y_{it} - \hat{\mu}_{it}}{(v(\hat{\mu}_{it}))^{\frac{1}{2}}}.$$

The dispersion parameter is estimated consistently by

$$\hat{\phi} = \frac{1}{N-p} \sum_{i=1}^{n} \sum_{t=1}^{T_i} \hat{r}_{it}^2, \qquad N = \sum_{i=1}^{n} T_i.$$

The estimation of $\boldsymbol{\alpha}$ depends on the choice of $\boldsymbol{R}_i(\boldsymbol{\alpha})$. For equicorrelation,

$$\hat{\alpha} = \frac{1}{\hat{\phi}\{\sum_{i=1}^{n} \frac{1}{2} T_i(T_i - 1) - p\}} \sum_{i=1}^{n} \sum_{k>j} \hat{r}_{ik} \hat{r}_{ij}.$$

If the cluster sizes are all equal to m and small compared to n, an unspecified $\boldsymbol{R} = \boldsymbol{R}(\boldsymbol{\alpha})$ can be estimated by

$$\hat{\boldsymbol{R}} = \frac{1}{n\hat{\phi}} \sum_{i=1}^{n} \boldsymbol{V}_i^{-\frac{1}{2}} (\boldsymbol{y}_i - \hat{\boldsymbol{\mu}}_i)(\boldsymbol{y}_i - \hat{\boldsymbol{\mu}}_i)^T \boldsymbol{V}_i^{-\frac{1}{2}},$$

where $\boldsymbol{V}_i$ is a diagonal matrix containing σ_{is}^2. Cycling between Fisher scoring steps for $\boldsymbol{\beta}$ and estimation of $\boldsymbol{\alpha}$ and ϕ leads to a consistent estimation of $\boldsymbol{\beta}$.

More generally, one may define for each cluster the vector $\boldsymbol{w}_i^T = (w_{i12}, w_{i13}, \ldots, w_{iT_i-1T_i})$ of products $w_{ijk} = (y_{ij} - \mu_{ij})(y_{ik} - \mu_{ik})$ and the vector $\boldsymbol{n}_i = E(\boldsymbol{w}_i)$ of corresponding expectations. Then $\boldsymbol{\alpha}$ is estimated by the additional estimation equation

$$s_{\boldsymbol{\alpha}}(\boldsymbol{\beta}, \boldsymbol{\alpha}) = \sum_{i=1}^{n} \frac{\partial \boldsymbol{n}_i}{\partial \boldsymbol{\alpha}} cov(\boldsymbol{w}_i)^{-1} (\boldsymbol{w}_i - \boldsymbol{n}_i) = \boldsymbol{0}. \tag{13.13}$$

Solving (13.12) and (13.13) yields estimates of $\hat{\boldsymbol{\beta}}, \hat{\boldsymbol{\alpha}}$. This procedure has been called GEE1 by Liang et al. (1992) and Liang and Zeger (1993).

Asymptotic Properties and Extensions

The strength of the GEE1 approach is that $\boldsymbol{\beta}$ may be estimated consistently even if the working covariance is not equal to the driving covariance matrix. One only has to assume that $\hat{\boldsymbol{\alpha}}$ is consistent for some $\boldsymbol{\alpha}^o$ (see Liang and Zeger, 1986). Asymptotically, $n \to \infty$, one obtains

$$\hat{\boldsymbol{\beta}} \sim N(\boldsymbol{\beta}, \boldsymbol{V}_{\boldsymbol{W}}^{-1}\boldsymbol{V}_{\boldsymbol{\Sigma}}\boldsymbol{V}_{\boldsymbol{W}}^{-1}),$$

where the sandwich matrix $\boldsymbol{V}_{\boldsymbol{W}}^{-1}\boldsymbol{V}_{\boldsymbol{\Sigma}}\boldsymbol{V}_{\boldsymbol{W}}^{-1}$ has the components

$$\begin{aligned}
\boldsymbol{V}_{\boldsymbol{W}} &= \sum_{i=1}^{n} \frac{\partial \boldsymbol{\mu}_i(\boldsymbol{\beta})}{\partial \boldsymbol{\beta}} \boldsymbol{W}_i(\boldsymbol{\beta}, \boldsymbol{\alpha})^{-1} \frac{\partial \boldsymbol{\mu}_i(\boldsymbol{\beta})}{\partial \boldsymbol{\beta}^T} \\
&= \sum_{i=1}^{n} \boldsymbol{X}_i^T \boldsymbol{D}_i(\boldsymbol{\beta}) \boldsymbol{W}_i(\boldsymbol{\beta}, \boldsymbol{\alpha})^{-1} \boldsymbol{D}_i(\boldsymbol{\beta}) \boldsymbol{X}_i, \\
\boldsymbol{V}_{\boldsymbol{\Sigma}} &= \sum_{i=1}^{n} \frac{\partial \boldsymbol{\mu}_i(\boldsymbol{\beta})}{\partial \boldsymbol{\beta}} \boldsymbol{W}_i(\boldsymbol{\beta}, \boldsymbol{\alpha})^{-1} \boldsymbol{\Sigma}_i(\boldsymbol{\beta}, \boldsymbol{\alpha}) \boldsymbol{W}_i(\boldsymbol{\beta}, \boldsymbol{\alpha})^{-1} \frac{\partial \boldsymbol{\mu}_i(\boldsymbol{\beta})}{\partial \boldsymbol{\beta}^T} \\
&= \sum_{i=1}^{n} \boldsymbol{X}_i^T \boldsymbol{D}_i(\boldsymbol{\beta}) \boldsymbol{W}_i(\boldsymbol{\beta}, \boldsymbol{\alpha})^{-1} \boldsymbol{\Sigma}_i(\boldsymbol{\beta}, \boldsymbol{\alpha}) \boldsymbol{W}_i(\boldsymbol{\beta}, \boldsymbol{\alpha})^{-1} \boldsymbol{D}_i(\boldsymbol{\beta}) \boldsymbol{X}_i.
\end{aligned}$$

Some motivation for the asymptotic behavior is obtained by considering a Taylor approximation. Omitting arguments one has $\boldsymbol{s}_{\boldsymbol{\beta}}(\boldsymbol{\beta}, \boldsymbol{\alpha}) = \sum_{i=1}^{n} \boldsymbol{X}_i^T \boldsymbol{D}_i \boldsymbol{W}_i^{-1}(\boldsymbol{y}_i - \boldsymbol{\mu}_i)$ and $\hat{\boldsymbol{\beta}}$ solves $s_{\boldsymbol{\beta}}(\hat{\boldsymbol{\beta}}, \boldsymbol{\alpha}) = \boldsymbol{0}$. By Taylor approximation one obtains

$$\boldsymbol{0} = s_{\boldsymbol{\beta}}(\hat{\boldsymbol{\beta}}, \boldsymbol{\alpha}) \approx \boldsymbol{s}_{\boldsymbol{\beta}}(\boldsymbol{\beta}, \boldsymbol{\alpha}) + \frac{\partial \boldsymbol{s}_{\boldsymbol{\beta}}}{\partial \boldsymbol{\beta}^T}(\hat{\boldsymbol{\beta}} - \boldsymbol{\beta}),$$

which yields

$$\hat{\boldsymbol{\beta}} - \boldsymbol{\beta} \approx (-\frac{\partial \boldsymbol{s}_{\boldsymbol{\beta}}}{\partial \boldsymbol{\beta}^T})^{-1} \boldsymbol{s}_{\boldsymbol{\beta}}(\boldsymbol{\beta}, \boldsymbol{\alpha}).$$

If $\partial \boldsymbol{s}_{\boldsymbol{\beta}} / \partial \boldsymbol{\beta}^T$ is replaced by its expectation, one obtains the sandwich

$$\operatorname{cov}(\hat{\boldsymbol{\beta}}) \approx \mathrm{E}(-\frac{\partial \boldsymbol{s}_{\boldsymbol{\beta}}}{\partial \boldsymbol{\beta}^T})^{-1} \operatorname{cov}(\boldsymbol{s}_{\boldsymbol{\beta}}(\boldsymbol{\beta}, \boldsymbol{\alpha}))\, \mathrm{E}(-\frac{\partial \boldsymbol{s}_{\boldsymbol{\beta}}}{\partial \boldsymbol{\beta}^T})^{-1},$$

where $\operatorname{cov}(\boldsymbol{s}_{\boldsymbol{\beta}}(\boldsymbol{\beta}, \boldsymbol{\alpha})) = \boldsymbol{X}_i^T \boldsymbol{D}_i \boldsymbol{W}_i^{-1} \boldsymbol{\Sigma}_i \boldsymbol{W}_i^{-1} \boldsymbol{D}_i \boldsymbol{X}_i$.

One drawback of the GEE1 approach is that in estimating $\boldsymbol{\beta}$ and $\boldsymbol{\alpha}$ it acts as if they were independent of each other. Therefore, little information from $\boldsymbol{\beta}$ is used when estimating $\boldsymbol{\alpha}$, which may lead to significant loss of $\boldsymbol{\alpha}$ information. As a remedy, Zhao and Prentice (1990) and Liang et al. (1992) discuss estimating the total parameter vector $\boldsymbol{\delta}^T = (\boldsymbol{\beta}^T, \boldsymbol{\alpha}^T)$ jointly by solving

$$\boldsymbol{s}(\boldsymbol{\alpha}, \boldsymbol{\beta}) = \sum_{i=1}^{n} \frac{\partial(\boldsymbol{\mu}_i, \boldsymbol{n}_i)}{\partial \boldsymbol{\delta}} cov(\boldsymbol{y}_i, \boldsymbol{w}_i)^{-1} \begin{pmatrix} \boldsymbol{y}_i - \boldsymbol{\mu}_i \\ \boldsymbol{w}_i - \boldsymbol{n}_i \end{pmatrix} = \boldsymbol{0}. \tag{13.14}$$

This expanded procedure, which estimates $\boldsymbol{\beta}$ and $\boldsymbol{\alpha}$ simultaneously, has been termed GEE2. The parameter $\boldsymbol{\beta}$ is estimated consistently with asymptotic normal distribution with mean $\boldsymbol{0}$ and a covariance that can be consistently estimated by

$$\operatorname{cov}(\hat{\boldsymbol{\beta}}) = \boldsymbol{V}_{\boldsymbol{W}}^{-1}\boldsymbol{V}_{\boldsymbol{\Sigma}}\boldsymbol{V}_{\boldsymbol{W}}^{-1}$$

Asymptotics GEE1

Approximation $n \to \infty$

$$\hat{\boldsymbol{\beta}} \sim N(\boldsymbol{\beta}, \boldsymbol{V}_{\boldsymbol{W}}^{-1}\boldsymbol{V}_{\boldsymbol{\Sigma}}\boldsymbol{V}_{\boldsymbol{W}}^{-1})$$

Sandwich

$$\boldsymbol{V}_{\boldsymbol{W}} = \sum_{i=1}^{n} \boldsymbol{X}_i^T \boldsymbol{D}_i(\boldsymbol{\beta})\boldsymbol{W}_i(\boldsymbol{\beta},\boldsymbol{\alpha})^{-1}\boldsymbol{D}_i(\boldsymbol{\beta})\boldsymbol{X}_i,$$

$$\boldsymbol{V}_{\boldsymbol{\Sigma}} = \sum_{i=1}^{n} \boldsymbol{X}_i^T \boldsymbol{D}_i(\boldsymbol{\beta})\boldsymbol{W}_i(\boldsymbol{\beta},\boldsymbol{\alpha})^{-1}\boldsymbol{\Sigma}_i(\boldsymbol{\beta},\boldsymbol{\alpha})\boldsymbol{W}_i(\boldsymbol{\beta},\boldsymbol{\alpha})^{-1}\boldsymbol{D}_i(\boldsymbol{\beta})\boldsymbol{X}_i$$

with

$$\boldsymbol{V}_{\boldsymbol{W}} = \sum_{i=1}^{n} \frac{\partial \boldsymbol{\mu}_i(\boldsymbol{\beta})}{\partial \boldsymbol{\beta}} \operatorname{cov}(\boldsymbol{y}_i, \boldsymbol{w}_i)^{-1} \frac{\partial \boldsymbol{\mu}_i(\boldsymbol{\beta})}{\partial \boldsymbol{\beta}^T},$$

$$\boldsymbol{V}_{\boldsymbol{\Sigma}} = \sum_{i=1}^{n} \frac{\partial \boldsymbol{\mu}_i(\boldsymbol{\beta})}{\partial \boldsymbol{\beta}} \operatorname{cov}(\boldsymbol{y}_i, \boldsymbol{w}_i)^{-1} \begin{pmatrix} \boldsymbol{y}_i - \hat{\boldsymbol{\mu}}_i \\ \boldsymbol{w}_i - \hat{\boldsymbol{\eta}}_i \end{pmatrix} \begin{pmatrix} \boldsymbol{y}_i - \hat{\boldsymbol{\mu}}_i \\ \boldsymbol{w}_i - \hat{\boldsymbol{\eta}}_i \end{pmatrix}^T \operatorname{cov}(\boldsymbol{y}_i, \boldsymbol{w}_i)^{-1} \frac{\partial \boldsymbol{\mu}_i(\boldsymbol{\beta})}{\partial \boldsymbol{\beta}^T},$$

evaluated at $\hat{\boldsymbol{\delta}}$; see Liang et al. (1992). One drawback of the GEE2 approach is that the consistency derived from (13.14) depends on the correct specifications of both $\boldsymbol{\mu}_i$ and $\operatorname{cov}(\boldsymbol{y}_i)$. Liang et al. (1992) discuss extensions where $\operatorname{cov}(\boldsymbol{y}_i, \boldsymbol{w}_i)$ is replaced by a working covariance that depends on higher order parameters obtaining consistency of $\boldsymbol{\delta}$ if the higher order parameters are estimated consistently.

Rigorous investigations of the asymptotic behavior of GGE estimates were given by Xie and Yang (2003) and Wang (2011).

Example 13.4: Knee Injuries

In the knee injuries study (Example 1.4) pain was recorded for each subject after 3, 7, and 10 days of treatment. The effect of treatment and the covariates gender and age (centered around 30) may be investigated by use of a marginal model. To keep the model simple, the pain level was dichotomized by grouping the two lowest pain levels and the three higher pain levels. With $\pi_{it} = P(y_{it} = 1|\boldsymbol{x}_i)$ the marginal model has the form

$$\operatorname{logit}(\pi_{it}) = \beta_0 + x_{T,i}\beta_r + x_{G,i}\beta_G + \text{Age}_i\beta_A + \text{Age}_i^2\beta_{A^2},$$

where $x_{T,i} = 1$ for treatment, $x_{T,i} = 0$ for placebo and $x_{G,i} = 1$ for females, $x_{G,i} = 0$ for males. In Table 13.2 naive standard errors have been obtained by assuming the working correlation to be correct, whereas the robust standard errors are based on the sandwich matrix. It is seen that there is quite some difference between the naive and robust standard errors if independence is used as working covariance. When the working covariance gets more complex, possibly coming closer to the true underlying covariance structure, the difference between naive and robust estimates become rather small. In Table 13.3 estimates and standard errors are given by fitting a logistic model and simply ignoring that the responses are clustered. The standard errors in this case are rather small and should be severely biased. For independence and an exchangeable structure one obtains the effect of therapy by -0.673, which corresponds

TABLE 13.2: Estimated regression coefficients, and naive and robust standard errors for three different choices of $R(\alpha)$. Naive standard errors are obtained by assuming the working correlation matrix to be correct.

Correlation Structure: Independent			
	Estimate	Naive S.E.	Robust S.E.
Intercept	1.172	0.284	0.448
Therapy (treatment)	−0.673	0.223	0.334
Gender (female)	0.265	0.241	0.366
Age	0.013	0.012	0.017
Age^2	−0.006	0.001	0.017
Correlation Structure: Exchangeable			
	Estimate	Naive S.E.	Robust S.E.
Intercept	1.172	0.423	0.448
Therapy (treatment)	−0.673	0.333	0.334
Gender (female)	0.265	0.360	0.366
Age	0.013	0.012	0.017
Age^2	−0.006	0.002	0.002
Correlation Structure: Exponential Correlation			
	Estimate	Naive S.E.	Robust S.E.
Intercept	1.122	0.427	0.443
Therapy (treatment)	−0.748	0.333	0.328
Gender (female)	0.192	0.361	0.362
Age	0.012	0.018	0.017
Age^2	−0.006	0.002	0.002

TABLE 13.3: Estimated regression coefficients and standard errors for a GLM with logit link.

	Estimate	Std. Error
(Intercept)	1.172	0.282
Therapy (treatment)	−0.673	0.221
Gender (female)	0.265	0.239
Age	0.013	0.012
Age^2	−0.006	0.001

to the odds ratio 0.510, signaling a drastic reduction of pain when an active ingredient is used rather than a placebo. It is seen from Table 13.2 that the estimates and the robust standard errors are the same for the independence model and the exchangeable correlation structure. The effect holds more generally. If all the covariates are cluster-specific, it can be shown that the independence model and the exchangeable correlation model yield the same estimates (see Exercise 13.3). □

The marginal model considered in Example 13.4 has the simple structure $\pi_{it} = h(\boldsymbol{x}_i^T\boldsymbol{\beta})$, where the covariate $\boldsymbol{x}_i$ does not vary over repeated measurements and $\boldsymbol{\beta}$ is the same for all responses. But in general marginal models are more flexible. In the general form one has $\pi_{it} = h(\boldsymbol{x}_{it}^T\boldsymbol{\beta})$, with covariates that can vary over measurements. That can also be used to model that the effects are specific for the response. That is especially needed if the responses differ in substance, as in the following example, where the first response is Cesarean section and the second is treatment in intensive care. The used model with response-specific effects,

$$\pi_{i1} = h(\boldsymbol{x}_i^T\boldsymbol{\beta}_1), \quad \pi_{i2} = h(\boldsymbol{x}_i^T\boldsymbol{\beta}_2),$$

is of the general form $\pi_{it} = h(\boldsymbol{x}_{it}^T\boldsymbol{\beta})$. One simply has to define $\boldsymbol{x}_{i1}^T = (\boldsymbol{x}_i^T, \mathbf{0}^T)$, $\boldsymbol{x}_{i2}^T = (\mathbf{0}^T, \boldsymbol{x}_i^T)$, and $\boldsymbol{\beta}^T = (\boldsymbol{\beta}_1^T, \boldsymbol{\beta}_2^T)$, where $\mathbf{0}$ is a vector of zeros.

Example 13.5: Birth Data

Table 13.4 shows the fitted values of a marginal logit model (independent correlation structure) with response-specific parameters $\pi_{it} = h(\boldsymbol{x}_i^T\boldsymbol{\beta}_t)$ for the birth study (Examples 13.2 and 13.3). The two response variables are Cesarean section (C; 1: yes; 0: no) and intensive care (IC; 1: yes; 0: no). The explanatory variables are the same as in Example 13.3 (gender of child, 1:male; 2:female), weight, age of mother, number of previous pregnancies, (1: no previous pregnancy; 2: one pregnancy; 3: more than 2 pregnancies). It is seen that the effects are about the same as for the model that includes the effects on odds ratios (Table 13.1). Also, the similarity of naive values and values that are based on the sandwich matrix suggests that correlation between responses given covariates is weak. □

TABLE 13.4: Estimated regression coefficients for birth data with responses Ceaserian section and intensive care; marginal model with independent working correlation, estimated scale parameter: 1.216.

	Estimate	Naive z	Robust z
InterceptInt	4.161	3.498	3.789
InterceptCes	−0.992	−1.109	−1.097
WeightInt	−0.002	−8.163	−8.334
WeightCes	−0.001	−4.095	−4.019
AgeMotherInt	0.007	0.215	0.233
AgeMotherCes	0.079	3.096	3.319
SexInt	−0.208	−0.751	−0.842
SexCes	−0.309	−1.462	−1.638
PreviousInt1	−0.457	−1.111	−1.268
PreviousCes1	−0.595	−2.097	−2.249
PreviousInt2	0.636	1.169	1.080
PreviousCes2	−2.136	−2.576	−2.684

Types of Covariates and Loss of Efficiency

In applications one often has two types of variables, *cluster-specific* or *cluster-level* covariates that vary across clusters but are constant within one cluster, and *within-cluster* covariates that vary across the measurements taken within one cluster. In treatment studies dosage may often be a within-cluster covariate when it varies across repeated measurements. A simple example of a cluster-specific covariate in treatment studies is gender. If $y_{i1}, \dots, y_{iT_i}$ represent repeated measurements, within-cluster covariates are time-varying covariates in contrast to cluster-specific covariates, which are time-invariant. In experimental designs within-cluster covariates may be the same for all of the clusters, for example, if a treatment varies across time but in the same way for all individuals. To distinguish it from the usual within-cluster covariate, it will be called a fixed (by design) within-cluster covariate.

Responses and covariates for these differing types of covariates are given in the form

$y_{i1}, \dots, y_{iT_i}$	responses
$x_i, \dots, x_i$	cluster-specific (for example, gender)
$x_{i1}, \dots, x_{iT_i}$	within-cluster (time-varying subject-specific covariate)
$x_{i1}, \dots, x_{iT_i}$	fixed within-cluster (fixed treatment plan).

If the working covariance is not equal to the underlying true covariance, usually some efficiency is lost. The loss of efficiency depends on the covariance and the discrepancy between the covariance and the working covariance as well as on the design. Fitzmaurice (1995) showed that the assumption of independence can lead to a considerable loss of efficiency when the responses are strongly correlated and the design includes a within-cluster covariate that is not invariant across clusters. In particular the coefficient associated with that covariate may be estimated rather inefficiently. Mancl and Leroux (1996) showed that asymptotic efficiency may be low even for little within-cluster variation (see also Wang and Carey, 2003). On the other hand, it has been demonstrated for cluster-level covariates and fixed within-cluster covariates that asymptotic relative efficiency is near unity when the GEE with independence or equicorrelation working correlation is compared to the maximum likelihood estimates (see Fitzmaurice, 1995). Mancl and Leroux (1996) demonstrated that asymptotic efficiency is one if all the covariates are mean-balanced in the sense that the cluster means are constant across clusters.

13.3.2 Marginal Models for Multinomial Responses

Let $Y_{i1}, \dots, Y_{iT_i}$ denote measurements on the ith unit with Y_{it} taking values from $\{1, \dots, m\}$. The marginal probability of responses, depending on the covariates x_{it}, is denoted by $\pi_{itr} = P(Y_{it} = r|\boldsymbol{x}_{it}), r = 1, \dots, m$. The underlying multinomial distribution uses dummy variables $(y_{it1}, \dots, y_{itq}), q = m - 1$, where $y_{itr} = 1$ if $Y_{it} = r$ and $y_{itr} = 0$ otherwise. It is given by

$$\boldsymbol{y}_{it}^T = (y_{it1}, \dots, y_{itq}) \sim M(n_i, \boldsymbol{\pi}_{it}^T = (\pi_{it1}, \dots, \pi_{itq})).$$

In the same way as for univariate marginal models, one specifies the marginal response probabilities together with a working covariance matrix. Candidates for marginal models are the nominal logit model (Chapter 8) and the ordinal models (Chapter 9), depending on the response. In detail, the specification is given by:

(1) The *marginal* vector of probabilities is assumed to be correctly specified by

$$\boldsymbol{\pi}_{it} = h(\boldsymbol{X}_{it}^T\boldsymbol{\beta}), \tag{13.15}$$

where $\boldsymbol{X}_{it}$ is composed from the covariates $\boldsymbol{x}_{it}$.

(2) The *marginal covariance* of response vector $\boldsymbol{y}_{it}$ has the usual form of multinomial covariances:

$$\boldsymbol{V}_{it} = \text{cov}(\boldsymbol{y}_{it}|\boldsymbol{x}_{it}) = \text{diag}(\boldsymbol{\pi}_{it}) - \boldsymbol{\pi}_{it}\boldsymbol{\pi}_{it}^T. \tag{13.16}$$

(3) The covariance between vectors $\boldsymbol{y}_{it}$ and $\boldsymbol{y}_{is}$ is specified as a *working covariance matrix* $\boldsymbol{W}_{ist} = \boldsymbol{W}_{ist}(\boldsymbol{\beta}, \boldsymbol{\alpha})$, where $\boldsymbol{\alpha}$ denotes a vector of association parameters.

For the total covariance of the measurements on unit i one obtains the block matrix

$$\boldsymbol{W}_i = \begin{pmatrix} \boldsymbol{V}_{i1} & \boldsymbol{W}_{i12} & \cdots & \boldsymbol{W}_{i1T_i} \\ \vdots & \ddots & & \\ \boldsymbol{W}_{iT_i1} & \cdots & & \boldsymbol{V}_{iT_i} \end{pmatrix}.$$

By collecting the components in matrices one obtains with $\boldsymbol{X}_i^T = (\boldsymbol{X}_{i1}^T, \dots, \boldsymbol{X}_{iT_i}^T)$ the same generalized estimation equation as in Section 13.3.1,

$$\sum_{i=1}^{n} \boldsymbol{X}_i^T \boldsymbol{D}_i(\boldsymbol{\beta}) \boldsymbol{W}_i^{-1}(\boldsymbol{\beta}, \boldsymbol{\alpha})(\boldsymbol{y}_i - \boldsymbol{\mu}_i(\boldsymbol{\beta})) = \boldsymbol{0}, \tag{13.17}$$

where $\boldsymbol{D}_i(\boldsymbol{\beta})$ is a block-diagonal matrix with blocks $\boldsymbol{D}_{i1}(\boldsymbol{\beta}), \dots, \boldsymbol{D}_{iT_i}(\boldsymbol{\beta})$, which contain the derivatives. The variances of estimates are again approximated by the sandwich matrix

$$\text{cov}(\hat{\boldsymbol{\beta}}) = \boldsymbol{V}_{\boldsymbol{W}}^{-1} \boldsymbol{V}_{\Sigma} \boldsymbol{V}_{\boldsymbol{W}}^{-1}.$$

As in the binary response case, specification of the working covariance typically uses odds ratios. But in the multinomial case one has to specify the covariance $\text{cov}(\boldsymbol{y}_{is}, \boldsymbol{y}_{it})$, which is a $(q \times q)$-matrix with elements $\text{cov}(y_{isl}, y_{itm})$. The corresponding odds ratio is

$$\gamma_{ist}(l, m) = \frac{P(y_{isl} = 1, y_{itm} = 1)/P(y_{isl} = 0, y_{itm} = 1)}{P(y_{isl} = 1, y_{itm} = 0)/P(y_{isl} = 0, y_{itm} = 0)}.$$

The choice $\gamma_{ist}(l, m) = 1, s \neq t$, corresponds to uncorrelated responses, whereas $\gamma_{ist}(l, m) = \gamma_{lm}, s \neq t$, corresponds to the equicorrelation structure.

For ordered response categories, in particular when the cumulative model is used, it is advantageous to use a different representation of the multinomial responses. Let $y_{it(l)} = I(Y_{it} \leq l), l = 1, \dots, m-1$ denote a dummy variable that codes if the response is below or in category l and $\pi_{it(l)} = P(Y_{it} \leq l | \boldsymbol{x}_{it})$ denote the corresponding probability. Then the cumulative model has the simple form $\text{logit}(\pi_{it(l)}) = \boldsymbol{x}_{itl}^T \boldsymbol{\beta}$, where $\boldsymbol{x}_{itl}$ contains the category-specific intercept. The covariance of the vector $\boldsymbol{y}_{it}^T = (y_{it(1)}, \dots, y_{it(m-1)})$ is given by $\boldsymbol{V}_{it} = \text{cov}(\boldsymbol{y}_{it} | \boldsymbol{x}_{it}) = \text{diag}(\boldsymbol{\pi}_{it}) - \boldsymbol{\pi}_{it} \boldsymbol{\pi}_{it}^T$, where $\boldsymbol{\pi}_{it}^T = (\pi_{it(1)}, \dots, \pi_{it(m-1)})$. The corresponding odds ratios that specify the working covariance are the so-called *global odds ratios*:

$$\gamma_{ist}(l, m) = \frac{P(Y_{is} \leq l, Y_{it} \leq m)/P(Y_{is} > l, Y_{it} \leq m)}{P(Y_{is} \leq l, Y_{it} > m)/P(Y_{is} > l, Y_{it} > m)}.$$

Marginal GEE models based on global odds ratios were considered by Williamson et al. (1995), Fahrmeir and Pritscher (1996), and Heagerty and Zeger (1996). Miller et al. (1993) used pairwise correlations for specifying the working covariance.

Models for ordered categorical outcomes with time-dependent parameters have been proposed by Stram et al. (1988) and Stram and Wei (1988). They consider the marginal cumulative response model $\pi_{it} = h(\boldsymbol{X}_{it} \boldsymbol{\beta}_t)$, in which the parameters vary over t. The combined estimate $\hat{\boldsymbol{\beta}} = (\hat{\boldsymbol{\beta}}_1, \dots, \hat{\boldsymbol{\beta}}_T)$ becomes asymptotically normal ($n \to \infty$), and its asymptotic covariance matrix can be estimated empirically. Zeger (1988) showed that $\hat{\boldsymbol{\beta}}$ can be viewed as the solution of a GEE with working correlation matrix $\boldsymbol{R}(\alpha) = \boldsymbol{I}$, and that the covariance matrix is identical to the one obtained from the GEE approach. In a similar approach, Moulton and Zeger (1989) combine the estimated coefficients at each time point by using bootstrap methods or weighted least-squares methods (see also Davis, 1991).

13.3.3 Penalized GEE Approaches

If covariates are collinear, the usual GEE methodology fails because estimates may not exist. However, by using shrinkage methods, estimates can be stabilized.

Penalized estimates as considered in Chapter 6 are based on the the penalized maximum likelihood $l_p = \sum_i l_i(\boldsymbol{\beta}) - \lambda \Sigma_j p(\beta_j)$, where $p(\beta_j)$ is a penalty function, for example, $p(\beta_j) = |\beta_j|^\gamma$. The corresponding estimation equation is defined by setting the penalized score function $\boldsymbol{s}_p(\boldsymbol{\beta}) = \partial l_p / \partial \boldsymbol{\beta}$ equal to zero.

For marginal models the maximum likelihood is no longer available. But considering $\boldsymbol{s}_{\boldsymbol{\beta}}(\boldsymbol{\beta}, \boldsymbol{\alpha})$ as a quasi-score function, a corresponding GEE may be derived. When setting $\boldsymbol{s}_{\boldsymbol{\beta}}(\boldsymbol{\beta}) = \boldsymbol{0}$ the corresponding penalty matrix is given by $\lambda \boldsymbol{P}$, where $\boldsymbol{P} =$ Diag $(\partial p(\beta_1)/\partial \beta, \dots,$

$\partial p(\beta_p)/\partial \beta)$. For $p(\beta_j) = |\beta_j|^\gamma$ one obtains $\partial p(\beta_j)/\partial \beta = \gamma sign(\beta_j)|\beta_j|^{\gamma-1}$. The corresponding *penalized GEE* is given by

$$\sum_{i=1}^{n} \boldsymbol{X}_i^T \boldsymbol{D}_i(\boldsymbol{\beta}) \boldsymbol{W}_i^{-1}(\boldsymbol{\beta}, \boldsymbol{\alpha})(\boldsymbol{y}_i - \boldsymbol{\mu}_i(\boldsymbol{\beta})) - \lambda \boldsymbol{P} = \boldsymbol{0}, \tag{13.18}$$

where $\lambda > 0$ is a fixed smoothing parameter and γ determines the type of penalization. The choice $\gamma = 2$ yields the penalty term $\gamma sign(\beta_j)|\beta_j|^{\gamma-1} = 2\beta_j$ and the simple *ridge-type penalized GEE*

$$\sum_{i=1}^{n} \boldsymbol{X}_i^T \boldsymbol{D}_i(\boldsymbol{\beta}) \boldsymbol{W}_i^{-1}(\boldsymbol{\beta}, \boldsymbol{\alpha})(\boldsymbol{y}_i - \boldsymbol{\mu}_i(\boldsymbol{\beta})) - 2\lambda \boldsymbol{\beta} = \boldsymbol{0}.$$

The solution of equation (13.18) is computed in rather the same way as in a GEE. It only has to take care of the penalization term. For example, an iterative computation of $\hat{\boldsymbol{\beta}}$ for fixed $\boldsymbol{\alpha}, \phi$ is given by

$$\hat{\boldsymbol{\beta}}^{(k+1)} = \hat{\boldsymbol{\beta}}^{(k)} + \hat{\boldsymbol{F}}_p(\hat{\boldsymbol{\beta}}^{(k)}, \boldsymbol{\alpha})^{-1} \boldsymbol{s}_{p,\boldsymbol{\beta}}(\hat{\boldsymbol{\beta}}^{(k)}, \boldsymbol{\alpha}),$$

where

$$\begin{aligned}
\hat{\boldsymbol{F}}_p(\hat{\boldsymbol{\beta}}, \boldsymbol{\alpha}) &= \textstyle\sum_i \boldsymbol{X}_i^T \boldsymbol{D}_i(\hat{\boldsymbol{\beta}}) \boldsymbol{W}_i^{-1}(\hat{\boldsymbol{\beta}}, \boldsymbol{\alpha}) \boldsymbol{D}_i(\hat{\boldsymbol{\beta}})^T \boldsymbol{X}_i + \partial \boldsymbol{P}, \\
\partial \boldsymbol{P} &= Diag(\partial^2 p(\beta_1)/\partial \beta^2, \dots, \partial^2 p(\beta_p)/\partial \beta^2), \\
\boldsymbol{s}_{p,\boldsymbol{\beta}}(\hat{\boldsymbol{\beta}}, \boldsymbol{\alpha}) &= \boldsymbol{X}_i^T \boldsymbol{D}_i(\hat{\boldsymbol{\beta}}) \hat{\boldsymbol{W}}_i(\hat{\boldsymbol{\beta}}, \boldsymbol{\alpha})(\boldsymbol{y}_i - \boldsymbol{\mu}_i(\hat{\boldsymbol{\beta}})) - \lambda \boldsymbol{P}.
\end{aligned}$$

For $\lambda > 0, \gamma \geq 1$, the resulting estimator is unique under weak conditions; for $\lambda = o(\sqrt{n})$, it is consistent and asymptotically normally distributed (Fu, 2003). An approximation to the covariance of $\hat{\boldsymbol{\beta}}$ is given by $\boldsymbol{V}_{\boldsymbol{W}}^{-1} \boldsymbol{V}_{\boldsymbol{\Sigma}} \boldsymbol{V}_{\boldsymbol{W}}^{-1}$, where

$$\begin{aligned}
\boldsymbol{V}_{\boldsymbol{W}} &= \textstyle\sum_i \boldsymbol{X}_i^T \boldsymbol{D}_i \boldsymbol{W}_i^{-1} \boldsymbol{D}_i \boldsymbol{X}_i \quad + \quad \lambda \partial \boldsymbol{P}, \\
\boldsymbol{V}_{\boldsymbol{\Sigma}} &= \textstyle\sum_i \boldsymbol{X}_i^T \boldsymbol{D}_i \boldsymbol{\Sigma}_i^{-1} \boldsymbol{D}_i \boldsymbol{X}_i.
\end{aligned}$$

13.3.4 Generalized Additive Marginal Models

The assumption of linear predictors in marginal modeling can be weakened by including additive functions in the predictor. Instead of the linear predictor $\eta_{it} = \boldsymbol{x}_{it}^T \boldsymbol{\beta}$ one assumes the additive structure

$$\eta_{it} = \beta_0 + f_{(1)}(x_{it1}) + \cdots + f_{(p)}(x_{itp}),$$

where the $f_{(j)}(.)$ are unspecified, unknown functions and $\boldsymbol{x}_{it}^T = (x_{it1}, \dots, x_{itp})$. For the unknown functions smoothing methods like localization or splines as considered in Chapter 10 can be applied. When the functions are expanded in basis functions, $f_{(j)}(x_{itj}) = \sum_{s=1}^{m_j} \beta_{js} \phi_{js}(x_{itj})$, one obtains the linear predictor

$$\eta_{it} = \beta_0 + \boldsymbol{\phi}_{it1}^T \boldsymbol{\beta}_1 + \cdots + \boldsymbol{\phi}_{itp}^T \boldsymbol{\beta}_p = \boldsymbol{\phi}_{it}^T \boldsymbol{\beta},$$

where $\boldsymbol{\phi}_{itj}^T = (\phi_{j1}(x_{itj}), \dots, \phi_{jm_i}(x_{itj}))$ are the evaluations of the basis functions for the jth variable at x_{itj}. With evaluations collected in $\boldsymbol{\phi}_{it}$ and $\boldsymbol{\beta}$ denoting the whole parameter vector one has a linearly structured predictor. After specifying the marginal variance and the working covariance as in (13.8) and (13.9) one can use the penalized GEE:

$$\sum_{i=1}^{n} \boldsymbol{\Phi}_i^T \boldsymbol{D}_i(\boldsymbol{\beta}) \boldsymbol{W}_i^{-1}(\boldsymbol{\beta}, \boldsymbol{\alpha})(\boldsymbol{y}_i - \boldsymbol{\mu}_i(\boldsymbol{\beta})) - \frac{1}{2} \sum_{j=1}^{p} \lambda_j \boldsymbol{P}_j \boldsymbol{\beta}_j = \boldsymbol{0},$$

where $\Phi_i^T = (\phi_{i1}, \ldots, \phi_{iT})$, and D_i, W_i denote the matrix of derivatives and the working covariance, respectively. The penalty term $(1/2)\sum_j \lambda_j P_j \beta_j$ can be seen as the derivative of the penalty $(1/2)\sum_j \lambda_j \beta_j^T P_j \beta_j$, which is typically used to penalize the log-likelihood, if it is available. The choice of the penalty matrix P_j depends on the used basis functions and the type of penalization one wants to impose (see Chapter 10). The use of individual smoothing parameters can raise severe selection problems, and it might be useful to assume that $\lambda_j = \lambda$, $j = 1, \ldots, p$.

For the extension of GEE approaches to contain additive components see Wild and Yee (1996). Non-parametric modeling of predictors in GEEs based on local regression techniques were given by Carroll et al. (1998); for longitudinal data with ordinal responses see Kauermann (2000) and Heagerty and Zeger (1998).

13.4 Marginal Homogeneity

When the same variable is measured repeatedly at different times or under different conditions one often wants to investigate if the distribution has changed over measurements or conditions. In the simplest case of two binary measurements (y_{i1}, y_{i2}) on subject i the two distributions are the same if the hypothesis of *marginal homogeneity* $P(y_{i1} = 1) = P(y_{i2} = 1)$ holds. A simple example is given in Table 13.5, which shows the measured pain levels before the beginning of treatment and after 10 days of treatment for the placebo group (compare Example 13.4). Of course one has to assume that the two responses are correlated, but correlation is of minor interest. The more interesting question is if the pain level is the same for both measurements although there was placebo treatment only. An example with differing conditions for measurements is given in Table 13.6. The data were collected in a psychiatric ward at the University of Regensburg; 177 members of the nursing staff were asked if patients talk about problems with their partner to them. If the hypothesis of marginal homogeneity holds, the probability of talking about problems does not depend on the gender of the patient. Comparisons of the response rates between correlated proportions are also found in the comparison of treatments or in problems of establishing equivalence or noninferiority between two medical diagnostic procedures (Berger and Sidik, 2003).

TABLE 13.5: Measured pain levels before beginning of treatment (measurement 1) and after 10 days (measurement 2) for placebo group.

	Measurement 2		
Measurement 1	Low	High	
Low	12	0	12
High	13	37	50
	25	37	62

TABLE 13.6: Talking about problems with partner depending on gender of patient.

	Male Patient		
Female Patient	Yes	No	
Yes	39	62	101
No	13	63	76
	52	125	177

The methods considered in the following do not apply only in cases where the same subject is measured repeatedly. It is only needed that the observations are linked, forming so-called matched pairs. *Matched pairs* arise when each observation in one sample pairs with an observation in an other sample. For example, in case control studies, frequently a population is stratified with respect to some control variable and one individual per stratum is chosen randomly to be given the treatment and one is given the control. The responses on the two individuals, matched, for example, according to gender and age, cannot be considered as independent. Therefore, as in longitudinal studies, where the responses of the same individual are observed repeatedly, one has to account for the dependence of the responses.

When investigating marginal distributions for correlated measurements, marginal models are a natural instrument. But it can be useful to model the heterogeneity by including subject-specific parameters.

13.4.1 Marginal Homogeneity for Dichotomous Outcome

The general form of repeated measurements or matched pairs with a binary response is determined by the bivariate variables (y_1, y_2) measured on each individual or stratum. One obtains a (2×2)-table containing the number of individuals with responses from $\{(0,0), (0,1), (1,0), (1,1)\}$.

Let $\pi_{ij} = P(y_1 = i, y_2 = j)$ denote the probability of outcomes (i, j) with $i, j \in \{0, 1\}$, n_{ij} denote the number of pairs with outcomes (i, j), and $p_{ij} = n_{ij}/n$ the sample proportion. The counts $\{n_{ij}\}$ are treated as a sample from the multinomial distribution $M(n, \{\pi_{ij}\})$. As usual, the subscript "+" denotes the sum over that index. The table of probabilities and the table of counts are shown in Table 13.7.

TABLE 13.7: Table of probabilities and table of counts.

		y_2: 1	y_2: 0	
y_1	1	π_{11}	π_{10}	π_{1+}
	0	π_{01}	π_{00}	π_{0+}
		π_{+1}	π_{+0}	1

		y_2: 1	y_2: 0	
y_1	1	n_{11}	n_{10}	n_{1+}
	0	n_{01}	n_{00}	n_{0+}
		n_{+1}	n_{+0}	n

Comparison of the marginal distribution focuses on the difference:

$$\delta = \pi_{+1} - \pi_{1+}.$$

When $\pi_{+1} = \pi_{1+}$, then $\pi_{+0} = \pi_{0+}$ also, and the table shows *marginal homogeneity*. In (2×2)-tables marginal homogeneity is equivalent to $\pi_{10} = \pi_{01}$, since

$$\delta = \pi_{+1} - \pi_{1+} = \pi_{11} + \pi_{01} - \pi_{11} - \pi_{10} = \pi_{01} - \pi_{10}.$$

The hypothesis of marginal homogeneity,

$$H_0 : \pi_{+1} = \pi_{1+} \quad (\text{or } \delta = 0),$$

may be tested by using *McNemar's test* (McNemar, 1947)

$$M = \frac{(n_{01} - n_{10})^2}{n_{01} + n_{10}},$$

which in large samples is approximately χ^2-distributed with $df = 1$, if H_0 holds. McNemar's test is based on the ML estimate of δ and has the form $M = \hat{\delta}^2/\hat{\sigma}(\hat{\delta})$, where

$$\hat{\delta} = \frac{n_{+1}}{n} - \frac{n_{1+}}{n} = \frac{n_{01} - n_{10}}{n}$$

is the ML estimate of δ and $\hat{\sigma}(\hat{\delta}) = (n_{01} + n_{10})/n^2$ is an estimate of the standard deviation derived under the null hypothesis.

An alternative test that suggests itself is the *exact sign test.* It is based on the equivalence of the hypotheses $H_0 : \pi_{+1} = \pi_{1+}$ and $H_0 : \pi_{01} = \pi_{10}$. By conditioning on observations $n_{01} + n_{10}$ one has to test the hypothesis $H_0 : \pi_{01} = 1/2$. Conditioning on observations $n_{01} + n_{10}$ may be seen as using the variables

$$\tilde{y}_i = \begin{cases} 1 & (y_{i1}, y_{i2}) = (0,1) \\ 0 & (y_{i1}, y_{i2}) = (1,0), \end{cases}$$

where y_{i1}, y_{i2} denote the original observations. Then the hypothesis $H_0 : \pi_{01} = 1/2$ corresponds to $H_0 : E(\tilde{y}_i) = 0.5$ and the conditional sign test, based on the binomial distribution of $\sum_i \tilde{y}_i$, applies. The corresponding p-value is given by $P_{cs} = 2\sum_{i=0}^{min(n_{01},n_{10})} B(i; n_{01} + n_{10}, 1/2)$, where $B(.; t, \pi)$ denotes the cumulative distribution function of the binomial distribution with parameters t and π. In both the examples given in Tables 13.5 and 13.6, the exact test as well as McNemar's test reject the null hypothesis of homogeneity ($M = 13$ for Table 13.5 and $M = 32.01$ for Table 13.6).

Both test statistics may also be used in one-sided test problems. For McNemar's test statistic one uses that the signed square root of M has approximately a standard normal distribution. The exact test uses the p-value $P_{cs} = \sum_{i=0}^{n_{01}} B(i, n_{01} + n_{10}, 1/2)$ for $H_0 : \pi_{+1} \leq \pi_{1+}$ and $P_{cs} = \sum_{i=n_{01}}^{n_{01}+n_{10}} B(i; n_{01} + n_{10}, 1/2)$ for $H_0 : \pi_{+1} \geq \pi_{1+}$. Both test statistics depend only on so-called "discordant pairs," which means on cases classified in different categories for the two observations. The observations $n_{00} + n_{11}$ are considered as irrelevant for inference (see also next section).

Likelihood Ratio Statistic and Alternatives

The counts $\{n_{ij}\}$ are tested as a sample from the multinomial distribution $M(n, \{\pi_{ij}\})$. When investigating the underlying probabilities it is helpful to use that the multinomial distribution can be expressed as a product of the binomial distributions (see also Lloyd, 2008). This means that the probability mass function of the multinomial, denoted by $m(n_{11}, n_{10}, n_{01}, n_{11})$, may be expressed as

$$b(t; n, \phi) b(n_{01}; t, \eta) b(n_{11}; n - t, \psi), \tag{13.19}$$

where $b(x; n, \pi)$ denotes the probability of outcomes x in a binomial distribution $B(n, \pi)$, and

$t = n_{01} + n_{10}$ is the number of discordant responses;

$\phi = \pi_{01} + \pi_{10}$ is the probability of discordant responses, $P(y_1 \neq y_2)$;

$\eta = (\pi_{01})/(\pi_{01} + \pi_{10})$ is the probability that a discordant response favors $y_2 = 1, P(y_2 = 1|y_1 \neq y_2)$;

$\psi = (\pi_{11})/(\pi_{11} + \pi_{00})$ is the probability that response 1 is favored when the responses are equal, $P(y_1 = y_2 = 1|y_1 = y_2)$.

The first term in (13.19) corresponds to the occurrence of discordant pairs. Conditional on this, the second and third terms correspond to counts n_{01} within the t discordant pairs and counts n_{11} within the $n - t$ concordant pairs. The decomposition (13.19) is a representation that turns the original parameters $\{\pi_{ij}\}$ into the parameters ϕ, η, ψ. To test the hypothesis $H_0 : \delta = 0$ it is important that the parameter η may also be written as

$$\eta = \frac{\delta + \phi}{2\phi}.$$

With $t = n_{01} + n_{10}$, the likelihood in this parameterization is given by

$$L(\phi, \eta, \psi) = \phi^t (1-\phi)^{n-t} \eta^{n_{01}} (1-\eta)^{t-n_{01}} \psi^{n_{11}} (1-\psi)^{n-t-n_{11}}.$$

The first two blocks contain the parameters ϕ and η, or, equivalently δ and ϕ, while the last block contains only $\psi = P(y_1 = y_2 = 1 | y_1 = y_2)$, which does not contribute to differences in the marginal distributions. Therefore one considers the reduced likelihood by omitting the last term $\psi^{n_{11}}(1-\psi)^{n-t-n_{11}}$. By using δ, ϕ rather than η, ϕ one obtains the likelihood

$$\begin{aligned} L(\phi, \delta) &= \phi^t (1-\phi)^{n-t} (\delta + \phi)^{n_{01}} (\phi - \delta)^{t-n_{01}} (2\phi)^{-t} \\ &= 2^{-t} (1-\phi)^{n-t} (\delta + \phi)^{n_{01}} (\phi - \delta)^{t-n_{01}}. \end{aligned}$$

It is easily shown that maximization yields the estimates $\hat{\delta} = (2n_{01} - t)/n = (n_{01} - n_{10})/n$, $\hat{\phi} = t/n$ (Exercise 13.6). The likelihood ratio test

$$\lambda = 2(l(\hat{\delta}, \hat{\phi}) - l(0, \hat{\phi}_0))$$

uses the estimate $\hat{\phi}_0$, which is obtained under the restriction $\delta = 0$. Simple computation shows that $\hat{\phi} = \hat{\phi}_0$. One obtains

$$\begin{aligned} \lambda &= 2n_{01} \log(n_{01}) + 2n_{10} \log(n_{10}) - 2t \log(t) + t \log(2) \\ &= 2n_{01} \log\left(\frac{2n_{01}}{n_{01} + n_{10}}\right) + 2n_{10} \log\left(\frac{2n_{10}}{n_{01} + n_{10}}\right), \end{aligned}$$

which may be compared to a $\chi^2(1)$-distribution.

McNemar's test statistic and the likelihood ratio statistic are asymptotically equivalent, although they may differ for small samples. For example, one obtains $M = 32.01$ and $\lambda = 34.80$ for Table 13.6. But for Table 13.5, which contains fewer observations, the statistics are not comparable since λ does not exist. Both test statistics approximate the p-value by using the asymptotic distribution. There has been some effort to obtain more concise p-values. Approaches differ in the way they handle the nuisance parameter ϕ. When the tests are viewed unconditionally, the distributions of the test statistics depend on ϕ. One strategy is to maximize the p-value over ϕ (Suissa and Shuster, 1991). Alternatively, one may use partial maximization over ϕ with a penalty on the range of maximization (Berger and Sidik, 2003). Methods that replace ϕ by the maximum likelihood estimate under the null hypothesis $\hat{\phi}$ and then maximize the p-value have been proposed by Lloyd (2008), who also gives an overview of exact methods.

13.4.2 Regression Approach to Marginal Homogeneity

Let y_1, y_2 denote the pair of binary measurements with $y_t \in \{0, 1\}$ that produces the counts in the (2×2)-table. A marginal regression model for the two measurements is given by

$$P(y_t = 1) = \delta_0 + x_t \delta, \tag{13.20}$$

where $x_1 = 0, x_2 = 1$. Simple computation shows that

$$\delta = P(y_2 = 1) - P(y_1 = 1) = \pi_{+1} - \pi_{1+}$$

is equivalent to the distance considered in Section 13.4.1. The parameter δ_0 is determined as $\delta_0 = P(y_1 = 1) = \pi_{+1}$, yielding $\delta_0 - \delta = \pi_{1+}$.

Of course, for dependent variables, the regression model is not fully specified by δ_0 and δ since δ_0 and δ determine only the marginal distributions $P(y_1 = 1)$ and $P(y_2 = 1)$, and not

$P(y_1 = y_2 = 1)$. The additional parameter $\phi = \pi_{01} + \pi_{10}$ completes the specification. The model may be represented by δ_0, δ, ϕ or ϕ, η, ψ from the previous section. Model (13.20) is a linear model for probabilities. Although the parameters have a simple interpretation in terms of probabilities, it is often preferable to use alternative parameterizations. Instead of a linear model one can consider the *marginal logistic model*

$$\text{logit}(P(y_t = 1)) = \beta_0 + x_t\beta.$$

Then the parameter β has the form

$$\beta = \log\left(\frac{P(y_2 = 1)/(1 - P(y_2 = 1))}{P(y_1 = 1)/(1 - P(y_1 = 1))}\right),$$

which represents odds ratios built from marginal distributions. Although the parameterization is different from the linear model, the hypothesis of marginal homogeneity in both cases is equivalent to the vanishing of the parameter that is connected to x_t ($H_0 : \delta = 0$ for the linear, $H_0 : \beta = 0$ for the logit model). A full parameterization is determined by specifying appropriately an additional parameter. Then the likelihood estimation would refer to an alternative set of parameters, but essentially the inference would be the same as for the linear model. The advantage of the logistic model is a different one. It is better suited for conditional approaches, which are considered in the next section.

Conditional Logit Models

Although it is not directly observed, in general one has to assume that individuals differ in more aspects than are modeled in a marginal model, where it is only distinguished between the first and second responses, that is, $x_t = 0$ or $x_t = 1$. The natural heterogeneity may be modeled explicitly by assuming that each observation has its own *subject-specific* parameter. Models of this type were considered by Cox (1958) for matched pairs and in psychometrics (Rasch (1961)).

Now let (y_{i1}, y_{i2}) denote the pair of observations collected for subject i. A model that allows for heterogeneity among subjects is the subject-specific conditional model:

$$\text{logit}(P(y_{it} = 1)) = \beta_{0i} + x_t\beta,$$

where β_{0i} is a subject-specific parameter and $P(y_{it} = 1)$ is the conditional probability given subject i. One obtains

$$\beta = \log\left(\frac{P(y_{i2} = 1)/(1 - P(y_{i2} = 1))}{P(y_{i1} = 1)/(1 - P(y_{i1} = 1))}\right),$$

which does not depend on i, and

$$\beta_{0i} = \log\left(\frac{P(y_{i1} = 1)}{1 - P(y_{i1} = 1)}\right),$$

which depends on i. The parameter β, which quantifies the effect of the measurement, is a common effect and does not vary over observations. The hypothesis of marginal homogeneity is equivalent to $\beta = 0$. If $\beta = 0$ holds, one has

$$\beta_{0i} = \log\left(\frac{P(y_{i1} = 1)}{1 - P(y_{i1} = 1)}\right) = \log\left(\frac{P(y_{i2} = 1)}{1 - P(y_{i2} = 1)}\right).$$

This means that the marginals for the two measurements may vary across subjects but are identical given the subject.

Parameter estimation for the subject-specific model is difficult because the number of parameters can be huge. Before considering fitting procedures, let us extend the model slightly by including the effects of additional covariates. The subject-specific model with covariates has the form

$$\text{logit}(P(y_{it} = 1)) = \beta_{0i} + \boldsymbol{x}_{it}^T\boldsymbol{\beta}, \tag{13.21}$$

where $\boldsymbol{x}_{it}$ is a vector of predictors that varies across subjects and measurements.

In the following we will consider the *conditional maximum likelihood.* The alternative estimation procedure that is based on the assumption of *random effects* will be considered later, in Chapter 14.

Conditional Maximum Likelihood Estimation

Given the parameters, a common assumption is that the two measurements are independent. This means that the association between the measurements that is found on the population level is due to the heterogeneity of the observations. On the subject level, given the parameters, the observations are independent. Then the probability for the occurrence of measurements y_{i1}, y_{i2} is

$$\begin{aligned} P(y_{i1}, y_{i2}) &= \left(\frac{e^{\beta_{0i}+\boldsymbol{x}_{i1}^T\boldsymbol{\beta}}}{1+e^{\beta_{0i}+\boldsymbol{x}_{i1}^T\boldsymbol{\beta}}}\right)^{y_{i1}} \left(\frac{1}{1+e^{\beta_{0i}+\boldsymbol{x}_{i1}^T\boldsymbol{\beta}}}\right)^{1-y_{i1}} \\ &\times \left(\frac{e^{\beta_{0i}+\boldsymbol{x}_{i2}^T\boldsymbol{\beta}}}{1+e^{\beta_{0i}+\boldsymbol{x}_{i2}^T\boldsymbol{\beta}}}\right)^{y_{i2}} \left(\frac{1}{1+e^{\beta_{0i}+\boldsymbol{x}_{i2}^T\boldsymbol{\beta}}}\right)^{1-y_{i2}} \\ &= c_i \quad e^{(y_{i1}+y_{i2})\beta_{0i}+(y_{i1}\boldsymbol{x}_{i1}^T+y_{i2}\boldsymbol{x}_{i2}^T)\boldsymbol{\beta}}, \end{aligned}$$

where $c_i = (1+e^{\beta_{0i}+\boldsymbol{x}_{i1}^T\boldsymbol{\beta}})^{-1}(1+e^{\beta_{0i}+\boldsymbol{x}_{i2}^T\boldsymbol{\beta}})^{-1}$. Conditional maximum likelihood estimation uses that the parameters β_{0i} may be eliminated by conditioning on the sufficient statistic $t_i = y_{i1}+y_{i2}$, which is on the subject level. There are two uninteresting cases, namely, $t_i = 0$, since then $P(y_{i1} = y_{i2} = 0|y_{i1} + y_{i2} = 0) = 1$, and $t_i = 2$, since then $P(y_{i1} = y_{i2} = 1|y_{i1} + y_{i2} = 2) = 1$. For the case $y_{i1} + y_{i2} = 1$ one derives

$$P(y_{i1} + y_{i2} = 1) = P(y_{i1} = 1, y_{i2} = 0) + P(y_{i1} = 0, y_{i2} = 1) = c_i e^{\beta_{0i}}(e^{\boldsymbol{x}_{i1}^T\boldsymbol{\beta}} + e^{\boldsymbol{x}_{i2}^T\boldsymbol{\beta}}),$$

yielding

$$\begin{aligned} P((y_{i1}, y_{i2}) = (0,1)|y_{i1} + y_{i2} = 1) &= \exp(\boldsymbol{x}_{i2}^T\boldsymbol{\beta})/[\exp(\boldsymbol{x}_{i1}^T\boldsymbol{\beta}) + \exp(\boldsymbol{x}_{i2}^T\boldsymbol{\beta})], \\ P((y_{i1}, y_{i2}) = (1,0)|y_{i1} + y_{i2} = 1) &= \exp(\boldsymbol{x}_{i1}^T\boldsymbol{\beta})/[\exp(\boldsymbol{x}_{i1}^T\boldsymbol{\beta}) + \exp(\boldsymbol{x}_{i2}^T\boldsymbol{\beta})]. \end{aligned}$$

The resulting conditional probabilities specify a binary logit model with outcomes $(1,0)$ and $(0,1)$. By defining $y_i^* = 1$ if $(y_{i1}, y_{i2}) = (0,1)$ and $y_i^* = 0$ if $(y_{i1}, y_{i2}) = (1,0)$, one obtains

$$P(y_i^* = 1|y_{i1} + y_{i2} = 1) = \frac{\exp((\boldsymbol{x}_{i2} - \boldsymbol{x}_{i1})^T\boldsymbol{\beta})}{1+\exp((\boldsymbol{x}_{i2} - \boldsymbol{x}_{i1})^T\boldsymbol{\beta})},$$

which is a binary logit model with predictor $\boldsymbol{x}_{i2} - \boldsymbol{x}_{i1}$. The conditional likelihood, given $t_i = y_{i1} + y_{i2} = 1$, has the form

$$L_c(\boldsymbol{\beta}) = \prod_{(y_{i1},y_{i2})=(0,1)} \frac{\exp(\boldsymbol{x}_{i2}^T\boldsymbol{\beta})}{\exp(\boldsymbol{x}_{i1}^T\boldsymbol{\beta}) + \exp(\boldsymbol{x}_{i2}^T\boldsymbol{\beta})} \prod_{(y_{i1},y_{i2})=(1,0)} \frac{\exp(\boldsymbol{x}_{i1}^T\boldsymbol{\beta})}{\exp(\boldsymbol{x}_{i1}^T\boldsymbol{\beta}) + \exp(\boldsymbol{x}_{i2}^T\boldsymbol{\beta})}$$

$$= \prod_{i:t_i=1} \left(\frac{\exp((\boldsymbol{x}_{i2}-\boldsymbol{x}_{i1})^T\boldsymbol{\beta})}{1+\exp((\boldsymbol{x}_{i2}-\boldsymbol{x}_{i1})^T\boldsymbol{\beta})}\right)^{y_i^*} \left(\frac{1}{1+\exp((\boldsymbol{x}_{i2}-\boldsymbol{x}_{i1})^T\boldsymbol{\beta}}\right)^{1-y_i^*}.$$

Thus *conditional maximum likelihood estimates* may be obtained by fitting a logistic model with an artificial response y_i^* and a predictor given by $\boldsymbol{x}_{i2} - \boldsymbol{x}_{i1}$ without an intercept (see also Breslow et al., 1978, Agresti, 2002).

It should be mentioned that it is essential that the covariates $\boldsymbol{x}_{it}$ vary with t. For covariates that characterize only the subject but not the measurement one obtains $\boldsymbol{x}_{is} - \boldsymbol{x}_{it} = \boldsymbol{0}$ and no inference on $\boldsymbol{\beta}$ is possible. When estimating $\boldsymbol{\beta}$ by conditioning on sufficient statistics, variables that depend only on i are eliminated together with the nuisance parameters β_{0i}. The effect may also be seen by considering two sets of variables, $\tilde{\boldsymbol{x}}_i$, which does not depend on i, and $\boldsymbol{x}_{it}$, which varies across measurements. Then the corresponding logit model $\text{logit}(P(y_{it} = 1)) = \beta_{0i} + \tilde{\boldsymbol{x}}_i^T\tilde{\boldsymbol{\beta}} + \boldsymbol{x}_{it}^T\boldsymbol{\beta}$ can be reparameterized to $\text{logit}(P(y_{it} = 1)) = \tilde{\beta}_{0i} + \boldsymbol{x}_{it}^T$, where $\tilde{\beta}_{0i} + \tilde{\boldsymbol{x}}_i^T\tilde{\boldsymbol{\beta}}$ is the subject-specific effect that includes the subject-specific effects $\tilde{\boldsymbol{x}}_i^T\tilde{\boldsymbol{\beta}}$, which do not vary across measurements.

For the simple case, where $x_{i1} = 0$ for measurement 1 and $x_{i2} = 1$ for measurement 2, the conditional likelihood, conditional on $t_i = 1$, simplifies to

$$L_c(\beta) = \prod_{i:t_i=1} \left(\frac{\exp(\beta)}{1+\exp(\beta)}\right)^{y_{i2}} \left(\frac{1}{1+\exp(\beta)}\right)^{y_{i1}} = \frac{\exp(\beta)^{n_{01}}}{(1+\exp(\beta))^{n_{10}+n_{01}}},$$

yielding the conditional ML estimate and standard error

$$\hat{\beta} = \log\left(\frac{n_{01}}{n_{10}}\right), \qquad s(\hat{\beta}) = \sqrt{1/n_{01} + 1/n_{10}}.$$

For Table 13.6 one obtains $\hat{\beta} = 0.67$ with a standard error 0.305. The estimated marginal odds ratio $\exp(\hat{\beta}) = 4.7$ suggests that the odds for talking about problems with a partner are much higher for female patients.

The advantage of the conditional approach to estimating parameters is that simple logit models may be used for estimation; the drawback is that only within-cluster effects can be estimated. Between-cluster effects, which vary across clusters, are lost by conditioning on the subject. An alternative approach that allows one to estimate within-cluster effects and models heterogeneity over subjects is provided by random effects models, which are considered in Chaper 14.

13.4.3 Marginal Homogeneity for Multicategorical Outcome

Let the two measurements taken from the same individual or a matched pair be given by (Y_1, Y_2), where $Y_t \in \{1, \ldots, k\}$. The observations may be summarized in a square $(k \times k)$-table containing the number of individuals with responses from $\{(1,1), \ldots, (k,k)\}$. Let n_{ij} denote the number of pairs with outcomes (i,j), $p_{ij} = n_{ij}/n$ the sample proportion, and $\pi_{ij} = P(y_1 = i, y_2 = j)$ the underlying probability of outcomes (i,j). Then the counts $\{n_{ij}\}$, which form the contingency table, are a sample from the multinomial distribution $M(n, \{\pi_{ij}\})$.

Marginal homogeneity holds if

$$P(Y_1 = r) = P(Y_2 = r) \qquad \text{for all} \quad r.$$

By use of logit models for the marginal distributions,

$$\log(P(Y_t = r)/P(Y_t = k)) = \beta_{0r} + x_t\beta_r, \qquad \text{for all} \quad r, \tag{13.22}$$

where $x_1 = 0, x_2 = 1$, and the marginal homogeneity hypothesis corresponds to

$$H_0 : \beta_1 = \cdots = \beta_k.$$

Fitting of the marginal homogeneity model is obtained by maximizing the multinomial likelihood subject to the constraints (13.22). The full model may be compared to the constrained model by the use of likelihood ratio tests (see Lipsitz et al., 1990). For alternative test strategies see Bhapkar (1966).

For ordered response categories a sparser representation is obtained by using cumulative logit models for the marginals. Also, tests may profit from using the order information in the marginals, resulting in more powerful tests (see Agresti, 2009). The conditional ML approach for multinomial margins was discussed by Conaway (1989).

Symmetry and Quasi-Symmetry

Marginal homogeneity without covariates can also be tested within the log-linear models framework by using *quasi-symmetry log-linear models*. With $\pi_{rs} = P(Y_1 = r, Y_2 = s)$ denoting the probability for an observation in cell (r, s) of a square contingency table, the quasi-symmetry model has the form

$$\log(\pi_{rs}) = \lambda + \lambda_{1(r)} + \lambda_{2(s)} + \lambda_{12(rs)},$$

where $\lambda_{12(rs)} = \lambda_{12(sr)}$. When, in addition, $\lambda_{1(r)} = \lambda_{2(r)}$ holds, one obtains symmetry, that is, $P(Y_1 = r, Y_2 = s) = P(Y_1 = s, Y_2 = r)$. The corresponding *symmetry log-linear model* may also be given in the simpler form

$$\log(\pi_{rs}) = \lambda + \lambda_r + \lambda_s + \lambda_{12(rs)}.$$

It can be shown that symmetry is equivalent to quasi-symmetry and marginal homogeneity holding simultaneously (see Agresti, 2002). Therefore, given that quasi-symmetry holds, a test of symmetry may be composed by comparing the fit of the quasi-symmetry model with the fit of the symmetry model. The difference of deviances,

$$D(\text{Symmetry}|\text{Quasi-symmetry}) = D(\text{Symmetry}) - D(\text{Quasi-symmetry}),$$

follows asymptotically a χ^2-distribution with $k - 1$ degrees of freedom.

Quasi-symmetry models have been further investigated by Conaway (1989), Agresti (1997), Tjur (1982), and Goodman (1968). Ordinal quasi-symmetry models were discussed by Agresti (1993b). For a concise overview, see Agresti (2002).

13.5 Further Reading

Surveys on Multivariate Models. Molenberghs and Verbeke (2005) discuss various models for discrete longitudinal data. Bergsma et al. (2009) gave an account of the full maximum likelihood approach to marginal models. Matched samples and correlated binary data are also considered in Fleiss et al. (2003). As always, the books of Agresti (2002, 2009) give an excellent overview.

Further Marginal Models for Ordered Responses. Dale (1986) proposed a marginal regression model for bivariate ordered responses that is based on the Plackett distribution. The model was extended to the multi-dimensional case by Molenberghs and Lesaffre (1994).

Semiparametric Marginal Models. Semiparametric approaches to the modeling of longitudinal data based on kernel methods was considered by Lin and Carroll (2006); Welsh et al. (2002); and Zhu, Fung, and He (2008).

Variable Selection for Marginal Models. Cantoni et al. (2005) proposed a variable selection procedure that is based on a generalized version of Mallow's C_p.

R Packages. Simple marginal binary regression with two responses and a specified odds-ratio can be fitted by using the R-function *vglm* from the versatile package *VGAM.* Marginal models based on GEEs are available in the library *gee.* Alternatively, one can use the package *geepack,* which provides functions for the fitting of binary marginal models (function *geeglm*) and ordinal responses (*ordgee*).

13.6 Exercises

13.1 A simple (2×2)-contingency table can be parameterized in various ways. One possibility is in probabilities $\pi_{11}, \pi_{10}, \pi_{01}, \pi_{00}$, where $\pi_{ij} = P(y_1 = i, y_2 = j)$. Alternatively, one can use the marginals together with the odds ratio $\{\pi_{1+}, \pi_{+1}, \gamma\}$ or the parameters $\lambda_1, \lambda_2, \lambda_{12}$ of the log-linear model. Yet another parameterization uses $\phi = \pi_{01} + \pi_{10}$, $\eta = \frac{\pi_{01}}{\pi_{01}+\pi_{10}}$, $\psi = \frac{\pi_{11}}{\pi_{11}+\pi_{00}}$.

(a) Discuss the use of these parametrizations with respect to modeling problems. What parameterization is to be preferred in which application?

(b) Show how the parameterizations transform into each other.

13.2 For three binary variables in (0-1)-coding the logit model with y_1, y_2 as the response variables and y_3 as the covariate can be given in the form

$$\log\left(\frac{P(y_1, y_2|y_3)}{P(y_1 = 0, y_2 = 0|y_3)}\right) = y_1\lambda_1 + y_2\lambda_2 + y_1y_2\lambda_{12} + y_1y_3\lambda_{13} + y_2y_3\lambda_{23} + y_1y_2y_3\lambda_{123}.$$

(a) Give the parameters λ_1, λ_2, and λ_{13} as functions of probabilities.

(b) Give an alternative model that uses marginal parameterization and compare the interpretation of the parameters of the two models.

13.3 Show that the estimates based on the generalized estimation equation (13.12) are equal for the independence and the equicorrelation model if all the covariates are cluster-specific and the number of observations is the same for all clusters.

13.4 The epilepsy dataset (Example 13.1) is available at http://www.biostat.harvard.edu/ fitzmaur/ala.

(a) Show the number of seizures for the four two-week periods in a boxplot.

(b) Fit marginal models with alternative correlation structures and investigate the effect of treatment.

13.5 The R package *Fahrmeir* contains the dataset ohio, which is a subset from the Harvard Study of Air Pollution and Health (Laird et al., 1984). For 537 children from Ohio, examined annually from ages 7 to 10, binary responses, with $y_{it} = 1$ for the presence and $y_{it} = 0$ for the absence of respiratory infection, are given. Fit marginal models that model the influence of mother's smoking status and of age on children's respiratory infection. Try several working covariances and investigate if an interaction effect of mother's smoking status and age is needed.

13.6 Let $m(n_{11}, n_{10}, n_{01}, n_{11})$ denote the probability mass function of the multinomial distribution for a (2×2)-contingency table, $M(n, \{\pi_{ij}\})$.

(a) Show that the probability mass function of the multinomial distribution can be expressed as

$$b(t; n, \phi)b(n_{01}; t, \eta)b(n_{11}; n - t, \psi),$$

where $b(x; n, \pi)$ denotes the probability of outcomes x in a binomial distribution $B(n, \pi)$, $t = n_{01} + n_{10}$, $\phi = \pi_{01} + \pi_{10}$, $\eta = (\pi_{01})/(\pi_{01} + \pi_{10})$, $\psi = (\pi_{11})/(\pi_{11} + \pi_{00})$.

(b) Show that maximization of the reduced likelihood $L(\phi, \delta) = 2^{-t}(1-\phi)^{n-t}(\delta+\phi)^{n_{01}}(\phi-\delta)^{t-n_{01}}$ yields the ML estimates $\hat{\delta} = (n_{01} - n_{10})/n$, $\psi = (\pi_{11})/(\pi_{11} + \pi_{00})$.

Chapter 14

Random Effects Models and Finite Mixtures

In Chapter 13 the marginal modeling approach has been used to model observations that occur in clusters. An alternative approach to dealing with repeated measurements is by modeling explicitly the heterogeneity of the clustered responses. By postulating the existence of unobserved latent variables, the so-called random effects, which are shared by the measurement within a cluster, one introduces correlation between the measurements within clusters.

The introduction of cluster-specific parameters has consequences on the interpretation of parameters. Responses are modeled given covariates and cluster-specific terms. Therefore, interpretation is *subject-specific*, in contrast to marginal models, which have *population-averaged* interpretations. For illustration let us consider Example 13.4, where a binary response indicating pain depending on treatment and other covariates is measured repeatedly. When each individual has its own parameter, which represents the individual's sensitivity to pain, modeling of the response given the covariates *and* the individual level means that effects are measured on the individual level. For non-linear models, which are the standard in categorical regressions, the effect strength will differ from the effect strength found in marginal modeling without subject-specific parameters. The difference will be discussed in more detail in Section 14.2.1 for the simple case of binary response models with random intercepts.

Explicit modeling of heterogeneity by random effects is typically found in repeated measurements, as in the pain study. In the following we give two more examples.

Example 14.1: AIDS Study

The data were collected within the Multicenter AIDS Cohort Study (MACS), which has followed nearly 5000 gay or bisexual men from Baltimore, Pittsburgh, Chicago, and Los Angeles since 1984 (see Kaslow et al., 1987; Zeger and Diggle, 1994). The study includes 1809 men who were infected with HIV when the study began and another 371 men who were seronegative at entry and seroconverted during the follow-up. In our application we use 369 seroconverters with 2376 measurements over time. The interesting response variable is the number or percent of CD4 cells, by which progression of disease may be assessed. The covariates include years since seroconversion, packs of cigarettes a day, recreational drug use (yes/no), number of sexual partners, age, and a mental illness score. The main interest is in the typical time course of CD4 cell decay and the variability across subjects (see also Zeger and Diggle, 1994). Figure 14.1 shows the data together with an estimated overall smooth effect of time on CD4 cell decay. For more details see Example 14.6. □

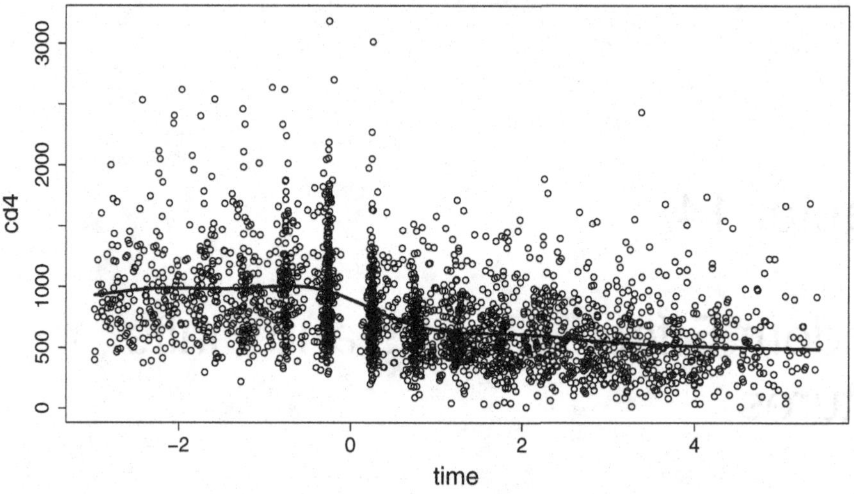

FIGURE 14.1: Data from Multicenter AIDS Cohort Study (MACS) and smoothed time effect.

Example 14.2: Recovery Scores
Davis (1991) considered post-surgery recovery data. In a randomized study, 60 children undergoing surgery were treated with one of four dosages of an anaesthetic. Upon admission to the recovery room and at minutes 5, 15, and 30 following admission, recovery scores were assigned on a categorical scale ranging from 1 (least favorable) to 7 (most favorable). Therefore, one has four repetitions of a variable having 7 categories. One wants to model how recovery scores depend on covariables as dosage of the anaesthetic (four levels), duration of surgery (in minutes), and age of the child (in months). □

Random effects models are a flexible tool for modeling correlated data. However, one consequence of flexibility is that additional choices are necessary. In addition to the usual modeling issues concerning the choice of the link function and the specifications of the explanatory variables, one has to specify which effects have to be considered random and how they are distributed.

Although we are focussing on discrete responses, it is helpful to consider first the special case of a linear mixed model with normally distributed responses. Some of the estimation concepts derived for the classical normal distribution case can be used as building blocks for the estimation of generalized mixed models.

14.1 Linear Random Effects Models for Gaussian Data

14.1.1 Random Effects for Clustered Data

Let $\boldsymbol{y}_i^T = (y_{i1,\dots,}y_{iT_i})$ denote the vector of observations on unit i $(i = 1, \dots, n)$ and $\boldsymbol{x}_{it}, \boldsymbol{z}_{it}$ denote the covariates associated with response y_{it}.

A linear random effects model can be defined as a two-stage model. At the *first stage* one assumes that the normally distributed response is specified by

$$y_{it} = \boldsymbol{x}_{it}^T \boldsymbol{\beta} + \boldsymbol{z}_{it}^T \boldsymbol{b}_i + \varepsilon_{it}, \tag{14.1}$$

where $\boldsymbol{x}_{it}, \boldsymbol{z}_{it}$ are design vectors built from covariates, $\boldsymbol{\beta}$ is a population-specific parameter, and $\boldsymbol{b}_i$ is a cluster-specific effect. It is assumed that the noise variables are centered, $E(\varepsilon_{it}) = 0$, homogeneous, $var(\varepsilon_{it}) = \sigma^2$, and uncorrelated $cov(\varepsilon_{is}, \varepsilon_{it}) = 0, s \neq t$. Moreover, ε_{it} is normally distributed. More precisely, in (14.1) it is assumed that y_{it} given $\boldsymbol{x}_{it}, \boldsymbol{z}_{it}, \boldsymbol{b}_i$ is normally distributed, that is,

$$y_{it}|\boldsymbol{x}_{it}, \boldsymbol{z}_{it}, \boldsymbol{b}_i \sim N(\mu_{it}, \sigma^2),$$

where $\mu_{it} = \boldsymbol{x}_{it}^T\boldsymbol{\beta} + \boldsymbol{z}_{it}^T\boldsymbol{b}_i$ denotes the structural term. In matrix form one obtains

$$\boldsymbol{y}_i = \boldsymbol{X}_i\boldsymbol{\beta} + \boldsymbol{Z}_i\boldsymbol{b}_i + \boldsymbol{\varepsilon}_i, \tag{14.2}$$

where $\boldsymbol{y}_i^T = (y_{i1}, \dots, y_{iT_i}), \boldsymbol{X}_i^T = (\boldsymbol{x}_{i1}, \dots, \boldsymbol{x}_{iT_i}), \boldsymbol{Z}_i^T = (\boldsymbol{z}_{i1}, \dots, \boldsymbol{z}_{iT_i})$ and $\boldsymbol{\varepsilon}_i^T = (\varepsilon_{i1}, \dots, \varepsilon_{iT_i})$ is the vector of within-cluster errors, $\boldsymbol{\varepsilon}_i \sim N(\boldsymbol{0}, \sigma^2\boldsymbol{I})$.

At the *second stage* it is assumed that the cluster-specific effects $\boldsymbol{b}_i$ vary independently across clusters, where the common assumption is the normal distribution:

$$\boldsymbol{b}_i \sim N(\boldsymbol{0}, \boldsymbol{Q}). \tag{14.3}$$

In addition it is assumed that $\boldsymbol{\varepsilon}_i$ and $\boldsymbol{b}_i$ are uncorrelated.

Since the conditional distribution of $\boldsymbol{y}_i$ and the random effect $\boldsymbol{b}_i$ are normally distributed, one obtains for the distribution of $\boldsymbol{y}_i$ (given $\boldsymbol{X}_i, \boldsymbol{Z}_i$) the *marginal version* of the random effects model:

$$\boldsymbol{y}_i = \boldsymbol{X}_i\boldsymbol{\beta} + \boldsymbol{\varepsilon}_i^*, \tag{14.4}$$

where $\boldsymbol{\varepsilon}_i^* \sim N(\boldsymbol{0}, \sigma^2\boldsymbol{I} + \boldsymbol{Z}_i\boldsymbol{Q}\boldsymbol{Z}_i^T)$. The corresponding linear regression model is heteroscedastic with a specific covariance structure $\sigma^2\boldsymbol{I} + \boldsymbol{Z}_i\boldsymbol{Q}\boldsymbol{Z}_i^T$ (Exercise 14.1).

Gaussian Random Effects Model for Clustered Data

$$\boldsymbol{y}_i = \boldsymbol{X}_i\boldsymbol{\beta} + \boldsymbol{Z}_i\boldsymbol{b}_i + \boldsymbol{\varepsilon}_i, \boldsymbol{b}_i \sim N(\boldsymbol{0}, \boldsymbol{Q}), \boldsymbol{\varepsilon}_i \sim N(\boldsymbol{0}, \sigma^2\boldsymbol{I})$$

Marginal Version

$$\boldsymbol{y}_i = \boldsymbol{X}_i\boldsymbol{\beta} + \boldsymbol{\varepsilon}_i^*, \boldsymbol{\varepsilon}_i^* \sim N(\boldsymbol{0}, \boldsymbol{V}_i), \boldsymbol{V}_i = \sigma^2\boldsymbol{I} + \boldsymbol{Z}_i\boldsymbol{Q}\boldsymbol{Z}_i^T$$

Random Intercept Model

A simple case is the random effects analysis of covariance model, where the only random effects are the random intercepts. Therefore, each cluster/individual has its own response level. The model is given by

$$y_{it} = \beta_0 + \boldsymbol{x}_{it}^T\boldsymbol{\gamma} + b_i + \varepsilon_{it} = (1, \boldsymbol{x}_{it})^T\boldsymbol{\beta} + b_i + \varepsilon_{it},$$

with $b_i \sim N(0, \sigma_b^2)$ independent of $\varepsilon_{it} \sim N(0, \sigma^2)$. Then the cluster-specific intercept $\beta_0 + b_i$ has mean β_0 and variance σ_b^2. It is easily derived that the covariance of $\boldsymbol{y}_i$(given $\boldsymbol{X}_i$) is

$$cov(\boldsymbol{y}_i) = \sigma^2\boldsymbol{I} + \sigma_b^2\boldsymbol{1}\boldsymbol{1}^T = \begin{pmatrix} \sigma^2 + \sigma_b^2 & \sigma_b^2 & \cdots & \sigma_b^2 \\ \sigma_b^2 & \sigma^2 + \sigma_b^2 & \ddots & \\ & & & \sigma^2 + \sigma_b^2. \end{pmatrix}.$$

Thus the observations in the same cluster are correlated. One has

$$var(y_{it}) = \sigma^2 + \sigma_b^2,\ cov(y_{is}, y_{it}) = \sigma_b^2 \quad \text{for} \quad s \neq t.$$

The variances σ^2 and σ_b^2 are called elementary- and cluster-level variance components, respectively. For $s \neq t$ one obtains the *intraclass correlation* coefficient $\rho(y_{is}, y_{it}) = \sigma_b^2/(\sigma^2 + \sigma_b^2)$, which is also called the variance components ratio.

14.1.2 General Linear Mixed Model

The model for clustered data is a special case of the more general *linear mixed model* (LMM):

$$\boldsymbol{y} = \boldsymbol{X\beta} + \boldsymbol{Zb} + \boldsymbol{\varepsilon}, \begin{bmatrix} \boldsymbol{b} \\ \boldsymbol{\varepsilon} \end{bmatrix} \sim N\left(\begin{bmatrix} \mathbf{0} \\ \mathbf{0} \end{bmatrix}, \begin{bmatrix} \boldsymbol{Q}_b & \mathbf{0} \\ \mathbf{0} & \boldsymbol{R} \end{bmatrix}\right), \tag{14.5}$$

where the covariances of the random effect vector and the noise are given by $\text{cov}(\boldsymbol{b}) = \boldsymbol{Q}_b$ and $\text{cov}(\boldsymbol{\varepsilon}) = \boldsymbol{R}$. In (14.5), the vector $\boldsymbol{y}$ has length N with $\boldsymbol{X}$ and $\boldsymbol{Z}$ having proper dimensions. The corresponding *marginal* version of the model, which shows how the responses are correlated, has the form

$$\boldsymbol{y} = \boldsymbol{X\beta} + \boldsymbol{\varepsilon}^*,$$

where $\boldsymbol{\varepsilon}^* \sim N(\mathbf{0}, \boldsymbol{V})$, with covariance matrix $\boldsymbol{V} = \boldsymbol{ZQ}_b\boldsymbol{Z}^T + \boldsymbol{R}$.

It is easily seen that the random effects model (14.2) is a special mixed effects model. By defining $\boldsymbol{y}^T = (\boldsymbol{y}_1^T, \dots, \boldsymbol{y}_n^T), \boldsymbol{b}^T = (\boldsymbol{b}_1^T, \dots, \boldsymbol{b}_n^T)$, $\boldsymbol{X}^T = (\boldsymbol{X}_1^T, \dots, \boldsymbol{X}_n^T)$, $\boldsymbol{Z} = Diag(\boldsymbol{Z}_1, \dots, \boldsymbol{Z}_n)$, $\boldsymbol{\varepsilon}^T = (\boldsymbol{\varepsilon}_1^T, \dots, \boldsymbol{\varepsilon}_n^T)$, $\boldsymbol{Q}_b = Diag(\boldsymbol{Q}, \dots, \boldsymbol{Q})$, $\boldsymbol{R} = Diag(\sigma^2\boldsymbol{I}, \dots, \sigma^2\boldsymbol{I})$, one obtains the general form (14.5) of the mixed model. The number of observations in (14.5) is $N = T_1 + \dots + T_n$.

The general linear model includes multilevel models that are needed when clustering occurs on more than one level and the clusters are hierarchically nested. For example, in educational studies on the first level the classes form clusters, and on the second levels the students form clusters (within classes). Repeated measurements taken on the student level can include class-specific and student-specific effects. The general model also includes crossed effects models. For example, in experimental studies where a sample of subjects has to solve a set of tasks repeatedly, it can be appropriate to include random effects for the subjects and the task, obtaining an additive term that includes two random effects. Then, with a multiple index, one row of the general model has the form $y_{ij} = \boldsymbol{x}_{ij} + b_{1i} + b_{2j} + \varepsilon_{ij}$, where b_{1i} refers to the subjects and b_{2j} refers to the tasks. Here the focus is on clustered data, but the general model is useful for a closed representation.

Linear Mixed Model (LMM)

$$\boldsymbol{y} = \boldsymbol{X\beta} + \boldsymbol{Zb} + \boldsymbol{\varepsilon}, \quad \begin{bmatrix} \boldsymbol{b} \\ \boldsymbol{\varepsilon} \end{bmatrix} \sim N\left(\begin{bmatrix} \mathbf{0} \\ \mathbf{0} \end{bmatrix}, \begin{bmatrix} \boldsymbol{Q}_b & \mathbf{0} \\ \mathbf{0} & \boldsymbol{R} \end{bmatrix}\right)$$

Marginal Version

$$\boldsymbol{y} = \boldsymbol{X\beta} + \boldsymbol{\varepsilon}^*, \boldsymbol{\varepsilon}^* \sim N(\mathbf{0}, \boldsymbol{V}), \boldsymbol{V} = \boldsymbol{ZQ}_b\boldsymbol{Z}^T + \boldsymbol{R}$$

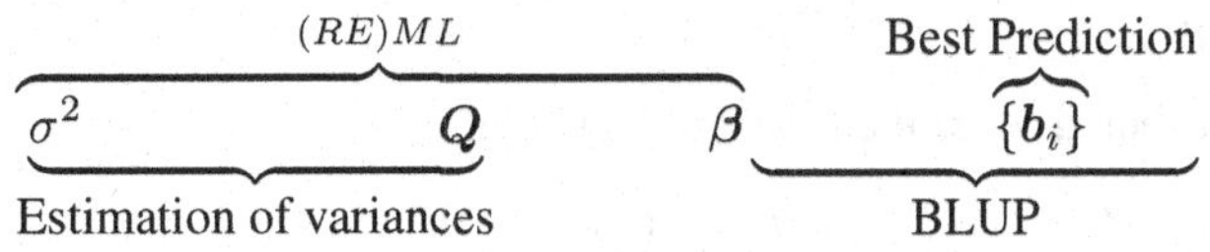

FIGURE 14.2: Fitting of random effects models.

14.1.3 Inference for Gaussian Response

One approach to estimating parameters distinguishes between the structural parameters $\boldsymbol{\beta}, \sigma^2, \boldsymbol{Q}$ and the random effects $\boldsymbol{b}_1, \ldots, \boldsymbol{b}_n$. Estimation is obtained by

- Estimation of fixed effects $\boldsymbol{\beta}$ and covariances $\sigma^2, \boldsymbol{Q}$ (more general $\boldsymbol{Q}_b$ and $\boldsymbol{R}$) using maximum likelihood (ML) or restricted maximum likelihood (REML)
- Prediction of random effects $\{\boldsymbol{b}_i\}$ using best prediction $\hat{\boldsymbol{b}}_i = E(\boldsymbol{b}_i|\boldsymbol{y}_i)$

An alternative approach is to find estimates of $\boldsymbol{\beta}$ and $\boldsymbol{b}_1, \ldots, \boldsymbol{b}_n$ simultaneously and estimate the variances separately. Both approaches are visualized in Figure 14.2. In the following the underlying estimation methods are discussed briefly.

Maximum Likelihood

From the marginal model (14.4) one obtains up to constants the log-likelihood

$$l(\boldsymbol{\beta}, \sigma^2, \boldsymbol{Q}) = \sum_{i=1}^{n} -\frac{1}{2}\log|\boldsymbol{V}_i| - \frac{1}{2}(\boldsymbol{y}_i - \boldsymbol{X}_i\boldsymbol{\beta})^T \boldsymbol{V}_i^{-1}(\boldsymbol{y}_i - \boldsymbol{X}_i\boldsymbol{\beta}),$$

where $\boldsymbol{V}_i = \sigma^2\boldsymbol{I} + \boldsymbol{Z}_i\boldsymbol{Q}\boldsymbol{Z}_i^T$. The computation of ML estimates is based on solving the score equations $\partial l/\partial\boldsymbol{\beta} = \boldsymbol{0}, \partial l/\partial\sigma^2 = 0, \partial l/\partial\boldsymbol{Q} = \boldsymbol{0}$. The first equation yields

$$\sum_{i=1}^{n} \boldsymbol{X}_i^T \hat{\boldsymbol{V}}_i^{-1}(\boldsymbol{y}_i - \boldsymbol{X}_i\hat{\boldsymbol{\beta}}) = \boldsymbol{0},$$

and therefore

$$\hat{\boldsymbol{\beta}} = \left\{\sum_{i=1}^{n}(\boldsymbol{X}_i^T \hat{\boldsymbol{V}}_i^{-1}\boldsymbol{X}_i)\right\}^{-1}\left\{\sum_{i=1}^{n}\boldsymbol{X}_i^T \hat{\boldsymbol{V}}_i^{-1}\boldsymbol{y}_i\right\}, \tag{14.6}$$

or, with $\hat{\boldsymbol{V}} = Diag(\hat{\boldsymbol{V}}_1, \ldots, \hat{\boldsymbol{V}}_n)$,

$$\hat{\boldsymbol{\beta}} = (\boldsymbol{X}^T\hat{\boldsymbol{V}}^{-1}\boldsymbol{X})^{-1}\boldsymbol{X}^T\hat{\boldsymbol{V}}^{-1}\boldsymbol{y}. \tag{14.7}$$

Thus, if $\hat{\boldsymbol{V}}_i$ is found, $\hat{\boldsymbol{\beta}}$ has the familiar form of a weighted least-squares estimate. If $\boldsymbol{V}_i$ is known, by setting $\hat{\boldsymbol{V}}_i = \boldsymbol{V}_i$ the generalized least-squares estimator $\hat{\boldsymbol{\beta}}$ is unbiased. The solutions of the second and third score equations $\partial l/\partial\sigma^2 = 0, \partial l/\partial\boldsymbol{Q} = \boldsymbol{0}$ usually make iterative procedures necessary (see, e.g., Longford, 1993, Chapter 2.3). For given $\boldsymbol{V}$ the estimate $\hat{\boldsymbol{\beta}}$ can be justified as a best linear unbiased estimator (BLUE) for $\boldsymbol{\beta}$. For a normally distributed $\boldsymbol{y}$, it is also the uniformly minimum variance unbiased estimator (UMVUE).

Best Prediction

For a normally distributed $\boldsymbol{b}_i, \boldsymbol{\varepsilon}_i$ the model (14.2) yields

$$\begin{pmatrix} \boldsymbol{y}_i \\ \boldsymbol{b}_i \end{pmatrix} \sim N\left(\begin{pmatrix} \boldsymbol{X\beta} \\ \boldsymbol{0} \end{pmatrix}, \begin{pmatrix} \boldsymbol{V}_i & \boldsymbol{Z}_i\boldsymbol{Q} \\ \boldsymbol{QZ}_i^T & \boldsymbol{Q} \end{pmatrix} \right)$$

and the posterior has the form

$$\boldsymbol{b}_i|\boldsymbol{y}_i \sim N(\boldsymbol{QZ}_i^T\boldsymbol{V}_i^{-1}(\boldsymbol{y}_i - \boldsymbol{X}_i\boldsymbol{\beta}), \boldsymbol{Q} - \boldsymbol{QZ}_i^T\boldsymbol{Q}^{-1}\boldsymbol{Z}_i\boldsymbol{Q}).$$

Thus, for known $\boldsymbol{Q}, \boldsymbol{V}_i, \boldsymbol{\beta}$, the best predictor is given by the posterior mean:

$$\hat{\boldsymbol{b}}_i = E(\boldsymbol{b}_i|\boldsymbol{y}_i) = \boldsymbol{QZ}_i^T\boldsymbol{V}_i^{-1}(\boldsymbol{y}_i - \boldsymbol{X}_i\boldsymbol{\beta}), \qquad (14.8)$$

which for normally distributed responses coincides with the posterior mode.

Estimation of Variances

In general, ML estimates of $\boldsymbol{V}$ can be found by maximizing the profile-likelihood. By substituting (14.6) into the log-likelihood, the profile-likelihood for $\boldsymbol{V}_i(\boldsymbol{V})$ is given by

$$l_{pr}(\boldsymbol{V}) = -\frac{1}{2}\log|\boldsymbol{V}| - \frac{1}{2}(\boldsymbol{y} - \boldsymbol{X}\hat{\boldsymbol{\beta}})^T\boldsymbol{V}^{-1}(\boldsymbol{y} - \boldsymbol{X}\hat{\boldsymbol{\beta}})$$

$$= -\frac{1}{2}\log|\boldsymbol{V}| - \frac{1}{2}(\boldsymbol{y}^T\boldsymbol{V}^{-1}\{\boldsymbol{I} - \boldsymbol{X}(\boldsymbol{X}^T\boldsymbol{V}^{-1}\boldsymbol{X})^{-1}\boldsymbol{X}^T\boldsymbol{V}^{-1}\}\boldsymbol{y}).$$

Maximization with respect to the parameters that specify $\boldsymbol{V}_i(\boldsymbol{V})$ yields the ML estimates.

A criticism of the maximum likelihood estimation in regression models is that the estimator of variances is biased. Even in a simple linear regression with $\boldsymbol{y} = \boldsymbol{X\beta}+\boldsymbol{\varepsilon}, \boldsymbol{\varepsilon} \sim N(\boldsymbol{0}, \sigma^2\boldsymbol{I})$ one obtains the ML estimator $\hat{\sigma}^2 = (\boldsymbol{y}-\boldsymbol{X}\hat{\boldsymbol{\beta}})^T(\boldsymbol{y}-\boldsymbol{X}\hat{\boldsymbol{\beta}})/n$, which has mean $E(\hat{\sigma}^2) = \sigma^2(n-p)/n$ and therefore has a downward bias that increases with the number of estimated parameters p. The reason is that the ML estimator fails to take into account the uncertainty about the regression parameters: $\hat{\sigma}^2$ would be unbiased if $\hat{\boldsymbol{\beta}}$ could be replaced by $\boldsymbol{\beta}$. The same is to be expected from ML estimators in random effects models.

A solution to the bias problem is to estimate variances without reference to $\hat{\boldsymbol{\beta}}$. A *restricted maximum likelihood* (Patterson and Thomson, 1971) is based on the construction of a likelihood that depends only on a complete set of error contrasts and therefore does not depend on $\boldsymbol{\beta}$. An error contrast is a linear combination of $\boldsymbol{y}$ that has zero expectation, that is, $E(\boldsymbol{u}^T\boldsymbol{y}) = 0$. In a simple linear regression a set of linearly independent error contrasts is $\boldsymbol{I} - \boldsymbol{X}(\boldsymbol{X}^T\boldsymbol{X})^{-1}\boldsymbol{X}$. For random effects models Harville (1976) derived the likelihood of error contrasts; see also Searle et al. (1992). The resulting criterion is the restricted log-likelihood:

$$l_R(\boldsymbol{V}) = l_{pr}(\boldsymbol{V}) - \frac{1}{2}\log|\boldsymbol{X}^T\boldsymbol{V}^{-1}\boldsymbol{X}|,$$

which yields restricted maximum likelihood (REML) estimates. Computations of ML and REML estimates are based on iterative procedures.

Best Linear Unbiased Prediction (BLUP)

An approach that aims at estimating of parameters and random effects simultaneously (separated from the estimation of variances) is based on the joint density of $\boldsymbol{y}_i$ and $\boldsymbol{b}_i$, which is maximized with respect to $\boldsymbol{\beta}$ and $\boldsymbol{b}_1, \dots, \boldsymbol{b}_n$. One uses the closed form,

$$\boldsymbol{y} = \boldsymbol{X\beta} + \boldsymbol{Zb} + \boldsymbol{\varepsilon},$$

with covariance given by

$$\text{cov}\begin{pmatrix} b \\ \varepsilon \end{pmatrix} = \begin{pmatrix} Q_b & 0 \\ 0 & R \end{pmatrix} = \begin{pmatrix} \tilde{Q}_b & 0 \\ 0 & \tilde{R} \end{pmatrix} \sigma^2,$$

where $\tilde{Q}_b$ and $\tilde{R}$ are known. Thus it is assumed that the structure of $\text{cov}(b_i)$ is known up to a constant. This is the case, for example, for random intercepts where $\tilde{Q}_b = (\sigma_b^2/\sigma^2)I$ and $\tilde{R} = I$. The joint density of y and b is normal and maximization with respect to β and b requires minimizing:

$$(y - X\beta - Zb)^T R^{-1}(y - X\beta - Zb) + b^T Q_b^{-1} b. \tag{14.9}$$

Minimization corresponds to a generalized least-squares with a penalty term.

Differentiating the joint density with respect to β and b yields Henderson's "mixed model equations" for β and b:

$$X^T R^{-1} X\hat{\beta} + X^T R^{-1} Z\hat{b} = X^T R^{-1} y,$$

$$Z^T R^{-1} X\hat{\beta} + (Z^T R^{-1} Z + Q_b^{-1})\hat{b} = Z^T R^{-1} y.$$

Maximization of the joint density of y and b is not a common maximum likelihood estimate since b is not a parameter. Therefore, estimates of b that result from solving these equations are often called predictions, while estimations of fixed effects are called estimations. It should be noted that the mixed model equations and the other formulas in this section may be written with Q_b, R, *or* $\tilde{Q}_b, \tilde{R}$. Some matrix algebra (Robinson, 1991) shows that the mixed model equations yield

$$\hat{\beta} = [X^T V^{-1} X]^{-1} X^T V^{-1} y,$$

where $V = R + ZQ_b Z^T$. Thus $\hat{\beta}$ is equivalent to the generalized least-squares estimator (14.7) for a known covariance matrix. For $\hat{b}$ one obtains $\hat{b} = V^{-1}Z^T R^{-1}(y - X\hat{\beta})$, which is equivalent to

$$\hat{b} = Q_b Z^T V^{-1}(y - X\hat{\beta}).$$

With $V = Diag(V_1, \dots, V_n)$ and $Q_b = Diag(Q, \dots, Q)$ one obtains $\hat{b}_i = QZ_i^T V_i^{-1}$ $(y_i - X_i\hat{\beta})$, which is equivalent to (14.8). Therefore, the best linear unbiased prediction yields an estimate $\hat{\beta}$ that is equivalent to the ML estimate for a known covariance and predictions $\hat{b}_i$ that correspond to the posterior mean. By using (14.9), the estimates $\hat{\beta}, \hat{b}$ can be written in closed form as

$$\begin{bmatrix} \hat{\beta} \\ \hat{b} \end{bmatrix} = (C^T R^{-1} C + B)^{-1} C^T R^{-1} y,$$

where $C = [X\ Z]$ and

$$B = \begin{bmatrix} 0 & 0 \\ 0 & Q_b^{-1} \end{bmatrix},$$

and one obtains the fitted values

$$\hat{y} = \text{BLUP}(y) = X\hat{\beta} + Z\hat{b} = C(C^T R^{-1} C + B)^{-1} C^T R^{-1} y. \tag{14.10}$$

Ruppert et al. (2003) call (14.10) the "ridge regression" formulation of BLUP. It shows that $\hat{\beta}$ and $\hat{b}$ are estimated by weighted least squares with a penalty term that penalizes b.

$\hat{\beta} = (X^T V^{-1} X)^{-1} X^T V^{-1} y$	ML estimator (if V is ML estimator) BLUP (for given V)
$\hat{b}_i = Q Z_i^T V_i^{-1}(y_i - X_i\beta)$	BLUP (for given V)
Variance V_i	ML or REML

14.2 Generalized Linear Mixed Models

Generalized linear mixed models extend generalized models to permit random effects as well as fixed effects in the linear predictor. We will first consider simple models that contain only a random intercept and discuss the differences between random effects models and marginal models.

14.2.1 Binary Response Models with Random Intercepts

Logistic-Normal Model

Let $y_i^T = (y_{i1}, \dots, y_{iT_i})$ denote the observation of one cluster, where $y_{it} \in \{0, 1\}$ is the binary response for cluster i and measurement t. The simple logistic-normal random intercept model with covariates x_{it} is given by

$$\text{logit}(P(y_{it} = 1|x_{it}, b_i)) = x_{it}^T\beta + b_i, \quad i = 1, \dots, n, t = 1, \dots, T_i. \tag{14.11}$$

While β is a fixed effect, b_i is considered to be a cluster-specific random effect. It is assumed that the $\{b_i\}$ are independent $N(0, \sigma_b^2)$ variates and the $y_{i1}, \dots, y_{iT_i}$ are conditionally independent given b_i, therefore the name *logistic-normal model*. In (14.11), heterogeneity across clusters is explicitly modeled by allowing each cluster to have its own intercept. Since b_i has mean $\text{E}(b_i) = 0$, it is natural to incorporate a fixed intercept in the design vector x_{it}.

In the conditional model (14.11) the regression parameter β measures the change in logits per unit change in a covariate, controlling for all other variables, including the random effect b_i. Interpretation of β is always *conditional on* b_i.

Model (14.11) is a non-linear mixed model since the linear predictor $\eta_{it} = x_{it}^T\beta + b_i$ effects upon the logit transformed mean $\mu_{it} = \text{E}(y_{it}|x_{it}, b_i)$. Non-linearity has severe consequences if one wants to interpret the effect of x_{it} on the marginal probability $P(y_{it} = 1|x_{it})$. Consider a *linear* random effects model $\text{E}(y_{it}|x_{it}, b_i) = x_{it}^T\beta + b_i$, where $b_i \sim N(0, \sigma_b^2)$. Then the marginal mean $\text{E}(y_{it}|x_{it}) = x_{it}^T\beta$ is again specified in a linear way and β is the parameter that determines the response. The crucial point is that for model (14.11) the marginal probability is

$$P(y_{it} = 1|x_{it}) = \int P(y_{it} = 1|x_{it}, b_i)p(b_i)db_i,$$

where $p(b_i)$ denotes the density of b_i. With $P(y_{it} = 1|x_{it}, b_i)$ given by the conditional logit model

$$P(y_{it} = 1|x_{it}, b_i) = \frac{\exp(x_{it}^T\beta + b_i)}{1 + \exp(x_{it}^T\beta + b_i)},$$

the marginal probability $P(y_{it} = 1|x_{it})$ is *not* a linear logistic model. A simple approximation of the marginal probability may be derived from a Taylor expansion (compare Cramer, 1991).

With F denoting the logistic distribution function $F(\eta) = \exp(\eta)/(1+\exp(\eta))$, the conditional model is given by

$$P(y_{it} = 1|\boldsymbol{x}_{it}, b_i) = F(\boldsymbol{x}_{it}^T\boldsymbol{\beta} + b_i).$$

A Taylor approximation of second order (Appendix B) at $\eta_i = \boldsymbol{x}_{it}^T\boldsymbol{\beta}$ yields $F(\eta_i + b_i) \approx F(\eta_i) + F'(\eta_i)b_i + \frac{1}{2}F''(\eta_i)b_i^2$, where $F'(\eta_i) = \partial F(\eta_i)/\partial\eta$, $F''(\eta_i) = \partial^2 F(\eta_i)/\partial\eta^2$. One obtains with $\sigma_b^2 = \text{var}(b_i)$

$$\begin{aligned} P(y_{it} = 1|\boldsymbol{x}_{it}) &= \mathrm{E}_{b_i}\, P(y_{it} = 1|\boldsymbol{x}_{it}, b_i) \approx F(\eta_i) + \frac{1}{2}F''(\eta_i)var(b_i) \\ &= F(\eta_i) + \frac{1}{2}F(\eta_i)(1 - F(\eta_i))(1 - 2F(\eta_i))\sigma_b^2. \end{aligned}$$

Figure 14.3 shows the (approximative) dependence of the marginal response on $F(\eta_i)$ for several values of σ_b^2. It is seen that for large values of η_i (corresponding to $F(\eta_i) > 0.5$) the marginal probability is shrunk toward 0.5, whereas for small values of η_i (corresponding to $F(\eta_i) < 0.5$) the marginal probability is larger than $F(\eta_i)$. This means that marginal probabilities are always closer to 0.5 when compared to $F(\eta_i) = F(\boldsymbol{x}_{it}^T\boldsymbol{\beta})$. Consequently, if one fits the marginal model

$$P(y_{it} = 1|\boldsymbol{x}_{it}) = F(\boldsymbol{x}_{it}^T\boldsymbol{\beta}^m),$$

the parameter estimates $\hat{\boldsymbol{\beta}}^m$ are attenuated as compared to $\boldsymbol{\beta}$. If one ignores heterogeneity and fits a marginal model, the effect strength $\boldsymbol{\beta}$ can be strongly underestimated.

Figure 14.4 visualizes the effect in an alternative way. The panels show the response curves for the conditional probabilities $F(x\beta + b_i)$ as functions in x for various fixed intercepts b_i and the marginal response averaged over the conditional responses, $P(y_{it} = 1|x) = \int F(x\beta + b_i)p(b_i)db_i$. The latter is given as a superimposed thick curve. It is seen that larger values of the variance σ_b^2 yield flatter marginal response curves. The marginal response curves are also sigmoidal but not logistic functions. Therefore, if heterogeneity is in the data and the conditional logistic-normal model holds, ignoring the heterogeneity by fitting a marginal logistic model means misspecification, that is, one is fitting a model that does not hold. As is shown in the following the connection between conditional and marginal models is slightly different in probit-normal models.

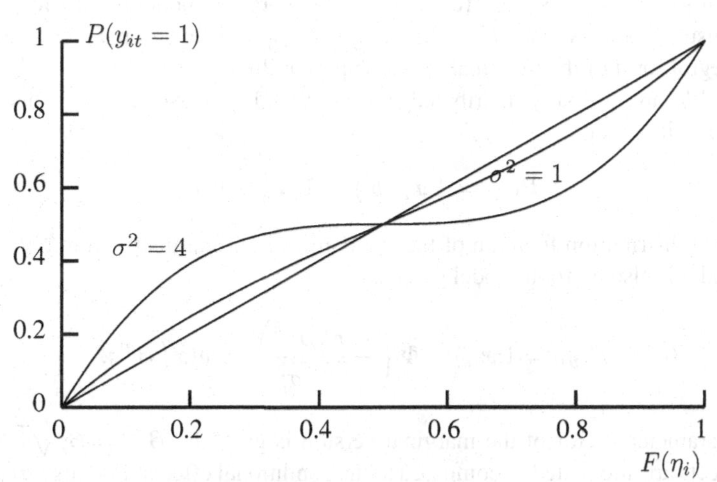

FIGURE 14.3: Dependence of marginal probability on $F(\eta_i)$.

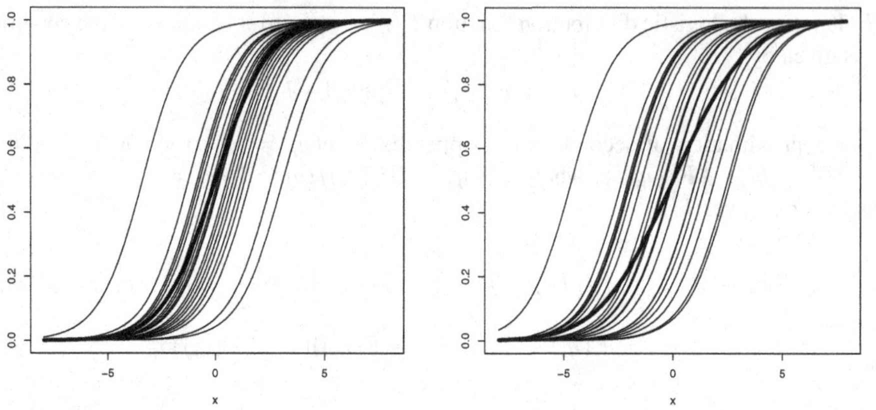

FIGURE 14.4: Mixture of logistic responses for $\sigma_b^2 = 1$ (left panel) and $\sigma_b^2 = 4$ (right panel).

Probit-Normal Model and Alternatives

As shown in Chapter 2, binary response models may be derived from latent regression models. Let a latent response $\tilde{y}_{it}$ be given by $\tilde{y}_{it} = \boldsymbol{x}_{it}^T\boldsymbol{\beta} + \varepsilon_{it} + b_i$, where $-\varepsilon_{it}$ and $-b_i$ are independent and have the distribution function F_ε, F_b. By assuming that the observed response $y_{it} = 1$ occurs if $\tilde{y}_{it} > 0$, one obtains the conditional model

$$P(y_{it} = 1|\boldsymbol{x}_{it}, b_i) = P(-\varepsilon_{it} \leq \boldsymbol{x}_{it}^T\boldsymbol{\beta} + b_i) = F_\varepsilon(\boldsymbol{x}_{it}^T\boldsymbol{\beta} + b_i).$$

The corresponding marginal model is given by

$$P(y_{it} = 1|\boldsymbol{x}_{it}) = \int F_\varepsilon(\boldsymbol{x}_{it}^T\boldsymbol{\beta} + b_i)p(b_i)db_i$$

or

$$P(y_{it} = 1|\boldsymbol{x}_{it}) = P(-\varepsilon_{it} - b_i \leq \boldsymbol{x}_{it}^T\boldsymbol{\beta}) = F_m(\boldsymbol{x}_{it}^T\boldsymbol{\beta}),$$

where, for symmetrically distributed b_i, F_m is the distribution function of $-\varepsilon_{it} - b_i$. While the conditional model has the response function F_ε, the marginal model has the response function F_m, which usually differs from F_ε. The marginal response function F_ε can also be derived from the integral form of the marginal response probability.

If ε_{it} *and* b_i are normally distributed, $\varepsilon_{it} \sim N(0, 1^2), b_i \sim N(0, \sigma_b^2)$, one assumes the conditional probit model

$$P(y_{it} = 1|\boldsymbol{x}_{it}, b_i) = \Phi(\boldsymbol{x}_{it}^T\boldsymbol{\beta} + b_i),$$

where Φ is the distribution function of the standardized normal distribution $\mathrm{N}(0, 1)$. Then the marginal model is also a probit model given by

$$P(y_{it} = 1|\boldsymbol{x}_{it}) = \Phi\left(\frac{\boldsymbol{x}_{it}^T\boldsymbol{\beta}}{\sqrt{1 + \sigma_b^2}}\right) = \Phi(\boldsymbol{x}_{it}^T\boldsymbol{\beta}^m),$$

where the parameter vector of the marginal version is given by $\boldsymbol{\beta}^m = \boldsymbol{\beta}/\sqrt{1 + \sigma_b^2}$. Again, marginal effects are attenuated as compared to the conditional effects. For large σ_b^2, the marginal effects $\boldsymbol{\beta}^m$ are much closer to zero than the conditional effects $\boldsymbol{\beta}$. Although marginal and conditional effects usually differ, for probit models the model structure is the same for the marginal and the conditional model. The model is closed in the sense that the distributions associated with the marginal and conditional link functions are of the same family. More generally, one also gets closed models for mixtures of normals; see Caffo et al. (2007).

14.2.2 Generalized Linear Mixed Models Approach

A general form of the model is obtained by considering the response variables y_i and covariates $\boldsymbol{x}_i, \boldsymbol{z}_i, i = 1, \dots, N$. One assumes for the mean response, conditionally on the random effect $\boldsymbol{b}$, denoted by $\mu_i = \mathrm{E}(y_i|\boldsymbol{x}_i, \boldsymbol{z}_i, \boldsymbol{b})$, that the link to the predictor is given by

$$g(\mu_i) = \boldsymbol{x}_i^T\boldsymbol{\beta} + \boldsymbol{z}_i^T\boldsymbol{b} \quad \text{or} \quad \mu_i = h(\boldsymbol{x}_i^T\boldsymbol{\beta} + \boldsymbol{z}_i^T\boldsymbol{b}),$$

where g represents the link function and $h = g^{-1}$ is the response function. While the $\boldsymbol{\beta}$'s are the *fixed effects model parameters*, $\boldsymbol{b}$ is a *random effect*, for which a distribution is assumed. Most often one assumes $\boldsymbol{b} \sim N(\boldsymbol{0}, \boldsymbol{Q}_b)$ for an unknown covariance matrix $\boldsymbol{Q}_b$. In addition, it is assumed that $y_1, y_2, \dots, y_N$ are independent *given* $\boldsymbol{b}$. Hence one assumes that the correlation between observations is due to the common latent variable $\boldsymbol{b}$. Considering the model as an extension of GLMs, it is assumed that the distribution of y_i is in the exponential family. The structural part of the model is often given in matrix form as

$$g(\boldsymbol{\mu}) = \boldsymbol{X\beta} + \boldsymbol{Zb} \quad \text{or} \quad \boldsymbol{\mu} = h(\boldsymbol{X\beta} + \boldsymbol{Zb}), \tag{14.12}$$

where $\boldsymbol{\mu}^T = (\mu_1, \dots, \mu_N)$, $\boldsymbol{X}$ and $\boldsymbol{Z}$ have rows $\boldsymbol{x}_i$ and $\boldsymbol{z}_i$, respectively, and g and h are understood componentwise.

Generalized Linear Mixed Model (GLMM)

$$g(\boldsymbol{\mu}) = \boldsymbol{X\beta} + \boldsymbol{Zb}$$

with components

$$g(\mu_i) = \boldsymbol{x}_i^T\boldsymbol{\beta} + \boldsymbol{z}_i^T\boldsymbol{b}$$

and

$$\mathrm{E}(\boldsymbol{b}) = \boldsymbol{0}, \ \mathrm{cov}(\boldsymbol{b}) = \boldsymbol{Q}_b,$$

$$y_1, \dots, y_N \quad \text{independent given } \boldsymbol{b}$$

14.2.3 Generalized Linear Mixed Models for Clustered Data

An important special case of the generalized linear mixed model occurs when data are collected in clusters. A typical example is repeated measurements, where the response variable is observed repeatedly for each unit. A unit may be an individual for which a response is observed under different conditions or a family with the measurements taken on the members of the family. For clustered data the random effects usually are cluster-level terms; the single random effects are specific for the individuals. As in Section 14.2.1, in the notation a double index is used to represent the cluster and the observation within the cluster. Let y_{it} denote observation t in cluster $i, t = 1, \dots, T_i$, collected in $\boldsymbol{y}_i^T = (y_{i1}, \dots, y_{iT_i})$.

With explanatory variables $\boldsymbol{x}_{it}, \boldsymbol{z}_{it}$ that may depend on i and t, one considers the conditional means $\mu_{it} = E(y_{it}|\boldsymbol{x}_{it}, \boldsymbol{z}_{it}, \boldsymbol{b}_i)$, where $\boldsymbol{b}_i$ is the cluster-specific random effect. The structural assumption is

$$g(\mu_{it}) = \boldsymbol{x}_{it}^T\boldsymbol{\beta} + \boldsymbol{z}_{it}^T\boldsymbol{b}_i \quad \text{or} \quad \mu_{it} = h(\boldsymbol{x}_{it}^T\boldsymbol{\beta} + \boldsymbol{z}_{it}^T\boldsymbol{b}_i), \tag{14.13}$$

or, simpler, $g(\mu_{it}) = \eta_{it}$, where $\eta_{it} = \boldsymbol{x}_{it}^T\boldsymbol{\beta} + \boldsymbol{z}_{it}^T\boldsymbol{b}_i$ is the linear predictor. Moreover, it is assumed that the random effects $\boldsymbol{b}_1, \dots, \boldsymbol{b}_n$ are independent with $\mathrm{E}(\boldsymbol{b}_i) = \boldsymbol{0}$ and have density $p(\boldsymbol{b}_i, \boldsymbol{Q})$, where $\boldsymbol{Q}$ represents the covariance matrix of a single random effect, $\mathrm{cov}(\boldsymbol{b}_i) = \boldsymbol{Q}$. In the following we give some examples for the specification of the explanatory variables and the random effects.

Random Intercept Models

In Section 14.2.1, the random intercept model for binary observations was considered. In general, for random intercept models, the linear predictor of observation t in cluster i is given by

$$\eta_{it} = b_i + \boldsymbol{x}_{it}^T\boldsymbol{\beta},$$

where b_i is the cluster-specific intercept. Thus the effect of the covariates is determined by the fixed effects $\boldsymbol{\beta}$, but the response strength may vary across clusters. By specifying $z_{it} = 1$ one obtains the random intercept model for clustered data from the general form (14.13).

Random Slopes

The random intercept model assumes that only the response level varies across individuals (or, more general, clusters). However, for example, the effect of a drug may also vary across individuals. The potential heterogeneity of this sort may be modeled by allowing for random slopes of the explanatory variable. The linear predictor of the model, where *all* variables have cluster-specific slopes, has the form

$$\eta_{it} = \boldsymbol{x}_{it}^T\boldsymbol{\beta} + \boldsymbol{x}_{it}^T\boldsymbol{b}_i,$$

where $\boldsymbol{\beta}$ represents the common effect of covariates and $\boldsymbol{b}_i$ represents the cluster-specific deviations from $\boldsymbol{\beta}$. Since one assumes $\mathrm{E}(\boldsymbol{b}_i) = \boldsymbol{0}$, the mean predictor is given by $\mathrm{E}_{\boldsymbol{b}_i}(\boldsymbol{x}_{it}^T(\boldsymbol{\beta}+\boldsymbol{b}_i)) = \boldsymbol{x}_{it}^T\boldsymbol{\beta}$, where $\mathrm{E}_{\boldsymbol{b}_i}$ denotes expectation with respect to $\boldsymbol{b}_i$. Most often, only part of the explanatory variables are assumed to have cluster-specific slopes. Then the variables $\boldsymbol{z}_{it}$ represent a subvector of $\boldsymbol{x}_{it}$. Let $\boldsymbol{x}_{it}$ be decomposed into $\boldsymbol{x}_{it}^T = (\boldsymbol{x}_{it(1)}^T, \boldsymbol{x}_{it(2)}^T)$ and $\boldsymbol{\beta}$ into $\boldsymbol{\beta}^T = (\boldsymbol{\beta}_{(1)}^T, \boldsymbol{\beta}_{(2)}^T)$. By setting $\boldsymbol{z}_{it} = \boldsymbol{x}_{it(2)}$ and assuming

$$\eta_{it} = \boldsymbol{x}_{it}^T\boldsymbol{\beta} + \boldsymbol{z}_{it}^T\boldsymbol{b}_i = \boldsymbol{x}_{it(1)}^T\boldsymbol{\beta}_{(1)} + \boldsymbol{x}_{it(2)}^T(\boldsymbol{\beta}_{(2)} + \boldsymbol{b}_i),$$

the variables $\boldsymbol{x}_{it(1)}$ have fixed effects whereas the variables $\boldsymbol{x}_{it(2)}$ have effects that vary across clusters.

Generalized Linear Mixed Model for Clustered Data

$$g(\mu_{it}) = \boldsymbol{x}_{it}^T\boldsymbol{\beta} + \boldsymbol{z}_{it}^T\boldsymbol{b}_i$$

with

$$\mathrm{E}(\boldsymbol{b}_i) = \boldsymbol{0}, \ \mathrm{cov}(\boldsymbol{b}_i) = \boldsymbol{Q}$$

$$y_{i1}, \dots, y_{nT_i} \quad \text{independent given } \{\boldsymbol{b}_i\}$$

It is easily seen that the model for clustered data has the form of the general model (14.12). By collecting the cluster-specific random effects in one vector, $\boldsymbol{b}^T = (\boldsymbol{b}_1^T, \dots, \boldsymbol{b}_n^T)$, one has for single observations the form

$$g(\mu_{it}) = \boldsymbol{x}_{it}^T\boldsymbol{\beta} + \tilde{\boldsymbol{z}}_{it}^T\boldsymbol{b},$$

where $\tilde{\boldsymbol{z}}_{it}^T = (\boldsymbol{0}^T, \dots, \boldsymbol{z}_{it}^T, \dots, \boldsymbol{0}^T)$. The matrix $\boldsymbol{X}$ in (14.12) contains the vectors $\boldsymbol{x}_{it}$ as rows; the rows of $\boldsymbol{Z}$ are given by the vectors $\tilde{\boldsymbol{z}}_{it}$. The covariance of the total random effects vector $\boldsymbol{b}$ is given as a block-diagonal matrix $\boldsymbol{Q}_b = Diag(\boldsymbol{Q}, \dots, \boldsymbol{Q})$. By collecting the observations of one cluster in a vector one obtains the matrix form $g(\boldsymbol{\mu}_i) = \boldsymbol{X}_i\boldsymbol{\beta} + \boldsymbol{Z}_i\boldsymbol{b}_i$, with $\boldsymbol{X}_i$ and $\boldsymbol{Z}_i$ containing observations $\boldsymbol{x}_{it}, \boldsymbol{z}_{it}$, $t = 1, \dots, T_i$. Stacking the matrices together yields the general form $g(\boldsymbol{\mu}) = \boldsymbol{X}\boldsymbol{\beta} + \boldsymbol{Z}\boldsymbol{b}$, where $\boldsymbol{X}^T = (\boldsymbol{X}_1^T, \dots, \boldsymbol{X}_n^T)$ and $\boldsymbol{Z}$ is a block-diagonal matrix with blocks $\boldsymbol{Z}_i$.

For clustered data the assumption of conditional independence given the random effects simplifies to

$$f(\boldsymbol{y}|\boldsymbol{b}; \boldsymbol{\beta}) = \prod_{i=1}^{n} f(\boldsymbol{y}_i|\boldsymbol{b}_i; \boldsymbol{\beta}) \text{ with } f(\boldsymbol{y}_i|\boldsymbol{b}_i; \boldsymbol{\beta}) = \prod_{t=1}^{T_i} f(y_{it}|\boldsymbol{b}_i; \boldsymbol{\beta}),$$

where $\boldsymbol{y}^T = (\boldsymbol{y}_1^T, \dots, \boldsymbol{y}_n^T)$, $\boldsymbol{y}_i^T = (y_{i1}, \dots, y_{iT_i})$, represents the whole set of responses.

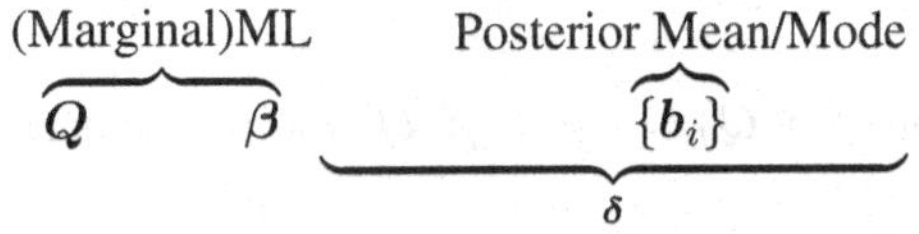

FIGURE 14.5: Fitting of generalized linear models with random effects.

14.3 Estimation Methods for Generalized Mixed Models

Several approaches to fitting generalized linear models may be distinguished. Maximum likelihood approaches aim at estimating the structural parameters $\boldsymbol{\beta}$ and $\boldsymbol{Q}$ by maximizing the marginal log-likelihood. Predictors for the random effects $\boldsymbol{b}_i$ are then derived as posterior mean estimators (see Sections 14.3.1 and 14.3.2). Alternatively, one may separate the estimation of the variance $\boldsymbol{Q}$ from the estimation/prediction of $\boldsymbol{\beta}$ and $\boldsymbol{b}_i$, which are collected in $\boldsymbol{\delta}^T = (\boldsymbol{\beta}^T, \boldsymbol{b}_1^T, \dots, \boldsymbol{b}_n^T)$ (see Section 14.3.3).

14.3.1 Marginal Maximum Likelihood Estimation by Integration Techniques

The fixed parameters $\boldsymbol{\beta}$ and $\boldsymbol{Q}_b$ can be estimated by maximization of the *marginal* log-likelihood

$$l(\boldsymbol{\beta}, \boldsymbol{Q}_b) = \log(\int f(\boldsymbol{y}|\boldsymbol{b}; \boldsymbol{Q}_b)p(\boldsymbol{b}; \boldsymbol{Q}_b)d\boldsymbol{b}).$$

In the case of crossed random effects, numerical maximizing is very cumbersome because the integral does not simplify. But it works relatively well when the random effects come in clusters (model (14.13)). Then the log-likelihood takes the form

$$l(\boldsymbol{\beta}, \boldsymbol{Q}) = \sum_{i=1}^{n} l_i(\boldsymbol{\beta}, \boldsymbol{Q}), \tag{14.14}$$

where

$$l_i(\boldsymbol{\beta}, \boldsymbol{Q}) = \log(\int f(\boldsymbol{y}_i|\boldsymbol{b}_i; \boldsymbol{\beta})p(\boldsymbol{b}_i; \boldsymbol{Q})d\boldsymbol{b}_i) = \log(\int \prod_{t=1}^{T_i} f(y_{it}|\boldsymbol{b}_i; \boldsymbol{\beta})p(\boldsymbol{b}_i; \boldsymbol{Q})d\boldsymbol{b}_i)$$

is the contribution of observation $\boldsymbol{y}_i^T = (y_{i1}, \ldots, y_{iT_i})$ that results from integration with respect to $\boldsymbol{b}_i$ by use of $p(\boldsymbol{b}_i; \boldsymbol{Q})$, the density of random effects within clusters.

One approach that works for low-dimensional $\boldsymbol{b}_i$ is to approximate the integral in $l_i(\boldsymbol{\beta}, \boldsymbol{Q})$ by numerical or Monte Carlo quadrature techniques. For estimating it is often useful to consider the standardized random effects

$$\boldsymbol{a}_i = \boldsymbol{Q}^{-1/2}\boldsymbol{b}_i,$$

where $\boldsymbol{Q}^{1/2}$ denotes the left Cholesky factor, which is a lower triangular matrix, so that $\boldsymbol{Q} = \boldsymbol{Q}^{1/2}\boldsymbol{Q}^{T/2}$, where ^{T} denotes the transpose. With $\boldsymbol{Q}^{-1/2}$ denoting the inverse of $\boldsymbol{Q}^{1/2}$, one obtains $\text{cov}(\boldsymbol{a}_i) = \boldsymbol{I}$. By using matrix algebra (e.g., Magnus and Neudecker, 1988, p. 30) one obtains the linear predictor

$$\eta_{it} = \boldsymbol{x}_{it}^T\boldsymbol{\beta} + \boldsymbol{z}_{it}^T\boldsymbol{Q}^{1/2}\boldsymbol{a}_i = \boldsymbol{x}_{it}^T\boldsymbol{\beta} + (\boldsymbol{a}_i^T \otimes \boldsymbol{z}_{it}^T)\boldsymbol{\theta},$$

where $\otimes$ is the Kronecker product and $\boldsymbol{\theta}$ denotes the vectorization of $\boldsymbol{Q}^{1/2}$, $\boldsymbol{\theta} = vec(\boldsymbol{Q}^{1/2})$. For the simple case of a scalar random effect one has $\text{var}(b_i) = \sigma_b^2$ and a_i is given by $a_i = b_i/\sigma_b$. With $\theta = \sigma_b$ the linear predictor simplifies to

$$\eta_{it} = \boldsymbol{x}_{it}^T\boldsymbol{\beta} + a_i\sigma_b.$$

The likelihood contribution $l_i(\boldsymbol{\beta}, \boldsymbol{Q}) = \log L_i(\boldsymbol{\beta}, \boldsymbol{Q})$ for the standardized random effects $\boldsymbol{a}_i$ is given by

$$L_i(\boldsymbol{\beta}, \boldsymbol{Q}) = \int f(\boldsymbol{y}_i|\boldsymbol{a}_i; \boldsymbol{\beta}, \boldsymbol{Q})p(\boldsymbol{a}_i)d\boldsymbol{a}_i = \int \prod_{t=1}^{T_i} f(y_{it}|\boldsymbol{a}_i; \boldsymbol{\beta}, \boldsymbol{Q})p(\boldsymbol{a}_i)d\boldsymbol{a}_i,$$

and $p(\boldsymbol{a}_i)$ denotes the (standardized) density of $\boldsymbol{a}_i$, which has zero mean and covariance matrix $\boldsymbol{I}$.

For a low dimension of $\boldsymbol{a}_i$, the integral may be approximated by integration techniques like Gauss-Hermite (for normally distributed $\boldsymbol{a}_i$) or Monte Carlo techniques. The Gauss-Hermite approximation has the form

$$L_i^{GH}(\boldsymbol{\beta}, \boldsymbol{Q}) = \sum_j v_j f(\boldsymbol{y}_i|\boldsymbol{d}_j; \boldsymbol{\beta}, \boldsymbol{Q}),$$

where $\boldsymbol{d}_j$ denotes fixed quadrature points and v_j denotes fixed weights that are associated with $\boldsymbol{d}_j$. Quadrature points and weights are given, for example, in Stroud and Secrest (1966) and Abramowitz and Stegun (1972). For more details see Appendix E.1.

A simple Monte Carlo technique approximates the likelihood by

$$L_i^{MC}(\boldsymbol{\beta}, \boldsymbol{Q}) = \frac{1}{m}\sum_{j=1}^{m} f(\boldsymbol{y}_i|\boldsymbol{d}_{ij}; \boldsymbol{\beta}, \boldsymbol{Q}),$$

where $\boldsymbol{d}_{ij}$ are m iid drawings from the mixing density $p(\boldsymbol{a}_i)$. In both, Gauss-Hermite and Monte Carlo approximations, the random effects $\boldsymbol{a}_i$ are replaced by known values, quadrature points, or random drawings. That makes it possible to compute estimates of $\boldsymbol{\beta}$ and $\boldsymbol{Q}$ within the generalized linear model framework with predictors $\eta_{itj} = \boldsymbol{x}_{it}^T\boldsymbol{\beta} + (\boldsymbol{d}_j^T \otimes \boldsymbol{z}_{it}^T)\boldsymbol{\theta}$, $j = 1, \ldots, m$, where $\boldsymbol{\theta} = vec(\boldsymbol{Q}^{1/2})$. The embedding into the GLM framework is given in the next section for the Gauss-Hermite procedure.

Direct Maximization by Using GLM Methodology

The Gauss-Hermite approximation $L_i^{GH}(\boldsymbol{\beta}, \boldsymbol{Q})$ is a function of $\boldsymbol{\alpha}^T = (\boldsymbol{\beta}^T, \boldsymbol{\theta}^T)$ with the predictor having the form $\eta_{it} = \boldsymbol{x}_{it}^T\boldsymbol{\beta} + (\boldsymbol{a}_i^T \otimes \boldsymbol{z}_{it}^T)\boldsymbol{\theta}$. By using

$$\frac{\partial f(\boldsymbol{y}_i|\boldsymbol{d}_j;\boldsymbol{\alpha})}{\partial \boldsymbol{\alpha}} = f(\boldsymbol{y}_i|\boldsymbol{d}_j;\boldsymbol{\alpha})\frac{\partial \log f(\boldsymbol{y}_i|\boldsymbol{d}_j;\boldsymbol{\alpha})}{\partial \boldsymbol{\alpha}},$$

one obtains the score approximation

$$s_i^{GH}(\boldsymbol{\alpha}) = \frac{\partial \log L_i^{GH}(\boldsymbol{\alpha})}{\partial \boldsymbol{\alpha}} = \sum_{j=1}^{m} c_{ij}^{GH}(\boldsymbol{\alpha})\frac{\partial \log f(\boldsymbol{y}_i|\boldsymbol{d}_j;\boldsymbol{\alpha})}{\partial \boldsymbol{\alpha}}, \tag{14.15}$$

where

$$c_{ij}^{GH}(\boldsymbol{\alpha}) = \frac{v_j f(y_i|\boldsymbol{d}_j;\boldsymbol{\alpha})}{\sum_k v_k f(\boldsymbol{y}_i|\boldsymbol{d}_k;\boldsymbol{\alpha})}, \quad \text{with} \quad \sum_j c_{ij}^{GH}(\boldsymbol{\alpha}) = 1, \tag{14.16}$$

denote weight factors that depend on the parameters $\boldsymbol{\alpha}$. The derivative

$$\partial \log f(\boldsymbol{y}_i|\boldsymbol{d}_j, \boldsymbol{\alpha})/\partial \boldsymbol{\alpha}^T = (\partial \log f(\boldsymbol{y}_i|\boldsymbol{d}_j, \boldsymbol{\alpha})/\partial \boldsymbol{\beta}^T,\ \partial \log f(\boldsymbol{y}_i|\boldsymbol{d}_j, \boldsymbol{\alpha})/\partial \boldsymbol{\theta}^T)$$

corresponds to the score function of the GLM $E(\tilde{y}_{itj}) = h(\eta_{itj})$ with the predictors

$$\eta_{itj} = \boldsymbol{x}_{it}^T\boldsymbol{\beta} + (\boldsymbol{d}_j^T \otimes \boldsymbol{z}_{it}^T)\boldsymbol{\theta} = \tilde{\boldsymbol{x}}_{itj}^T\boldsymbol{\alpha} \tag{14.17}$$

for observations $\tilde{y}_{itj}, t = 1, \ldots, T_i,\ j = 1, \ldots, m$, where m is the number of quadrature points and $\tilde{y}_{itj} = y_{it}$. This means that the original T_i observations for cluster i become $T_i m$ observations in the corresponding GLM. The essential point in (14.17) is that the unknown $\boldsymbol{a}_i$ is replaced by the known quadrature point $\boldsymbol{d}_j$, yielding a weighted score function of GLM type.

MLEs for $\boldsymbol{\alpha}$ have to be computed by an iterative procedure such as Newton-Raphson or Fisher scoring. Both algorithms imply the calculation of the observed or expected information matrix. Since the weights c_{ij}^{GH} depend on the parameters to be estimated, the analytical derivation of information matrices is cumbersome. An alternative is to calculate the observed information matrix by numerical differentiation of s^{GH}. A direct maximization procedure for cumulative logit and probit models is considered by Hedeker and Gibbons (1994). For more details see Hinde (1982), Anderson and Aitkin (1985), and Fahrmeir and Tutz (2001, Section 7.4).

If the number of quadrature points in a Gauss-Hermite quadrature is large enough, the approximation of the likelihood becomes sufficiently accurate. Thus, as n and the number of quadrature points tend to infinity, the MLEs for $\boldsymbol{\beta}$ will be consistent and asymptotically normal under the usual regularity conditions. A procedure that may reduce the number of quadrature points is the adaptive Gauss-Hermite quadrature (Liu and Pierce, 1994; Pinheiro and Bates, 1995; Hartzel et al., 2001). An adaptive quadrature is based on the log-likelihood (14.14); it first centers the modes with respect to the mode of the function being integrated and in addition scales them according to the curvature (see also Appendix E.1).

Indirect Maximization Based on the EM Algorithm

Indirect maximization of the log-likelihood (14.14) can be obtained by use of an EM algorithm (for the general form see Appendix B). The EM algorithm distinguishes between the observable data, given by the response vector $\boldsymbol{y}$, and the unobservable data, given by $\boldsymbol{a}^T = (\boldsymbol{a}_1^T, \ldots, \boldsymbol{a}_n^T)$. The complete data log-density is

$$\log f(\boldsymbol{y}, \boldsymbol{a}; \boldsymbol{\alpha}) = \sum_{i=1}^{n} \log f(\boldsymbol{y}_i|\boldsymbol{a}_i; \boldsymbol{\alpha}) + \sum_{i=1}^{n} \log p(\boldsymbol{a}_i). \tag{14.18}$$

In the E-step of the $(s+1)$th EM cycle, one determines the conditional expectation, given the data $\boldsymbol{y}$ and an estimate $\boldsymbol{\alpha}^{(s)}$ from the previous EM cycle:

$$M(\boldsymbol{\alpha}|\boldsymbol{\alpha}^{(s)}) = E\{\log f(\boldsymbol{y}, \boldsymbol{a}; \boldsymbol{\alpha})|\boldsymbol{y}; \boldsymbol{\alpha}^{(s)}\} = \int \log(f(\boldsymbol{y}, \boldsymbol{a}; \boldsymbol{\alpha}))f(\boldsymbol{a}|\boldsymbol{y}; \boldsymbol{\alpha}^{(s)})d\boldsymbol{a},$$

where the density $f(\boldsymbol{a}|\boldsymbol{y}; \boldsymbol{\alpha}^{(s)})$ denotes the posterior

$$f(\boldsymbol{a}|\boldsymbol{y}; \boldsymbol{\alpha}^{(s)}) = \frac{\prod_{i=1}^{n} f(\boldsymbol{y}_i|\boldsymbol{a}_i; \boldsymbol{\alpha}^{(s)}) \prod_{i=1}^{n} p(\boldsymbol{a}_i)}{\prod_{i=1}^{n} \int f(\boldsymbol{y}_i|\boldsymbol{a}_i; \boldsymbol{\alpha}^{(s)})p(\boldsymbol{a}_i)d\boldsymbol{a}_i}, \tag{14.19}$$

which is obtained from Bayes' theorem and the conditional independence given the random effects. One obtains

$$M(\boldsymbol{\alpha}|\boldsymbol{\alpha}^{(s)}) = \sum_{i=1}^{n} k_i^{-1} \int [\log f(\boldsymbol{y}_i|\boldsymbol{a}_i; \boldsymbol{\alpha}) + \log g(\boldsymbol{a}_i)] f(\boldsymbol{y}_i|\boldsymbol{a}_i; \boldsymbol{\alpha}^{(s)})p(\boldsymbol{a}_i)d\boldsymbol{a}_i,$$

where the factor $k_i = \int f(\boldsymbol{y}_i|\boldsymbol{a}_i; \boldsymbol{\alpha}^{(s)})g(\boldsymbol{a}_i)d\boldsymbol{a}_i$ does not depend on the parameters $\boldsymbol{\alpha}$ and the re-parameterized random effects $\boldsymbol{a}_i$. The integral in $M(\boldsymbol{\alpha}|\boldsymbol{\alpha}^{(s)})$ has to be approximated, for example, by Gauss-Hermite or Monte Carlo integration. When assuming a normal distribution for the random effects, using the Gauss-Hermite integration yields the approximation

$$M^{GH}(\boldsymbol{\alpha}|\boldsymbol{\alpha}^{(s)}) = \sum_{i=1}^{n} \sum_{j} c_{ij}^{GH} [\log f(\boldsymbol{y}_i|\boldsymbol{d}_j; \boldsymbol{\alpha}) + \log p(\boldsymbol{d}_j)] \tag{14.20}$$

with the weight factors

$$c_{ij}^{GH} = \frac{v_j f(\boldsymbol{y}_i|\boldsymbol{d}_j; \boldsymbol{\alpha}^{(s)})}{\sum_k v_k f(\boldsymbol{y}_i|\boldsymbol{d}_k; \boldsymbol{\alpha}^{(s)})}, \quad \sum_j c_{ij}^{GH} = 1,$$

where $\boldsymbol{d}_j$ denotes the fixed quadrature points and v_j denotes fixed weights associated with $\boldsymbol{d}_j$. The essential point, which makes maximization easier, is that the weights c_{ij}^{GH} do not depend on $\boldsymbol{\alpha}$, in contrast to the weights in direct maximization given in (14.16).

In the M-step the function $M(\boldsymbol{\alpha}|\boldsymbol{\alpha}^{(p)})$ is maximized with respect to $\boldsymbol{\alpha}$. For the Gauss-Hermite approximation one obtains the derivative

$$\frac{\partial M^{GH}(\boldsymbol{\alpha}|\boldsymbol{\alpha}^{(p)})}{\partial \boldsymbol{\alpha}} = \sum_{i=1}^{n} \sum_{j=1}^{m} c_{ij}^{GH} \frac{\partial \log f(\boldsymbol{y}_i|\boldsymbol{d}_{ij}; \boldsymbol{\alpha})}{\partial \boldsymbol{\alpha}}, \tag{14.21}$$

which has the same form as (14.15) but with weights that do not depend on $\boldsymbol{\alpha}$. Solving the equation $\partial M^{GH}(\boldsymbol{\alpha}|\boldsymbol{\alpha}^{(s)})/\partial\boldsymbol{\alpha} = \boldsymbol{0}$ uses that $\partial M^{GH}(\boldsymbol{\alpha}|\boldsymbol{\alpha}^{(s)})/\partial\boldsymbol{\alpha}$ corresponds to the weighted score function of the GLM $E(\tilde{y}_{itj}) = h(\eta_{itj})$ with predictors

$$\eta_{itj} = \boldsymbol{x}_{it}^T\boldsymbol{\beta} + (\boldsymbol{d}_j^T \otimes \boldsymbol{z}_{it}^T)\boldsymbol{\theta} = \tilde{\boldsymbol{x}}_{itj}^T\boldsymbol{\alpha} \tag{14.22}$$

for observations $\tilde{y}_{itj}, t = 1, \ldots, T_i,\ j = 1, \ldots, m$, where m is the number of quadrature points and $\tilde{y}_{itj} = y_{it}$. The resulting EM algorithm is often slow but rather stable and simple to implement.

The EM algorithm has been used by Hinde (1982), Brillinger and Preisler (1983), Anderson and Hinde (1988), and Jansen (1990) for one-dimensional random effects; Anderson and Aitkin (1985) and Im and Gianola (1988) for bivariate random effects; and Tutz and Hennevogl (1996) for ordinal models. An elaborate Monte Carlo technique was proposed by Booth and Hobert (1999).

14.3.2 Posterior Mean Estimation of Random Effects

Prediction of random effects $\boldsymbol{b}_i$ may be based on the posterior density of $\boldsymbol{b}_i$ given the observations $\boldsymbol{y}^T = (\boldsymbol{y}_1^T, \ldots, \boldsymbol{y}_n^T)$. Due to the independence assumptions, the posterior of $\boldsymbol{b}_i$ depends only on $\boldsymbol{y}_i$ and one obtains

$$f(\boldsymbol{b}_i|\boldsymbol{y}_i;\boldsymbol{\beta},\boldsymbol{Q}) = \frac{f(\boldsymbol{y}_i|\boldsymbol{b}_i;\boldsymbol{\beta})p(\boldsymbol{b}_i;\boldsymbol{Q})}{\int f(\boldsymbol{y}_i|\boldsymbol{b}_i;\boldsymbol{\beta})p(\boldsymbol{b}_i;\boldsymbol{Q})d\boldsymbol{b}_i}.$$

After replacing $\boldsymbol{\beta}$ and $\boldsymbol{Q}$ by estimates $\hat{\boldsymbol{\beta}}, \hat{\boldsymbol{Q}}$ one obtains the posterior mean estimate:

$$\hat{\boldsymbol{b}}_i = \int \boldsymbol{b}_i f(\boldsymbol{b}_i|\boldsymbol{y}_i, \hat{\boldsymbol{\beta}}, \hat{\boldsymbol{Q}})d\boldsymbol{b}_i.$$

Evaluation of integrals again makes approximation techniques, for example, numerical or Monte Carlo techniques, necessary.

14.3.3 Penalized Quasi-Likelihood Estimation for Given Variance

Alternative approaches to fitting random effects models yield a penalized log-likelihood (or quasi-log-likelihood) for the estimation of $\boldsymbol{\beta}$ and $\boldsymbol{b}_1, \ldots, \boldsymbol{b}_n$. Let $\boldsymbol{\beta}$ and $\boldsymbol{b}_1, \ldots, \boldsymbol{b}_n$ be collected in $\boldsymbol{\delta}^T = (\boldsymbol{\beta}^T, \boldsymbol{b}_1^T, \ldots, \boldsymbol{b}_n^T)$ and $\boldsymbol{Q}$ be known. The "joint maximization" approaches aim at estimating $\boldsymbol{\beta}$ and $\boldsymbol{b}_1, \ldots, \boldsymbol{b}_n$ together.

Motivation as Posterior Mode Estimator

Since $\boldsymbol{b}_1, \ldots, \boldsymbol{b}_n$ are independent, the assumption of a flat prior on $\boldsymbol{\beta}$ ($\text{cov}(\boldsymbol{\beta}) \to \infty$) yields

$$p(\boldsymbol{\delta};\boldsymbol{Q}) \propto \prod_{i=1}^{n} p(\boldsymbol{b}_i;\boldsymbol{Q}),$$

which depends only on the covariance $cov(\boldsymbol{b}_i) = \boldsymbol{Q}$. Bayes' theorem yields

$$f(\boldsymbol{\delta}|\{\boldsymbol{y}_i\};\boldsymbol{Q}) = \frac{\prod_{i=1}^{n} f(\boldsymbol{y}_i|\boldsymbol{b}_i,\boldsymbol{\beta}) \prod_{i=1}^{n} p(\boldsymbol{b}_i,\boldsymbol{Q})}{\int \prod_{i=1}^{n} f(\boldsymbol{y}_i|\boldsymbol{b}_i,\boldsymbol{\beta})p(\boldsymbol{b}_i;\boldsymbol{Q})d\boldsymbol{b}_1 \ldots d\boldsymbol{b}_n d\boldsymbol{\beta}}. \tag{14.23}$$

A posterior mean implies a heavy computational burden. However, posterior mode estimation turns out to be a feasible alternative. By using only the nominator of (14.23), maximization of (14.23) is equivalent to maximizing the log-posterior

$$\sum_{i=1}^{n} \log(f(\boldsymbol{y}_i|\boldsymbol{b}_i;\boldsymbol{\beta})) + \sum_{i=1}^{n} \log\ p(\boldsymbol{b}_i;\boldsymbol{Q})$$

with respect to $\boldsymbol{\delta}$. For normally distributed random effects one obtains (after dropping irrelevant terms)

$$l_p(\boldsymbol{\delta}) = \sum_{i=1}^{n} \log\left(f(\boldsymbol{y}_i|\boldsymbol{b}_i,\boldsymbol{\beta})\right) - \frac{1}{2}\sum_{i=1}^{n} \boldsymbol{b}_i^T \boldsymbol{Q}^{-1}\boldsymbol{b}_i. \tag{14.24}$$

Motivation by Laplace's Approximation

For the general mixed model (14.12) the marginal likelihood has the form

$$L(\boldsymbol{\beta}, \boldsymbol{Q_b}) = \int f(\boldsymbol{y}|\boldsymbol{b}; \boldsymbol{\beta}) \;\; p(\boldsymbol{b}; \boldsymbol{Q_b}) d\boldsymbol{b}.$$

For normally distributed $\boldsymbol{b} \sim N(\boldsymbol{0}, \boldsymbol{Q_b})$ one obtains

$$L(\boldsymbol{\beta}, \boldsymbol{Q_b}) = |\boldsymbol{Q_b}|^{-1/2}(2\pi)^{-q/2} \int \exp\{\log(f(\boldsymbol{y}|\boldsymbol{b}; \boldsymbol{\beta})) - \frac{1}{2}\boldsymbol{b}^T \boldsymbol{Q_b}^{-1} \boldsymbol{b}\} d\boldsymbol{b},$$

$q = dim(\boldsymbol{b})$. With $\kappa_{\boldsymbol{\beta}}(\boldsymbol{b}) = -\log f(\boldsymbol{y}|\boldsymbol{b}; \boldsymbol{\beta}) + \frac{1}{2}\boldsymbol{b}^T \boldsymbol{Q_b}^{-1} \boldsymbol{b}$, the integrand is $\exp(-\kappa_{\boldsymbol{\beta}}(\boldsymbol{b}))$ and Laplace approximation (see Appendix E.1) yields

$$L(\boldsymbol{\beta}, \boldsymbol{Q_b}) \approx |\boldsymbol{Q_b}|^{-1/2} \exp(-\kappa_{\boldsymbol{\beta}}(\tilde{\boldsymbol{b}})) \; |\partial^2 \kappa_{\boldsymbol{\beta}}(\tilde{\boldsymbol{b}})/\partial \boldsymbol{b} \partial \boldsymbol{b}^T)|^{-1/2},$$

where $\tilde{\boldsymbol{b}}$ minimizes $\kappa_{\boldsymbol{\beta}}(\boldsymbol{b})$. One obtains the approximative log-likelihood

$$l(\boldsymbol{\beta}, \boldsymbol{Q_b}) \approx -\kappa_{\boldsymbol{\beta}}(\tilde{\boldsymbol{b}}) - \frac{1}{2}\log(|\boldsymbol{Q_b}|) - \frac{1}{2}\log|\partial^2 \kappa(\tilde{\boldsymbol{b}})/\partial \boldsymbol{b} \partial \boldsymbol{b}^T|.$$

The second derivative $\partial^2 \kappa_{\boldsymbol{\beta}}(\boldsymbol{b})/\partial \boldsymbol{b} \partial \boldsymbol{b}^T$ has the form

$$\frac{\partial^2 \kappa(\boldsymbol{b})}{\partial \boldsymbol{b} \partial \boldsymbol{b}^T} = \boldsymbol{Z}^T \boldsymbol{D} \boldsymbol{\Sigma}^{-1} \boldsymbol{D}^T \boldsymbol{Z} + \boldsymbol{Q_b}^{-1} + \boldsymbol{M},$$

where $\boldsymbol{D} = diag(\partial h(\eta_1)/\partial \eta, \dots, \partial h(\eta_N)/\partial \eta)$, $\boldsymbol{\Sigma} = diag(\sigma_1^2, \dots, \sigma_N^2), \sigma_i^2 = \text{var}(b_i)$, and the remainder term $\boldsymbol{M}$ has expectation zero. Thus, by ignoring $\boldsymbol{M}$ and inserting $\kappa_{\boldsymbol{\beta}}(\tilde{\boldsymbol{b}})$, the approximative log-likelihood has the form

$$\begin{aligned} l(\boldsymbol{\beta}, \boldsymbol{Q_b}) &\approx \log f(\boldsymbol{y}|\tilde{\boldsymbol{b}}, \boldsymbol{\beta}) - \frac{1}{2}\tilde{\boldsymbol{b}}^T \boldsymbol{Q_b}^{-1} \tilde{\boldsymbol{b}} - \frac{1}{2}\log(|\boldsymbol{Z}^T \boldsymbol{D} \boldsymbol{\Sigma}^{-1} \boldsymbol{D} \boldsymbol{Z} + \boldsymbol{Q_b}^{-1}|) - \frac{1}{2}\log(|\boldsymbol{Q_b}|)) \\ &= \log f(\boldsymbol{y}|\tilde{\boldsymbol{b}}, \boldsymbol{\beta}) - \frac{1}{2}\tilde{\boldsymbol{b}}^T \boldsymbol{Q_b}^{-1} \tilde{\boldsymbol{b}} - \frac{1}{2}\log|\boldsymbol{Z}^T \boldsymbol{D} \boldsymbol{\Sigma}^{-1} \boldsymbol{D} \boldsymbol{Z} \boldsymbol{Q_b} + \boldsymbol{I}|. \end{aligned}$$

Breslow and Clayton (1993) also ignore the last term. Since $\tilde{\boldsymbol{b}}$ is the minimum of $\kappa_{\boldsymbol{\beta}}(\boldsymbol{b})$, by definition of $\kappa_{\boldsymbol{\beta}}(\boldsymbol{b})$ it may also be seen as the maximum of the penalized likelihood:

$$l_p(\boldsymbol{\delta}) = \log(f(\boldsymbol{y}|\boldsymbol{b}, \boldsymbol{\beta})) - \frac{1}{2}\boldsymbol{b}^T \boldsymbol{Q_b}^{-1} \boldsymbol{b}, \tag{14.25}$$

which is the general form of (14.24).

Solution of the Penalized Likelihood Problem

$l_p(\boldsymbol{\delta})$ may be maximized by solving $s_p(\boldsymbol{\delta}) = (\partial l_p(\boldsymbol{\delta})/\partial \boldsymbol{\beta}^T, \partial l_p(\boldsymbol{\delta})/\partial \boldsymbol{b}^T)^T = \boldsymbol{0}$, where the first term is given by

$$\frac{\partial l_p(\boldsymbol{\delta})}{\partial \boldsymbol{\beta}} = \boldsymbol{X}^T \boldsymbol{D}(\boldsymbol{\delta}) \boldsymbol{\Sigma}^{-1}(\boldsymbol{y} - \boldsymbol{\mu}(\boldsymbol{\delta})),$$

and the second is

$$\frac{\partial l_p(\boldsymbol{\delta})}{\partial \boldsymbol{b}} = \boldsymbol{Z}^T \boldsymbol{D}(\boldsymbol{\delta}) \boldsymbol{\Sigma}^{-1}(\boldsymbol{\delta})(\boldsymbol{y} - \boldsymbol{\mu}(\boldsymbol{\delta})) - \boldsymbol{Q_b}^{-1} \boldsymbol{b}.$$

In closed form the penalized likelihood and the corresponding score function, which use predictor $\eta = X\beta + Zb$, are given as

$$l_p(\delta) = \log(f(y|b, \beta)) - \frac{1}{2}\delta^T K\delta,$$

$$s_p(\delta) = \tilde{X}^T D\Sigma^{-1}(y - \mu) - K\delta,$$

where $\tilde{X} = [X|Z]$ with

$$K = \begin{pmatrix} 0 & 0 \\ 0 & Q_b^{-1} \end{pmatrix}.$$

Iterative pseudo-Fisher scoring has the form

$$\hat{\delta}^{(k+1)} = \hat{\delta}^{(k)} + F_p^{-1}(\hat{\delta}^{(k)})s_p(\hat{\delta}^{(k)}), \tag{14.26}$$

with $F_p(\delta) = \tilde{X}^T W\tilde{X} + K$, $W(\delta) = D(\delta)\Sigma^{-1}(\delta)D(\delta)^T$. An alternative form is

$$\hat{\delta}^{(k+1)} = (\tilde{X}^T W(\hat{\delta}^{(k)})\tilde{X} + K)^{-1}\tilde{X}^T W(\hat{\delta}^{(k)})\tilde{\eta}(\hat{\delta}^{(k)}),$$

with pseudo-observations $\tilde{\eta}(\delta) = \tilde{X}\delta + D^{-1}(\delta)(y - \mu(\delta))$. Therefore, one step of pseudo-Fisher scoring solves the system of equations for the best linear unbiased estimation in normal response models (Harville, 1977):

$$\begin{bmatrix} X^T WX & X^T WZ \\ Z^T WX & Q_b^{-1} + Z^T WZ \end{bmatrix} \hat{\delta}^{(k+1)} = \begin{bmatrix} X^T W \\ Z^T W \end{bmatrix} \tilde{\eta}(\hat{\delta}^{(k)}),$$

where the dependence of matrices W and D on δ is suppressed. The solutions have the form

$$\hat{\beta}^{(k+1)} = [X^T V^{-1} X]^{-1} X^T V^{-1}\tilde{\eta}(\hat{\delta}^{(k)}),$$

$$\hat{b}^{(k+1)} = Q_b Z^T V^{-1}(y - X\hat{\beta}^{(k)}),$$

with $V = W^{-1} + ZQ_b Z^T$.

For clustered data y_{it} with predictor $\eta_{it} = x_{it}^T\beta + z_{it}^T b_i$ the score functions simplify to

$$\frac{\partial l_p(\delta)}{\partial b_i} = Z_i^T D_i(\delta)\Sigma_i^{-1}(y_i - \mu_i) - Q^{-1}b_i,$$

$$\frac{\partial l_p(\delta)}{\partial \beta} = \sum_{i=1}^{n} X_i^T D_i(\delta)\Sigma_i^{-1}(y_i - \mu_i),$$

where $D_i(\delta) = \text{diag}(\partial h(\eta_{i1})/\partial\eta, \partial h(\eta_{i2})/\partial\eta, \dots)$, $\Sigma_i = \text{diag}(\sigma_{i1}, \sigma_{i2}, \dots)$. Iterative solutions for single effects are given as

$$\hat{b}_i^{(k+1)} = QZ_i^T V_i^{-1}(y_i - X_i\hat{\beta}^{(k)}).$$

Details for the inversion of the pseudo-Fisher matrix $F_p(\delta)$ are given in the Appendix (see E.1).

14.3.4 Estimation of Variances

The penalized log-likelihood approach considered in Section 14.3.3 yields estimates of $\delta^T = (\beta^T, b_1^T, \dots, b_n^T)$ under the assumption that Q is known. In the following we consider estimation methods for Q.

REML-Type Estimates

For the estimation of variances Breslow and Clayton (1993) maximize the profile likelihood that is associated with the normal theory model. With $\boldsymbol{\beta}$ replaced by $\hat{\boldsymbol{\beta}}$ one maximizes

$$l(\hat{\boldsymbol{\beta}}, \boldsymbol{Q}_b) = -\frac{1}{2}\log(|\boldsymbol{V}|) - \frac{1}{2}\log(|\boldsymbol{X}^T\boldsymbol{V}^{-1}\boldsymbol{X}|) - \frac{1}{2}(\tilde{\boldsymbol{\eta}}(\hat{\boldsymbol{\delta}}) - \boldsymbol{X}\hat{\boldsymbol{\beta}})^T\boldsymbol{V}^{-1}(\tilde{\boldsymbol{\eta}}(\hat{\boldsymbol{\delta}}) - \boldsymbol{X}\hat{\boldsymbol{\beta}})$$

with respect to $\boldsymbol{Q}_b$, with pseudo-observations $\tilde{\boldsymbol{\eta}}(\boldsymbol{\delta}) = \tilde{\boldsymbol{X}}\boldsymbol{\delta} + \boldsymbol{D}^{-1}(\boldsymbol{\delta})(\boldsymbol{y} - \boldsymbol{\mu}(\boldsymbol{\delta}))$. Typically, the unknown matrix $\boldsymbol{Q}_b$ is parameterized, $\boldsymbol{Q}_b = \boldsymbol{Q}_b(\gamma)$, where some structure of $\boldsymbol{Q}_b$ is assumed, and maximization refers to the parameter γ. In practice, one iterates between one step of Fisher scoring (yielding $\hat{\boldsymbol{\delta}}^{(k)}$) and one step of maximizing $l(\hat{\boldsymbol{\beta}}, \boldsymbol{Q}_b)$ (yielding $\hat{\boldsymbol{Q}}_b^{(k)}$).

Alternative Estimates

Estimates obtained by iteratively improving the estimates of $\boldsymbol{\delta}$ and $\boldsymbol{Q}$ may also be based on an alternative estimation of variances. A simple estimate, which can be derived as an approximate EM algorithm, uses the posterior mode estimates $\hat{\boldsymbol{\delta}}^{(k)}$ and posterior curvatures $\hat{\boldsymbol{V}}_{ii}^{(k)}$ (evaluated at $\hat{\boldsymbol{\delta}}^{(k)}$) by computing

$$\hat{\boldsymbol{Q}}^{(k)} = \frac{1}{n}\sum_{i=1}^{n}(\hat{\boldsymbol{V}}_{ii}^{(k)} + \hat{\boldsymbol{b}}_i^{(k)}(\hat{\boldsymbol{b}}_i^{(k)})^T).$$

Joint maximization of a penalized log-likelihood with respect to parameters and random effects appended by estimation of the variance of random effects can be justified in various ways (see also Schall, 1991; Wolfinger, 1994; McCulloch and Searle, 2001). The derivation of Breslow and Clayton (1993) is often referred to as a penalized quasi-likelihood (PQL) because it uses the more general concept of quasi-likelihood. Although modifications were proposed (Breslow and Lin, 1995; Lin and Breslow, 1996) joint maximization algorithms tend to underestimate the variance and therefore the true values of the random effects (see, for example, McCulloch, 1997). In particular, for binary data in small clusters performance might be poor. Similar approaches have been used by Stiratelli et al. (1984) for binary logistic models, Harville and Mee (1984) for cumulative models, and Wong and Mason (1985) for multilevel analysis.

Error Approximation

If the cluster sizes are large enough, one can use the normal approximation

$$\hat{\boldsymbol{\delta}} \sim N(\boldsymbol{\delta}, \boldsymbol{F}_p(\boldsymbol{\delta})^{-1})$$

to evaluate standard errors of $\hat{\boldsymbol{\delta}}$.

Example 14.3: Knee Injuries

In the knee injuries study pain was recorded for each subject after 3, 7, and 10 days of treatment. The marginal effects of treatment and the covariates gender and age were evaluated in Example 13.4. Again we consider the dichotomized response and fit the random effects logit model:

$$\text{logit}(P(y_{it} = 1|\boldsymbol{x}_i)) = b_i + x_{T,i}\beta_r + x_{G,i}\beta_G + Age_i\beta_A + Age_i^2\beta_{A^2},$$

where $x_{T,i} = 1$ for treatment, $x_{T,i} = 0$ for placebo, and $x_{G,i} = 1$ for females, $x_{T,i} = 0$ for males. Table 14.1 shows the estimated parameters resulting from a Gauss-Hermite quadrature with 20 quadrature points (GH(20)), a penalized quasi-likelihood (PQL), and the marginal model with an exchangeable correlation structure. For ease of interpretation the variable age has been centered around 30. The estimated standard deviation of the mixing distribution is 3.621 for GH(20) and 2.706 for PQL. Since PQL tends to underestimate the variation of random effects, it was to be expected that the value is smaller than for a Gauss-Hermite quadrature. Consequently, the estimated fixed effects are also smaller for PQL. However, both procedures yield distinctly larger parameter estimates than the marginal model. Of course the parameters of the marginal model are population-averaged effects, whereas the parameters of random effects models are measured on the individual level and therefore the estimates cannot be compared directly (see also the introductory remark to this chapter and Section 14.2.1).

□

TABLE 14.1: Knee data binary; Gauss-Hermite, penalized quasi-likelihood, and marginal models.

	Gauss-Hermite		PQL		Marginal	
	Coef	Pr > $\|z\|$	Coef	Pr> $\|z\|$	Coef	Robust z
Intercept	3.054	0.005	2.142	0.003	1.172	2.617
factor(Th)2	−1.861	0.027	−1.294	0.023	−0.673	−2.016
factor(Sex)1	0.607	0.494	0.414	0.506	0.265	0.724
Age	0.032	0.460	0.022	0.463	0.013	0.788
Age2	−0.015	0.003	−0.010	0.001	−0.006	−3.085

14.3.5 Bayesian Approaches

In Bayesian approaches all the unknown parameters are considered as random variables steming from prior distributions. Thus the structural assumption is considered as conditional on $\boldsymbol{\beta}$ and $\boldsymbol{b}_i$ and has the form

$$E(y_{it}|\boldsymbol{\beta}, \boldsymbol{b}_i) = h(\boldsymbol{X}_{it}\boldsymbol{\beta} + \boldsymbol{Z}_{it}\boldsymbol{b}_i). \tag{14.27}$$

In addition to a mixing distribution for random effects $p(\boldsymbol{b}_i)$ one has to specify a prior distribution $p(\boldsymbol{\beta})$. This can be done in a hierarchical way by specifying, for example, a normal distribution $p(\boldsymbol{b}_i|\boldsymbol{Q}) \sim N(\boldsymbol{0}, \boldsymbol{Q})$ together with a prior for the hyperparameter $\boldsymbol{Q}$. A better choice than Jeffreys prior, which may be improper, is the assumption of a highly dispersed inverse Wishart distribution $\boldsymbol{Q} \sim IW_r(\xi, \Psi)$ with density

$$p(\boldsymbol{Q}) \propto |\boldsymbol{Q}|^{-(\xi+m+1)/2} \exp(-tr(\Psi \boldsymbol{Q}^{-1})/2),$$

where m is the dimension of $\boldsymbol{b}_i$ and ξ and Ψ are hyperparameters that have to be fixed (see Besag et al., 1995). In the one-dimensional case the distribution reduces to the inverse gamma distribution

$$p(\sigma^2|\alpha, \beta) = \frac{\beta^\alpha \sigma^{2(-\alpha-1)} exp(-\beta/\sigma^2)}{\Gamma(\alpha)}$$

with shape parameter $\alpha > 0$ and scale parameter $\beta > 0$.

If one assumes conditional independence among the response variables $y_{it}|\boldsymbol{b}_i, \boldsymbol{\beta}$, the random effects $\boldsymbol{b}_i|\boldsymbol{Q}$, the regression parameters $\boldsymbol{\beta}$, and the hyperparameter $\boldsymbol{Q}$, one obtains for the posterior distribution

$$p(\boldsymbol{\beta}, \boldsymbol{b}_1, \dots, \boldsymbol{b}_n, \boldsymbol{Q}|data) \propto \prod_{i=1}^{n}\prod_{t=1}^{T_i} f(y_{it}|\boldsymbol{\beta}, \boldsymbol{b}_i), p(\boldsymbol{\beta}) \prod_{i=1}^{n} p(\boldsymbol{b}_i|\boldsymbol{Q}) p(\boldsymbol{Q}),$$

and the full conditionals $p(\boldsymbol{\beta}|\cdot), p(\boldsymbol{b}_i|\cdot), p(\boldsymbol{Q}|\cdot)$, given the data and the rest of parameters, simplify to

$$
\begin{aligned}
p(\boldsymbol{\beta}|\cdot) &\propto \prod_{i=1}^{n}\prod_{t=1}^{T_i} f(y_{it}|\boldsymbol{\beta}, \boldsymbol{b}_i)p(\boldsymbol{\beta}),\\
p(\boldsymbol{b}_i|\cdot) &\propto \prod_{t=1}^{T_i} f(y_{it}|\boldsymbol{\beta}, \boldsymbol{b}_i)p(\boldsymbol{b}_i|\boldsymbol{Q}),\\
p(\boldsymbol{Q}|\cdot) &\propto \prod_{i=1}^{n} f(\boldsymbol{b}_i|\boldsymbol{Q})p(\boldsymbol{Q}).
\end{aligned}
$$

The latter conditional $p(\boldsymbol{Q}|\cdot)$ is again an inverse Wishart with updated parameters $\xi + n/2$ and $\Psi + \frac{1}{2}\sum_{i=1}^{n} \boldsymbol{b}_i\boldsymbol{b}_i^T$. For computational details see Zeger and Karim (1991) and Gamerman (1997). An extensive treatment of Bayesian mixed models is found in Fahrmeir and Kneib (2010); see also Chib and Carlin (1999), Kinney and Dunson (2007), and Tüchler (2008).

14.4 Multicategorical Response Models

In the knee injury example pain was originally measured on a five-point scale. With ordered categories like that, which are subjective judgements, heterogeneity is to be expected since each individual has its own sensitivity for pain. Therefore, modeling a subject-specific parameter seems warranted. In the following we consider the extension to multicategorical responses.

Ordered Response Categories

Let Y_{it} denote observation t on unit i ($i = 1, \dots, n, t = 1, \dots, T_i$), where $Y_{it} \in \{1, \dots, k\}$. Then a simple ordinal mixed model of the cumulative type is

$$P(Y_{it} \le r|\boldsymbol{x}_{it}) = F(b_i + \gamma_{0r} + \boldsymbol{x}_{it}^T\boldsymbol{\gamma}),$$

where the $\boldsymbol{x}_{it}$'s are design vectors built from covariates, $\boldsymbol{\gamma}$ is a population-specific parameter, and b_i is a cluster-specific random effect, which follows a mixing distribution. The corresponding sequential model has the form

$$P(Y_{it} = r|Y_{it} \ge r, \boldsymbol{x}) = F(b_i + \gamma_{0r} + \boldsymbol{x}_{it}^T\boldsymbol{\gamma}).$$

In both models, the random effect b_i represents an individual response level that differs across subjects.

As shown in Section 9.5, both models can be written as multivariate GLMs if no random effect is present. With random effects one obtains the general form

$$g(\boldsymbol{\pi}_{it}) = \boldsymbol{X}_{it}\boldsymbol{\beta} + \boldsymbol{Z}_{it}\boldsymbol{b}_i \quad \text{or} \quad \boldsymbol{\pi}_{it} = h(\boldsymbol{X}_{it}\boldsymbol{\beta} + \boldsymbol{Z}_{it}\boldsymbol{b}_i),$$

where $\boldsymbol{\pi}_{it}^T = (\pi_{it1}, \dots, \pi_{itq})$, $q = k - 1$, is the vector of response probabilities, g is the (multivariate) link function, and $h = g^{-1}$ is the inverse link function. The matrix $\boldsymbol{Z}_{it}$ specifies the structure of the random effects. For simple random intercepts the matrix is $\boldsymbol{Z}_{it}^T = (1, \dots, 1)$. In the case of random slopes, the matrix contains the corresponding variables. In the more general case with category-specific random effects b_{ir}, problems may occur when using the cumulative model since the intercepts have to be ordered. However, the assumption $\gamma_{01} + b_{i1} \le \gamma_{02} + b_{i2} \le \dots$ cannot hold if a normal distribution for b_{ir} is assumed. The problem can be overcome by reparameterizing the thresholds (see Tutz and Hennevogl, 1996). An advantage of the sequential model is that random intercepts are not restricted and binary mixed model software can be used to fit the model (see Problem 14.3).

Example 14.4: Knee Injuries

In the knee injuries study pain was recorded for each subject after 3, 7, and 10 days of treatment. A simple binary model with a random intercept and dichotomized responses was investigated in Example 14.3. Now we fit the sequential random effects logit model:

$$\text{logit}(P(Y_{it} = r|Y_{it} \geq r, \boldsymbol{x}_i)) = b_i + \beta_{0r} + x_{T,i}\beta_T + Age_i\beta_A + Age_i^2\beta_{A^2},$$

where $x_{T,i} = 1$ for treatment, $x_{T,i} = 0$ for placebo, and age (centered around 30) is included as a linear and quadratic term. Gender has been omitted because it turned out not to be influential. Table 14.2 shows the estimated parameters resulting from a Gauss-Hermite quadrature with 25 quadrature points (R function *lmer*) and penalized quasi-likelihood (PQL; R function *glmmPQL*). The standard deviation of the random effect is 5.425, with a standard error 0.630 for PQL and 5.910 for Gauss-Hermite. In contrast to the dichotomized response now age should not be neglected. The effect of therapy now refers to the transition between categories of pain with a distinct effect of therapy. An alternative is the cumulative model

$$\text{logit}(P(Y_{it} \leq r|\boldsymbol{x}_i)) = b_i + \beta_{0r} + x_{T,i}\beta_r + Age_i\beta_A + Age_i^2\beta_{A^2}.$$

Because of restrictions on the parameters the model is harder to fit. The R function *clmm* yields estimates for a Laplace approximation and a Gauss-Hermite approximation, but no standard errors are available. With standard deviation 6.37 for the Laplace approximation and 6.25 for the Gauss-Hermite, heterogeneity seems to be even stronger than for the sequential model. □

TABLE 14.2: Knee data sequential model; Gauss-Hermite and penalized quasi-likelihood.

	Gauss-Hermite		PQL	
	Coef	p-Value	Coef	p-Value
Therapy	2.402	0.031	2.113	0.037
Age	0.002	0.976	0.0003	0.996
Age2	0.023	0.001	0.017	0.006

Example 14.5: Recovery Scores

In the recovery data (Example 14.2) recovery scores were measured repeatedly on a categorical scale immediately after admission to the recovery room and at minutes 5, 15, and 30 following admission. The covariates are dosage of the anaesthetic (four levels), duration of surgery (in minutes), and age of the child (in months). The cumulative model that is considered is

$$\begin{aligned}\text{logit}(P(Y_{it} \leq r|\boldsymbol{x}_i)) =& b_i + \beta_{0r} + \text{Dose1}_i\beta_{D1} + \text{Dose2}_i\beta_{D2} + \text{Dose3}_i\beta_{D3} + \text{Duration}_i\beta_{Du} \\ &+ \text{Age}_i\beta_A + \text{Age}_i^2\beta_{A^2} + \text{Rep1}_i\beta_{R1} + \text{Rep2}_i\beta_{R2} + \text{Rep3}_i\beta_{R3},\end{aligned}$$

where Dose_i and Rep_i are (0-1)-dummies for dose and repetition, respectively. Instead of fitting the whole model we used a boosting algorithm that successively includes the relevant variables. Figure 14.6 shows the coefficient buildups for the standardized predictors. It is seen that the dominant influence comes from the replications; all other predictors are neglected (see also Tutz and Groll, 2010a, and Exercise 14.4) □

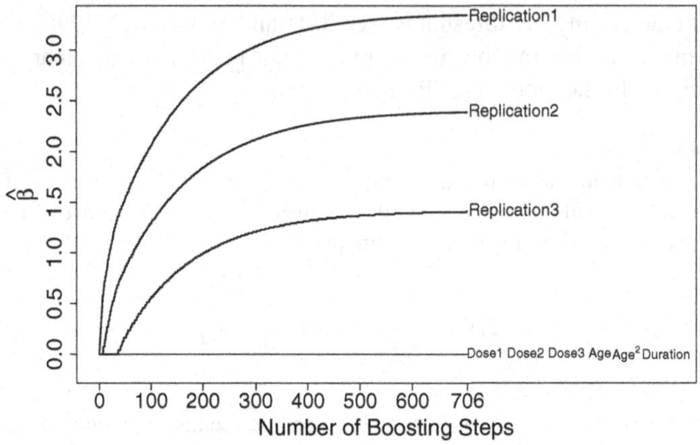

FIGURE 14.6: Coefficient buildup for recovery data.

Mixed Multinomial Logit Model

In the multinomial logit model (see Chapter 8),

$$P(Y_{it} = r|\boldsymbol{x}) = \frac{\exp(\eta_{itr})}{1 + \sum_{s=1}^{k-1} \exp(\eta_{its})}, \qquad r = 1, \ldots, k-1,$$

random intercepts are included by assuming $\eta_{itr} = b_{ir} + \boldsymbol{x}^T \boldsymbol{\beta}_r$, where the vector of random intercepts, $\boldsymbol{b}_i = (b_{i1}, \ldots, b_{i,k-1})$, is normally distributed, $\mathrm{N}(\boldsymbol{0}, \boldsymbol{Q})$. Then the probability of response in one of the categories $\{1, \ldots, k\}$ may vary across individuals.

Generalized Linear Mixed Model for Clustered Data

$$g(\boldsymbol{\pi}_{it}) = \boldsymbol{X}_{it}^T \boldsymbol{\beta} + \boldsymbol{Z}_{it}^T \boldsymbol{b}_i$$

with

$$\mathrm{E}(\boldsymbol{b}_i) = \boldsymbol{0}, \ \mathrm{cov}(\boldsymbol{b}_i) = \boldsymbol{Q}$$

In the same way as for univariate responses one can find closed forms by collecting the observations of one cluster in a vector. Let the underlying observations that have mean $\boldsymbol{\pi}_{it}$ be given as $\boldsymbol{y}_{it}^T = (y_{it1}, \ldots, y_{itq})$, where $y_{its} = 1$ if $Y_{it} = s$. Then the whole set of responses is $\boldsymbol{y}^T = (\boldsymbol{y}_1^T, \ldots, \boldsymbol{y}_n^T)$, $\boldsymbol{y}_i^T = (\boldsymbol{y}_{i1}^T, \ldots, \boldsymbol{y}_{iT_i}^T)$, and $\boldsymbol{\pi}_i = \mathrm{E}(\boldsymbol{y}_i)$ has the matrix form $g(\boldsymbol{\pi}_i) = \boldsymbol{X}_i \boldsymbol{\beta} + \boldsymbol{Z}_i \boldsymbol{b}_i$, with $\boldsymbol{X}_i$ and $\boldsymbol{Z}_i$ containing matrices $\boldsymbol{X}_{it}, \boldsymbol{Z}_{it}, t = 1, \ldots, T_i$. Stacking matrices together yields the general form $g(\boldsymbol{\mu}) = \boldsymbol{X}\boldsymbol{\beta} + \boldsymbol{Z}\boldsymbol{b}$, where $\boldsymbol{X}^T = (\boldsymbol{X}_1^T, \ldots, \boldsymbol{X}_n^T)$ and $\boldsymbol{Z}$ is a block-diagonal matrix with blocks $\boldsymbol{Z}_i$, $\boldsymbol{b}^T = (\boldsymbol{b}_1^T, \ldots, \boldsymbol{b}_n^T)$. With that notation and small adaptations one may use the estimation concepts for univariate GLMMs.

Cumulative type random effects models were considered by Harville and Mee (1984), Jansen (1990), and Tutz and Hennevogl (1996); adjacent categories type models were considered by Hartzel et al. (2001). Hartzel, Agresti, and Caffo (2001) investigated multinomial random effects models.

14.5 The Marginalized Random Effects Model

Random effects models parameterize the mean response conditional on covariates and random effects. The interpretation of parameters is subject-specific and refers to the change in the mean response having accounted for all the conditioning variables including the random effects. Since it is implicitly assumed that the distribution of the random effects is correct, violation of the random effects distribution will affect the estimation and interpretation of parameters. Marginal models as considered in Chapter 13 do not depend on the random effects distribution because they do not assume random effects. Marginal models that include random effects but are less sensitive to the assumed distribution are the *marginalized random effects models* or the *marginalized latent variable models*.

Marginalized random effects models specify a marginal model that is connected to a latent variable model (see Fitzmaurice and Laird, 1993; Azzalini, 1994; Heagerty, 1999; Heagerty and Zeger, 1996). Let μ_{it} denote the marginal mean, $\mu_{it} = \mathrm{E}(y_{it}|\boldsymbol{x}_{it})$. One specifies a regression structure for the marginal mean by

$$\mu_{it} = h(\boldsymbol{x}_{it}^T\boldsymbol{\beta}) \quad \text{or} \quad g(\mu_{it}) = \boldsymbol{x}_{it}^T\boldsymbol{\beta}, \tag{14.28}$$

where $g = h^{-1}$ is the link function. The second component of the model describes the dependence among measurements within a cluster by conditioning on the random effect $\boldsymbol{b}_i$, assuming

$$\mu_{it}^c = h(\Delta_{it} + b_{it}) \quad \text{or} \quad g(\mu_{it}^c) = \Delta_{it} + b_{it}, \tag{14.29}$$

where $\mu_{it}^c = \mathrm{E}(y_{it}|\boldsymbol{x}_i, \boldsymbol{b}_i)$ with $\boldsymbol{x}_i^T = (\boldsymbol{x}_{i1}, \ldots, \boldsymbol{x}_{iT_i})$ is the conditional mean given $\boldsymbol{x}_i$ *and* $\boldsymbol{b}_i$. Moreover, one assumes that $y_{i1}, \ldots y_{iT_i}$ are conditionally independent given $\boldsymbol{b}_i$ and the covariates $\boldsymbol{x}_i$. For $\boldsymbol{b}_i$ given the covariates a distribution is assumed, for example,

$$\boldsymbol{b}_i \sim N(\boldsymbol{0}, \boldsymbol{Q}_i),$$

where $\boldsymbol{Q}_i$ is a function of a parameter vector $\boldsymbol{\alpha}$. When g is the logit function one obtains the *marginally specified logistic-normal model*.

The value Δ_{it} in the conditional model (**??**) is implicitly determined by the integral that links the marginal mean and the conditional distribution, $\mu_{it} = \int h(\Delta_{it} + b) p_{0,\sigma_{it}^2}(b) db$, where p_{0,σ_{it}^2} is the normal density function with mean zero and variance $\sigma_{it}^2 = \mathrm{var}(b_{it})$. Thus, Δ_{it} is a function of μ_{it} and σ_{it}^2. Given these values, the integral can be solved for Δ_{it} using numerical integration.

The marginalized model specified by (14.28) and (14.29) parameterizes the marginal response, which determines the interpretation of $\boldsymbol{\beta}$, but accounts for heterogeneity and the resulting correlation by additionally assuming a random effects model. The parameter $\boldsymbol{\beta}$ measures the change in the marginal response due to a change in covariates, that is, the effect of covariates averaged over the distribution within subgroups that are defined by the covariates. Heagerty (1999) illustrates the interpretation of the heterogeneity parameter by considering the marginally specified logistic-normal model with $b_{it} = b_{i0}$ and $b_{i0} \sim N(0, \sigma^2)$. Then the individual variation is seen from the model representation $\mathrm{logit}\, \mathrm{E}(y_{it}|\boldsymbol{x}_{it}, z_i) = \Delta_{it} + \sigma z_i$. Thus σ measures the variation between individuals within a group that is defined by the measured covariates $\boldsymbol{x}_i$, which also determine Δ_{it}. In the more general case of b_{it} varying over t one assumes, for example, an autoregressive model. Then the parameters in $\boldsymbol{Q}_i$ are measures of variation across t and individuals. The regression coefficients have interpretations that refer to the mean change given the observed covariates, averaged over unobserved latent variables.

For ML estimation and estimation-equation approaches to estimation and inference see Heagerty (1999). A generalization to multilevel models is given by Heagerty and Zeger (2000).

14.6 Latent Trait Models and Conditional ML

The modeling and measurement of heterogeneity has a long tradition in psychometrics. In item response theory latent traits that are to be measured often refer to abilities like "intelligence." The famous Rasch model (Rasch, 1961) assumes that the probability of participant i solving test item t follows a logit model,

$$\text{logit}(P(y_{it} = 1|\alpha_i, \beta_j)) = \alpha_i + \beta_j, \tag{14.30}$$

where α_i is a person parameter that represents the ability and β_j is an item parameter that represents the easiness of an item. Both are measured on the same latent scale. With $\delta_j = -\beta_j$ denoting the difficulty of an item one obtains the more intuitive form $\text{logit}(P(y_{it}|\alpha_i, \beta_j)) = \alpha_i - \delta_j$, which shows that the probability of solving the item grows with the difference between the ability of the person and the difficulty of the item, $\alpha_i - \delta_j$.

For n persons trying to solve T items, model (14.30) is a model for repeated binary measurements with fixed effects. Each item has its own effect and the heterogeneity of the population is represented by the person parameters. Fixed person parameters have the effect that usual ML estimates are not consistent for growing n since the number of parameters increases with the number of persons. An alternative is *conditional maximum likelihood estimation*. By conditioning on the sum of solved items $\sum_t y_{it}, i = 1, \dots, n$, which are sufficient statistics, the dependence on the person parameters α_i is eliminated. Although conditional ML has the advantage that no distributional assumption is needed, a drawback is that between-cluster effects are conditioned away and cannot be estimated (compare the conditional ML estimates in Section 13.4.2, where only two binary measurements are assumed). In addition, the sum of solved items forms a set of sufficient statistics only for the logit link. An alternative that is used in modern item response theory is the modeling of the person parameters as random effects. Then item parameters and the effect of explanatory variables that explain the item difficulty can be estimated by marginal ML estimates.

Extensions of the Rasch model to the proportional odds model and corresponding conditional ML estimates were proposed by Agresti and Lang (1993) and Agresti (1993a, 1993d). For an overview on generalized versions of the Rasch model that include explanatory variables see De Boeck and Wilson (2004).

14.7 Semiparametric Mixed Models

In the generalized linear mixed models as considered in the previous sections, the effect of covariates has been modeled linearly. However, the assumption of a strictly parametric form is frequently too restrictive when modeling longitudinal data. An alternative is to use smoothing methods as considered in Chapter 10.

Generalized linear mixed models for clustered data assume $g(\mu_{it}) = \eta_{it}$, where the linear predictor has the form $\eta_{it} = \boldsymbol{x}_{it}^T\boldsymbol{\beta} + \boldsymbol{z}_{it}^T\boldsymbol{b}_i$. For a more general additive predictor let the data be given by $(y_{it}, \boldsymbol{x}_{it}, \boldsymbol{u}_{it}, \boldsymbol{z}_{it})$, $i = 1, \dots, n$, $t = 1, \dots, T_i$, where y_{it} is the response for observation t within cluster i and $\boldsymbol{x}_{it}^T = (x_{it1}, \dots, x_{itp})$, $\boldsymbol{u}_{it}^T = (u_{it1}, \dots, u_{itm})$, $\boldsymbol{z}_{it}^T = (z_{it1}, \dots, z_{its})$ are vectors of covariates, which may vary across clusters and observations. Then the additive semiparametric mixed model has the predictor

$$\eta_{it} = \boldsymbol{x}_{it}^T\boldsymbol{\beta} + \sum_{j=1}^{m} \alpha_{(j)}(u_{itj}) + \boldsymbol{z}_{it}^T\boldsymbol{b}_i = \mu_{it}^{par} + \mu_{it}^{add} + \mu_{it}^{rand},$$

where

$\mu_{it}^{par} = \boldsymbol{x}_{it}^T\boldsymbol{\beta}$ is a linear parametric term;

$\mu_{it}^{add} = \sum_{j=1}^{m} \alpha_{(j)}(u_{itj})$ is an additive term with unspecified influence functions $\alpha_{(1)}, \ldots, \alpha_{(m)}$;

$\mu_{it}^{rand} = \boldsymbol{z}_{it}^T\boldsymbol{b}_i$ contains the cluster-specific random effect $\boldsymbol{b}_i$, $\boldsymbol{b}_i \sim N(\boldsymbol{0}, \boldsymbol{Q}(\boldsymbol{\rho}))$, where $\boldsymbol{Q}(\boldsymbol{\rho})$ is a parameterized covariance matrix.

Let the unknown function be expanded in basis functions $\alpha_{(j)}(u) = \sum_{s=1}^{m} \alpha_s^{(j)}\phi_s^{(j)}(u) = \boldsymbol{\alpha}_j^T\boldsymbol{\phi}_j(u)$ by using the basis functions given in Chapter 10. With $\boldsymbol{\alpha}^T = (\boldsymbol{\alpha}_1^T, \ldots, \boldsymbol{\alpha}_m^T)$ and $\boldsymbol{\phi}_{it}^T = (\boldsymbol{\phi}_1(u_{it1})^T, \ldots, \boldsymbol{\phi}_m(u_{itm})^T)$ one obtains the linear predictor

$$\eta_{it} = \boldsymbol{x}_{it}^T\boldsymbol{\beta} + \boldsymbol{\phi}_{it}^T\boldsymbol{\alpha} + \boldsymbol{z}_{it}^T\boldsymbol{b}_i,$$

which corresponds to a GLMM with fixed parameters $\boldsymbol{\beta}, \boldsymbol{\alpha}$ and random effects $\boldsymbol{b}_i$. In matrix notation one obtains for one cluster $\boldsymbol{\eta}_i = \boldsymbol{X}_i\boldsymbol{\beta} + \boldsymbol{\Phi}_i\boldsymbol{\alpha} + \boldsymbol{Z}_i\boldsymbol{b}_i$, and $\boldsymbol{\eta} = \boldsymbol{X}\boldsymbol{\beta} + \boldsymbol{\Phi}\boldsymbol{\alpha} + \boldsymbol{Z}\boldsymbol{b}$ for the whole predictor.

Simultaneous estimations of fixed parameters and random effects, collected in $\boldsymbol{\delta}^T = (\boldsymbol{\beta}^T, \boldsymbol{\alpha}^T, \boldsymbol{b}_1^T, \ldots, \boldsymbol{b}_n^T)$ can be based on the methods given in Section 14.3.3. When many basis functions are used, the coefficients that refer to basis functions should be penalized. By using a penalized likelihood one obtains the double penalized log-likelihood:

$$l_p(\boldsymbol{\delta}) = \log(f(\boldsymbol{y}|\boldsymbol{b}, \boldsymbol{\beta})) - \frac{1}{2}\sum_{j=1}^{m} \lambda_j\boldsymbol{\alpha}_j^T\boldsymbol{K}_j\boldsymbol{\alpha}_j - \frac{1}{2}\boldsymbol{b}^T\boldsymbol{Q}_b^{-1}\boldsymbol{b}, \tag{14.31}$$

where $\boldsymbol{K}_j$ penalizes the coefficients $\boldsymbol{\alpha}_j$. In closed form the penalty has the form $-\frac{1}{2}\boldsymbol{\delta}^T\boldsymbol{K}\boldsymbol{\delta}$, with block-diagonal matrix $\boldsymbol{K} = Block(\boldsymbol{0}, \lambda_1\boldsymbol{K}_1, \ldots, \lambda_m\boldsymbol{K}_m, \boldsymbol{Q}_b^{-1})$. Typically, the variance of random effects depends on the parameters, $\boldsymbol{Q}_b = \boldsymbol{Q}_b(\boldsymbol{\rho})$. Considering the penalties on coefficients $\boldsymbol{\alpha}_j$ as functions of λ_j, $\boldsymbol{K}_j(\lambda_j) = \lambda_j\boldsymbol{K}_j$, one obtains the penalized log-likelihood of a GLMM with variance parameters $\lambda_1, \ldots, \lambda_m, \boldsymbol{\rho}$, which can be estimated simultaneously by ML or REML methods. The approach to consider the smoothing parameters as variance components in a mixed model was already helpful in the selection of smoothing parameters (Section 10.1.5). The penalized log-likelihood (14.31) just includes a variance component that actually refers to a random effect. Therefore, models with additive terms and random effects are embedded quite naturally within the mixed modeling representation of smoothing (see also Wood, 2006a; Ruppert et al., 2003; Lin and Zhang, 1999). A procedure that automatically selects relevant functions and performs well in high dimensions was given by Tutz and Reithinger (2007).

Example 14.6: AIDS Study

For the AIDS Cohort Study MACS (Example 14.1) the covariates were years since seroconversion, recreational drug use (yes/no), number of sexual partners, age, and a mental illness score (cesd). Since the forms of the effects are not known, time since seroconversion, age, and the mental illness score may be considered as unspecified additive effects. We consider the semiparametric mixed model with Poisson distribution, log-link, and the linear predictor $\eta_{it} = \mu_{it}^{par} + \mu_{it}^{add} + b_{it}$, where

$$\mu_{it}^{\text{par}} = \beta_0 + \text{drugs}_{it}\beta_D + \text{partners}_{it}\beta_P, \quad \mu_{it}^{\text{add}} = \alpha_T(\text{time}_{it}) + \alpha_A(\text{age}_i) + \alpha_C(\text{cesd}_{it}).$$

Fitting was performed using function *gamm* from the R package *mgcv*. Figure 14.7 shows the smooth effects of time and the mental illness score. The corresponding p-values were below 0.002. The effect of age is not shown because it is linear with slope and very close to zero (p-value 0.66). It is seen that there is a slight decrease in CD4 cells for increasing values of the mental illness score. The coefficients of number of partners and drugs were 0.036 and 0.003, both of them having p-value larger than 0.1. When additive components are selected by an appropriate boosting procedure, age is not included (see Groll and Tutz, 2011a). Slightly different functions are found when the response, after some transformation, is treated as approximately normally distributed (see Tutz and Reithinger, 2007). □

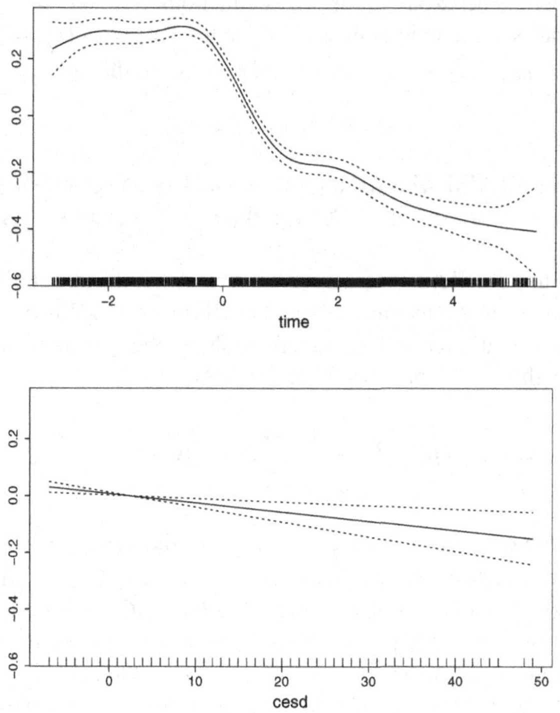

FIGURE 14.7: Estimated effect of time and mental illness score for AIDs study.

14.8 Finite Mixture Models

In cross-sectional data as well as in repeated measurements, a bad fit of a model can be due to unobserved heterogeneity. When an important categorical covariate has not been observed and therefore not included in the predictor, one observes a mixture of responses that follow different models. Instead of assuming a smooth distribution for the unobserved part heterogeneity can be modeled by allowing a finite mixture of models. In finite mixtures of generalized linear models it is assumed that the density or mass function of observation y given $\boldsymbol{x}$ is a mixture:

$$f(y|\boldsymbol{x}) = \sum_{j=1}^{m} \pi_j f_j(y|\boldsymbol{x}, \boldsymbol{\beta}_j, \phi_j), \tag{14.32}$$

where $f_j(y|\boldsymbol{x}, \boldsymbol{\beta}_j, \phi_j)$ represents the jth component of the mixture that follows a simple exponential family parameterized by the parameter vector from the model $\mu_j = \mathrm{E}(y|\boldsymbol{x}, j) = h(\boldsymbol{x}^T\boldsymbol{\beta}_j)$ and dispersion parameter ϕ_j. The unknown component weights follow $\sum_{j=1}^{m} \pi_j = 1, \pi_j > 0, j = 1, \ldots, m$.

Rather than allowing all the parameter vectors to be varying across components, it is often more appropriate to let only part of the variables, for example, the intercepts, vary across components. A simple example is the finite mixture of two logit models, where one assumes

$$P(y=1|\boldsymbol{x}) = \pi_1 P_1(y=1|\boldsymbol{x}) + (1-\pi_1)P_2(y=1|\boldsymbol{x}).$$

For the components one can assume that all the weights differ:

$$\text{logit}(P_j(y=1|\boldsymbol{x})) = \beta_{j0} + \boldsymbol{x}^T\boldsymbol{\beta}_j,$$

or one can assume that only intercepts are component-specific:

$$\text{logit}(P_j(y=1|\boldsymbol{x})) = \beta_{j0} + \boldsymbol{x}^T\boldsymbol{\beta}.$$

The latter is a random intercept model that uses finite mixtures. In particular, the case where only intercepts are component-specific yields a flexible model with a straightforward interpretation. Let the means be specified by $\mu_j = h(\beta_{j0} + \boldsymbol{x}^T\boldsymbol{\beta})$ and σ_j^2 denote the variance of the jth component. Since the mean and variance of the response are given by

$$\mu = \sum_{j=1}^{m} \pi_j \mu_j, \quad \sigma^2 = \sum_{j=1}^{m} \pi_j \sigma_j^2 - \mu^2 + \sum_{j=1}^{m} \pi_j \mu_j^2$$

(Exercise 14.6), the mean structure may also be given as

$$\mu = h_0(\boldsymbol{x}^T\boldsymbol{\beta}),$$

where $h_0(\eta) = \sum_j \pi_j h(\beta_{j0} + \eta)$ is the response function with linear predictor $\eta = \boldsymbol{x}^T\boldsymbol{\beta}$. One obtains a model for which the link function is determined by the data. This is similar to the estimation of link functions considered in Section 5.2 with basis functions given by $h(\beta_{10} + \eta), \dots, h(\beta_{m0} + \eta)$. The link $\mu = h_0(\boldsymbol{x}^T\boldsymbol{\beta})$ allows for easy interpretation of the parameters that has to refer to the estimated link function $h_0(.)$. Moreover, the link function $h_0(.)$ automatically maps the predictor into the admissible range if $h(.)$ does. If the model contains more than one component and the variances of the components are equal, the model implies a larger variance than for the single components, which makes it a tool for modeling overdispersion (see also Section 5.3).

Models of that type are considered by Follmann and Lambert (1989), and Aitkin (1999). Follmann and Lambert (1989) investigated the identifiability of finite mixtures of binomial regression models and gave sufficient identifiability conditions for mixing at the binary and the binomial level. Grün and Leisch (2008) consider identifiability for mixtures of multinomial logit models.

When modeling repeated measurements the mixture distribution $\{\pi_1, \dots, \pi_m\}$ replaces the assumption of a continuous distribution of random intercepts. With large m it can be seen as a non-parametric approach to approximate the unknown mixture distribution of the parameters. More general models in which the mixing distribution also depends on covariates are given by McLachlan and Peel (2000).

Estimation

For given data $y_{it}|\boldsymbol{x}_{it}$, $i = 1, \dots, n$, $t = 1, \dots, T_i$, the log-likelihood to be maximized is

$$l(\{\boldsymbol{\beta}_j, \phi_j\}) = \sum_{i=1}^{n} \log\left(\sum_{j=1}^{m} \pi_j f_j(\boldsymbol{y}_i|\boldsymbol{x}_i)\right),$$

where $f_j(\boldsymbol{y}_i|\boldsymbol{x}_i) = \prod_{t=1}^{T_i} f_j(y_{it}|\boldsymbol{x}_{it}, \boldsymbol{\beta}_j, \phi_j)$. Direct maximization is tedious; therefore indirect maximization based on the EM algorithm is often used. When using the EM algorithm (Appendix B) one distiguishes between the observable data and the unobservable data. In the case of finite mixtures, $\boldsymbol{y}_i^T = (y_{i1}, \dots, y_{iT_i})$ are observable whereas the mixture component is unobservable. Let the latter be given by 0-1 variables $z_{i1}, \dots, z_{im}$, where $z_{ij} = 1$ denotes that the ith observation is from the jth mixture component. By collecting all parameters in the parameter $\boldsymbol{\theta}$ one has

$$f(\boldsymbol{y}_i|z_{ir} = 1, \boldsymbol{x}_i, \boldsymbol{\theta}) = f_r(\boldsymbol{y}_i|\boldsymbol{x}_i, \boldsymbol{\beta}_r) = \prod_{j=1}^{m} f_j(\boldsymbol{y}_i|\boldsymbol{x}_i, \boldsymbol{\beta}_j)^{z_{ij}}.$$

Since $\boldsymbol{z}_i^T = (z_{i1}, \dots, z_{im})$ is multinomially distributed with probability vector $\pi^T = (\pi_1, \dots, \pi_m)$ the complete density for $\boldsymbol{y}_i, \boldsymbol{z}_i$ is

$$f(\boldsymbol{y}_i, \boldsymbol{z}_i|\boldsymbol{x}_i, \boldsymbol{\theta}) = f(\boldsymbol{y}_i|\boldsymbol{z}_i, \boldsymbol{x}_i, \boldsymbol{\theta}) f(\boldsymbol{z}_i|\boldsymbol{\theta}) = \prod_{j=1}^{m} f_j(\boldsymbol{y}_i|\boldsymbol{z}_i, \boldsymbol{x}_i, \boldsymbol{\beta}_j)^{z_{ij}} \prod_{j=1}^{m} \pi_j^{z_{ij}}$$

yielding the complete log-likelihood

$$l_c(\boldsymbol{\theta}) = \sum_{i=1}^{n} \log(f(\boldsymbol{y}_i, \boldsymbol{z}_i|\boldsymbol{x}_i, \boldsymbol{\theta})) = \sum_{i=1}^{n}\sum_{j=1}^{m} z_{ij}(\log(\pi_j) + \log(f_j(\boldsymbol{y}_i|\boldsymbol{z}_i, \boldsymbol{x}_i, \boldsymbol{\beta}_j)).$$

Within the iterative EM algorithm an estimate $\boldsymbol{\theta}^{(s)}$ is improved by computing the expectation $M(\boldsymbol{\theta}|\boldsymbol{\theta}^{(s)}) = \mathrm{E}\, l_c(\boldsymbol{\theta})$ with respect to the conditional density $f(\boldsymbol{z}|\boldsymbol{y}, \boldsymbol{\theta}^{(s)})$. With

$$f(z_{ir} = 1|\boldsymbol{y}_i, \boldsymbol{x}_i) = f(\boldsymbol{y}_i|z_{ir} = 1, \boldsymbol{x}_i) f(z_{ir} = 1)/f(\boldsymbol{y}_i|\boldsymbol{x}_i) = \pi_r f_r(\boldsymbol{y}_i|\boldsymbol{x}_i)/f(\boldsymbol{y}_i|\boldsymbol{x}_i)$$

one obtains

$$M(\boldsymbol{\theta}|\boldsymbol{\theta}^{(s)}) = \sum_{i=1}^{n}\sum_{j=1}^{m} \pi_{ij}^{(s)}(\log(\pi_j) + \log(f_j(\boldsymbol{y}_i|\boldsymbol{z}_i, \boldsymbol{x}_i, \boldsymbol{\beta}_j)),$$

where $\pi_{ij}^{(s)} = \pi_j^{(s)} f_j(\boldsymbol{y}_i|\boldsymbol{x}_i, \boldsymbol{\beta}_j^{(s)})/(\sum_{r=1}^{m} \pi_r^{(s)} f_r(\boldsymbol{y}_i|\boldsymbol{x}_i, \boldsymbol{\beta}_r^{(s)}))$. Thus, for given $\boldsymbol{\theta}^{(s)}$ one computes in the E-step the weights $\pi_{ij}^{(s)}$ and in the M-step maximizes $M(\boldsymbol{\theta}|\boldsymbol{\theta}^{(s)})$, which yields the new estimates

$$\pi_j^{(s+1)} = \frac{1}{n}\sum_{i=1}^{n} \pi_{ij}^{(s)} \quad \boldsymbol{\beta}_j^{(s+1)} = \mathrm{argmax}_{\boldsymbol{\beta}_j} \sum_{i=1}^{n} \pi_{ij}^{(s)} \log(f_j(\boldsymbol{y}_i|\boldsymbol{x}_i, \boldsymbol{\beta}_j)).$$

Computation of $\boldsymbol{\beta}_j^{(s+1)}$ can be based on familiar maximization tools because one maximizes a weighted log-likelihood with known weights. In the case where only intercepts are component-specific, the derivatives are very similar to the score function used in a Gauss-Hermite quadrature and a similar EM algorithm applies with an additional calculation of the mixing distribution $\{\pi_1, \dots, \pi_m\}$ (see Aitkin, 1999).

Maximization is more difficult in the general case, where one has to distinguish carefully between fixed effects, which do not depend on the mixture component, and component-specific effects. Then part of the variables have fixed effects and part have component-specific effects. In the case of repeated measurements the component can also be linked to the units or clusters. Let $C = \{1, \dots, n\}$ denote the set of units that are observed. Then one specifies in a random intercept model for the jth component

$$\mathrm{logit}(P_j(y_{it} = 1|\boldsymbol{x}_{it})) = \beta_{j(i)} + \boldsymbol{x}_{it}^T\boldsymbol{\beta},$$

where $\beta_{j(i)}$ denotes that component membership is fixed for each unit, that is, $\beta_{j(i)} = \beta_j$ for all $i \in C_j$, where $C_1, \ldots, C_m$ is a disjunct partition of C. Therefore the units are clustered into subsets with identical intercepts. Grün and Leisch (2007) describe how to use their R package *flexmix* and give various applications. An extensive treatment of mixture models is given by Frühwirth-Schnatter (2006).

Example 14.7: Beta-Blockers

The dataset is from a 22-center clinical trial of beta-blockers for reducing mortality after myocardial infarction (see also Aitkin, 1999; McLachlan and Peel, 2000; Grün and Leisch, 2007). In addition to the centers there is only one explanatory variable, treatment, coded as 0 for control and 1 for beta-blocker treatment. Therefore one has 44 binomial observations. The estimated treatment effects for various models are given in Table 14.3. Fitting of a simple logit model with treatment as a fixed effect and ignoring the effects of the center (GLM Treatment) yields a deviance of 305.76 with 42 df, which reduces to 23.62 with 21 df if the center is included as a factor (GLM Treatment + Center). Therefore, some heterogeneity across hospitals is to be suspected. A Gauss-Hermite quadrature yields a comparable effect that is quite stable over quadrature points (GLMM GH(quadrature points)). Estimates for 4 and 20 quadrature points are given in Table 14.3. For the standard deviation one obtains in both cases 0.513, with standard error 0.085. The fitting of discrete mixtures uses an intercept that is linked to the centers. The estimates are given for three and four components, which have BIC 341.42 and 346.76, respectively. Therefore, in terms of BIC, three components are to be preferred. Discrete mixture models yield about the same treatment effect as the other procedures. For the discrete mixture model with three components, the estimated component weights were 0.512, 0.249, 0.239 with corresponding sizes 24, 10, 10. Table 14.4 shows the estimates. □

TABLE 14.3: Estimates and standard errors for several logit models fitted to beta-blocker data.

	Treatment Effect	Std. Error
GLM Treatment	−0.257	0.049
GLM Treatment + Center	−0.261	0.049
GLMM GH(4) Treatment	−0.252	0.057
GLMM GH(20) Treatment	−0.252	0.057
Discrete(3)	−0.258	0.049
Discrete(4)	−0.260	0.049

TABLE 14.4: Estimates and standard errors for mixture model with three components fitted to beta-blocker data.

	Estimate	Std. Error	z-Value	Pr(>|z|)
Treatment	−0.258	0.049	−5.17	0.0
Intercept comp 1	−2.250	0.040	−55.52	0.0
Intercept comp 2	−1.609	0.055	−28.88	0.0
Intercept comp 3	−2.833	0.075	−37.74	0.0

Example 14.8: Knee Injuries

For the knee injuries study with dichotomized responses (Example 14.3) discrete mixture models with random intercepts were fitted with the intercepts being linked to the subjects. It turns out that BIC is the smallest for the two component mixture models for which estimates are given in Table 14.5. Comparison to Table 14.1 shows that the significant coefficients are quite similar to the coefficients obtained for a Gauss-Hermite quadrature. □

TABLE 14.5: Knee data binary, discrete mixture (2).

	Coef	Pr(>$\|z\|$)
factor(Th)2	−1.841	0.0003
factor(Sex)1	−0.101	0.798
Age	0.020	0.330
Age2	−0.006	0.000

For repeated measurements, finite mixture models provide a non-parametric alternative to random effects models with a specified distribution of random effects. Frequently only a few components are needed, and the estimates of explanatory variables do not change much when the number of components is increased. An advantage of non-parametric approaches is that one does not have to choose a fixed distribution, a choice that might affect the estimates (for references on misspecification see Section 14.9). However, when random effects are multivariate, random effects models with normally distributed effects have the advantage that the distribution is described by relatively few (correlation) parameters.

14.9 Further Reading

Linear and Generalized Linear Mixed Models. Detailed expositions of linear mixed models are found in Hsiao (1986), Lindsey (1993), and Jones (1993). A book with many applications and references to the use of SAS as software is Verbeke and Molenberghs (2000). Longford (1993) and McCulloch and Searle (2001) discuss linear as well as generalized mixed models. General treatments of longitudinal data including mixed models were given by Diggle et al. (2002) and Molenberghs and Verbeke (2005).

Random Effects Distribution. Misspecification of the model for the random effects can affect the likelihood-based inference. Although earlier papers found that ML estimates are not severely biased (Neuhaus, Hauck, and Kalbfleisch, 1992), more recently it has been shown that substantial bias can result; see Heagerty and Kurland (2001), Agresti et al. (2004), and Neuhaus et al. (1992). Non-parametric approaches were proposed by Chen and Davidian (2002) and Magder and Zeger (1996). Huang (2009) proposed diagnostic methods for random-effect misspecification. Claeskens and Hart (2009) proposed tests for the assumption of the normal distribution.

Variable Selection in Random Effects and Mixture Models. For the selection of variables in random effects models L_1-penalty terms can be included in the marginal likelihood or the penalized quasi-likelihood. For linear mixed models procedures of that type were proposed by Ni et al. (2010). Alternatively, one can use boosting techniques; see Tutz and Reithinger (2007), Tutz and Groll (2010a), Tutz and Groll (2010b). Groll and Tutz (2011b) used lasso type penalties in generalized linear mixed models; Ibrahim, Zhu, Garcia, and Guo (2011) also include the selection of random effects. For mixture models Khalili and Chen (2007) proposed penalty-driven selection procedures that select single coefficients.

R Packages. The package *glmmML* allows one to fit GLMs with random intercepts by maximum likelihood and numerical integration via a Gauss-Hermite quadrature. Estimation of random effects models with ordinal responses is performed by the function *clmm* from the package *ordinal*. The function *glmmPQL* from the package *MASS* fits GLMMs, using the penalized quasi-likelihood. The package *flexmix* provides procedures for fitting finite mixture models including binomial, Poisson, and Gaussian mixtures. The function *glmer* from the package *lme4* fits by using the adaptive Gauss-Hermite approximation proposed by Liu and Pierce (1994). The versatile package *mgcv* contains the function *gamm*, which allows one to fit GLMMs that contain smooth functions.

14.10 Exercises

14.1 The Gaussian random effects model for clustered data is given by $\boldsymbol{y}_i = \boldsymbol{X}_i\boldsymbol{\beta} + \boldsymbol{Z}_i\boldsymbol{b}_i + \boldsymbol{\varepsilon}_i$, where $\boldsymbol{b}_i \sim N(\boldsymbol{0}, \boldsymbol{Q})$, $\boldsymbol{\varepsilon}_i \sim N(\boldsymbol{0}, \sigma^2\boldsymbol{I})$ and $\boldsymbol{b}_i$ and $\boldsymbol{\varepsilon}_i$ are independent. Derive the distribution of $\boldsymbol{\varepsilon}_i^*$ in the marginal representation $\boldsymbol{y}_i = \boldsymbol{X}_i\boldsymbol{\beta} + \boldsymbol{\varepsilon}_i^*, \boldsymbol{\varepsilon}_i^* \sim N(\boldsymbol{0}, \boldsymbol{V}_i)$.

14.2 The R package *flexmix* provides the dataset *betablocker*.

(a) Use descriptive tools to learn about the data.

(b) Fit a GLMM by using a Gauss-Hermit quadrature. Choose an appropriate number of quadrature points.

(c) Fit a GLMM by using a Laplace approximation and compare to (b).

(d) Fit a discrete mixture model with random intercepts linked to centers. Choose an appropriate number of mixing components.

(e) Fit a discrete mixture model with random intercepts and treatment effect linked to centers. Choose an appropriate number of mixing components and compare with (d).

14.3 Consider the sequential mixed model with category-specific random intercepts

$$P(Y_{it} = r | Y_{it} \geq r, \boldsymbol{x}) = F(\gamma_{0r} + b_{ir} + \boldsymbol{x}_{it}^T\boldsymbol{\gamma}), \quad , r = 1, \ldots, k-1,$$

where $\boldsymbol{b}_i^T = (b_{i1}, \ldots, b_{i,k-1})$ is normally distributed.

(a) Show how the model can be fitted by using binary mixed model methodology (see also Section 9.5.2).

(b) Fit a sequential model with random intercepts for the recovery score data (Example 14.2).

(c) Fit a cumulative model with random intercepts for the recovery score data and compare with the results from (b).

14.4 Consider the cumulative model from Example 14.2. If estimation becomes unstable, also consider a rougher response that is obtained by collapsing of the categories.

(a) Fit cumulative models for the separate measurements.

(b) Fit a GLMM by Gauss-Hermite integration and penalized quasi-likelihood methods.

(c) Use in the GLMM a linear trend over repetitions instead of dummy variables.

14.5 Consider again the epilepsy data from Example 13.1 (available at http://biosun1.harvard.edu/ fitzmaur/ala).

(a) Fit a model with random intercepts and investigate the effect of treatment.

(b) Compare the estimates to the marginal model fits from Exercise 13.4.

14.6 Let a finite mixture model be given by $f(y) = \sum_{j=1}^{m} \pi_j f_j(y)$, and let μ_j, σ_j^2 denote the mean and variance of the jth component.

(a) Show that the mean and variance of the mixture are given by $\mu = \sum_{j=1}^{m} \pi_j\mu_j$, $\sigma^2 = \sum_j \pi_j\sigma_j^2 - \mu^2 + \sum_j \pi_j\mu_j^2$. (The variance can be computed directly or by using the variance decomposition.)

(b) Show that the variance of the mixture is at least as large as the mean variance $\sum_j \pi_j\sigma_j^2$. Is the variance of the mixture also larger than the variance of any single component, that is, $\sigma^2 \geq \sigma_j^2$?

Chapter 15

Prediction and Classification

In prediction problems one considers a new observation (y, x). While the predictor value x is observed, y is unknown and is to be predicted. In general, the unknown y may be from any distribution, continuous or discrete, depending on the prediction problem. When the unknown value is categorical we will often denote it by Y, with Y taking values from $\{1, \ldots, k\}$. Then prediction means to find the true underlying value from the set $\{1, \ldots, k\}$. The problem is strongly related to the common classification problem where one wants to find the true class from which the observation stems. When the numbers $1, \ldots, k$ denote the underlying classes, the classification problem has the same structure as the prediction problem. Classification problems are basically diagnostic problems. In medical applications one wants to identify the type of disease, in pattern recognition one might aim at recognizing handwritten characters, and in credit scoring (Example 1.7) one wants to identify risk clients. Sometimes the distinction between prediction and classification is philosophical. In credit scoring, where one wants to find out if a client is a risk client, one might argue that it is a prediction problem since the classification lies in the future. Nevertheless, it is mostly seen as a classification problem, implying that the client is already a risk client or not. The following example illustrates the uncertainty of classification rules by giving the true distribution of a simple indicator given the class. Usually the distribution is not known and one has to start from data (Example 15.2).

Example 15.1: Drug Use

In some companies it is not unusual that applicants for a job are tested for drug use. Marylin vos Savant, known as the person with the highest IQ, discussed in the *Gainesville Sun* the use of a diagnostic test that is .95 accurate if the person is a user or a non-user. The probabilities of test results given the class are given in the following table.

	Test Positive	Test Negative
User	0.95	0.05
Non-user	0.05	0.95

Tests are usually not 100% reliable. One distinguishes between sensitivity, which is the probability of a positive test, given the signal (user) is present, and specificity, which is the probability of a negative test, given no signal (non-user) is present. In the table above sensitivity as well as specificity are .95. If a test like that is used to infer on the drug use of a person selected at random, the question is how well it performs. □

Example 15.2: Glass Identification

The identification of types of glass can be very important in criminological investigations since at the scene of the crime, the glass left can be used as evidence. A dataset coming from USA Forensic Science Service distinguishes between seven types of glass (four types of window glass and three types of non-window); predictors are the refractive index and the oxide content of various minerals like Na, Fe, and K. The data have been used by Ripley (1996) and others. It is available from the UCI Machine Learning Repository. □

In the last decade in particular, the analysis of genetic data has become an interesting field of application for classification techniques. For example, gene expression data may be used to distinguish between tumor classes and to predict responses to treatment (e.g., Golub et al., 1999b). The challenge in gene expression data is in the dimension of the datasets. With about $30,000$ genes in the human genome, the data to be analyzed have the unusual feature that the number of variables (genes) is much higher than the number of cases (tumor samples), which is referred to as the "$p > n$" case. Standard classification procedures fail in the "$p > n$" case. Therefore, one of the main goals with this type of data is variable selection, which will be considered in a separate section.

Classification methods run under several names; they are also referred to as *pattern recognition* in technical applications, *discriminant analysis* in statistics, and *supervised learning* in the machine learning community. Statistical approaches are well covered in McLachlan (1992) and Ripley (1996). Ripley also includes methods developed in the machine learning community. Bishop (2006) is guided by pattern recognition and machine learning and Hastie et al. (2009) consider regression and classification from a general statistical learning viewpoint.

The main objective of prediction is high accuracy, where of course accuracy has to be defined, for example, by the use of loss functions. In some applications, for instance, in recognizing handwritten characters, accuracy is indeed prevailing. In other fields, like credit scoring, it is often found unsatisfying to use a black box for prediction purposes. Users prefer to know which variables, in what form, determine the prediction, and in particular how trustworthy the prediction is. Thus, users often favor simple rules in prediction, for instance, linear discrimination, which captures the importance of predictors by parameters. What users usually want is a combination; they want a simple model of the underlying structure that in addition has good prediction properties. A simple model will usually yield good prediction performance only if it is a fair approximation of the underlying structure. If that structure is complicated, no simple model will provide a good approximation and performance will be poor. Then, approaches that are highly efficient but do not identify an interpretable structure should be chosen. In the following we will consider several simple prediction rules that are based on statistical models and therefore are user friendly but also non-parametric approaches that do not imply an easy-to-interpret structure but also show excellent performance for complicated distributions.

We will first consider the basic concepts of prediction that apply in general prediction problems. If one is interested in particular in classification rules, one can also skip the next section and start with Section 15.2.

15.1 Basic Concepts of Prediction

Let a new observation $(y, \boldsymbol{x})$ be randomly drawn from the distribution. While $\boldsymbol{x}$ is observed, y is unknown and is to be predicted. The prediction is denoted by $\hat{y}(\boldsymbol{x})$, which is based on a mapping $\hat{y} : \boldsymbol{x} \mapsto \hat{y}(\boldsymbol{x})$. Let the performance of a prediction rule be measured by a loss function $L(y, \hat{y}(\boldsymbol{x}))$. Classical losses for metric responses are the squared loss $L_2(y, \hat{y}) = (y - \hat{y})^2$ and the L_1 norm $L_1(y, \hat{y}) = |y - \hat{y}|$. It is useful to distinguish between two cases: the case where the prediction rule $\hat{y}$ is considered as given and the case where $\hat{y}$ is an estimated rule. We will first consider fixed rules and how they may be chosen optimally.

For randomly drawn $(y, \boldsymbol{x})$ the *actual prediction error* is given by

$$\mathrm{E}_{y,\boldsymbol{x}} \; L(y, \hat{y}(\boldsymbol{x})) = \mathrm{E}_{\boldsymbol{x}} \, \mathrm{E}_{y|\boldsymbol{x}} \; L(y, \hat{y}(\boldsymbol{x})) = \int \mathrm{E}_{y|\boldsymbol{x}} \; L(y, \hat{y}(\boldsymbol{x})) f(\boldsymbol{x}) d\boldsymbol{x}, \tag{15.1}$$

where $f(\boldsymbol{x})$ is the marginal distribution of $\boldsymbol{x}$. $E_{y,\boldsymbol{x}} \; L(y, \hat{y}(\boldsymbol{x}))$ is a measure for the mean prediction error averaged across all possible predictors $\boldsymbol{x}$. For $\boldsymbol{x}$ fixed it reduces to $\mathrm{E}_{y|\boldsymbol{x}} \; L(y, \hat{y}(\boldsymbol{x}))$.

Optimal Prediction

When minimization of the actual prediction error is considered as the criterion for choosing $\hat{y}(\boldsymbol{x})$, one may ask what the optimal prediction rule is. The answer, of course, depends on the loss function. Since it suffices to minimize the conditional loss, given $\boldsymbol{x}$, the optimal prediction at value $\boldsymbol{x}$ is determined by

$$\hat{y}_{opt}(\boldsymbol{x}) = \mathrm{argmin}_c L(y, c).$$

For the quadratic loss one obtains as a minimizer

$$\hat{y}_{opt}(\boldsymbol{x}) = \mathrm{E}(y|\boldsymbol{x}),$$

which is also called a *regression function*. When using the L_1-norm one obtains the median

$$\hat{y}_{opt}(\boldsymbol{x}) = \mathrm{med}(y|\boldsymbol{x}).$$

The loss function $L_p(y, \hat{y}) = |y - \hat{y}| + (2p - 1)(y - \hat{y})$, with $p \in (0, 1)$, yields the p-quantile

$$\hat{y}_{opt}(\boldsymbol{x}) = \inf\{y_0 | P(y \le y_0 | \boldsymbol{x}) \ge p\},$$

also known as a *quantile regression*. Therefore, the loss determines the functional of the conditional distribution $y|\boldsymbol{x}$ that is specified. The mean, the median, and the quantile represent different forms of regression.

Estimated Prediction Rule

Most often $\hat{y}(\boldsymbol{x})$ is based on a fitted model. For example, one fits a model for the conditional mean $\mu(\boldsymbol{x}) = E(y|\boldsymbol{x}) = h(\boldsymbol{x}, \theta)$ and then uses the estimated mean at $\boldsymbol{x}$, $\hat{\mu}(\boldsymbol{x}) = h(\boldsymbol{x}, \hat{\theta})$, for prediction. Then the actual prediction error is a random variable that depends on the sample $S = \{(y_i, \boldsymbol{x}_i), i = 1, \dots, n\}$ that was used to obtain the estimate. The dependence on the sample can be made explicit by the denotation $\hat{y}_S(\boldsymbol{x}) = \hat{\mu}(\boldsymbol{x})$. Different models should be compared by the *mean actual prediction error*, also known as the *test* or *generalization error*:

$$\mathrm{E} \; \mathrm{E}_{y,\boldsymbol{x}} \; L(y, \hat{y}_S(\boldsymbol{x})),$$

where E denotes the expectation over the learning sample. The mean actual prediction error also reflects the uncertainty, which stems from the estimation of the prediction rule. Thus it refers to the uncertainty of the total system and not only to the data at hand. For fixed $\boldsymbol{x}$ the corresponding generalization error is $\mathrm{E} \; \mathrm{E}_{y|\boldsymbol{x}} \; L(y, \hat{y}_S(\boldsymbol{x}))$.

Having fitted a model, one also wants to estimate its performance in future samples. Simple estimates that apply the predictor retrospectively to the sample, obtaining the so-called apparent error rate, tend to be overly optimistic and have to be bias corrected.

15.1.1 Squared Error Loss

In the following we will first consider basic concepts in the well-investigated case of a squared error loss. For a squared error the *predictive mean squared error* (PMSE; actual prediction error) of the prediction $\hat{\mu}(\boldsymbol{x})$ is given by

$$\begin{aligned}\text{PMSE} &= \mathrm{E}_{y,\boldsymbol{x}}(y-\hat{\mu}(\boldsymbol{x}))^2 = \mathrm{E}_{\boldsymbol{x}}\, E_{y|\boldsymbol{x}}(y-\mu(\boldsymbol{x}))^2 + \mathrm{E}_{\boldsymbol{x}}\,\mathrm{E}_{y|\boldsymbol{x}}(\mu(\boldsymbol{x})-\hat{\mu}(\boldsymbol{x}))^2\\ &= \mathrm{E}_{\boldsymbol{x}}\,\sigma^2_{y|\boldsymbol{x}} + \mathrm{E}_{\boldsymbol{x}}(\mu(\boldsymbol{x})-\hat{\mu}(\boldsymbol{x}))^2,\end{aligned}$$

where $\sigma^2_{y|\boldsymbol{x}} = \mathrm{E}_{y|\boldsymbol{x}}(y-\mu(\boldsymbol{x}))^2$ is the conditional variance of $y|\boldsymbol{x}$. For fixed $\boldsymbol{x}$ PMSE reduces to $\text{PMSE}(\boldsymbol{x}) = \sigma^2_{y|\boldsymbol{x}} + (\mu(\boldsymbol{x})-\hat{\mu}(\boldsymbol{x}))^2$. It is seen that the precision of the prediction at $\boldsymbol{x}$ depends on the variance at $\boldsymbol{x}$ and the precision of the (model-based) estimate $\hat{\mu}(\boldsymbol{x})$. The latter term depends on the sample; better estimates will result in a smaller term. In contrast, the conditional variance does not depend on the sample and reflects the basic variability of $y|\boldsymbol{x}$, which cannot be reduced by increasing the sample size or using better estimates.

Let us consider a familiar example, namely, a prediction within the classical linear model. Under the assumption of the classical linear model $y_i = \boldsymbol{x}_i^T\boldsymbol{\beta} + \epsilon_i,\ \epsilon_i \sim N(0,\sigma^2)$ one obtains

$$\text{PMSE}(\boldsymbol{x}) = \sigma^2 + (\mu(\boldsymbol{x}) - \boldsymbol{x}^T\hat{\boldsymbol{\beta}})^2,$$

where $\mu(\boldsymbol{x}) = \boldsymbol{x}^T\boldsymbol{\beta}$. Let $\hat{\boldsymbol{\beta}}$ denote the least-squares estimate. If the classical linear model is the operating model, one obtains, conditionally on the observations $\boldsymbol{x}_1,\dots,\boldsymbol{x}_n$, the mean actual prediction error:

$$\begin{aligned}\mathrm{E}\ \text{PMSE}(\boldsymbol{x}) &= \sigma^2 + \mathrm{E}(\boldsymbol{x}^T\boldsymbol{\beta} - \boldsymbol{x}^T\hat{\boldsymbol{\beta}})^2 = \sigma^2 + \mathrm{E}(\boldsymbol{x}^T(\hat{\boldsymbol{\beta}}-\boldsymbol{\beta})(\hat{\boldsymbol{\beta}}-\boldsymbol{\beta})^T\boldsymbol{x})\\ &= \sigma^2(1+\boldsymbol{x}^T(\boldsymbol{X}^T\boldsymbol{X})^{-1}\boldsymbol{x}).\end{aligned}$$

While the uncertainty contained in σ^2 is fixed by the underlying model, the uncertainty of the prediction depends on the predictor value $\boldsymbol{x}$ as well as on the design matrix. The actual predicton error $\text{PMSE}(\boldsymbol{x})$ is a random variable, since $\hat{\boldsymbol{\beta}}$ is. In contrast, the mean actual prediction is not random and includes the variability due to the random responses in the learning sample.

For the assessment of prediction accuracy let the model be $y_i = \mu_i + \varepsilon_i, \varepsilon_i \sim N(0,\sigma_i^2)$. A simple estimate of the generalization error uses the fit retrospectively on the data that generated the fit, yielding the *apparent error rate*:

$$e_{\text{app}} = \frac{1}{n}\sum_{i=1}^{n}(y_i - \hat{\mu}_i)^2.$$

For the linear model with homogeneous variances it is well known that e_{app} is a biased estimate of the variance. Correcting for the number of estimated parameters, p, yields the unbiased estimate $\{n/(n-p)\}e_{\text{app}}$.

The apparent error rate estimates the error at the predictor values given in the sample. Let $\boldsymbol{y}_i^T = (y_1,\dots,y_n)$ collect the observations at predictor values $\boldsymbol{x}_1,\dots,\boldsymbol{x}_n$ and let $\boldsymbol{y}_0^T = (y_{01},\dots,y_{0n})$ denote a vector of new, independent observations, where y_{0i} is taken at predictor value $\boldsymbol{x}_i$. Then $e_i = (y_i - \hat{\mu}_i)^2$ may be seen as an estimate of $\mathrm{E}\,\mathrm{E}_0(y_{0i}-\hat{\mu}_i)^2$, where E_0 refers to the expectation of y_{0i} and E to the expectation over the sample. For the mean at $\boldsymbol{x}_i$ one obtains

$$\mathrm{E}\,\mathrm{E}_0(y_{0i}-\hat{\mu}_i)^2 = \mathrm{E}\{(y_i-\hat{\mu}_i)^2 + 2\,\mathrm{cov}(\hat{\mu}_i, y_i)\}, \tag{15.2}$$

where one uses $\mathrm{E}(y_i-\mu_i)^2 = \mathrm{E}_0(y_{0i}-\mu_i)^2$, yielding $\mathrm{E}\,\mathrm{E}_0(y_{0i}-\hat{\mu}_i)^2 = \mathrm{E}(y_i-\mu_i)^2 + \mathrm{E}(\mu_i - \hat{\mu}_i)^2$ and the expectations in the decomposition

$$(y_i-\hat{\mu}_i)^2 = (y_i-\mu_i)^2 + (\hat{\mu}_i-\mu_i)^2 - 2(y_i-\mu_i)(\hat{\mu}_i-\mu_i)$$

(compare Efron, 2004). Thus the apparent error term $(y_i - \hat{\mu}_i)^2$ has to be corrected by the covariance between the estimate at $\boldsymbol{x}_i$ and the observation at $\boldsymbol{x}_i$.

For a homogeneous covariance and a linear estimate $\hat{\boldsymbol{\mu}} = \boldsymbol{M}\boldsymbol{y}$ one obtains from (15.2)

$$\mathrm{E}(\sum_{i=1}^{n} E_0(y_{0i} - \hat{\mu}_i)^2) = \mathrm{E}(ne_{\text{app}} + 2\sigma^2 \text{ trace } (\boldsymbol{M})).$$

The right-hand side corresponds to Mallow's C_p , which may be given as

$$C_p = \sum_{i=1}^{n} (y_i - \hat{\mu}_i)^2 + 2\hat{\sigma}^2 \text{ trace } (\boldsymbol{M}),$$

where for a linear model with p parameters one has trace $(\boldsymbol{M}) = p$ and σ^2 has to be estimated. The term $2\sigma^2$ trace $(\boldsymbol{M})$ may be considered the optimism of the apparent error rate. In general, one has to add the (unobservable) term $\Sigma_i 2\,\mathrm{cov}(\hat{\mu}_i, y_i)$ to the apparent error term to obtain an unbiased estimate. Efron (2004) calls the additional term a covariance penalty.

In the Gaussian case, the $\mathrm{cov}(\hat{\mu}_i, y_i)$ has the form $\sigma^2\,\mathrm{E}(\partial\hat{\mu}_i/\partial y_i)$, yielding Stein's unbiased risk estimate (SURE) for the total prediction error, which uses the correction term $2\sigma^2\Sigma_i\partial\hat{\mu}_i/\partial y_i$ (Stein, 1981). An alternative way to estimate the covariance $\mathrm{cov}(\hat{\mu}_i, y_i)$ is based on bootstrap techniques (see Efron, 2004; Ye, 1998).

The penalty correction term may also be used to define the degrees of freedom of a fit. For the linear model with fit $\hat{\boldsymbol{\mu}} = \boldsymbol{M}\boldsymbol{y}$, where $\boldsymbol{M}$ is a projection matrix, trace $(\boldsymbol{M})$ is the dimension of the projected space. More generally, one may consider the degrees of freedom

$$df = \sum_{i=1}^{n} \mathrm{cov}(\hat{\mu}_i, y_i)/\sigma^2$$

(compare Ye, 1998).

Predicting a Distribution

From a more general point of view, if $\boldsymbol{x}$ is observed, prediction may be understood as finding the density of y given $\boldsymbol{x}$, $\hat{f}(y|\boldsymbol{x})$. The prediction error then is measured as a discrepancy between the true density $f(.\,|\boldsymbol{x})$ and the estimated density $\hat{f}(.\,|\boldsymbol{x})$. A measure that applies is the Kullback-Leibler discrepancy:

$$KL(f(.\,|\boldsymbol{x}), \hat{f}(.\,|\boldsymbol{x})) = E_f\ \log(f(.\,|\boldsymbol{x})/\hat{f}(.\,|\boldsymbol{x})),$$

where expectation is based on the underlying density f. If one assumes that the underlying density is a normal distribution $N(\mu(\boldsymbol{x}), \sigma^2)$ and the estimated density is $N(\hat{\mu}(\boldsymbol{x}), \hat{\sigma}^2)$, it is easily shown that one obtains

$$KL(f, \hat{f}) = \{\sigma^2 + (\mu(\boldsymbol{x}) - \hat{\mu}(\boldsymbol{x}))^2\}/(2\hat{\sigma}^2) + \log\frac{\hat{\sigma}}{\sigma} - \frac{1}{2},$$

which, for given $\hat{\sigma}$, is a scaled version of $\mathrm{PMSE}(\boldsymbol{x}) = \sigma^2 + (\mu(\boldsymbol{x}) - \hat{\mu}(\boldsymbol{x}))^2$. Thus, for normal distributions, the Kullback-Leibler discrepancy is strongly linked to the mean squared error.

15.1.2 Discrete Data

In discrete data let the new observation be given by $(Y, \boldsymbol{x})$, with Y taking values from $\{1, \ldots, k\}$. When $\boldsymbol{x}$ is observed, prediction is often understood as finding a prediction rule $\hat{Y} = \hat{Y}(\boldsymbol{x})$ with

$\hat{Y}$ taking values from $\{1,\ldots,k\}$. Prediction in this sense is often called *classification* since one wants to find the true underlying class. Y is the indicator for the underlying class and $\hat{Y}$ is the diagnosis. In the following, finding the point prediction $\hat{Y}$ will be referred to as a *direct prediction* or *classification*.

When a statistical model is used for prediction one often first estimates the probability $P(Y = r|\boldsymbol{x})$ and then derives a classification rule from these probabilities. But then more information is available than a simple assignment to classes. The conditional probabilities contain additional information about the precision of the classification rule derived from it. Therefore, in a more general sense one might view the estimation of the conditional distribution given $\boldsymbol{x}$, which is identical to the estimated probabilities $P(Y = r|\boldsymbol{x}), r = 1,\ldots,k$, as the prediction rule. Depending on the type of prediction one has in mind, different loss functions apply. In the following we will first consider loss functions for direct predictions and then loss functions for the case where the prediction is based on an estimate of the underlying probability. Since the actual or expected loss is given by (15.1), it again suffices to minimize the conditional loss given $\boldsymbol{x}$. Therefore, in the notation, for the most part, $\boldsymbol{x}$ is dropped.

Direct Prediction

Loss functions for direct predictions come in two forms. One may use Y and $\hat{Y}$ as arguments or one may refer to the multivariate nature of the response by using vector-valued arguments. The latter form is often useful to see the connection to common loss functions like the quadratic loss.

Let $Y \in \{1,\ldots,k\}$ denote the categorical response and $\boldsymbol{y}^T = (y_1,\ldots,y_k)$ denote the corresponding vector-valued response, where $y_i = 1$ if $Y = i$ and $y_i = 0$ otherwise. Let $(\boldsymbol{y},\boldsymbol{x})$ with $\boldsymbol{y}^T = (y_1,\ldots,y_k)$ denote a new observation and $\hat{\boldsymbol{y}}^T = (\hat{y}_1,\ldots,\hat{y}_k)$ the prediction for the given value $\boldsymbol{x}$, where $\hat{y}_r \in \{0,1\}$ and $\sum_r \hat{y}_r = 1$.

One may again consider functions like the quadratic or L_1 loss, which are in common use for continuous response variables. However, for categorical responses with observations and direct predictions as arguments, the quadratic loss and the L_1-norm loss

$$L_2(\boldsymbol{y},\hat{\boldsymbol{y}}) = \sum_{r=1}^{k}(y_r - \hat{y}_r)^2, \quad L_1(\boldsymbol{y},\hat{\boldsymbol{y}}) = \sum_{r=1}^{k}|y_r - \hat{y}_r|$$

reduce to twice the simple 0-1 loss $2L_{01}(\boldsymbol{y},\hat{\boldsymbol{y}})$, where

$$L_{01}(\boldsymbol{y},\hat{\boldsymbol{y}}) = \begin{cases} 1 & \boldsymbol{y} \neq \hat{\boldsymbol{y}} \\ 0 & \boldsymbol{y} = \hat{\boldsymbol{y}}. \end{cases}$$

L_{01} has value 0 if the prediction is correct and 1 if it is incorrect. With arguments $Y,\hat{Y} \in \{1,\ldots,k\}$ the 0-1 loss has the form

$$L_{01}(Y,\hat{Y}) = \begin{cases} 0 & Y = \hat{Y} \\ 1 & Y \neq \hat{Y}. \end{cases}$$

The 0-1 function plays a central role in classification since the corresponding actual prediction error at a fixed value $\boldsymbol{x}$ is given by

$$\mathrm{E}\ L_{01}(\boldsymbol{y},\hat{\boldsymbol{y}}) = P(\boldsymbol{y} \neq \hat{\boldsymbol{y}}|\boldsymbol{x}) = 1 - P(\boldsymbol{y} = \hat{\boldsymbol{y}}|\boldsymbol{x}),$$

which is the *actual probability of misclassification*. The optimal prediction rule obtained by minimizing $E\ L_{01}(\boldsymbol{y},\hat{\boldsymbol{y}})$ is known as the *Bayes classifier* and is given by

$$\hat{Y}(\boldsymbol{x}) = r \quad \text{if} \quad P(Y = r|\boldsymbol{x}) = \max_{i=1,\ldots,k} P(Y = i|\boldsymbol{x}).$$

In the case of two classes it reduces to the intuitively appealing rule

$$\hat{Y}(\boldsymbol{x}) = 1 \quad \text{if} \quad P(Y = 1|\boldsymbol{x}) \geq 0.5.$$

The Bayes classifier minimizes the probability of misclassification by minimizing $P(\boldsymbol{y} \neq \hat{\boldsymbol{y}}|\boldsymbol{x})$. It is the central issue of Section 15.2, where classification methods that minimize the probability of misclassification are considered extensively. The Bayes rule results from minimizing the expected 0-1 loss. Therefore, in direct predictions all loss functions that reduce to 0-1 loss when given in the form $L_{01}(\boldsymbol{y}, \hat{\boldsymbol{y}})$ yield the Bayes classifier. The reason is that minimization of $\mathrm{E}\ L_{01}(\boldsymbol{y}, \hat{\boldsymbol{y}})$ or $\mathrm{E}\ L_{01}(Y, \hat{Y})$ is over discrete values, and in the latter form over $\hat{Y} \in \{1, \ldots, k\}$. If minimization is over real-valued approximations of the true vector $\boldsymbol{y}$, the quadratic loss or L_1 loss will yield different expected losses and therefore different allocation rules. The effect is investigated in the next section.

Prediction Based on Estimated Probabilities

For categorical data prediction may also be understood as finding the best real-valued approximation to the unknown $\boldsymbol{y}$, which usually is an estimate $\hat{\boldsymbol{\pi}}^T = (\hat{\pi}_1, \ldots, \hat{\pi}_k)$ of the underlying probabilities. Then the corresponding loss functions have the form $L(\boldsymbol{y}, \hat{\boldsymbol{\pi}})$. Slightly more generally, one can also consider the discrepancy between the true probability function represented by $\boldsymbol{\pi}^T = (\pi_1, \ldots, \pi_k)$ and the estimated probability function $\hat{\boldsymbol{\pi}}^T = (\hat{\pi}_1, \ldots, \hat{\pi}_k)$, and derive the loss $L(\boldsymbol{y}, \hat{\boldsymbol{\pi}})$ from it. Possible discrepancy measures are the squared loss:

$$L_2(\boldsymbol{\pi}, \hat{\boldsymbol{\pi}}) = \sum\nolimits_r (\pi_r - \hat{\pi}_r)^2,$$

the L_1-norm:

$$L_1(\boldsymbol{\pi}, \hat{\boldsymbol{\pi}}) = \sum\nolimits_r |\pi_r - \hat{\pi}_r|,$$

and the Kullback-Leibler discrepancy:

$$L_{KL}(\boldsymbol{\pi}, \hat{\boldsymbol{\pi}}) = \sum\nolimits_r \pi_r \log(\pi_r / \hat{\pi}_r).$$

The Kullback-Leibler discrepancy uses explicitly that the underlying distribution is the multinomial distribution with parameter $\boldsymbol{\pi}$. Further possible measures are the Pearson chi-squared discrepancy $P(\boldsymbol{\pi}, \hat{\boldsymbol{\pi}}) = \sum_r (\pi_r - \hat{\pi}_r)^2 / \hat{\pi}_r$ and the Neyman-Pearson discrepancy $\mathrm{NP}(\boldsymbol{\pi}, \hat{\boldsymbol{\pi}}) = \sum_r (\pi_r - \hat{\pi}_r)^2 / \pi_r$. When $\hat{\boldsymbol{\pi}}$ is considered as an estimate of a new observation $\boldsymbol{y}$, loss functions like L_2 and L_1 do not reduce to the 0-1 loss. Let us consider a new observation $Y \in \{1, \ldots, k\}$ or $\boldsymbol{y}^T = (y_1, \ldots, y_k)$, $y_r \in \{0, 1\}$, and the corresponding loss $L(\boldsymbol{y}, \hat{\boldsymbol{\pi}})$ and actual prediction error $\mathrm{E}\, L(\boldsymbol{y}, \hat{\boldsymbol{\pi}})$. For the quadratic loss one obtains the *quadratic score*, which is also known as *Brier score*:

$$L_2(\boldsymbol{y}, \hat{\boldsymbol{\pi}}) = (1 - \hat{\pi}_Y)^2 + \sum_{r \neq Y} \hat{\pi}_r^2.$$

In the literature on scoring rules usually the reward for estimating an outcome $\boldsymbol{y}$ by the distribution $\hat{\boldsymbol{\pi}}$ is measured and the corresponding Brier score is considered to be $-L_2(\boldsymbol{y}, \hat{\boldsymbol{\pi}})$. However, here the loss function perspective is preferred. When using the quadratic score one obtains for the actual prediction error

$$\mathrm{E}(L_2(\boldsymbol{y}, \hat{\boldsymbol{\pi}})) = \mathrm{E}\left\{\sum_{r=1}^{k} (y_r - \hat{\pi}_r)^2\right\} = L_2(\boldsymbol{\pi}, \hat{\boldsymbol{\pi}}) + \sum_{r=1}^{k} \mathrm{var}(y_r)$$

(Exercise 15.2). Therefore, the prediction error is determined by the distance between the true probability $\boldsymbol{\pi}$ and the estimate $\hat{\boldsymbol{\pi}}$ and a term that depends on the true probability only. $E(L_2(\boldsymbol{y}, \boldsymbol{\pi}))$ obtains its minimum value if $\boldsymbol{\pi} = \hat{\boldsymbol{\pi}}$, yielding $L_2(\boldsymbol{\pi}, \hat{\boldsymbol{\pi}}) = 0$.

For the Kullback-Leibler discrepancy one obtains the distance between a new observation $\boldsymbol{y}$ and its estimate $\hat{\boldsymbol{\pi}}$ as

$$L_{KL}(\boldsymbol{y}, \hat{\boldsymbol{\pi}}) = -\log(\hat{\pi}_Y) = -\sum_{r=1}^{k} y_r \log(\hat{\pi}_r),$$

which is called the *logarithmic score*. It is strongly linked to the deviance since for multinomial distribution the deviance is twice the Kullback-Leibler distance. For example, for a binary distributions the deviance in its usual form is given as $D(y, \hat{\pi}) = -2\log(\hat{\pi})$ if $y = 1$ and $D(y, \hat{\pi}) = -2\log(1-\hat{\pi})$ if $y = 0$. Therefore, for the vector $\boldsymbol{y}^T = (y_1, y_2)$ encoding the two classes one obtains $D(y, \hat{\pi}) = 2L_{KL}((y_1, y_2), (\hat{\pi}_1, \hat{\pi}_2))$. The relation is the reason why, for multinomial distributions, the logarithmic score when applied to new observations is also called the *predictive deviance*. As for the squared loss, a decomposition into a term that is determined by the distance between the true probability $\boldsymbol{\pi}$ and the estimate $\hat{\boldsymbol{\pi}}$ and a term that does depend on the true probability only can be found. One obtains $\mathrm{E}(L_{KL}(\boldsymbol{y}, \hat{\boldsymbol{\pi}})) = L_{KL}(\boldsymbol{\pi}, \hat{\boldsymbol{\pi}}) + \mathrm{Entr}(\boldsymbol{\pi})$, with $\mathrm{Entr}(\boldsymbol{\pi}) = -\sum_{r=1}^{k} \pi_r \log(\pi_r)$ denoting the *entropy* of probability vector $\boldsymbol{\pi}$.

The link between a direct prediction and a prediction as an estimation of probabilities is found by considering how the estimation of probabilities turns into a prediction of the new observation. Usually the prediction of $\boldsymbol{y}$ or Y is derived from $\hat{\boldsymbol{\pi}}$ by setting $\hat{Y} = r$ if $\hat{\pi}_r = \max_{i=1,\dots,k} \hat{\pi}_i$. Uniqueness can be obtained by assigning an observation to the class with the smallest number if the estimated probabilities are equal. The maximization procedure can be included into the loss function by using

$$L_B(\boldsymbol{\pi}, \hat{\boldsymbol{\pi}}) = \sum_{r=1}^{k} \pi_r (1 - \mathrm{Ind}_r(\hat{\boldsymbol{\pi}})),$$

where

$$\mathrm{Ind}_r(\boldsymbol{\pi}) = \begin{cases} 1 & \pi_r = \max_{i=1,\dots,k} \pi_i,\ \pi_r > \pi_i \text{ for } i < r \\ 0 & \text{otherwise} \end{cases}$$

is an indicator function for optimal classification. The corresponding loss,

$$L_B(\boldsymbol{y}, \hat{\boldsymbol{\pi}}) = \sum_{r=1}^{k} y_r (1 - \mathrm{Ind}_r(\hat{\boldsymbol{\pi}})) = \mathrm{Ind}_Y(\hat{\boldsymbol{\pi}}),$$

is equivalent to a 0-1 loss, but built with argument $\hat{\boldsymbol{\pi}}$. For the underlying function one obtains

$$L_B(\boldsymbol{\pi}, \hat{\boldsymbol{\pi}}) = \sum_{r=1}^{k} \pi_r (1 - \mathrm{Ind}_r(\hat{\boldsymbol{\pi}})) = \sum_{r:\hat{\pi}_r < \max_i \hat{\pi}_i} \pi_r = 1 - \pi_{\hat{Y}},$$

which is the probability of class $\hat{Y}$ that has been determined as an estimated Bayes classifier. Since $\mathrm{E}(L_B(\boldsymbol{y}, \hat{\boldsymbol{\pi}})) = L_B(\boldsymbol{\pi}, \hat{\boldsymbol{\pi}})$ holds, $L_B(\boldsymbol{\pi}, \hat{\boldsymbol{\pi}})$ can be considered the expected loss or risk of the estimated Bayes classifier.

There is a strong difference between using $L_B(\boldsymbol{y}, \hat{\boldsymbol{\pi}})$ or a loss like the quadratic loss $L_2(\boldsymbol{y}, \hat{\boldsymbol{\pi}})$. The expectation of the quadratic loss is minimal only if $\boldsymbol{\pi} = \hat{\boldsymbol{\pi}}$, whereas minimization of the expectation of $L_B(\boldsymbol{y}, \hat{\boldsymbol{\pi}})$ is also obtained for estimates $\hat{\boldsymbol{\pi}}$ that have the same indicator function as $\boldsymbol{\pi}$. While the quadratic loss aims at exact estimations of probabilities, 0-1 loss ignores the precision of these estimates and focuses solely on misclassification. Therefore, when prediction is based on estimated probabilities, the losses $L(\boldsymbol{y}, \hat{\boldsymbol{\pi}})$ may be used as empirical measures for the precision of the predictions. For the quadratic loss one obtains the Brier score, and for the Kullback-Leibler loss the logarithmic score. Both give more precise measures for the accuracy of prediction than the simple 0-1 loss.

Within the framework of scoring rules one considers the reward of a prediction in the form of scores instead of loss functions. However, by using the negative values of scores one obtains loss functions. A rigorous treatment of score functions is given by Gneiting and Raftery (2007), who also consider alternative scoring rules like the spherical score and the beta family score, which are proposed by Buja et al. (2005). In particular, one distinguishes between proper, strictly proper, and improper scores. In terms of loss functions, a function is proper if $\mathrm{E}\, L_{\boldsymbol{\pi}}(\boldsymbol{y}, \boldsymbol{\pi}) \le \mathrm{E}\, L_{\boldsymbol{\pi}}(\boldsymbol{y}, \tilde{\boldsymbol{\pi}})$ for all distributions $\boldsymbol{\pi}, \tilde{\boldsymbol{\pi}}$. It is strictly proper if that holds with equality if and only if $\boldsymbol{\pi} = \tilde{\boldsymbol{\pi}}$. Brier scores and logarithmic scores are strictly proper but the 0-1-loss is merely proper.

TABLE 15.1: Loss functions, scores, and actual prediction error.

$L_2(\boldsymbol{\pi}, \hat{\boldsymbol{\pi}}) = \sum_{r=1}^{k} (\pi_r - \hat{\pi}_r)^2$	quadratic loss
$L_2(\boldsymbol{y}, \hat{\boldsymbol{\pi}}) = (1 - \hat{\pi}_Y)^2 + \sum_{r \ne Y} \hat{\pi}_r^2$	quadratic or Brier score
$E(L_2(\boldsymbol{y}, \hat{\boldsymbol{\pi}})) = L_2(\boldsymbol{\pi}, \hat{\boldsymbol{\pi}}) + \sum_{r=1}^{k} \pi_r (1 - \pi_r)$	actual prediction error
$L_B(\boldsymbol{\pi}, \hat{\boldsymbol{\pi}}) = \sum_{r=1}^{k} \pi_r (1 - \mathsf{Ind}_r(\hat{\boldsymbol{\pi}})) = \varepsilon(\boldsymbol{x})$	Bayes loss
$L_B(\boldsymbol{y}, \hat{\boldsymbol{\pi}}) = 1 - \mathsf{Ind}_Y(\hat{\boldsymbol{\pi}}) = \frac{1}{2} \sum_{r=1}^{k} \lvert y_r - \mathsf{Ind}_r(\hat{\boldsymbol{\pi}}) \rvert$	0–1-loss
$E(L_B(\boldsymbol{y}, \hat{\boldsymbol{\pi}})) = L_B(\boldsymbol{\pi}, \hat{\boldsymbol{\pi}})$	actual prediction error
$L_{KL}(\boldsymbol{\pi}, \hat{\boldsymbol{\pi}}) = \sum_{r=1}^{k} \pi_r \log(\pi_r / \hat{\pi}_r)$	Kullback-Leibler discrepancy
$L_{KL}(\boldsymbol{y}, \hat{\boldsymbol{\pi}}) = -\log(\hat{\pi}_Y) = -\sum_{r=1}^{k} y_r \log(\hat{\pi}_r)$	logarithmic score
$E(L_{KL}(\boldsymbol{y}, \hat{\boldsymbol{\pi}})) = L_{KL}(\boldsymbol{\pi}, \hat{\boldsymbol{\pi}}) + \sum_{r=1}^{k} -\pi_r \log(\pi_r)$	actual prediction error

A direct prediction of a class in the form $\hat{\boldsymbol{y}}^T = (\hat{y}_1, \dots, \hat{y}_k)$ if $\boldsymbol{y}^T = (y_1, \dots, y_k)$ is observed may be seen as a special case of a degenerate distribution. For the Bayes loss, as well as for $L_2(\boldsymbol{y}, \hat{\boldsymbol{y}})$ and $L_1(\boldsymbol{y}, \hat{\boldsymbol{y}})$, one obtains the 0-1 loss L_{01} and, therefore, $\mathrm{E}(L(\boldsymbol{y}, \hat{\boldsymbol{y}}))$ is the probability of misclassification. For the Kullback-Leibler discrepancy, however, one has $L_{KL}(\boldsymbol{y}, \hat{\boldsymbol{y}}) = KL(Y, \hat{Y}) = -\log(\hat{\pi}_Y)$, where $Y \in \{1, \dots, k\}$ is the observation. Alternative loss functions that are usefull in classification are the hinge loss and the exponential loss, which will be considered later.

Loss for Univariate Discrete Response

In the preceding sections implicitly the response was assumed to be multinomial. For univariate discrete responses, for example, count data with $y \in \{0, 1, \dots\}$ alternative loss functions can be useful. A distance measure that explicitly depends on the assumed distribution may be derived from the deviance, which in generalized linear models measures the discrepancy between the data and the fit. It may also be used for future observations in the form of a *predictive deviance* that measures the discrepancy between the prediction and the observation. The deviance is an appropriate measure since its scaling is based on the underlying distribution. For a Poisson distributed response, it is given by $D(y, \hat{\mu}) = 2y\log(y/\hat{\mu})$, where $\hat{\mu}$ is the estimated mean. A disadvantage of the predictive deviance is that it is sensible only if the assumed distribution assumption holds.

In this respect, the Kullback-Leibler distance is more general. For a discrete response it has the general form $L_{\text{KL}}(\boldsymbol{\pi}, \hat{\boldsymbol{\pi}}) = \sum_i \pi_i \log(\pi_i/\hat{\pi}_i)$, where $\boldsymbol{\pi}^T = (\pi_1, \pi_2, \dots)$ is the vector of the true probabilities and $\hat{\boldsymbol{\pi}}^T = (\hat{\pi}_1, \hat{\pi}_2, \dots)$ is the estimate. Instead of using dummy variables, a new response $y \in \{0, 1, \dots\}$ may be identified with the degenerate distribution that puts weight 1 at y, represented by the vector $\boldsymbol{\delta}_y^T = (\delta_1(y), \delta_2(y), \dots)$, $\delta_y(y) = 1$, $\delta_y(z) = 0$, $z \neq y$. The corresponding Kullback-Leibler distance is again the logarithmic score $L_{\text{KL}}(\boldsymbol{\delta}_y, \hat{\boldsymbol{\pi}}) = -\log(\hat{\pi}_y)$. It measures the distance between y and the estimate $\hat{\boldsymbol{\pi}}$ simply by the negative log-transformed probability at the new observation y. The distance does not depend on the assumed distribution and therefore may be used quite generally without reference to a specific discrete distribution (although the range of the distance depends on the underlying distribution). The decomposition considered previously also holds for the case with infinite support. One obtains the actual prediction error in the form $\text{E}(L_{KL}(\boldsymbol{\delta}_y, \hat{\boldsymbol{\pi}})) = L_{KL}(\boldsymbol{\pi}, \hat{\boldsymbol{\pi}}) + \sum_r -\pi_r \log(\pi_r)$. The first term in the equation is the error due to estimation of the underlying probability distribution, whereas the second term gives the error that cannot be avoided and which is a function of the true distribution.

A criticism of scores like the logarithmic score is that the predictive distribution $\hat{\boldsymbol{\pi}}$ is only evaluated at the observation y. Therefore, it does not take the whole predictive distribution into account. As Gneiting and Raftery (2007) postulate, a desirable predictive distribution should be as sharp as possible and well calibrated. Sharpness refers to the concentration of the distribution and calibration to the agreement between the distribution and the observation. For nominal data, quantification of the distance tends to become trivial. But if y represents count data, a more appropriate loss function derived from the continuous ranked probability score (Gneiting and Raftery, 2007) is

$$L_{RPS}(y, \hat{\boldsymbol{\pi}}) = \sum_r (\hat{\pi}(r) - Ind(y \leq r))^2, \tag{15.3}$$

where $\hat{\pi}(r) = \hat{\pi}_1 + \dots + \hat{\pi}_r$. For binary data, it is a sum over quadratic (or Brier) scores and takes the closeness between the whole distribution and the observed value into account. It may also be used for categorial ordered responses, which actually are not uni-dimensional but contain at least order information. For further measures see Gneiting and Raftery (2007).

15.2 Methods for Optimal Classification

In this section we consider how the predictor space that contains the observations $\boldsymbol{x}$ can be partitioned to obtain optimal classification rules. The objective is a direct prediction, that is, one wants to find the the true class $Y \in \{1, \dots, k\}$ when only $\boldsymbol{x}$ from a new observation $(Y, \boldsymbol{x})$ is observed.

15.2.1 Bayes Rule and the Minimization of the Rate of Misclassification

In the following we consider direct prediction rules that usually are derived within the framework of discriminant analysis. Therefore we focus on the classification problem with a finite number of classes, $Y \in \{1, \dots, k\}$. Some denotations that are used in the following are

- $P(r) = P(Y = r)$, $r = 1, \dots, k$, for the *prior*.
- $P(r|\boldsymbol{x}) = P(Y = r|\boldsymbol{x})$, $r = 1, \dots, k$, for the conditional probability for class r given $\boldsymbol{x}$, also called *posterior probability*.

- $f(\boldsymbol{x}|1), \ldots, f(\boldsymbol{x}|k)$ for the densities within classes, which may refer to continuous or discrete distributions.
- $f(\boldsymbol{x}) = P(1)f(\boldsymbol{x}|1) + \cdots + P(k)f(\boldsymbol{x}|k)$ for the mixture density.

Let a classifier or classification rule be defined as a mapping

$$\begin{aligned} \delta : \mathbb{R}^p &\mapsto \{1, \ldots, k\} \\ \boldsymbol{x} &\mapsto \delta(\boldsymbol{x}). \end{aligned}$$

A basic classification rule that can be derived from decision theory arguments (see next section) and has been already considered briefly in the preceding section is the *Bayes rule*. The Bayes rule has the simple form

$$\delta^*(\boldsymbol{x}) = r \quad \Longleftrightarrow \quad P(r|\boldsymbol{x}) = \max_{i=1,\ldots,k} P(i|\boldsymbol{x}). \tag{15.4}$$

The Bayes rule explains itself. For a given $\boldsymbol{x}$ one chooses the class for which the posterior probability $P(r|\boldsymbol{x})$ takes its maximal value. The rule is strongly connected to 0-1 loss $L_{01}(Y, \hat{Y}) = 0$ if $Y = \hat{Y}$ and $L_{01}(Y, \hat{Y}) = 1$ if $Y \neq \hat{Y}$. When considering the 0-1 loss one may distinguish between several error rates that are connected to the rule. For a fixed classifier δ the prediction error for randomly drawn $(Y, \boldsymbol{x})$ has the form

$$\varepsilon = \mathrm{E}\ L_{01}(Y, \delta(\boldsymbol{x})) = P(Y \neq \delta(\boldsymbol{x})) = 1 - P(Y = \delta(\boldsymbol{x}))$$

and is also known as the *global rate of misclassification* or the *global probability of misclassification*. From $P(Y \neq \delta(\boldsymbol{x}))$ it is seen that ε is the probability that an observation $(Y, \boldsymbol{x})$ is misclassified. Conditioning on $\boldsymbol{x}$ yields the *error rate of misclassification conditional on* $\boldsymbol{x}$:

$$\begin{aligned} \varepsilon(\boldsymbol{x}) &= \mathrm{E}_{y|\boldsymbol{x}}\ L_{01}(Y, \delta(\boldsymbol{x})) = P(Y \neq \delta(\boldsymbol{x})|\boldsymbol{x}) = 1 - P(Y = \delta(\boldsymbol{x})|\boldsymbol{x}) = \\ &= 1 - P(\delta(\boldsymbol{x})|\boldsymbol{x}). \end{aligned}$$

The connection between these error rates is given by

$$\varepsilon = \mathrm{E}_{\boldsymbol{x}}\ \varepsilon(\boldsymbol{x}) = \int \varepsilon(\boldsymbol{x}) f(\boldsymbol{x}) d\boldsymbol{x}.$$

It is obvious from $\varepsilon(\boldsymbol{x}) = 1 - P(\delta(\boldsymbol{x})|\boldsymbol{x})$ how to choose the best prediction with reference to the 0-1 loss for a fixed $\boldsymbol{x}$. One just has to choose $\delta(\boldsymbol{x})$ such that $P(\delta(\boldsymbol{x})|\boldsymbol{x})$ takes its maximal value. Since ε is an integral across $\varepsilon(\boldsymbol{x})$, the optimal error rate is obtained by choosing the optimal rule for each value $\boldsymbol{x}$. Therefore, the Bayes rule is optimal when the 0-1 loss is considered and one obtains:

> The Bayes rule
>
> $$\delta^*(\boldsymbol{x}) = r \quad \Longleftrightarrow \quad P(r|\boldsymbol{x}) = \max_{i=1,\ldots,k} P(i|\boldsymbol{x})$$
>
> minimizes the global probability of misclassification

The Bayes rule obtains the *optimal error rate*:

$$\varepsilon_{opt} = 1 - \int \max_{r=1,\ldots,k} P(r|\boldsymbol{x}) f(\boldsymbol{x}) d\boldsymbol{x},$$

with the optimal value at $\boldsymbol{x}$ given as $\epsilon_{opt}(\boldsymbol{x}) = 1 - \max_{r=1,\ldots,k} p(r|\boldsymbol{x})$.

Further error rates are the *individual errors*:

$$\varepsilon_{rs} = P(\delta(\boldsymbol{x}) = s|Y = r) = \int_{\boldsymbol{x}:\delta(\boldsymbol{x})=s} f(\boldsymbol{x}|r)d\boldsymbol{x},$$

which represent the probability of classifying into class s if the underlying true class is r. Individual error rates are linked to the global error rate by

$$\varepsilon = P(\delta(\boldsymbol{x}) \neq Y) = \sum_{r=1}^{k} P(\delta(\boldsymbol{x}) \neq r|Y = r)P(r) = \sum_{r=1}^{k}\sum_{s\neq r} \varepsilon_{rs}\, P(r). \tag{15.5}$$

When looking for good classifiers that minimize the global error, the main problem is that $P(r|\boldsymbol{x})$ usually is unknown and has to be estimated. For the estimated rate one obtains for a given rule $\hat{Y} = \delta(\boldsymbol{x})$ the actual error rate ε. Of course for the actual error rate $\varepsilon_{opt} \leq \varepsilon$ holds.

Error Rates

Global rate of misclassification

$$\varepsilon = P(Y \neq \delta(\boldsymbol{x}))$$

Conditional rate of misclassification (conditional on $\boldsymbol{x}$)

$$\varepsilon(\boldsymbol{x}) = 1 - P(\delta(\boldsymbol{x})|\boldsymbol{x})$$

Individual error rates

$$\varepsilon_{rs} = P(\delta(x) = s|Y = r)$$

15.2.2 Classification with Discriminant Functions

The Bayes rule as given above uses directly the posterior probabilities $P(Y = r|\boldsymbol{x})$. Alternative forms of the classification rule can be formulated by using the prior and the densities of the predictors given the class. It is helpful to consider the Bayes rule as a maximizer of so-called *discriminant functions*. For each $\boldsymbol{x}$ let the discriminant functions $d_r(\boldsymbol{x})$, $r = 1, \ldots, k$, contain some measure for the plausibility that observation $\boldsymbol{x}$ comes from class r. If one defines $d_r(\boldsymbol{x}) = P(r|\boldsymbol{x})$, one obtains the Bayes rule in the form

$$\delta^*(\boldsymbol{x}) = r \quad \Longleftrightarrow \quad d_r(\boldsymbol{x}) = \max_{i=1,\ldots,k} d_i(\boldsymbol{x}). \tag{15.6}$$

Alternative forms that yield the same classification rule can be obtained by using Bayes' theorem

$$P(r|\boldsymbol{x}) = \frac{f(\boldsymbol{x}|r)P(r)}{f(\boldsymbol{x})} = \frac{f(\boldsymbol{x}|r)P(r)}{\sum_{i=1}^{k} P(i)f(\boldsymbol{x}|i)}.$$

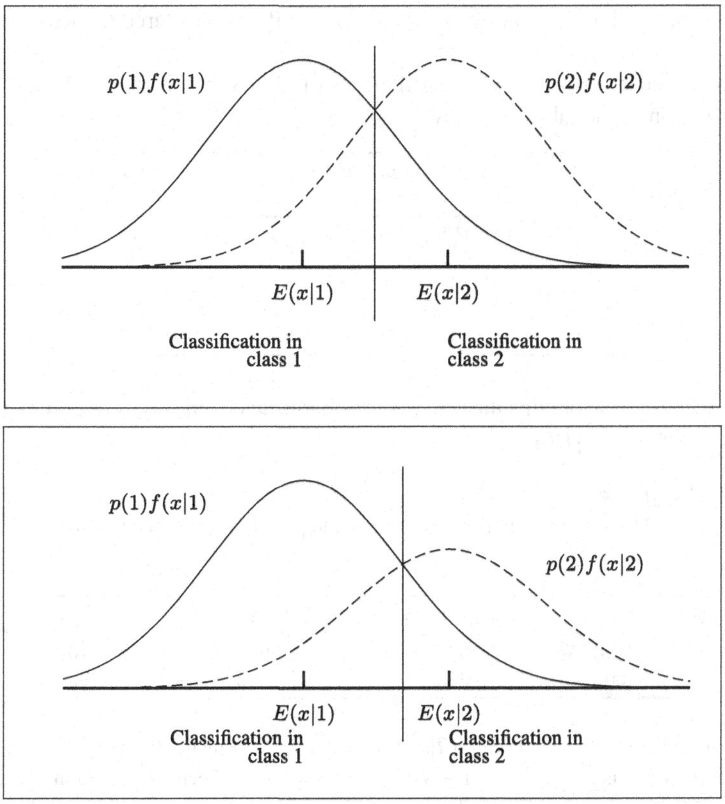

FIGURE 15.1: Bayes classifications for $p(1) = p(2) = 0.5$ (upper panel) and $p(1) = 0.6$, $p(2) = 0.4$ (lower panel).

Comparison of two classes by use of the Bayes rule has the form

$$\begin{aligned} P(r|\boldsymbol{x}) \geq P(s|\boldsymbol{x}) \iff & \frac{f(\boldsymbol{x}|r)P(r)}{f(\boldsymbol{x})} \geq \frac{f(\boldsymbol{x}|s)P(s)}{f(\boldsymbol{x})} \\ \iff & f(\boldsymbol{x}|r)P(r) \geq f(\boldsymbol{x}|s)P(s) \\ \iff & \log(f(\boldsymbol{x}|r)) + \log(P(r)) \geq \log(f(\boldsymbol{x}|s)) + \log(P(s)). \end{aligned}$$

Thus the maximization of $P(r|\boldsymbol{x})$ over the classes is also obtained by using a maximization of discriminant functions with the functions given in one of the following forms:

$$d_r(\boldsymbol{x}) = P(r|\boldsymbol{x}), \quad d_r(\boldsymbol{x}) = f(\boldsymbol{x}|r)P(r)/f(\boldsymbol{x}),$$
$$d_r(\boldsymbol{x}) = f(\boldsymbol{x}|r)P(r), \quad d_r(\boldsymbol{x}) = \log(f(\boldsymbol{x}|r)) + \log(P(r)).$$

The different forms may be used to show different aspects of the Bayes rule and to simplify the classification rule. The form $d_r(\boldsymbol{x}) = f(\boldsymbol{x}|r)P(r)$ shows that the Bayes classification is equivalent to classifying an observation into class r if the weighted density of the class is the maximal one. Figure 15.1 shows the classification rule for a one-dimensional predictor with equal priors $P(1) = P(2)$ and with priors $P(1) = 0.6$, $P(2) = 0.4$. It is seen that the increase in prior probability from $P(1) = 0.5$ to $P(1) = 0.6$ shifgts the cut-off to the right. Therefore, the area where classification is into class 1 increases. The logarithmic form of the density may be used to obtain a simple form of the Bayes rule when the distribution of the predictor is normal. The case of normal distributions will be considered in the next section as an example.

For discrete predictors with given distributions of indicators given the classes, one can collect the information in the table as follows:

		Indicator Values			Priors
		x_1	...	x_m	
Classes	1	$P(x_1\|1)$	...	$P(x_m\|1)$	$P(1)$
	.	.		.	.
	.			.	.
	.	.		.	.
	k	$P(x_1\|k)$	...	$P(x_m\|k)$	$P(k)$

For fixed value $\boldsymbol{x}_i$ one obtains the posterior probabilities from Bayes' theorem as $P(r|\boldsymbol{x}_i) = P(\boldsymbol{x}_i|r)P(r)/(\sum_{i=1}^{k} P(i)P(\boldsymbol{x}_i|i))$.

Example 15.3: Drug Use

If one assumes that 10 % of the population uses drugs, the probabilities of the test from Example 15.1 are given by

	x = 1 (positive)	x = 0 (negative)	Priors
1 (User)	0.95	0.05	0.10
2 (Non-user)	0.05	0.95	0.90

One computes $P(1|x = 1) = 0.68$, $P(2|x = 1) = 0.32$, yielding the Bayes rule $\delta^*(x) = 1$ if $x = 1$. The corresponding error is $\epsilon(x = 1) = 1 - P(1|x = 1) = 0.32$. Therefore, 32% of the applicants that test positive will be falsely classified as drug users. Thus, for $x = 1$, the classification rule is not very convincing. However, for $x = 0$, the results differ. One computes $P(1|x = 0) = 0.006$, $P(2|x = 0) = 0.994$, yielding the Bayes rule $\delta^*(x) = 2$ if $x = 0$ with error $\epsilon(x = 0) = 1 - P(2|x = 0) = 0.006$. It is obvious that the accuracy of the prediction strongly depends on the observation that is considered. The global error rate, which is 0.05 in the present case, is just an overall measure. The variation of the accuracy of the prediction across values of the predictor is a general property of classification rules (see also Exercise 15.4). □

15.2.3 Discrimination with Normally Distributed Predictors

Let the predictors be normally distributed within classes, that is, one assumes $\boldsymbol{x}|Y = r \sim N(\boldsymbol{\mu}_r, \boldsymbol{\Sigma}_r)$. A simplifying assumption that is often used is that the covariances are the same in each class. In this *homogeneous case* one assumes $\boldsymbol{\Sigma} = \boldsymbol{\Sigma}_1 = \ldots = \boldsymbol{\Sigma}_k$. Then the discriminant function $d_r(\boldsymbol{x}) = \log(f(\boldsymbol{x}|r)) + \log(P(\boldsymbol{x}))$ takes the form

$$d_r(\boldsymbol{x}) = -\frac{1}{2}(\boldsymbol{x} - \boldsymbol{\mu}_r)^T \boldsymbol{\Sigma}^{-1}(\boldsymbol{x} - \boldsymbol{\mu}_r) - \frac{p}{2}\log(2\pi) - \frac{1}{2}\log(|\boldsymbol{\Sigma}|) + \log(P(r)).$$

Let us investigate how the maximization of $d_r(\boldsymbol{x})$ discriminates between two classes, namely, classes r and s. It is easily seen that a comparison of $d_r(\boldsymbol{x})$ and $d_s(\boldsymbol{x})$ yields a minimum distance classifier with a shift. If $P(r) = P(s)$, one prefers class r over class s if

$$(\boldsymbol{x} - \boldsymbol{\mu}_r)^T \boldsymbol{\Sigma}^{-1}(\boldsymbol{x} - \boldsymbol{\mu}_r) \leq (\boldsymbol{x} - \boldsymbol{\mu}_s)^T \boldsymbol{\Sigma}^{-1}(\boldsymbol{x} - \boldsymbol{\mu}_s);$$

this means that the Mahalanobis distance between $\boldsymbol{x}$ and the center of the rth class $\boldsymbol{\mu}_r$ is smaller than the Mahalanobis distance between $\boldsymbol{x}$ and $\boldsymbol{\mu}_s$. For unequal priors one has to include a shift determined by $\log(P(r)/P(s))$.

By building differences between $d_r(\boldsymbol{x})$ and $d_s(\boldsymbol{x})$ one obtains after simple computation

$$d_r(\boldsymbol{x}) - d_s(\boldsymbol{x}) = \beta_{0rs} + \boldsymbol{x}^T \boldsymbol{\beta}_{rs},$$

where

$$\beta_{0rs} = -\frac{1}{2}\boldsymbol{\mu}_r^T \boldsymbol{\Sigma}^{-1} \boldsymbol{\mu}_r + \frac{1}{2}\boldsymbol{\mu}_s^T \boldsymbol{\Sigma}^{-1} \boldsymbol{\mu}_s + \log\left(\frac{P(r)}{P(s)}\right),$$

$$\boldsymbol{\beta}_{rs} = \boldsymbol{\Sigma}^{-1}(\boldsymbol{\mu}_r - \boldsymbol{\mu}_s). \tag{15.7}$$

This means that the classifier is linear with the rule:

$$\text{Prefer class } r \text{ over } s \text{ if } \beta_{0rs} + \boldsymbol{x}^T \boldsymbol{\beta}_{rs} \geq 0. \tag{15.8}$$

Geometrically, the two classes r and s are separated by a hyperplane. For the simple two-class case one obtains

$$\delta(\boldsymbol{x}) = 1 \quad \text{if} \quad \beta_0 + \boldsymbol{x}^T \boldsymbol{\beta} \geq 0,$$

where $\beta_0 = \beta_{012}$, $\boldsymbol{\beta} = \boldsymbol{\beta}_{12}$. The function $s(\boldsymbol{x}) = \beta_0 + \boldsymbol{x}^T \boldsymbol{\beta}$ is sometimes called a *score* that distinguishes between classes 1 and 2.

Estimated versions of the classifier are obtained by replacing the class centers $\boldsymbol{\mu}_r$ and $\boldsymbol{\Sigma}$ by estimates from a sample (see Section 15.4). When the number of predictors is not too large, linear classifiers of the form (15.8) are attractive because only a few parameters have to be estimated. On the other hand, they are only appropriate if the classes can be separated linearly.

It is interesting to consider an alternative representation of the linear classification rule. For the case of two classes, the posterior is given by

$$P(1|\boldsymbol{x}) = \frac{f(\boldsymbol{x}|1)P(1)}{f(\boldsymbol{x}|1)P(1) + f(\boldsymbol{x}|2)P(2)} = \frac{\exp(a)}{1 + \exp(a)}, \tag{15.9}$$

where $a = \log\{[f(\boldsymbol{x}|1)P(1)]/[f(\boldsymbol{x}|2)P(2)]\} = d_1(\boldsymbol{x}) - d_2(\boldsymbol{x})$. Therefore, one has a linear logistic regression model:

$$P(1|x) = \frac{\exp(\beta_0 + \boldsymbol{x}^T \boldsymbol{\beta})}{1 + \exp(\beta_0 + \boldsymbol{x}^T \boldsymbol{\beta})},$$

with the parameters $\beta_0, \boldsymbol{\beta}$ given by $\beta_0 = \beta_{012}, \boldsymbol{\beta} = \boldsymbol{\beta}_{12}$ from (15.7). Although the linear logistic regression model derives from the assumption of normally distributed predictors, $\boldsymbol{x}|Y = r \sim N(\boldsymbol{\mu}_r, \boldsymbol{\Sigma})$, the model holds under more general assumptions.

The linear multinomial logistic model for k classes,

$$P(r|\boldsymbol{x}) = \frac{\exp(\beta_{0r} + \boldsymbol{x}^T \boldsymbol{\beta}_r)}{1 + \sum_{j=1}^{k-1} \exp(\beta_{0j} + \boldsymbol{x}^T \boldsymbol{\beta}_j)},$$

may be derived in the same way by assuming $\boldsymbol{x}|Y = r \sim N(\boldsymbol{\mu}_r, \boldsymbol{\Sigma})$, yielding $\beta_{0r} = \beta_{0rk}$, $\boldsymbol{\beta}_r = \boldsymbol{\beta}_{rk}$ from (15.7) (Exercise 15.5).

In the more general case, with potentially differing covariances, $\boldsymbol{x}|Y = r \sim N(\boldsymbol{\mu}_r, \boldsymbol{\Sigma}_r)$, the classification rule does not simplify to a linear form. Then, the difference between discriminant functions $d_r(\boldsymbol{x}) - d_s(\boldsymbol{x})$ contains quadratic terms $x_1^2, \dots, x_p^2$ as well as interaction terms $x_i x_j$. When the terms $\boldsymbol{\mu}_r$ and $\boldsymbol{\Sigma}_r$ are replaced by estimates one refers to the method as *quadratic discrimination*. The rule is more flexible, but many more parameters have to be estimated, which for small sample sizes often yields inferior results than simple linear discrimination.

15.2.4 Bayes Rule for General Loss Functions

Classification rules may be considered within the more general framework of decision theory. On the basis of an observed feature vector $\boldsymbol{x}$, the decision $\delta(\boldsymbol{x}) = r$ corresponds to claiming that $\boldsymbol{x}$ is from class r. Decision theory requires a loss function $L(Y, \delta(\boldsymbol{x}))$ for penalizing errors in classification. Typically one assumes

$$L(r,s) = \begin{cases} 0 & \text{if} \quad r = s \quad \text{(correct decision)} \\ >0 & \text{if} \quad r \neq s \quad \text{(wrong decision).} \end{cases}$$

The criterion for the performance of a classification rule δ is the *expected loss* or the *total Bayes risk*:

$$R = \mathrm{E}(L(Y, \delta(\boldsymbol{x}))),$$

where $(Y, \boldsymbol{x})$ is a randomly drawn observation. By conditioning on $\boldsymbol{x}$ the risk can be written as

$$\begin{aligned} R &= \mathrm{E}_{\boldsymbol{x}}\, \mathrm{E}_{Y|\boldsymbol{x}}(L(Y, \delta(\boldsymbol{x}))) = \mathrm{E}_{\boldsymbol{x}} \sum_{i=1}^{k} L(i, \delta(\boldsymbol{x})) P(i|\boldsymbol{x}) \\ &= \int \sum_{i=1}^{k} L(i, \delta(\boldsymbol{x})) P(i|\boldsymbol{x}) f(\boldsymbol{x}) d\boldsymbol{x}. \end{aligned}$$

Therefore, the total risk is a weighted average across the *conditional risk* (conditional on $\boldsymbol{x}$):

$$r(\boldsymbol{x}) = \sum_{i=1}^{k} L(i, \delta(\boldsymbol{x})) P(i|\boldsymbol{x}).$$

That shows how the total risk may be minimized. The optimal classifier (with loss L), called the *Bayes rule (with loss L)*, is given by

$$\delta(\boldsymbol{x}) = r \Leftrightarrow \sum_{i=1}^{k} L(i, r) P(i|\boldsymbol{x}) = \min_{j=1,\ldots,k} \sum_{i=1}^{k} L(i, j) P(i|\boldsymbol{x}). \tag{15.10}$$

It minimizes the total Bayes risk R. When ties occur they can be broken arbitrarily. The value R for the Bayes rule is called the *minimum Bayes risk*. Some authors (for example, Ripley, 1996) call it the minimal R Bayes risk while Fukunaga (1990) calls it the Bayes error. The minimum Bayes risk obtained by (15.10) is the best one can achieve if the posterior probabilities are known.

It is immediately seen that the Bayes rule with loss function may be written as a maximizer of discriminant functions in the form (15.6) by using the discriminant functions

$$d_r(\boldsymbol{x}) = -\sum_{i=1}^{k} L(i, r) P(i|\boldsymbol{x}).$$

It is easy to show that for the symmetrical loss function

$$L(r,s) = \begin{cases} 0 & \text{if} \quad r = s \\ c & \text{if} \quad r \neq s, \end{cases}$$

which differs from the 0-1 loss just by a constant c, the Bayes rule is given by maximizing the posterior probability, $\delta^*(\boldsymbol{x}) = r \quad \Leftrightarrow \quad P(r|\boldsymbol{x}) = \max_{j=1,\ldots,k} P(j|\boldsymbol{x})$. For the loss function

$$L(r,s) = \begin{cases} 0 & \text{if} \quad r = s \\ c/P(r) & \text{if} \quad r \neq s, \end{cases}$$

which specifies that the loss is proportional to $1/P(r)$, one obtains the *maximum likelihood* (ML) rule:

$$\delta(\boldsymbol{x}) = r \quad \Leftrightarrow \quad f(\boldsymbol{x}|r) = \max_{j=1,\dots,k} f(\boldsymbol{x}|j).$$

It is also possible to extend the decisions to incorporate the decision D, which means "being in doubt" (e.g., Ripley, 1996). When the corresponding loss function is defined by

$$L(r,s) = \begin{cases} 0 & \text{if} \quad r = s \\ 1 & \text{if} \quad r \neq s, \ s \in \{1,\dots,k\} \\ d & \text{if} \quad s = D \end{cases}$$

the classifier that minimizes the total risk is given by

$$\delta(\boldsymbol{x}) = \begin{cases} r & \text{if} \quad P(r|\boldsymbol{x}) = \max_{j=1,\dots,k} P(j|\boldsymbol{x}) \quad \text{and} \quad P(r|\boldsymbol{x}) > 1-d \\ D & \text{if} \quad P(j|\boldsymbol{x}) \leq 1-d \quad \text{for all} \quad j. \end{cases}$$

15.3 Basics of Estimated Classification Rules

15.3.1 Samples and Error Rates

The optimal classification rule is easily derived when the class densities (and priors) or the posterior probabilities are known. In practice, the classification rule has to be estimated and δ is replaced by $\hat{\delta}$. Therefore one needs a *training set* or *learning sample* that is used to obtain the estimated classifier $\hat{\delta}$. There are different forms of learning sets. One may distinguish them as follows:

- *total sample*, where $(Y_i, \boldsymbol{x}_i), i = 1,\dots,n$ are iid variables;
- *stratified sample, conditional on classes*, where $\boldsymbol{x}_{ir}|Y = r, i = 1,\dots,n_r$, are iid variables;
- *stratified sample, conditional on $\boldsymbol{x}$*, where $Y_i^{(\boldsymbol{x})}, i = 1,\dots,n(\boldsymbol{x})$, are iid variables.

The type of learning sample determines how a classification rule may be estimated. Based on the optimal Bayes rule (for L_{01}-loss) one classifies by

$$\hat{\delta}(\boldsymbol{x}) = r \Leftrightarrow \hat{d}_r(\boldsymbol{x}) = \max_{i=1,\dots,k} \hat{d}_i(\boldsymbol{x}), \tag{15.11}$$

where the functions $\hat{d}_r(\boldsymbol{x})$ are estimated discriminant functions. If one has a total sample, the discriminant functions $\hat{d}_r(\boldsymbol{x}) = \hat{P}(r|\boldsymbol{x})$ or $\hat{d}_r(\boldsymbol{x}) = \hat{f}(\boldsymbol{x}|r)P(r)$ can be used. The priors $P(r)$ in the latter form may also be replaced by the proportion $\hat{P}(r) = n_r/n$. When one has a stratified sample, conditional on classes, $\hat{d}_r(\boldsymbol{x}) = \hat{f}(\boldsymbol{x}|r)P(r)$ may be estimated directly, whereas $\hat{P}(r|\boldsymbol{x})$ is harder to obtain. The form $\hat{P}(r|\boldsymbol{x})$ is more suitable for the stratified sample, conditional on $\boldsymbol{x}$. Unless stated otherwise, in the following we will assume that the learning sample is a total sample.

When using estimated classification rules the error (for L_{01}-loss) becomes

$$\varepsilon(\hat{\delta}) = P(\hat{\delta}(\boldsymbol{x}) \neq Y) = \sum_{r=1}^{k} \int_{\hat{\delta}(\boldsymbol{x}) \neq r} f(\boldsymbol{x}|r)P(r)d\boldsymbol{x},$$

which is a random variable because it depends on the learning sample. The corresponding error rate when a new sample $(Y, \boldsymbol{x})$ is drawn is the *expected actual rate of misclassification* $\mathrm{E}_L\, \varepsilon(\hat{\delta})$, where E_L denotes that expectation is built over the learning sample.

15.3.2 Prediction Measures

In the following we consider prediction measures for classification. We focus on measures that apply for arbitrary classification rules rather than giving measures that depend on specific assumptions, for example, a normal distribution of predictors. When investigating the performance of a classifier it is essential to distinguish between measures that are computed by use of the learning sample or by use of new observations. Therefore one distinguishes several types of samples:

- The *learning* or *training set* $L = \{(Y_i, \boldsymbol{x}_i), i = 1, \dots, n\}$, from which the classifier is derived;
- the *test set*, which is a sample of new observations $T = \{(Y_i, \boldsymbol{x}_i), i = 1, \dots, n_T\}$, set aside to assess the performance of the predictor;
- a *validation set* $V = \{(Y_i, \boldsymbol{x}_i), i = 1, \dots, n_V\}$, which is a sample of new observations used to choose hyperparameters.

For the assessment of performance typically the test set is used. The validation set is an additional set that is used to optimize the classification rule if it contains additional parameters, for example, the number of neighbors used in the nearest neighbors method. The terminology is not standardized. Some authors use the term "validation set" for the test set.

Empirical Error Rates

The apparent error rate that directly derives from the learning data is the *reclassification* or *resubstitution error rate*, which for the L_{01} loss is given by

$$\hat{\varepsilon}_{rs}(\hat{\delta}) = \frac{1}{n} \sum_{(Y_i, \boldsymbol{x}_i) \in L} I(\hat{\delta}(\boldsymbol{x}_i) \neq Y_i),$$

where $I(a) = 1$ if a is true and zero otherwise. The resubstitution error is the apparent error for the 0-1 loss. It tends to underestimate the expected actual error rate, since the learning set is used twice, once for deriving the classification rule and once for evaluating of the accuracy. It is to be expected that the rule performs better in the sample from which the rule is derived than in future samples. Especially if one uses complicated classification rules, one often may reduce the error in the learning sample to zero but naturally the perfect separation in the training sample cannot be expected to hold in the future (see also Section 1.1).

A better criterion is the test error. If the data are split from the beginning into a learning set and a training set, one computes the *empirical test error*:

$$\hat{\varepsilon}_{Test}(\hat{\delta}) = \frac{1}{n_T} \sum_{(Y_i, \boldsymbol{x}_i) \in T} I(\hat{\delta}(\boldsymbol{x}_i) \neq Y_i),$$

where n_T is the number of observations in the test set. It is an unbiased estimate of the expected actual error rate (for sample size n) but wastes data since only part of the data determines the classifier.

If one wants to not waste data by setting aside a test set, one can try to obtain better estimates from the learning set by dividing it several times. A strategy of this type is K-fold cross-validation. In *K-fold cross-validation* the data of the learning set is split into K roughly equal-sized parts. Let $T_1, \dots, T_K$, where $T_1 \cup \dots \cup T_K = L$, denote the partition of the learning sample. Now for the part T_r the classifier is derived from $\cup_{j \neq r} T_j$ and the error rates are computed in T_r. Therefore, T_r are new data that have not been used for the estimation of the classification rule. This is done for $r = 1, \dots, K$, yielding the K-fold cross-validation error

$$\hat{\varepsilon}_{K-CV}(\hat{\delta}) = \frac{1}{n}\sum_{r=1}^{K}\sum_{(Y_i,\boldsymbol{x}_i)\in T_r} I(Y_i \neq \hat{\delta}_{\backslash r}(\boldsymbol{x}_i)),$$

where $\hat{\delta}_{\backslash r}$ denotes that the classification rule is estimated from $\cup_{j\neq r}T_j$.

The extreme case $K = n$, where one observation at a time is left out, is known as *leave-one-out* cross-validation or *leave-one-out error rate*:

$$\hat{\varepsilon}_{loo}(\hat{\delta}) = \frac{1}{n}\sum_{(Y_i,\boldsymbol{x}_i)\in L} I(Y_i \neq \hat{\delta}_{\backslash i}(\boldsymbol{x}_i)),$$

where $\hat{\delta}_{\backslash i}$ is estimated from $L \setminus (Y_i, \boldsymbol{x}_i)$. The leaving-one-out error rate is approximately unbiased as an estimate of the expected actual error rate.

An alternative to classical K-fold cross-validation is *Monte Carlo cross-validation* or *subsampling*. In contrast to classical cross-validation, subsampling does not use a fixed partition of the original dataset. The test sets are generated by several random splittings of the original dataset into learning sets and test sets. Let $(L_r, T_r), r = 1, \dots, K$, denote the pairs of learning samples L_r and test sample T_r with fixed samples sizes n_L and n_T, respectively. The common size ratios are n_L/n_T and are 4/1 and 9/1. The resulting subsampling estimator has the form

$$\hat{\varepsilon}_{sub}(\hat{\delta}) = \frac{1}{K}\sum_{r=1}^{K}\frac{1}{n_T}\sum_{(Y_i,\boldsymbol{x}_i)\in T_r} I(Y_i \neq \hat{\delta}_{L_r}(\boldsymbol{x}_i)),$$

where $\hat{\delta}_{L_r}$ denotes the classifier based on the learning set L_r. In contrast to classical cross-validation, the number of splittings K and the used sample sizes n_L, n_V are not linked; therefore, the number of splittings may be chosen very large.

An interesting, cleverly designed error rate is based on Efron's bootstrap methodology. Starting from a dataset with observation size n, a bootstrap sample L_r^* of size n is drawn randomly with replacement. Thus, in L_r^* single observations may be represented several times. L_r^* forms the learning set and the corresponding test set T_r^* is built from the observations that were selected when constructing L_r^*. The observations of the latter set, T_r^*, are usually called "out-of-bag" observations. When the classifier is derived from observations L_r^* the classifier has not "seen" the out-of-bag observations, and therefore these are perfect for evaluating the performance in future samples. While the sample size of the learning sets L_r^* has been fixed at n, due to the random drawing, the number of out-of-bag observations varies. Therefore, the sample size n_r of T_r^* is a random variable. The estimator of the error rate typically uses a large number of bootstrap samples. Let $(L_r^*, T_r^*), r = 1, \dots, B$, denote the bootstrap samples and $\hat{\delta}_{L_r^*}$ denote the estimated classifier based on learning sample L_r^*. The bootstrap estimators in common use are

$$\hat{\varepsilon}_{b1}(\hat{\delta}) = \frac{\sum_{r=1}^{B}\sum_{i=1}^{n} I(Y_i \in T_r^*) I(Y_i \neq \hat{\delta}_{L_r^*}(\boldsymbol{x}_i))}{\sum_{r=1}^{B}\sum_{i=1}^{n} I(Y_i \in T_r^*)}$$

and

$$\hat{\varepsilon}_{b2}(\hat{\delta}) = \frac{1}{n}\sum_{i=1}^{n}\frac{\sum_{r=1}^{B} I(Y_i \in T_r^*) I(Y_i \neq \hat{\delta}_{L_r^*}(\boldsymbol{x}_i))}{\sum_{r=1}^{B} I(Y_i \in T_r^*)}.$$

The first estimator uses all observations simultaneously, whereas the second estimator within the sum computes the individual errors for single observations. For large B the estimators produce nearly identical results (compare Efron and Tibshirani, 1997).

By construction bootstrap estimators are upwardly biased estimates of the mean actual error rate, since only a fraction of the available data size n is used in deriving the classifier. Efron (1983) proposed a modification that corrects for this bias. It is based on the observation that the probability that a data point appears in the bootstrap sample is $1-(1-1/n)^n$, which for $n > 40$ may be approximated by .632. Therefore, Efron (1983) proposed a combination of the bootstrap estimator and the resubstitution error rate of the form

$$\hat{\varepsilon}_{.632}(\hat{\delta}) = 0.368\hat{\varepsilon}_{rs}(\hat{\delta}) + 0.632\hat{\varepsilon}_{b1}(\hat{\delta}).$$

The estimator corrects for bias if the resubstitution error is moderately biased. However, for highly overfitting classification rules the resubstitution error becomes very small and then $\hat{\varepsilon}_{.632}(\hat{\delta})$ may become overly optimistic. In cases of strong overfitting one should use a modification of $\hat{\varepsilon}_{.632}(\hat{\delta})$, proposed by Efron and Tibshirani (1997) and called the .632+ estimator. The modification puts more weight on the bootstrap error rate by use of a relative overfitting rate (for details see Efron and Tibshirani, 1997).

When comparing estimates of prediction error the central issues are bias and computational effort. Although the resubstitution error rate is easily computed, it cannot be recommended because of its optimistic bias. Test errors make it necessary to set aside the test data. Therefore, the dataset used to compute the classifier is smaller than the available data, resulting in pessimistic estimates, since more data could be used to compute the classifier. In small sample settings classical cross-validation techniques have been critized because of their high variability (Braga-Neto and Dougherty, 2004). In contrast, Molinaro et al. (2005) report small mean squared errors also in small data settings. Subsampling and bootstrap errors are stable alternatives but need computational power and are slightly pessimistic.

Receiver Operating Characteristic Curves (ROC Curves)

Receiver operating characteristic curves (ROC) are a device for measuring the predictive power in two class problems. Typically the two classes are considered in an asymmetrical way. One class is considered as the signal to be detected while the other class represents the no-signal case. In medical classification problems, the signal frequently refers to a disease, while in credit risk modeling one focusses on default events as the signal. For simplicity we will denote the two classes by $y = 1$ for the signal to be detected and $y = 0$ for the no-signal class.

One often distinguishes between *sensitivity* and *specificity*, which are defined as follows:

$$\text{Sensitivity}: \; P(\hat{y} = 1|y = 1), \quad \text{Specificity}: \; P(\hat{y} = 0|y = 0).$$

Sensitivity is the probability of detecting the signal if it is present. In medical applications sensitivity may represent the probability of predicting disease given the true state is disease, while in credit risks it means that the borrower is correctly classified as a defaulter. Specificity, on the other hand, describes the probability of correctly identifying the no-signal case, namely, predicting non-disease (non-defaulter) given the true state is non-disease (non-defaulter).

Sensitivity and specificity characterize the correct classification, which may also be described as "hit" and "correct rejection." Complementary events are characterized as "miss" when the signal has falsely not been detected and "false alarm" when the signal has not been present but has been diagnosed. Borrowing terminology from hypothesis testing, missing the signal may be seen as a type II error while a false alarm corresponds to a type I error. Table 15.2 summarizes the events.

In many applications one is not only interested in the predictive power of a fixed rule that gives prediction $\hat{y} \in \{0, 1\}$. Rather than investigating the preference in terms of one hit rate and false alarm rate, one wants to evaluate how well a classification rule performs for different

TABLE 15.2: Classification results given y.

			$\hat{y}$	
			Positive 1	Negative 0
y	Signal	1	hit/sensitivity	miss / type II
	Noise	0	false alarm / type I	correct rejection/specifity

cut-offs. Typically classification rules for two-class problems are based on an underlying score that determines how classification is done. With the score representing the preference of class $y = 1$ over class $y = 0$, one considers the classification rule

$$\text{Classify } \hat{y} = 1 \text{ if the score } s(\boldsymbol{x}) \text{ is above threshold } c.$$

The ROC curves give sensitivity and specificity for varying cut-offs. More concise, the ROC curve is a plot of sensitivity as a function of 1-specificity; it plots the hit probability against the false alarm rate. With score $s(\boldsymbol{x})$ and cut-off value c one obtains the hit rate and the false alarm rate as

$$\begin{aligned}\text{HR}(c) =& P(\hat{y} = 1 \text{ based on cut-off } c|y = 1) = P(s(\boldsymbol{x}) > c|y = 1),\\ \text{FAR}(c) =& P(\hat{y} = 1 \text{ based on cut-off } c|y = 0) = P(s(\boldsymbol{x}) > c|y = 0).\end{aligned}$$

ROC curves are plots of $(\text{FAR}(c), \text{HR}(c))$ for varying cut-off values c. They are monotone increasing functions in the positive quadrant. Typically the curve has a concave shape connecting the points $(0, 0)$ and $(1, 1)$. It may be given as $\{(\text{FAR}(c), \text{HR}(c)), c \in (-\infty, \infty)\}$ or $\{(t, \text{ROC}(t)), t \in (0, 1)\}$, where $\text{ROC}(t)$ maps t to $\text{HR}(c)$, with c being the value for which $\text{FAR}(c) = t$. It has some nice properties; for example, it is invariant to strictly increasing transformations of the classification score $s(\boldsymbol{x})$.

The curve is much more informative than a simple classification rule, which is based on just one cut-off value. A classification score that perfectly separates two classes has for some value c $\text{HR}(c) = 1$ and $\text{FAR}(c) = 0$ and the ROC curve is along the axes connecting $(0, 0)$, $(0, 1)$, and $(1, 1)$. If the score is non-informative, one has $\text{ROC}(t) = t$. Discrimination is better the closer the curve is to the left upper point $(0, 1)$. For illustration, in Figure 15.2 the empirical ROC curves for two classes of the glass identification data (Example 15.2) are given. It shows the curves for three classification rules, namely, linear, quadratic, and logistic discrimination,

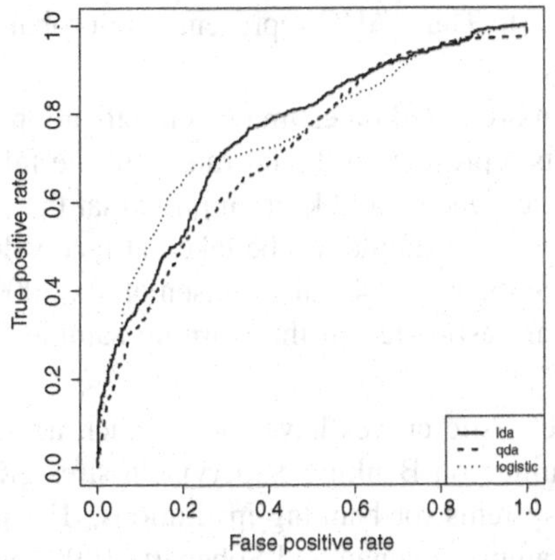

FIGURE 15.2: ROC curves for two classes of glass identification data

which will be considered in the next sections. It is seen that there is no strict ordering of classification rules; one does not dominate the others over the whole range.

In applications, scores are found in various ways. With the optimal Bayes classification in mind, the scores may be represented as differences or properties of discriminant functions. With $s(\boldsymbol{x}) = d_1(\boldsymbol{x})/d_2(\boldsymbol{x})$ and $d_r(\boldsymbol{x}) = f(\boldsymbol{x}|r)P(r)$ one obtains

$$s(\boldsymbol{x}) > c \Leftrightarrow \frac{f(\boldsymbol{x}|1)P(1)}{f(\boldsymbol{x}|2)P(2)} \geq c.$$

The optimal Bayes classification uses $c = 1$, yielding the minimal probability of misclassification. By rewriting the allocation rule as

$$s(\boldsymbol{x}) > c \Leftrightarrow \frac{f(\boldsymbol{x}|1)}{f(\boldsymbol{x}|2)} \geq c\frac{P(2)}{P(1)},$$

one sees that varying c corresponds to varying the cut-off values for the proportion of densities. For continuous $\boldsymbol{x}$ one obtains a closed curve, whereas for discrete $\boldsymbol{x}$ one obtains single points in the unit square that have to be connected to give a curve.

As a curve, ROC shows how the classification rate works for differing cut-offs. The higher the area under the curve, the better the scoring or classification rule. Therefore, the area under the ROC curve can be used as an indicator for the performance of a classification rule. Given that the curve is concave with end points (0,0) and (1,1), one often considers the area under the ROC curve (AUC):

$$\text{AUC} = \int \text{HR } d(\text{FAR}) = \int_0^1 \text{ROC}(t)dt$$

or the area without the lower triangle, which gives the *ROC accuracy ratio*:

$$\text{ROC accuracy ratio} = 2\left(\int \text{HR } d(\text{FAR}) - \frac{1}{2}\right).$$

Perfect discrimination rules have AUC $= 1$; if the score has the same distribution in both classes and therefore is uninformative, one obtains AUC $= 0.5$. It can be shown that AUC is equivalent to $P(s_1 > s_0)$, where s_1, s_0 are scores for randomly drawn scores from classes 1 and 0, respectively (see Pepe, 2003). Thus, AUC represents the probability of correctly ordering the classes.

Empirical ROC curves that are based on estimated classification rules result from connecting the finite number of points representing the hit rates and the false alarm rates for varying cut-off values c within a sample. One should keep in mind that the estimated curves are based on a random sample. Therefore they should not be taken at face value. Campbell (1994) proposed simultaneous confidence intervals (see also Jensen et al., 2000). In particular, when hit rates and false alarm values are estimated in the learning sample biased estimates are to be expected.

Receiver operating characteristic curves have found much attention in default risk modelling, since the Basel Committee on Banking Supervision stressed the supervisory risks assessment and early warning systems for banking institutions. For performance measures for credit risk models see, for example, Keenan and Sobehart (1999) and Sobehart et al. (2000). The evaluation of tests in medicine has been considered by Pepe (2003), Zweig and Campbell

(1993), Lee and Hsiao (1996), and Campbell (1994). In applications it can be seen as a disadvantage of ROC curves and derived measures that performance is summarized over regions of the curve that are not of interest. Therefore, Dodd and Pepe (2003) proposed a partial AUC, which restricts the curve to a relevant range of false alarm rates and gave a regression framework for making inferences about covariate effects on the partial AUC. An extension to ordinal classes was given by Toledano and Gatsanis (1996).

15.4 Parametric Classification Methods

In this section methods for obtaining estimated classification rules by use of parameterized classifiers are considered. Let $\boldsymbol{x}_{r1}, \dots, \boldsymbol{x}_{rn_r}, r = 1, \dots, k$, denote the samples from classes $1, \dots, k$. If not stated otherwise, they stem from iid samples of $(Y, \boldsymbol{x})$ with sample size $n = n_1 + \dots + n_k$.

15.4.1 Linear and Quadratic Discriminations

Plug-In Rules

It was shown in Section 15.2 that if one assumes that predictors are normal with mean $\boldsymbol{\mu}_r$ and common covariances matrix $\boldsymbol{\Sigma}$, the optimal Bayes rule prefers class r over class s if

$$d_r(\boldsymbol{x}) - d_s(\boldsymbol{x}) = \beta_{0rs} + \boldsymbol{x}^T \boldsymbol{\beta}_{rs} \geq 0,$$

where $\boldsymbol{\beta}_{rs} = \boldsymbol{\Sigma}^{-1}(\boldsymbol{\mu}_r - \boldsymbol{\mu}_s)$ and β_{0rs} contains $\boldsymbol{\mu}_r, \boldsymbol{\mu}_s$ and $p(r), p(s)$ (see equation (15.7)). In estimated rules one replaces $\boldsymbol{\mu}_r$ by $\bar{\boldsymbol{x}}_r = \frac{1}{n_r}\sum_{i=1}^{n_r} \boldsymbol{x}_{ri}$ and $\boldsymbol{\Sigma}$ by the pooled empirical covariance matrix

$$\boldsymbol{S} = \frac{1}{n-k} \sum_{r=1}^{k} \sum_{i=1}^{n_r} (\boldsymbol{x}_{ri} - \bar{\boldsymbol{x}}_r)(\boldsymbol{x}_{ri} - \bar{\boldsymbol{x}}_r)^T$$

and thereby obtains $\hat{\boldsymbol{\beta}}_{rs}$ and $\hat{\beta}_{0rs}$. The corresponding linear discrimination rule prefers class r over class s if $\hat{\beta}_{0rs} + \boldsymbol{x}^T \hat{\boldsymbol{\beta}}_{rs} \geq 0$. Figure 15.3 shows the linear separation of observations from three classes that were generated from normal distributions.

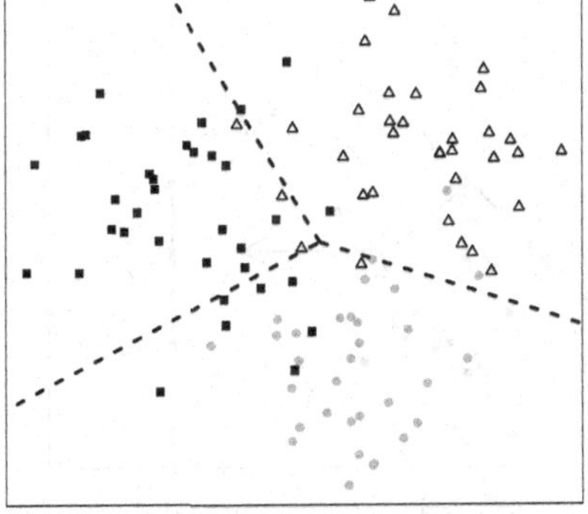

FIGURE 15.3: Linear separation of observations from three classes.

Fisher's Discriminant Analysis

The linear classification rule obtained for normally distributed predictors with common covariance can be motivated in a quite different way. Fisher (1936) motivated linear discriminant analysis by considering a criterion that separates two groups in a linear way. He proposed looking for a linear combination $y = \boldsymbol{a}^T\boldsymbol{x}$ of predictors such that the data are well separated in y-values. Geometrically, the linear combination $\boldsymbol{a}^T\boldsymbol{x}$ may be seen as a projection of observations $\boldsymbol{x}$ on vector $\boldsymbol{a}$. Thus one is looking for directions that promise separation of classes. Figure 15.4 shows two projections of data, one that yields a perfect separation of classes and one that does not separate classes perfectly. Mathematically, the criterion to be maximized is

$$Q(\boldsymbol{a}) = \frac{(\bar{y}_1 - \bar{y}_2)^2}{w_1^2 + w_2^2}$$

under a side constraint that fixes the length of $\boldsymbol{a}$. Typically one uses $||\boldsymbol{a}|| = 1$. In $Q(\boldsymbol{a})$ the values $\bar{y}_1, \bar{y}_2$ represent the means after projecting the observations from the two classes on $\boldsymbol{a}$. With $y_{ri} = \boldsymbol{a}^T\boldsymbol{x}_{ri}$ being the linear combination of observation $\boldsymbol{x}_{ri}$, one has

$$\bar{y}_r = \frac{1}{n_r}\sum_{i=1}^{n_r} y_{ri} = \frac{1}{n_r}\sum_{i=1}^{n_r} \boldsymbol{a}^T\boldsymbol{x}_{ri} = \boldsymbol{a}^T\bar{\boldsymbol{x}}_r.$$

Therefore $\bar{y}_r$ is identical to the projection of the mean $\bar{\boldsymbol{x}}_r$ in class r. The values in the denominator represent the empirical within-group variances after projection, with w_r^2 given by

$$w_r^2 = \sum_{i=1}^{n_r}(y_{ri} - \bar{y}_r)^2 = \sum_{i=1}^{n_r}(\boldsymbol{a}^T\boldsymbol{x}_{ri} - \boldsymbol{a}^T\bar{\boldsymbol{x}}_r)^2.$$

The nominator of $Q(\boldsymbol{a})$ aims at separating the mean values of the classes, whereas the denominator represents the variability within classes. Thus two components are contained: separation of mean values and minimization of variability within classes. Maximization of $Q(\boldsymbol{a})$ yields up to constants the vector $\boldsymbol{a} = \boldsymbol{S}^{-1}(\bar{\boldsymbol{x}}_1 - \bar{\boldsymbol{x}}_2)$, which is the same as the weight vector in the normal distribution case with a common covariance matrix $\boldsymbol{\Sigma}$ (Exercise 15.6). However, Fisher's

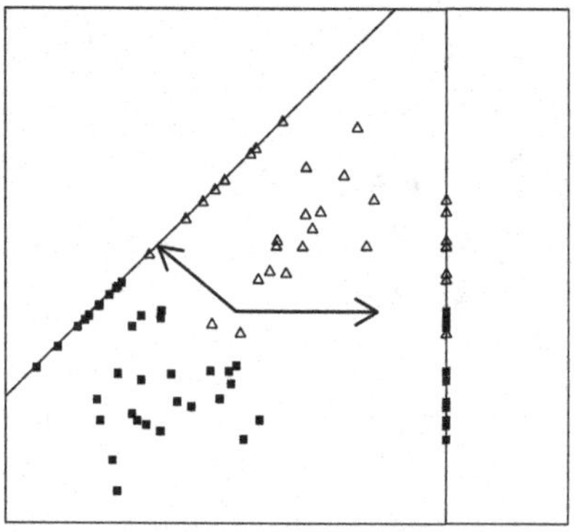

FIGURE 15.4: Two projections of data, one with a perfect separation of classes and one with an imperfect separation.

approach does not assume normality, but starts from a plausible empirical criterion. It may also be seen as motivation for the more general use of linear discriminant analysis, which is not restricted to normal distributions. Since the criterion is chosen in a sensible way, it also applies in cases where the distribution is non-normal, for example, in the mixed variables case, where metric and binary predictors are included. Indeed, linear discriminant analysis has been shown to be robust against deviations from the normal distribution.

In the more general case with k classes, one maximizes the criterion

$$Q(\boldsymbol{a}) = \frac{\sum_{r=1}^{k} n_r(\bar{y}_r - \bar{y})^2}{\sum_{r=1}^{k} w_r^2},$$

where $\bar{y} = \frac{1}{n}\sum_{r=1}^{k}\sum_{i=1}^{n_r} \boldsymbol{a}^T\boldsymbol{x}_{ri} = \sum_{r=1}^{k}(n_r/n)\boldsymbol{a}^T\bar{\boldsymbol{x}}_r$ is the mean across all observations. An alternative way of presenting the criterion is

$$Q(\boldsymbol{a}) = \frac{\boldsymbol{a}^T\boldsymbol{B}\boldsymbol{a}}{\boldsymbol{a}^T\boldsymbol{W}\boldsymbol{a}},$$

where $\boldsymbol{B} = \sum_{r=1}^{k} n_r(\bar{\boldsymbol{x}}_r - \bar{\boldsymbol{x}})(\bar{\boldsymbol{x}}_r - \bar{\boldsymbol{x}})^T$ and $\boldsymbol{W} = \sum_{r=1}^{k}\sum_{i=1}^{n_r}(\boldsymbol{x}_{ri} - \bar{\boldsymbol{x}}_r)(\boldsymbol{x}_{ri} - \bar{\boldsymbol{x}}_r)^T$ are the between-group and within-group sums of squares. As a side constraint on $\boldsymbol{a}$ one typically uses $\boldsymbol{a}^T\boldsymbol{W}\boldsymbol{a} = 1$. The extreme values are obtained by solving the generalized eigenvalue problem $\boldsymbol{B}\boldsymbol{a} = \lambda\boldsymbol{W}\boldsymbol{a}$, which yields the largest eigenvalue of the not-necessarily-symmetric matrix $\boldsymbol{W}^{-1}\boldsymbol{B}$. One obtains at most $m = \min\{k-1, p\}$ non-zero eigenvalues $\lambda_1 \geq \cdots \geq \lambda_m$ with corresponding eigenvalues $\boldsymbol{a}_1, \ldots, \boldsymbol{a}_m$, for which $\boldsymbol{a}_i^T\boldsymbol{W}\boldsymbol{a}_j = 0$, $i \neq j$ holds. Therefore, they are not orthogonal in the common sense but for a generalized metric that uses $\boldsymbol{W}$ (Exercise 15.7). In particular, $\boldsymbol{a}_1$ maximizes $Q(\boldsymbol{a})$. The *canonical discriminant variables* defined as $u_r = \boldsymbol{a}_r^T\boldsymbol{x}$ are used in classification by using the Euclidean distance after projection. One classifies into class r if

$$\sum_{s=1}^{m}\{\boldsymbol{a}_s^T(\boldsymbol{x} - \bar{\boldsymbol{x}}_r)\}^2 = \min_{j=1,\ldots,k}\sum_{s=1}^{m}\{\boldsymbol{a}_s^T(\boldsymbol{x} - \bar{\boldsymbol{x}}_j)\}^2.$$

Thus one chooses the class whose mean vector is closest to $\boldsymbol{x}$ in the space of canonical discriminant variables.

Quadratic Discrimination

For normally distributed predictors $\boldsymbol{x}|Y = r \sim N(\boldsymbol{\mu}_r, \boldsymbol{\Sigma}_r)$ the discriminant function $d_r(\boldsymbol{x}) = \log(f(\boldsymbol{x}|r)) + \log(p(\boldsymbol{x}))$ has the form

$$d_r(\boldsymbol{x}) = -\frac{1}{2}(\boldsymbol{x} - \boldsymbol{\mu}_r)^T\boldsymbol{\Sigma}_r^{-1}(\boldsymbol{x} - \boldsymbol{\mu}_r) - \frac{p}{2}\log(2\pi) - \frac{1}{2}\log(|\boldsymbol{\Sigma}_r|) + \log(p(r)).$$

Estimated versions are obtained by using class-specific estimates $\hat{\boldsymbol{\Sigma}}_r$ (and estimates of $\boldsymbol{\mu}_r$). The resulting classification rule is referred to as a quadratic discrimination since quadratic terms are contained in differences between estimated disciminant functions. Figure 15.5 shows data generated from underlying normal distributions with different covariance matrices. The left panel shows the separation obtained from a linear discrimination rule, and the right panel shows the estimated and the optimal classification regions for the quadratic discrimination rule.

Although quadratic discrimination is more flexible than linear discrimination, if the learning sample is not large, and the true variances are not very different, the linear rule often outperforms the quadratic rule (e.g. Marks and Dunn, 1974). Several approaches have been suggested to obtain a compromise between equal and unequal covariances by means of regularization; see Campbell (1980), Friedman (1989), Rayens and Greene (1991), and Flury (1986).

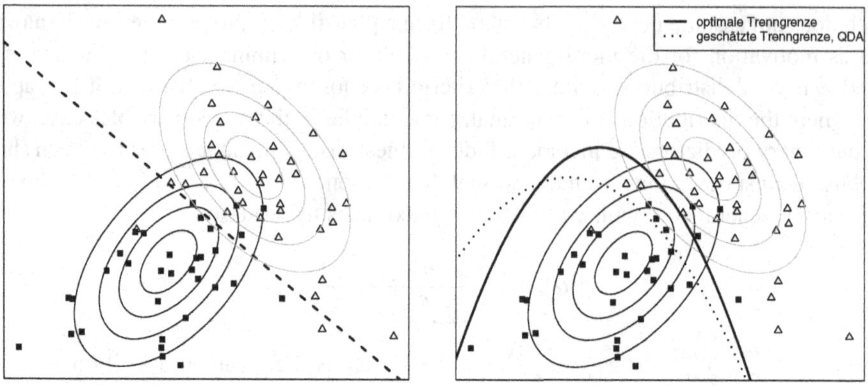

FIGURE 15.5: Linear and quadratic separations of observations from two classes.

Diagonal Discriminant Analysis

Linear discrimination based on estimates of $\boldsymbol{\Sigma}$ and $\boldsymbol{\mu}_1, \dots, \boldsymbol{\mu}_k$, as well as on Fisher's approach, require the computation of the inverse (estimated) covariance $\boldsymbol{S}^{-1}$, which might be quite unstable when the number of predictors is large. Simpler rules are obtained by assuming a more restrictive distribution of $\boldsymbol{x}|Y = r$. By assuming $\boldsymbol{x}|Y = r \sim N(\boldsymbol{\mu}_r, \boldsymbol{\Sigma}_d)$, where $\boldsymbol{\Sigma}_d$ is a diagonal matrix, $\boldsymbol{\Sigma}_d = diag(\sigma_1^2, \dots, \sigma_p^2)$, the optimal separation is linear, with differences of discriminant functions $d_r(\boldsymbol{x}) - d_s(\boldsymbol{x}) = \beta_{0rs} + \boldsymbol{x}^T \boldsymbol{\beta}_{rs}$ having the parameters

$$\beta_{0rs} = -\frac{1}{2}\sum_{j=1}^{p}(\mu_{rj}/\sigma_j)^2 + \frac{1}{2}\sum_{j=1}^{p}(\mu_{sj}/\sigma_j)^2 + \log(p(r)/p(s)),$$

$$\beta_{rsj} = (\mu_{rj} - \mu_{sj})^2/\sigma_j^2.$$

Estimation of the parameters requires only that one obtains estimates of $\boldsymbol{\mu}_r, \dots, \boldsymbol{\mu}_k$ and $\sigma_1^2, \dots,$ σ_p^2. Although the underlying assumption seems artificial and overly simplistic, for classification purposes it turns out to be very stable in high-dimensional settings. Dudoit et al. (2002) used the term *diagonal discriminant analysis* and demonstrated that in gene expression data, which are notoriously high-dimensional, large gains in accuracy may be obtained by ignoring correlations between genes.

Regularized Discriminant Analysis

Stability of the estimation of covariance matrices (or inverse covariance matrices) can also be obtained by regularized estimates. One version of linear discriminant analysis uses ridge-type estimates $\boldsymbol{S} + \lambda \boldsymbol{I}$ with tuning parameter λ. A compromise between linear and quadratic discriminant analyses uses the regularized estimates of covariance matrices

$$(1-\lambda)\hat{\boldsymbol{\Sigma}}_r + \lambda\hat{\boldsymbol{\Sigma}},$$

where $\hat{\boldsymbol{\Sigma}}$ is the pooled covariance matrix and $\hat{\boldsymbol{\Sigma}}_r$ are estimated class-specific covariances (Friedman, 1989). In alternative approaches the pooled covariance matrix is replaced by a scaled version of the identity matrix. Guo, Hastie, and Tibshirani (2007) considered the regularized estimate $(1-\lambda)\hat{\boldsymbol{\Sigma}} + \lambda\hat{\boldsymbol{I}}$ and a version in which the correlation matrix is regularized. A specific algorithm that uses the nearest shrunken centroids method (Tibshirani, Hastie,

Narasimhan, and Chu (2003)) allows one to select variables. An overview of methods including direct regularization of inverse covariance matrices is given by Mkhadri et al. (1997).

15.4.2 Logistic Discrimination

It was shown in Section 15.2 that the linear logistic model may be derived from the assumption of normally distributed predictors with a common covariance matrix. As in linear discrimination, for example, by use of Fisher's criterion, the separation of classes is based on a linear form of the predictors, although a non-linear transformation is used to transform the linear predictors into probabilities. The logistic representation has several advantages. It holds not only when predictors are normal, but also if binary predictors follow a log-linear model and for other distributions (Day and Kerridge, 1967; Anderson, 1972). Moreover, it offers a more direct way of estimating posterior probabilities and is easier to generalize.

Rather than replacing unknown covariances and means by estimates, logistic discrimination typically is based on conditional maximum likelihood estimations, as outlined in Section 4.1. Therefore one estimates parameters conditionally on predictors, which yields estimates that only assumes that the logistic regression model holds, but not that the predictors are necessarily normal. ML estimates are appropriate when the sample is drawn conditionally on the predictor or in total samples by conditioning on predictors. When the sample is conditional on classes, one has to be careful with the intercepts. Since priors cannot be estimated from samples that are stratified by classes, intercepts have to be corrected by known priors; see Section 4.1 for estimating in stratified samples.

A problem that arises with maximum likelihood estimates is that they may become infinite. That happens in particular if observations in some classes can be completely separated from others. What comes as a problem when investigating the effect strength of predictors may be seen as an advantage when the objective is the separation of classes. However, the question arises of whether perfect separation obtained in the learning sample generalizes to future data. By using shrinkage estimators (Chapter 6), the problem of infinite estimates is easily avoided; for example, Zhu and Hastie (2004) used a quadratically regularized estimate for the classification of gene microarrays. For a discussion of the problem of $\boldsymbol{\beta} \to \infty$, see also Albert and Lesaffre (1986) and Lesaffre and Albert (1989). An early comparison of the efficiency of logistic discrimination compared to normal discriminant analysis is found in Efron (1975). McLachlan (1992) gave a concise summary of logistic discrimination methods.

Logistic discrimination may be extended to include more general forms by using the model $P(r|x) = \exp(\eta_r)/(1 + \sum_{j=1}^{k-1} \exp(\eta_j))$, with η_r having a more general form. The predictors may be specified as a parametric family of functions or may contain additive terms. The estimation of parameters (or functions of predictors) can be obtained by maximum likelihood methods only when there are not too many predictors. An alternative approach that allows one to estimate parameters and simultaneously select variables is based on boosting techniques. Although these methods apply to parametric models, they are considered within a more general setting in Section 15.5.3.

15.4.3 Linear Separation and Support Vector Classifiers

In the case of two classes, linear scoring rules have the form $s(\boldsymbol{x}) = \beta_0 + \boldsymbol{x}^T\boldsymbol{\beta}$, with classification in class 1 if $s(\boldsymbol{x}) \geq 0$ and in class 2 otherwise. Geometrically, the decision boundary $H = \{s(\boldsymbol{x}) = 0\}$ forms a $(p-1)$-dimensional separating hyperplane within the p-dimensional predictor space. For $p = 2$ this is a simple line. For two points $\boldsymbol{x}_0$ $\tilde{\boldsymbol{x}}_0$ from hyperplane H one obtains $(\boldsymbol{x}_0 - \tilde{\boldsymbol{x}}_0)^T\boldsymbol{\beta} = 0$, and therefore $\boldsymbol{\beta}$ is orthogonal to every vector in the hyperplane.

Thus $\boldsymbol{\beta}$ (or, equivalently, the standardized value $\boldsymbol{\beta}^* = \boldsymbol{\beta}/||\boldsymbol{\beta}||$) determines the orientation of the hyperplane. For $\boldsymbol{x}_0 \in H$ one obtains $\boldsymbol{x}_0^T\boldsymbol{\beta}^* = \boldsymbol{x}_0^T\boldsymbol{\beta}/||\boldsymbol{\beta}|| = -\beta_0/||\boldsymbol{\beta}||$, which shows that the intercept determines the distance between the hyperplane and the origin. The orthogonal distance γ between any point $\boldsymbol{x}_i$ and the hyperplane can be found by considering the decomposition $\boldsymbol{x}_i = \boldsymbol{x}_i^{ort} + \gamma_i\boldsymbol{\beta}/||\boldsymbol{\beta}||$, where $\boldsymbol{x}_i^{ort}$ is the orthogonal projection of $\boldsymbol{x}_i$ onto the hyperplane. Multiplying both sides by $\boldsymbol{\beta}^T$ and using $\boldsymbol{\beta}^T\boldsymbol{x}_i^{ort} = -\beta_0$ yields the distance $\gamma_i = (\beta_0 + \boldsymbol{x}_i^T\boldsymbol{\beta})/||\boldsymbol{\beta}|| = s(\boldsymbol{x}_i)/||\boldsymbol{\beta}||$. Therefore, the scoring rule $s(\boldsymbol{x}_i)$ is proportional to the distance between the point $\boldsymbol{x}_i$ and the separating hyperplane. The distance is signed, taking positive and negative values depending on which side of the hyperplane the point is located.

If classes 1 and 2 are represented by 1 and -1 the classification based on score $s(\boldsymbol{x}) = \beta_0 + \boldsymbol{x}^T\boldsymbol{\beta}$ can be given as $\delta(\boldsymbol{x}) = \text{sign}(\beta_0 + \boldsymbol{x}^T\boldsymbol{\beta})$ and an observation $y_i \in \{1, -1\}$ is classified correctly if $y_i s(\boldsymbol{x}_i) = y_i(\beta_0 + \boldsymbol{x}_i^T\boldsymbol{\beta}) \geq 0$, where the value $y_i s(\boldsymbol{x}_i)$ measures the distinctness of the classification. In the case where the training data $(y_i, \boldsymbol{x}_i)$, $y_i \in \{1, -1\}$, $i = 1, \dots, n$, are separable, that is, all of them can be classified correctly, one may choose the orientation of $\boldsymbol{\beta}$ such that the distinctness (or *margin*) is maximized. One chooses $\beta_0, \boldsymbol{\beta}$ with $||\boldsymbol{\beta}|| = 1$ such that M is maximized subject to $y_i(\beta_0 + \boldsymbol{x}_i^T\boldsymbol{\beta}) \geq M$. Therefore, among the $\boldsymbol{\beta}$s, for which $y_i s(\boldsymbol{x}_i) > 0$, $i = 1, \dots, n$, that one is chosen for which the distance between observations and the hyperplane is at least M. A reformulation of the problem is

$$\min_{\beta_0,\boldsymbol{\beta}} ||\boldsymbol{\beta}|| \quad \text{subject to} \quad y_i(\beta_0 + \boldsymbol{x}_i^T\boldsymbol{\beta}) \geq 1, i = 1, \dots, n \tag{15.12}$$

(Exercise 15.10). If the observations are not separable in the feature space, the procedure has to be modified by introducing so-called slack variables $\xi_1, \dots, \xi_n$ and considering minimization of $||\boldsymbol{\beta}||$ subject to

$$y_i(\beta_0 + \boldsymbol{x}_i^T\boldsymbol{\beta}) \geq 1 - \xi_i, i = 1, \dots, n, \quad \xi_i \geq 0, \sum_i \xi_i \leq \text{constant}. \tag{15.13}$$

The resulting classifier is a so-called *support vector* classifier. The solution has the form $\hat{\boldsymbol{\beta}} = \sum_i \hat{\alpha}_i y_i \boldsymbol{x}_i$, where the estimated coefficients $\hat{\alpha}_i$ are non-zero for only those observations for which the constraints are exactly met. These observations constitute the support vectors that give the procedure its name. For details on how to find the solution, see, for example, Hastie et al. (2009).

It is noteworthy that the support vector classifier uses a specific loss function. If $y_i s(\boldsymbol{x}_i) < 0$, the observation is misclassified. The misclassification is more distinct if the value $y_i s(\boldsymbol{x}_i)$ is small. Therefore, the problem is to maximize $y_i s(\boldsymbol{x}_i)$ or to minimize $-y_i s(\boldsymbol{x}_i)$, which is equivalent to minimizing the so-called hinge loss $L_h(y, s(\boldsymbol{x})) = 1 - ys(\boldsymbol{x})$. Thus, the hinge loss is implicitly used in the minimization problem (15.12) that aims at maximizing the margin to obtain separated observations. For true values coded by $\{-1, 1\}$, the *hinge* or *soft margin loss* has the general form

$$L_h(y, \hat{y}) = 1 - y\hat{y}I(y\hat{y} < 1).$$

When the expectation of the hinge loss is maximized over values $\hat{y} \in [-1, 1]$ one obtains the optimal prediction $\hat{y} = \text{sign}(P(y = 1|\boldsymbol{x}) - 1/2)$, which is equivalent to the Bayes classifier (Exercise 15.11). Therefore, for this loss function the Bayes classifier can also be derived from real-valued optimal predictions (see also Section 15.5.3).

In practice, one is not restricted to using the linear classifier $s(\boldsymbol{x}) = \beta_0 + \boldsymbol{x}^T\boldsymbol{\beta}$, but can enlarge the feature space using basis transformations. But perfect separation in constructed spaces may not generalize well. Thus, even in enlarged spaces, the relaxed optimization problem (15.13) is helpful. It can also be given in the form of a penalized estimate with some constant C, $\min_{\beta_0,\boldsymbol{\beta}} ||\boldsymbol{\beta}||^2 + C\sum_{i=1}^n \xi_i$, subject to $y_i(\beta_0 + \boldsymbol{x}_i^T\boldsymbol{\beta}) \geq 1 - \xi_i, i = 1, \dots, n, \xi_i \geq 0$.

Since for fixed $\boldsymbol{\beta}$ the optimal choice of ξ is $\xi = L_h(y_i, s(\boldsymbol{x}_i))$, the minimization can also be formulated as the minimization of the hinge loss under regularization. More concretely, one formulates it as a minimization of a regularized functional with solutions in a reproducing Hilbert space, which is beyond the scope of this book (see, for example, Blanchard et al., 2008). Although the general support vector classifier is not a parametric classifier, it is included here because of its strong connection to linear separation. Support vector classifiers have been extensively studied in the machine learning community. Steinwart and Christmann (2008) collected their results in a book, and a short introduction from a statistical perspective is given by Blanchard et al. (2008).

15.5 Non-Parametric Methods

In the previous sections mostly classification rules were obtained by estimating parameters. Typically the parametrization is motivated by underlying distribution assumptions. Non-parametric methods that do not rely on specific distributions provide more flexible classification rules that frequently show good performance if the complexity of the rule is restricted adequately.

15.5.1 Nearest Neighborhood Methods

Nearest neighborhood methods are non-parametric methods that require no model to be fit. The basic concept uses only distances of observations in feature space and is due to Fix and Hodges (1951) (reprinted in Agrawala, 1977).

Based on a sample $\{(Y_i, \boldsymbol{x}_i), i = 1, \dots, n\}$ and a distance in feature space $d(\boldsymbol{x}, \tilde{\boldsymbol{x}}), \boldsymbol{x}, \tilde{\boldsymbol{x}} \in \mathbb{R}^p$, one determines for a new observation $\boldsymbol{x}$ the K observation points that are closest in distance to $\boldsymbol{x}$. This means that one seeks the nearest neighbors $\boldsymbol{x}_{(1)}, \dots, \boldsymbol{x}_{(K)}$ with

$$d(\boldsymbol{x}, \boldsymbol{x}_{(1)}) \leq \dots \leq d(\boldsymbol{x}, \boldsymbol{x}_{(K)}),$$

where $\boldsymbol{x}_{(1)}, \dots, \boldsymbol{x}_{(K)}$ are values from the learning sample. With $Y_{(1)}, \dots, Y_{(K)}$ denoting the corresponding classes, one classifies, using the majority vote rule, by

$$\hat{d}(\boldsymbol{x}) = r \Leftrightarrow \text{class r is the most frequent class in } \{Y_{(1)}, \dots, Y_{(K)}\}.$$

In effect, for a given $\boldsymbol{x}$, one looks within the learning sample for almost perfect matches of $\boldsymbol{x}$ and then classifies by using the class labels of these observations. In the extreme case $K = 1$, one simply chooses the class from which the observation stems that is most similar to the value to be classified. If ties occur, they are broken at random. The resulting classifier is called the K-nearest-neighbor (K-NN) classifier.

The most simple and often surprisingly successful classification rule is 1-NN, where one chooses the class of the nearest neighbor. It has been shown that, asymptotically, the rule is suboptimal but with specific upper bounds. Cover and Hart (1967) gave an often referred to result on asymptotic behavior. They show that the asymptotic error ($n \to \infty$) of 1-NN, denoted by $\varepsilon_{as}(\hat{\delta}_1)$, and the optimal Bayes error, denoted by ε_{opt}, fulfill

$$\varepsilon_{opt} \leq \varepsilon_{as}(\hat{\delta}_1) \leq \varepsilon_{opt}(2 - \frac{k}{k-1}\varepsilon_{opt}),$$

where k is the number of classes. The first inequality simply reflects that no classification rule can improve on the Bayes error, while the second inequality gives an upper bound for the (asymptotic) error. For the two-class problems the upper bound is $2\varepsilon_{opt}(1-\varepsilon_{opt})$, and therefore it is always smaller than twice the optimal error.

When using the nearest neighbor classification rule one has to choose K, the number of neighbors. One may expect that the rule improves with increasing K, although a very large K will result in a trivial classifier that does not depend on $\boldsymbol{x}$. An interesting case is 2-NN as compared to 1-NN. In the 2-NN case one either has a majority of two (which yields the same decision as 1-NN) or a tie, which will be broken at random. Thus one might expect no improvement. Indeed, the asymptotic error rates are the same. Ripley (1996) shows more generally that, for the asymptotic error of the K-NN rule, denoted by $\varepsilon_{as}(\hat{\delta}_k)$, one has

$$\varepsilon_{as}(\hat{\delta}_{2K}) = \varepsilon_{as}(\hat{\delta}_{2K-1}) \leq \cdots \leq \varepsilon_{as}(\hat{\delta}_2) = \varepsilon_{as}(\hat{\delta}_1).$$

Thus, increasing K can improve the error rate, and error rates for $2K$-rules and ($2K$-1)-rules are identical for $n \to \infty$. A further result on large sample behavior was given by Stone (1977). He showed that the risk for the K-NN rule converges in probability to the Bayes risk provided $K \to \infty$ and $K/n \to 0$.

The distance measure $d(.,.)$, which is used when computing the nearest neighbors, should be chosen carefully. For metric predictors, typically one uses the Euclidean distance, after the predictors have been standardized to have mean 0 and variance 1. For binary variables and a mixture of different types of variables, other measures will be more appropriate. An alternative choice of distance is based on the empirical correlation coefficient, used, for example, by Dudoit et al. (2002).

When compared to alternative classifiers, the nearest neighborhood methods often perform remarkably well, even in high dimensions. In a comparison of several methods Dudoit et al. (2002) used gene expression data with pre-selected variables ranging from $p = 30$ to $p = 50$. The nearest neighborhood methods with cross-validatory choices of K did rather well when compared to more sophisticated methods. For literature that deals with extended versions of nearest neighbor methods see Section 15.11.

15.5.2 Random Forests and Ensemble Methods

One of the most efficient classifiers that have been proposed in recent years are random forests (Breiman, 2001a). Random forests are an example of a wider group of methods, called *ensemble methods*. When using ensembles one fits several models, for example, several trees, and lets them vote for the most popular class. Ensemble methods can be divided into two groups. In the first group the distribution of the training set is changed adaptively based on the performances of previous classifiers, whereas in the second group the distribution is unchanged. Random forests are examples of the latter group. Boosting methods, which will be considered in the next section, are of the former type.

Breiman (2001a) defines, rather generally, a random forest as a classifier consisting of a collection of tree-structured classifiers where the parameters are iid random vectors. Let us consider some examples. In *bagging* (for bootstrap aggregation), proposed by Breiman (1998), trees are grown on bootstrap samples from the training set. A bootstrap sample is generated by uniform sampling from the training set with replacement. For each bootstrap sample one obtains a tree. An ensemble of trees may also be obtained by a random split selection (Dietterich, 2000). In a random split selection, at each node the split is selected from among the best splits, where a fixed number of n_{split} best splits is considered. For $n_{split} = 1$ the ensemble collapses and the original tree results. Alternatively, one may select the training sets for the trees by using random sets of weights on the observations in the training set or use a random selection of subsets of variables (e.g., Ho, 1998).

Breiman (1996a) favors random trees of the latter type, which use randomly selected inputs (or combinations of inputs) at each node to grow a new tree. The simplest version he considers, called Forest-RI for forests with random input, selects at random at each node a small group of input variables to split on. Trees are grown without pruning. He demonstrates that the method compares favorably to methods based on reweighting sets like AdaBoost (see next section). In particular, it was surprising that the selection of just one input variable at each node did very well. Improvement by selecting more variables was negligible. The more advanced version, called Forest-RC for forests with random combinations, uses a linear combination of variables at each node. At a given node n_I, combinations of combined input variables are selected where a combined input variable consists of n_v randomly selected variables, added together with coefficients that are uniform random numbers on $[-1, 1]$. The performance of the datasets considered by Breiman (1996a) improves on Forest-RI.

In his paper Breiman demonstrates not only that random forests work well; in addition he investigates the reason why that is so. In particular, he gives an upper bound for the generalization error. We will briefly summarize some of his results. Let the classifier for a new observation $\boldsymbol{x}$ be given by $\delta_{\boldsymbol{\theta}}(\boldsymbol{x})$, where the vector $\boldsymbol{\theta}$ contains the parameters that are randomly drawn when generating a tree. For instance, in a random split selection based on the n_{split} best splits, the parameter contains the randomly drawn splits. In random forests the parameter $\boldsymbol{\theta}$ is generated randomly, yielding a sequence $\boldsymbol{\theta}_1, \boldsymbol{\theta}_2, \ldots$ of iid parameters.

The margin function for a random forest is defined by

$$m_r(Y, \boldsymbol{x}) = P_{\boldsymbol{\theta}}(\delta_{\boldsymbol{\theta}}(\boldsymbol{x}) = Y) - P_{\boldsymbol{\theta}}(\delta_{\boldsymbol{\theta}}(\boldsymbol{x}) = \tilde{r}),$$

where $\tilde{r} = argmax_{r \neq Y} P_{\boldsymbol{\theta}}(\delta_{\boldsymbol{\theta}}(\boldsymbol{x}) = r)$ is the maximal probability for choosing a class that is not equal to Y. $m_r(Y, \boldsymbol{x})$ measures the extent to which the probability for the right class exceeds the probability for any other class. The strength of the random forest is the expectation $s = E_{Y,\boldsymbol{x}} m_r(Y, \boldsymbol{x})$. The margin function may also be written as

$$m_r(Y, \boldsymbol{x}) = E_{\boldsymbol{\theta}}\{I(\delta_{\boldsymbol{\theta}}(\boldsymbol{x}) = Y) - I(\delta_{\boldsymbol{\theta}}(\boldsymbol{x}) = \tilde{r})\},$$

where I is the indicator function. By defining $rmg_{\boldsymbol{\theta}}(Y, \boldsymbol{x}) = I(\delta_{\theta}(\boldsymbol{x}) = Y) - I(\delta_{\theta}(\boldsymbol{x}) = \tilde{r})$ as the raw margin function, the margin function is its expectation.

Breiman (1996a) showed that the generalization error converges to $P_{Y,\boldsymbol{x}}(m_r(Y, \boldsymbol{x}) < 0)$ for almost surely all sequences $\boldsymbol{\theta}_1, \boldsymbol{\theta}_2, \ldots$, which explains why random forests are not apt to overfit as more trees are added. Moreover, he showed that an upper bound for the generalization error is $\bar{\varrho}(1 - s^2)/s^2$, where $\bar{\varrho}$ is the mean correlation between the raw margin function of randomly generated trees indicated by $\theta, \tilde{\theta}$ and s is the strength of the random forest. Thus the bound of the generalization error is mainly determined by the strength of the classifier in the forest and the correlation between them measured is in terms of the raw margin function. Thus strength and correlation yield guidelines for highly accurate random forests.

In many studies random forests were shown to be among the best predictors. Comparisons for gene expression data were given by Diaz-Uriarte and de Andres (2006b) and Huang et al. (2005). Moreover, random forests have a built-in selection procedure that selects the most important predictors. Therefore, they can be also applied when the predictor space is high dimensional.

A disadvantage of random forests is that the contribution of single variables in the classification rule gets lost. Importance measures try to measure this contribution for random forests. A permutation accuracy importance measure, proposed by Breiman (2001a), evaluates the difference in prediction accuracy between the trees based on the original observations and trees that are built on data in which one predictor variable is randomly permuted, thereby breaking the original association with the response variable. A disadvantage of the method might be that the permutation also breaks the association with the other predictors. Conditional variable importance measures that try to avoid that effect were proposed by Strobl et al. (2008).

15.5.3 Boosting Methods

In several chapters of this book boosting methods have been used for modeling purposes. However, the roots of boosting techniques lie in the construction of accurate classification algorithms. Therefore, in the following we consider basic boosting algorithms for classification. The first well-established boosting algorithm, AdaBoost (Freund and Schapire, 1997), works in the spirit of ensemble schemes. In contrast to the simple bagging strategy, it uses an adaptive updating strategy for the weights in each step of the ensemble construction.

Let a classifier based on learning set L have the form

$$\begin{aligned} \delta(.,L): X &\longrightarrow \{1,\dots,k\}, \\ \boldsymbol{x} &\longrightarrow \delta(\boldsymbol{x},L), \end{aligned}$$

where X denotes the space of covariates and $\delta(\boldsymbol{x},L)$ is the predicted class for observation $\boldsymbol{x}$.

Early Boosting Approaches: AdaBoost

In *boosting* the data are resampled adaptively and the predictors are aggregated by weighted voting. The discrete AdaBoost procedure proposed by Freund and Schapire (1997) starts with weights $w_1 = \dots = w_{n_L} = 1/n_L$. In the following we give the mth step of the algorithm (for the case of two classes).

Discrete AdaBoost (mth step)

1. (a) The current weights $w_1,\dots,w_{n_L}$ form the resampling probabilities. Based on these probabilities, the learning set L_m is sampled from L with replacement.
   (b) The classifier $\delta(.,L_m)$ is built based on L_m.
2. The learning set is run through the classifier $\delta(.,L_m)$, yielding error indicators $\epsilon_i = 1$ if the ith observation is classified incorrectly and $\epsilon_i = 0$ otherwise.
3. With $e_m = \sum_{i=1}^{n_L} w_i\epsilon_i$, $b_m = (1-e_m)/e_m$ and $c_m = \log(b_m)$, the resampling weights are updated for the next step by

$$w_{i,new} = \frac{w_i b_m^{\epsilon_i}}{\sum_{j=1}^{n_L} w_j b_m^{\epsilon_j}} = \frac{w_i \exp(c_m\epsilon_i)}{\sum_{j=1}^{n_L} w_j \exp(c_m\epsilon_j)}.$$

After M steps, the aggregated voting for observation x is obtained by

$$\operatorname{argmax}_j \left(\sum_{m=1}^{M} c_m I(\delta(\boldsymbol{x},L_m) = j) \right).$$

While e_m is a weighted sum of errors, the parameters $c_m = \log((1-e_m)/e_m)$ are log-odds comparing weighted hits to errors. It is easily seen that for the new weighting scheme one has $\sum_{\epsilon_i=1} w_{i,new} = \sum_{\epsilon_i=0} w_{i,new} = 0.5$, and therefore in the next step the resampling probability put on the observations which have been misclassified in the mth step sums up to 0.5.

The algorithm as given above is based on weighted resampling. In alternative versions of boosting, the observations are not resampled. Instead, the classifier is computed by weighting the original observations by weights $w_1, \ldots, w_{n_L}$, which are updated iteratively. Then $\delta(., L_m)$ should be read as the classifier based on the current weights $w_1, \ldots, w_L$ (in the mth step).

In the case of two classes, it is more common to use binary observations $y_i \in \{1, 0\}$ or $\tilde{y}_i \in \{1, -1\}$ rather than class labels 1 and 2 as indicators. The coding $\tilde{y}_i \in \{1, -1\}$ is used especially in the machine learning community. The class indicator $Y \in \{1, 2\}$ transforms into the binary case by using $y_i = 1$ if $Y_i = 1$ and $y_i = 0$ if $Y_i = 2$. The version $\tilde{y}_i \in \{1, -1\}$ is obtained from $\tilde{y}_i = 2y_i - 1$.

Real AdaBoost (Friedman et al., 2000) uses real-valued classifier functions $f(\boldsymbol{x}, L)$ instead of $\delta(\boldsymbol{x}, L)$, with the convention that $f(\boldsymbol{x}, L) \geq 0$ corresponds to $\delta(\boldsymbol{x}, L) = 1$ and $f(\boldsymbol{x}, L) < 0$ corresponds to $\delta(\boldsymbol{x}, L) = 2$.

Real AdaBoost (mth step)

1. Based on weights $w_1, \ldots, w_{n_L}$ the classifier $\delta(., L_m)$ is built.
2. The learning set is run through the classifier $\delta(., L_m)$, yielding estimated class probabilities $p(\boldsymbol{x}_i) = \hat{P}(\tilde{y}_i = 1|\boldsymbol{x}_i)$.
3. Based on these probabilities a real-valued classifier is built by

$$f(\boldsymbol{x}_i, L_m) = 0.5 \cdot \log \frac{p(\boldsymbol{x}_i)}{1 - p(\boldsymbol{x}_i)},$$

and the weights are updated for the next step by

$$w_{i,new} = \frac{w_i \exp(-\tilde{y}_i f(\boldsymbol{x}_i, L_m))}{\sum_{j=1}^{n_L} w_j \exp(-\tilde{y}_j f(\boldsymbol{x}_j, L_m))}.$$

After M steps the aggregated voting for observation $\boldsymbol{x}$ (for coding {1, –1}) is obtained by $\text{sign}(\sum_{m=1}^{M} f(\boldsymbol{x}, L_m))$. In this version of Real AdaBoost, either resampled observations or weighted observations may be used. The essential term in the updating is $w_i \exp(-\tilde{y}_i f(\boldsymbol{x}_i, L_m))$, which, depending on hits ($\epsilon_i = 0$) and misclassifications ($\epsilon_i = 1$), has the form

$$w_i \exp(-\tilde{y}_i f(\boldsymbol{x}_i, L_m)) = \begin{cases} w_i \exp(-|f(\boldsymbol{x}_i, L_m)|) & \epsilon_i = 0 \\ w_i \exp(|f(\boldsymbol{x}_i, L_m)|) & \epsilon_i = 1. \end{cases}$$

It is seen that for misclassified observations the weight w_i is increased, whereas for correctly classified observations w_i is decreased. In order to ensure the existence of $f(\boldsymbol{x}_i, L_m)$, $1/n_L$ is added to the numerator and denominator of the fraction, yielding

$$f(\boldsymbol{x}_i, L_m) = 0.5 \cdot \log \frac{p(\boldsymbol{x}_i) + 1/n_L}{1 - p(\boldsymbol{x}_i) + 1/n_L}.$$

Functional Gradient Descent Boosting

Breiman (1999) examined boosting from a game theoretical point of view and established for the first time a connection between boosting and numerical optimization techniques. Friedman et al. (2000) elaborated these connections in more detail and established a close relationship between boosting algorithms and logistic regression. In Friedman (2001), Bühlmann and Yu

(2003), and Bühlmann and Hothorn (2007), the idea of boosting as an optimization technique in function space was investigated further. In the following we give the basic minimization concept.

For a new observation $(y, \boldsymbol{x})$ one considers the problem of minimizing $E[L(y, G(\boldsymbol{x}))]$ with respect to $G(\boldsymbol{x})$ for some specified loss function $L(.,.)$ and an appropriately chosen value $G(\boldsymbol{x})$. A simple example is the squared error loss $L_2(y, G(\boldsymbol{x})) = (y - G(\boldsymbol{x}))^2$, where $G(\boldsymbol{x})$ is considered to represent an approximation of y. To obtain a practical implementation for finite datasets, one minimizes the empirical version of the expected loss, $\sum_i L(y_i, G(\mathbf{x}_i))/n_L$. Minimization is obtained iteratively by utilizing a *steepest gradient descent approach*. An essential ingredient of the method is the fitting of a structured function as an approximation to $G(\boldsymbol{x})$. This fitting may be seen as a *base procedure*. Depending on the modeling problem, one may fit a regression tree or (in smoothing problems) a regression spline function. Thus, in each iteration step $G(\mathbf{x})$ is approximated by an (parameterized) estimate $\hat{g}(\boldsymbol{x}, \{u_i, \boldsymbol{x}_i\})$ that is based on input data $\{u_i, \boldsymbol{x}_i\}$. The input data are not the original data but are generated during the fitting process by computing the derivatives

$$u_i = -\frac{\partial L(y_i, G(\mathbf{x}))}{\partial G(\mathbf{x})}|_{G(\mathbf{x})=G(\mathbf{x}_i)}, \quad i = 1, \ldots, n. \tag{15.14}$$

Basically the functional gradient descent algorithm consists of iteratively refitting these pseudo-response values. The final solution is attained in a stage-wise manner. We give in the following a version of this algorithm, which is close to the gradient boost algorithm given in Friedman (2001).

Let $\hat{g}(\boldsymbol{x}, \{u_i, \boldsymbol{x}_i\})$ denote the base procedure at value $\boldsymbol{x}$ based on input data $\{u_i, \boldsymbol{x}_i\}$, which are not necessarily the original data $\{y_i, \boldsymbol{x}_i\}$.

Functional Gradient Descent Boosting

Step 1 (Initialization)

Given the data $\{y_i, \boldsymbol{x}_i\}$, fit a base procedure for initialization that yields the function estimate $G^{(0)}(\boldsymbol{x}) = \hat{g}_0(., \{y_i, \boldsymbol{x}_i\})$. For example, a constant c is fitted, yielding $G^{(0)}(\mathbf{x}) = \text{argmin}_c \frac{1}{n} \sum_{i=1}^{n} L(y_i, c)$.

Step 2 (Iteration) For $l = 0, 1, \ldots$

1. *Fitting step*

   Compute the values of the negative gradient u_i as given in (15.14), evaluated at $G^{(l)}(\mathbf{x}_i)$. Fit a base procedure to the current data $\{u_i, \boldsymbol{x}_i\}$. The fit $\hat{g}(., \{u_i, \boldsymbol{x}_i\})$ is an estimate based on the original predictor variables and the current negative gradient vector.

2. *Update step*

   The improved fit is obtained by the update

   $$G^{(l+1)}(.) = G^{(l)}(.) + \nu \hat{g}(., \{u_i, \boldsymbol{x}_i\}),$$

   where $\nu \in (0, 1]$ is a shrinkage parameter that should be sufficiently small.

Step 3 (Final estimator)

Obtain the final estimator after an optimized number of iterations l_{opt}, that is, $\hat{G}^{(l_{\text{opt}})}(\mathbf{x})$.

In contrast to the original version of GradientBoost, this algorithm renounces an additional line search step between the fitting and the update step, which calibrates the shrinkage parameter ν within each step. For obtaining an accurate estimate $\hat{G}^{(l_{\text{opt}})}(\mathbf{x})$ the constant ν seems to do well (see Bühlmann and Hothorn, 2007). However, the shrinkage parameter ν plays an important role in boosting algorithms. It makes the learner *weak* in the sense that within one step not a perfect fit but a slightly better fit structured by the parametric learner is intended. Small values of ν have been shown to avoid early overfitting of the procedure. Therefore, ν should be chosen rather small (e.g., $\nu = 0.1$). But its choice has to be balanced with the number of iterations, because for very small values a large number of iterations is needed to obtain the best results.

An example for which the negative gradient has a very simple form is the squared error loss $L_2(y, G(\boldsymbol{x})) = (y - G(\boldsymbol{x}))^2$. For the squared error loss the negative gradient vector consists of the simple residuals $u_i = -\partial L(y_i, G_i)/\partial G = 2(y_i - G_i)$. Therefore, in the fitting step, a model is fit with the original responses replaced by the current residuals. Essentially the same procedure with just one boosting step was proposed already by Tukey (1977). Friedman (2001) noted that boosting by gradient descent is a stagewise strategy that is different from a stepwise approach that readjusts previously entered terms. Since previously entered terms are not adjusted, it is a *greedy function approximation* that performs *forward stagewise additive modeling*. The additive fit becomes obvious from the final estimator, which has the form

$$\hat{G}^{(l_{\text{opt}})}(.) = \hat{g}_0(., \{y_i, \boldsymbol{x}_i\}) + \sum_{j=1}^{l_{\text{opt}}} \nu \hat{g}(., \{u_i^{(j-1)}, \boldsymbol{x}_i\}),$$

where $u_i^{(j)} = -\partial L(y_i, G^{(j)}(\mathbf{x}_i))/\partial G(\mathbf{x})$ are the negative gradient values from step j.

For the squared error loss, $G(\boldsymbol{x}_i)$ is estimated as a direct approximation to $\mathrm{E}(y_i|\boldsymbol{x}_i)$. However, for a binary response the approximating function $G(\boldsymbol{x}_i)$ may also denote an approximation to a transformation of $\mathrm{E}(y_i|\boldsymbol{x}_i)$. For binary responses, $y_i \in \{0, 1\}$, one often uses the logit transformation. Let $\eta_i = \eta(\boldsymbol{x}_i) = \log(\pi_i/(1 - \pi_i))$ denote the logits, let $\tilde{\eta}_i = \eta_i/2$ denote the half-logits, and let $\tilde{y}_i = 2y_i - 1$ be the rescaled response ($\tilde{y}_i \in \{-1, 1\}$). The corresponding approximation of y_i is the probability

$$\pi_i = \pi(\boldsymbol{x}_i) = \frac{\exp(\eta_i)}{1 + \exp(\eta_i)} = \frac{\exp(\tilde{\eta}_i)}{\exp(-\tilde{\eta}_i) + \exp(\tilde{\eta}_i)},$$

where π_i denotes the probability of response one. Then the usual negative binary log-likelihood for binary data (or the half-deviance) $-l_i = -\{y_i \log(\pi_i) + (1 - y_i)\log(1 - \pi_i)\}$ can be given as $-l_i = \log(1 + e^{-(2y_i - 1)\eta_i}) = \log(1 + e^{-2\tilde{y}_i\tilde{\eta}_i})$. It can be expressed as a loss function with arguments y_i and $\pi(\boldsymbol{x}_i)$ as

$$L_{lik}(y_i, \pi(\boldsymbol{x}_i)) = -\{y_i \log(\pi(\boldsymbol{x}_i)) + (1 - y_i)\log(1 - \pi(\boldsymbol{x}_i))\},$$

or with $G(\boldsymbol{x}_i)$ set to be $\tilde{\eta}_i$ in the form of a loss function with arguments $\tilde{y}_i, G(\boldsymbol{x}_i)$:

$$L^{lik}(\tilde{y}_i, G(\boldsymbol{x}_i)) = \log(1 + e^{-2\tilde{y}_i G(\boldsymbol{x}_i)}).$$

In the last form, which uses the superscript lik, the gradient descent algorithm implicitly fits a logit model and $G(\boldsymbol{x}_i)$ approximates the half-logits $\log(\pi_i/(1 - \pi_i))/2$. Therefore, the approximation is on the level of the predictor after the logit transformation has been applied. When a structured response is fitted to the gradient, for example, a linear term $\eta_i = \boldsymbol{x}_i^T\boldsymbol{\beta}$, the gradient descent algorithm fits a linear logit model. The use of $\tilde{\eta}_i$ instead of η_i and $\tilde{y}_i \in \{-1, 1\}$ instead of $y_i \in \{0, 1\}$ is a convention that comes from the machine learning community and is useful to show the connection to other loss functions. It should be noted that for binary responses the likelihood-based loss $L_{lik}(y_i, \pi(\boldsymbol{x}_i))$ is equivalent to the logarithmic score, which can be seen as an empirical version of the Kullback-Leibler distance (Section 15.1). An alternative loss function is the *exponential loss*,

$$L^{\exp}(y_i, G_i) = \exp(-\tilde{y}_i G_i) = \exp(-(2y_i - 1)G_i).$$

The role of G_i in the exponential loss may be derived from looking at the population minimizer, which is given by

$$G^*(\boldsymbol{x}) = \text{argmin}_{G(\boldsymbol{x})} E_{y|\boldsymbol{x}}(\exp(-\tilde{y}G(\boldsymbol{x}))) = \frac{1}{2}\log(\frac{\pi(\boldsymbol{x})}{1-\pi(\boldsymbol{x})}) = \tilde{\eta}(\boldsymbol{x}).$$

Therefore, G_i corresponds again to the half-logits $\tilde{\eta}_i$ as in the log-likelihood loss function L^{lik} and is on the level of the predictor in a logit model. It should be noted that the population minimizer of the likelihood-based loss L^{lik}, $\text{argmin}_{G(\boldsymbol{x})} = E_{y|\boldsymbol{x}} L(y, G(\boldsymbol{x}))$, is also $\tilde{\eta}(\boldsymbol{x})$. The exponential loss is just an alternative loss function that has the same population minimizer but different computational properties. Exponential loss is in particular of interest since it fills the gap between reweighting algorithms developed in the machine learning community and the gradient descent boosting algorithms, because it may be shown that AdaBoost is a boosting algorithm that minimizes exponential loss (see Friedman et al., 2000).

TABLE 15.3: Losses as functions of $y \in \{0,1\}$ and the approximating π and functions of $\tilde{y} \in \{-1,1\}$ and the approximating $G(\boldsymbol{x})$.

$L^{lik}(\tilde{y}, G(\boldsymbol{x})) = \log(1 + e^{-2\tilde{y}\, G(\boldsymbol{x})})$
$L_{lik}(y, \pi(\boldsymbol{x})) = -\{y\ \log(\pi(\boldsymbol{x})) + (1-y)\ \log(1-\pi(\boldsymbol{x}))\}$
$L^{exp}(\tilde{y}, G(\boldsymbol{x})) = \exp(-\tilde{y}\ G(\boldsymbol{x}))$
$L_{exp}(y, \pi(\boldsymbol{x})) = (\frac{1-\pi(\boldsymbol{x})}{\pi(\boldsymbol{x})})^{(2y-1)/2}$
$L^{01}(\tilde{y}, G(\boldsymbol{x})) = I\{\tilde{y}\ G(\boldsymbol{x}) \le 0\} = I\{\text{sign}\, G(\boldsymbol{x}) \ne \text{sign}\, \tilde{y}\}$
$L_{01}(y, \pi(\boldsymbol{x})) = I\{\lvert y - \pi(\boldsymbol{x})\rvert < 0.5\}$

When the focus is on classification, the loss functions should be evaluated with respect to the underlying classification rule. Thereby one should distinguish between $y \in \{0,1\}$ and $\tilde{y} \in \{-1,1\}$, which represent different scalings of the response. The approximation of $\tilde{y}$ may be measured by $G(\boldsymbol{x})$ while the approximation of y is usually measured by the corresponding probability $\pi(\boldsymbol{x}) = \exp(G(\boldsymbol{x})/(\exp(G(\boldsymbol{x})) + \exp(-G(\boldsymbol{x}))))$. The classification rule for 0-1 responses is $\delta(\boldsymbol{x}) = 1$ if $\pi(\boldsymbol{x}) > 0.5$, and $\delta(\boldsymbol{x}) = 0$ if $\pi(\boldsymbol{x}) < 0.5$ and the corresponding 0-1 loss is

$$L_{01}(y, \pi(\boldsymbol{x})) = I\{\delta(\boldsymbol{x}) \ne y\} = I\{\lvert y - \pi(\boldsymbol{x})\rvert < 0,5\}.$$

For the scaling $\tilde{y} \in \{-1,1\}$, the classification rule may be given in the form $\delta(\boldsymbol{x}) = 1$ if $G(\boldsymbol{x}) > 0$, and $\delta(\boldsymbol{x}) = -1$ if $G(\boldsymbol{x}) < 0$ and the corresponding 0-1 loss is

$$L^{01}(\tilde{y}, G(\boldsymbol{x})) = I\{\tilde{y}G(\boldsymbol{x}) \le 0\} = I\{\text{sign}\, G(\boldsymbol{x}) \ne \text{sign}\, \tilde{y}\}.$$

Therefore, misclassification occurs only if one has $\tilde{y}G(\boldsymbol{x}) \le 0$, where $\tilde{y}G(\boldsymbol{x})$ is called a "margin." For the $\tilde{y}$-scaling, the margin plays an important role for the loss functions, since likelihood-based loss as well as exponential loss are monotone decreasing functions of the margin. Figure 15.6 shows the likelihood, exponential, and the 0-1 loss as a function of the margin $\tilde{y}G(\boldsymbol{x})$. In addition, we also show the hinge loss $(1 - \tilde{y}G(\boldsymbol{x}))I(\tilde{y}G(\boldsymbol{x}) < 1)$ and the squared loss $(\tilde{y} - G(\boldsymbol{x}))^2$ for $\tilde{y} = 1$, since the losses in Figure 15.6 may be seen as function of $G(\boldsymbol{x})$ when $\tilde{y} = 1$. Table 15.3 collects the loss functions, given in the boosting form with $\tilde{y}$-scaling, and the regression form with arguments $y \in \{0, 1\}$ and the approximating probability $\pi(\boldsymbol{x})$.

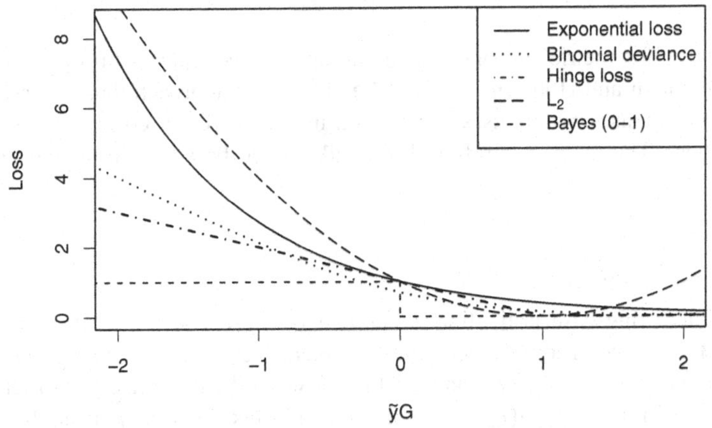

FIGURE 15.6: Loss functions $L(\tilde{y}, G(\boldsymbol{x}))$ as functions of $\tilde{y}G(\boldsymbol{x})$, $\tilde{y} \in \{-1, 1\}$. For $\tilde{y} = 1$ they show the dependence on $G(\boldsymbol{x})$.

Friedman (2001) gave algorithms developed from gradient boosting for several specific loss functions. They include loss functions that are commonly used in the regression setting, like L_2-loss, which means iteratively fitting the residuals, or Huber-loss for robustification,

$$L(y, G(\boldsymbol{x})) = \begin{cases} |y - G(\boldsymbol{x})|^2 & \text{for } |y - G(\boldsymbol{x})| \leq \delta \\ 2\delta(|y - G(\boldsymbol{x})| - \delta/2) & \text{otherwise.} \end{cases}$$

Figure 15.7 shows the L_2-loss, the L_1-loss, and the robust Huber-loss.

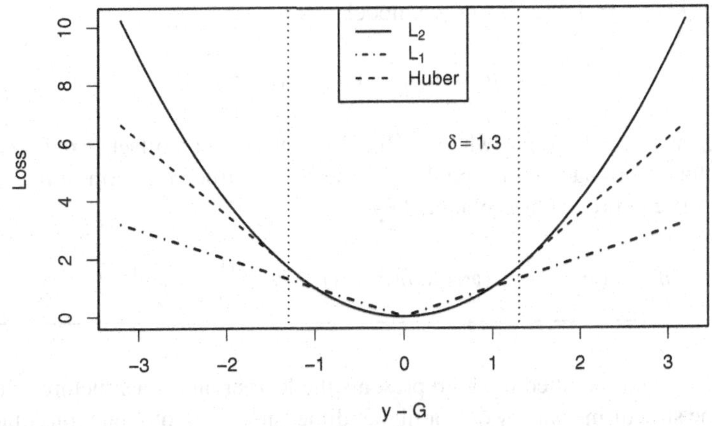

FIGURE 15.7: Loss functions $L(y, G(\boldsymbol{x}))$ as functions of $y - G(\boldsymbol{x})$.

Up until now it has not yet been specified *how* the improvement $\hat{g}(., \{u_i, \boldsymbol{x}_i\})$ is fitted within the functional gradient descent approach. In the regression setting with the squared error loss, the u_i-values are the residuals $2(y_i - G_i)$, and it is straightforward to fit a regression model by using least-squares fitting where u_i are the responses and $\boldsymbol{x}_i$ are the covariates. For binary observations the negative derivatives of the likelihood loss, $-\partial L^{lik}(\tilde{y}_i, \tilde{\eta}_i)/\partial\eta$, are again residuals of the form

$$2\left\{\frac{1}{2}(\tilde{y}_i + 1) - \frac{e^{\tilde{\eta}_i}}{e^{\tilde{\eta}_i} + e^{-\tilde{\eta}_i}}\right\} = 2\{y_i - \pi_i\}.$$

In the same way as for metric responses, one can apply least-squares fitting with responses $2\{y_i - \hat{\pi}_i\}$ (Bühlmann and Hothorn, 2007). An alternative approach that generalizes to exponential family response models uses a weighted update step that corresponds to a shrunk Fisher-scoring step. The method is referred to as likelihood-based boosting and given in the next section.

Likelihood-Based Boosting

A particularly interesting case is likelihood-based boosting because it is strongly related to likelihood-based inference as considered in most chapters. Let y_i be from an exponential family distribution with mean $\mu_i = E(y_i|\boldsymbol{x}_i)$ and the link between the mean and the structuring term specified in the usual form $\mu_i = h(\eta_i)$ or $g(\mu_i) = \eta_i$ with link function g. Then the *likelihood-based generalized model boosting* (GenBoost) tries to improve the predictor $\eta(\boldsymbol{x})$ by greedy forward additive fitting. It has essentially the same form as GAMBoost for structured additive regressions as considered in Section 10.3.4, but in the fitting step more general models can be used. The basic form of the algorithm (for simplicity without selection step) is

---

GenBoost

Step 1 (Initialization)

For the given data $(y_i, \boldsymbol{x}_i), i = 1, \dots, n$, fit the intercept model $\mu^{(0)}(\boldsymbol{x}) = h(\eta_0)$ by maximizing the likelihood, yielding $\eta^{(0)} = \hat{\eta}_0, \hat{\mu}^{(0)} = h(\hat{\eta}_0)$.

Step 1 (Iteration) For $l = 0, 1, 2, \dots$ fit the model

$$\mu_i = h(\hat{\eta}^{(l)}(\boldsymbol{x}_i) + \eta(\boldsymbol{x}_i, \gamma))$$

to data $(y_i, \boldsymbol{x}_i), i = 1, \dots, n$, where $\hat{\eta}^{(l)}(\boldsymbol{x}_i)$ is treated as an offset (fixed constant) and the predictor is estimated by fitting the parametrically structured term $\eta(\boldsymbol{x}_i, \gamma)$, thus obtaining $\hat{\gamma}_l$. The improved fit is obtained by

$$\hat{\eta}^{(l+1)}(\boldsymbol{x}_i) = \hat{\eta}^{(l)}(\boldsymbol{x}_i) + \hat{\eta}(\boldsymbol{x}_i, \hat{\gamma}_l), \quad \hat{\mu}_i^{(l+1)} = h(\hat{\eta}^{(l+1)}(\boldsymbol{x}_i)).$$

The term $\eta(\boldsymbol{x}_i, \gamma)$ in the fitted model represents the learner and is a structured function. In Section 10.3.4 the structuring was by one or higher dimensional smooth functions of predictors, and in Section 6.3.2 linear functions were used. More generally, $\eta(\boldsymbol{x}_i, \gamma)$ can represent any structure one wants to fit, including, for example, trees. One more advantage of the general algorithm is that alternative link functions can be used (not only the logit link as in *LogitBoost* proposed by Friedman et al., 2000). Moreover, the basic boosting algorithm also works in the same way if the response is multivariate, but, of course, learners for the predictors have to be adapted.

The results of boosting depend in particular on the structuring of the predictor. In earlier statistical boosting literature (Breiman, 1998; Friedman, Hastie, and Tibshirani, 2000; Friedman, 2001) CARTs were recommended. Then the performance may depend on the tree depth and higher tree depths may yield superior error rates; however, the results are hardly interpretable. By using stumps (trees with two terminal nodes), used by Friedman et al. (2000) a decomposition of $G(\boldsymbol{x})$ into univariate functions of each covariate is obtained. The advantage is that the result can be visualized in a similar way as GAM results. Bühlmann and Yu (2003) use componentwise smoothing splines as the base procedure for fitting and doing variable selection in additive models. Tutz and Binder (2006) extend a similar approach to GAMs.

In recent years, many properties of boosting algorithms have been analyzed theoretically by both machine learning theorists and statisticians. Among the most important results from a statistical point of view are those of Bühlmann and Yu, 2003 for functional gradient descent boosting using the L_2-loss. They verify a bias-variance trade-off, with the variance exponentially small increasing with the number of iterations l (Bühlmann and Yu, 2003). In addition, they show that when a smoothing spline is used as the base procedure, the algorithm reaches the optimal rate of convergence for a one-dimensional function estimation. Furthermore, it is capable of capturing a higher degree of smoothness than the smoothing spline (Bühlmann and Yu, 2003). Later, Bühlmann (2006) investigated the aforementioned functional gradient descent approach using the L_2-loss and componentwise least-squares estimates as base procedures. He showed that this algorithm provides a consistent estimator for high-dimensional models in which the number of covariates is allowed to grow exponentially with the sample size under some sparseness assumptions.

Boosting approaches were shown to perform well in high-dimensional classification problems where different phenotypes, mostly cancer types, are classified using microarray gene expression data. While AdaBoost did not show superior performance in the comparison of Dudoit et al. (2002), LogitBoost with simple modifications was very successful in the comparison study of Dettling and Bühlmann (2003), where stumps were used as learners after the preselection of genes based on Wilcoxon's test statistic.

Multiple Category Case

The boosting approaches considered in the preceding sections mostly use a univariate response, which in classification means two classes. When the classes are given in $k > 2$ categories, a one-against-all approach can be used (for example, Dettling and Bühlmann, 2003). Let the dichotomous variable that distinguishes between category r and the other categories be given by $y^{(r)} = 1$ if $Y = r$ and $y^{(r)} = 0$ otherwise. The fitting procedures are applied to the binary response data $(y_i^{(r)}, \boldsymbol{x}_i)$, yielding estimates of probabilities $\hat{P}^{(r)}(r|\boldsymbol{x}) = P^{(r)}(y^{(r)} = 1|\boldsymbol{x})$ or dichotomous classifiers $\hat{y}^{(r)}(\boldsymbol{x})$. Classification rules are found by aggregating over splits. For the fixed split that distinguishes between r and the rest, estimates of probabilities are

$$\hat{P}^{(r)}(Y = r|\boldsymbol{x}) = \hat{P}^{(r)}(r|\boldsymbol{x}), \quad \hat{P}^{(r)}(Y = j|\boldsymbol{x}) = (1 - \hat{P}^{(r)}(r|\boldsymbol{x}))/(k-1),$$

which are combined to

$$\hat{P}(Y = r|\boldsymbol{x}) = \sum_{j=1}^{k} \hat{P}^{(j)}(Y = r|\boldsymbol{x})/k.$$

The Bayes rule applied to $\hat{P}(Y = r|\boldsymbol{x})$ yields the classifier. When dichotomous classifiers are given one can use the majority vote

$$\hat{Y} = r_0, \quad \text{if} \quad r_0 = \text{argmax}_r\{\hat{y}_r^{(1)} + \cdots + \hat{y}_r^{(k)}\} \tag{15.15}$$

over dummy variables

$$\hat{y}_r^{(r)}(\boldsymbol{x}) = \hat{y}^{(r)}(\boldsymbol{x}), \quad \hat{y}_j^{(r)}(\boldsymbol{x}) = 1 - \hat{y}^{(r)}(\boldsymbol{x}), \quad j \neq r.$$

The rules are referred to as nominal aggregation (ensemble) rules, in contrast to the ordinal aggregation rules considered in Section 15.9. Of course, if a multinomial or ordinal model is used in GenBoost, aggregation is not needed.

15.6 Neural Networks

Neural networks were developed mainly in the machine learning community with some input from statisticians. They encompass many techniques, only part of them biologically motivated by the information processing in organisms. Their motivation by biology and their performance in classification problems made them an attractive and somehow fancy tool. From a modeling viewpoint neural networks are just non-linear models. Non-linearity is obtained by extracting linear combinations of the inputs, which are then processed by use of non-linear transformations. In that respect they are close to non-linear modeling by expansion in basis functions as considered in Chapter 10. In this section only the basic concepts of neural networks, namely, feed-forward networks and radial basis function networks, are considered. An extensive treatment of neural networks in pattern recognition is found in Ripley (1996) and Bishop (2006).

15.6.1 Feed-Forward Networks

Feed-forward networks may be seen as a general framework for non-linear functional mappings between the set of input variables $x_1, \ldots, x_p$ and the set of output variables $y_1, \ldots, y_q$. Feed-forward networks consist of successive layers of processing units where connections run only from every unit in one layer to every unit in the next layer. A simple example with only one intermediate layer of hidden units between input and output variables is given in Figure 15.8.

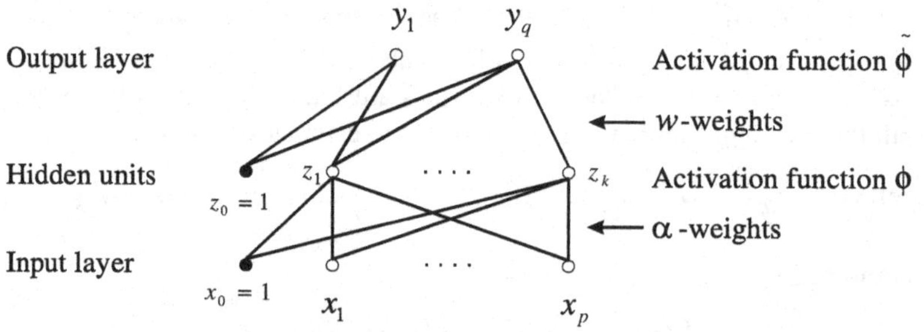

FIGURE 15.8: Feed-forward network with one hidden layer.

The network has p inputs, k hidden units, and q output units. In addition, there is an input variable $x_0 = 1$ and a hidden unit $z_0 = 1$ that represent the intercepts. In the context of neural networks these intercepts are called biases. The flow of information is from input to output units, where the units are connected by specific transformations of linear terms. The basic principle with one hidden layer is given by the following two steps:

1. *Connection between input and hidden units*
   The input of the hidden unit l is given as a linear combination of the input variables and an intercept (also called the bias). The output is obtained by a transformation of these linear combinations given by

$$z_l = \phi(\alpha_{l0} + \sum_{j=1}^{p} \alpha_{lj} x_j),$$

   where ϕ is a fixed transformation function called the activation function. The output z_l of unit l is considered as the activation of this unit.

2. *Connection between hidden and output units*
   The input of the output unit y_t is given in the same form by considering linear combinations of the outputs $z_1, \ldots, z_k$ of the hidden layer and an intercept. The output is obtained by using a fixed transformation or activation function $\tilde{\phi}$ that yields

$$y_t = \tilde{\phi}(w_{t0} + \sum_{l=1}^{k} w_{tl} z_l).$$

Thus two sorts of weights and two activation functions are involved. The α-weights determine the input for the hidden units and are a combination of the input variables $x_1, \ldots, x_p$. The w-weights determine the linear combination of the activations of the hidden units $z_1, \ldots, z_k$. The activation functions ϕ and $\tilde{\phi}$ transform the input within the hidden and output layers, respectively. Combining the two steps yields the form

$$y_t(\boldsymbol{x}) = \tilde{\phi}(w_{t0} + \sum_{l=1}^{k} w_{tl} \phi(\alpha_{l0} + \sum_{j=1}^{p} \alpha_{lj} x_j)), \quad (15.16)$$

where $y_t(\boldsymbol{x}), \boldsymbol{x}^T = (x_1, \ldots, x_p)$, is the output resulting from input vector $\boldsymbol{x}$.

In the special case of a univariate response y ($q = 1$) in the weights $w_{t0}, \ldots, w_{tk}$, the index t is omitted and one obtains

$$y(\boldsymbol{x}) = \tilde{\phi}(w_0 + \sum_{l=1}^{k} w_l \phi(\alpha_{l0} + \sum_{j=1}^{p} \alpha_{lj} x_j)). \quad (15.17)$$

Up until now the activation functions ϕ and $\tilde{\phi}$ have not been specified. One possibility for ϕ is the *Heaviside function*:

$$\phi(a) = \begin{cases} 1 & a \geq 0 \\ 0 & a < 0. \end{cases}$$

If it is used for the hidden units, these units behave as *threshold units* that send a signal or not. As the activation function in the output units, it makes sense only if the response variables are dichotomous with $y_i \in \{0, 1\}$. For continuous responses, sigmoid functions like the *logistic function*,

$$\phi_{\log}(a) = \frac{1}{1 + \exp(-a)} = \frac{\exp(a)}{1 + \exp(a)},$$

or the "tanh" activation function,

$$\phi_{\tan}(a) = \tanh(a) = \frac{\exp(a) - \exp(-a)}{\exp(a) + \exp(-a)},$$

are more appropriate. While $\phi_{\log}(a) \in [0, 1)$ one has $\phi_{\tan}(a) \in [-1, 1]$. However, $\phi_{\log}(a)$ and $\phi_{\tan}(a)$ differ only through a linear transformation; more precisely, one has $\phi_{\tan}(a/2) = \phi_{\log}(a) - 1$. Therefore, networks based on $\phi_{\log}$ or $\phi_{\tan}$ are equivalent but have different values for weights and biases. It should be noted that the activation functions are essential for making the system identifiable. For example, if ϕ and $\tilde{\phi}$ are chosen as linear functions or, simpler, as $\phi(a) = \tilde{\phi}(a) = a$, one obtains $y(x) = w_0 + \sum_{l=1}^{h}(w_l \alpha_{l0} + \sum_{j=1}^{p} w_l \alpha_{lj} x_j)$, where the weights on $x_1, \ldots, x_p$ are not unique since one may always choose, for example, $\tilde{w}_l = 1, \tilde{\alpha}_{lj} = w_l \alpha_{lj}$, and obtain $w_l \alpha_{lj} = \tilde{w}_l \tilde{\alpha}_{lj}$. Moreover, if the activation functions are identity functions, the model collapses to a simple linear model.

For feed-forward networks, the estimation of parameters, also called the learning or training of the neural network, is usually based on a steepest gradient descent, which is called back-propagation in this setting. A detailed discussion of algorithms is given by Ripley (1996), Chapter 5.

15.6.2 Radial Basis Function Networks

Radial basis function networks have the same architecture as feed-forward networks with one hidden layer, with the modification that no bias term is needed as input to the hidden units (see Figure 15.9). However, there are several modifications concerning the functional relationship between the layers. The essential difference is that the output of the hidden layers is not given as transformations of a linear input for unit l but has the form

$$z_l(\boldsymbol{x}) = \phi_l(\boldsymbol{x}),$$

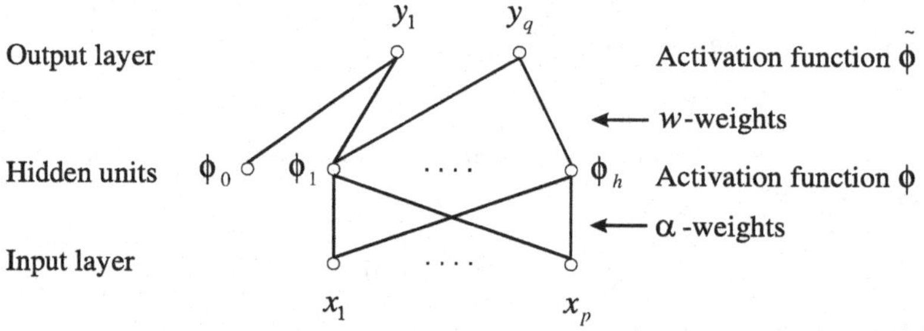

FIGURE 15.9: Radial basis function network.

where ϕ_l is a radial basis function that is connected to unit l. The function ϕ_l is specific for unit l by using (unknown) parameters that characterize this unit. $\phi_l(x)$ has the general form $\phi_l(\boldsymbol{x}) = \phi(\boldsymbol{x}; \boldsymbol{\theta}_l)$, where ϕ is a function in $\boldsymbol{x}$ depending on the parameter vector $\boldsymbol{\theta}_l$. Examples are the Gaussian function

$$\phi(\boldsymbol{x}; \boldsymbol{\mu}_l, \sigma_l) = \exp\left(-\frac{\| \boldsymbol{x} - \boldsymbol{\mu}_l \|^2}{2\sigma_l^2}\right)$$

and localized basis functions of the type

$$\phi(\boldsymbol{x}; \boldsymbol{\theta}_l) = \bar{\phi}(\| \boldsymbol{x} - \boldsymbol{\theta}_l \|),$$

where $\bar{\phi}$ is chosen by thin plate splines $\bar{\phi}(z) = z^2 \log(z)$, the linear function $\bar{\phi}(z) = z$, or $\bar{\phi}(z) = (z^2 + \sigma^2)^\beta, 0 < \beta < 1$, which, for $\beta = 1/2$, yields the multi-quadratic function. Based on these basis functions the output is given as a linear combination:

$$y_t(\boldsymbol{x}) = w_{t0} + \sum_{l=1}^{h} w_{tl}\phi_l(\boldsymbol{x}).$$

Thus the functional form of $y_t(\boldsymbol{x})$ is given as a linear weighted sum of basis functions, which is quite similar to the methods considered in Chapter 10.

Radial basis functions may be motivated in different ways that yield specific basis functions. A powerful approach is based on the theory of Poggio and Girosi (1990). An alternative motivation is by representing the distribution within classes as mixtures (see Bishop, 2006, Chapter 5).

Neural networks are strong competitors in classification problems. They are quite flexible and able to find complex classification regions, in contrast to simple rules like linear and quadratic discriminations. However, flexibility needs regularization; otherwise, overfitting with bad generalization errors can be expected. The training of neural networks is an art in itself. With all the tuning parameters, like the number of hidden units, the architecture of the network itself (one or more layers), the choice of the activation function, and the placement of basis functions, neural networks become very flexible but also unstable. A more severe problem in applications is that they work like a black box. However, even though the prediction may be excellent, practioners often do not like black boxes. They prefer to know how predictors enter the prediction rule and how strong their effect is within the prediction rule. The implicit feature selection that is at work in neural networks makes interpreting of the impact of single variables a challenge.

15.7 Examples

In this section we give some applications of the methods that were discussed in the preceding sections. When comparing methods, fine tuning can often improve the performance of specific methods considerably. In the following we use cross-validation for the selection of tuning parameters but otherwise use the default values of the used program packages. Performance is evaluated in the test data; 50 random splits into training and test data were used.

Classification typically focusses on misclassification error, although other loss functions may be used in estimating the best empirical classifier. For example, support vector classifiers implicitly use the hinge loss and AdaBoost minimizes exponential loss. When evaluating the performance of classifiers, in particular if estimated probabilities are available, one should not restrict consideration to simple misclassification errors but use alternative loss functions that include the precision of classification. In Section 15.1, several loss functions were considered, in particular, the Brier or quadratic score $L_2(\boldsymbol{y}, \hat{\boldsymbol{\pi}}) = (1-\hat{\pi}_Y)^2 + \sum_{i \neq Y} \hat{\pi}_i^2$, and the logarithmic score $L_{KL}(\boldsymbol{y}, \hat{\boldsymbol{\pi}}) = -\log(\hat{\pi}_Y)$. The former uses estimation of all the probabilities, whereas the latter uses only the probability at the target value. But both use more information than the simple classification decision.

Example 15.4: Glass Identification

In Example 15.2 the objective was the identification of types of glass based on the refractive index and oxide content of various minerals like Na, Fe, K. Figure 15.10 shows the performance of classifiers for two classes (originally coded as types 1 and 2) and Figure 15.11 for three classes (originally coded as types 1, 2, and 7) in terms of misclassification errors and quadratic scores. The errors were computed for several splits into learning and test sample. The functions *lda* and *qda* from the package *MASS* were used for linear and quadratic discriminations, *glmnet* was used to fit the lasso for the logit model (log), *party* was used to fit random forests (rf), *mboost* was used for boosting methods (bst), *svm* from *e1071* was used for support vector machines (sv), and *nnet* was used for feed-forward networks. It is seen that in both cases the quadratic discriminant analysis, although more flexible than the linear discriminant analysis, performs worse. The effect is even more distinct when the quadratic score is used to evaluate performance. The lasso

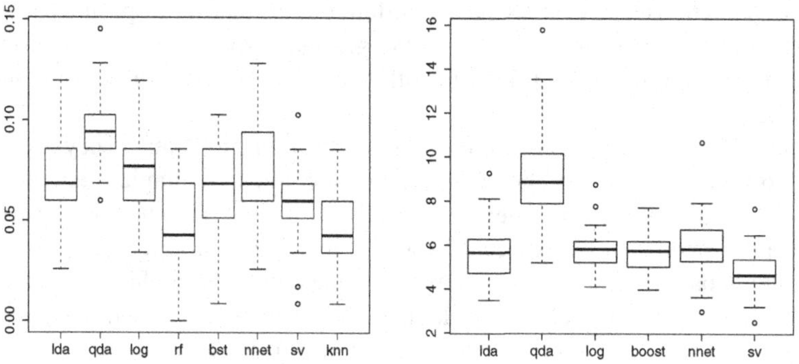

FIGURE 15.10: Misclassification rate (left) and quadratic score (right) in test samples for two classes of glass identification data.

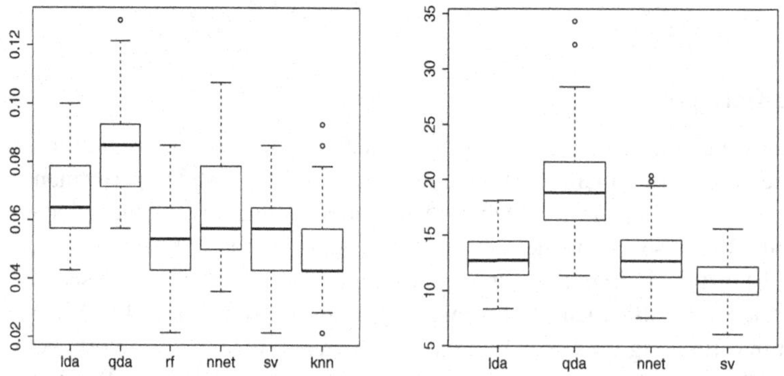

FIGURE 15.11: Misclassification rate (left) and quadratic score (right) in test samples for three classes of glass identification data.

(log) and boosting (bst) for the logistic model with linear predictor are comparable to linear discriminant analysis. The non-parametric methods, support vector machines (with Gaussian kernel), random forests, and nearest neigbors ($K = 5$) show the best performance in terms of misclassification. The support vector machine has the best performance in terms of squared error. □

Example 15.5: DLBCL Data

For the illustration of high-dimensional classification problems we will use the DLBCL data available from the website http://www.gems-system.org/; see Shipp et al. (2002). Diffuse large B-cell lymphoma (DLBCL) is a lymphoid malignancy that is curable in less than 50% of patients. As predictors the expression of 6,817 genes in diagnostic tumor specimens from patients under treatment is available. The prediction target is to identify cured versus fatal or refractory disease. Simple procedures, like linear or quadratic discriminant analysis, fail in high-dimensional datasets like this. Therefore, regularized estimates and methods for high-dimensional classifications were applied by use of the Bioconductor package *CMA* (Slawski et al. (2008)) and *rda* for regularized discriminant analysis. Figure 15.12 shows the misclassification error for 25 random splits into training and test data. The methods used were lasso (las), ridge, elastic net (en), regularized discriminant analysis with covariance matrix (rda), regularized discriminant analysis with correlation matrix (rdak), boosting (bst), nearest neighbors (knn), and random forests (rf). For this dataset interaction seems to be less important; knn and random forests, which are able to model interactions, perform worse than regularized discriminant analysis and ridge regression. Classification procedures that select variables like elastic net and boosting show a slightly worse performance but are helpful to identify the genes that have discriminatory power. □

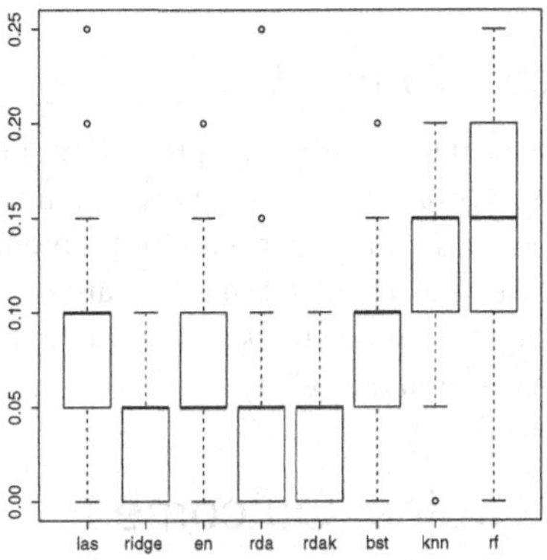

FIGURE 15.12: Misclassification rate in test sample for DLBCL data.

15.8 Variable Selection in Classification

In particular in high-dimensional settings, for example, in microarray-based prediction, where simple classification rules fail, the selection of predictors is an important task. In microarray data the focus is often on the selection of variables, which in that case correspond to genes. When one wants to identify the role of genes in distinguishing between diseases, prediction accuracy is often of minor importance. But careful selection of predictors is to be recommended not only in the $p > n$ case; for moderate numbers of predictors, too, the selection of relevant variables tends to reduce noise and therefore improves prediction accuracy.

Kohavi and John (1998) distinguish between the *wrapper* and the *filter approaches* for selecting predictors. Algorithms from the first class judge the usefulness of predictors for classification based on the algorithm used for classification, whereas in the filter approach predictors are judged irrespectively of the classification algorithm. This means that the first variables are selected and then the selected variables are used for prediction.

Preselection of Predictors

In applications with some 10,000 variables it is common to do a preliminary selection of predictors, in order to reduce the predictors to a number that can be handled by the classification procedure. This filter approach is standard procedure in high-dimensional gene expression data. Most procedures are univariate, which means that the predictive strength of each variable is determined by using that variable only. Variables are ranked according to some criterion and the best variables on the list are used in further analysis.

Several criteria have been proposed for continuous predictors. Dudoit et al. (2002) used the ratio of between-group to within-group sums of squares,

$$\frac{BSS_j}{WSS_j} = \frac{\sum_{i=1}^{n}\sum_{r=1}^{k} I(Y_i = r)(\bar{x}_{rj} - \bar{x}_{.j})}{\sum_{i=1}^{n}\sum_{r=1}^{k} I(Y_i = r)(x_{rj} - \bar{x}_{rj})},$$

where $\bar{x}_{.j}$ denotes the average of variable j across all samples and $\bar{x}_{rj}$ denotes the average of variable j across samples belonging to class r. Alternative criteria are the t-statistic and regularized variants, Wilcoxon's rank sum statistic, the Mann-Whitney statistic, and heuristic signal-to-noise ratios (Golub et al., 1999b). For an overview on non-parametric criteria used in gene expression see Troyanskaya et al. (2002). For an overview on multivariate selection criteria for microarray-based classifiers see Boulesteix et al. ().

Classifier-Based Selection of Predictors

Some classification methods have a built-in selection procedure. For example, trees and random forests select predictors by construction. For other classification methods, like logistic discrimination, one has to decide which predictors contribute to the prediction. For many procedures stepwise procedures based on test statistics (used in a forward or backward manner) have been proposed (see Section 6.1 for stepwise procedures). Better results are often obtained by using regularization methods like penalization (see Chapter 6).

15.9 Prediction of Ordinal Outcomes

When the categorical responses are ordinal the usual 0-1 loss, which is connected to the minimization of the misclassification error, is not appropriate because misclassification into adjacent categories should be considered more acceptable than misclassification into more remote categories. Loss functions should take the ordinal nature of the response into account. When scores $s_1, \dots, s_k$ can be assigned to the response categories $1, \dots, k$ the performance of classification rules can be evaluated by using loss functions for metric responses like the squared error in the form

$$L(Y, \hat{Y}) = (s_Y - s_{\hat{Y}})^2,$$

where Y denotes the true class and $\hat{Y}$ the estimated class. The simple score $s_r = r$ yields $L(Y, \hat{Y}) = (Y - \hat{Y})^2$. Then $\mathrm{E}(Y|\boldsymbol{x})$ would be the best approximation to the response. However, it is obvious that by computing the mean one uses more than the ordinal scale level; the responses Y are treated as metric, interval-scaled responses. Therefore, if no scores are available, the simple scores $s_r = r$ cannot be considered as appropriate for the ordinal scale level.

Using misclassification errors and avoiding assigned scores does not solve the problem. A simple example shows that misclassification cannot be the main objective when classes are ordered. If one has ten categories with probabilities given by $P(Y = 1|\boldsymbol{x}) = 0.28$, $P(Y = 8|\boldsymbol{x}) = P(Y = 9|\boldsymbol{x}) = P(Y = 10|\boldsymbol{x}) = 0.24$, misclassification is minimized by the Bayes rule $\hat{Y} = 1$. But the probability of a response in categories 8, 9 and 10 sums up to 0.72. Of course, for unordered categories, adding of probabilities over classes would make no sense. This is different for ordered classes. For ordered classes it is obvious that $\hat{Y} = 1$ is a bad allocation rule; any value from 8, 9 and 10 would be more appropriate because the probability is concentrated in these categories. What may be learned from this simple example is that classification should not be based on the Bayes rule if the categories are ordered.

A more natural choice that takes the ordering of classes into account is the median of the distribution of Y over the ordered categories $1, \dots, k$. In the example the median is class 8, which is close to the concentration of probability. In general, the median is strongly connected to the L_1-norm $L(Y, \hat{Y}) = |Y - \hat{Y}|$. The minimizer

$$r_0 = \text{argmin}_a \, \mathrm{E}\, L(Y, a) = \text{argmin}_a \sum_{r=1}^{k} \pi_r |r - a|$$

yields the median of the distribution of Y, which may be given as

$$r_0 = \text{med}_Y = \text{argmin}_r \{\pi(r) | \pi(r) \geq 0.5\},$$

where $\pi(r)$ denotes the cumulative probability $\pi_1 + \dots + \pi_r$.

15.9.1 Ordinal Response Models

A simple way of utilizing the ordinal nature of the response is to use a parametric model for ordinal responses (see Chapter 9). A candidate is the cumulative-type model, which has the form $P(Y \leq r|\boldsymbol{x}) = F(\gamma_{0r} + \boldsymbol{x}^T\boldsymbol{\gamma})$, where F is a fixed distribution function. The probabilities for the response categories $P(Y = r|\boldsymbol{x}) = F(\gamma_{0r} + \boldsymbol{x}^T\boldsymbol{\gamma}) - F(\gamma_{0,r-1} + \boldsymbol{x}^T\boldsymbol{\gamma})$ could be used to derive the Bayes rule. Better rules that take the ordering of categories into account can be derived from the underlying metric response. As shown in Chapter 9, the cumulative model can be derived from the assumption of an underlying latent variable that follows the regression model $\tilde{Y} = -\boldsymbol{x}^T\boldsymbol{\gamma} + \varepsilon$, where ε is a noise variable with a continuous distribution function F. Then the responses are modeled by assuming

$$Y = r \Leftrightarrow \gamma_{0,r-1} < \tilde{Y} \leq \gamma_{0r}.$$

When $\mathrm{E}(\varepsilon) = 0$ a simple classification rule that is obtained after replacing the parameters by estimates is

$$\hat{Y} = r \Leftrightarrow \hat{\gamma}_{0,r-1} < -\boldsymbol{x}^T\hat{\boldsymbol{\gamma}} \leq \hat{\gamma}_{0r}.$$

The rule implicitly uses the prediction of the latent variable $-\boldsymbol{x}^T\hat{\boldsymbol{\gamma}}$ to derive where on the latent scale the response is located. Although it does not necessarily yield the median, it will frequently be close to it. If $\mathrm{E}(\varepsilon) = 0$ does not hold, one has to include the mean into the predicted latent variable by using the estimate $-\boldsymbol{x}^T\hat{\boldsymbol{\gamma}} + E(\varepsilon)$. For the logistic (and probit) cumulative model $E(\varepsilon) = 0$ holds, but not for models like the proportional hazards model. The use of the logit model has been propagated by Anderson and Phillips (1981). Of course alternative ordinal regression models like the sequential model can be be used to find estimates of probabilities that then are used to find the estimated median.

15.9.2 Aggregation over Binary Splits

To avoid the use of artificially assigned scores but still use the ordering of responses one can combine predictions that are obtained for splits within the categories $1, \dots, k$. Let the response be split at category r into the sets $\{Y \leq r\}$ and $\{Y > r\}$, $r \in \{1, \dots, q = k-1\}$. The corresponding dichotomous variables are

$$y(r) = \begin{cases} 1 & Y \leq r \\ 0 & Y > r, \end{cases}$$

with the underlying probabilities given by $\pi(r) = \pi_1 + \dots + \pi_r$. For a *fixed* split at category r the Bayes rule is given by

$$\hat{y}(r) = 1 \quad \Longleftrightarrow \quad \pi(r) \geq 1 - \pi(r).$$

The assignments resulting from the Bayes rule applied to fixed splits can be combined into one decision: Select the category that is preferred for all splits. The resulting rule can be seen as a majority vote over all splits. For the split at category r let the indicator functions be given by

$$y_s^{(r)} = 1, \quad s = 1, \dots, r, \quad y_s^{(r)} = 0, \quad s = r+1, \dots, k,$$

if $\pi(r) \geq 0.5$, which corresponds to prediction into categories $\{1, \dots, r\}$, and

$$y_s^{(r)} = 0, \quad s = 1, \dots, r, \quad y_s^{(r)} = 1, \quad s = r+1, \dots, k,$$

if $\pi(r) < 0.5$, which corresponds to prediction into categories $\{r+1, \dots, k\}$. Then the majority vote combines the decisions over all splits in the allocation rule

$$\hat{Y} = r_0, \quad \text{if} \quad r_0 = \text{argmax}_s\{y_s^{(1)} + \dots + y_s^{(k-1)}\}. \tag{15.18}$$

In the ideal case, when the probabilities are known, the assignment resulting from the majority vote is into the median $r_0 = \text{med}_Y = \text{argmin}_r\{\pi(r) | \pi(r) \geq 0.5\}$.

However, for estimated classification rules derived from dichotomization, the majority rule is an ensemble method, which only approximates the underlying median. Then the probability $\pi(r)$ is replaced by an estimate $\hat{\pi}(r)$, which is obtained by fitting a binary response model to the data $(y_i(r), \boldsymbol{x}_i)$, $i = 1, \dots, n$. Simple parametric models like the logit model can be used as binary response models. But non-parametric approaches like binary trees, random forests, and nearest neighborhood methods may also be applied. Non-parametric methods do not necessarily use the Bayes rule based on estimates $\hat{\pi}(r)$ but will directly yield assignments into the split classes $\{1, \dots, r\}$, $\{r+1, \dots, k\}$, which yields the indicator functions $\hat{y}_s^{(r)}$. The corresponding ensemble is formed by the majority vote,

$$\hat{Y} = r_0, \quad \text{if} \quad r_0 = \text{argmax}_s\{\hat{y}_s^{(1)} + \dots + \hat{y}_s^{(k-1)}\}. \tag{15.19}$$

When applying binary models one relies on the majority vote over binary splits. For fixed splits, the allocation may be derived from the Bayes rule based on the estimated probabilities $\hat{\pi}(r)$. Since the estimated probabilities $\hat{\pi}(r)$ are based on the split at category r without using that the original response was in k categories, the resulting estimates do not have to fulfill the natural order restriction $\hat{\pi}(r) \leq \hat{\pi}(r+1)$. Therefore, there is no estimated vector of response probabilities for all categories and one does not necessarily assign into the class that corresponds to the (estimated) median.

Performance may improve by weighting the classification rules that result from dichotomization by taking into account where the split was made. For a fixed split at category r, one can use

$$\hat{y}_s^{(r)} = 1/r, \quad s = 1, \dots, r, \quad \hat{y}_s^{(r)} = 0, \quad s = r+1, \dots, k, \tag{15.20}$$

if the prediction is into categories $\{1, \dots, r\}$ and

$$y_s^{(r)} = 0, \quad s = 1, \dots, r, \quad y_s^{(r)} = 1/(k-r), \quad s = r+1, \dots, k, \tag{15.21}$$

if the prediction is into categories $\{r+1, \dots, k\}$. The weighting smoothes the 0-1-decision over the categories that are involved in the dichotomization. If estimates $\hat{\pi}(r)$ are available, $y_s^{(r)}$ in (15.20) and (15.21) can be replaced by the estimated probabilities

$$\hat{\pi}_s^{(r)} = \hat{\pi}(r)/r, \quad s = 1, \dots, r, \quad \hat{\pi}_s^{(r)} = (1 - \hat{\pi}(r))/(k-r), \quad s = r+1, \dots, k, \tag{15.22}$$

yielding the majority vote $r_0 = \text{argmax}_s\{\pi_s^{(1)} + \dots + \pi_s^{(k-1)}\}$ and also an ensemble-based estimate of probabilities $\hat{\pi}_s = \sum_{r=1}^{k-1} \hat{\pi}_s^{(r)}$.

An alternative scheme to obtain estimated probabilities uses the estimates $\hat{\pi}(1), \dots, \hat{\pi}(k-1)$ obtained for the $k-1$ splits. After transforming them such that $\bar{\pi}(1) \le \dots \le \bar{\pi}(k-1)$, for example, by monotone regression tools, one obtains estimated probabilities $\hat{\pi}_s = \bar{\pi}(s) - \bar{\pi}(s-1)$, which can be directly used to predict and to compute the precision of the estimates. We will refer to it as the monotonized ensemble method.

As in unordered regression parametric regression models have the advantage that simple parameter estimates show how the classification rule is obtained. However, parametric approaches are restricted to low dimensions. In terms of prediction, power non-parametric approaches often prevail over parametric models. The dichotomization approach has the advantage that all binary classifiers are potential candidates. Boosting methods, for example, a boosted logit model that automatically selects predictors, can also be used. The latter has the advantage that it also applies to high-dimensional data because of its built-in selection procedure. In most applications the selected variables will vary across splits since the discriminatory power of variables usually varies across splits.

For illustration several methods were applied to real datasets. We found that ensembles based on (15.22) or the monotonized version did not differ strongly in terms of misclassification error. When performance is measured by taking the estimated probabilities into account the monotonized version frequently shows better performance. Therefore, in the following only the results for the latter ensemble method are given. The classification methods considered comprise parametric and non-parametric approaches. With the understanding that the ending e in the notation refers to an ensemble that uses monotonized probabilities, the methods were CUM: cumulative logit model (function *polr* from package *MASS*); GLMe: binary logit model ensemble (function *glm*); LASSOe: lasso ensemble (function *penalized* from package *penalized*); BOOSTe: logit boost ensemble (function *glmboost* from package *mboost*); WNN: weighted nearest neighbors (K = 7, R function kknn); WNNe: weighted nearest neighbors ensemble; LDA: linear discriminant analysis (function *lda* from package *MASS*); LDAe: linear discriminant analysis ensemble; RF: random forests (R function *randomForest*); and RFe: random forests ensemble.

The datasets were repeatedly split into learning and test set, and classification based on the median was applied to the learning set and accuracy of prediction evaluated in the test set. An important measure of performance is misclassification error. Since responses are ordered, one can also consider the distance between the prediction and the true value, although one implicitly uses a higher scale level than ordinal. A measure that only uses the ordinal scale level is the ranked probability score given in (15.3). The major advantage of the latter measure is that

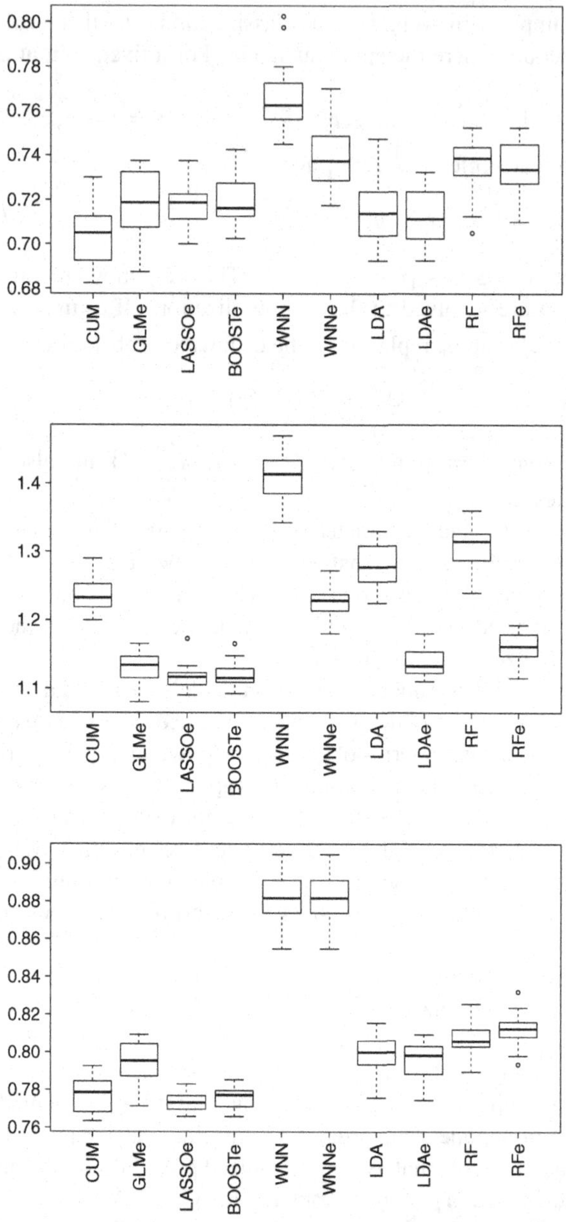

FIGURE 15.13: Misclassifications, absolute differences, and ranked probability scores in test data for the satisfaction data.

it not only measures the correctness of the prediction but, by using the distance between the prediction and the estimated distribution over categories, it also takes the distinctness of the prediction into account.

Example 15.6: Satisfaction with Life

The data stem from a household panel (German Socio-Economic Panel, 2007). The response is the satisfaction with life in six ordered categories; 17 predictors were used, 4 of which were metrically scaled. Twenty splits into learning and test data were performed, and the sample size of the learning data was 400. Figure 15.13 shows the misclassification rates, the absolute differences between prediction and true value, and the ranked probability scores. It is seen that the absolute difference distinguishes stronger between classification methods than the misclassification rate. Classifiers that are not designed for ordinal responses, like nearest neighbors, linear discriminant analysis, and random forests, improve strongly in terms of the absolute differences if the ensemble method is used. Interaction seems not to be of major importance because linear approaches perform quite well. Approaches that include variable selection, like the lasso and the boosted logit models, do not outperform other procedures in terms of misclassification but show better performance in absolute differences. In terms of the distinctness of the predicted distribution, which is measured by the ranked probability score, nearest neighbor approaches in particular show poor performance. By construction, nearest neighbors focus on exact predictions but yield inferior distributions, and therefore poor performance is to be expected. □

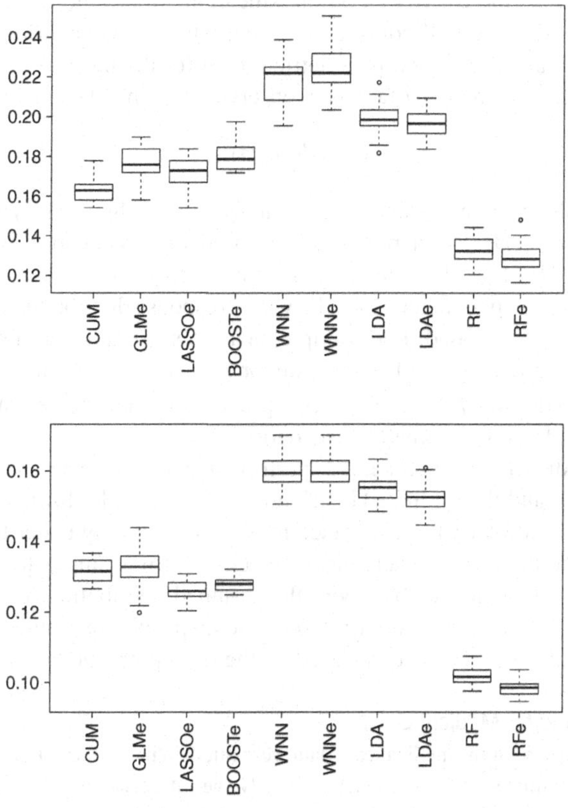

FIGURE 15.14: Absolute differences and ranked probability scores in test data for housing data.

Example 15.7: Housing Data

In the housing data, one wants to predict the price by using continuous and categorical predictors. The dataset is available from the UCI repository (http://archive.ics.uci.edu/ml). Four ordered categories of prices are considered with predictors CRIM: per capita crime rate by town, ZN: proportion of residential land zoned for lots over 25,000 sq. ft., INDUS: proportion of non-retail business acres per town, CHAS: Charles River dummy variable, NOX: nitric oxides concentration (parts per 10 million), RM: average number of rooms per dwelling, AGE: proportion of owner-occupied units built prior to 1940, DIS: weighted distances to five Boston employment centres, RAD: index of accessibility to radial highways, TAX: full-value property-tax rate per 10, 000, PTRATIO: pupil-teacher ratio by town 12, B: $1000(Bk - 0.63)^2$, where Bk is the proportion of blacks by town, and LSTAT: percent lower status of the population. The dataset, which comprises 506 observations, was split such that the training data have 400 observations. Figure 15.13 shows the performance of the classifiers. We omit the error rate because the results are very similar to the results for the absolute differences. For this dataset random forests are dominating, which may be due to interaction effects. The transition to ensemble methods shows only slight improvement, and for nearest neighbors there is even a small increase in terms of absolute error. □

15.10 Model-Based Prediction

In the preceding sections the focus was on classification, where the response is in one of k categories. In this section we will consider briefly the use of parametric models for general responses. When a parametric or non-parametric model for the mean $\mu = h(\boldsymbol{x}; \boldsymbol{\beta})$ is fit, prediction of the (univariate) response y_0 at a new observation $\boldsymbol{x}$ may be constructed as

$$\hat{y}_0 = h(\boldsymbol{x}; \hat{\boldsymbol{\beta}}).$$

When $\hat{y}_0$ is not a valid response value, that is, not integer-valued, one can select the value $\hat{y}_0$ that is closest to $h(\boldsymbol{x}; \hat{\boldsymbol{\beta}})$. That rule has been used in classification but also applies for count data, where $y \in \{0, 1, \dots\}$. If one assumes for the response y given $\boldsymbol{x}$ a specific density $f(y; \gamma(\boldsymbol{x}))$ that depends on parameter $\gamma(\boldsymbol{x})$, the whole response distribution is determined when a parameter estimate $\hat{\gamma}(\boldsymbol{x})$ is found. For example, in a generalized linear model, the parameters that determine $\gamma(\boldsymbol{x})$ are $\boldsymbol{\beta}$ from the linear predictor $\eta = \boldsymbol{x}^T\boldsymbol{\beta}$ and a dispersion parameter ϕ, yielding the estimated density $f(y; \hat{\boldsymbol{\beta}}, \hat{\phi})$. Then a prediction $\hat{y}_0$ can be determined by the mean, the median, or the mode of the estimated distribution.

When a specific distribution is assumed, performance can be measured by distance measures between the true and the estimated distributions, for example, Kullback-Leibler loss and likelihood-based loss. However, these distance measures work only under the assumption that the underlying distribution is true. Then, they can be used, for example, to evaluate the impact of predictors. If one has major doubts about the assumed distribution, it is preferable to use measures that can be interpreted without reference to a specific distribution, for example, the squared loss, the L_1-norm, and measures based on the ranked probability score.

Example 15.8: Demand for Medical Care

In Chapter 7, several models for the medical care data were fitted, in particular, a Poisson model, a negative binomial model, a zero-inflated Poisson model with only the intercept for the mixture (zero1), a zero-inflated Poisson model with predictors in the mixture (zero2), a hurdle model with only the intercept for the mixture (hurdle1), and a hurdle model with predictors in the mixture (hurdle2). Here, only males with responses smaller than 30 are included. The dataset was split various times with 600 men in the learning sample; prediction was investigated in the test sample. Figure 15.15 shows the results in terms of absolute values of differences between the medians of the estimated distributions and the actual observation, and the ranked probability score. It is seen that the negative binomial model dominates the other procedures in terms of the absolute difference; the zero inflation and hurdle models improve when covariates that determine the mixture are included. In terms of the ranked probability score, which also includes the concentration of the predicted distribution, the negative binomial model dominates the Poisson model more distinctly, but now the hurdle models are the best performers. In the whole dataset the negative binomial model has the smallest AIC (value of 9291) followed by the zero inflation model and the hurdle model with covariates (value of 11 014). □

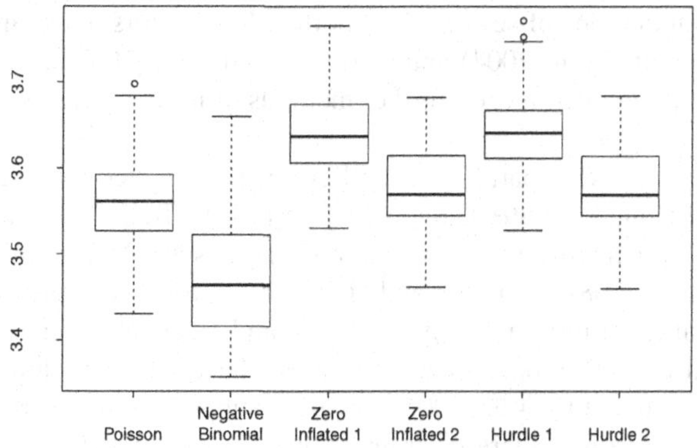

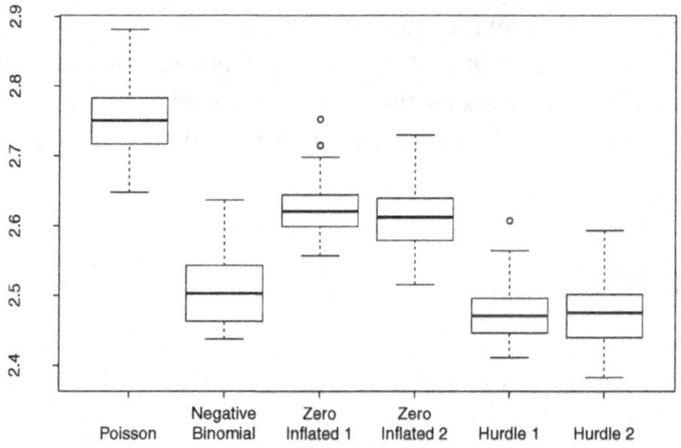

FIGURE 15.15: Absolute differences and ranked probability scores in test data for medical care data.

15.11 Further Reading

Books. Statistical approaches to classification are well covered in McLachlan (1992) and Ripley (1996). Bishop (2006) is guided by pattern recognition and machine learning, Hastie et al. (2009) considered regression and classification from a general statistical learning viewpoint.

Surveys on High-Dimensional Classifications. Classification methods in high-dimensional settings were compared by Dudoit et al. (2002), Romualdi et al. (2003), Lee et al. (2005), Statnikov et al. (2005), and Diaz-Uriarte and de Andres (2006b). An overview on the evaluation of microarray-based classifiers were given by Boulesteix et al. ().

Nearest Neighborhood Methods. Neighborhood methods with weighted neighbors were proposed by Morin and Raeside (1981), Parthasarthy and Chatterji (1990), and Silverman and Jones (1989). Paik and Yang (2004) introduced a method called adaptive classification by mixing (ACM). They used combinations of many K-NN classifiers with different values for K and different subsets of predictors to improve the results of one single K-NN prediction. Friedman (1994) proposed local flexible weights for the predictors to account for their local relevance that is estimated by recursive partitioning techniques. Alternative approaches which also choose the metric adaptively were given by Hastie and Tibshirani (1996), and Domeniconi et al. (2002). A connection between these adaptively chosen metrics and random forests was derived by Lin and Jeon (2006). Combinations of several neighborhood estimators in an ensemble were considered by Domeniconi and Yan (2004) and Gertheiss and Tutz (2009a). More detailed large sample results as well as an overview on further methods of nearest neighborhoods were given in Ripley (1996).

Ordinal Prediction The use of parametric models was investigated by Anderson and Phillips (1981), Campbell and Donner (1989), Campbell et al. (1991), Bell (1992), and Rudolfer et al. (1995). Comparisons of methods are also found in Demoraes and Dunsmore (1995) and Coste et al. (1997). Piccarreta (2008) and Frank and Hall (2001) proposed using classification trees. Ensemble methods came up more recently; see Tutz and Hechenbichler (2005) for boosting approaches in ordinal classification. Some approaches have been developed in the machine learning community; Chu and Keerthi (2005) proposed support vector classification tools, and Herbrich et al. (1999) investigated large margin boundaries for ordinal regression.

Classifiers Using R. The package *MASS* contains the functions *lda* and *qda* for linear and quadratic discriminations, *party* provides random forests, and *mboost* boosting methods; support vector algorithms are available in *e1071*. Feed-forward networks can be fitted by use of *nnet*, regularized discriminant analysis by use of *rda*. Most of the methods considered in this chapter are also accessible in an R Bioconductor package called *CMA* (Slawski, Daumer, and Boulesteix, 2008).

15.12 Exercises

15.1 Let the optimal prediction be defined by $\hat{y}_{opt} = argmin_c L(y, c)$ for the loss function $L(.,.)$. Show that one obtains as minimizers

(a) $\hat{y}_{opt} = \mathrm{E}(y|\boldsymbol{x})$ for the quadratic loss,

(b) $\hat{y}_{opt} = med(y|\boldsymbol{x})$ for the L_1-norm,

(c) the p-quantile $\hat{y}_{opt} = inf\{y_0|P(y \leq y_0|\boldsymbol{x}) \geq p\}$ for the loss function $L_p(y, \hat{y}) = |y - \hat{y}| + (2p-1)(y - \hat{y})$, with $p \in (0, 1)$.

15.2 Let $\hat{\boldsymbol{\pi}}^T = (\hat{\pi}_1, \dots, \hat{\pi}_k)$ denote an estimated vector of probabilities and $\boldsymbol{y}^T = (y_1, \dots, y_k)$, $y_r \in \{0, 1\}$ a new observation. Show that for the actual prediction error of the Brier score

$$\mathrm{E}(L_2(\boldsymbol{y}, \hat{\boldsymbol{\pi}})) = \mathrm{E}\left\{\sum_{r=1}^{k} (y_r - \hat{\pi}_r)^2\right\} = L_2(\boldsymbol{\pi}, \hat{\boldsymbol{\pi}}) + \sum_{r=1}^{k} \mathrm{var}(y_r)$$

holds.

15.3 For fixed classifier δ the prediction error for randomly drawn $(Y, \boldsymbol{x})$ with 0-1 loss is defined as $\varepsilon = \mathrm{E}\ L_{01}(Y, \delta(\boldsymbol{x}))$. Derive the connection to the error rate of misclassification conditional on $\boldsymbol{x}$ defined as $\varepsilon(\boldsymbol{x}) = \mathrm{E}_{y|\boldsymbol{x}}\ L_{01}(Y, \delta(\boldsymbol{x}))$ and the connection to the individual errors $\varepsilon_{rs} = p(\delta(\boldsymbol{x}) = s|Y = r)$.

15.4 In Example 15.3 the probabilities are given by $P(x = 1|Y = 1) = 0.95$, $P(x = 1|Y = 2) = 0.05$, where $Y = 1$ refers to user, $Y = 2$ to non-user, and $x = 1$ to positive test results and $x = 0$ to negative test results.

(a) Now let the prior be given by $P(1) = 0.05$. Compute the Bayes rule, the conditional error rates $\varepsilon(x)$, ε_{12}, ε_{21}, and the global error rate $\varepsilon(x)$. Discuss the accuracy of the predictions in terms of overall accuracy and conditional accuracy.

(b) What are the results for prior probability $P(1) < 0.05$?

15.5 There is a strong connection between the logit model and normally distributed predictors.

(a) Show that under the assumption of normally distributed predictors, $\boldsymbol{x}|Y = r \sim N(\boldsymbol{\mu}_r, \boldsymbol{\Sigma})$, one obtains for the posterior probabilities the linear multinomial logistic model for k classes $P(r|\boldsymbol{x}) = \exp(\beta_{0r} + \boldsymbol{x}^T\boldsymbol{\beta}_r)/1 + \sum_{j=1}^{k-1} \exp(\beta_{0j} + \boldsymbol{x}^T\boldsymbol{\beta}_j)$. Give the parameters $\beta_{0r}, \boldsymbol{\beta}_r$ as functions of $\boldsymbol{\mu}_r, \boldsymbol{\Sigma}$.

(b) Derive the logistic model for $P(r|\boldsymbol{x})$ that holds if the predictors follow $\boldsymbol{x}|Y = r \sim N(\boldsymbol{\mu}_r, \boldsymbol{\Sigma}_r)$.

15.6 Fisher's discriminant analysis for two classes uses the projections $y_{ri} = \boldsymbol{a}^T\boldsymbol{x}_{ri}$ and maximizes the criterion $Q(\boldsymbol{a}) = (\bar{y}_1 - \bar{y}_2)^2/(w_1^2 + w_2^2)$ with $\bar{y}_r = \sum_{i=1}^{n_r} y_{ri}/n_r$, $w_r^2 = \sum_{i=1}^{n_r}(y_{ri} - \bar{y}_r)^2$.

(a) Show that maximization of $Q(\boldsymbol{a})$ yields a solution that is proportional to $\boldsymbol{a} = \boldsymbol{S}^{-1}(\bar{\boldsymbol{x}}_1 - \bar{\boldsymbol{x}}_2)$.

(b) Show that the criterion for two classes is equivalent to the general criterion $Q(\boldsymbol{a}) = \boldsymbol{a}^T\boldsymbol{B}\boldsymbol{a}/\boldsymbol{a}^T\boldsymbol{W}\boldsymbol{a}$, where $\boldsymbol{B} = \sum_{r=1}^{k} n_r(\bar{\boldsymbol{x}}_r - \bar{\boldsymbol{x}})(\bar{\boldsymbol{x}}_r - \bar{\boldsymbol{x}})^T$, $\boldsymbol{W} = \sum_{r=1}^{k}\sum_{i=1}^{n_r}(\boldsymbol{x}_{ri} - \bar{\boldsymbol{x}}_r)(\boldsymbol{x}_{ri} - \bar{\boldsymbol{x}}_r)^T$.

15.7 The Fisher criterion that is to be maximized in the general case of k classes has the form $Q(\boldsymbol{a}) = \boldsymbol{a}^T\boldsymbol{B}\boldsymbol{a}/\boldsymbol{a}^T\boldsymbol{W}\boldsymbol{a}$, under the side constraint $||\boldsymbol{a}^T\boldsymbol{W}\boldsymbol{a}|| = 1$.

(a) Why is a side constraint needed? What are alternative choices for the side constraint and how does the choice affect the maximization procedure?

(b) Use that there exists an invertible matrix $\boldsymbol{H}$ such that $\boldsymbol{W} = \boldsymbol{H}\boldsymbol{H}^T$, $\boldsymbol{B} = \boldsymbol{H}\boldsymbol{\Omega}\boldsymbol{H}^T$, with $\boldsymbol{\Omega} = \text{diag}(\lambda_1, \ldots \lambda_p)$, $\lambda_1 \geq \ldots \geq \lambda_p \geq 0$. Show that the first $m = \min\{k-1, p\}$ columns of $(\boldsymbol{H}^T)^{-1}$ are solutions of the generalized eigenvalue problem $Q(\boldsymbol{a}_i) = \max Q(\boldsymbol{a})$, where $\boldsymbol{a}$ is restricted by $\boldsymbol{a}^T\boldsymbol{W}\boldsymbol{a} = 1$, $\boldsymbol{a}_j^T\boldsymbol{W}\boldsymbol{a} = 0$, $j = 1, \ldots, i-1$ (for the existence of the matrices $\boldsymbol{H}$ and $\boldsymbol{\Omega}$ see Harville, 1997, who considers the generalized eigenvalue problem).

15.8 ROC curves are plots of (FAR(c), HR(c)) for varying cut-off values c, where FAR(c), HR(c) denote the hit rate $P(s(\boldsymbol{x}) > c|y = 1)$ and the false alarm rate $P(s(\boldsymbol{x}) > c|y = 0)$. Let the distribution of $\boldsymbol{x}$ in classes 1 and 2 be given.

(a) Draw the ROC curve if the distribution of $\boldsymbol{x}$ is the same in both classes.

(b) Draw the ROC curve if the distributions of $\boldsymbol{x}$ in classes are not overlapping.

(c) Compute the hit and false alarm rates in terms of the normal distribution function if the distribution of x in class 0 is N(0,1) and in class 1 N(1,1). Draw the approximate curve by computing some values.

15.9 The optimal Bayes classifier with loss function is given by equation (15.10).

(a) Show that the Bayes classifier with symmetrical loss function $L(r, s) = 0$ if $r = s$, $L(r, s) = c$, with constant c if $r \neq s$, yields the common Bayes classifier that minimizes the misclassification error.

(b) Show that the Bayes classifier with loss function $L(r, s) = 0$ if $r = s$, $L(r, s) = c/P(r)$ if $r \neq s$, yields the maximum likelihood rule.

15.10 Show that $\max_{\beta_0, \boldsymbol{\beta}, ||\boldsymbol{\beta}||=1} M$ subject to $y_i(\beta_0 + \boldsymbol{x}_i^T\boldsymbol{\beta}) \geq M$, $i = 1, \ldots, n$, is equivalent to $\min_{\beta_0, \boldsymbol{\beta}} ||\boldsymbol{\beta}||$ subject to $y_i(\beta_0 + \boldsymbol{x}_i^T\boldsymbol{\beta}) \geq 1$, $i = 1, \ldots, n$. Use that one can get rid of the constraint $||\boldsymbol{\beta}|| = 1$ by considering the constraint $y_i(\beta_0 + \boldsymbol{x}_i^T\boldsymbol{\beta}) \geq M||\boldsymbol{\beta}||$.

15.11 Derive for binary observations with $y \in \{-1, 1\}$ the optimal prediction under hinge loss $L_h(y, \hat{y}) = 1 - y\hat{y}$ by deriving the value c that minimizes $EL_h(y, c)$ over $c \in [-1, 1]$.

15.12 The R package *datasets* contains Anderson's well-known iris data. It contains four measurements on three groups of iris.

(a) Find the classification rules by using classifiers like linear, quadratic, logistic discrimination, nearest neighborhood methods, trees, and random forests. Compute the misclassification rate in the learning sample.

(b) Split the dataset several times and use 80% of the data as the learning sample and the rest as test data. Compute the average test error. Which classifiers perform best? Compare the test errors to the test errors from (a).

15.13 In credit scoring one wants to identify risk clients. The dataset credit contains a sample of 1000 consumer credit scores collected at a German bank with 20 predictors. The dataset is available from the UCI Machine Learning Repository or the R package *Fahrmeir*.

(a) Find the classification rules by using classifiers like linear, quadratic, logistic discrimination, nearest neighborhood methods, trees, and random forests. Investigate which predictors are useful and determine the misclassification rate in the learning sample.

(b) Split the dataset several times and use 80% of the data as the learning sample and the rest as test data. Compute the average test error. Which classifiers perform best? Compare the test errors to the test errors from (a).

Appendix A

Distributions

A.1 Discrete Distributions

A.1.1 Binomial Distribution

The variable Y is binomially distributed, $Y \sim B(n, \pi)$, if the probability mass function is given by

$$f(y) = \begin{cases} \binom{n}{y}\pi^y(1-\pi)^{n-y} & y \in \{0, 1, \ldots, n\} \\ 0 & \text{otherwise.} \end{cases}$$

The parameters are $n \in \mathbb{N}, \pi \in [0, 1]$. One obtains $E(Y) = n\pi$, $\text{var}(Y) = n\pi(1-\pi)$.

A.1.2 Poisson Distribution

A variable Y follows a Poisson distribution, $Y \sim P(\lambda)$, if the probability mass function is given by

$$f(y) = \begin{cases} \frac{y^\alpha}{y!}e^{-\lambda} & y = 0, 1, 2, \ldots \\ 0 & \text{otherwise.} \end{cases}$$

The parameter $\lambda > 0$ determines $E(Y) = \lambda$, $\text{var}(Y) = \lambda$.

A.1.3 Negative Binomial Distribution

The variable $Y \in \{0, 1, \ldots\}$ follows a negative binomial distribution, $NB(\nu, \mu)$, if the mass function is given by

$$f(y) = \frac{\Gamma(y+\nu)}{\Gamma(y+1)\Gamma(\nu)}\left(\frac{\mu}{\mu+\nu}\right)^y\left(\frac{\nu}{\mu+\nu}\right)^\nu, \quad y = 0, 1, \ldots$$

where $\nu, \mu > 0$ and $\Gamma(\nu) = \int_0^\infty t^{\nu-1}e^{-t}dt$. One has

$$E(Y) = \mu, \quad \text{var}(Y) = \mu + \mu^2/\nu.$$

An alternative parameterization uses $\pi = \nu/(\mu+\nu)$, yielding $NB(\nu, \nu(1-\pi)/\pi)$ with

$$f(y) = \frac{\Gamma(y+\nu)}{\Gamma(\nu+1)\Gamma(\nu)}\pi^\nu(1-\pi)^y.$$

The mean and variance are given by

$$E(Y) = \nu\frac{1-\pi}{\pi}, \quad \text{var}(Y) = \nu\frac{1-\pi}{\pi^2} = \mu/\pi.$$

For integer-valued ν's the random variable Y describes the number of trials needed (in addition to ν) until ν successes are observed. The underlying experiment consists of repeated trials, where it is assumed that one trial can result in just two possible outcomes (success or failure), the probability of success, π, is the same on every trial, and the trials are independent. The experiment continues until ν successes are observed.

A.1.4 Hypergeometric Distribution

The variable y follows a hypergeometric distribution, $Y \sim H(n, N, M)$, if the probability mass function is given by

$$f(y) = \begin{cases} \dfrac{\begin{pmatrix} M \\ y \end{pmatrix}\begin{pmatrix} N-M \\ y \end{pmatrix}}{\begin{pmatrix} N \\ n \end{pmatrix}} & \begin{matrix} max\{0, M+n-N\} \leq y \\ \leq min\{M, n\} \end{matrix} \\ 0 & \text{otherwise.} \end{cases}$$

The distribution arises if n marbles are drawn at random from a box that contains N marbles, M of which are "red" and $N - M$ are "white." Then Y counts the number of red marbles and one obtains

$$E(y) = n\frac{M}{N}, \; var(y) = \frac{N-n}{N-1}n(1-\frac{M}{N})\frac{M}{N}.$$

A.1.5 Beta-Binomial Distribution

The variable $Y \in \{0, 1, \ldots, n\}$ follows a beta-binomial distribution, $Y \sim BetaBin(n, \alpha, \beta)$, if the mass function is given by

$$f(y) = \begin{cases} \binom{n}{y}\frac{B(\alpha+y,\beta+n-y)}{B(\alpha,\beta)} & y \in \{0, 1, \ldots, n\} \\ 0 & \text{otherwise,} \end{cases}$$

where $\alpha, \beta > 0$ and $B(\alpha, \beta)$ is the beta function defined as

$$B(\alpha, \beta) = \Gamma(\alpha)\Gamma(\beta)/\Gamma(\alpha+\beta) = \int_0^1 t^{\alpha-1}(1-t)^{\beta-1}dt.$$

One has with $\mu = \alpha/(\alpha+\beta)$, $\delta = 1/(\alpha+\beta+1)$

$$E(Y) = n\mu, \quad \text{var}(Y) = n\mu(1-\mu)[1+(n-1)\delta].$$

As $\delta \to 0$, the beta-binomial distribution converges to the binomial distribution $B(n, \mu)$.

A.1.6 Multinomial Distribution

The vector $\boldsymbol{y}^T = (y_1, \ldots, y_k)$ is multinomially distributed, $\boldsymbol{y} \sim M(n, (\pi_1, \ldots, \pi_k))$, if the probability mass function is given by

$$f(y_1, \ldots, y_k) = \begin{cases} \frac{n!}{y_1! \ldots y_k!} \quad \pi_1^{y_1} \ldots \pi_k^{y_k} & \begin{matrix} y_i \in \{0, \ldots, n\}, \\ \sum_i y_i = n \end{matrix} \\ 0 & \text{otherwise,} \end{cases}$$

where $\boldsymbol{\pi}^T = (\pi_1, \ldots, \pi_k)$ is a probability vector, that is, $\pi_i \in [0, 1]$, $\sum_i \pi_i = 1$.
One has

$$E(y_i) = n\pi_i, \; \text{var}(y_i) = n\pi_i(1-\pi_i),$$
$$\text{cov}(y_i, y_j) = -n\pi_i\pi_j, i \neq j.$$

A.2 Continuous Distributions

A.2.1 Normal Distribution

The normal distribution, $Y \sim N(\mu, \sigma^2)$, has density

$$f(y) = \frac{1}{\sqrt{2\pi}\sigma} \exp\left(-\frac{1}{2}\left(\frac{y-\mu}{\sigma}\right)^2\right)$$

with $E(Y) = \mu$, $\text{var}(Y) = \sigma^2$.

A.2.2 Logistic Distribution

A variable Y has a logistic distribution, $Y \sim \text{Logistic}(\mu, \beta)$, if the density is given by

$$f(y) = \frac{1}{\beta} \frac{\exp(-\frac{y-\mu}{\beta})}{(1+\exp(-\frac{y-\mu}{\beta}))^2}$$

with distribution function

$$F(y) = \frac{1}{1+\exp(-\frac{y-\mu}{\beta})} = \frac{\exp(\frac{y-\mu}{\beta})}{1+\exp(\frac{y-\mu}{\beta})}.$$

One has $E(y) = \mu$, $\text{var}(y) = \beta^2\pi^2/3$.

A.2.3 Gumbel or Maximum Extreme Value Distribution

The Gumbel distribution, $Y \sim \text{Gu}(\alpha, \beta)$, has density

$$f(y) = \frac{1}{\beta} \exp\left(-\frac{y-\alpha}{\beta}\right) \exp\left(-\exp\left(-\frac{y-\alpha}{\beta}\right)\right)$$

and distribution function

$$F(y) = \exp\left(-\exp\left(-\left(\frac{y-\alpha}{\beta}\right)\right)\right),$$

with

$$E(Y) = \alpha + \beta\gamma \quad (\gamma = 0.5772), \quad \text{var}(Y) = \beta^2\pi^2/6.$$

A.2.4 Gompertz or Minimum Extreme Value Distribution

The density of the Gompertz distribution, $Y \sim \text{Go}(\tilde{\alpha}, \beta)$, is given by

$$f(y) = \frac{1}{\beta} \exp\left(\frac{y-\tilde{\alpha}}{\beta}\right) \exp\left(-\exp\left(\frac{y-\tilde{\alpha}}{\beta}\right)\right)$$

with the distribution function

$$F(y) = 1 - \exp\left(-\exp\left(\frac{y-\tilde{\alpha}}{\beta}\right)\right)$$

and

$$\text{E}(y) = \tilde{\alpha} - \beta\alpha, \quad \text{var}(y) = \beta^2\pi^2/6.$$

There is a strong connection to the maximum extreme value distribution since y has a minimum extreme value distribution $\text{Go}(\tilde{\alpha}, \beta)$ if $-y$ has a maximum extreme value distribution Gu (α, β), where $\tilde{\alpha} = -\alpha$.

A.2.5 Exponential Distribution

A random variable Y has an exponential distribution, $Y \sim E(\alpha)$, if the density function is given by

$$f(y) = \begin{cases} 0 & y < 0 \\ \lambda e^{-\lambda y} & y \geq 0, \end{cases}$$

where $\lambda > 0$. One has

$$E(Y) = \frac{1}{\lambda}, \text{ var}(Y) = \frac{1}{\lambda^2}.$$

A.2.6 Gamma-Distribution

A random variable Y is gamma-distributed, $Y \sim \Gamma(\nu, \alpha)$, if it has the density function

$$f(y) = \begin{cases} 0 & y \leq 0 \\ \frac{\alpha^\nu}{\Gamma(\nu)} y^{\nu-1} e^{-\alpha y} & y > 0, \end{cases}$$

where, for $\nu > 0, \Gamma(\nu)$ is defined by

$$\Gamma(\nu) = \int_0^\infty t^{\nu-1} e^{-t} dt.$$

One obtains $E(Y) = \nu/\alpha$, var$(Y) = \nu/\alpha^2$. For $\nu = 1$, the exponential distribution is a special case.

A.2.7 Inverse Gaussian Distribution

The density of the inverse Gaussian distribution, $Y \sim IG(\mu, \lambda)$, is given by

$$f(y) = \left(\frac{\lambda}{2\pi y^3}\right)^{1/2} \quad exp\left\{-\frac{\lambda}{2\pi^2 y}(y-\mu)^2\right\}, y > 0,$$

where $\mu, \lambda > 0$. One obtains

$$E(Y) = \mu, \quad var(Y) = \mu^3/\lambda.$$

For $\mu = 1$, the distribution is also called the Wald distribution.

A.2.8 Dirichlet Distribution

The vector $\boldsymbol{Z}^T = (Z_1, \ldots, Z_k)$, $k \geq 2$, is Dirichlet distributed, $\boldsymbol{Z} \sim D(\boldsymbol{\alpha})$, with $\boldsymbol{\alpha}^T = (\alpha_1, \ldots \alpha_k)$ if the density on the simplex $\{(z_1, \ldots, z_k) : \sum_{i=1}^k z_i = 1, z_i > 0\}$ is given by

$$f(z_1, \ldots, z_k) = \frac{\Gamma\left(\sum_{i=1}^k \alpha_i\right)}{\prod_{i=1}^k \Gamma(\alpha_i)} \prod_{i=1}^k z_i^{\alpha_i - 1},$$

where $\Gamma()$ is the Gamma function. An often-used reparameterization is obtained with

$$\mu_i = \frac{\alpha_i}{\sum_{i=1}^k \alpha_i}, \quad i = 1, \ldots, k, \quad K = \sum_{i=1}^k \alpha_i.$$

With vectors $\boldsymbol{\mu}^T = (\mu_1, \ldots, \mu_k) = \boldsymbol{\alpha}^T/K$ one obtains

$$\text{E}(\boldsymbol{Z}) = \boldsymbol{\mu}, \quad \text{var}(Z_i) = \frac{\mu_i(1-\mu_i)}{K+1}.$$

A.2.9 Beta Distribution

The Beta distribution, $(\text{Beta}(\alpha_1, \alpha_2))$, is a special case of the Dirichlet distribution where $k = 2$. It is enough to consider the first component, which has the density

$$f(z) = \frac{\Gamma(\alpha_1 + \alpha_2)}{\Gamma(\alpha_1)\Gamma(\alpha_2)} z^{\alpha_1 - 1}(1 - z)^{\alpha_2 - 1}.$$

Reparameterization yields

$$\mu_i = \frac{\alpha_i}{\alpha_1 + \alpha_2} \quad i = 1, 2 \quad K = \alpha_1 + \alpha_2$$

with density

$$f(z) = \frac{\Gamma(K)}{\Gamma(K\mu_1)\Gamma(K\mu_2)} z^{\mu_1 K - 1}(1 - z)^{\mu_2 K - 1}.$$

One has for $Z = Z_1$

$$E(Z) = \mu_1, \quad \text{var}(Z) = \frac{\mu_1(1 - \mu_1)}{K + 1}.$$

Appendix B

Some Basic Tools

B.1 Linear Algebra

Derivatives

The log-likelihood is typically a real-valued function of a p-dimensional parameter. When maximizing the likelihood one seeks parameter values where the derivative becomes zero.

Let $f : \mathbb{R}^p \to \mathbb{R}$ be a function of the vector $\boldsymbol{x}$, that is, $f(\boldsymbol{x}) = f(x_1, \dots, x_p)$. The *derivative of f with respect to $\boldsymbol{x}$* is determined by the vector of partial derivatives:

$$\frac{\partial f(\boldsymbol{x})}{\partial \boldsymbol{x}} = \begin{pmatrix} \frac{\partial f(\boldsymbol{x})}{\partial x_1} \\ \vdots \\ \frac{\partial f(\boldsymbol{x})}{\partial x_p} \end{pmatrix}.$$

The transposed vector is denoted by

$$\left[\frac{\partial f(\boldsymbol{x})}{\partial \boldsymbol{x}}\right]^T = \frac{\partial f(\boldsymbol{x})}{\partial \boldsymbol{x}^T} = \left(\frac{\partial f(\boldsymbol{x})}{\partial x_1}, \dots, \frac{\partial f(\boldsymbol{x})}{\partial x_p}\right).$$

The (asymptotic) covariance of the ML estimator is linked to the Fisher matrix, which contains the second derivative of the log-likelihood computed at the ML estimate. The matrix of second derivatives of $f : \mathbb{R}^p \to \mathbb{R}$ is determined by

$$\frac{\partial f(\boldsymbol{x})}{\partial \boldsymbol{x} \partial \boldsymbol{x}^T} = \frac{\partial}{\partial \boldsymbol{x}^T}\left(\frac{\partial f}{\partial \boldsymbol{x}}\right) = \begin{pmatrix} \frac{\partial f(\boldsymbol{x})}{\partial x_1^2} & \frac{\partial f(\boldsymbol{x})}{\partial x_1 \partial x_2} & \cdots & \frac{\partial f(\boldsymbol{x})}{\partial x_1 \partial x_p} \\ \frac{\partial f(\boldsymbol{x})}{\partial x_2 \partial x_1} & \ddots & & \vdots \\ \vdots & & & \\ \frac{\partial f(\boldsymbol{x})}{\partial x_p \partial x_1} & & \cdots & \frac{\partial f(\boldsymbol{x})}{\partial x_p \partial x_p} \end{pmatrix}.$$

This so-called *Hesse matrix* is symmetric if the function f has continuous second partial derivatives.

For a vector-valued function $f : \mathbb{R}^p \to \mathbb{R}^q$, where $f(\boldsymbol{x}) = (f_1(\boldsymbol{x}), \dots, f_q(\boldsymbol{x}))$, the derivative is

$$\frac{\partial f(\boldsymbol{x})}{\partial \boldsymbol{x}} = \begin{pmatrix} \frac{\partial f_1(\boldsymbol{x})}{\partial x_1} & \cdots & \frac{\partial f_q(\boldsymbol{x})}{\partial x_1} \\ \vdots & \ddots & \vdots \\ \frac{\partial f_1(\boldsymbol{x})}{\partial x_p} & \cdots & \frac{\partial f_q(\boldsymbol{x})}{\partial x_p} \end{pmatrix} = \left(\frac{\partial f_1(\boldsymbol{x})}{\partial \boldsymbol{x}}, \dots, \frac{\partial f_q(\boldsymbol{x})}{\partial \boldsymbol{x}}\right).$$

Covariance

The covariance (matrix) of a random vector $\boldsymbol{x}^T = (x_1, \ldots, x_k)$ is given by the symmetric matrix

$$\operatorname{cov}(\boldsymbol{x}) = \begin{pmatrix} \operatorname{var}(x_1) & \operatorname{cov}(x_1, x_2) & \ldots & \operatorname{cov}(x_1, x_k) \\ \operatorname{cov}(x_2, x_1) & & \ddots & \vdots \\ \vdots & & & \\ \operatorname{cov}(x_k, x_1) & & \ldots & \operatorname{var}(x_k) \end{pmatrix}.$$

For the transformed random vector $\boldsymbol{A}\boldsymbol{x}$ one obtains

$$\operatorname{cov}(\boldsymbol{A}\boldsymbol{x}) = \boldsymbol{A}\operatorname{cov}(\boldsymbol{x})\boldsymbol{A}^T, \qquad \text{where } \boldsymbol{A} \text{ is any } (n \times k)\text{-matrix.}$$

Square Root of a Matrix

Let $\boldsymbol{A}$ be a symmetric and positive definite matrix. For a positive definite matrix $\boldsymbol{x}^T\boldsymbol{A}\boldsymbol{x} > 0$ holds for all $\boldsymbol{x} \neq \boldsymbol{0}$. Then there exists a matrix $\boldsymbol{A}^{1/2}$ such that

$$\boldsymbol{A} = \boldsymbol{A}^{1/2}\boldsymbol{A}^{T/2}$$

holds, where $\boldsymbol{A}^{T/2} = (\boldsymbol{A}^{1/2})^T$ denotes the transposed of $\boldsymbol{A}^{1/2}$. The matrix $\boldsymbol{A}^{1/2}$ is the left and $\boldsymbol{A}^{T/2}$ is the right square root of $\boldsymbol{A}$. Both are not singular, that is, $\det(\boldsymbol{A}^{1/2}) \neq 0$.

In general, the square root of a matrix is not unique. A unique decomposition is based on the Cholesky decomposition, which in addition postulates that $\boldsymbol{A}^{1/2}$ is a lower triangular matrix and the conjugate transpose $\boldsymbol{A}^{T/2}$ is an upper triangular matrix.

Frequently one uses the square root of an inverse matrix $\boldsymbol{A}^{-1}$. From $\boldsymbol{A} = \boldsymbol{A}^{1/2}\boldsymbol{A}^{T/2}$ one obtains $\boldsymbol{A}^{-1} = (\boldsymbol{A}^{T/2})^{-1}(\boldsymbol{A}^{1/2})^{-1}$. With the denotation $\boldsymbol{A}^{-T/2} = (\boldsymbol{A}^{T/2})^{-1}$, $\boldsymbol{A}^{-1/2} = (\boldsymbol{A}^{1/2})^{-1}$ one obtains the decomposition of $\boldsymbol{A}^{-1}$ as

$$\boldsymbol{A}^{-1} = \boldsymbol{A}^{-T/2}\boldsymbol{A}^{-1/2}.$$

The square root of an inverse matrix is used, for example, if one wants to standardize a random vector. Let $\boldsymbol{x}$ have variance $\boldsymbol{\Sigma}$. It is easily derived that the transformed vector $\boldsymbol{\Sigma}^{-1/2}\boldsymbol{x}$ has variance $\boldsymbol{I}$, that is, $\operatorname{cov}(\boldsymbol{\Sigma}^{-1/2}\boldsymbol{x}) = \boldsymbol{I}$.

B.2 Taylor Approximation

Taylor's theorem asserts that any sufficiently smooth function can be locally approximated by polynomials. It is in particular useful for the motivation of local polynomial fits and the derivation of asymptotic results.

Univariate Version

Let f be a function that is n-times continuously differentiable on a closed interval and $(n+1)$-times differentiable on the open interval. Then one obtains for an inner point x_0

$$f(x) = f(x_0) + \sum_{r=1}^{n} \frac{f^{(r)}(x_0)}{r!}(x - x_0)^r + f^{(n+1)}(x_0 + \vartheta(x - x_0))\frac{(x - x_0)^{n+1}}{(n+1)!},$$

where x is from the interval, ϑ is a value for which $0 < \vartheta < 1$ holds, and $f^{(r)} = d^r f/dx^r$ are the r-order derivatives.

Multivariate Version

Let f be a function that is n-times continuously differentiable on a ball in $\mathbb{R}^p$ and $(n+1)$-times differentiable on the closure. Then one obtains for any $\boldsymbol{x}$

$$f(\boldsymbol{x}) = \sum_{r=0}^{n} \sum_{r_1+\cdots+r_p=r} \frac{1}{r_1!\ldots r_p!}(x_1 - x_{01})^{r_1}\ldots(x_1 - x_{0p})^{r_p}\frac{\partial^r f(\boldsymbol{x}_0)}{\partial x_1^{r_1}\ldots\partial x_p^{r_p}}$$
$$+ \sum_{r_1+\cdots+r_p=n+1} \frac{1}{r_1!\ldots r_p!}(x_1 - x_{01})^{r_1}\ldots(x_1 - x_{0p})^{r_p}\frac{\partial^r f(\boldsymbol{x}_0 + \vartheta(\boldsymbol{x} - \boldsymbol{x}_0))}{\partial x_1^{r_1}\ldots\partial x_p^{r_p}},$$

where $\boldsymbol{x}_0^T = (x_{01}, \ldots, x_{0p}), \boldsymbol{x}^T = (x_1, \ldots, x_p)$. The second-order approximation (without remainder) is

$$f(\boldsymbol{x}) \approx f(\boldsymbol{x}_0) + \sum_{i=1}^{p}\frac{\partial f(\boldsymbol{x}_0)}{\partial x_i}(x_i - x_{0i}) + \frac{1}{2}\sum_{i=1}^{p}\frac{\partial^2 f(\boldsymbol{x}_0)}{\partial x_i^2}(x_i - x_{0i})^2$$
$$+ \sum_{i \neq j}\frac{\partial^2 f(\boldsymbol{x}_0)}{\partial x_i \partial x_j}(x_i - x_{0i})(x_j - x_{0j})$$
$$= f(\boldsymbol{x}_0) + (\boldsymbol{x} - \boldsymbol{x}_0)^T\frac{\partial f(\boldsymbol{x}_0)}{\partial \boldsymbol{x}} + \frac{1}{2}(\boldsymbol{x} - \boldsymbol{x}_0)^T\frac{\partial f(\boldsymbol{x}_0)}{\partial \boldsymbol{x}\partial \boldsymbol{x}^T}(\boldsymbol{x} - \boldsymbol{x}_0).$$

For a vector-valued function $f = (f_1, \ldots, f_s) : \mathbb{R}^p \to \mathbb{R}^s$ one obtains for each component the first-order approximation

$$f_i(\boldsymbol{x}) \approx f_i(\boldsymbol{x}_0) + (\boldsymbol{x} - \boldsymbol{x}_0)^T\frac{\partial f_i(\boldsymbol{x}_0)}{\partial \boldsymbol{x}} = f_i(\boldsymbol{x}_0) + \frac{\partial f(\boldsymbol{x}_0)}{\partial \boldsymbol{x}^T}(\boldsymbol{x} - \boldsymbol{x}_0).$$

The function itself can be approximated through

$$f(\boldsymbol{x}) \approx f(\boldsymbol{x}_0) + \frac{\partial f(\boldsymbol{x}_0)}{\partial \boldsymbol{x}^T}(\boldsymbol{x} - \boldsymbol{x}_0).$$

Taylor Approximation and the Asymptotic Covariance of ML Estimates

The Taylor approximation is often used to derive asymptotic properties of the ML estimate $\hat{\boldsymbol{\beta}}$. In the following we just give some motivation for the asymptotic covariance. For GLMs the first-order Taylor approximation of the score function $\boldsymbol{s}(\boldsymbol{\beta}) = \partial l(\boldsymbol{\beta})/\partial\boldsymbol{\beta}$ at $\boldsymbol{\beta}$ yields

$$\boldsymbol{s}(\hat{\boldsymbol{\beta}}) \approx \boldsymbol{s}(\boldsymbol{\beta}) + \frac{\partial \boldsymbol{s}(\boldsymbol{\beta})}{\partial \boldsymbol{\beta}^T}(\hat{\boldsymbol{\beta}} - \boldsymbol{\beta}).$$

Since $\boldsymbol{s}(\hat{\boldsymbol{\beta}}) = \boldsymbol{0}$ one obtains the approximation $\hat{\boldsymbol{\beta}} - \boldsymbol{\beta} \approx (-\partial \boldsymbol{s}(\boldsymbol{\beta})/\partial\boldsymbol{\beta}^T)^{-1}\boldsymbol{s}(\boldsymbol{\beta})$. With $-\partial \boldsymbol{s}(\boldsymbol{\beta})/\partial\boldsymbol{\beta}^T = -\partial^2 l(\boldsymbol{\beta})/\partial\boldsymbol{\beta}\partial\boldsymbol{\beta}^T = \boldsymbol{F}_{obs}(\boldsymbol{\beta})$ one obtains $\hat{\boldsymbol{\beta}} - \boldsymbol{\beta} = \boldsymbol{F}_{obs}(\boldsymbol{\beta})^{-1}\boldsymbol{s}(\boldsymbol{\beta})$. If the observed Fisher matrix $\boldsymbol{F}_{obs}(\boldsymbol{\beta})$ is replaced by the Fisher matrix $\boldsymbol{F}(\boldsymbol{\beta}) = \boldsymbol{X}^T\boldsymbol{D}\boldsymbol{\Sigma}^{-1}\boldsymbol{D}\boldsymbol{X}$, one obtains the approximate covariance matrix of the ML estimate $\boldsymbol{F}(\boldsymbol{\beta})^{-1}\operatorname{cov}(\boldsymbol{s}(\boldsymbol{\beta})\boldsymbol{F}(\boldsymbol{\beta})^{-1}$. Since $\operatorname{cov}(\boldsymbol{s}(\boldsymbol{\beta}) = \mathrm{E}(\boldsymbol{s}(\boldsymbol{\beta})\boldsymbol{s}(\boldsymbol{\beta})^T) = \boldsymbol{F}(\boldsymbol{\beta})$ the approximate covariance reduces to $\boldsymbol{F}(\boldsymbol{\beta})^{-1}$. To make the approximation work asymptotically regularity conditions are needed, in particular, one assumes that $\boldsymbol{F}(\hat{\boldsymbol{\beta}})/n$ converges to a limit; for details see Fahrmeir and Kaufmann (1985).

B.3 Conditional Expectation, Distribution

Let X, Y be random variables with the joint density function $f(x, y)$. Then the *conditional density* of Y given $X = x$ is given by

$$f_{Y|X}(y|x) = \frac{f(y,x)}{f_X(x)}$$

for any x such that $f_X(x) = \int f(x,y)dy > 0$. For discrete random variables X, Y the analog is the *conditional probability mass function*:

$$f_{Y|X}(y|x) = P(Y = y|X = x),$$

where $P(X = x) > 0$.

The *conditional distribution function* of Y given $X = x$ is defined to be

$$F_{Y|X}(y|x) = P(Y \leq y|X = x),$$

which has the form

$$F_{Y|X}(y|x) = \int_{-\infty}^{x} f_{Y|X}(y|x)dx$$

for variables with conditional density $f_{Y|X}$ and

$$F_{Y|X}(x|y) = \sum_{y_i \leq y} P(y = y_i|X = x)$$

for a discrete-valued Y.

The *conditional expectation* of Y given X, denoted by $\mathrm{E}(Y|X)$, *is a random variable* that varies with the values of X. If X takes the value x, the conditional expectation takes the value

$$\mathrm{E}(Y|x) = \begin{cases} \sum_{y_i} y_i P(Y = y_i|X = x) & \text{for discrete variables} \\ \int y f_{Y|X}(y|x)dy & \text{for continuous-type variables.} \end{cases}$$

In a slightly more general form, one may consider the transformation $h(Y)$ with the conditional expectation of $h(Y)$ given X taking the value

$$\mathrm{E}(h(Y)|x) = \begin{cases} \sum_{y_i} h(y_i) P(Y = y_i|X = x) & \text{for discrete variables} \\ \int h(y) f_{Y|X}(y|x)dy & \text{for continuous-type variables.} \end{cases}$$

if X takes the value x.

The higher moments follow from this definition. $\mathrm{E}(Y^r|x)$ is based on the transformation $h(y) = y^r$. In particular, one has

$$\begin{aligned} \mathrm{var}(Y|x) &= \mathrm{E}(\{Y - \mathrm{E}(Y|X = x)\}^2|X = x) \\ &= \mathrm{E}(Y^2|X = x) - \mathrm{E}(Y|X = x)^2. \end{aligned}$$

One obtains for the random variable $\mathrm{E}(Y|X)$:

(1) $\mathrm{E}(\mathrm{E}(Y|X)) = \mathrm{E}(Y)$.

(2) If $\mathrm{E}(Y^2) < \infty$, one has

$$\mathrm{var}(Y) = \mathrm{var}(\mathrm{E}(Y|X)) + \mathrm{E}(\mathrm{var}(Y|X)),$$

where $\mathrm{var}(Y|X)$ is given by $\mathrm{var}(Y|X) = \mathrm{E}(Y^2|X) - \mathrm{E}(Y|X)^2$.

The decomposition of variance also holds for vectors $\boldsymbol{y}, \boldsymbol{x}$:

$$\mathrm{cov}(\boldsymbol{y}) = \mathrm{cov}_{\boldsymbol{x}}(\mathrm{E}(\boldsymbol{y}|\boldsymbol{x})) + \mathrm{E}_{\boldsymbol{x}}\,\mathrm{cov}(\boldsymbol{y}|\boldsymbol{x})$$

with $\mathrm{cov}(\boldsymbol{y}|\boldsymbol{x}) = \mathrm{E}(\boldsymbol{y}\boldsymbol{y}^T|\boldsymbol{x}) + \mathrm{E}(\boldsymbol{y}|\boldsymbol{x})\,\mathrm{E}(\boldsymbol{y}|\boldsymbol{x})^T$.

B.4 EM Algorithm

The basic EM algorithm was proposed by Dempster et al. (1977). The algorithm provides a general device to obtain maximum likelihood estimators in incomplete data situations. Let $\boldsymbol{y} \in \mathbb{R}^n$ denote a vector of observed data and $\boldsymbol{z} \in \mathbb{R}^m$ a vector of unobservable data. The hypothetical complete data are $(\boldsymbol{y}, \boldsymbol{z})$ and the incomplete data that were observed are $\boldsymbol{y}$. Let the joint density of the complete data, $f(\boldsymbol{y}, \boldsymbol{z}; \boldsymbol{\theta})$, depend on an unknown parameter vector $\boldsymbol{\theta} \in \boldsymbol{\Theta}$. Then the maximum likelihood estimator for $\boldsymbol{\theta}$ can be obtained by maximizing the marginal log-likelihood:

$$l(\boldsymbol{\theta}) = \log \int f(\boldsymbol{y}, \boldsymbol{z}; \boldsymbol{\theta}) d\boldsymbol{z}. \tag{B.1}$$

However, maximization my involve difficult integration procedures. Indirect maximization by the EM algorithm helps to avoid numerical evaluations of the integral. With $g(\boldsymbol{z}|\boldsymbol{y}; \boldsymbol{\theta})$ denoting the conditional density of the unobservable data $\boldsymbol{z}$, given the observed data $\boldsymbol{y}$, one obtains for the marginal density $f(\boldsymbol{y}; \boldsymbol{\theta}) = f(\boldsymbol{y}, \boldsymbol{z}; \boldsymbol{\theta})/g(\boldsymbol{z}|\boldsymbol{y}; \boldsymbol{\theta})$. Therefore, the marginal log-likelihood has the form

$$l(\boldsymbol{\theta}) = \log f(\boldsymbol{y}, \boldsymbol{z}; \boldsymbol{\theta}) - \log g(\boldsymbol{z}|\boldsymbol{y}; \boldsymbol{\theta}). \tag{B.2}$$

Since $\boldsymbol{z}$ is unobservable, expectations are taken on both sides of (B.2) with respect to the conditional density $g(\boldsymbol{z}|\boldsymbol{y}; \boldsymbol{\theta}_0)$ for fixed $\boldsymbol{\theta}_0 \in \boldsymbol{\Theta}$, obtaining

$$\begin{aligned} l(\boldsymbol{\theta}) &= \mathrm{E}_{\boldsymbol{\theta}_0}\{\log f(\boldsymbol{y}, \boldsymbol{z}; \boldsymbol{\theta})\} - \mathrm{E}_{\boldsymbol{\theta}_0}\{\log g(\boldsymbol{z}|\boldsymbol{y}; \boldsymbol{\theta})\} \\ &= M(\boldsymbol{\theta}|\boldsymbol{\theta}_0) - H(\boldsymbol{\theta}|\boldsymbol{\theta}_0), \end{aligned}$$

where

$$M(\boldsymbol{\theta}|\boldsymbol{\theta}_0) = \int \log f(\boldsymbol{y}, \boldsymbol{z}; \boldsymbol{\theta})\, g(\boldsymbol{z}|\boldsymbol{y}; \boldsymbol{\theta}_0)\, d\boldsymbol{z}, \quad H(\boldsymbol{\theta}|\boldsymbol{\theta}_0) = \int \log g(\boldsymbol{z}|\boldsymbol{y}; \boldsymbol{\theta})\, g(\boldsymbol{z}|\boldsymbol{y}; \boldsymbol{\theta}_0)\, d\boldsymbol{z}.$$

The EM algorithm maximizes $l(\boldsymbol{\theta})$ iteratively by maximizing $M(\boldsymbol{\theta}|\boldsymbol{\theta}_0)$ with respect to $\boldsymbol{\theta}$, where $\boldsymbol{\theta}_0$ is given at each cycle of the iteration. In contrast to the integral in (B.1), evaluation of the integral in $M(\boldsymbol{\theta}|\boldsymbol{\theta}_0)$ is straightforward for many applications. With $\hat{\boldsymbol{\theta}}^{(0)}$ denoting a starting value for $\boldsymbol{\theta}$, the $(p+1)$-th cycle of the EM algorithm consists of the following two steps:

E(xpectation) step: Compute the expectation $M(\boldsymbol{\theta}|\hat{\boldsymbol{\theta}}^{(p)})$.

M(aximizing) step: The improved estimate $\hat{\boldsymbol{\theta}}^{(p+1)}$ is obtained by maximizing $M(\boldsymbol{\theta}|\hat{\boldsymbol{\theta}}^{(p)})$ as a function in $\boldsymbol{\theta}$.

The EM algorithm has the desirable property that the log-likelihood l always increases or remains at least constant at each cycle. If $\hat{\boldsymbol{\theta}}$ maximizes $M(\boldsymbol{\theta}|\boldsymbol{\theta}_0)$ for fixed $\boldsymbol{\theta}_0$, one has $M(\hat{\boldsymbol{\theta}}|\boldsymbol{\theta}_0) \geq M(\boldsymbol{\theta}|\boldsymbol{\theta}_0)$ for all $\boldsymbol{\theta}$'s by definition and $H(\boldsymbol{\theta}|\boldsymbol{\theta}_0) \leq H(\boldsymbol{\theta}_0|\boldsymbol{\theta}_0)$ for all $\boldsymbol{\theta}$'s by Jensen's inequality, so that $l(\hat{\boldsymbol{\theta}}) \geq l(\boldsymbol{\theta}_0)$ holds.

Convergence of the log-likelihood sequence $l(\boldsymbol{\theta}^{(p)})$, $p = 0, 1, 2 \ldots$, against a global or local maximum or a stationary point l_* is ensured under weak regularity conditions concerning $\boldsymbol{\Theta}$ and $l(\boldsymbol{\theta})$ (see, e.g., Dempster et al., 1977). However, if more than one maximum or stationary point exists, convergence against one of these points depends on the starting value. Moreover, convergence of the log-likelihood sequence $l(\boldsymbol{\theta}^{(p)})$, $p = 0, 1, 2, \ldots$, against l_* does not imply the convergence of $(\boldsymbol{\theta}^{(p)})$ against a point $\boldsymbol{\theta}_*$ (Wu, 1983; Boyles, 1983). In general, convergence of $(\boldsymbol{\theta}^{(p)})$ requires stronger regularity conditions, which are ensured in particular for complete data densities $f(\boldsymbol{y}, \boldsymbol{z}; \boldsymbol{\theta})$ of the simple or curved exponential family. The rate of convergence depends on the relative size of the unobservable information on $\boldsymbol{\theta}$. If the information loss due to the missing $\boldsymbol{z}$ is a small fraction of the information in the complete data $(\boldsymbol{y}, \boldsymbol{z})$, the algorithm converges rapidly. But the rate of convergence becomes rather slow for parameters $\boldsymbol{\theta}$ near the boundary of $\boldsymbol{\Theta}$.

An estimator for the variance-covariance matrix of the MLE for $\boldsymbol{\theta}$, for example, the observed or expected information on $\boldsymbol{\theta}$ in the observed data $\boldsymbol{y}$, is not provided by the EM algorithm. Newton-Raphson or other gradient methods that maximize (B.1) directly are generally faster and yield an estimator for the variance-covariance matrix of the MLE. However, the EM algorithm is simpler to implement and numerically more stable. An estimate for the variance-covariance matrix of the MLE is obtained if an additional analysis is applied after the last cycle of the EM algorithm (see Louis, 1982). The method can also be used to speed up the EM algorithm (see also Meilijson, 1989).

Many extensions of the EM algorithm have been proposed; see, for example, McLachlan and Krishnan (1997). In statistical versions of the EM algorithm, the E-step consists in simulating the missing data from the conditional distribution; see Celeux and Diebolt (1985). In Monte Carlo EM (MCEM), this step is replaced by Monte Carlo approximations (Wei and Tanner, 1990). Quasi–Monte Carlo methods have been proposed by Jank (2004). To reduce the computational costs Delyon et al. (1999) proposed SAEM, a stochastic approximation version of the EM algorithm.

Appendix C

Constrained Estimation

C.1 Simplification of Penalties

A common penalty that is easy to handle has the form

$$J(\boldsymbol{\delta}) = \boldsymbol{\delta}^T \boldsymbol{K} \boldsymbol{\delta},$$

where $\boldsymbol{\delta}^T = (\delta_1, \ldots, \delta_m)$ and $\boldsymbol{K}$ is a symmetric $(m \times m)$-matrix. Typically $\boldsymbol{K}$ has the form $\boldsymbol{K} = \boldsymbol{L}^T \boldsymbol{L}$, where $\boldsymbol{L}$ is an $(r \times m)$-matrix. A very simple form results when $\boldsymbol{K}$ is chosen as the identity matrix. Then one obtains the penalty term $J(\boldsymbol{\delta}) = \sum_{j=1}^{m} \delta_j^2$, which is used in ridge regressions.

When basis functions are defined on an equally spaced grid, as, for example, in P-splines, one uses difference penalties. First-order differences use the $((m-1) \times m)$-matrix

$$\boldsymbol{L} = \begin{pmatrix} -1 & 1 & & & \\ & -1 & 1 & & \\ & & \ddots & \ddots & \\ & & & -1 & 1 \end{pmatrix}, \tag{C.1}$$

which yields

$$J(\boldsymbol{\delta}) = \boldsymbol{\delta}^T \boldsymbol{L}^T \boldsymbol{L} \boldsymbol{\delta} = \sum_{j=1}^{m-1} (\delta_{j+1} - \delta_j)^2$$

and the corresponding matrix

$$\boldsymbol{K} = \begin{pmatrix} 1 & -1 & & & \\ -1 & 2 & -1 & & \\ \ddots & \ddots & \ddots & & \\ & -1 & 2 & -1 & \\ & & -1 & 1 & \end{pmatrix},$$

which has rank $m-1$. Second-order differences are specified by

$$\boldsymbol{L} = \begin{pmatrix} 1 & -2 & 1 & & & \\ & 1 & -2 & 1 & & \\ & & \ddots & \ddots & & \\ & & & 1 & -2 & 1 \end{pmatrix},$$

yielding

$$J(\boldsymbol{\delta}) = \sum_{j=1}^{m-2} \{(\delta_{j+2} - \delta_{j+1}) - (\delta_{j+1} - \delta_j)\}^2$$

and the corresponding matrix

$$\boldsymbol{K} = \begin{pmatrix} 1 & -2 & 1 & & & & \\ -2 & 5 & -4 & 1 & & & \\ 1 & -4 & 6 & -4 & 1 & & \\ & & \ddots & \ddots & \ddots & & \\ & & 1 & -4 & 6 & -4 & 1 \\ & & & 1 & -4 & 5 & -2 \\ & & & & 1 & -2 & 1 \end{pmatrix},$$

which has rank $m - 2$.

Simplifying the Penalty by Reparameterization

Let the penalty have the general form $J(\boldsymbol{\delta}) = \boldsymbol{\delta}^T \boldsymbol{K} \boldsymbol{\delta}$, where $\boldsymbol{K}$ is a symmetric, non-negative definite $(m \times m)$-matrix.

When $\boldsymbol{K}$ has full rank one obtains by defining $\boldsymbol{\delta}_p = \boldsymbol{K}^{1/2}\boldsymbol{\delta}$ the simple reparameterized penalty

$$\boldsymbol{\delta}^T \boldsymbol{K} \boldsymbol{\delta} = \boldsymbol{\delta}_p^T \boldsymbol{\delta}_p = \|\boldsymbol{\delta}_p\|^2.$$

In the more general case, let $\boldsymbol{K}$ have rank r, $r < m$. Then one wants a decomposition of $\boldsymbol{\delta}$ into an unpenalized part and a penalized part of the form

$$\boldsymbol{\delta} = \boldsymbol{T}\boldsymbol{\delta}_0 + \boldsymbol{W}\boldsymbol{\delta}_p, \tag{C.2}$$

where $\boldsymbol{T}$ is an $m \times (m - r)$-matrix and $\boldsymbol{W}$ is a $m \times r$-matrix. $\boldsymbol{\delta}_0$ represents the unpenalized part and $\boldsymbol{\delta}_p$ the penalized part. Given the decomposition one obtains by simple matrix algebra

$$\boldsymbol{\delta}^T \boldsymbol{K} \boldsymbol{\delta} = \boldsymbol{\delta}_0^T \boldsymbol{T}^T \boldsymbol{K} \boldsymbol{T} \boldsymbol{\delta}_0 + \boldsymbol{\delta}_0^T \boldsymbol{T}^T \boldsymbol{K} \boldsymbol{W} \boldsymbol{\delta}_p + \boldsymbol{\delta}_p^T \boldsymbol{W}^T \boldsymbol{K} \boldsymbol{T} \boldsymbol{\delta}_0 + \boldsymbol{\delta}_p^T \boldsymbol{W}^T \boldsymbol{K} \boldsymbol{W} \boldsymbol{\delta}_p. \tag{C.3}$$

If $\boldsymbol{T}$ and $\boldsymbol{W}$ are chosen such that $\boldsymbol{T}^T\boldsymbol{K}\boldsymbol{T} = \boldsymbol{0}$, $\boldsymbol{T}^T\boldsymbol{K}\boldsymbol{W} = \boldsymbol{0}$, and $\boldsymbol{W}^T\boldsymbol{K}\boldsymbol{W} = \boldsymbol{I}$, one obtains the simple form

$$\boldsymbol{\delta}^T \boldsymbol{K} \boldsymbol{\delta} = \boldsymbol{\delta}_p^T \boldsymbol{\delta}_p = \|\boldsymbol{\delta}_p\|^2,$$

which contains only the penalized parameters. In addition, it is required that the composed matrix $[\boldsymbol{T}|\boldsymbol{W}]$ has full rank, in order to obtain a one-to-one transformation between the parameterizations $\boldsymbol{\delta}$ and $(\boldsymbol{\delta}_0, \boldsymbol{\delta}_p)$. The effect of the reparameterization becomes especially obvious when a smoothing parameter is added. Then the penalty becomes

$$\lambda\boldsymbol{\delta}^T \boldsymbol{K} \boldsymbol{\delta} = \lambda\boldsymbol{\delta}_p^T \boldsymbol{\delta}_p = \lambda\|\boldsymbol{\delta}_p\|^2.$$

While $\boldsymbol{\delta}_p$ is penalized with the effect that $\boldsymbol{\delta}_p$ becomes zero for increasing λ, the parameters in $\boldsymbol{\delta}_0$ remain unaffected by λ.

For the construction of the matrices $\boldsymbol{T}$ and $\boldsymbol{W}$ one starts by choosing $\boldsymbol{W} = \boldsymbol{L}^T(\boldsymbol{L}\boldsymbol{L}^T)^{-1}$ from some factorization of the penalty matrix into $\boldsymbol{K} = \boldsymbol{L}^T\boldsymbol{L}$. Then one has $\boldsymbol{W}^T\boldsymbol{K}\boldsymbol{W} = \boldsymbol{I}$. The matrix $\boldsymbol{T}$ is chosen as an $(m-r)$-dimensional basis of the null space of $\boldsymbol{L}$, that is, $\boldsymbol{L}\boldsymbol{T} = \boldsymbol{0}$ holds. Then one has $\boldsymbol{T}^T\boldsymbol{K}\boldsymbol{T} = \boldsymbol{T}^T\boldsymbol{L}^T\boldsymbol{L}\boldsymbol{T} = \boldsymbol{0}$ and the first term in (C.3) can be omitted. Moreover, one obtains $\boldsymbol{W}^T\boldsymbol{K}\boldsymbol{T} = \boldsymbol{W}^T\boldsymbol{L}^T\boldsymbol{L}\boldsymbol{T} = \boldsymbol{0}$ and, in addition, the second and third terms in (C.3) can be omitted. Moreover, $[\boldsymbol{L}^T\boldsymbol{T}]$ is non-singular and the transformation is one-to-one.

It should be noted that the decomposition $\boldsymbol{K} = \boldsymbol{L}^T\boldsymbol{L}$ is not unique. A factorization of the matrix $\boldsymbol{K}$ may be based on the spectral decomposition $\boldsymbol{K} = \boldsymbol{\Gamma}\boldsymbol{D}\boldsymbol{\Gamma}^T$, where $\boldsymbol{D}$ is a diagonal matrix containing the eigenvalues in descending order. Then $\boldsymbol{L} = \boldsymbol{D}^{1/2}\boldsymbol{\Gamma}^T$ yields $\boldsymbol{K} = \boldsymbol{L}^T\boldsymbol{L}$. Alternatively, the matrix $\boldsymbol{L}$ can be used that defines a penalty. For example, with P-splines, the matrix $\boldsymbol{L}$ is given as a matrix containing differences of the first or higher order.

As examples let us consider the first- and second-order difference penalties for equally spaced P-splines. When using first differences it is obvious that adding a constraint c to the parameters, obtaining $\delta_j + c$, does not change the penatly. This means that the overall level is not penalized. Consequently, the matrix $\boldsymbol{T}$ that fulfills $\boldsymbol{L}\boldsymbol{T} = \boldsymbol{0}$ is $\boldsymbol{T}^T = (1, \dots, 1)$ and $[\boldsymbol{L}^T, \boldsymbol{T}]$ is non-singular. In the corresponding penalty, the overall level δ_0 of the components in $\boldsymbol{\delta}$ remains unpenalized. When second-order differences are used one obtains the matrix

$$\boldsymbol{T} = \begin{pmatrix} 1 & \xi_1 \\ 1 & \xi_2 \\ \vdots & \\ 1 & \xi_m \end{pmatrix},$$

with $\xi_j = a + j/\Delta, j = 1, \dots, m$ representing equidistant values (knots). The effect is that the level as well as the linear trend in the components δ_i from $\boldsymbol{\delta}^T = (\delta_1, \dots, \delta_m)$ are not penalized.

More generally, for B-splines with equidistant knots, the differences penalty of order d is determined by an $((m-d) \times m)$-matrix $\boldsymbol{L}$ and $\boldsymbol{L}\boldsymbol{L}^T$ has rank $m - d$. The corresponding $(m \times d)$-matrix $\boldsymbol{T}$ can be chosen by

$$\boldsymbol{T} = \begin{pmatrix} 1 & \xi_1 & \dots & \xi_1^{d-1} \\ 1 & \vdots & & \vdots \\ \vdots & \vdots & & \vdots \\ 1 & \xi_m & \dots & \xi_m^{d-1} \end{pmatrix},$$

which is spanned by a polynomial of degree $d - 1$ defined by the knots of the B-spline. An early reference to decompositions of the form given here is Green (1987).

C.2 Linear Constraints

Frequently, it is necessary to impose constraints on the parameters in the linear predictor of the form

$$\boldsymbol{C}\boldsymbol{\beta} = \boldsymbol{0},$$

where $\boldsymbol{C}$ is an $m \times p$-matrix of known coefficients. One approach to fitting subject to linear constraints is to rewrite the model in terms of $p - m$ unconstrained parameters by using the $\boldsymbol{QR}$ decomposition. According to the $\boldsymbol{QR}$ decomposition, any $n \times m$-matrix $(m \leq n)$ can be written as

$$\boldsymbol{M} = \boldsymbol{QR} = [\boldsymbol{Q}|\boldsymbol{Z}] \begin{bmatrix} \boldsymbol{R} \\ \boldsymbol{0} \end{bmatrix},$$

where $[\boldsymbol{Q}|\boldsymbol{Z}]$ is an $n \times n$ orthogonal matrix and $\boldsymbol{R}$ is an invertible $m \times m$ upper triangular matrix. The matrix $\boldsymbol{Q}$ is an $n \times m$-matrix with orthogonal columns. Let the $\boldsymbol{QR}$ decomposition of $\boldsymbol{C}^T$ be given by

$$\boldsymbol{C}^T = [\boldsymbol{Q}|\boldsymbol{Z}] \begin{bmatrix} \boldsymbol{R} \\ \boldsymbol{0} \end{bmatrix}.$$

It is easily seen that any vector of the form

$$\boldsymbol{\beta} = \boldsymbol{Z}\boldsymbol{\beta}_u$$

meets the constraints for any value of the unrestricted $p - m$-dimensional parameter $\boldsymbol{\beta}_u$.

Therefore, if one wants to fit a model with linear predictor $\boldsymbol{\eta} = \boldsymbol{X}\boldsymbol{\beta}$, subject to $\boldsymbol{C}\boldsymbol{\beta} = \mathbf{0}$, for example by minimizing $\|\boldsymbol{y} - \boldsymbol{X}\boldsymbol{\beta}\|^2$ subject to $\boldsymbol{C}\boldsymbol{\beta} = \mathbf{0}$, one reformulates the problem in terms of $\boldsymbol{\beta}_u$. First one has to find the $\boldsymbol{QR}$ decomposition of $\boldsymbol{C}^T$, in particular the matrix $\boldsymbol{Z}$. Then one minimize $\|\boldsymbol{y} - \boldsymbol{X}\boldsymbol{Z}\boldsymbol{\beta}_u\|$ w.r.t. $\boldsymbol{\beta}_u$, to obtain $\hat{\boldsymbol{\beta}}_u$, and then one sets $\hat{\boldsymbol{\beta}} = \boldsymbol{Z}\boldsymbol{\beta}_u$.

It is not necessary to compute $\boldsymbol{Z}$ explicitly because it is only necessary to compute $\boldsymbol{X}\boldsymbol{Z}$ and $\boldsymbol{Z}\boldsymbol{\beta}_u$, which may be done by simple matrix operations (Householder rotations; see, for example, Wood, 2006a, Appendix A5, A6).

When linear constraints $\boldsymbol{C}\boldsymbol{\beta} = \mathbf{0}$ are imposed in combination with a quadratic penalty $\boldsymbol{\beta}^T\boldsymbol{K}\boldsymbol{\beta}$, the quadratic penalty can be rewritten as

$$\boldsymbol{\beta}^T\boldsymbol{K}\boldsymbol{\beta} = \boldsymbol{\beta}_u^T\boldsymbol{Z}^T\boldsymbol{K}\boldsymbol{Z}\boldsymbol{\beta}_u.$$

Then, following (C.2), $\boldsymbol{\beta}_u$ is decomposed into $\boldsymbol{\beta}_u = \boldsymbol{T}\boldsymbol{\beta}_{u0} + \boldsymbol{W}\boldsymbol{\beta}_{up}$.

C.3 Fisher Scoring with Penalty Term

Consider the likelihood of a penalized GLM:

$$l_p(\boldsymbol{\beta}) = \sum_{i=1}^{p} l_i(\boldsymbol{\beta}) - \boldsymbol{\beta}^T\boldsymbol{\Lambda}\boldsymbol{\beta}.$$

The corresponding penalized score function and the pseudo-Fisher matrix have the form

$$\boldsymbol{s}_p(\boldsymbol{\beta}) = \boldsymbol{s}(\boldsymbol{\beta}) - \boldsymbol{\Lambda}\boldsymbol{\beta}, \boldsymbol{F}_p(\boldsymbol{\beta}) = \mathrm{E}(-\partial l_p/\partial\boldsymbol{\beta}\partial\boldsymbol{\beta}^T) = \boldsymbol{F}(\boldsymbol{\beta}) + \boldsymbol{\Lambda},$$

with score function $\boldsymbol{s}(\boldsymbol{\beta}) = \boldsymbol{X}^T\boldsymbol{D}(\boldsymbol{\beta})\boldsymbol{\Sigma}^{-1}(\boldsymbol{\beta})(\boldsymbol{y}-\boldsymbol{\mu})$ and Fisher matrix $\boldsymbol{F}(\boldsymbol{\beta}) = \boldsymbol{X}^T\boldsymbol{W}(\boldsymbol{\beta})\boldsymbol{X}$, $\boldsymbol{X}^T = (\boldsymbol{x}_1, \dots, \boldsymbol{x}_n)$, $\boldsymbol{D}(\boldsymbol{\beta}) = \mathrm{diag}((\partial h(\hat{\eta}_1)/\partial\eta)/\sigma_1^2, \dots))$, $\boldsymbol{W}(\boldsymbol{\beta}) = \boldsymbol{D}(\boldsymbol{\beta})\boldsymbol{\Sigma}^{-1}(\boldsymbol{\beta})\boldsymbol{D}(\boldsymbol{\beta})$. Then Fisher scoring for solving $s_p(\hat{\boldsymbol{\beta}}) = \mathbf{0}$, which has the form

$$\hat{\boldsymbol{\beta}}^{(k+1)} = \hat{\boldsymbol{\beta}}^{(k)} + \boldsymbol{F}_p(\hat{\boldsymbol{\beta}}^{(k)})^{-1}\boldsymbol{s}_p(\hat{\boldsymbol{\beta}}^{(k)}),$$

can also be given with pseudo-observations in the form

$$\hat{\boldsymbol{\beta}}^{(k+1)} = (\boldsymbol{F}(\hat{\boldsymbol{\beta}}^{(k)}) + \boldsymbol{\Lambda})^{-1}\boldsymbol{X}^T\boldsymbol{W}(\hat{\boldsymbol{\beta}}^{(k)})\tilde{\eta}(\hat{\boldsymbol{\beta}}^{(k)}),$$

where the pseudo-observations are $\tilde{\eta}(\boldsymbol{\beta}) = \boldsymbol{X}\boldsymbol{\beta} + \boldsymbol{D}(\boldsymbol{\beta})^{-1}(\boldsymbol{y} - \boldsymbol{\mu}(\boldsymbol{\beta}))$.

Appendix D

Kullback-Leibler Distance and Information-Based Criteria of Model Fit

D.1 Kullback-Leibler Distance

Consider two densities f^*, f, where f^* often denotes the true or operating model and $f = f(.;\theta)$ is a parameterized candidate model. E_* denotes expectation with respect to the true density f^*.

The *Kullback-Leibler distance* between two (continuous) densities f^*, f is defined as

$$\mathrm{KL}(f^*, f) = \mathrm{E}_* \log(\frac{f^*(\boldsymbol{z})}{f(\boldsymbol{z})}) = \int f^*(\boldsymbol{z}) \log(\frac{f^*(\boldsymbol{z})}{f(\boldsymbol{z})}) d\boldsymbol{z}.$$

Often one uses the notation $I(f^*, f)$, referring to the *information lost when f is used to approximate f^**. It should be noted that $\mathrm{KL}(f^*, f)$ is not a distance in the usual sense because $\mathrm{KL}(f^*, f) \neq KL(f, f^*)$. More precisely, KL may be considered a directed or oriented distance representing the distance *from f to f^**. The KL distance has also been called the *divergence* or *KL discrepancy* or *Kullback-Leibler information*. The KL distance is always positive, except when the two distributions f^* and f are identical. It is noteworthy that the KL distance is a distance between two statistical expectations. One obtains

$$\begin{aligned}\mathrm{KL}(f^*, f) = \mathrm{E}_* \log\left(\frac{f^*(\boldsymbol{z})}{f(\boldsymbol{z})}\right) &= \mathrm{E}_* \log(f^*(\boldsymbol{z})) - \mathrm{E}_* \log(f(\boldsymbol{z})) \\ &= \int \log(f^*(\boldsymbol{z})) f^*(\boldsymbol{z}) d\boldsymbol{z} - \int \log(f(\boldsymbol{z})) f^*(\boldsymbol{z}) d\boldsymbol{z}.\end{aligned}$$

The first expectation, $\mathrm{E}_* \log(f^*(\boldsymbol{z}))$, depends only on the unknown true distribution. Since in actual data analysis the true distribution is never known, it can be considered as a constant. What one obtains is a measure of the *relative distance*, which might be used to compare candidate models. For two candidate models f_1, f_2 one obtains $\mathrm{KL}(f^*, f_1) - \mathrm{KL}(f^*, f_1) = -\mathrm{E}_* \log(f_1(\boldsymbol{z})) + \mathrm{E}_* \log(f_2(\boldsymbol{z}))$, where the constant term has vanished. Since the constant is irrelevant for comparisons of models, some authors define $-\mathrm{E}_* \log(f(\boldsymbol{z}))$ as the Kullback-Leibler discrepancy.

Thus, for candidate model f the discrepancy $-\mathrm{E}_* \log(f(\boldsymbol{z}))$ should be small. The crucial point is that this discrepancy can only be computed if the operating model f^* is known and f is fully specified. If f is a parametric model $f(\boldsymbol{z}|\theta)$, then θ has to be estimated. Let $f(\boldsymbol{z}|\hat{\theta}(\boldsymbol{y}))$ denote the density, with the parameter $\hat{\theta}(\boldsymbol{y})$ being estimated from the sample $\boldsymbol{y}$.

In a series of papers Akaike (1973, 1974) proposed using for an applied KL model selection an estimate of

$$\mathrm{E}_{\boldsymbol{y}}\,\mathrm{E}_{\boldsymbol{z}}\ \log\ f(\boldsymbol{z}|\hat{\theta}(\boldsymbol{y})), \tag{D.1}$$

where $\boldsymbol{y}$ and $\boldsymbol{z}$ are independent random samples from the same distribution and both expectations are taken with respect to the true density f^*. Thus $\mathrm{E}_{\boldsymbol{y}}$ and $\mathrm{E}_{\boldsymbol{z}}$ represent E_*. The selection target $\mathrm{E}_{\boldsymbol{y}}\,\mathrm{E}_{\boldsymbol{z}}\ \log\ f(\boldsymbol{z}|\hat{\theta}(\boldsymbol{y}))$ represents the (negative) expected KL discrepancy in a future sample $\boldsymbol{z}$ where expectation is taken with respect to the parameter generating sample $\boldsymbol{y}$. Therefore it gives the expected KL discrepancy if estimation is taken into account. It is essential that $\boldsymbol{y}$ and $\boldsymbol{z}$ are samples of the same size and distribution.

An approximately unbiased estimation of (D.1) for large samples is the corrected log-likelihood

$$l(\hat{\boldsymbol{\theta}}) - dim(\boldsymbol{\theta}),$$

where $l(\hat{\boldsymbol{\theta}})$ is the log-likelihood evaluated at the estimate $\hat{\boldsymbol{\theta}}$ and $dim(\boldsymbol{\theta})$ is the dimensionality of the parameter $\boldsymbol{\theta}$. Akaike (1973) defined "an information criterion" (AIC) by multiplying the respected log-likelihood by -2 to obtain

$$AIC = -2(l(\hat{\boldsymbol{\theta}}) - dim(\boldsymbol{\theta})),$$

which estimates the expected KL discrepancy $-2\,\mathrm{E}_{\boldsymbol{y}}\,\mathrm{E}_{\boldsymbol{z}}\ \log\ f(\boldsymbol{z}|\hat{\theta}(\boldsymbol{y}))$. A bias-corrected AIC criterion that is more appropriate for small sample sizes has been suggested by Hurvich and Tsai (1989). It has with $d = dim(\boldsymbol{\theta})$ the form

$$AIC_c = -2(l(\hat{\boldsymbol{\theta}}) - dn/(n-d-1)) = AIC + \frac{2d(d+1)}{n-d-1}.$$

For an extensive treatment of AIC see Burnham and Anderson (2002).

D.1.1 Kullback-Leibler and ML Estimation

If $\delta_{\boldsymbol{y}}$ is considered a degenerate distribution with mass 1 on the data $\boldsymbol{y}$, that is, $\delta_{\boldsymbol{y}}(\boldsymbol{y}) = 1,\ \delta_{\boldsymbol{y}}(\boldsymbol{z}) = 0,\ \boldsymbol{z} \neq \boldsymbol{y}$ and $f = f(.;\boldsymbol{\theta})$ is a parameterized density, one obtains

$$\mathrm{KL}(\delta_{\boldsymbol{y}}, f(.,\boldsymbol{\theta})) = -\log(f(\boldsymbol{y};\boldsymbol{\theta})).$$

Thus the log-likelihood may be considered as the Kullback-Leibler discrepancy between $f_{\boldsymbol{\theta}}$ and the "data" $\delta_{\boldsymbol{y}}$. The log-likelihood takes its maximal value with respect to $\boldsymbol{\theta}$ if $\mathrm{KL}(\delta_{\boldsymbol{y}}, f_{\boldsymbol{\theta}})$ is minimal.

If $\boldsymbol{y}$ represents a vector $\boldsymbol{y}^T = (y_1, \dots, y_n)$ containing independent observations, the distribution representing the data is given by $\delta_{\boldsymbol{y}}(\boldsymbol{z}) = \delta_{y_1}(z_1) \cdot \ldots \cdot \delta_{y_n}(z_n)$, where $\boldsymbol{z}^T = (z_1, \dots, z_n)$. With f also denoting the distribution of $\boldsymbol{y}$ one obtains

$$\mathrm{KL}(\delta_{\boldsymbol{y}}, f(.,\boldsymbol{\theta})) = E_{\delta_{\boldsymbol{y}}} \log\left(\frac{\delta_{\boldsymbol{y}}(\boldsymbol{z})}{f(\boldsymbol{z};\boldsymbol{\theta})}\right) = -\log(f(y_1, \dots, y_n;\boldsymbol{\theta})) = -\sum_{i=1}^{n} \log(f(y_i;\boldsymbol{\theta})),$$

which equals the negative likelihood of observations $y_1, \dots, y_n$.

D.1.2 Kullback-Leibler and Discrete Distributions

For discrete distributions with support $\{w_i, i = 1, 2, \dots\}$, f^* and f are mass functions and one has the Kullback-Leibler distance

$$\mathrm{KL}\,(f^*,\ f) = E_* \log\left(\frac{f^*(z)}{f(z)}\right) = \sum_{w_i} f^*(w_i) \log\left(\frac{f^*(w_i)}{f(w_i)}\right).$$

The true mass function f^* may be replaced by the vector $\boldsymbol{\pi}^T = (\pi_1, \pi_2, \dots)$, $\pi_i = f(w_i)$, of the true probabilities and if f is the corresponding estimate $\hat{\boldsymbol{\pi}}^T = (\hat{\pi}_1, \hat{\pi}_2, \dots)$, $\hat{\pi}_i = f(w_i)$, one has

$$\mathrm{KL}(\boldsymbol{\pi}, \hat{\boldsymbol{\pi}}) = \sum_i \pi_i \log\left(\frac{\pi_i}{\hat{\pi}_i}\right).$$

If $\boldsymbol{\pi}$ is replaced by the data in the form of relative frequencies $\boldsymbol{p}^T = (p_1, p_2, \dots)$, which means independent observations have been assumed, one has

$$\mathrm{KL}(\boldsymbol{p}, \hat{\boldsymbol{\pi}}) = \sum_i p_i \log(p_i) - \sum_i p_i \log(\hat{\pi}_i) = l(\boldsymbol{p}; \boldsymbol{p}) - l(\boldsymbol{p}; \hat{\boldsymbol{\pi}}_i),$$

where $l(\boldsymbol{p}; \hat{\boldsymbol{\pi}}_i) = \sum_i p_i \log(\hat{\pi}_i)$ is the grouped form of the log-likelihood with $\boldsymbol{p}$ representing the data. $l(\boldsymbol{p}, \boldsymbol{p})$ corresponds to the saturated model and does not depend on the parameters that determine the candidate model and therefore the estimates $\hat{\pi}_i$. For the multinomial distribution and $\hat{\boldsymbol{\pi}}$ the maximum likelihood estimate $2\,\mathrm{KL}(\boldsymbol{p}, \hat{\boldsymbol{\pi}})$ is called the deviance.

D.1.3 Kullback-Leibler in Generalized Linear Models

Let f^*, f be replaced by the densities depending on the parameters $f(\boldsymbol{y}; \boldsymbol{\theta}^*)$, $f(\boldsymbol{y}; \boldsymbol{\theta})$, where f is from a natural exponential family:

$$f(\boldsymbol{y}|\boldsymbol{\theta}) = \exp\left\{(\boldsymbol{y}^T\boldsymbol{\theta} - b(\boldsymbol{\theta}))/\phi + c(\boldsymbol{y}, \phi)\right\}$$

with ϕ being an additional dispersion parameter that is omitted in the notation. Then one has

$$\mathrm{KL}(f(.,\boldsymbol{\theta}^*), f(.,\boldsymbol{\theta})) = E_{\boldsymbol{\theta}^*} \log\left(\frac{f(\boldsymbol{y},\boldsymbol{\theta}^*)}{f(\boldsymbol{y},\boldsymbol{\theta})}\right) = \frac{1}{\phi} E_{\boldsymbol{\theta}^*}\left\{(\boldsymbol{y}^T\boldsymbol{\theta}^* - b(\boldsymbol{\theta}^*)) - (\boldsymbol{y}^T\boldsymbol{\theta} - b(\boldsymbol{\theta}))\right\}.$$

Considering the dependence of $\boldsymbol{\theta}$ on the mean $\boldsymbol{\mu}$ in the form $\boldsymbol{\theta} = \boldsymbol{\theta}(\boldsymbol{\mu})$, one may consider the Kullback-Leibler distance between the distribution in which the mean is given as the observation and the underlying mean:

$$\mathrm{KL}(f(.,\boldsymbol{\theta}(\boldsymbol{y})), f(.,\boldsymbol{\theta}(\boldsymbol{\mu}))) = \frac{1}{\phi} E_{\boldsymbol{\theta}(\boldsymbol{y})}\left\{(\boldsymbol{y}^T\boldsymbol{\theta}(\boldsymbol{y}) - b(\boldsymbol{\theta}(\boldsymbol{y})) - (\boldsymbol{y}^T\boldsymbol{\theta}(\boldsymbol{\mu}) - b(\boldsymbol{\theta}(\boldsymbol{\mu}))\right\}.$$

This form of Kullback-Leibler is appropriate only in the non-degenerate case where $\boldsymbol{\theta}(\boldsymbol{y})$ is well defined. For a single observation from the binomial distribution with $y \in \{0, 1\}$, one has a degenerated distribution with $\theta(y) \in \{\infty, -\infty\}$. However, in this case one may consider that $f(y, \theta(y)) = 1$ and therefore obtains

$$\mathrm{KL}(f(.,\boldsymbol{\theta}(\boldsymbol{y})), f(.,\boldsymbol{\theta}(\boldsymbol{\mu}))) = \mathrm{KL}(\delta_{\boldsymbol{y}}, f(.,\boldsymbol{\theta}(\boldsymbol{\mu})) = -\log(f(\boldsymbol{y}; \boldsymbol{\theta}(\boldsymbol{\mu})),$$

which is the negative log-likelihood contribution of observation $\boldsymbol{y}$. It is equal to the Kullback-Leibler discrepancy where $\delta_{\boldsymbol{y}}$ puts mass 1 on the observation $\boldsymbol{y}$.

For independent observations $\boldsymbol{y}_{i1}, \dots, \boldsymbol{y}_{in_i}$ at design point $\boldsymbol{x}_i$, one may consider the vector $\bar{\boldsymbol{y}}_i = \sum_j \boldsymbol{y}_{ij}/n$ as the mean observation. One obtains for the log-likelihood contribution of $\boldsymbol{y}_{i1}, \dots, \boldsymbol{y}_{in_i}$

$$l_i(\hat{\boldsymbol{\mu}}_i) = \sum_{j=1}^{n_i} \left\{\boldsymbol{y}_{ij}^T\boldsymbol{\theta}(\hat{\boldsymbol{\mu}}_i) - b(\boldsymbol{\theta}(\hat{\boldsymbol{\mu}}_i))\right\}/\phi = n_i\left\{\bar{\boldsymbol{y}}_i^T\boldsymbol{\theta}(\hat{\boldsymbol{\mu}}_i) - b(\boldsymbol{\theta}(\hat{\boldsymbol{\mu}}_i))\right\}/\phi.$$

This is equivalent to $l_i(\hat{\boldsymbol{\mu}}_i) = \mathrm{KL}(\delta_{\boldsymbol{y}_i}, f(.,\boldsymbol{\theta}(\hat{\boldsymbol{\mu}}_i)) = n_i \mathrm{KL}(\delta_{\bar{\boldsymbol{y}}_i}, f(.,\boldsymbol{\theta}(\hat{\boldsymbol{\mu}}_i))$. Thus the deviance based on grouped observations has the form

$$\begin{aligned} D &= -2\phi\left\{\sum_{i=1}^{g} l_i(\hat{\boldsymbol{\mu}}_i) - l_i(\bar{\boldsymbol{y}}_i)\right\} \\ &= -2\phi\left\{\sum_{i=1}^{g} n_i\left\{\mathrm{KL}(\delta_{\bar{\boldsymbol{y}}_i}, f(.,\boldsymbol{\theta}(\hat{\boldsymbol{\mu}}_i))) - \mathrm{KL}(\delta_{\bar{\boldsymbol{y}}_i}, f(.,\boldsymbol{\theta}(\bar{\boldsymbol{y}}_i)))\right\}\right\}. \end{aligned}$$

D.1.4 Decomposition

Let $\boldsymbol{y}^*(\boldsymbol{x}) = E(\boldsymbol{y}|\boldsymbol{x})$ be the true expectation with parameter $\boldsymbol{\theta}^*$ and let $f(.,\boldsymbol{\theta})$ be a candidate model where $\boldsymbol{\theta} = \boldsymbol{\theta}(\boldsymbol{\mu}(\boldsymbol{x}))$. Then one obtains for the new observation $(\boldsymbol{y},\ \boldsymbol{x})$

$$\begin{aligned} &\mathrm{E}_{\boldsymbol{yx}}\,\mathrm{KL}(\delta_{\boldsymbol{y}}, f(.,\boldsymbol{\theta})) \\ &\quad= \mathrm{E}_{\boldsymbol{yx}}\,\mathrm{KL}(\delta_{\boldsymbol{y}}, f(.,\boldsymbol{\theta}^*)) + \mathrm{E}_{\boldsymbol{x}}\,\mathrm{KL}(f(.,\boldsymbol{\theta}^*), f(.,\boldsymbol{\theta})) \end{aligned} \tag{D.2}$$

Derivation omitting $\boldsymbol{x}$ as an argument yields

$$\begin{aligned} &\mathrm{E}_{\boldsymbol{yx}}\,\mathrm{KL}(\delta_{\boldsymbol{y}}, f(.,\boldsymbol{\theta})) = \\ &\quad= \mathrm{E}_{\boldsymbol{yx}}\,\mathrm{E}_{\delta_{\boldsymbol{y}}} \log\frac{\delta_{\boldsymbol{y}}(z)/f(z,\boldsymbol{\theta}^*)}{f(z,\boldsymbol{\theta})/f(z,\boldsymbol{\theta}^*)} \\ &\quad= \mathrm{E}_{\boldsymbol{yx}}\,\mathrm{KL}(\delta_{\boldsymbol{y}}, f(.,\boldsymbol{\theta}^*)) + \mathrm{E}_{\boldsymbol{y},\boldsymbol{x}}\,\mathrm{E}_{\delta_{\boldsymbol{y}}} \log(f(z,\boldsymbol{\theta}^*)/f(z,\boldsymbol{\theta})). \end{aligned}$$

The last term equals

$$\mathrm{E}_{\boldsymbol{yx}} \log(f(\boldsymbol{y},\boldsymbol{\theta}^*)/f(\boldsymbol{y},\boldsymbol{\theta})) = \mathrm{E}_{\boldsymbol{x}}\,\mathrm{E}_{\boldsymbol{y}} \log(f(\boldsymbol{y},\boldsymbol{\theta}^*)/f(\boldsymbol{y},\boldsymbol{\theta})) = \mathrm{E}_{\boldsymbol{x}}\,\mathrm{KL}(f(.,\boldsymbol{\theta}^*), f(.,\boldsymbol{\theta})).$$

The first term in (D.2) depends only on the distribution. It is easy to show that minimization of $E_{\boldsymbol{yx}}\,\mathrm{KL}(\delta_{\boldsymbol{y}}, f(.,|\boldsymbol{\theta}))$ yields expectation $\boldsymbol{\mu}(\boldsymbol{\theta})$, which is the closest to $\boldsymbol{\mu}(\boldsymbol{\theta}^*)$ (see also Hastie and Tibshirani, 1986).

Appendix E

Numerical Integration and Tools for Random Effects Modeling

E.1 Laplace Approximation

The Laplace approximation provides an approximation for integrals of the form $\int e^{nl(\theta)} d\theta$ when n is large (for example, De Bruijn, 1981). It states that for a unidimensional θ one has

$$\int e^{nl(\theta)} d\theta \approx \exp(nl(\hat{\theta})) \frac{\hat{\sigma}}{n^{1/2}} \sqrt{2\pi},$$

where $\hat{\theta}$ is the unique maximum of $l(\theta)$ and $\hat{\sigma}^2 = -1/(\partial^2 l(\hat{\theta})/\partial\theta^2)$. The result is easily derived by using the Taylor approximation of second order, $l(\theta) \approx l(\hat{\theta}) + l(\hat{\theta})(\theta - \hat{\theta}) + \frac{1}{2}(\partial^2 l(\hat{\theta})/\partial\theta^2)(\theta - \hat{\theta})^2$.

For a p-dimensional $\boldsymbol{\theta}$ one obtains by the Taylor approximation

$$\int e^{nl(\boldsymbol{\theta})} d\boldsymbol{\theta} \approx e^{nl(\hat{\boldsymbol{\theta}})} \frac{\sqrt{|\hat{\boldsymbol{\Sigma}}|}}{n^{1/2}} (2\pi)^{p/2},$$

where $\hat{\boldsymbol{\theta}}$ is the unique maximum of $l(\boldsymbol{\theta})$ and

$$\hat{\boldsymbol{\Sigma}} = (-\partial l(\hat{\boldsymbol{\theta}})/\partial\boldsymbol{\theta}\partial\boldsymbol{\theta}^T)^{-1}.$$

By substituting $g(\theta) = e^{nl(\theta)}$ one obtains in the unidimensional case

$$\int g(\theta) d\theta \approx g(\hat{\theta}) \hat{\sigma}_g \sqrt{2\pi},$$

where $\hat{\sigma}_g^2 = \hat{\sigma}^2/n = (-\frac{\partial^2 \log g(\hat{\theta})}{\partial\theta^2})^{-1}$ and $\hat{\theta}$ is the unique maximum of $g(\theta)$. For the multivariate case one has

$$\int g(\boldsymbol{\theta}) d\boldsymbol{\theta} \approx g(\hat{\boldsymbol{\theta}}) \hat{\sigma}_g (2\pi)^{p/2},$$

where $\hat{\sigma}_g^2 = |-\partial^2 \log\ g(\hat{\boldsymbol{\theta}})/\partial\boldsymbol{\theta}\partial\boldsymbol{\theta}^T|^{-1}$.

E.2 Gauss-Hermite Integration

For a unidimensional function $g(x)$, the Gauss-Hermite rule approximates integrals of $h(x)\exp(-x^2)$ by

$$\int_{-\infty}^{\infty} h(x)\exp(-x^2)dx \approx \sum_{i=1}^{k} h(\xi_i)w_i,$$

where the node ξ_i is the ith zero of the Hermite polynomial having degree k, and w_i represents fixed weights depending on k. Tables for nodes and weights are given, for example, in Abramowitz and Stegun (1972). The approximation is exact if $h(x)$ is a polynomial of degree $2k-1$. Thus, by increasing k the approximation improves.

An often-used integral is based on the normal density. By simply substitutiing $x = \sqrt{2\pi}\sigma z + \mu$ one obtains

$$\int h(x)\frac{1}{\sqrt{2\pi}\sigma}\exp(-\frac{(x-\mu)^2}{2\sigma^2})dx = \pi^{-1/2}\int h(\sqrt{2}\sigma z+\mu)\exp(-z^2)dz$$

and therefore the *Gauss-Hermite approximation*

$$\int h(x)\phi_{\mu,\sigma}(x) \approx \sum_{i=1}^{k} h(\sqrt{2}\sigma\xi_i+\mu)v_i, \tag{E.1}$$

where $v_i = w_i/\sqrt{\pi}$ is the transformed weight, ξ_i is the tabulated ith zero of the Hermite polynomial, and $\phi_{\mu,\sigma}(x) = (2\pi)^{-1/2}\sigma^{-1/2}\exp(-(x-\mu)^2/(2\sigma^2))$ is the Gaussian density with mean μ and variance σ^2.

The adaptive Gauss-Hermite quadrature aims at sampling in an appropriate region. Thus μ and σ are chosen deliberately. Consider the integral for function g:

$$\int_{-\infty}^{\infty} g(t)dt.$$

By choosing

$$\hat{\mu} = \arg\max_x\ g(x) \quad \hat{\sigma} = (-\frac{\partial^2 \log g(\hat{\mu})}{\partial\mu^2})^{-1},$$

one obtains from (E.1) the *adaptive Gauss-Hermite approximation*:

$$\begin{aligned}\int_{-\infty}^{\infty} g(t)dt &= \int_{-\infty}^{\infty}\frac{g(t)}{\phi_{\hat{\mu},\hat{\sigma}}(t)}\phi_{\hat{\mu},\hat{\sigma}}(t)dt\\ &\approx \sum_i h(\sqrt{2}\hat{\sigma}\xi_i+\hat{\mu})v_i\\ &= \sqrt{2\pi}\hat{\sigma}\sum_i \exp(\xi_i^2)g(\sqrt{2}\hat{\sigma}\xi_i+\hat{\mu})v_i,\end{aligned}$$

where $h(t) = g(t)/\phi_{\hat{\mu},\hat{\sigma}}(t)$. For only one node one obtains the Laplace approximation by inserting the node and weight of the corresponding Hermite polynomial:

$$\int_{-\infty}^{\infty} g(t)dt \approx \sqrt{2\pi}\hat{\sigma}_1\exp(\xi^2)g(\sqrt{2}\hat{\sigma}_1\xi+\hat{\mu})v_1 = \sqrt{2\pi}\hat{\sigma}g(\hat{\mu}).$$

Since the Laplace approximation turns out to be the special case with one node, Liu and Pierce (1994) call (E.1) the k-order Laplace approximation.

E.2.1 Multivariate Gauss-Hermite Integration

The m-dimensional Gauss-Hermite approximation has the form

$$\int h(\boldsymbol{x}) \exp(-\boldsymbol{x}'\boldsymbol{x}) dx = \sum_{i_1=1}^{k_1} \cdots \sum_{i_m=1}^{k_m} w_{i_1}^{(1)} \cdot \ldots \cdot w_{i_m}^{(m)} h(\xi_{i_1}^{(1)}, \ldots \xi_{i_m}^{(m)}),$$

where $w_{i_s}^{(s)}$ are the weights and $\xi_{i_s}^{(s)}$ are the nodes of the ith variable. With multiple index i for $(i_1, \ldots, i_m)$, $\boldsymbol{\xi}_i^T = (\xi_{i_1}^{(1)}, \ldots \xi_{i_m}^{(m)})$, and $w_i = w_{i_1}^{(1)} \cdot \ldots \cdot w_{im}^{(m)}$, one obtains the form

$$\int h(\boldsymbol{x}) \exp(-\boldsymbol{x}^T \boldsymbol{x}) = \sum_i w_i h(\boldsymbol{\xi}_i).$$

The number of terms within the sum of the left-hand side can become rather large. If one uses $k = k_1 = \ldots, k_m$ quadrature points in each component, the sum includes m^k terms. Therefore, m has to be rather small in order to reduce the computational task.

For the adaptive Gauss-Hermite integration one considers the general integral $\int g(\boldsymbol{t}) d\boldsymbol{t}$. By using the density $\phi_{\boldsymbol{\mu}, \boldsymbol{\Sigma}}$ of a normal distribution $N(\boldsymbol{\mu}, \boldsymbol{\Sigma})$, it may be written as

$$\begin{aligned} \int g(\boldsymbol{t}) d\boldsymbol{t} &= \int \{\frac{g(\boldsymbol{t})}{\phi_{\boldsymbol{\mu}, \boldsymbol{\Sigma}}(\boldsymbol{t})}\} \phi_{\boldsymbol{\mu}, \boldsymbol{\Sigma}}(\boldsymbol{t}) d\boldsymbol{t} \\ &= \int \frac{g(\boldsymbol{t})}{\phi_{\boldsymbol{\mu}, \boldsymbol{\Sigma}}(\boldsymbol{t})} |2\pi|^{-m/2} |\boldsymbol{\Sigma}|^{-1/2} \exp(-\frac{1}{2}(\boldsymbol{t} - \boldsymbol{\mu})^T \boldsymbol{\Sigma}^{-1} (\boldsymbol{t} - \boldsymbol{\mu})) d\boldsymbol{t}. \end{aligned}$$

With $\tilde{\boldsymbol{t}} = \frac{1}{\sqrt{2}} \boldsymbol{\Sigma}^{-1/2} (\boldsymbol{t} - \boldsymbol{\mu}) \Leftrightarrow \boldsymbol{t} = \sqrt{2} \boldsymbol{\Sigma}^{1/2} \tilde{\boldsymbol{t}} + \boldsymbol{\mu}$ and $h(\boldsymbol{t}) = g(\boldsymbol{t}) / \phi_{\boldsymbol{\mu}, \boldsymbol{\Sigma}}(\boldsymbol{t})$, one obtains the usual form of Gauss-Hermite integration:

$$\int g(\boldsymbol{t}) d\boldsymbol{t} = \int h(\sqrt{2} \boldsymbol{\Sigma}^{1/2} \tilde{\boldsymbol{t}} + \boldsymbol{\mu}) (2\pi)^{-m/2} \sqrt{2} \exp(\tilde{\boldsymbol{t}}' \tilde{\boldsymbol{t}}) d\tilde{\boldsymbol{t}}.$$

With $w_{i_s}^{(s)}$ and $\xi_{i_s}^{(s)}$ denoting the weights and nodes of the sth variable, one obtains

$$\int g(\boldsymbol{t}) d\boldsymbol{t} \approx 2^{1/2} (2\pi)^{-m/2} \sum_i w_{i_1}^{(1)} \ldots w_{i_m}^{(m)} h(\sqrt{2} \boldsymbol{\Sigma}^{1/2} \xi_i + \boldsymbol{\mu}),$$

where $\boldsymbol{\xi}_i^T = (\xi_{i_1}^{(1)}; \ldots \xi_{i_m}^{(m)})$ and i denotes the multiple index $(i_1, \ldots i_m)$. Therefore,

$$\begin{aligned} \int g(\boldsymbol{t}) d\boldsymbol{t} &\approx 2^{1/2} (2\pi)^{-m/2} \sum_i \frac{w_{i_1}^{(1)} \ldots w_{i_m}^{(m)} g(\sqrt{2} \boldsymbol{\Sigma}^{1/2} \xi_i + \boldsymbol{\mu})}{|2\pi|^{-m/2} |\boldsymbol{\Sigma}|^{-1/2} \exp(-\boldsymbol{\xi}_i^T \boldsymbol{\xi}_i)} \\ &= \sqrt{2} \sum_i |\boldsymbol{\Sigma}|^{1/2} w_{i_1}^{(1)} \ldots w_{i_m}^{(m)} g(\sqrt{2} \boldsymbol{\Sigma}^{1/2} \boldsymbol{\xi}_i + \boldsymbol{\mu}) \exp(\boldsymbol{\xi}_i^T \boldsymbol{\xi}_i). \end{aligned}$$

Up until now $\boldsymbol{\mu}$ and $\boldsymbol{\Sigma}$ have not been specified. To sample $g(t)$ in a suitable range one chooses $\boldsymbol{\mu}$ as the mode of $g(t)$ and $\boldsymbol{\Sigma} = (-\partial^2 / \partial \log\ g(\boldsymbol{t}) \boldsymbol{t} \partial \boldsymbol{t}^T)^{1/2}$.

For the log-likelihood of a mixed model one obtains

$$\begin{aligned} l_i(\boldsymbol{\beta}, \boldsymbol{Q}) &= \int f(\boldsymbol{y}_i, \boldsymbol{b}_i, \boldsymbol{\beta}) \phi_{\boldsymbol{O}, \boldsymbol{Q}}(\boldsymbol{b}_i) d\boldsymbol{b}_i \\ &\approx \sqrt{2} \sum_i |\boldsymbol{\Sigma}|^{1/2} w_{i_1}^{(1)} \ldots w_{i_m}^{(m)} f(\boldsymbol{y}_i, \sqrt{2} \boldsymbol{\Sigma}^{1/2} \boldsymbol{\xi}_i + \boldsymbol{\mu}) \phi_{\boldsymbol{O}, \boldsymbol{Q}}(\sqrt{2} \boldsymbol{\Sigma}^{1/2} \boldsymbol{\xi}_i + \boldsymbol{\mu}) \exp(\boldsymbol{\xi}_i^T \boldsymbol{\xi}_i). \end{aligned}$$

E.3 Inversion of Pseudo-Fisher Matrix

The pseudo-Fisher matrix $\boldsymbol{F}_p(\delta)$ from Section 14.3.3 can be inverted by using the partitioning into

$$\boldsymbol{F}_p(\boldsymbol{\delta}) = \begin{bmatrix} \boldsymbol{F}_{\beta\beta} & \boldsymbol{F}_{\beta 1} & \boldsymbol{F}_{\beta 2} & \cdots & \boldsymbol{F}_{\beta n} \\ \boldsymbol{F}_{1\beta} & \boldsymbol{F}_{11} & & & 0 \\ \boldsymbol{F}_{2\beta} & & \boldsymbol{F}_{22} & & \\ \vdots & & & \ddots & \\ \boldsymbol{F}_{n\beta} & 0 & & & \boldsymbol{F}_{nn} \end{bmatrix},$$

with

$$\begin{aligned} \boldsymbol{F}_{\beta\beta} &= -E\left(\frac{\partial^2 l(\boldsymbol{\delta})}{\partial\boldsymbol{\beta}\partial\boldsymbol{\beta}^T}\right) = \sum_{i=1}^{n} \boldsymbol{X}_i^T \boldsymbol{D}_i(\boldsymbol{\delta})\boldsymbol{\Sigma}_i^{-1}(\boldsymbol{\delta})\boldsymbol{D}_i(\boldsymbol{\delta})\boldsymbol{X}_i, \\ \boldsymbol{F}_{\beta i} &= \boldsymbol{F}_{i\beta}^T = -E\left(\frac{\partial^2 l(\boldsymbol{\delta})}{\partial\boldsymbol{\beta}\partial\boldsymbol{b}_i^T}\right) = \boldsymbol{X}_i^T \boldsymbol{D}_i(\boldsymbol{\delta})\boldsymbol{\Sigma}_i^{-1}(\boldsymbol{\delta})\boldsymbol{D}_i(\boldsymbol{\delta})\boldsymbol{Z}_i, \\ \boldsymbol{F}_{ii} &= -E\left(\frac{\partial^2 l(\boldsymbol{\delta})}{\partial\boldsymbol{b}_i\partial\boldsymbol{b}_i^T}\right) = \boldsymbol{Z}_i^T \boldsymbol{D}_i(\boldsymbol{\delta})\boldsymbol{\Sigma}_i^{-1}(\boldsymbol{\delta})\boldsymbol{D}_i(\boldsymbol{\delta})\boldsymbol{Z}_i + \boldsymbol{Q}^{-1}. \end{aligned}$$

By using standard results for inverting partitioned matrices one obtains for the inverse the easy-to-compute form

$$\boldsymbol{F}^{-1}(\boldsymbol{\delta}) = \begin{bmatrix} \boldsymbol{V}_{\beta\beta} & \boldsymbol{V}_{\beta 1} & \boldsymbol{V}_{\beta 2} & \cdots & \boldsymbol{V}_{\beta n} \\ \boldsymbol{V}_{1\beta} & \boldsymbol{V}_{11} & \boldsymbol{V}_{12} & \cdots & \boldsymbol{V}_{1n} \\ \boldsymbol{V}_{2\beta} & \boldsymbol{V}_{21} & \boldsymbol{V}_{22} & \cdots & \boldsymbol{V}_{2n} \\ \vdots & \vdots & & & \vdots \\ \boldsymbol{V}_{n\beta} & \boldsymbol{V}_{n1} & \cdots & \cdots & \boldsymbol{V}_{nn} \end{bmatrix},$$

with

$$V_{\beta\beta} = (\boldsymbol{F}_{\beta\beta} - \sum_{i=1}^{n} \boldsymbol{F}_{\beta i}\boldsymbol{F}_{ii}^{-1}\boldsymbol{F}_{i\beta})^{-1}, \quad \boldsymbol{V}_{\beta i} = \boldsymbol{V}_{i\beta}^T = -\boldsymbol{V}_{\beta\beta}\boldsymbol{F}_{\beta i}\boldsymbol{F}_{ii}^{-1},$$

$$\boldsymbol{V}_{ii} = \boldsymbol{F}_{ii}^{-1} + \boldsymbol{F}_{ii}^{-1}\boldsymbol{F}_{i\beta}\boldsymbol{V}_{\beta\beta}\boldsymbol{F}_{\beta i}\boldsymbol{F}_{ii}^{-1}, \quad \boldsymbol{V}_{ij} = \boldsymbol{V}_{ji}^T = \boldsymbol{F}_{ii}^{-1}\boldsymbol{F}_{i\beta}\boldsymbol{V}_{\beta\beta}\boldsymbol{F}_{\beta j}\boldsymbol{F}_{jj}^{-1}, \; i \neq j.$$

List of Examples

Bibliography

Abe, M. (1999). A generalized additive model for discrete-choice data. *Journal of Business & Economic Statistics 17*, 271–284.

Abramowitz, M. and I. Stegun (1972). *Handbook of Mathematical Functions*. New York: Dover.

Agrawala, A. K. (1977). *Machine Recognition of Patterns*. New York: IEEE Press.

Agresti, A. (1986). Applying R^2-type measures to ordered categorical data. *Technometrics 28*, 133–138.

Agresti, A. (1992a). *Analysis of Ordinal Categorical Data*. New York: Wiley.

Agresti, A. (1992b). A survey of exact inference for contingency tables. *Statistical Science 7*, 131–153.

Agresti, A. (1993a). Computing Conditional Maximum-Likehood-Estimates for Generalized Rasch Models using simple Loglinear Models with Diagonals Parameters. *Scandinavian Journal of Statistics 20*, 63–71.

Agresti, A. (1993b). Computing conditional maximum likelihood estimates for generalized Rasch models using simple loglinear models with diagonal parameters. *Scandinavian Journal of Statistics 20*, 63–72.

Agresti, A. (1993c). Distribution-free fitting of logit models with random effects for repeated categorical responses. *Statistics in Medicine 12*, 1969–1988.

Agresti, A. (1993d). Distribution-Free Fitting of Logit-Models with Random Effects for Repeated Categorical Responses. *Statistics in Medicine 12*, 1969–1987.

Agresti, A. (1997). A model for repeated measurements of a multivariate binary response. *Journal of the American Statistical Association 92*, 315–321.

Agresti, A. (2001). Exact inference for categorical data: recent advances and continuing controversies. *Statistics in Medicine 20*(17-18), 2709–2722.

Agresti, A. (2002). *Categorical Data Analysis*. New York: Wiley.

Agresti, A. (2009). *Analysis of Ordinal Categorical Data, 2nd Edition*. New York: Wiley.

Agresti, A., B. Caffo, and P. Ohman-Strickland (2004). Examples in which misspecification of a random effects distribution reduces efficiency, and possible remedies. *Computational Statistics and Data Analysis 47*, 639–653.

Agresti, A. and A. Kezouh (1983). Association models for multi-dimensional cross-classifications of ordinal variables. *Communications in Statistics, Part A – Theory Meth. 12*, 1261–1276.

Agresti, A. and J. Lang (1993). A proportional odds model with subject-specific effects for repeated ordered categorical responses. *Biometrika 80*, 527.

Aitkin, M. (1979). A simultaneous test procedure for contingency table models. *Journal of Applied Statistics 28*, 233–242.

Aitkin, M. (1980). A note on the selection of log-linear models. *Biometrics 36*, 173–178.

Aitkin, M. (1999). A general maximum likelihood analysis of variance components in generalized linear models. *Biometrics 55*, 117–128.

Akaike, H. (1973). In B. Petrov and F. Caski (Eds.), *Information Theory and the Extension of the Maximum Likelihood Principle*, Second International Symposium on Information Theory. Akademia Kiado.

Akaike, H. (1974). A new look at statistical model identification. *IEEE Transactions on Automatic Control 19*, 716–723.

Albert, A. and J. A. Anderson (1984). On the existence of maximum likelihood estimates in logistic regression models. *Biometrika 71*, 1–10.

Albert, A. and E. Lesaffre (1986). Multiple group logistic discrimination. *Computers and Mathematics with Applications 12*, 209–224.

Albert, J. H. and S. Chib (2001). Sequential ordinal modelling with applications to survival data. *Biometrics 57*, 829–836.

Amemiya, T. (1978). On two-step estimation of a multivariate logit model. *Journal of Econometrics 19*, 13–21.

Amemiya, T. (1981). Qualitative response models: A survey. *Journal of Economic Literature XIX*, 1483–1536.

Ananth, C. V. and D. G. Kleinbaum (1997). Regression models for ordinal responses: A review of methods and applications. *International Journal of Epidemiology 26*, 1323–1333.

Anbari, M. E. and A. Mkhadri (2008). Penalized regression combining the l1 norm and a correlation based penalty. Technical Report 6746, Institut National de recherche en informatique et en automatique.

Anderson, D. A. and M. Aitkin (1985). Variance component models with binary response: Interviewer variability. *Journal of the Royal Statistical Society Series B 47*, 203–210.

Anderson, D. A. and J. P. Hinde (1988). Random effects in generalized linear models and the EM algorithm. *Communications in Statistics A – Theory and Methods 17*, 3847–3856.

Anderson, J. A. (1972). Separate sample logistic discrimination. *Biometrika 59*, 19–35.

Anderson, J. A. (1984). Regression and ordered categorical variables. *Journal of the Royal Statistical Society B 46*, 1–30.

Anderson, J. A. and V. Blair (1982). Penalized maximum likelihood estimation in logistic regression and discrimination. *Biometrika 69*, 123–136.

Anderson, J. A. and R. R. Phillips (1981). Regression, discrimination and measurement models for ordered categorical variables. *Applied Statistics 30*, 22–31.

Aranda-Ordaz, F. J. (1983). An extension of the proportional-hazard-model for grouped data. *Biometrics 39*, 109–118.

Armstrong, B. and M. Sloan (1989). Ordinal regression models for epidemiologic data. *American Journal of Epidemiology 129*, 191–204.

Atkinson, A. and M. Riani (2000). *Robust Diagnostic Regression Analysis*. New York: Springer-Verlag.

Avalos, M., Y. Grandvalet, and C. Ambroise (2007). Parsimonious additive models. *Computational Statistics & Data Analysis 51*(6), 2851–2870.

Azzalini, A. (1994). Logistic regression for autocorrelated data with application to repeated measures. *Biometrika 81*, 767–775.

Azzalini, A., A. W. Bowman, and W. Härdle (1989). On the use of nonparametric regression for linear models. *Biometrika 76*, 1–11.

Barla, A., G. Jurman, S. Riccadonna, S. Merler, M. Chierici, and C. Furlanello (2008). Machine learning methods for predictive proteomics. *Briefings in Bioinformatics 9*, 119–128.

Baumgarten, M., P. Seliske, and M. S. Goldberg (1989). Warning re. the use of GLIM macros for the estimation of risk ratio. *American Journal of Epidemiology 130*, 1065.

Begg, C. and R. Gray (1984). Calculation of polytomous logistic regression parameters using individualized regressions. *Biometrika 71*, 11–18.

Belitz, C. and S. Lang (2008). Simultaneous selection of variables and smoothing parameters in structured additive regression models. *Computational Statistics and Data Analysis 51*, 6044–6059.

Bell, R. (1992). Are ordinal models useful for classification? *Statistics in Medicine 11*(1), 133–134.

Bellman, R. (1961). *Adaptive Control Processes*. Princeton University Press.

Ben-Akiva, M. E. and S. R. Lerman (1985). *Discrete Choice Analysis: Theory and Application to Travel Demand*. Cambridge, MA: MIT Press.

Bender, R. and U. Grouven (1998). Using binary logistic regression models for ordinal data with non–proportional odds. *Journal of Clinical Epidemiology 51*, 809–816.

Benedetti, J. K. and M. B. Brown (1978). Strategies for the selection of loglinear models. *Biometrics 34*, 680–686.

Berger, R. L. and K. Sidik (2003). Exact unconditional tests for a 2×2 matched pairs design. *Statistical Methods in Medical Research 12*, 91–108.

Bergsma, W., M. Croon, and J. Hagenaars (2009). *Marginal Models*. New York: Springer–Verlag.

Berkson, J. (1994). Application of the logistic function to bio-assay. *Journal of the American Statistical Association 9*, 357–365.

Besag, J., P. J. Green, D. Higdon, and K. Mengersen (1995). Bayesian computation and stochastic systems. *Statistical Science 10*, 3–66.

Bhapkar, V. P. (1966). A note on the equivalence of two test criteria for hypotheses in categorical data. *Journal of the American Statistical Association 61*, 228–235.

Binder, H. and G. Tutz (2008). A comparison of methods for the fitting of generalized additive models. *Statistics and Computing 18*, 87–99.

Bishop, C. M. (2006). *Pattern Recognition and Machine Learning*. New York: Springer–Verlag.

Bishop, Y., S. Fienberg, and P. Holland (1975). *Discrete Multivariate Analysis*. Cambridge, MA: MIT Press.

Bishop, Y., S. Fienberg, and P. Holland (2007). *Discrete Multivariate Analysis*. New York: Springer–Verlag.

Blanchard, G., O. Bousquet, and P. Massart (2008). Statistical performance of support vector machines. *Annals of Statistics 36*(2), 489–531.

Bliss, C. I. (1934). The method of probits. *Science 79*, 38–39.

Böckenholt, U. and W. R. Dillon (1997). Modelling within – subject dependencies in ordinal paired comparison data. *Psychometrika 62*, 412–434.

Bondell, H. D. and B. J. Reich (2008). Simultaneous regression shrinkage, variable selection and clustering of predictors with oscar. *Biometrics 64*, 115–123.

Bondell, H. D. and B. J. Reich (2009). Simultaneous factor selection and collapsing levels in anova. *Biometrics 65*, 169–177.

Bonney, G. E. (1987). Logistic regression for dependent binary observations. *Biometrics 43*, 951–973.

Booth, J. G. and J. P. Hobert (1999). Maximizing generalized linear mixed model likelihoods with an automated Monte Carlo EM algorithm. *Journal of the Royal Statistical Society B 61*, 265–285.

Börsch-Supan, A. (1987). *Econometric Analysis of Discrete Choice, with Applications on the Demand for Housing in the U.S. and West-Germany*. Berlin: Springer-Verlag.

Boulesteix, A., C. Strobl, T. Augustin, and M. Daumer. Evaluating microarray-based classifiers: an overview. *Cancer Informatics 6*, 77–97.

Boulesteix, A.-L. (2006). Maximally selected chi-squared statistics for ordinal variables. *Biometrical Journal 48*, 451–462.

Boyles, R. A. (1983). On the covergence of the EM algorithm. *Journal of the Royal Statistical Society B 45*, 47–50.

Bradley, R. A. (1976). Science, statistics, and paired comparison. *Biometrics 32*, 213–232.

Bradley, R. A. (1984). Paired comparisons: Some basic procedures and examples. In P. Krishnaiah and P. R. Sen (Eds.), *Handbook of Statistics*, Volume 4, pp. 299–326. Elsevier.

Bradley, R. A. and M. E. Terry (1952). Rank analysis of incomplete block designs, I: The method of pair comparisons. *Biometrika 39*, 324–345.

Braga-Neto, U. and E. R. Dougherty (2004). Is cross-validation valid for small-samplemicroarray classification? *Bioinformatics 20*, 374–380.

Brant, R. (1990). Assessing proportionality in the proportional odds model for ordinal logistic regression. *Biometrics 46*, 1171–1178.

Breiman, L. (1995). Better subset regression using the nonnegative garrotte. *Technometrics 37*, 373–384.

Breiman, L. (1996a). Bagging predictors. *Machine Learning 24*, 123–140.

Breiman, L. (1996b). Heuristics of instability and stabilisation in model selection. *Annals of Statistics 24*, 2350–2383.

Breiman, L. (1998). Arcing classifiers. *Annals of Statistics 26*, 801–849.

Breiman, L. (1999). Prediction games and arcing algorithms. *Neural Computation 11*, 1493–1517.

Breiman, L. (2001a). Random forests. *Machine Learning 45*, 5–32.

Breiman, L. (2001b). Statistical modelling. The two cultures. *Statistical Science 16*, 199–231.

Breiman, L., J. H. Friedman, R. A. Olshen, and J. C. Stone (1984). *Classification and Regression Trees*. Monterey, CA: Wadsworth.

Breslow, N. E. (1984). Extra-poisson variation in log-linear models. *Applied Statistics 33*, 38–44.

Breslow, N. E. and D. G. Clayton (1993). Approximate inference in generalized linear mixed model. *Journal of the American Statistical Association 88*, 9–25.

Breslow, N. E., K. Halvorsen, R. Prentice, and C. Sabai (1978). Estimation of multiple relative risk functions in matched case-control studies. *American Journal of Epidemiology 108*, 299–307.

Breslow, N. E. and X. Lin (1995). Bias correction in generalized linear mixed models with a single component of dispersion. *Biometrika 82*, 81–91.

Brezger, A. and S. Lang (2006). Generalized additive regression based on Bayesian p-splines. *Computational Statistics and Data Analysis 50*, 967–991.

Brillinger, D. R. and M. K. Preisler (1983). Maximum likelihood estimation in a latent variable problem. In T. Amemiya, S. Karlin, and T. Goodman (Eds.), *Studies in Econometrics, Time Series, and Multivariate Statistics*, pp. 31–65. New York: Academic Press.

Brown, C. C. (1982). On a goodness-of-fit test for the logistic models based on score statistics. *Communicattions in Statistics: Theory and Methods 11*, 1087–1105.

Brown, M. B. (1976). Screening effects in multidimensional contingency tables. *Journal of Applied Statistics 25*, 37–46.

Brownstone, D. and K. Small (1989). Efficient estimation of nested logit models. *Journal of Business & Economic Statistics 7*, 67–74.

Bühlmann, P. (2006). Boosting for high-dimensional linear models. *Annals of Statistics 34*, 559–583.

Bühlmann, P. and T. Hothorn (2007). Boosting algorithms: regularization, prediction and model fitting (with discussion). *Statistical Science 22*, 477–505.

Bühlmann, P. and S. Van De Geer (2011). *Statistics for High-Dimensional Data: Methods, Theory and Applications*. Springer-Verlag New York.

Bühlmann, P. and B. Yu (2003). Boosting with the L2 loss: Regression and classification. *Journal of the American Statistical Association 98*, 324–339.

Buja, A., T. Hastie, and R. Tibshirani (1989). Linear smoothers and additive models. *Annals of Statistics 17*, 453–510.

Buja, A., W. Stuetzle, and Y. Shen (2005). Loss functions for binary class probability estimation and classification: Structure and applications. Manuscript, Department of Statistics, University of Pennsylvania, Philadelphia.

Burnham, K. and D. Anderson (2004). Multimodel inference – understanding AIC and BIC in model selection. *Sociological Methods & Research 33*, 261–304.

Burnham, K. P. and D. R. Anderson (2002). *Model Selection and Multimodel Inference: A Practical Information–Theoretic Approach*. New York: Springer–Verlag.

Caffo, B., M.-W. An, and C. Rhode (2007). Flexible random intercept models for binary outcomes using mixtures of normals. *Computational Statistics & Data Analysis 51*, 5220–5235.

Cameron, A. C. and P. K. Trivedi (1998). *Regression Analysis of Count Data. Econometric Society Monographs No. 30*. Cambridge: Cambridge University Press.

Campbell, G. (1980). Shrunken estimators in discriminant and canonical variate analysis. *Applied Statistics 29*, 5–14.

Campbell, G. (1994). Advances in statistical methodology for the evaluation of diagnostic and laboratory tests. *Statistics in Medicine 13*, 499–508.

Campbell, M. K. and A. P. Donner (1989). Classification efficiency of multinomial logistic-regression relative to ordinal logistic-regression. *Journal of the American Statistical Association 84*(406), 587–591.

Campbell, M. K., A. P. Donner, and K. M. Webster (1991). Are ordinal models useful for classification? *Statistics in Medicine 10*, 383–394.

Candes, E. and T. Tao (2007). The Dantzig selector: Statistical estimation when p is much larger than n. *Annals of Statistics 35*(6), 2313–2351.

Cantoni, E., J. Flemming, and E. Ronchetti (2005). Variable selection for marginal longitudinal generalized linear models. *Biometrics 61*, 507–514.

Carroll, R., S. Wang, and C. Wang (1995). Prospective Analysis of Logistic Case-Control Studies. *Journal of the American Statistical Association 90*(429).

Carroll, R. J., J. Fan, I. Gijbels, and M. P. Wand (1997). Generalized partially linear single-index models. *Journal of the American Statistical Association 92*, 477–489.

Carroll, R. J. and S. Pederson (1993). On robustness in the logistic regression model. *Journal of the Royal Statistical Society B 55*, 693–706.

Carroll, R. J., D. Ruppert, and A. H. Welsh (1998). Local estimating equations. *Journal of the American Statistical Association 93*, 214–227.

Celeux, G. and J. Diebolt (1985). The SEM algorithm: A probabilistic teacher algorithm derived fom EM algorithm for the mixture Problem. *Computational. Statistics 2*, 73–82.

Chaganty, N. and H. Joe (2004). Efficiency of generalized estimation equations for binary responses. *Journal of the Royal Statistical Society B 66*, 851–860.

Chambers, J. M. and T. J. Hastie (1992). *Statistical Models in S.* Pacific Grove, CA: Wadsworth Brooks/Cole.

Chen, J. and M. Davidian (2002). A monte carlo EM algorithm for generalized linear models with flexible random effects distribution. *Biostatistics 3*, 347–360.

Chen, S. S., D. L. Donoho, and M. A. Saunders (2001). Atomic decomposition by basis pursuit. *Siam Review 43*(1), 129–159.

Chib, S. and B. Carlin (1999). On mcmc sampling in hierarchical longitudinal models. *Statistics and Computing 9*, 17–26.

Christensen, R. (1997). *Log-linear Models and Logistic Regression.* New York: Springer-Verlag.

Christmann, A. and P. J. Rousseeuw (2001). Measuring overlap in binary regression. *Computational Statistics and Data Analysis 37*, 65–75.

Chu, W. and S. Keerthi (2005). New approaches to support vector ordinal regression. In *Proceedings of the 22nd international conference on Machine learning*, pp. 145–152. ACM.

Ciampi, A., C.-H. Chang, S. Hogg, and S. McKinney (1987). Recursive partitioning: A versatile method for exploratory data analysis in biostatistics. In I. McNeil and G. Umphrey (Eds.), *Biostatistics.* New York: D. Reidel Publishing.

Claeskens, G. and J. D. Hart (2009). Goodness-of-fit tests in mixed models. *TEST 18*, 213–239.

Claeskens, G. and N. Hjort (2008). *Model Selection and Model Averaging.* Cambridge University Press.

Claeskens, G., T. Krivobokova, and J. D. Opsomer (2009). Asymptotic properties of penalized spline estimators. *Biometrika 96*(3), 529–544.

Clark, L. and D. Pregibon (1992). Tree-based models. In J. Chambers and T. Hastie (Eds.), *Statistical Models in S*, pp. 377–420. Pacific Grove, California: Wadsworth & Brooks.

Cleveland, W. S. and C. Loader (1996). Smoothing by local regression: Principles and methods. In W. Härdle and M. Schimek (Eds.), *Statistical Theory and Computational Aspects of Smoothing*, pp. 10–49. Heidelberg: Physica-Verlag.

Cochran, W. (1954). Some methods for strengthening the common χ 2 tests. *Biometrics 10*, 417–451.

Colonius, H. (1980). Representation and uniqueness of the Bradley-Terry-Luce model for paired comparisons. *British Journal of Mathematical & Statistical Psychology 33*, 99–103.

Conaway, M. R. (1989). Analysis of repeated categorical measurements with conditional likelihood methods. *Journal of the American Statistical Association 84*, 53–62.

Conolly, M. A. and K. Y. Liang (1988). Conditional logistic regression models for correlated binary data. *Biometrika 75*, 501–506.

Consul, P. C. (1998). *Generalized Poisson distributions.* New York: Marcel Dekker.

Cooil, B. and R. T. Rust (1994). Reliability and Expected Loss: A Unifying Principle. *Psychometrika 59*, 203–216.

Cook, R. D. (1977). Detection of influential observations in linear regression. *Technometrics 19*, 15–18.

Cook, R. D. and S. Weisberg (1982). *Residuals and Influence in Regression.* London: Chapman & Hall.

Copas, J. B. (1988). Binary regression models for contaminated data (with discussion). *Journal of the Royal Statistical Society B 50*, 225–265.

Cordeiro, G. and P. McCullagh (1991). Bias correction in generalized linear models. *Journal of the Royal Statistical Society. Series B (Methodological) 53*, 629–643.

Cornell, R. G. and J. A. Speckman (1967). Estimation for a simple exponential model. *Biometrics 23*, 717–737.

Coste, J., E. Walter, D. Wasserman, and A. Venot (1997). Optimal discriminant analysis for ordinal responses. *Statistics in Medicine 16*(5), 561–569.

Cover, T. M. and P. E. Hart (1967). Nearest neighbor pattern classification. *IEEE Transactions on Information Theory 13*, 21–27.

Cox, C. (1995). Location-scale cumulative odds models for ordinal data: A generalized nonlinear model approach. *Statistics in Medicine 14*, 1191–1203.

Cox, D. (1958). Two further applications of a model for binary regression. *Biometrika 45*, 562–565.

Cox, D. R. and D. V. Hinkley (1974). *Theoretical Statistics.* London: Chapman & Hall.

Cox, D. R. and N. Reid (1987). Approximations to noncentral distributions. *Canadian Journal of Statistics 15*, 105–114.

Cox, D. R. and E. J. Snell (1989). *Analysis of Binary Data (Second Edition).* London; New York: Chapman & Hall.

Cox, D. W. J. and N. Wermuth (1992). A comment on the coefficient of determination for binary responses. *American Statistician 46*, 1–4.

Cramer, J. S. (1991). *The Logit Model.* New York: Routhedge, Chapman & Hall.

Cramer, J. S. (2003). The origins and development of the logit model. Manuscript, University of Amsterdam and Tinbergen Institute.

Creel, M. D. and J. B. Loomis (1990). Theoretical and empirical advantages of truncated count data estimators for analysis of deer hunting in California. *Journal of Agricultural Economics 72*, 434–441.

Crowder, M. (1995). On the use of working correlation matrix in using generalized linear models for repeated measuresh. *Biometrika 82*, 407–410.

Crowder, M. J. (1987). Beta-binomial ANOVA for proportions. *Journal of the Royal Statistical Society 27*, 34–37.

Currie, I., M. Durban, and H. P. Eilers (2004). Smoothing and forecasting mortality rates. *Statistical Modelling 4*, 279–298.

Czado, C. (1992). On link selection in generalized linear models. In S. L. N. in Statistics (Ed.), *Advances in GLIM and Statistical Modelling*. New York: Springer–Verlag. 78, 60–65.

Czado, C. (1997). On selecting parametric link transformation families in generalized linear models. *Journal of Statistical Planning and Inference 61*(1), 125–139.

Czado, C., V. Erhardt, A. Min, and S. Wagner (2007). Zero-inflated generalized poisson models with regression effects on the mean. dispersion and zero-inflation level applied to patent outsourcing rates. *Statistical Modelling 7(2)*, 125–153.

Czado, C. and A. Munk (2000). Noncanonical links in generalized linear models – When is the effort justified? *Journal of Statistical Planning and Inference 87*(2), 317–345.

Czado, C. and T. Santner (1992). The effect of link misspecification on binary regression inference. *Journal of Statistical Planning and Inference 33*(2), 213–231.

Dahinden, C., G. Parmigiani, M. C. Emerick, and P. Bühlmann (2007). Penalized likelihood for sparse contingency tables with application to full-length cDNA libraries. *BMC Bioinformatics 8*, 476.

Dale, J. R. (1986). Global cross-ratio models for bivariate, discrete, ordered responses. *Biometrics 42*, 909–917.

Darroch, J. N., S. L. Lauritzen, and T. P. Speed (1980). Markov fields and log-linear interaction models for contingency tables. *Annals of Statistics 8(3)*, 522–539.

Davidson, R. (1970). On extending the Bradley-Terry model to accommodate ties in paired comparison experiments. *Journal of the American Statistical Association 65*, 317–328.

Davis, C. S. (1991). Semi-parametric and non-parametric methods for the analysis of repeated measurements with applications to clinical trials. *Statistics in Medicine 10*, 1959–1980.

Day, N. and D. Kerridge (1967). A general maximum likelihood discriminant. *Biometrics 23*, 313–323.

Daye, Z. and X. Jeng (2009). Shrinkage and model selection weighted fusion. *Computational Statistics and Data Analysis 53*, 1284–1298.

De Boeck, P. and M. Wilson (2004). *Explanatory item response models: A generalized linear and nonlinear approach.* Springer Verlag.

De Bruijn, N. G. (1981). *Asymptotic Methods in Analysis.* Dover.

Dean, C., J. F. Lawless, and G. E. Willmot (1989). A mixed poisson-inverse Gaussian regression model. *The Canadian Journal of Statistics 17*, 171–181.

Deb, P. and P. K. Trivedi (1997). Demand for medical care by the elderly: A finite mixture approach. *Journal of Applied Econometrics 12*(3), 313–336.

Delyon, B., M. Lavielle, and E. Moulines (1999). Convergence of a stochastic approximation version of the EM algorithm. *Annals of Statistics 27*, 94–128.

Deming, W. E. and F. F. Stephan (1940). On a least squares adjustment of a sampled frequency table when the expected marginal totals are known. *Annals of Mathematical Statistics 11*, 427–444.

Demoraes, A. R. and I. R. Dunsmore (1995). Predictive comparisons in ordinal models. *Communications in Statistics – Theory and Methods 24*(8), 2145–2164.

Dempster, A. P., N. M. Laird, and D. B. Rubin (1977). Maximum likelihood from incomplete data via the EM algorithm. *Journal of the Royal Statistical Society B 39*, 1–38.

Deng, P. and S. R. Paul (2005). Score tests for zero-inflation and over-dispersion in generalized linear models. *Statistica Sinica 15(1)*, 257–276.

Denison, D. G. T., B. K. Mallick, and A. F. M. Smith (1998). Automatic Bayesian curve fitting. *Journal of the Royal Statistical Society B 60*, 333–350.

Dettling, M. and P. Bühlmann (2003). Boosting for tumor classification with gene expression data. *Bioinformatics 19*, 1061–1069.

Dey, D. K., S. K. Ghosh, and B. K. Mallick (2000). *Generalized Linear Models: A Bayesian Perspective*. New York: Marcel Dekker.

Diaz-Uriarte, R. and S. de Andres (2006a). Gene selection and classification of microarray data using random forest. *BMC Bioinformatics 7*, 3.

Diaz-Uriarte, R. and S. A. de Andres (2006b). Gene selection and classification of microarray data using random forest. *Bioinformatics 7*, 3.

Dierckx, P. (1993). *Curve and Surface Fitting with Splines*. Oxford: Oxford Science Publications.

Dietterich, T. (2000). An experimental comparison of three methods for constructing ensembles of decision trees: Bagging boosting and randomization. *Machine Learning 40*(2), 139–157.

Diggle, P. J., P. J. Heagerty, K. Y. Liang, and S. L. Zeger (2002). *Analysis of Longitudinal Data (Second Edition)*. London: Chapman & Hall.

Dillon, W. R., A. Kumar, and M. de Borrero (1993). Capturing individual differences in paired comparisons: An extended BTL model incorporating descriptor variables. *Journal of Marketing Research 30*, 42–51.

Dittrich, R., R. Hatzinger, and W. Katzenbeisser (1998). Modelling the effect of subject-specific covariates in paired comparison studies with an application to university rankings. *Applied Statistics 47*, 511–525.

Dobson, A. J. (1989). *Introduction to Statistical Modelling*. London: Chapman & Hall.

Dodd, L. and M. Pepe (2003). Partial AUC estimation and regression. *Biometrics 59*(3), 614–623.

Domeniconi, C., J. Peng, and D. Gunopulos (2002). Locally adaptive metric nearest-neighbor classification. *IEEE Transactions on Pattern Analysis and Machine Intelligence 24*, 1281–1285.

Domeniconi, C. and B. Yan (2004). Nearest neighbor ensemble. In *Proc. of the 17th International Conference on Pattern Recognition*, Volume 1, pp. 228–231.

Duchon, J. (1977). Splines minimizing rotation-invariant semi-norms in solobev spaces. In W. Schemp and K. Zeller (Eds.), *Construction Theory of Functions of Several Variables*, pp. 85–100. Berlin: Springer-Verlag.

Dudoit, S., J. Fridlyand, and T. P. Speed (2002). Comparison of discrimination methods for the classification of tumors using gene expression data. *Journal of the American Statistical Association 97*, 77–87.

Duffy, D. E. and T. J. Santner (1989). On the small sample properties of restricted maximum likelihood estimators for logistic regression models. *Communication in Statistics – Theory & Methods 18*, 959–989.

Edwards, D. and T. Havranek (1987). A fast model selection procedure for large family of models. *Journal of the American Statistical Association 82*, 205–213.

Efron, B. (1975). The efficiency of logistic regression compared to normal discriminant analysis. *Journal of the American Statistical Association 70*, 892–898.

Efron, B. (1978). Regression and ANOVA with zero–one data: Measures of residual variation. *Journal of the American Statistical Association 73*, 113–121.

Efron, B. (1983). Estimating the error rate of a prediction rule: improvement on cross-validation. *Journal of the American Statistical Association 78*(382), 316–331.

Efron, B. (1986). Double exponential families and their use in generalized linear regression. *Journal of the American Statistical Association 81*, 709–721.

Efron, B. (2004). The estimation of prediction error: Covariance penalties and cross-validation. *Journal of the American Statistical Association 99*, 619–632.

Efron, B., T. Hastie, I. Johnstone, and R. Tibshirani (2004). Least angle regression. *Annals of Statistics 32*, 407–499.

Efron, B. and R. Tibshirani (1997). Improvements on cross-validation: The .632+ bootstrap method. *Journal of the American Statistical Association 92*, 548–60.

Eilers, P. H. C. and B. D. Marx (2003). Multivariate calibration with temperature interaction using two-dimensional penalized signal regression. *Chemometrics and intelligent laboratory systems 66*, 159–174.

Everitt, B. and T. Hothorn (2006). *A Handbook of Statistical Analyses Using R*. New York: Chapman & Hall.

Fahrmeir, L. (1987). Asymptotic likelihood inference for nonhomogeneous observations. *Statistische Hefte (N.F.) 28*, 81–116.

Fahrmeir, L. (1994). Dynamic modelling and penalized likelihood estimation for discrete time survival data. *Biometrika 81*(2), 317.

Fahrmeir, L. and H. Frost (1992). On stepwise variable selection in generalized linear regression and time series models. *Computational Statistics 7*, 137–154.

Fahrmeir, L. and A. Hamerle (1984). *Multivariate statistische Verfahren*. Berlin / New York: de Gruyter.

Fahrmeir, L. and H. Kaufmann (1985). Consistency and asymptotic normality of the maximum likelihood estimator in generalized linear models. *Annals of Statistics 13*, 342–368.

Fahrmeir, L. and H. Kaufmann (1987). Regression model for nonstationary categorical time series. *Journal of Time Series Analysis 8*, 147–160.

Fahrmeir, L. and T. Kneib (2009). Bayesian regularisation in structured additive regression: A unifying perspective on shrinkage, smoothing and predictor selection. *Statistics and Computing 2*, 203–219.

Fahrmeir, L. and T. Kneib (2010). *Bayesian Smoothing and Regression for Longitudinal, Spatial and Event History Data*. Oxford: Clarendon Press.

Fahrmeir, L., T. Kneib, and S. Lang (2004). Penalized structured additive regression for space-time data: a Bayesian perspective . *Statistica Sinica 14*, 715–745.

Fahrmeir, L., T. Kneib, S. Lang, and B. Marx (2011). *Regression. Models, Methods and Applications*. Berlin: Springer Verlag.

Fahrmeir, L. and S. Lang (2001). Bayesian inference for generalized additive mixed models based on Markov random field priors. *Applied Statistics 50*(2), 201–220.

Fahrmeir, L. and L. Pritscher (1996). Regression analysis of forest damage by marginal models for correlated ordinal responses. *Journal of Environmental and Ecological Statistics 3*, 257–268.

Fahrmeir, L. and G. Tutz (2001). *Multivariate Statistical Modelling based on Generalized Linear Models*. New York: Springer.

Famoye, F. and K. P. Singh (2003). On inflated generalized Poisson regression models. *Advanced Applied Statistics 3(2)*, 145–158.

Famoye, F. and K. P. Singh (2006). Zero-inflated generalized Poisson model with an application to domestic violence data. *Journal of Data Science 4(1)*, 117–130.

Fan, J. and I. Gijbels (1996). *Local Polynomial Modelling and Its Applications*. London: Chapman & Hall.

Fan, J. and R. Li (2001). Variable selection via nonconcave penalize likelihood and its oracle properties. *Journal of the American Statistical Association 96*, 1348–1360.

Fan, J. and W. Zhang (1999). Statistical estimation in varying coefficient models. *Annals of Statistics 27*(5), 1491–1518.

Faraway, J. (2006). *Extending the Linear Model with R*. London: Chapman & Hall.

Fienberg, S. E. (1980). *The Analysis of Cross-classified Categorical Data*. Cambridge: MIT Press.

Finney, D. (1947). The estimation from individual records of the relationship between dose and quantal response. *Biometrika 34*, 320–334.

Firth, D. (1987). On the efficiency of quasi-likelihood estimation. *Biometrika 74*, 233–245.

Firth, D. (1991). Generalized linear models. In D. V. Hinkley, N. Reid, and E. J. Snell (Eds.), *Statistical Theory and Modelling*. London: Chapman & Hall.

Firth, D. (1993). Bias reduction of maximum likelihood estimates. *Biometrika 80*(1), 27–38.

Firth, D. and R. De Menezes (2004). Quasi-variances. *Biometrika 91*, 65.

Fitzmaurice, G. M. (1995). A caveat concerning independence estimating equations with multivariate binary data. *Biometrics 51*, 309–317.

Fitzmaurice, G. M. and N. M. Laird (1993). A likelihood-based method for analysing longitudinal binary responses. *Biometrika 80*, 141–151.

Fitzmaurice, G. M., N. M. Laird, and J. H.Ware (2004). *Applied Longitudinal Analysis*. New York: Wiley.

Fix, E. and J. L. Hodges (1951). Discriminatory analysis-nonparametric discrimination: Consistency properties. US Air Force School of Aviation Medicine, Randolph Field, Texas.

Fleiss, J., B. Levin, and C. Paik (2003). *Statistical Methods for Rates and Proportions*. New York: Wiley.

Flury, B. (1986). Proportionality of k covariance matrices. *Statistics and Probability Letters 4*, 29–33.

Folks, J. L. and R. S. Chhikara (1978). The inverse Gaussian distribution and its statistical application, a review (with discussion). *Journal of the Royal Statistical Society B 40*, 263–289.

Follmann, D. and D. Lambert (1989). Generalizing logistic regression by non-parametric mixing. *Journal of the American Statistical Association 84*, 295–300.

Fowlkes, E. B. (1987). Some diagnosties for binary logistic regression via smoothing. *Biometrika 74*, 503–515.

Frank, E. and M. Hall (2001). A simple approach to ordinal classification. *Machine Learning: ECML 2001*, 145–156.

Frank, I. E. and J. H. Friedman (1993). A statistical view of some chemometrics regression tools (with discussion). *Technometrics 35*, 109–148.

Freund, Y. and R. E. Schapire (1997). A decision-theoretic generalization of on-line learning and an application to boosting. *Journal of Computer and System Sciences 55*, 119–139.

Frühwirth-Schnatter, S. (2006). *Finite mixture and Markov switching models*. New York: Springer–Verlag.

Friedman, J., T. Hastie, and R. Tibshirani (2008). *glmnet: Lasso and elastic-net regularized generalized linear models*. R package version 1.1.

Friedman, J. H. (1989). Regularized discriminant analysis. *Journal of the American Statistical Association 84*, 165–175.

Friedman, J. H. (1991). Multivariate adaptive regression splines (with discussion). *Annals of Statistics 19*, 1–67.

Friedman, J. H. (1994). Flexible metric nearest neighbor classification. Technical Report 113, Stanford University, Statistics Department.

Friedman, J. H. (2001). Greedy function approximation: a gradient boosting machine. *Annals of Statistics 29*, 1189–1232.

Friedman, J. H., T. Hastie, H. Höfling, and T. Tibshirani (2007). Pathwise coordinate optimization. *Applied Statistics 1*(2), 302–332.

Friedman, J. H., T. Hastie, and R. Tibshirani (2000). Additive logistic regression: A statistical view of boosting. *Annals of Statistics 28*, 337–407.

Friedman, J. H., T. Hastie, and R. Tibshirani (2010). Regularization paths for generalized linear models via coordinate descent. *Journal of Statistical Software 33*(1), 1–22.

Friedman, J. H. and W. Stützle (1981). Projection pursuit regression. *Journal of the American Statistical Association 76*, 817–823.

Fu, W. J. (1998). Penalized regression: the bridge versus the lasso. *Journal of Computational and Graphical Statistics 7*, 397–416.

Fu, W. J. (2003). Penalized estimation equations. *Biometrics 59*, 126–132.

Fukunaga, K. (1990). *Introduction to Statistical Pattern Recognition.* San Diego, California: Academic Press.

Furnival, G. M. and R. W. Wilson (1974). Regression by leaps and bounds. *Technometrics 16*, 499–511.

Gamerman, D. (1997). Efficient sampling from the posterior distribution in generalized linear mixed models. *Statistics and Computing 7*, 57–68.

Gay, D. M. and R. E. Welsch (1988). Maximum likelihood and quasi-likelihood for nonlinear exponential family regression models. *Journal of the American Statistical Association 83*, 990–998.

Genkin, A., D. Lewis, and D. Madigan (2004). Large-scale Bayesian logistic regression for text categorization. Technical report, Rutgers University.

Genter, F. C. and V. T. Farewell (1985). Goodness-of-link testing in ordinal regression models. *Canadian Journal of Statistics 13*, 37–44.

Gertheiss, J. (2011). *Feature Extraction in Regression and Classification with Structured Predictors.* Cuvillier Verlag.

Gertheiss, J., S. Hogger, C. Oberhauser, and G. Tutz (2011). Selection of ordinally scaled independent variables with applications to international classification of functioning core sets. *Journal of the Royal Statistical Society: Series C*, 377–396.

Gertheiss, J. and G. Tutz (2009a). Feature Selection and Weighting by Nearest Neighbor Ensembles. *Chemometrics and Intelligent Laboratory Systems 99*, 30–38.

Gertheiss, J. and G. Tutz (2009b). Penalized Regression with Ordinal Predictors. *International Statistical Review 77*, 345–365.

Gertheiss, J. and G. Tutz (2009c). Supervised feature selection in mass spectrometry based proteomic profiling by blockwise boosting. *Bioinformatics 8*, 1076–1077.

Gertheiss, J. and G. Tutz (2010). Sparse modeling of categorial explanatory variables. *Annals of Applied Statistics 4*, 2150–2180.

Gertheiss, J. and G. Tutz (2011). Regularization and model selection with categorial effect modifiers. *Statistica Sinica (to appear).*

Gijbels, I. and A. Verhaselt (2010). P-splines regression smoothing and difference type of penalty. *Statistics and Computing 4*, 499–511.

Glonek, G. F. V. and P. McCullagh (1995). Multivariate logistic models. *Journal of the Royal Statistical Society 57*, 533–546.

Glonek, G. V. F. (1996). A class of regression models for multivariate categorical responses. *Biometrika 83*, 15–28.

Gneiting, T. and A. Raftery (2007). Strictly proper scoring rules, prediction, and estimation. *Journal of the American Statistical Association 102*(477), 359–376.

Goeman, J. and S. le Cessie (2006). A goodness-of-fit test for multinomial logistic regression. *Biometrics 62*, 980–985.

Goeman, J. J. (2010). L_1 penalized estimation in the Cox proportional hazards model. *Biometrical Journal 52*, 70–84.

Golub, T., D. Slonim, P. Tamayo, C. Huard, M. Gaasenbeek, J. Mesirov, H. Coller, M. Loh, J. Downing, M. Caligiuri, C. Bloomfield, and E. Lander (1999a). Molecular classification of cancer: Class discovery and class prediction by gene expression monitoring. *Science 286*(5439), 531–537.

Golub, T. R., D. K. Slonim, P. Tamayo, C. Huard, M. Gaasenbeek, J. P. Mesirov, H. Coller, M. L. Loh, J. R. Downing, M. A. Caligiuri, C. D. Bloomfield, and E. S. Lander (1999b). Molecular classification of cancer: class discovery and class prediction by gene expression monitoring. *Science 286*, 531–537.

Goodman, L. A. (1968). The analysis of crossclassified data: Independence, quasi-independence and interaction in contingency tables with or without missing cells. *Journal of the American Statistical Association 63*, 1091–1131.

Goodman, L. A. (1971). The analysis of multidimensional contingency tables: Stepwise procedures and direct estimation methods for building models for multiple classifications. *Technometrics 13*, 33–61.

Goodman, L. A. (1979). Simple models for the analysis of association in cross-classification having ordered categories. *Journal of the American Statistical Society 74*, 537–552.

Goodman, L. A. (1981a). Association models and canonical correlation in the analysis of cross-classification having ordered categories. *Journal of the American Statistical Association 76*, 320–334.

Goodman, L. A. (1981b). Association models and the bivariate normal for contingency tables with ordered categories. *Biometrika 68*, 347–355.

Goodman, L. A. (1983). The analysis of dependence in cross-classification having ordered categories, using log-linear models for frequencies and log-linear models for odds. *Biometrika 39*, 149–160.

Goodman, L. A. (1985). The analysis of cross-classified data having ordered and/or ordered categories. *Annals of Statistics 13*(1), 10–69.

Goodman, L. A. and W. H. Kruskal (1954). Measures of associaton for cross classifications. *Journal of the American Statistical Association 49*, 732–764.

Gourieroux, C., A. Monfort, and A. Trognon (1984). Pseudo maximum likelihood methods: Theory. *Econometrica 52*, 681–700.

Green, D. J. and B. W. Silverman (1994). *Nonparametric Regression and Generalized Linear Models: A Roughness Penalty Approach*. London: Chapman & Hall.

Green, P. J. (1987). Penalized likelihood for general semi-parametric regression models. *International Statistical Review 55*, 245–259.

Greene, W. (2003). *Econometric Analysis*. New Jersey: Prentice Hall.

Greenland, S. (1994). Alternative models for ordinal logistic regression. *Statistics in Medicine 13*, 1665–1677.

Grizzle, J. E., C. F. Starmer, and G. G. Koch (1969). Analysis of categorical data by linear models. *Biometrika 28*, 137–156.

Grün, B. and F. Leisch (2007). Fitting finite mixtures of generalized linear regressions in R. *Computational Statistics & Data Analysis 51*(11), 5247–5252.

Grün, B. and F. Leisch (2008). Identifiability of finite mixtures of multinomial logit models with varying and fixed effects. *Journal of Classification 25*(2), 225–247.

Groll, A. and G. Tutz (2011a). Regularization for generalized additive mixed models by likelihood-based boosting. Technical Report 110, LMU, Department of Statistics.

Groll, A. and G. Tutz (2011b). Variable selection for generalized linear mixed models by l_1-penalized estimation. Technical Report 108, LMU, Department of Statistics.

Gschoessl, S. and C. Czado (2006). Modelling count data with overdispersion and spatial effects. *Statistical Papers 49(3)*, 531–552.

Gu, C. (2002). *Smoothing Splines ANOVA Models*. New York: Springer–Verlag.

Gu, C. and G. Wahba (1991). Minimizing GCV/GML Scores with Multiple Smoothing Parameters via the Newton Method. *SIAM Journal on Scientific and Statistical Computing 12*(2), 383–398.

Gu, C. and G. Wahba (1993). Semiparametric analysis of variance with tensor product thin plate splines. *Journal of the Royal Statistical Society Series B – Methodological 55*(2), 353–368.

Guess, H. A. and K. S. Crump (1978). Maximum likelihood estimation of dose-response functions subject to absolutely monotonic constraints. *Annals of Statistics 6*, 101–111.

Guo, Y., T. Hastie, and R. Tibshirani (2007). Regularized linear discriminant analysis and its application in microarrays. *Biostatistics 8*(1), 86.

Gupta, P. L., R. C. Gupta, and R. C. Tripathi (2004). Score test for zero inflated generalized Poisson regression model. *Communications in Statistics – Theory and Methods 33(1)*, 47–64.

Haberman, S. J. (1974). Loglinear models for frequency tables with ordered classifications. *Biometrics 30*, 589–600.

Haberman, S. J. (1977). Maximum likelihood estimates in exponential response models. *Annals of Statistics 5*, 815–841.

Haberman, S. J. (1982). Analysis of dispersion of multinomial responses. *Journal of the American Statistical Association 77*, 568–580.

Hamada, M. and C. F. J. Wu (1996). A critical look at accumulation analysis and related methods. *Technometrics 32*, 119–130.

Hamerle, A. K. P. and G. Tutz (1980). Kategoriale Reaktionen in multifaktoriellen Versuchsplänen und mehrdimensionale Zusammenhangsanalysen. *Archiv für Psychologie*, 53–68.

Hans, C. (2009). Bayesian lasso regression. *Biometrika 96*(1), 835–845.

Härdle, W., P. Hall, and H. Ichimura (1993). Optimal smoothing in single-index models. *Annals of Statistics 21*, 157–178.

Harrell, F. (2001). *Regression Modeling Strategies*. New York: Springer–Verlag.

Hartzel, J., A. Agresti, and B. Caffo (2001). Multinomial logit random effects models. *Statistical Modelling 1*, 81–102.

Hartzel, J., I. Liu, and A. Agresti (2001). Describing heterogenous effects in stratified ordinal contingency tables, with applications to multi-center clinical trials. *Computational Statistics & Data Analysis 35*(4), 429–449.

Harville, D. (1997). *Matrix Algebra from a Statistician's Perspective*. New York: Springer–Verlag.

Harville, D. A. (1976). Extension of the Gauss-Markov theorem to include the estimation of random effects. *Annals of Statistics 4*, 384–395.

Harville, D. A. (1977). Maximum likelihood approaches to variance component estimation and to related problems. *Journal of the American Statistical Association 72*, 320–338.

Harville, D. A. and R. W. Mee (1984). A mixed-model procedure for analyzing ordered categorical data. *Biometrics 40*, 393–408.

Hastie, T. (1996). Pseudosplines. *JRSS, Series B 58*, 379–396.

Hastie, T. and C. Loader (1993). Local regression: Automatic kernel carpentry. *Statistical Science 8*, 120–143.

Hastie, T. and R. Tibshirani (1986). Generalized additive models (c/r: p. 310–318). *Statist. Sci. 1*, 297–310.

Hastie, T. and R. Tibshirani (1990). *Generalized Additive Models*. London: Chapman & Hall.

Hastie, T. and R. Tibshirani (1993). Varying-coefficient models. *Journal of the Royal Statistical Society B 55*, 757–796.

Hastie, T. and R. Tibshirani (1996). Discriminant adaptive nearest-neighbor classification. *IEEE Transactions on Pattern Analysis and Machine Intelligence 18*, 607–616.

Hastie, T., R. Tibshirani, and J. H. Friedman (2001). *The Elements of Statistical Learning*. New York: Springer-Verlag.

Hastie, T., R. Tibshirani, and J. H. Friedman (2009). *The Elements of Statistical Learning (Second Edition)*. New York: Springer-Verlag.

Hausman, J. A. and D. McFadden (1984). Specification tests for the multinomial logit model. *Econometrika 52*, 1219–1240.

Hausman, J. A. and D. A. Wise (1978). A conditional probit model for qualitative choice: Discrete decisions recognizing interdependence and heterogeneous preference. *Econometrica 46*, 403–426.

He, X. and P. Ng (1999). COBS: Qualitatively constrained smoothing via linear programming. *Computational Statistics 14*, 315–337.

Heagerty, P. and B. F. Kurland (2001). Misspecified maximum likelihood estimates and generalised linear mixed models. *Biometrika 88*, 973–985.

Heagerty, P. and S. Zeger (2000). Marginalized multilevel models and likelihood inference. *Statistical Science 15*(1), 1–19.

Heagerty, P. J. (1999). Marginally specified logistic-normal models for longitudinal binary data. *Biometrics 55*, 688–698.

Heagerty, P. J. and S. Zeger (1998). Lorelogram: A regression approach to exploring dependence in longitudinal categorical responses. *Journal of the American Statistical Association 93*(441), 150–162.

Heagerty, P. J. and S. L. Zeger (1996). Marginal regression models for clustered ordinal measurements. *Journal of the American Statistical Association 91*, 1024–1036.

Hedeker, D. and R. B. Gibbons (1994). A random-effects ordinal regression model for multilevel analysis. *Biometrics 50*, 933–944.

Heim, A. (1970). *Intelligence and Personality*. Harmondsworth: Penguin.

Herbrich, R., T. Graepel, and K. Obermayer (1999). Large margin rank boundaries for ordinal regression. *Advances in neural information processing systems*, 115–132.

Heyde, C. C. (1997). *Quasi-likelihood and Its Applications*. New York: Springer–Verlag.

Hilbe, J. (2011). *Negative binomial regression*. Cambridge University Press.

Hinde, J. (1982). Compound poisson regression models. In R. Gilchrist (Ed.), *GLIM 1982 International Conference on Generalized Linear Models*, pp. 109–121. New York: Springer-Verlag.

Hinde, J. and C. Démetrio (1998). Overdispersion: Models and estimation. *Computational Statistics & Data Analysis 27*, 151–170.

Ho, T. K. (1998). The random subspace method for constructing decision forests. *IEEE Trans. on Pattern Analysis and Machine Intelligence 20*, 832–844.

Hoaglin, D. and R. Welsch (1978). The hat matrix in regression and ANOVA. *American Statistician 32*, 17–22.

Hoefsloot, H. C. J., S. Smit, and A. K. Smilde (2008). A classification model for the Leiden proteomics competition. *Statistical Applications in Genetics and Molecular Biology 7*, Article 8.

Hoerl, A. E. and R. W. Kennard (1970). Ridge regression: Bias estimation for nonorthogonal problems. *Technometrics 12*, 55–67.

Holtbrügge, W. and M. Schuhmacher (1991). A comparison of regression models for the analysis of ordered categorical data. *Applied Statistics 40*, 249–259.

Horowitz, J. and W. Härdle (1996). Direct semiparametric estimation of sngle-index models with discrete covariates. *Journal of the American Statistical Association 91*, 1623–9.

Hosmer, D. H. and S. Lemeshow (1980). Goodness-of-fit tests for the multiple logistic regression model. *Communications in Statistics – Theory & Methods 9*, 1043–1069.

Hosmer, D. H. and S. Lemeshow (1989). *Applied Logistic Regression*. New York: Wiley.

Hothorn, T., P. Bühlmann, T. Kneib, M. Schmid, and B. Hofner (2009). *mboost: Model-Based Boosting*. R package version 2.0-0.

Hothorn, T., K. Hornik, and A. Zeileis (2006). Unbiased recursive partitioning: A conditional inference framework. *Journal of Computational and Graphical Statistics 15*, 651–674.

Hothorn, T. and B. Lausen (2003). On the exact distribution of maximally selected rank statistics. *Computational Statistics and Data Analysis 43*, 121–137.

Hristache, M., A. Juditsky, and V. Spokoiny (2001). Direct estimation of the index coefficient in a single-index model. *Annals of Statistics 29*, 595–623.

Hsiao, C. (1986). *Analysis of Panel Data*. Cambridge: Cambridge University Press.

Huang, X. (2009). Diagnosis of random-effect model misspecification in generalized linear mixed models for binary response. *Biometrics 65*, 361–368.

Huang, X., W. Pan, S. Grindle, X. Han, Y. Chen, S. J. Park, I. W. Miller, and J. Hall (2005). A comparative study of discriminating human heart failure etiology using gene expression profiles. *Bioinformatics 6*, 205.

Hurvich, C. M. and C.-L. Tsai (1989). Regression and time series model selection in small samples. *BMA 76*, 297–307.

Ibrahim, J., H. Zhu, R. Garcia, and R. Guo (2011). Fixed and random effects selection in mixed effects models. *Biometrics 67*, 495–503.

Im, S. and D. Gianola (1988). Mixed models for bionomial data with an application to lamb mortality. *Applied Statistics 37*, 196–204.

James, G. (2002). Generalized linear models with functional predictors. *Journal of the Royal Statistical Society B 64*, 411–432.

James, G. M. and P. Radchenko (2008). A generalized Dantzig selector with shrinkage tuning. *Biometrika* , 127–142.

Jank, W. (2004). Quasi-Monte Carlo sampling to improve the efficiency of Monte Carlo EM. *Computational Statistics & Data Analysis 48*, 685–701.

Jansen, J. (1990). On the statistical analysis of ordinal data when extravariation is present. *Applied Statistics 39*, 74–85.

Jensen, K., H. Müller, and H. Schäfer (2000). Regional confidence bands for ROC curves. *Statistics in Medicine 19*(4), 493–509.

Joe, H. (1989). Relative entropy measures of multivariate dependence. *Journal of the American Statistical Association 84*, 157–164.

Joe, H. and R. Zhu (2005). Generalized Poisson distribution: the property of mixture of Poisson and comparison with negative binomial distribution. *Biometrical Journal 47(2)*, 219–229.

Jones, R. H. (1993). *Longitudinal Data with Serial Correlation: A State-Space Approach.* London: Chapman & Hall.

Jorgenson, B. (1987). Exponential dispersion models. *J. Roy. Stat. Soc. Ser. B 49*, 127–162.

Karimi, A., A. Windorfer, and J. Dreesman (1998). Vorkommen von zentralnervösen Infektionen in europäischen Ländern. Technical report, Schriften des Niedersächsischen Landesgesundheitsamtes.

Kaslow, R. A., D. G. Ostrow, R. Detels, J. P. Phair, B. F. Polk, and C. R. Rinaldo (1987). The multicenter aids cohort study: Rationale, organization and selected characteristic of the participiants. *American Journal of Epidemiology 126*, 310–318.

Kauermann, G. (2000). Modelling longitudinal data with ordinal response by varying coefficients. *Biometrics 56*, 692–698.

Kauermann, G., T. Krivobokova, and L. Fahrmeir (2009). Some asymptotic results on generalized penalized spline smoothing. *Journal of the Royal Statistical Society Series B – Statistical Methodology 71*(Part 2), 487–503.

Kauermann, G. and J. Opsomer (2004). Generalized cross-validation for bandwidth selection of backfitting estimates in generalized additive models. *Journal of Computational Comutational and Graphical Statistics 13*(1), 66–89.

Kauermann, G. and G. Tutz (2001). Testing generalized and semiparametric models against smooth alternatives. *Journal of the Royal Statistical Society B 63*, 147–166.

Kauermann, G. and G. Tutz (2003). Semi- and nonparametric modeling of ordinal data. *Journal of Computational and Graphical Statistics 12*, 176–196.

Kaufmann, H. (1987). Regression models for nonstationary categorical time series: Asymptotic estimation theory. *Annals of Statistics 15*, 79–98.

Kedem, b. and K. Fokianos (2002). *Regression Models for Time Series Analysis.* New York: Wiley.

Keenan, S. C. and J. R. Sobehart (1999). Performance measures for credit risk models. Research Report 13, Moody's Risk Management Services.

Khalili, A. and J. Chen (2007). Variable selection in finite mixture of regression models. *Journal of the American Statistical Association 102*(479), 1025–1038.

Kinney, S. K. and D. B. Dunson (2007). Fixed and random effects selection in linear and logistic models. *Biometrics 63*, 690–698.

Kleiber, C. and A. Zeileis (2008). *Applied Econometrics with R*. New York: Springer–Verlag.

Klein, R. L. and R. H. Spady (1993). An efficient semiparametric estimator for binary response models. *Econometrica 61*, 387–421.

Kneib, T. and L. Fahrmeir (2006). Structured additive regression for categorical space-time data: A mixed model approach. *Biometrics 62*, 109–118.

Kneib, T. and L. Fahrmeir (2008). A space-time study on forest health. In R. Chandler and M. Scott (Eds.), *Statistical Methods for Trend Detection and Analysis in the Environmental Sciences*. New York: Wiley.

Kneib, T., T. Hothorn, and G. Tutz (2009). Variable selection and model choice in geoadditive regression models. *Biometrics 65*, 626–634.

Kockelkorn, U. (2000). *Lineare Modelle*. Oldenbourg Verlag.

Koenker, R., P. Ng, and S. Portnoy (1994). Quantile smoothing splines. *Biometrika 81*, 673–680.

Kohavi, R. and G. H. John (1998). The wrapper approach. In H. Liu and H. Motoda (Eds.), *Feature Extraction, Construction and Selection. A Data Mining Perspective*. Dordrecht: Kluwer.

Krantz, D. H. (1964). Conjoint measurement: The Luce-Tukey axiomatization and some extentions. *Journal of Mathematical Psychology 1*, 248–277.

Krantz, D. H., R. D. Luce, P. Suppes, and A. Tversky (1971). *Foundations of Measurement*, Volume 1. New York: Academic Press.

Krishnapuram, B., L. Carin, M. A. Figueiredo, and A. J. Hartemink (2005). Sparse multinomial logistic regression: Fast algorithms and generalization bounds. *IEEE Transactions on Pattern Analysis and Machine Intelligence 27*, 957–968.

Krivobokova, T., C. Crainiceanu, and G. Kauermann (2008). Fast adaptive penalized splines. *Journal of Computational and Graphical Statistics 17*, 1–20.

Küchenhoff, H. and K. Ulm (1997). Comparison of statistical methods for assessing threshold limiting values in occupational epidemiology. *Computational Statistics 12*, 249–264.

Künsch, H. R., L. A. Stefanski, and R. J. Carroll (1989). Conditionally unbiased bounded-influence estimation in general regression models, with applications to generalized linear models. *Journal of the American Statistical Association 84*, 460–466.

Kuss, O. (2002). Global goodness-of-fit tests in logistic regression with sparse data. *Statistics in Medicine 21*, 3789–3801.

Laara, E. and J. N. Matthews (1985). The equivalence of two models for ordinal data. *Biometrika 72*, 206–207.

Laird, N. M., G. J. Beck, and J. H. Ware (1984). Mixed models for serial categorical response. Quoted in A. Eckholm (1991). Maximum Likelihood for Many Short Binary Time Series (preprint).

Lambert, D. (1992). Zero-inflated poisson regression with an application to defects in manufacturing. *Technometrics 34*, 1–14.

Lambert, D. and K. Roeder (1995). Overdispersion diagnostics for generalized linear models. *Journal of the American Statistical Association 90*, 1225–1236.

Land, S. R. and J. H. Friedman (1997). Variable fusion: A new adaptive signal regression method. Discussion paper 656, Department of Statistics, Carnegie Mellon University, Pittsburg.

Landwehr, J. M., D. Pregibon, and A. C. Shoemaker (1984). Graphical methods for assessing logistic regression models. *Journal of the American Statistical Association 79*, 61–71.

Lang, J. (1996a). On the comparison of multinomial and Poisson log-linear models. *Journal of the Royal Statistical Society B*, 253–266.

Lang, J. B. (1996b). Maximum likelihood methods for a generalized class of log-linear models. *Annals of Statistics 24*, 726–752.

Lang, J. B. and A. Agresti (1994). Simultaneous modelling joint and marginal distributions of multivariate categorical responses. *Journal of the American Statistical Association 89*, 625–632.

Lang, S. and A. Brezger (2004a, MAR). Bayesian P-splines. *Journal of Computational and Graphical Statistics 13*(1), 183–212.

Lang, S. and A. Brezger (2004b). Bayesian P-splines. *Journal of Computational and Graphical Statistics 13*, 183–212.

Lauritzen, S. (1996). *Graphical Models*. New York: Oxford University Press.

Lawless, J. F. and K. Singhal (1978). Efficient screening of nonnormal regression models. *Biometrics 34*, 318–327.

Lawless, J. F. and K. Singhal (1987). ISMOD: An all-subsets regression program for generalized linear models. *Computer Methods and Programs in Biomedicine 24*, 117–134.

LeCessie (1992). Ridge estimators in logistic regression. *Applied Statistics 41*(1), 191–201.

LeCessie, S. and J. C. van Houwelingen (1991). A goodness-of-fit test for binary regression models, based on smoothing methods. *Biometrics 47*, 1267–1282.

LeCessie, S. and J. C. van Houwelingen (1995). Goodness-of-fit tests for generalized linear models based on random effect models. *Biometrics 51*, 600–614.

Lee, J., M. Park, and S. Song (2005). An extensive comparison of recent classification tools applied to microarray data. *Computational Statistics and Data Analysis 48*, 869–885.

Lee, W.-C. and C. K. Hsiao (1996). Alternative summary indices for the receiver operating characteristic curve. *Epidemiology 7*, 605–611.

Leitenstorfer, F. and G. Tutz (2007). Generalized monotonic regression based on B-splines with an application to air pollution data. *Biostatistics 8*, 654–673.

Leitenstorfer, F. and G. Tutz (2011). Estimation of single-index models based on boosting techniques. *Statistical Modelling*.

Leng, C. (2009). A simple approach for varying-coefficient model selection. *Journal of Statistical Planning and Inference 139*(7), 2138–2146.

Lesaffre, E. and A. Albert (1989). Multiple-group logistic regression diagnostics. *Applied Statistics 38*, 425–440.

Li, Y. and D. Ruppert (2008). On the asymptotics of penalized splines. *Biometrika 95*, 415–436.

Liang, K.-Y. and P. McCullagh (1993). Case studies in binary dispersion. *Biometrics 49*, 623–630.

Liang, K.-Y. and S. Zeger (1986). Longitudinal data analysis using generalized linear models. *Biometrika 73*, 13–22.

Liang, K.-Y. and S. Zeger (1993). Regression analysis for correlated data. *Annual Reviews Public Health 14*, 43–68.

Liang, K.-Y., S. Zeger, and B. Qaqish (1992). Multivariate regression analysis for categorical data (with discussion). *Journal of the Royal Statistical Society B 54*, 3–40.

Lin, X. and N. E. Breslow (1996). Bias correction in generalized linear mixed models with multiple components of dispersion. *Journal of the American Statistical Association 91*, 1007–1016.

Lin, X. and R. Carroll (2006). Semiparametric estimation in general repeated measures problems. *Journal of the Royal Statistical Society, B, 68*, 69–88.

Lin, X. and D. Zhang (1999). Inference in generalized additive mixed models by using smoothing splines. *Journal of the Royal Statistical Society. Series B (Statistical Methodology) 61*, 381–400.

Lin, Y. and Y. Jeon (2006). Random forests and adaptive nearest neighbors. *Journal of the American Statistical Association 101*, 578–590.

Lindsey, J. J. (1993). *Models for Repeated Measurements*. Oxford: Oxford University Press.

Linton, O. B. and W. Härdle (1996). Estimation of additive regression models with known links. *Biometrika 83*, 529–540.

Lipsitz, S., G. Fitzmaurice, and G. Molenberghs (1996). Goodness-of-fit tests for ordinal response regression models. *Applied Statistics 45*, 175–190.

Lipsitz, S., N. Laird, and D. Harrington (1990). Finding the design matrix for the marginal homogeneity model. *Biometrika 77*, 353–358.

Lipsitz, S., N. Laird, and D. Harrington (1991). Generalized estimation equations for correlated binary data: Using the odds ratio as a measure of association in unbalanced mixed models with nested random effects. *Biometrika 78*, 153–160.

Liu, Q. and A. Agresti (2005). The analysis of ordinal categorical data: An overview and a survey of recent developments. *Test 14*, 1–73.

Liu, Q. and D. A. Pierce (1994). A note on Gauss-Hermite quadrature. *Biometrika 81*, 624–629.

Lloyd, C. J. (2008). A new exact and more powerful unconditional test of no treatment effect from binary matched pairs. *Biometrics 64*(3), 716–723.

Loader, C. (1999). *Local Regression and Likelihood*. New York: Springer-Verlag.

Loh, W. and Y. Shih (1997). Split selection methods for classification trees. *Statistica Sinica 7*, 815–840.

Longford, N. L. (1993). *Random Effect Models*. New York: Oxford University Press.

Louis, T. A. (1982). Finding the observed information matrix when using the EM algorithm. *Journal of the Royal Statistical Society B 44*, 226–233.

Luce, R. D. (1959). *Individual Choice Behaviour*. New York: Wiley.

Lunetta, K., L. Hayward, J. Segal, and P. Eerdewegh (2004). Screening Large-Scale Association Study Data: Exploiting Interactions Using Random Forests. *BMC Genetics 5*(1), 32.

Maddala, G. S. (1983). *Limited-Dependent and Qualitative Variables in Econometrics*. Cambridge: Cambridge University Press.

Magder, L. and S. Zeger (1996). A smooth nonparametric estimate of a mixing distribution using mixtures of gaussians. *Journal of the American Statistical Association 91*, 1141–1151.

Magnus, J. R. and H. Neudecker (1988). *Matrix Differential Calculus with Applications in Statistics and Econometrics*. London: Wiley.

Mancl, L. A. and B. G. Leroux (1996). Efficiency of regression estimates for clustered data. *Biometrics 52*, 500–511.

Mantel, N. and W. Haenszel (1959). Statistical aspects of the analysis of data from retrospective studies. *J. Natl. Cancer Inst. 22*, 719–48.

Marks, S. and O. J. Dunn (1974). Discriminant functions when covariance matrices are unequal. *Journal of the American Statistical Association 69*.

Marra, G. and S. Wood (2011). Practical variable selection for generalized additive models. *Computational Statistics and Data Analysis 55*, 2372–2387.

Marx, B. D. and P. H. C. Eilers (1998). Direct generalized additive modelling with penalized likelihood. *Computational Statistics & Data Analysis 28*, 193–209.

Marx, B. D. and P. H. C. Eilers (1999). Generalized linear regression on sampled signals and curves: A p-spline approach. *Technometrics 41*, 1–13.

Marx, B. D. and P. H. C. Eilers (2005). Multidimensional penalized signal regression. *Technometrics 47*, 13–22.

Masters, G. N. (1982). A Rasch model for partial credit scoring. *Psychometrika 47*, 149–174.

McCullagh, P. (1980). Regression model for ordinal data (with discussion). *Journal of the Royal Statistical Society B 42*, 109–127.

McCullagh, P. (1983). Quasi-likelihood functions. *Annals of Statistics 11*, 59–67.

McCullagh, P. and J. A. Nelder (1989). *Generalized Linear Models (Second Edition)*. New York: Chapman & Hall.

McCulloch, C. and S. Searle (2001). *Generalized, Linear, and Mixed Models*. New York: Wiley.

McCulloch, C. E. (1997). Maximum likelihood algorithms for generalized linear mixed models. *Journal of the American Statistical Association 92*, 162–170.

McDonald, B. W. (1993). Estimating logistic regression parameters for bivariate binary data. *Journal of the Royal Statistical Society B 55*, 391–397.

McFadden, D. (1973). Conditional logit analysis of qualitative choice behaviour. In P. Zarembka (Ed.), *Frontiers in Econometrics*. New York: Academic Press.

McFadden, D. (1978). Modelling the choice of residential location. In A. Karlquist et al. (Eds.), *Spatial Interaction Theory and Residential Location*. Amsterdam: North-Holland.

McFadden, D. (1981). Econometric models of probabilistic choice. In C. F. Manski and D. McFadden (Eds.), *Structural Analysis of Discrete Data with Econometric Applications*, pp. 198–272. Cambridge, MA: MIT Press.

McFadden, D. (1986). The choice theory approach to market research. *Marketing Science 5*, 275–297.

McLachlan, G. and T. Krishnan (1997). *The EM Algorithm and Extensions*. New York: Wiley.

McLachlan, G. J. (1992). *Discriminant Analysis and Statistical Pattern Recognition*. New York: Wiley.

McLachlan, G. J. and D. Peel (2000). *Finite Mixture Models*. New York: Wiley.

McNemar, Q. (1947). Note on the sampling error of the difference between correlated proportions or percentages. *Psychometrika 12*, 153–157.

Mehta, C. R., N. R. Patel, and A. A. Tsiatis (1984). Exact significance testing to establish treatment equivalence with ordered categorical data. *Biometrics 40*, 819–825.

Meier, L., S. van de Geer, and P. Bühlmann (2008). The group lasso for logistic regression. *Journal of the Royal Statistical Society, Series B 70*, 53–71.

Meier, L., S. Van De Geer, and P. Bühlmann (2009). High-dimensional additive modeling. *The Annals of Statistics 37*, 3779–3821.

Meilijson, I. (1989). A fast improvement to the EM-algorithm on its own terms. *Journal of the Royal Statistical Society B 51*, 127–138.

Miller, A. J. (1989). *Subset Selection in Regression*. London: Chapman & Hall.

Miller, M. E., C. S. Davis, and R. J. Landis (1993). The analysis of longitudinal polytomous data: Generalized estimated equations and connections with weighted least squares. *Biometrics 49*, 1033–1044.

Miller, R. and D. Siegmund (1982). Maximally selected chi-square statistics. *Biometrics 38*, 1011–1016.

Min, A. and C. Czado (2010). Testing for zero-modification in count regression models. *Statistica Sinica 20*, 323–341.

Mittlböck, M. and M. Schemper (1996). Explained variation for logistic regression. *Statistic in Medicine 15*, 1987–1997.

Mkhadri, A., G. Celeux, and A. Nasroallah (1997). Regularization in discriminant analysis: An overview. *Computational Statistics & Data analysis 23*, 403–423.

Molenberghs, G. and E. Lesaffre (1994). Marginal modelling of correlated ordinal data using a multivariate Plackett distribution. *Journal of the American Statistical Association 89*, 633–644.

Molenberghs, G. and G. Verbeke (2005). *Models for Discrete Longitdinal Data*. New York: Springer–Verlag.

Molinaro, A., R. Simon, and R. M. Pfeiffer (2005). Predition error estimation: a comparison of resampling methods. *Bioinformatics 21*, 3301–3307.

Moore, D. F. and A. Tsiatis (1991). Robust estimation of the variance in moment methods for extra-binomial and extra-poisson variation. *Biometrics 47*, 383–401.

Morgan, B. J. T. (1985). The cubic logistic model for quantal assay data. *Applied Statistics 34*, 105–113.

Morgan, B. J. T. and D. M. Smith (1993). A note on Wadley's problem with overdispersion. *Applied Statistics 41*, 349–354.

Morgan, J. N. and J. A. Sonquist (1963). Problems in the analysis of survey data, and a proposal. *Journal of the American Statistical Association 58*, 415–435.

Morin, R. L. and D. E. Raeside (1981). A reappraisal of distance-weighted k-nearest neighbor classification for pattern recognition with missing data. *IEEE Transactions on Systems, Man and Cybernetics 11*, 241–243.

Moulton, L. and S. Zeger (1989). Analysing repeated measures in generalized linear models via the bootstrap. *Biometrics 45*, 381–394.

Muggeo, V. M. R. and G. Ferrara (2008). Fitting generalized linear models with unspecified link function: A P-spline approach. *Computational Statistics & Data Analysis 52*(5), 2529–2537.

Mullahy, J. (1986). Specification and testing of some modified count data models. *Journal of Econometrics 33*, 341–365.

Nadaraya, E. A. (1964). On estimating regression. *Theory of Probability and Applications 10*, 186–190.

Nagelkerke, N. J. D. (1991). A note on a general definition of the coefficient of determination. *Biometrika 78*, 691–692.

Naik, P. A. and C. Tsai (2001). Single-index model selections. *Biometrika 88*, 821–832.

Nair, V. N. (1987). Chi-squared-type tests for ordered alternatives in contingency tables. *Journal of the American Statistical Association 82*, 283–291.

Nelder, J. A. (1992). Joint modelling of mean and dispersion. In P. van der Heijden, W. Jansen, B. Francis, and G. Seeber (Eds.), *Statistical Modelling*. Amsterdam: North-Holland.

Nelder, J. A. and D. Pregibon (1987). An extended quasi-likelihood function. *Biometrika 74*, 221–232.

Nelder, J. A. and R. W. M. Wedderburn (1972). Generalized linear models. *Journal of the Royal Statistical Society A 135*, 370–384.

Neuhaus, J., W. Hauck, and J. Kalbfleisch (1992). The effects of mixture distribution. misspecification when fitting mixed effect logistic models. *Biometrika 79*(4), 755–762.

Neuhaus, J. M., J. D. Kalbfleisch, and W. W. Hauck (1991). A comparison of cluster-specific and population-averaged approaches for analyzing correlated binary data. *International Statistical Review 59*, 25–35.

Newson, R. (2002). Parameters behind "nonparametric" statistics: Kendall's tau, Somers' D and median differences. *The Stata Journal 2*, 45–64.

Ni, X., D. Zhang, and H. H. Zhang (2010). Variable selection for semiparametric mixed models in longitudinal studies. *Biometrics 66*, 79–88.

Nyquist, H. (1991). Restricted estimation of generalized linear models. *Applied Statistics 40*, 133–141.

Ogden, R. T. (1997). *Essential Wavelets for Statistical Applications and Data Analysis*. Boston: Birkhäuser.

Opsomer, J. D. (2000). Asymptotic properties of backfitting estimators. *Journal of Multivariate Analysis 73*, 166–179.

Opsomer, J. D. and D. Ruppert (1997). Fitting a bivariate additive model by local polynomial regression. *Annals of Statistics 25*, 186–211.

Osborne, M., B. Presnell, and B. Turlach (2000). On the lasso and its dual. *Journal of Computational and Graphical Statistics 9*(2), 319–337.

Osborne, M. R., B. Presnell, and B. A. Turlach (1998). Knot selection for regression splines via the lasso. In S. Weisberg (Ed.), *Dimension Reduction, Computational Complexity, and Information*, Volume 30 of *Computing Science and Statistics*, pp. 44–49.

Osius, G. (2004). The association between two random elements: A complete characterization and odds ratio models. *Metrika 60*, 261–277.

Osius, G. and D. Rojek (1992). Normal goodness-of-fit tests for parametric multinomial models with large degrees of freedom. *Journal of the American Statistical Association 87*, 1145–1152.

Paik, M. and Y. Yang (2004). Combining nearest neighbor classifiers versus cross-validation selection. *Statistical Applications in Genetics and Molecular Biology 3*(12).

Palmgren, J. (1981). The Fisher information matrix for log-linear models arguing conditionally in the observed explanatory variables. *Biometrika 68*, 563–566.

Palmgren, J. (1989). Regression models for bivariate binary responses. *UW Biostatistics Working Paper Series*.

Park, M. Y. and T. Hastie (2007). An l1 regularization-path algorithm for generalized linear models. *Journal of the Royal Statistical Society B 69*, 659–677.

Park, T. and G. Casella (2008). The Bayesian lasso. *Journal of the American Statistical Association 103*, 681–686.

Parthasarthy, G. and B. N. Chatterji (1990). A class of new knn methods for low sample problems. *IEEE Transactions on systems, man and Cybernetics 20*, 715–718.

Patterson, H. and R. Thomson (1971). Recovery of inter-block information when block sizes are unequal. *Biometrika 58*, 545–554.

Pepe, M. S. (2003). *The Statistical Evaluation of Medical Tests for Classification and Prediction*. New York: Chapman & Hall.

Peterson, B. and F. E. Harrell (1990). Partial proportional odds models for ordinal response variables. *Applied Statistics 39*, 205–217.

Petricoin, E. F., D. K. Ornstein, C. P. Paweletz, A. M. Ardekani, P. S. Hackett, B. A. Hitt, A. Velassco, C. Trucco, L. Wiegand, K. Wood, C. B. Simone, P. J. Levine, W. M. Lineham, M. R. Emmert-Buck, S. M. Steinberg, E. C. Kohn, and L. A. Liotta (2002). Serum proteomic patterns for detection of prostate cancer. *Journal of the National Cancer Institute 94*, 1576–1578.

Petry, S. and G. Tutz (2011). The oscar for generalized linear models. Technical Report 112, LMU, Department of Statistics.

Petry, S. and G. Tutz (2012). Shrinkage and variable selection by polytopes. *Journal of Statistical Planning and Inference 142*, 48–64.

Petry, S., G. Tutz, and C. Flexeder (2011). Pairwise fused lasso. Technical Report 102, LMU, Department of Statistics.

Piccarreta, R. (2008). Classification trees for ordinal variables. *Computational Statistics 23*(3), 407–427.

Piegorsch, W. (1992). Complementary log regression for generalized linear models. *The American Statistician 46*, 94–99.

Piegorsch, W. W., C. R. Weinberg, and B. H. Margolin (1988). Exploring simple independent action in multifactor tables of proportions. *Biometrics 44*, 595–603.

Pierce, D. A. and D. W. Schafer (1986). Residuals in generalized linear models. *Journal of the American Statistical Association 81*, 977–986.

Pigeon, J. and J. Heyse (1999). An improved goodness-of-fit statistic for probability prediction models. *Biometrical Journal 41*, 71–82.

Pinheiro, J. C. and D. M. Bates (1995). Approximations to the log-likelihood function in the nonlinear mixed-effects model. *Journal of Computational and Graphical Statistics 4*, 12–35.

Poggio, T. and F. Girosi (1990). Regularization algorithms for learning that are equivalent to multilayer networks. *Science 247*, 978–982.

Pohlmeier, W. and V. Ulrich (1995). An econometric model of the two-part decisionmaking process in the demand for health care. *Journal of Human Resources 30*, 339–361.

Poortema, K. L. (1999). On modelling overdispersion of counts. *Statistica Neerlandica 53*, 5–20.

Powell, J. L., J. H. Stock, and T. M. Stoker (1989). Semiparametric estimation of index coefficients. *Econometrica 57*, 1403–1430.

Pregibon, D. (1980). Goodness of link tests for generalized linear models. *Applied Statistics 29*, 15–24.

Pregibon, D. (1981). Logistic regression diagnostics. *Annals of Statistics 9*, 705–724.

Pregibon, D. (1982). Resistant fits for some commonly used logistic models with medical applications. *Biometrics 38*, 485–498.

Pregibon, D. (1984). Review of generalized linear models by mccullagh and nelder. *American Statistician 12*, 1589–1596.

Prentice, R. and R. Pyke (1979). Logistic disease incidence models and case-control studies. *Biometrika 66*, 403.

Prentice, R. L. (1976). A generalization of the probit and logit methods for close response curves. *Biometrics 32*, 761–768.

Prentice, R. L. (1986). Binary regression using an extended beta-binomial distribution, with discussion of correlation induced by covariate measurement errors. *Journal of the American Statistical Association 81*, 321–327.

Prentice, R. L. (1988). Correlated binary regression with covariates specific to each binary observation. *Biometrics 44*, 1033–1084.

Pulkstenis, E. and T. J. Robinson (2002). Two goodness-of-fit tests for logistic regression models with continuous covariates. *Statistics in Medicine 21*, 79–93.

Pulkstenis, E. and T. J. Robinson (2004). Goodness-of-fit tests for ordinal response regression models. *Statistics in Medicine 23*, 999–1014.

Qu, Y., G. W. Williams, G. J. Beck, and M. Goormastic (1987). A generalized model of logistic regression for clustered data. *Communications in Statistics – Theory and Methods 16*, 3447–3476.

Quinlan, J. R. (1986). Industion of decision trees. *Machine Learning 1*, 81–106.

Quinlan, J. R. (1993). *Programs for Machine Learning.* San Francisco: Morgan Kaufmann PublisherInc.

R Development Core Team (2010). *R: A Language and Environment for Statistical Computing.* Vienna, Austria: R Foundation for Statistical Computing. ISBN 3-900051-07-0.

Radelet, M. and G. Pierce (1991). Choosing those who will die: Race and the death penalty in florida. *Florida Law Review 43*, 1–34.

Ramsey, J. O. and B. W. Silverman (2005). *Functional Data Analysis.* New York: Springer–Verlag.

Rao, P. and L. Kupper (1967). Ties in paired-comparison experiments: A generalization of the Bradley-Terry model. *Journal of the American Statistical Association 62*, 194–204.

Rasch, G. (1961). On general laws and the meaning of measurement in psychology. In J. Neyman (Ed.), *Proceedings of the Fourth Berkeley Symposium on Mathematical Statistics and Probability*, Berkeley.

Ravikumar, P., M. Wainwright, and J. Lafferty (2009). High-dimensional graphical model selection using l_1-regularized logistic regression. *Annals of Statistics 3*, 1287–1319.

Rawlings, J., S. Pantula, and D. Dickey (1998). *Applied Regression Analysis.* New York: Springer–Verlag.

Rayens, W. and T. Greene (1991). Covariance pooling and stabilization for classification. *Computational Statistics and Data Analysis 11*, 17–42.

Read, I. and N. Cressie (1988). *Goodness-of-Fit Statistics for Discrete Multivariate Data.* New York: Springer-Verlag.

Reinsch, C. (1967). Smoothing by spline functions. *Numerische Mathematik 10*, 177–183.

Ridgeway, G. (1999). *Generalization of boosting algorithms and applications of bayesian inference for massive datasets.* Ph. D. thesis, University of Washington.

Ripley, B. D. (1996). *Pattern Recognition and Neural Networks.* Cambridge: Cambridge University Press.

Robinson, G. K. (1991). That blup is a good thing. the estimation of random effects. *Statistical Science 6*, 15–51.

Romualdi, C., S. Campanaro, D. Campagna, B. Celegato, N. Cannata, S. Toppo, G. Valle, and G. Lanfranchi (2003). Pattern recognition in gene expression profiling using dna array: A comparison study of different statistical methods applied to cancer classification. *Human Molecular Genetics 12*, 823–836.

Rosenstone, S., D. Kinder, and W. Miller (1997). *American National Election Study*. MI: Inter–University Consortium for Political and Social Research.

Rosner, B. (1984). Multivariate methods in orphthalmology with applications to other paired-data situations. *Biometrics 40*, 1025–1035.

Rosset, S. (2004). Tracking curved regularized optimization solution paths. In *Advances in Neural Information Processing Systems*, Cambridge. MIT Press.

Rousseeuw, P. J. and A. Christmann (2003). Robustness against separation and outliers in logistic regression. *Computational Statistics and Data Analysis 43*, 315–332.

Ruckstuhl, A. and A. Welsh (1999). Reference bands for nonparametrically estimated link functions. *Journal of Computational and Graphical Statistics 8*(4), 699–714.

Rudolfer, S. M., P. C. Watson, and E. Lesaffre (1995). Are ordinal models useful for classification? a revised analysis. *Journal of Statistical Computation Simulation 52*(2), 105–132.

Rue, H. and L. Held (2005). *Gaussian Markov Random Fields.Theory and Applications*. London: CRC / Chapman & Hall.

Rumelhart, D. L. and J. G. Greeno (1971). Similarity between stimuli: An experimental test of the Luce and restle choice methods. *Journal of Mathematical Psychology 8*, 370–381.

Ruppert, D. (2002). Selecting the number of knots for penalized splines. *Journal of Computational and Graphical Statistics 11*, 735–757.

Ruppert, D., M. P. Wand, and R. J. Carroll (2003). *Semiparametric Regression*. Cambridge: Cambridge University Press.

Ruppert, D., M. P. Wand, and R. J. Carroll (2009). Semiparametric regression during 2003 – 2007. *Electronic Journal of Statistics 3*, 1193–1256.

Ryan, T. (1997). *Modern Regression Methods*. New York: Wiley.

Sampson, A. and H. Singh (2002). Min and max scorings for two sample partially ordered categorical data. *Journal of statistical Planning and Inference 107*, 219–236.

Santner, T. J. and D. E. Duffy (1986). A note on A. Albert and J. A. Anderson's conditions for the existence of maximum likelihood estimates regression models. *Biometrika 73*, 755–758.

Santner, T. J. and D. E. Duffy (1989). *The Statistical Analysis of Discrete Data*. New York: Springer-Verlag.

Schaefer, R. L., L. D. Roi, and R. A. Wolfe (1984). A ridge logistic estimate. *Communication in Statistics, Theory & Methods 13*, 99–113.

Schall, R. (1991). Estimation in generalised linear models with random effects. *Biometrika 78*, 719–727.

Schwarz, G. (1978). Estimating the dimension of a model. *Annals of Statistics 6*, 461–464.

Scott, A. and C. Wild (1986). Fitting logistic models under case-control or choice based sampling. *Journal of the Royal Statistical Society. Series B (Methodological) 48*(2), 170–182.

Searle, S., G. Casella, and C. McCulloch (1992). *Variance Components*. New York: Wiley.

Seeber, G. (1977). *Linear Regression Analysis*. New York: Wiley.

Segerstedt, B. (1992). On ordinary ridge regression in generalized linear models. *Communications in Statistics – Theory and Methods 21*, 2227–2246.

Shapire, R. E. (1990). The strength of weak learnability. *Machine Learning 5*, 197–227.

Shih, Y.-S. (2004). A note on split selection bias in classification trees. *Computational Statistics and Data Analysis 45*, 457–466.

Shih, Y.-S. and H. Tsai (2004). Variable selection bias in regression trees with constant fits. *Computational Statistics and Data Analysis 45*, 595–607.

Shipp, M., K. Ross, P. Tamayo, A. Weng, J. Kutok, R. Aguiar, M. Gaasenbeek, M. Angelo, M. Reich, G. Pinkus, et al. (2002). Diffuse large B-cell lymphoma outcome prediction by gene-expression profiling and supervised machine learning. *Nature medicine 8*(1), 68–74.

Silverman, B. W. and M. C. Jones (1989). Commentary on Fix and Hodges (1951): An important contribution to nonparametric discriminant analysis and density estimation. *International Statistical Review 57*, 233–238.

Simonoff, J. (1995). Smoothing categorical data. *Journal of Statistical Planning and Inference 47*, 41–69.

Simonoff, J. S. (1983). A penalty function approach to smoothing large sparse contingency tables. *Annals of Statistics 11*, 208–218.

Simonoff, J. S. (1996). *Smoothing Methods in Statistics*. New York: Springer-Verlag.

Simonoff, J. S. and G. Tutz (2000). Smoothing methods for discrete data. In M. Schimek (Ed.), *Smoothing and Regression. Approaches, Computation and Application*. New York: Wiley.

Slawski, M. (2010). The structured elastic net for quantile regression and support vector classification. *Statistics and Computing*.

Slawski, M., M. Daumer, and A.-L. Boulesteix (2008). CMA – A comprehensive bioconductor package for supervised classification with high dimensional data. *BMC Bioinformatics 9*, 439.

Smith, M. and R. Kohn (1996). Nonparametric regression using Bayesian variable selection. *Journal of Econometrics 75*, 317–343.

Smith, P. L. (1982). Curve fitting and modeling with splines using statistical variable selection techniques. Report 166034, NASA.

Snell, E. J. (1964). A scaling procedure for ordered categorical data. *Biometrics 20*, 592–607.

Sobehart, J., S. Keenan, and R. Stein (2000). Validation methodologies for default risk models. *Credit*, 51–56.

Soofi, E. S., J. J. Retzer, and M. Yasai-Ardekani (2000). A framework for measuring the importance of variables with applications to management research and decision models. *Decision Sciences 31*, 595–625.

Statnikov, A., C. F. Aliferis, I. Tsamardinos, D. Hardin, and S. Levy (2005). A comprehensive evaluation of multicategory classification methods for microarray gene expression cancer diagnosis. *Bioinformatics 21*, 631–643.

Steadman, S. and L. Weissfeld (1998). A study of the effect of dichotomizing ordinal data upon modelling. *Communications in Statistics – Simulation and Computation 27(4)*, 871–887.

Stein, C. (1981). Estimation of the mean of a multivariate normal distribution. *Annals of Statistics 9*, 1135–1151.

Steinwart, I. and A. Christmann (2008). *Support vector machines*. Springer Verlag.

Stiratelli, R., N. Laird, and J. H. Ware (1984). Random-effects models for serial observation with binary response. *Biometrics 40*, 961–971.

Stone, C., M. Hansen, C. Kooperberg, and Y. Truong (1997). Polynomial splines and their tensor products in extended linear modeling. *The Annals of Statistics 25*, 1371–1470.

Stone, C. J. (1977). Consistent nonparametric regression (with discussion). *Annals of Statistics 5*, 595–645.

Stram, D. O. and L. J. Wei (1988). Analyzing repeated measurements with possibly missing observations by modelling marginal distributions. *Statistics in Medicine 7*, 139–148.

Stram, D. O., L. J. Wei, and J. H. Ware (1988). Analysis of repeated categorical outcomes with possibly missing observations and time-dependent covariates. *Journal of the American Statistical Association 83*, 631–637.

Strobl, C., A.-L. Boulesteix, and T. Augustin (2007). Unbiased split selection for classification trees based on the gini index. *Computational Statistics & Data Analysis 52*, 483–501.

Strobl, C., A.-L. Boulesteix, T. Kneib, T. Augustin, and A. Zeileis (2008). Conditional variable importance for random forests. *BMC Bioinformatics 9*(1), 307.

Strobl, C., J. Malley, and G. Tutz (2009). An Introduction to Recursive Partitioning: Rationale, Application and Characteristics of Classification and Regression Trees, Bagging and Random Forests. *Psychological Methods 14*, 323–348.

Stroud, A. H. and D. Secrest (1966). *Gaussian Quadrature Formulas*. Englewood Cliffs, NJ: Prentice-Hall.

Stukel, T. A. (1988). Generalized logistic models. *Journal of the American Statistical Association 83*(402), 426–431.

Suissa, S. and J. J. Shuster (1991). The 2×2 method-pairs trial: Exact unconditional design and analysis. *Biometrics 47*, 361–372.

Tüchler, R. (2008). Bayesian variable selection for logistic models using auxiliary mixture sampling. *Journal of Computational and Graphical Statistics 17*, 76–94.

Thall, P. F. and S. C. Vail (1990). Some covariance models for longitudinal count data with overdispersion. *Biometrics 46*, 657–671.

Theil, H. (1970). On the estimation of relationships involving qualitative variables. *American Journal of Sociology 76(1)*, 103–154.

Thurner, P. and A. Eymann (2000). Policy-specific alienation and indifference in the calculus of voting: A simultaneous model of party choice and abstention. *Public Choice 102*, 49–75.

Thurstone, L. L. (1927). A law of comparative judgement. *Psychological Review 34*, 273–286.

Tibshirani, R. (1996). Regression shrinkage and selection via the lasso. *Journal of the Royal Statistical Society B 58*, 267–288.

Tibshirani, R. and T. Hastie (1987). Local likelihood estimation. *Journal of the American Statistical Association 82*, 559–568.

Tibshirani, R., T. Hastie, B. Narasimhan, and G. Chu (2003). Class prediction by nearest shrunken centroids, with applications to DNA microarrays. *Statistical Science*, 104–117.

Tibshirani, R., T. Hastie, B. Narasimhan, S. Soltys, G. Shi, A. Koong, and Q.-T. Le (2004). Sample classification from protein mass spectrometry, by "'peak probability contrasts". *Bioinformatics 20*, 3034–3044.

Tibshirani, R., M. Saunders, S. Rosset, J. Zhu, and K. Kneight (2005). Sparsity and smoothness via the fused lasso. *Journal of the Royal Statistical Society B 67*, 91–108.

Tjur, T. (1982). A connection between Rasch's item analysis model and a multiplicative Poisson modelb. *Scandinavian Journal of Statistics 9*, 23–30.

Toledano, A. and C. Gatsanis (1996). Ordinal regression methodology for ROC curves derived from correlated data. *Statistics in Medicine 15*, 1807–1826.

Troyanskaya, O. G., M. E. Garber, P. O. Brown, D. Botstein, and R. B. Altman (2002). Nonparametric methods for identifying differentially expressed genes in microarray data. *Bioinformatics 18*, 1454–1461.

Tsiatis, A. A. (1980). A note on a goodness-of-fit test for the logistic regression model. *Biometrika 67*, 250–251.

Tukey, J. (1977). *Exploratory Data Analysis*. Reading, Pennsylvania: Addison Wesley.

Tutz, G. (1986). Bradley-Terry-Luce models with an ordered response. *Journal of Mathematical Psychology 30*, 306–316.

Tutz, G. (1991). Sequential models in ordinal regression. *Computational Statistics & Data Analysis 11*, 275–295.

Tutz, G. (2003). Generalized semiparametrically structured ordinal models. *Biometrics 59*, 263–273.

Tutz, G. (2005). Modelling of repeated ordered measurements by isotonic sequential regression. *Statistical Modelling 5*(4), 269–287.

Tutz, G. and H. Binder (2004). Flexible modelling of discrete failure time including time-varying smooth effects. *Statistics in Medicine 23*(15), 2445–2461.

Tutz, G. and H. Binder (2006). Generalized additive modeling with implicit variable selection by likelihood-based boosting. *Biometrics 62*, 961–971.

Tutz, G. and H. Binder (2007). Boosting ridge regression. *Computational Statistics & Data Analysis 51*, 6044–6059.

Tutz, G. and J. Gertheiss (2010). Feature extraction in signal regression: A boosting technique for functional data regression. *Journal of Computational and Graphical Statistics 19*, 154–174.

Tutz, G. and A. Groll (2010a). Binary and ordinal random effects models including variable selection. Technical Report 97, LMU, Department of Statistics.

Tutz, G. and A. Groll (2010b). Generalized linear mixed models based on boosting. In T. Kneib and G. Tutz (Eds.), *Statistical Modelling and Regression Structures - Festschrift in the Honour of Ludwig Fahrmeir*, pp. 197–215. Physica.

Tutz, G. and K. Hechenbichler (2005). Aggregating classifiers with ordinal response structure. *Journal of Statistical Computation and Simulation 75*(5), 391–408.

Tutz, G. and W. Hennevogl (1996). Random effects in ordinal regression models. *Computational Statistics and Data Analysis 22*, 537–557.

Tutz, G. and G. Kauermann (1997). Local estimators in multivariate generalized linear models with varying coefficients. *Computational Statistics 12*, 193–208.

Tutz, G. and F. Leitenstorfer (2006). Response shrinkage estimators in binary regression. *Computational Statistics and Data Analysis 50*, 2878–2901.

Tutz, G. and S. Petry (2011). Nonparametric estimation of the link function including variable selection. *Statistics and Computing, to appear*.

Tutz, G. and F. Reithinger (2007). A boosting approach to flexible semiparametric mixed models. *Statistics in Medicine 26*, 2872–2900.

Tutz, G. and T. Scholz (2004). Semiparametric modelling of multicategorial data. *Journal of Statistical Computation & Simulation 74*, 183–200.

Tutz, G. and J. Ulbricht (2009). Penalized regression with correlation based penalty. *Statistics and Computing 19*, 239–253.

Tversky, A. (1972). Elimination by aspects: A theory of choice. *Psychological Review 79*, 281–299.

Tweedie, M. C. K. (1957). Statistical properties of inverse Gaussian distributions. I. *Annals of Mathematical Statistics 28*(2), 362–377.

Ulbricht, J. and G. Tutz (2008). Boosting correlation based penalization in generalized linear models. In Shalabh and C. Heumann (Eds.), *Recent Advances In Linear Models and Related Areas*. New York: Springer–Verlag.

Ulm, K. (1991). A statistical method for assessing a threshold in epidemiological studies. *Statistics in Medicine 10*, 341–348.

van den Broek, J. (1995). A score test for zero inflation in a Poisson distribution. *Biometrics 51(2)*, 738–743.

Van der Linde, A. and G. Tutz (2008). On association in regression: the coefficient of determination revisited. *Statistics 42*, 1–24.

van Houwelingen, J. C. and S. L. Cessie (1990). Predictive value of statistical models. *Statistics in Medicine 9*, 1303–1325.

Venables, W. N. and B. D. Ripley (2002). *Modern Applied Statistics with S. Fourth edition.* New York: Springer–Verlag.

Verbeke and G. Molenberghs (2000). *Linear Mixed Models for longitudinal data.* New York: Springer–Verlag.

Vidakovic (1999). *Statistical Modelling by Wavelets.* Wiley Series in Probability and Statistics. New York: Wiley.

Vuong, Q. (1989). Likelihood ratio tests for model selection and non-nested hypotheses. *Econometrica 2*, 307–333.

Wacholder, S. (1986). Binomial regression in GLIM: Estimation risk ratios and risk differences. *American Journal of Epidemiology 123*, 174–184.

Wahba, G. (1990). *Spline Models for Observational Data.* Philadelphia: Society for Industrial and Applied Mathematics.

Walker, S. H. and D. B. Duncan (1967). Estimation of the probability of an event as a function of several independent variables. *Biometrika 54*, 167–178.

Wand, M. P. (2000). A comparison of regression spline smoothing procedures. *Computational Statistics 15*, 443–462.

Wand, M. P. (2003). Smoothing and mixed models. *Computational Statistics 18*(2), 223–249.

Wang, H. and Y. Xia (2009). Shrinkage estimation of the varying coefficient model. *Journal of the American Statistical Association 104*(486), 747–757.

Wang, L. (2011). GEE analysis of clustered binary data with diverging number of covariates. *Ann. Statist. 39*, 389–417.

Wang, Y.-F. and V. Carey (2003). Working correlation structure missclassification, estimation and covariate design: Implicationa for generalized estimating equations. *Biometrika 90*, 29–41.

Watson, G. S. (1964). Smooth regression analysis. *Sankhyā, Series A, 26*, 359–372.

Wedderburn, R. W. M. (1974). Quasilikelihood functions, generalized linear models and the Gauss-Newton method. *Biometrika 61*, 439–447.

Wei, G. and M. Tanner (1990). A Monte Carlo implementation of the EM algorithm and the poor man's data augmentation algorithms. *Journal of the American Statistical Association 85*, 699–704.

Weisberg, S. and A. H. Welsh (1994). Adapting for the missing link. *Annals of Statistics 22*, 1674–1700.

Welsh, A., X. Lin, and R. Carroll (2002). Marginal longitudinal nonparametric regression. *Journal of the American Statistical Association 97*(458), 482–493.

Whittaker, J. (1990). *Graphical Models in Applied Multivariate Statistics.* Chichester: Wiley.

Whittaker, J. (2008). *Graphical Models in Applied Multivariate Statistics.* Wiley Publishing.

Whittemore, A. S. (1983). Transformations to linearity in binary regression. *SIAM Journal of Applied Mathematics 43*, 703–710.

Wild, C. J. and T. W. Yee (1996). Additive extensions to generalized estimating equation methods. *Journal of the Royal Statistical Society B58*, 711–725.

Wilkinson, G. N. and C. E. Rogers (1973). Symbolic description of factorial models for analysis of variance. *Applied Statistics 22*, 392–399.

Williams, D. A. (1982). Extra binomial variation in logistic linear models. *Applied Statistics 31*, 144–148.

Williams, O. D. and J. E. Grizzle (1972). Analysis of contingency tables having ordered response categories. *Journal of the American Statistical Association 67*, 55–63.

Williamson, J. M., K. Kim, and S. R. Lipsitz (1995). Analyzing bivariate ordinal data using a global odds ratio. *Journal of the American Statistical Association 90*, 1432–1437.

Winkelmann, R. (1997). *Count Data Models: Econometric Theory and Application to Labor Mobility (Second Edition)*. Berlin: Springer-Verlag.

Wolfinger, R. W. (1994). Laplace's approximation for nonlinear mixed models. *Biometrika 80*, 791–795.

Wong, G. Y. and W. M. Mason (1985). The hierarchical logistic regression model for multilevel analysis. *Journal of the American Statistical Association 80*, 513–524.

Wood, S. N. (2000). Modelling and smoothing parameter estimation with multiple quadratic penalties. *Journal of the Royal Statistical Society B 62*, 413–428.

Wood, S. N. (2004). Stable and efficient multiple smoothing parameter estimation for generalized additive models. *Journal of the American Statistical Association 99*, 673–686.

Wood, S. N. (2006a). *Generalized Additive Models: An Introduction with R*. London: Chapman & Hall/CRC.

Wood, S. N. (2006b). On confidence intervals for generalized additive models based on penalized regression splines. *Australian & New Zealand Journal of Statistics 48*, 445–464.

Wood, S. N. (2006c). Thin plate regression splines. *Journal of the Royal Statistical Society, Series B 65*, 95–114.

Wu, J. C. F. (1983). On the covergence properties of the EM-algorithm. *Annals of Statistics 11*, 95–103.

Xia, T., F. Kong, S. Wang, and X. Wang (2008). Asymptotic properties of the maximum quasi-likelihood estimator in quasi-likelihood nonlinear models. *Communications in Statistics – Theory and Methods 37*(15), 2358–2368.

Xia, Y., H. Tong, W. K. Li, and L. Zhu (2002). An adaptive estimation of dimension reduction. *Journal of the Royal Statistical Society B 64*, 363–410.

Xie, M. and Y. Yang (2003). Asymptotics for generalized estimating equations with large cluster sizes. *The Annals of Statistics 31*(1), 310–347.

Ye, J. M. (1998). On measuring and correcting the effects of data mining and model selection. *Journal of the American Statistical Association 93*(441), 120–131.

Yee, T. (2010). The VGAM package for categorical data analysis. *Journal of Statistical Software 32*(10), 1–34.

Yee, T. and T. Hastie (2003). Reduced-rank vector generalized linear models. *Statistical Modelling 3*, 15.

Yee, T. W. and C. J. Wild (1996). Vector generalized additive models. *Journal of the Royal Statistical Society B*, 481–493.

Yellott, J. I. (1977). The relationship between Luce's choice axiom, Thurstone's theory of comparative judgement, and the double exponential distribution. *Journal of Mathematical Psychology 15*, 109–144.

Yu, Y. and D. Ruppert (2002). Penalized spline estimation for partially linear single-index models. *Journal of the American Statistical Association 97*, 1042–1054.

Yuan, M. and Y. Lin (2006). Model selection and estimation in regression with grouped variables. *Journal of the Royal Statistical Society B 68*, 49–67.

Zahid, F. M. and G. Tutz (2009). Ridge estimation for multinomial logit models with symmetric side constraints. Technical Report 67, LMU, Department of Statistics.

Zahid, F. M. and G. Tutz (2010). Multinomial logit models with implicit variable selection. *Technical Report 89, Department of Statistics LMU*.

Zeger, S. L. (1988). Commentary. *Statistics in Medicine 7*, 161–168.

Zeger, S. L. and P. J. Diggle (1994). Semi-parametric models for longitudinal data with application to CD4 cell numbers in HIV seroconverters. *Biometrics 50*, 689–699.

Zeger, S. L. and M. R. Karim (1991). Generalized linear models with random effects; a Gibbs' sampling approach. *Journal of the American Statistical Association 86*, 79–95.

Zeileis, A., C. Kleiber, and S. Jackman (2008). Regression models for count data in R. *Journal of Statistical Software 27*.

Zhang, H. (1998). Classification trees for multiple binary responses. *Journal of the American Statistical Association 93*, 180–193.

Zhang, H. and B. Singer (1999). *Recursive Partitioning in the Health Sciences*. New York: Springer–Verlag.

Zhang, Q. and E. Ip (2011). Generalized linear model for partially ordered data. *Statistics in Medicine (to appear)*.

Zhao, L. P. and R. Prentice (1990). Correlated binary regression using a quadratic exponential model. *Biometrika 77*, 642–48.

Zhao, L. P., R. L. Prentice, and S. Self (1992). Multivariate mean parameter estimation by using a partly exponential model. *Journal of the Royal Statistical Society B 54*, 805–811.

Zhao, P., G. Rocha, and B. Yu (2009). The composite absolute penalties family for grouped and hierarchical variable selection. *Annals of Statistics 37*, 3468–3497.

Zhao, P. and B. Yu (2004). Boosted lasso. Technical report, University of California, Berkeley, USA.

Zheng, B. and A. Agresti (2000). Summarizing the predictive power of a generalized linear model. *Statistics in Medicine 19*, 1771–1781.

Zheng, S. (2008). Selection of components and degrees of smoothing via lasso in high dimensional nonparametric additive models. *Computational Statistics & Data Analysis 53*, 164–175.

Zhu, J. and T. Hastie (2004). Classification of gene microarrays by penalized logistic regression. *Biostatistics 5*, 427–443.

Zhu, Z., W. Fung, and X. He (2008). On the asymptotics of marginal regression splines with longitudinal data. *Biometrika 95*(4), 907.

Zou, H. (2006). The adaptive lasso and its oracle properties. *Journal of the American Statistical Association 101*(476), 1418–1429.

Zou, H. and T. Hastie (2005). Regularization and variable selection via the elastic net. *Journal of the Royal Statistical Society B 67*, 301–320.

Zweig, M. and G. Campbell (1993). Receiver-operating characteristic (ROC) plots: A fundamental evaluation tool in clinical medicine. *Clinical Chemistry 39*, 561–577.

Author Index

Subject Index

Printed in the United States
By Bookmasters